Principles of Surface-Enhanced Raman Spectroscopy

Principles of Surface-Enhanced Raman Spectroscopy

and related plasmonic effects

Eric C. Le Ru and Pablo G. Etchegoin

Elsevier
Radarweg 29, PO Box 211, 1000 AE Amsterdam, The Netherlands
Linacre House, Jordan Hill, Oxford OX2 8DP, UK

First edition 2009

British Library Cataloguing in Publication Data
A catalogue record for this book is available from the British Library

Library of Congress Cataloging-in-Publication Data
A catalog record for this book is available from the Library of Congress

ISBN: 978-0-444-52779-0

For information on all Elsevier publications
visit our website at elsevierdirect.com

Printed and bound in Great Britain
08 09 10 11 12 10 9 8 7 6 5 4 3 2 1

Principles of Surface-Enhanced Raman Spectroscopy

and related plasmonic effects

Eric C. Le Ru and Pablo G. Etchegoin
School of Chemical and Physical Sciences
Victoria University of Wellington
PO Box 600, Wellington
New Zealand

ELSEVIER

Amsterdam • Boston • Heidelberg • London • New York • Oxford
Paris • San Diego • San Francisco • Singapore • Sydney • Tokyo

Preface

Surface-enhanced Raman scattering (SERS) was discovered in 1974 [1] and correctly interpreted in 1977 [2,3]. Since then, the field has grown enormously in breadth, depth, and understanding. One of the major characteristics of SERS is its *interdisciplinary nature*. SERS exists at the boundaries shared among physics, chemistry, colloid science, plasmonics, technology, engineering, and biology. There are several review articles in the field [4–6] for the advanced researcher together with a recent book dedicated to surface-enhanced vibrational spectroscopy by Ricardo Aroca [7]. Still, we put ourselves in the situation of a graduate student in physics, chemistry, physical chemistry, or chemical physics, undertaking a Ph.D. project in the area of SERS or related subjects and not having an in-depth understanding of Raman spectroscopy itself, the theory of plasmon resonances, or elements of colloid science. By their very nature, it is difficult to find a textbook that will summarize the principles of these rather dissimilar and disconnected topics. It is even less likely that this collection of topics was touched upon as a coherent unit during most undergraduate studies in physics or chemistry. A similar situation can arise for established researchers, either chemists or physicists, who are newcomers to the field but might not have a background in Raman spectroscopy or the physics of plasmons. Yet, a basic understanding of these topics is desirable to start a research project in SERS, and as a stepping stone to tackle the more specialized literature. This book finds its justification in that fact, and will hopefully fill (at least) a fraction of what we feel is an existing gap in the literature.

The content of the book covers most of the topics related to SERS and presents them as a coherent study program that can be tackled at different levels of complexity depending on the individual needs of the reader. For the most important subjects, we have attempted in our presentation to provide a *graded approach*: starting with a simple explanation of the most relevant concepts, which is then developed into a more rigorous exposition, including the more advanced aspects. In this way, we hope that this book will cater to a variety of readers with different skills and scientific backgrounds; an intrinsic characteristic of the general SERS and plasmonics community. To help the reader find his/her way through the various topics and the different level of complexities, a detailed overview of the content of the book and a few suggested reading plans are provided at the end of the introductory chapter.

This book is about *principles* and therefore does not attempt to replace the many excellent reviews in the field, which are concentrated mainly on the exposition of the latest research results and their interpretations. Review articles tend to be too specialized to spend time on basic aspects of, for example, molecular Raman spectroscopy or the physics of plasmon resonances in metals. This book therefore attempts to make emphasis on these underlying concepts. The selection of topics is *not* intended as a detailed collection of results of the current literature and the accompanying bibliography is far from being exhaustive. Such an extensive review of the older and current literature of SERS is, in fact, largely provided already in Ref. [7]. The most important examples of the current literature are used, of course, to stress concepts or to make the explanation of certain topics clearer, but it is by no means exhaustive. Moreover, we emphasize concepts and principles that *we* judge important as a general background to SERS, but it does not represent a complete (and unbiased) list of topics. Both authors are physicists by training (and at heart...), and there is a natural emphasis on physical aspects of the problem in the presentation. We have in fact deliberately tried to avoid too much overlap in the selection of topics with the recent book by Ricardo Aroca [7]. Not only that Aroca's insight into the field, from a more chemical point of view, is excellent but also, in this manner, we hope that the books will complement each other. One aspect we do particularly emphasize is the intricate link between SERS and the wider research field of *plasmonics*, i.e. the study and applications of the optical properties of metals. SERS can, in fact, be viewed as a subfield of plasmonics. The relation between SERS and related plasmonics effects is, we believe, symbiotic, and we attempt to emphasize this aspect repeatedly.

To conclude this preface, a tradition that we shall not attempt to escape is to thank the many people and institutions that made the book (directly or indirectly) possible. First of all, we would like to thank the continuous support of the MacDiarmid Institute for Advanced Materials and Nanotechnology in New Zealand, and by the same token, Victoria University of Wellington (where part of the Institute is hosted). In particular, we would like to thank its founding director (Prof. Paul T. Callaghan) who has been a continuous source of inspiration and support (economic and personal) during the last few years. Without the financial support of the MacDiarmid Institute and Victoria University of Wellington, this book would not have been possible. The Royal Society of New Zealand is also gratefully acknowledged for financial support during this period. In addition, we would like to thank our direct collaborators (past and present), and our students (in particular Robert C. Maher from Imperial College London, and Matthias Meyer, Evan Blackie, and Chris Galloway from Victoria University of Wellington) who paid (and are still paying) the high price of long hours in the lab studying the SERS effect. Special thanks are also given to Prof. Lesley F. Cohen of Imperial College London, who, many years ago, proposed for the first time the subject of SERS as a possible research topic to one of the authors (PGE). For the many scientific discussions and the longstanding collaboration we are very grateful.

Last but not least, we would like to thank our respective family members (Nancy and little Noah!, Sofía, and Julián) for their understanding and support during the long period while the writing was under way.

Eric C. Le Ru, Pablo G. Etchegoin

Wellington, New Zealand

Notations, units and other conventions

We have made our best efforts to use notations, conventions, and units that are consistent throughout the book. We summarize here (for reference) our specific choices.

Units:

We use S.I. units throughout in all our expressions (except when discussing other units that are commonly used in the literature). These are, in our opinion, the more versatile choice for a subject spanning through such diverse areas of physics and chemistry. They are also more rigorous in many respects compared, for example, to Gaussian units.

We have also endeavored when possible to specify the units of the variables we define. This should help, we hope, in understanding the physical meaning of each variable. These are given in between brackets [...], using either:

- The basic S.I. units: kilogram [kg] for mass, meter [m] for length, second [s] for time, Ampere [A] for electric current, Kelvin [K] for temperature, and mole [mol] for amount of substance,
- Or commonly used *derived* S.I. units, such as Joule [J] = [m^2 kg s^{-2}] for energy, Watt [W] = [m^2 kg s^{-3}] = [J s^{-1}] for power, Coulomb [C] = [s A] for electric charge, or Volt [V] = [m^2 kg s^{-3} A^{-1}] for voltage.
- Or sometimes for simplicity in units of common physical constants, such as ϵ_0 [kg^{-1} m^{-3} s^4 A^2], the permittivity of vacuum. For example, polarizability is given in [ϵ_0 m^3] rather than the equivalent (but more cumbersome) S.I. expression [kg^{-1} s^4 A^2].
- Or common adimensional units to further clarify the meaning of the physical quantity. These include radians [rad] for angles or [rad s^{-1}] for angular frequency, and steradians [sr] for solid angles.

When relevant, we may also use "less rigorous", but "more conventional" units, such as electron-volt [eV] for energy, liter [L] for volume, or molar [M] = [mol L^{-1}] for concentration.

Mathematical notations:

Most mathematical notations we use are fairly standard. Variables are Greek or Roman letters in italics, such as a, A, or α. Vectors are represented by bold letters, such as $\mathbf{A}$. The unit vectors for a given coordinate frame are written as $\mathbf{e}_i$, where the subscript i refers to the corresponding axis. In Cartesian coordinates, where the vector position is $\mathbf{r} = (x, y, z)$, they are therefore $\mathbf{e}_x$, $\mathbf{e}_y$, $\mathbf{e}_z$. Spherical coordinates are denoted $\mathbf{r} = (r, \theta, \phi)$ and defined in Appendix H. The unit vectors are then $\mathbf{e}_r$, $\mathbf{e}_\theta$, $\mathbf{e}_\phi$ (and depend on position $\mathbf{r}$). Tensors are represented with a hat, such as $\hat{\alpha}$, or may be explicitly given as the tensorial product of two vectors, such as $\mathbf{e}_x \otimes \mathbf{e}_y$.

Variable names:

We have attempted to follow standard practices in terms of variable names, especially for common physical constants or quantities. All of them will be obvious within the context and in agreement with standard conventions in the literature.

Conventions:

We use a number of conventions that may differ from other treatments of the subject:

- A time dependence as $\exp(-i\omega t)$ is assumed for complex notations, which results in positive imaginary parts for response functions, such as the dielectric function ϵ or the polarizability α. This convention is commonly used in the *physics* literature, but is different from the convention normally used in *engineering*.
- Dielectric constants and dielectric functions are always *relative*. They are therefore adimensional quantities and should be multiplied by ϵ_0, the permittivity of vacuum, to obtain the absolute dielectric constant.

Moreover, as in many scientific publications, we make use of numerous acronyms, starting with SERS, the main subject of the book! These will be defined in the text as they are introduced, but in case of doubt, we have attempted to include them all in the index at the end of the book.

Computer codes:

Many of the most complicated equations given in this book are not given with the expectation that the reader will carry out further analytical studies from them. Rather, they are provided to be used for numerical calculations, thanks to which the reader may experiment at will, to understand the underlying physics or model problems adapted to his/her own specific needs.

To make this easier, we therefore also provide in some places a brief description of the actual numerical implementation (as Matlab scripts or functions). All the corresponding codes are available for download from the book's website: http://www.victoria.ac.nz/raman/book, and will be updated as required in the future. We have also included there (as examples) a number of Matlab scripts that can be used to reproduce (and adapt if necessary) many figures of the book. We hope that they will be easily usable by someone not familiar with the underlying mathematics or computer coding. A minimum knowledge of Matlab is, however, necessary and can be acquired quickly by browsing the Matlab help menu.

Book's website:

The book's website can be found at:

http://www.victoria.ac.nz/raman/book.

It contains an extensive section dedicated to Matlab computer codes relevant to SERS and plasmonics, many of which are based on the theory presented in the book and – in particular – on the material presented in the appendices. We will also attempt to update it regularly with various other information related to the book itself, and to SERS and plasmonics in general.

Chapter 1

A quick overview of surface-enhanced Raman spectroscopy

The technical complexity of this book will scale rapidly in the forthcoming chapters. Still, we try to imagine first a potential reader who might have heard about surface-enhanced Raman spectroscopy (SERS) only superficially (or somebody who has been asked to look at its potential for a specific application) and wants to have a bird's eye view about the general principles and applications of SERS. That includes somebody who might be curious or interested in how the technique actually works in practice (at a basic level). This introductory chapter is, therefore, not for the experienced scientist or student in the field, but rather for the complete newcomer looking for a broad map that will guide him/her toward more advanced studies and applications. Whether the technique will provide the ideal solution to the problem at hand (or not) will probably require the more in-depth analysis presented in the forthcoming chapters. This overview of the main characteristics of the effect and some of its applications, however, should certainly convey a general impression to the reader of how the technique actually works, and a flavor (without the technicalities) of its underlying principles. By the same token, we shall try to put the SERS effect into its historical context, and highlight its present status and future challenges. The chapter will finish with a brief overview of the content of the book (and how it addresses some of the issues raised in this chapter) and a suggested reading plan that (hopefully) will cater to a wide variety of potential readers with different needs.

1.1. WHAT IS SERS? – BASIC PRINCIPLES

In a nutshell, the SERS effect is about amplifying *Raman signals* (almost exclusively coming from molecules) by several orders of magnitude.

The amplification of the signals in SERS comes (mainly) through the electromagnetic interaction of light with metals, which produces large amplifications of the laser field through excitations generally known as *plasmon resonances*. To profit from these, the molecules must typically be adsorbed on the metal *surface*, or at least be very close to it (typically $\approx$10 nm maximum). The denomination surface-enhanced Raman scattering or SERS, summarizes particularly well these three cornerstones of the effect:

- Surface (S): SERS is a *surface spectroscopy* technique; the molecules must be on (or close to) the surface. This is a major point for applications of SERS. One must ensure that the molecules to be detected can *attach* to (or at least be in close proximity to) the surface of the metal substrate. The transfer of molecules from a volume to a surface is a recurrent theme (and problem) in practical implementations of SERS.

- Enhanced (E): The signal enhancement is provided by *plasmon resonances* in the metal substrate. The term 'plasmon resonances' is, in fact, a shorthand for a family of effects associated with the interaction of electromagnetic radiation with metals. A full description of the many different aspects of plasmon resonances and the way they influence SERS phenomena are given in Chapters 3–5. Also, metals appear in the SERS effect (more often than not) in the form of *metallic nano-structures*, which encompass a variety of different SERS substrates, from metallic colloids in solution (described in Chapter 7) to substrates fabricated by nano-lithography or self-organization (described in Chapter 8).

- Raman (R): The technique consists in measuring the *Raman signals* of molecules (the SERS probes or analytes). Raman spectroscopy is the study of *inelastic light scattering* and, when applied to molecules, it provides an insight into their chemical structure (in particular their vibrational structure). A detailed description of the Raman effect itself is given in Chapter 2, with a special emphasis on Raman scattering from molecules.

The final S in SERS can stand for Scattering or Spectroscopy, depending on whether one prefers to emphasize the optical effect (scattering) or the technique and its applications (spectroscopy).

This simple description of the effect should convey one particular interesting characteristic of SERS, namely: *its multi-disciplinary nature*. Although typically classified as a topic in 'chemical physics' or 'physical chemistry', some aspects of it – such as the electromagnetic theory of plasmon resonances – are very much *physical*, while others such as molecular adsorption on the surfaces are very much *chemical* in nature. To these, one may add engineering aspects of SERS substrate fabrication and biological aspects of many potential applications.

1.2. SERS PROBES AND SERS SUBSTRATES

Among the many parameters that can be varied in a SERS experiment, two stand out naturally: the molecular species to be detected (the probe), and the metallic structures onto which it adsorbs (the SERS substrate). These two aspects are to a large extent independent, although some degree of 'compatibility' is required: one must ultimately ensure that the probe goes onto the substrate to profit from the amplification of Raman signals by plasmon resonances.

1.2.1. SERS substrates

What is a good SERS substrate?

Good SERS substrates are in simple terms those that support the 'strongest' (in a sense that will be defined throughout this book) plasmon resonances; in other words, those that provide the largest enhancement or amplification. In this respect, one should in addition distinguish between those that provide a relatively uniform enhancement on the surface and those with large variations. The latter typically exhibit some highly localized positions of very high enhancement (hot-spots), particularly suited for single-molecule detection. Nevertheless, the former should be preferred for reproducibility in applications.

Moreover, because the SERS enhancements arise from a resonant response of the substrate, they are typically strongly wavelength dependent, i.e. they vary with the excitation wavelength (and to a lesser degree with the Raman shift of the modes, to be defined in Chapter 2). A given SERS substrate will, therefore, typically exhibit good enhancements in a limited excitation wavelength range. In fact, the optimum excitation wavelength could be viewed as part of the definition of the SERS substrate: a SERS substrate excited at the wrong wavelength is no longer a SERS substrate (or only a really bad one). Most SERS substrates are designed to operate with visible/near-infrared excitation (~400–1000 nm), which is the typical range of interest for molecular Raman scattering experiments.

As a rule of thumb, enhancements suitable for a successful implementation of the technique typically arise from:

- Structures made of *gold or silver*; the two metals most used for SERS and plasmonics in general. This is simply because they have the 'right' optical properties (see Chapter 3 and Appendix E) to sustain 'good' plasmon resonances in the visible/near-infrared range (~400–1000 nm), which is the most interesting range for SERS.

- Objects (or structures) with dimensions in the *sub-wavelength* range, and typically less than ∼100 nm. This requirement creates a strong connection between SERS and the general area of *nano-science* or *nano-technology*. There is, in principle, no limit on how small the metallic objects constituting a SERS substrate should be. As an example, a *rough* metallic surface can be used as a SERS substrate, and it typically has 'structures' on its surface that span a wide range of characteristic dimensions down to ∼1 nm.

These two conditions are not strictly exclusive; a simple *flat* metallic surface can already serve as an 'amplifier' of Raman signals for molecules deposited on it, albeit achieving a much lower level of amplification than that reached normally in metallic nano-structures. SERS has also been measured on structures made of a wide variety of metals, such as copper or platinum, but again with lower enhancements than those typically achieved with gold and silver.

Other considerations

Enhancements are not the only important characteristics of a SERS substrate. Among other aspects, let us mention here the *surface area*. SERS being a surface spectroscopy, the surface area of the substrate should obviously be an important parameter (the surface area should be understood here as the metallic surface area within the scattering volume of observation). A larger surface area increases the potential number of molecules that can produce SERS (for example the number of molecules in a monolayer). This does not improve the sensitivity, since at low concentrations (sub-monolayer coverage) we are mainly limited by the intrinsic 'strength' of the SERS signals of the molecule. There are, nonetheless, situations where molecules only attach to the substrate in the first layer (by direct contact on the metal). The maximum achievable SERS signal is then limited by the maximum number of molecules in this layer (the 'parking-problem'). If the molecule is a weak Raman scatterer, and the maximum achievable SERS signal is too low, then the SERS signal cannot be measured. The several possible options to avoid this issue are: (i) to use a substrate with a larger average enhancement (this increases the average SERS signal of individual molecules), (ii) to use a substrate with a larger surface area (this increases the maximum number of molecules producing the signal), (iii) to increase the laser power, and (iv) to increase the scattering volume (and therefore the probed surface area). This latter option is only worth if the power density is kept constant, which usually requires increasing the laser power as in (iii). The latter two options are therefore usually limited by instrumental considerations (available laser power).

In addition to these basic characteristics, one should also consider the ease and costs of fabrication and sample preparation. Finally, the substrate/probe interactions play a major role in SERS and some substrates may therefore be better suited (or even specifically designed) for use with a particular type of molecule(s). We will be more specific about this point in the forthcoming discussion of SERS probes.

Three main classes of SERS substrates

A 'SERS substrate' is therefore a general denomination for any plasmon-resonance-supporting structure that will produce suitable Raman amplifications. SERS substrates can be tentatively classified (only for the purpose of fixing ideas) into three main classes:

- Metallic particles (usually nano-particles) in solution, such as colloidal solutions.
- 'Planar' metallic structures, such as arrays of metallic nano-particles supported on a planar substrate (glass, silicon, or metallic, for example).
- Metallic electrodes.

Electrodes have played an important role in the historical development of SERS, including its discovery. It is fair to say that SERS started as a discipline in *electro-chemistry*. Its importance has however been decreasing substantially, mostly because of the relatively low enhancement factors typically achievable. We have accordingly chosen not to discuss them further in this book, and refer the reader to the specialized literature [4–6]. Let us note however that it remains an important approach, for example, for (i) SERS on metals other than (the most commonly-used) silver and gold, (ii) investigations of chemical enhancement mechanisms (discussed in Section 4.8), and (iii) SERS applications to electro-chemistry itself, as a tool to monitor specific aspects of electro-chemical reactions.

Among the other two classes of SERS substrates, solutions of metallic colloids (predominantly made of silver (Ag) or gold (Au)), occupy an unquestionable place of pre-eminence in SERS, both in early and more recent studies. One of the important applications of SERS is in the tracing of molecules in water, where Ag and Au colloids can exist and provide the necessary SERS enhancements. Such colloids can, moreover, be dried or attached to a suitable substrate as a simple means to fabricate the second type: planar metallic structures. Indeed, this approach and that of metal island films, have constituted for a long time the main examples of planar metallic structures. More recently, with the advent of nano-technologies, a whole new range of *ordered* planar metallic structures has appeared (some examples of which are discussed in Chapter 8). SERS substrates in general, and metallic colloids in solution in particular, will be discussed in detail in Chapters 7 and 8.

The major difference between particles in solution and planar substrates does not lie in the nature of the SERS enhancements but in the actual implementation of the technique. In the first case (SERS solutions), the SERS signal arises from a *3D volume* (defined by the experimental set-up). This volume is what is known in optics (and spectroscopy) as the *scattering volume*, and it is defined by the excitation/collecting optics of the spectrometer. In addition to the SERS enhancement of individual nano-particles, factors such as the nano-particle concentration and their dynamics (Brownian motion) can play a major role in SERS experiments in solution. In the second case (SERS on planar substrates), the SERS signal arises from a '*2D plane*' (although a plane with some 3D structure). The necessary transfer of the probe molecules to be detected (typically in solution) onto this 2D plane is then one of the most important aspects to be considered for the interpretation of results, or for practical applications.

1.2.2. SERS probes

What is a good SERS probe?

Not all molecules are good SERS probes, even though the technique can be used with a remarkable variety of analytes. The two major characteristics of a SERS probe are:

- Intrinsic Raman properties: The intensity of Raman scattering (characterized by the Raman cross-section, see Chapter 2) can vary by many orders of magnitude depending on the molecules under study and the incident laser wavelength. Raman scattering is, for example, particularly intense for molecules with electronic energies close to the exciting laser energy, for example *dyes*; this is then called *resonant Raman scattering* (RRS) . RRS intensities can be $\approx 10^6$ larger than normal (off-resonance) Raman intensities. As a rule of thumb, *good Raman scatterers (like dyes) make good SERS probes.* This is, in some ways, obvious: if the Raman signal is $\sim 10^6$ times stronger before amplification, it will still be (in general) $\sim 10^6$ times stronger after amplification! (i.e. in SERS conditions). Note in this context that when SERS is measured with a probe in RRS conditions, it is sometimes referred to as 'SERRS' or 'SE(R)RS', for surface-enhanced resonant Raman scattering. However, the main enhancement mechanisms are the same (only the molecules they apply to are different), and we shall not make this distinction in the rest of this book. Finally, other intrinsic Raman properties, such as the Raman mode symmetries also influence their SERS properties, but this is in most cases secondary.

- Probe/Metal interactions: The condition for a molecule to be a 'good Raman scatterer' is not enough to make it a good SERS probe. It must, in addition, be able *to adsorb efficiently on the SERS substrate* to be used, i.e. typically on gold or silver surfaces. Some molecules have a strong chemical affinity for such metal surfaces (for example forming strong covalent bonds) and are therefore easier to work with. Examples of the latter include molecules with thiol or triazole moieties in their structure (see Chapter 7 and Appendix A for specific examples), which display strong affinity for gold and silver. Other common mechanisms of probe attachment are through electrostatic interactions; but then only probes with the correct (opposite) charge will adsorb on a charged substrate. The probe and substrate can then no longer be considered as independent. This concept can be pushed even further in SERS substrates with *surface functionalization*: the metallic surface is then chemically prepared to allow (and ideally facilitate) binding of only one specific type of analyte. A typical implementation (in biological applications) is that of a metallic surface coated with antigens that would only bind to specific anti-bodies (serving as SERS probes here). Problems of probe/metal interactions are among the most important, and also most difficult, in SERS implementations.

Can any type of molecule be measured with SERS?

With the aforementioned considerations in mind, it is now worth recalling that, from a practical point of view, choosing a SERS probe is not always an option! For fundamental studies of the SERS process itself, one is typically free to choose the SERS probe that best suits the experimental needs. In this case, dyes are often preferred, simply because they produce larger signals. It is also possible to consider 'tagging' the target analyte with a good SERS probe. This is a common approach in biological studies, where proteins, anti-bodies, DNA-strands, etc., are 'tagged' with dyes, that can then be detected by fluorescence spectroscopy. These techniques can be readily transferred to SERS.

But for many applications, the probe is a *fixed* parameter of the problem and one must simply adapt to it by choosing the appropriate SERS substrate (and possibly excitation wavelength). One of the most important and basic question is therefore: can *any* molecule be measured with SERS? The answer to this question is most of the time 'yes', but not always with the maximum level of amplification or the most convenient experimental procedures.

Let us now try to be more precise. If a molecule produces a Raman signal, then it can in principle be amplified by interaction with plasmon resonances on a metallic substrate and, therefore, produce SERS. Two important questions arise then: can any molecule be attached (or at least brought close) to a metallic substrate? and, will the resulting SERS signal be sufficiently strong to be observed (and distinguished from any other unavoidable signal and noise)? The first part, attaching the molecule, is in general possible but may require

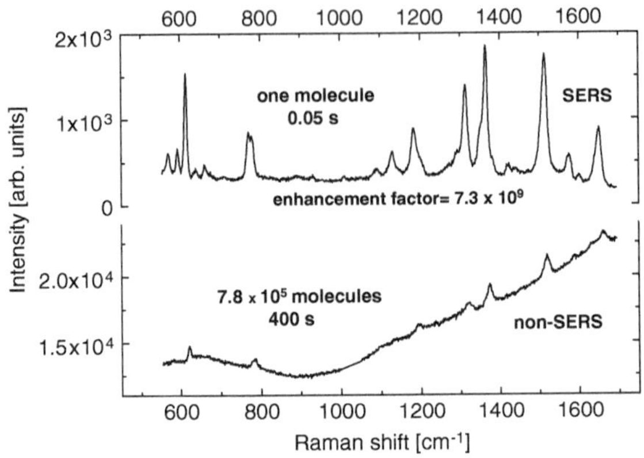

Figure 1.1. Raman (non-SERS) and SERS spectra at 633 nm laser excitation (3 mW) for rhodamine 6G molecules (RH6G). The (vertical) intensity axis is in arbitrary units but the same for both spectra. Bottom spectrum: signal of $\sim 7.8 \times 10^5$ RH6G molecules (100 μM solution in a 13 μm^3 scattering volume, ×100 immersion objective [8]) with 400 s integration time. Top: signal from a *single* RH6G molecule (isolated by the two-analyte method described later in Chapter 8) under the same experimental conditions but with 0.05 s integration time. More experimental details are reported in Ref. [8]. In order to go from the spectrum at the bottom to the one at the top, an amplification of the Raman signal by an enhancement factor of $\sim 7.3 \times 10^9$ is required.

chemical manipulations and may not be straightforward. The second part, obtaining a detectable signal, is in general easy with good Raman scatterers, but can be more challenging with weaker ones. It then comes down to a proper optimization of the various parameters to maximize the enhancement, the surface area, the number of adsorbed molecules, the optical set-up (detection, scattering volume, incident power), etc. Overall, it may not be straightforward and may require a lot of effort, but *any* molecule can in principle be observed in SERS.

1.2.3. Example

Before continuing, it is worth having a brief look at typical data to convey a direct visual impression of the amplification of Raman signals achieved in SERS. Strictly speaking, we have not introduced yet all the details of Raman spectra (treated in Chapter 2) and all the details of the enhancement factor (treated in Chapter 4), but we only need to accept for the moment that molecules produce a Raman spectrum with a series of peaks (which are fingerprints of a specific molecule), and that this spectrum may be amplified (in intensity) as a result of the interaction with the metal. Accordingly, we will be able to see a much smaller number of molecules under SERS conditions.

This is illustrated in Fig. 1.1. The Raman (non-SERS) and SERS spectra of a very common (and widely used) SERS dye, rhodamine 6G (RH6G), are shown for two different conditions. The spectrum at the bottom is obtained in a solution (100 μM) of RH6G in water (no SERS amplification) with an integration time of 400 s. The number of molecules contributing to the signal can be obtained from the concentration and the knowledge of the *scattering volume* of the system (which needs to be thoroughly characterized beforehand [8]). The signal can be compared to that of a *single* RH6G molecule under SERS conditions, here in an aggregated Ag colloidal solution (see Chapter 7), obtained in the same system, with the same experimental conditions but with an integration time of only 0.05 s. The method used to decide that we are are really measuring a single molecule in this second case is based on a two-analyte technique fully explained in Chapter 8, but we shall take it as a fact at this stage. If we now compare the signals of both cases and normalize for the different integration times and the number of molecules, we conclude that the signal of the single molecule in SERS conditions has been amplified by a factor $\sim 7 \times 10^9$ (the *enhancement factor*, see Chapter 4) compared to that of the same single molecule in normal Raman conditions. This figure should suffice to demonstrate visually why SERS is promising and why it is the subject of intense research; it can be exploited (under the appropriate conditions) as an analytical tool capable of boosting the sensitivity of Raman experiments to the point of detecting a single molecule.

A big fraction of the work in SERS, nonetheless, is dedicated to the 'taming of these large enhancement factors'. The example in Fig. 1.1 has swept under the carpet the important detail that we cannot usually control very well such large enhancement factors. As a matter of fact, there is some complementarity rule that applies in practice: the highest the enhancements (typical of single-molecule SERS conditions) the more 'uncontrollable' from the experimental point of view they become. This will be a recurrent theme in this book.

1.3. OTHER IMPORTANT ASPECTS OF SERS

Having laid out the scene, namely: (i) Raman scattering, (ii) plasmon resonances, and (iii) surface chemistry; as well as presented the main protagonists, *SERS probes* and *SERS substrates*, we will now review briefly some of the main characteristics of the SERS effect.

1.3.1. SERS enhancements

The magnitude of the enhancement factor (EF) (i.e. by how much the Raman signal is amplified with respect to normal conditions, as in Fig. 1.1) is one of the most crucial aspect of SERS. This is not only true for most applications, where the interest in the technique lies in its improved sensitivity, but also for understanding the origins of SERS and the physical mechanisms of this enhancement.

Unfortunately, this enhancement is not as straightforward to measure as it would appear at first sight. In fact, it has been (and still is) a subject of intense controversy and discussion in the SERS literature. The main difficulty lies in the estimation of the *number of molecules* producing the SERS signal, i.e. determining how many molecules there are on the SERS substrate surface. We will return in detail to SERS enhancements in Chapter 4, and give only a brief preliminary discussion here.

There are, broadly speaking, two main important characteristics for the SERS enhancement factor in a given SERS substrate: (i) the *maximum* SERS EF, and (ii) the *average* SERS EF.

Maximum SERS EF

The maximum SERS EF typically occurs at specific positions on the surface (so-called hot-spots) and only those molecules adsorbed there can profit from it. The maximum SERS EF may be of the order of $\sim 10^6$ on a spherical nano-particle, and be as high as $\sim 10^{10}$–10^{11}; for example at the apex of a metallic tip, or at a nano-meter gap between two nano-particles. Such large EFs are typically sufficient to detect the SERS signal from a *single molecule*, arguably the ultimate sensitivity in terms of analytical applications (as mentioned before). However, there is currently no real control on how to create such hot-spots at pre-determined locations; or – equivalently – on how to position a given molecule at a hot-spot. Note however, that enhancements of the order of $\sim 10^7$–10^8 can already be sufficient to detect single molecules in the case of *good SERS probes* (typically, resonant Raman dyes) [8]. The difficulties associated with single-molecule detection are further discussed in Section 8.1.

Average SERS EF

The average SERS EF is, as its name suggests, the SERS EF averaged over all possible positions on the metallic surface. It therefore corresponds to the enhancement in signal expected for molecules randomly adsorbed on the surface (as compared to the *same number of non-adsorbed molecules*). Average SERS EFs can be as low as ~ 10–10^3 for non-optimized conditions. More typical values are in the range $\sim 10^5$–10^6 and should be 'easily' achievable with 'standard' substrates. Values as large as $\sim 10^7$–10^8 are possible and should be considered as very good SERS substrates.

1.3.2. Sample preparation and metal/probe interaction

In order to exploit a given SERS EF at its best, the sample preparation (closely linked to the metal/probe interactions) is of paramount importance. Most analytes are first prepared in solution, and their transfer from a volume (in solution) to a surface (on the substrate) is a critical step. As a way of introduction to the more general aspects of how SERS works in practice, we shall provide a few comments on these issues here.

Adsorption efficiency

An extreme example of this aspect is a situation where the probe does not attach to the metallic surface (for example because of electrostatic repulsion). No SERS signal is then observed, regardless of how large the actual average SERS EF might be on the surface of the substrate. This is a common situation in SERS-active colloidal liquids, which are (typically) negatively charged. This prevents most negatively-charged species in solution from adsorbing on the surface and profiting from SERS enhancements. Even without going into these extreme cases, if, for some reason, only $\sim$10% of the molecules in solution adsorb onto the substrate, the SERS signal will obviously be 10 times less than predicted by the average SERS EF. Hence, the *adsorption efficiency* of the probe directly affects the SERS signal and, accordingly, the sensitivity.

Transfer onto 2D SERS substrates

Let us also consider the case of 2D SERS substrates. This corresponds to the common situation where the SERS substrate (metallic objects, rough metallic surface, etc...) is supported onto a single macroscopically flat surface (like a glass slide or a silicon substrate). To transfer a solution of analytes to be studied onto such a substrate, one can, for example, either dip the substrate into the solution, or dry a drop of solution onto the substrate, or deposit this drop by spin-coating. Note that in all these approaches, a precise estimate of how many molecules are transferred is extremely difficult, unless careful control and calibration of the dipping, drying, or spin-coating process is carried out. Even then, for dipping and drying, the molecular surface density tends to be non-uniform and affected by surface tension and 'edge effects'. Spin-coating provides typically a much more uniform density, but the 'transfer efficiency' is then lower. Moreover, if one was able to dry or spin-coat a drop that is 10$\times$ larger (in volume) onto the same surface area, then the molecule surface density would be, in principle, 10 times larger; and the SERS sensitivity would increase 10-fold (unless the surface is saturated, which depends on the concentration of the analyte solution).

These simple considerations emphasize further a fact already stressed: *the enhancement factor itself is not the only parameter determining the sensitivity of the SERS technique.* Sample preparation (and control) issues can also play an important, if not decisive, role. Among these, the adsorption efficiency, or more generally the transfer of the analyte from solution to the substrate, is a major issue to consider.

1.3.3. Main characteristics of the SERS signals

Let us now list a few important characteristics of SERS signals (spectra and intensities).

- SERS spectrum vs Raman spectrum: As a rule of thumb, most molecules exhibit a SERS spectrum that is very similar to their normal Raman spectrum (at the same excitation wavelength), and most of the fingerprint Raman peaks in particular are easily identifiable. This is evident for example in Fig. 1.1. However, some minor differences may arise, and they should be kept in mind because they are sometimes indicators of important characteristics of the SERS process. For a start, the Raman spectrum under SERS conditions can be affected by the fact that the plasmon resonances (producing the enhancement) are typically *wavelength dependent*. As a consequence, different parts of the spectrum can be amplified by different amounts, depending on the dispersion of the underlying resonance producing the enhancement. Even more subtler (and in general secondary) effects can arise from the particular orientation of the molecule on the surface and the specific Raman mode symmetry. These are called *surface selection rules* and are discussed in Section 4.5. Both effects may result in different relative intensities of the Raman peaks under SERS conditions. In addition, the molecule may change its 'identity' upon adsorption and become a *surface complex*. This may result in small shifts and/or broadening of the Raman peaks. In more extreme cases, Raman modes that are easily visible in the bare molecule can disappear upon interaction with the surface. By the same token, other modes can be 'activated' and even new modes may arise. The intrinsic Raman intensities (cross-sections) may also be modified upon adsorption. These latter effects are usually classified as the *chemical enhancement* effect, and are discussed in Section 4.8.

- Polarization effects: SERS signals can also differ from Raman signals in their polarization properties (see Chapter 2 and Chapter 4). This is a result of the polarization dependence of plasmon resonances.

- SERS continuum: SERS spectra are sometimes associated with a broad background. A background is also present in most Raman spectra, but attributed to impurities or residual intrinsic fluorescence. In the case of SERS, it is believed to have, at least in some cases, a real physical origin. This broad background is often called the *SERS continuum* , but its origin is still controversial [9]. The SERS continuum fluctuates like the SERS signal and has the same polarization properties.

- Photo-bleaching/photo-chemistry: Many SERS probes like dyes are known to photo-bleach under normal non-SERS conditions (at least at sufficiently high excitation powers). It is therefore not so surprising that photo-bleaching also occurs under SERS conditions; decays of the SERS signal because of photo-bleaching are indeed observed experimentally, and the photo-products may also sometimes appear in the SERS spectrum itself [10]. In addition, the electromagnetic enhancements that give rise to the SERS signal can also dramatically affect the photo-bleaching properties, possibly resulting in new photo-chemical phenomena, although the exact details are not yet fully understood. The photo-stability of the probe (and associated photo-chemical phenomena) may or may not be a problem depending on the specific probe, and this needs to be analyzed on a case-by-case basis. It should nevertheless be taken into account when analyzing SERS experiments, especially from dyes.

- Signal fluctuations: SERS signals can also show brusque intensity fluctuations which are not present at all in conventional Raman conditions. These may be linked to changes in SERS substrate configurations (e.g. Brownian motion of colloids), photo-bleaching, single-molecule sensitivity conditions (see Chapter 8), or combinations thereof. SERS fluctuations are further discussed in Chapter 7).

1.3.4. Related techniques

There are many related techniques with a natural link to SERS. The most obvious one is *fluorescence spectroscopy*. This is not a book about fluorescence spectroscopy but, nevertheless, we shall be explaining its basic concepts at an introductory level in Chapter 2. In fact, there are many aspects of fluorescence spectroscopy that do play a role in the framework of SERS, either as additional background signals or through other indirect effects. Moreover, the fluorescence signal, like the Raman signal, can be enhanced under appropriate conditions for molecules on metal surfaces; this effect is called *surface-enhanced fluorescence* (SEF). SEF has many similarities, but also important differences, with SERS. It will therefore be discussed in numerous instances in this book, and in particular described in detail in Chapter 4. It is, in fact, one of the 'related plasmonics effects' hinted at in the subtitle of this book. Other related plasmonics effects will be discussed in the broader context of Chapter 3.

Likewise, the variety of surface-enhanced spectroscopies does not stop at SERS and SEF. There is, in fact, a long list of surface-enhanced phenomena that can potentially play a role under the same conditions as those required for SERS. Many of these have a link with plasmon resonances and, hence, an indirect connection with SERS. These will only be mentioned in passing when relevant, and we list a few of them here for reference:

- Surface-enhanced infrared absorption spectroscopy (SEIRA) consists of IR absorption spectroscopy on surfaces. In its simplest form, for molecules on metals, plasmon resonances do not play a direct role in SEIRA (because they do not occur at far-infrared wavelengths). In this sense, SEIRA is only a distant cousin of SERS. However, there are crystals (like some semiconductors) that exhibit an optical response in the far-infrared that is very similar to that of metals in the UV/visible.

The resonance is then mediated by phonons rather than plasmons. It is possible in principle to exploit these phonon resonances for SEIRA in a similar fashion as plasmon resonances are used for SERS. See for example Chapter 7 in Ref. [7] for more information on SEIRA.

- Most nonlinear optical effects can also in principle profit from surface enhancements. Among these, let us mention surface-enhanced second harmonic generation (SESHG), surface-enhanced four wave mixing (SEFWM), and surface-enhanced hyper-Raman scattering (SEHRS); the latter being the closest to SERS itself. Some of these techniques are slowly growing into specific research areas in their own right at the moment.

1.3.5. Related areas

Because of the important role played by plasmon resonances, SERS is closely related to the problem of optics of metals and metallic nano-structures. From this standpoint, the field has a large overlap of interests with the fields of 'plasmonics' (or 'nano-plasmonics') [11] and nano-photonics (or nano-optics) [12]. It is possible to a large degree to consider SERS as a sub-field of plasmonics [11]. Nano-optics by itself is a field which is gaining a life of its own, with the advent of many international conferences dedicated to it and specialized books in the field [12]. The additional aspect of spectroscopy brought in by Raman scattering, along with the chemical variable related to molecular adsorption on metallic structures is what completes the SERS picture. In fact, from an entirely different point of view, SERS could be, for example, viewed as a sub-field of analytical chemistry [13]. Finally, by the very nature of the SERS probes (molecules) and SERS substrates, SERS is intrinsically part of, and has numerous connections with, the broader field of nano-science and nano-technology.

As pointed out already, SERS is intrinsically *multi-disciplinary*; and a big part of its attraction (and difficulty) stems, actually, from this fact.

1.4. APPLICATIONS OF SERS

The Raman signals from molecules reveal a distinct spectrum (as shown in Fig. 1.1), with characteristic peaks that can be observed with the appropriate lasers, spectrometers, detectors, and instrumentation. In the same sense that a given molecule will have a characteristic (and in many cases unique) infrared or nuclear magnetic resonance (NMR) spectrum, the Raman spectrum provides a 'fingerprint' of a molecule, which can be used for analytical purposes in a myriad of cases and combinations. This 'spectroscopic fingerprint' is a very valuable feature of Raman spectroscopy, which makes it a lot more specific (in terms of the information it provides) than other commonly-used techniques like *fluorescence spectroscopy*. For these reasons, the Raman effect is already exploited in a variety of applications spanning many scientific and industrial areas, see for example Refs. [14,15] for further examples.

The main obstacle against a much more widespread implementation of Raman spectroscopy in applications is that *the Raman effect is typically weak*, much weaker than fluorescence. In this context, the attraction of SERS is obvious: it promises to combine the high specificity and other advantages of the Raman technique with a much higher sensitivity, possibly comparable to that of fluorescence (and in some cases surpassing it [16]). With this in mind, the possible applications of SERS can therefore be classified into three categories, which we discuss sequentially.

1.4.1. Raman with improved sensitivity

These encompass any applications that already use Raman spectroscopy, or would use it if the signals were stronger. The gain in Raman intensity provided by SERS can then simply increase the sensitivity or detection limit of the technique. This could improve existing Raman applications, and even make it possible to apply it to systems that could not be envisaged with conventional Raman. This applies to most potential applications of SERS to analytical chemistry and biochemistry, forensic sciences, etc. and in particular to trace analysis (detection, identification, and quantification) of medicines, drugs, explosives, or, at higher concentrations, bio-fluids (e.g. glucose sensing or monitoring [17]). Molecules relevant to these applications such as glucose [17], proteins [18], DNA [19], a wide range of medicinal drugs [20–22], and substances for forensic science [23], have recently been characterized for their SERS activity. Another example is the detection and identification of dyestuffs from old artwork (paintings) and medieval manuscripts using SERS [24–26].

'Trace detection' is in fact one of the classical applications of SERS (and one of its main driving forces). In this case, the strategies are quite different from those used for fundamental studies, in the sense that we do not choose the optimum analyte to exploit the technique with its best performance, but rather the molecules are defined by the application itself (detection of specific illegal substances, for example). In this case, the science is concerned with the optimization of the experimental conditions (substrates, laser excitation, etc.) to optimize the signals and enhance the detection limits.

1.4.2. SERS vs fluorescence spectroscopy

A second broad class of applications comes from the use of SERS probes as (hopefully better) substitutes to fluorophores in many fluorescence-based applications, in particular in biology. In fact, as far as applications are concerned, fluorescence spectroscopy has always been looked at as the 'biggest competitor' to SERS. The weakness of the Raman effect is the main reason why Raman spectroscopy, despite its superior specificity, is much less widespread than fluorescence; SERS can in principle solve this problem [16]. Both techniques have actually advantages and shortcomings, but let us attempt a simple comparison here, starting with the advantages of fluorescence over SERS:

- Fluorescence from a good fluorophore is very efficient, and allows for routine single-molecule detection. Although the same can be achieved with SERS, it is not as straightforward in its practical implementation, and the necessary presence of the SERS substrate adds an additional potential complication.
- Fluorescence is currently a well-established technique with fully packaged instrumentation and countless available probes and chemical tools to attach them where needed. Compared to it, SERS appears *prima facie* at a much less developed level in these aspects.

These important advantages are counter-balanced by a number of additional features that SERS may offer:

- Probably the most attracting aspect of SERS is its high specificity, providing a unique 'fingerprint' of the molecule under study. This makes it easier to distinguish the SERS signature from any spurious background signals (a common problem in biological environments). It also allows for the possibility of high-level multiplexing (simultaneous monitoring of many different probes or tags).
- Another attractive feature of SERS is that it can be directly applied to any molecule, whereas fluorescence requires the presence of a fluorophore. This difficulty is usually overcome by 'tagging' various biomolecules with fluorescent tags. One could also envisage 'tagging' with good SERS probes, but avoiding this step altogether is obviously an attractive possibility.
- SERS can in principle work at any excitation wavelength (with the appropriate substrate), whereas fluorescence is typically limited to the visible. Near-infrared or even infrared excitation may in some situations be the only possibility (for example, because of the large optical absorption of living tissues in the visible). This limitation may be complicated or even impossible to handle with fluorescence-based techniques, but not with SERS.
- Finally, one well-known shortcoming of fluorescence is the problem associated with the fluorophore stability (i.e. photo-bleaching), especially in single-molecule applications. It is often argued that SERS would solve this issue, but this is debatable. Photo-bleaching does occur under SERS conditions too. Therefore, although there is scope for improved photo-stability using SERS, it remains an open issue.

Ultimately, the perceived competition between SERS and fluorescence-based techniques is to some degree artificial. The hope at the moment is that SERS will contribute to fill the gaps that fluorescence cannot easily cover, and expand into areas that have not been yet explored and where the unique characteristics of Raman spectroscopy can be exploited.

1.4.3. Applications specific to SERS

There is a number of additional possible applications of SERS that could not have been envisaged with conventional Raman spectroscopy or fluorescence. These include those related to the presence of the metal surface. The SERS signal emitted by the analytes can be used as an indirect probe of the surface properties, including any surface chemistry, or the analyte adsorption properties. Similarly, it can be used as a probe of the electromagnetic response of the metallic substrate, i.e. a tool to study plasmon resonances (in particular local field properties associated with plasmon resonances).

Another group of SERS applications is associated with the single-molecule detection capabilities of the technique. This for example opens up the possibility of detecting and identifying a single DNA base; a first step toward a potential single DNA strand sequencer. Single-molecule SERS detection is still the subject of intense fundamental research nonetheless, and such applications (although proposed in the literature) are at this stage only speculative.

1.5. THE CURRENT STATUS OF SERS

Before moving into the full description of the many aspects of SERS in the forthcoming chapters, we complete this general introductory material with a brief reflection on the current status of SERS (at the time of writing this book), starting by placing it into its historical context.

1.5.1. Brief history of SERS

The historical developments of the SERS effect have been described and reported many times [4,7,27,28]. We shall give here, accordingly, only a brief account of events and refer the interested reader to the original articles.

The discovery of SERS

In the early 70s, several research groups were trying to study possible ways to observe molecules on surfaces at the single monolayer coverage level. This objective had been already achieved at the time for other types of optical techniques (like infrared spectroscopy). Nonetheless, it was initially thought that observing Raman scattering from a monolayer of molecules on surfaces would not be a feasible option, due to the weakness of the Raman signal (and the limited sensitivity of the Raman systems at the time). Ref. [27] describes in some detail the original ideas at the time.

The work on pyridine on electro-chemical roughened silver electrodes by Fleischmann et al. [1] in 1974 (which is considered now to be the first ever reported observation of the SERS effect) was presented and interpreted as due to an increase in the effective surface area caused by the roughening of the electrodes. In historical terms, the concept of a surface enhancement other than a resonance (Raman) effect was not envisioned as possible. The original idea of an increase in the effective surface area due to roughness was appealing for its simplicity, and was also timely at a moment when theories of 'fractal effects' started to proliferate in many disciplines. The increased *number of molecules* resulting from the larger surface area seemed to be the most natural explanation for the observed signal enhancements.

However, there were many unanswered questions that could not be explained by this simple hypothesis. Among them, the striking observations were that the signal could actually weaken under increasing roughening in some circumstances and the lack of a capacitive effect in the electro-chemical electrodes. If the signal was purely due to an increased number of adsorbed molecules, they should have formed a layer on the electrodes that could have been detected easily as an additional capacitance in the system. Two independent (and almost simultaneous) papers by Jeanmaire and Van Duyne [2] on one side, and Albrecht and Creighton [3] on the other, provided a demonstration that the observed signal levels could *not* be accounted for by an increased surface area. These papers hence suggested that a different form of enhancement acting directly on the Raman intensities of the molecules had to be invoked. Several articles [27–29] from some of the pioneers in the field recount the early developments and the way the ideas developed from puzzling experimental observations to the present understanding (arguably, still incomplete in many aspects).

The following thirty years

The decade following the discovery of SERS was the subject of intense research activity. Actually, most of the important theoretical aspects of SERS, and to some degree the content of this book, find its origin in this 'early work'. The landmark review by Moskovits in 1985 [4] provides a great summary of the activity up to that time. Many aspects are also discussed from a more modern perspective in Ref. [7].

The excitement of the discovery appears to have faded in the following decade, with a more latent, but still persistent activity in the field. The interest in SERS has however been rekindled in the last decade, and several facts may have contributed to this:

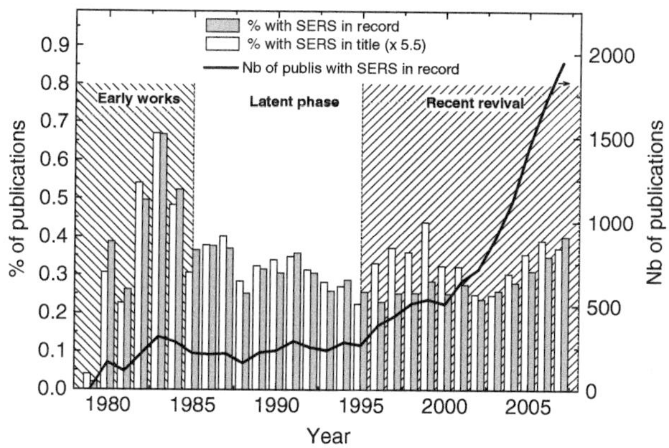

Figure 1.2. Bibliographic data for SERS publications since 1979. The data were compiled on 24/04/2008 from the Scopus database (http://www.scopus.com). The 'full-record' searches were carried out with the keyword 'surface-enhanced raman' or any of its hyphenated derivatives. For the 'title-only' searches, the acronym 'SERS' was also included. The percentage of publications was calculated with respect to the total number of publications in each year in the database. A multiplicative factor of 5.5 was applied to the 'title-only' searches to make it match the 'full-record' results. Finally, a multiplicative factor of 4 was applied to all 'full-record' searches before (and including) 1995, to account for the limited full-text abstract search capabilities of the database before 1995.

- Improvements in Raman instrumentation have made the technique much more easily available to a wider community of scientists from various backgrounds: physics, chemistry, biology, engineering, etc.
- The dramatic development of nano-science and nano-technologies has opened up many new possibilities, notably in terms of SERS substrate design and fabrication. Similarly, the increased research activity in plasmonics is indirectly connected to many fundamental and applied aspects of SERS.
- The two independent proposals [30,31] that SERS might be capable of single-molecule detection has also dramatically stimulated the field, both in terms of theoretical and experimental studies.

As a result, SERS has dramatically expanded its visibility across many disciplines. Although much work is still being carried out in terms of the fundamental understanding of the effect, many more publications are now entirely dedicated to its applications, either as a research tool or even toward its commercialization.

These phases are reflected in the number of SERS publications, as illustrated in Fig. 1.2. The data in this plot, explained in the caption, are interesting in many ways. Firstly, they show that in order to assess the importance of a field or technique, the number of publications in itself is not really meaningful, but must be renormalized to the total number of publications (in the same database, for example), a figure that is currently increasing every year. Secondly, the comparison between the appearance of SERS in the publication title or in the full publication record is instructive: it indicates in our opinion whether the publication is really dedicated to SERS or only makes use of the technique. With this in mind, the various phases discussed earlier, i.e. early works, latent period, and revival, are evident in the graph of Fig. 1.2. The last phase also exhibits an interesting feature: a short period of increased interest in dedicated studies about SERS itself appears around 1997 as seen in the spike in 'SERS in title' vs 'SERS in record'. The actual recent trend of revived interest in SERS and its application actually only started around 2003. Interestingly, in proportion of the total research output, it still remains smaller than the activity recorded in the early 80s; this fact is possibly a reflection of the different total number of research topics in the two different periods separated by three decades.

1.5.2. Where is SERS now?

With the benefit of hindsight (from what happened in other techniques), it is possible to draw some imperfect analogies and claim that SERS is perhaps today at the level of development that Nuclear Magnetic Resonance (NMR) was in the early 70s. At that stage, it was already obvious that NMR had a great potential as an analytical tool with many different ramifications (including imaging). Still, many basic fundamental aspects of the techniques had to be pinned down properly (in particular regarding the imaging capabilities), and some of the basic science and instrumentation was still under development. It would also have been very difficult to imagine thirty years ago a modern NMR spectrometer, like the ones available at present. If the analogy is valid at all, it is quite obvious today that SERS has a great potential to become an established analytical tool, with the desirable capability of analytical studies down to the single-molecule level. As with NMR, many of the achievements are strongly linked to improvements in instrumentation, that did not exist at the time when the effect was discovered for the first time. In the case of SERS, Raman instrumentation has already made great strides forward in the last few decades. A lot more advances are probably to be expected, for example in the area of substrate fabrication (a very active field in recent years). But it is interesting to note that some of the basic science of SERS is still under development, with many important issues needing to be resolved. Whether the historical parallel with more developed techniques like NMR will live to its promises remains to be seen, and we hope that the content of this book encourage new practitioners to use, improve, investigate, and further develop the technique in years to come.

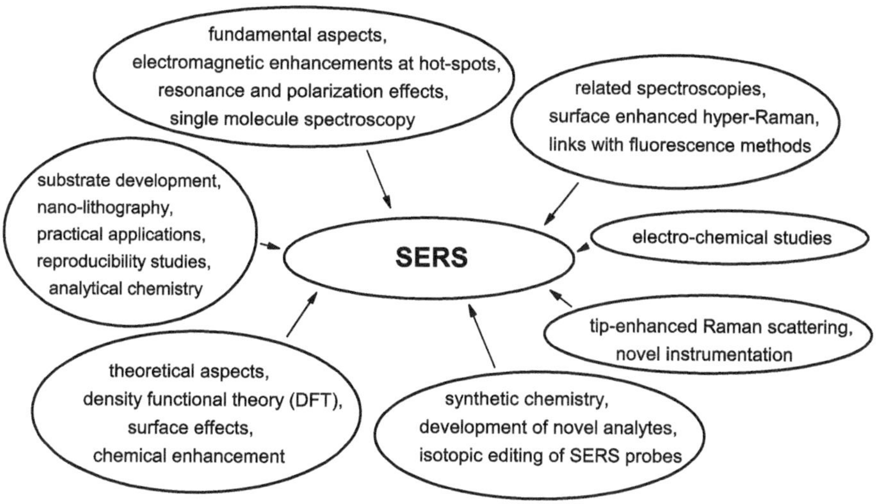

Figure 1.3. Schematic classification of some of the main areas contributing to SERS at present.

1.5.3. Current 'hot topics'

Chapter 8 is fully devoted to recent advances in fundamental research, instrumentation, and applications of SERS. However, in the general spirit of this introductory chapter, we shall make a few comments here on current 'hot topics'. Figure 1.3 shows a schematic (and self-explanatory) representation of some of the most important areas of research and applications that are contributing to SERS at the moment. The separation of topics in Fig. 1.3 is, of course, arbitrary and it is likely that many specific applications of the technique lie either at the boundaries of this classification, or belong to more than one of them. We provide in what follows a few additional comments to reinforce some of the content of Fig. 1.3 on selected current areas of research.

Fundamental studies

In terms of the study of the fundamental mechanisms of the effect, one of the areas that has showed the largest amount of progress in recent years is, undoubtedly, the problem of *single-molecule SERS* (SM-SERS). The single-molecule problem is interesting for both fundamental and applied reasons. It is only in the last ten years that a wealth of activity in SM-SERS has been developed, with many conflicting results and a lively scientific debate [30–36]. These are discussed in more detail in Section 8.1.

Without attempting to be exhaustive, we mention that other recent fundamental studies include:

- A number of experimental verifications of the SERS electromagnetic model (described in detail in Chapter 4) using various approaches [8,38–49].
- Ongoing theoretical and experimental investigations of the SERS chemical enhancement (discussed in Section 4.8) [50,51].
- Theoretical and experimental studies of polarization effects and surface selection rules in SERS [48,49,52–57].
- Studies of vibrational pumping under SERS conditions [58,59].

Other theoretical aspects

A brief outlook of theoretical activities can also be given at this stage to convey a general impression without going deep into the details. Theoretical aspects of SERS can be concentrated mainly into two categories: (i) calculations of the enhancement factors (which typically involve concepts from plasmonics), and (ii) quantum chemical calculations of the electronic and vibrational structure of molecules, by themselves or in the presence of metallic surfaces [60,61]. The latter topic has a strong link with the wider field of surface science and molecules at interfaces, as well as the understanding of topics that are directly linked to the SERS effect (like the chemical enhancement, discussed in Section 4.8).

Novel SERS substrates and reproducibility

A significant fraction of the recent advances in the field have been concentrated into the production of novel SERS substrates [13,17–23,62–66], with a particular emphasis on standardization, uniformity, and reproducibility. The problem of reproducibility has been for many years a major issue in SERS, and one that has hampered the transition from the 'potential' of the technique to 'real applications'. There are many possible approaches to this problem, and this is where the links between SERS and many areas of engineering are perhaps the strongest [62]. As mentioned already in the example in Section 1.2.3, most of the time there is a complementarity between 'reproducibility' and 'enhancement': the '*SERS uncertainty principle*' as dubbed recently [67]. In general, this means that the more reproducible the enhancement, the less the amplification (enhancement factor) of the Raman signals. In the opposite limit, large enhancement factors (required for single-molecule detection, as in Fig. 1.1) are in general a lot more difficult to reproduce, and one has to resort to statistical analysis to understand the meaning of the signals. It is important to stress, however, that even in the simplest cases of more reproducible 'low enhancement' situations, SERS can still improve the sensitivity of detection of certain analytes by several orders of magnitude compared to conventional Raman techniques.

Accordingly, the technique finds already many applications despite the fact that several fundamental aspects of the practical implementation have not been yet fully resolved.

Tip-enhanced Raman spectroscopy

Other techniques have been developed from SERS by (sometimes simple) instrumental variations of the effect. An example of this latter category is the emerging field of *tip-enhanced Raman spectroscopy* (TERS), which is basically identical to SERS except for the fact that the SERS substrate is brought to the analyte, rather than the conventional transfer of the analyte onto the substrate. This is achieved by creating a 'hot-spot' between a metallic tip and a substrate. The tip normally comes from a scanning tunneling microscope (STM), or an atomic force microscope (AFM) [68], which allows for an exquisite control over the hot-spot position and its characteristics (mainly defined by the distance between tip and substrate). Hence, this is more than just a different way of achieving the conditions for the SERS effect, for it could allow (in principle) the combination of spectroscopy (SERS) with microscopy (AFM or STM), possibly at a single-molecule level. TERS is a technique that is (relatively speaking) still in its infancy [69] but will surely see a wealth of activity in years to come.

1.6. OVERVIEW OF THE BOOK CONTENT

The brief overview of the technique presented so far was aimed at giving a flavor of the basic phenomenology and the diversity of subjects that are involved in SERS. We hope to have conveyed the impression that the technique is truly general and, at the same time, that there is still a lot of basic science that needs to be surveyed to improve our understanding (even on very elementary topics). SERS is expected to evolve into a standard technique for many applications and the links represented in Fig. 1.3 are likely to expand, and will surely have to be revisited in the future.

We now explain briefly the overall structure of the book and how the material was split into different chapters. After that, we shall suggest several possible *reading plans* of the book, depending on the different needs of potential readers, and the level of depth that is desired.

1.6.1. General outline of the book

The introduction to Raman spectroscopy in Chapter 2, together with the introduction to plasmonics in Chapter 3, lay out the most fundamental principles which define the SERS effect. These two ingredients – Raman scattering and plasmon resonances – are then connected in Chapter 4 to describe in detail the electromagnetic mechanisms giving rise to SERS and related plasmonic effects; such as surface-enhanced fluorescence (SEF). The rest of the book is then dedicated to a more practical understanding of these effects. In Chapter 5, we focus on approaches to model electromagnetic effects for SERS, and we then discuss in Chapter 6 their link to plasmon resonances and their many important properties in detail. Chapter 7 reviews a number of effects relevant to SERS substrates, with a special emphasis on colloidal solutions. Finally, in Chapter 8, we review some (but not all) of the most exciting recent developments in SERS, and discuss them in the light of the understanding gained in the previous chapters.

The appendices of the book are almost exclusively dedicated to more advanced topics. There are included to provide in-depth background material on specific topics. Appendix A (practical aspects of Density Functional Theory (DFT) for applications in Raman spectroscopy) and Appendix B (bond-polarizability theory) can be read in conjunction with Chapter 2 on Raman spectroscopy. Appendix C (a brief overview of Maxwell's equations), and Appendix D (elements of polarizability theory) provide an introduction to general electromagnetic theory. The other appendices are maybe more technical but also more practical. They summarize (and justify) all the necessary expressions for numerical implementations of various EM problems of relevance to SERS and plasmonics. They can be used in conjunction with the Matlab codes that can be found on the book's website. Appendix F (reflection/refraction of waves at planar interfaces) is particularly relevant to Chapter 3 on plasmonics and both can therefore be read in conjunction. Appendices E (model dielectric functions of silver and gold), G (optical properties of ellipsoids in the electrostatic approximation), and H (Mie theory) can be used as references for EM calculations.

1.6.2. General 'spirit' of the book

We have tried to present material in this book that *is not conventionally found in other sources*, especially regarding the background material necessary to understand SERS itself. In that respect, we have chosen to present the different subjects from alternative points of view, hopefully more directly suited to SERS itself. This is particularly challenging for well-established topics like Raman spectroscopy (Chapter 2); but we believe that the approach we followed produced a much more specialized and focused outlook of the field; which is more directly related to SERS and its applications.

Another aspect that we particularly emphasize throughout is the intricate link between SERS and the wider research field of *plasmonics*, i.e. the study and applications of the optical properties of metals. In fact, SERS uses the same raw ingredients of plasmonics (metals and light), requires similar substrates, etc. Some of the tools of plasmonics (substrate fabrication, electromagnetic modeling, etc.) are, in fact, directly relevant to SERS. Other sub-fields of plasmonics, such as surface-enhanced fluorescence (SEF), are based on exactly the same underlying principles. Moreover, SERS can be used by itself as an additional characterization tool of plasmonic structures. This symbiotic relation, therefore, is stressed in as many places as possible. This is also the reason why we dedicate a lengthy chapter to the principles of plasmons and plasmonics (Chapter 3).

As always happens in most books – and particularly so in a subject of such inter-disciplinary nature as SERS – it is a difficult exercise to assess the appropriate level of the material to be included in each chapter. Some chapters/sections may appear too trivial for some and too complicated for others. We have attempted when possible, and especially in the first few chapters on the fundamental principles, to adopt a *graded approach*: we start with a description of the concepts in simple terms, and then present the 'proper' (more complicated) approach. The detailed understanding of several important aspects of the electromagnetic theory of SERS does require some advanced electromagnetic skills and knowledge. Rather than ignoring them for the sake of simplicity, we have decided to include them in the appendices. The content of the appendices can be found in various physics textbooks, but we felt that a single 'self-contained' presentation, adapted to the needs of SERS and plasmonics (and in the context and framework of the rest of the book), will be valuable to some (and hopefully all) readers.

Finally, in many instances, and in particular in the appendices, complex analytical expressions are provided. The utility of these expressions is only realized if they are computed numerically to provide a meaningful result, such as the simulation of the outcome of an experiment. It is with this in mind that the expressions are provided, to spare the readers the need to carry out the analytical calculations, and focus on using them for their own problems. To this end, we also provide – when possible – Matlab implementations of the tools/methods encountered in this book. The corresponding source files can be found on the book's website: http://www.victoria.ac.nz/raman/book. Again, this should enable the reader who understands the concepts but does not want to do the maths, to carry out some basic EM modeling relevant to SERS and plasmonics.

1.6.3. Different reading plans

Obviously, different readers will have different needs/inclinations to cover different parts of the book and achieve different levels of depth. We suggest a few possible (but obviously not restrictive) reading plans of the material in the following:

- Basic reading: A reader willing to obtain a basic understanding of the main topics of SERS, while avoiding the more technical or difficult concepts, might consider the following reading plan. Regarding the information needed from Raman scattering, Sections 2.1, 2.2 and 2.4 of Chapter 2 contain most of the basic material required. Despite its apparent simplicity, this material covers the main aspects of the classical theory of Raman scattering needed for its generalization to the SERS case. This can be complemented with the beginning of Chapter 3 (up to Section 3.3.3), to gain a basic understanding of the optical properties of metals and what plasmon resonances are. After these two basic elements have been incorporated, the reader can proceed with the simplest examples of enhancement factors (EFs) in Section 4.1–4.3 and 4.8 of Chapter 4, which include a simple description of their physical origin and of the experimental aspects of the problem. Chapters 5 and 6 can be avoided in the first read, unless the reader is willing to explore the details of how the electromagnetic enhancements are calculated in practice and how they relate to plasmon resonances. The reading can then continue with the first two sections of Chapter 7, which are mainly descriptive and provide all the necessary information to gain a good understanding of metallic colloids and other SERS substrates. Finally, Chapter 8 provides an (optional) overview of interesting and important areas of current research. All the appendices can be avoided in this reading plan, and this would complete a basic understanding of SERS and its implementations.

- More advanced readings: By definition, everything that is added to the above 'basic plan' becomes a 'more advanced plan'. The question is then which aspect a potential reader wants to emphasize. The complementary sections 2.5–2.7 in Chapter 2, together with Appendix A, can greatly enhance the understanding of Raman spectroscopy on aspects that have direct applications to SERS (but are naturally more advanced). On the contrary, one might choose to deepen the 'electromagnetic enhancement' aspects of the problem. This can be achieved by adding to the basic reading list the rest of Chapter 4, and the whole of Chapters 5 and 6. A reader more interested in the connection to plasmonics in general can read the rest of Chapter 3, supplemented by Appendix F. Finally, advanced aspects of colloid science and its relevance to SERS are discussed in the second part of Chapter 7.
- Complementary information in the appendices: The appendices can be used in all cases as self-contained complementary 'advanced material' on specific topics. This applies in particular to the *in-depth tutorials* in Appendices A, C, F and H. Appendix C in particular refreshes the main concepts surrounding electromagnetic effects in media, from a point of view that is slightly different from the one found in conventional books on electrodynamics, and hopefully more suited to SERS. As mentioned earlier, some of these appendices (in particular E, F, G and H) will also be linked to material and computer codes that will be available from the book's website (http://www.victoria.ac.nz/raman/book), for direct use in advanced research applications.

These reading plans are obviously not exclusive, and it is perfectly possible that specific sections will cater to specific aspects of the problem. We have tried to cross-reference as many topics as possible in the *index* at the end of the book to facilitate the search for specific aspects of the problem, which can be studied as a separate unit from the rest of the material. The index is also used to compile (and define) most of the acronyms used in this book.

Chapter 2

Raman spectroscopy and related optical techniques

A book on surface-enhanced Raman scattering (SERS) cannot avoid a description of the Raman process itself, mainly because many newcomers to the field (driven by possible applications in chemistry, biology, forensic sciences, etc.) may encounter the Raman effect for the first time and may not have a comprehensive background on Raman spectroscopy. There are two possibilities at this stage and, arguably, none of them is fully satisfactory. One is to refer the reader to the tens (even hundreds!) of excellent references on Raman spectroscopy to gain a background in the field, and then try to discuss only those aspects that are relevant in a SERS context. Another possibility is to try to present a self-contained introduction to Raman spectroscopy and related optical techniques, which is more general than what is actually needed for SERS in several aspects, but covers the essential concepts and provides an unifying view of the basic material. We have chosen the second option, bearing in mind that it is probably impossible to satisfy the needs and tastes of all potential readers.

Needless to say, there have been tens of possible approaches to the field of Raman spectroscopy reported in the literature over the years [70–75]. All of them are equivalent at some level and all of them have advantages and disadvantages; here we try to highlight the aspects we judge to be more important for both molecular Raman spectroscopy and SERS. The level of presentation is always a compromise between clarity and depth. While avoiding sophisticated calculations throughout, we try nevertheless not to sacrifice depth, and we present and discuss the main formulas and their physical relevance where required.

We have attempted to provide a *gradual approach* to the subject with an underlying thread toward the ultimate goal of the book: explaining SERS and related effects. The presentation of the material in this chapter is therefore in many aspects recursive; in the sense that many times we shall come back to the same concepts but from a slightly different point of view. We start by introducing the Raman effect and other basic optical processes such as absorption and fluorescence from a relatively simple phenomenological point of view. In fact, this description – mostly contained in Sections 2.2 and 2.4 – is sufficient to move on to the description of SERS and related plasmonic effects. We do provide additional information on the same topics in the other sections, but on an increased level of complexity. The most advanced aspects of Raman spectroscopy are only treated at an introductory level in the last sub-sections of this chapter; with the practical use of density functional theory (DFT) for the prediction of Raman spectra and vibrational analysis only included as an appendix (Appendix A). These aspects are only needed for researchers in the field or for advanced readers, and none of the basic aspects will be missed if this information is skipped.

Arguably, one of the most complete descriptions of Raman spectroscopy in molecules is the one provided by D. A. Long in his book 'The Raman effect' [70]. In particular, all the detailed aspects of resonance Raman scattering and their quantum mechanical description are treated in full in Ref. [70], together with the connection to semi-classical approaches (like Placzek's polarizability theory). It would make no sense whatsoever to repeat that information in full here. We shall instead cite or make reference to results in more advanced textbooks on Raman spectroscopy when we avoid the derivation of the more sophisticated aspects of the theory. Readers with a strong desire for a more comprehensive quantum mechanical theory of the effect (in particular under resonance conditions) are strongly encouraged to consider the contents of Ref. [70], which is an excellent reference in any case as a long-standing source of in-depth material on molecular Raman spectroscopy.

Finally, SERS is about Raman scattering in the presence of metallic nanostructures that will produce an *enhancement* of the Raman signal. We have chosen to explain the Raman effect first, but in doing so, we will have to mention the existence of the 'enhancement' in SERS before we actually show its origin. The reader should take any reference to the 'enhancement' or 'enhancement factor' at this stage as a comment in passing for something that will be explained in detail in the following chapters.

2.1. A BRIEF INTRODUCTION

2.1.1. The discovery of the Raman effect

What we call today *Raman scattering* is an effect discovered by Chandrasekhara Venkata Raman [76] in 1921 in India[1]. It is interesting to

[1] The phenomenon had been predicted theoretically by Smekal; it was observed by C.V. Raman and K.S. Krishnan in India and almost simultaneously by G. Landesberg and L. Mandelstan in Russia. The succession of historical events that led to the discovery of the Raman effect has been described many times in the literature, for example in Ref. [77].

note that Raman himself observed the effect with very rudimentary optical tools (essentially color filters) by using the sun as a light source and his eyes as detector [76]. The discovery won him the Nobel prize in physics in 1930 'for his work on the scattering of light and for the discovery of the effect named after him'[2]. Moreover, what he observed is what we call today 'anti-Stokes Raman scattering', which is normally quite weak. For anti-Stokes Raman scattering, the scattered photons have an energy that is *larger* than the incoming ones (following interaction with an excitation already present in the sample). In fact, it was this up-conversion process in energy that convinced Raman that he was in the presence of a new optical scattering phenomenon. A down-conversion in energy (such as Stokes Raman scattering) could have been caused by and/or confused with other more efficient optical processes, such as fluorescence, already well-known at the time of the discovery. The 'up-conversion' Raman observed by eye in several organic liquids [76] implied an *inelastic* scattering process. Raman called it 'A New Type of Secondary Radiation' in his original paper [76] and this is what we now call *the Raman effect* or *Raman scattering*. The measurement and analysis of the signals (photons) arising from the Raman effect is called *Raman spectroscopy*.

2.1.2. Some applications of Raman spectroscopy

Raman spectroscopy is such a general technique that even reviewing briefly what can be done with it (and the different fields where it finds applications at present) would require a book by itself. Instead, we base the description here on summaries of recent advances that can be found in the literature (and will not be repeated here, accordingly). As an illustration, W. Kiefer has recently published a summary [78] of recent advances in Raman spectroscopy with over 300 references of key developments published *only* in the *Journal of Raman Spectroscopy* (Wiley) in the last few years. A quick scan of Ref. [78] shows that recent advances (other than SERS) in the application of Raman spectroscopy have been produced in the fields of: (i) art and archeology, (ii) biosciences, (iii) vibrational studies and analytical chemistry, (iv) solid state physics (minerals, crystals, glasses, ceramics, etc.), (v) liquids and liquid interactions, (vi) nano-materials (nano-tubes, nano-particles, etc.), (vii) phase transitions of various kinds, including liquid crystals (viii) resonant Raman scattering, (ix) pharmaceutical studies, (x) high-pressure physics and chemistry, (xi) forensic science, etc. All these topics are without including the additional degrees of freedom that the technique allows in some non-linear versions of Raman spectroscopy (like coherent anti-Stokes Raman scattering, CARS), or in time-resolved Raman spectroscopy. It is not an exaggeration to claim that Raman is a truly universal type of spectroscopy and most conceivable materials

[2] For more details, see for example http://nobelprize.org

will produce a Raman signal. Whether the information provided by Raman spectroscopy for a given sample is the most interesting one for a specific application is an altogether different question; but it remains true that Raman spectroscopy can *a priori* be applied to materials as diverse as foodstuffs, biomolecules, crystals, liquids, ceramics, gases, etc. It is also non-invasive and allows remote sensing through appropriate optics and instrumentation. It is timely perhaps to echo the words of W. Kiefer in the concluding remarks of Ref. [78] when he states: *sometimes I am asking myself what cannot be done with Raman spectroscopy.* The perspective at the time of writing this book is that of a field with an explosion of activity in many areas, and an ever increasing base of practitioners. As a matter of fact, SERS, as a technique intimately related to Raman spectroscopy, has a lot to do with this increase in activity.

2.1.3. Raman spectroscopy instrumentation

The dramatic development of Raman spectroscopy for research and industrial applications has been possible largely thanks to the progresses made in the instrumentation. C.V. Raman made his measurements on a number of highly-concentrated substances (mostly pure organic solvents), and he was therefore capable of seeing the effect by eye. However, in the vast majority of applications (where diluted solutions or solids are used), the Raman effect is either 'weak' or 'very weak'. Many applications were therefore delayed for several decades, until the discovery of the laser together with the development of more efficient detection systems substantially improved the sensitivity and scope of Raman spectroscopy. Modern implementations have replaced sunlight by lasers (with the advantages not only of very large intensities, but also of monochromaticity; two extremely important characteristics for the detection of Raman signals) and sophisticated detection systems with combinations of notch filters, holographic gratings, high throughput monochromators, photomultipliers and/or multi-channel charge-coupled devices (CCD) as detectors. It is now possible to observe and study Raman signals from materials that would have been completely out of reach by many orders of magnitude at the time Raman discovered the effect. Moreover, with the advent of commercial 'ready-to-use' Raman spectrometers, and even portable systems, the technique becomes increasingly available to a wider range of users [14]. We shall not further review any of the 'technical' aspects of the equipment. We refer instead the reader to, for example, Ref. [71] for more technical details on the Raman instrumentation, and Ref. [14] for a recent review of commercial Raman systems. Other resources from which to get a flavor of current techniques, methods, and available 'hardware' are specialized scientific journals such as (for example) *Vibrational Spectroscopy* (Elsevier) or the *Journal of Raman Spectroscopy* (Wiley).

2.2. OPTICAL SPECTROSCOPY OF MOLECULES

Optical spectroscopy broadly consists in the analysis of the optical spectrum generated by a sample under some experimental conditions. By generalization, it can be loosely defined in simple terms as the study of the interaction of light with matter. Many forms of spectroscopies exist, depending on the type of the sample under investigation, and the underlying physical processes that occur during the interaction of light with matter: absorption, emission, or scattering; linear or non-linear processes; etc. Entire textbooks are dedicated to molecular spectroscopy, or even to some of its sub-fields [71,79,80]. We attempt to provide in this section a brief overview, at a relatively simple level, of those aspects of molecular spectroscopy that are most relevant to SERS and plasmonics:

- We start with the energy levels in molecules (Section 2.2.1), with a brief digression about spectroscopic units (Section 2.2.2).
- We then describe the main optical processes: absorption (Section 2.2.3), luminescence (Section 2.2.4), and scattering (Section 2.2.5).
- To quantify these, we then introduce the concept of cross-section, focusing in particular on the rigorous definition of the Raman cross-section (Section 2.2.7), along with its experimental measurement (Section 2.2.8).
- We conclude by a discussion of a few simple mechanical analogs of these optical processes.

2.2.1. The energy levels of molecules

When dealing with molecules, the interaction with light is primarily determined by the energy levels of the degrees of freedom of the molecule. These may be either associated with the movement of the electrons (electronic energy levels, possibly further differentiated by their spin), or of the atoms in the molecule (vibrational, rotational, or translational energy levels, i.e. motional energy states).

The Jablonski diagram

A convenient way to visualize these energy levels, and the various transitions among them, is through the use of a Jablonski diagram [71,79,80] (after the Polish physicist Aleksander Jabłoński). Such diagrams are widely used in molecular spectroscopy. An example is shown in Fig. 2.1, and it will illustrate many aspects of the forthcoming discussion. The electronic states of the molecule are schematically represented as bold curves showing the energy of the state as a function of the nuclear (atomic) coordinates (which

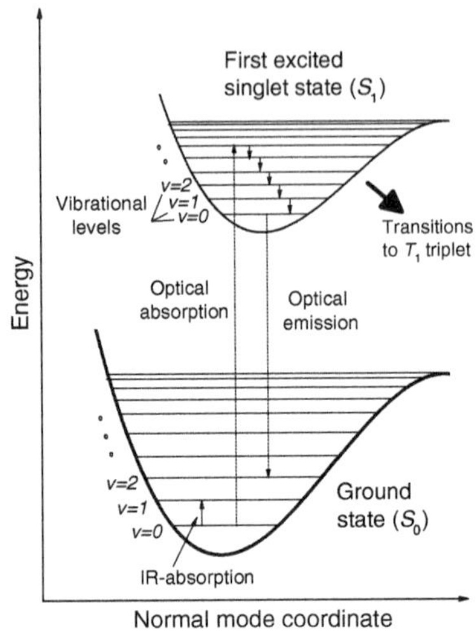

Figure 2.1. Schematic representation of electronic (bold lines) and rotational/vibrational (thin lines) energy levels of a molecule in a Jablonski diagram. Possible transitions between states are indicated by arrows, either as dotted lines for radiative (dipole-allowed) transitions or solid lines for non-radiative transitions.

is schematically represented in the figure as a single variable, the *normal mode coordinate*). The energy minima in these curves then correspond to the equilibrium position of the atoms. Motional states (typically only vibrational states) for each electronic state are shown as thin lines. Transitions between states are represented by arrows. In simplified versions of the Jablonski diagram, which we will use later, the atomic coordinates' dependence of the electronic energy is not shown and the equilibrium position is then assumed. All energy levels (electronic and vibrational) are then represented by horizontal lines.

Electronic states

For a molecule in its *ground electronic state*, or ground state, the electrons occupy their lowest energy state (as allowed by the Pauli exclusion principle). In general, this means that all electrons are paired (with two electrons of opposite spins[3]). The ground state is then called a *singlet state* (S_0) , to indicate that its total spin is zero. When one electron in a pair is transferred to its excited state, the Pauli exclusion principle no longer prevents the two

[3] There are a few prominent exceptions, notably molecular oxygen, which is a triplet in its ground state, i.e. the orbital wave-function of the ground state of O_2 is in the *anti-bonding* configuration. This allows the two electrons to be with the same spin, thus conferring O_2 with a triplet ground state. This is partly responsible for the high 'reactivity' of O_2.

electrons from having the same spin (because they are in different states). The *electronic excited state* is therefore usually four-times degenerate, because there are four possible spin states, which are commonly grouped as: a *singlet state* (S_1) corresponding to the only excited state configuration of total spin zero, and a *triplet state* (T_1) corresponding to a triply-degenerate excited state with a total spin of one. In the Jablonski diagram, the electronic states are arranged vertically by increasing energy, and grouped horizontally by spin multiplicity, i.e. the triplet state (T_1) is represented (when needed) to the right of the singlet state S_1, itself represented above S_0. Note that in many molecules, the triplet state T_1 is not perfectly degenerate with S_1, and typically lies below S_1 (in energy). Higher-energy states, denoted S_2, T_2, etc. may also be shown if necessary.

Motional states

The motional energy states correspond to the degrees of freedom associated with translations (usually irrelevant), rotations of the molecule as a whole, and vibrations of the constituent atoms (around their equilibrium positions). These internal atomic vibrations are, in a first approximation, a 'small' perturbation to the electronic state. This is the basis for the *Born–Oppenheimer approximation* , which allows for a certain 'separation' of the problem into its electronic and atomic contributions. Since electrons are much 'lighter' than nuclei, we can think of the electrons as reacting instantaneously to any change in the atomic positions.

The energy of the vibrational states is typically smaller than those of the electronic states and they can be considered as a sub-structure of the latter, as seen in Fig. 2.1. For organic molecules (except for the smallest), many vibrational states exist over a typical energy range of ≈ 1500 cm^{-1} $\sim$ 0.2 eV. To this vibrational energy level structure one can superimpose the rotational levels (if the molecule is free to move). The combination of rotational/vibrational sub-structure typically forms a quasi-continuum of energy states. In reality, vibrational (and rotational) states are usually (slightly) coupled to electronic states. They cannot therefore be (strictly speaking) separated, and constitute what is known as a *vibronic state*. The energy levels in Fig. 2.1 specify a particular excitation state of the molecule, and they are a combination of both an electronic and a vibrational state.

Transitions between molecular states

Various interactions can induce transitions between molecular states. Because energy must be conserved, every transition is associated with either absorption or loss of energy. If the transition occurs between two different electronic states, it is usually called an *electronic transition*. Transitions between two states in the sub-structure of the same electronic state may be called internal or vibrational transitions. These transitions can further be classified in our context into two groups: *radiative transitions* if they involve interaction with a photon, and *non-radiative transitions* otherwise:

- *Radiative* (i.e. dipole-allowed) transitions are indicated by straight vertical (dotted) arrows in Fig. 2.1 and they involve the absorption (if the transition occurs to a higher-energy level) or the emission (in the reverse case) of a photon. Radiative transitions are in principle 'forbidden' between singlet and triplet states because a photon cannot introduce a change in the total spin. Such spin-forbidden transitions may however become weakly allowed if a spin relaxation mechanism is present (for example spin–orbit coupling).

- *Non-radiative* transitions may be the result of interactions with the environment (solvent, other molecules) or of internal interactions. Examples of the latter are internal conversion (or relaxation) [79] or intra-molecular vibrational redistribution (IVR), which ensures relaxation from excited vibrational states to the ground vibrational state (i.e. the bottom of the electronic state). This occurs by redistributing energy into lower-energy vibrational or rotational states, on a typical timescale of $\sim 10^{-12}$ s. Non-radiative transitions can also occur to higher-energy levels, through thermal activation (by an amount comparable to the average thermal energy $\sim k_B T$). Finally, non-radiative transitions may also occur between singlet and triplet states.

A transition (radiative or non-radiative) between singlet and triplet states is usually referred to as *inter-system crossing* [71,79,80] (ISC), which represents more generally a transition to a state with a different spin multiplicity. In typical situations, ISC is much less likely than radiative transitions among singlet states, and the triplet states can often be 'ignored'. However, ISC plays a major role in some processes like phosphorescence or photo-bleaching[4].

2.2.2. Spectroscopic units and conversions

Several units are commonly used for the transition energy and wavelength, depending sometimes on the type of spectroscopy, but also on the tastes of the users. We summarize briefly here the most important ones for reference and convenience.

[4] Triplet states are typically unstable and can result in photo-dissociation or photo-bleaching [81–83]. This instability is often associated with a larger 'chemical reactivity', in particular to the presence of oxygen. Many SERS probes of interest, for example, do have markedly different photo-bleaching rates when they are exposed to an either oxygen-rich or oxygen-deficient atmosphere, respectively. We come back to this subject in Section 2.3.3.

A photon (or transition) of energy E [J] can otherwise be specified by:

- its angular frequency ω [rad s^{-1}] (often called frequency for short),
- its frequency $\nu = \omega/(2\pi)$ [Hz or s^{-1}],
- its wavelength $\lambda = c/\nu$ [m],
- its (absolute) wave-number $\bar{\nu} = 1/\lambda$ [m^{-1}],
- or (less often) an effective temperature T [K], i.e. the temperature that corresponds to the same thermal energy.

These are related by the standard expressions:

$$\boxed{E = h\nu = \hbar\omega = \frac{hc}{\lambda} = hc\bar{\nu} = k_B T,} \tag{2.1}$$

where:

- $h \approx 6.626 \times 10^{-34}$ J s^{-1} ($\hbar = h/(2\pi) \approx 1.0546 \times 10^{-34}$ J s^{-1}) is the (reduced) Planck constant,
- $c \approx 3 \times 10^{8}$ m s^{-1} is the speed of light,
- and $k_B \approx 1.381 \times 10^{-23}$ J K^{-1} is the Boltzmann constant.

Note that E is proportional to ω, ν, $\bar{\nu}$, and T, all of which can therefore be viewed as a measure of energy (with some appropriate units). This is not the case for λ, which is inversely proportional to E.

Other common units for these quantities are electron-volt (eV) or milli-eV (meV) for E, nano-meters (nm) for λ and cm^{-1} for wave-numbers $\bar{\nu}$. Here are some useful conversion expressions:

$$\bar{\nu}[\text{cm}^{-1}] = \frac{10^7}{\lambda[\text{nm}]}, \tag{2.2}$$

$$E[\text{eV}] \equiv \frac{1239.8}{\lambda[\text{nm}]} \equiv 1.2398 \times 10^{-4}(\bar{\nu}[\text{cm}^{-1}]) \equiv 0.025\frac{T[\text{K}]}{290}. \tag{2.3}$$

Useful figures to remember (and other expressions simply deduced by proportionality) are:

- 1 eV corresponds to $\bar{\nu} \approx 8065.5 \approx 8000$ cm^{-1}.
- 1 eV corresponds to $\nu \approx 2.418 \times 10^{14} \approx 2.4 \times 10^{14}$ Hz.
- 1 eV corresponds to $\omega \approx 1.519 \times 10^{15} \approx 1.5 \times 10^{15}$ rad s^{-1}.
- 1000 cm^{-1} correspond to $E \approx 123.98 \approx 124$ meV.

- 290 K ($\sim$ room temperature) correspond to 25 meV and to $201.56 \approx 200\ \text{cm}^{-1}$.

In addition, the common laser excitation line at 633 nm (He–Ne-laser) corresponds to ≈ 1.96 eV, while 514 nm (Ar^+-ion laser) to ≈ 2.41 eV.

In the following few sub-sections, we discuss in simple terms the nature of the simplest radiative transitions that may occur in a typical molecule (and in many other optically active systems, such as atoms or crystals). A more quantitative description of the most relevant processes will then be given in later sections.

2.2.3. Optical absorption

Perhaps the most intuitive optical process is that of *optical absorption*. A photon of energy E can excite a molecule from a level of energy E_a to a higher level of energy E_b. The photon energy is transferred to the molecule and the photon is therefore 'absorbed'. Moreover, energy conservation implies that $E = E_b - E_a$. Such transitions are possible only between states with specific properties that ensure that the photon can indeed couple to the transition (the so-called dipole-allowed transitions). Without going into more details, let us simply mention that these requirements represent the selection rules for optical absorption.

Let us mention two important types of optical absorption:

- Electronic absorption, whereby an electron is excited to a higher electronic state. In its simplest (and most common) form, a molecule in its ground electronic state S_0 is excited to its first excited state S_1. Such transitions typically occur in the UV ($\sim$200–400 nm) for small molecules, or in the visible ($\sim$400–800 nm) for dye molecules (this is directly linked to their color and that is why they are called dyes!). The study of electronic absorption is therefore typically called *UV/Visible (or UV/Vis) spectroscopy*. The resulting UV/Vis absorption spectrum (i.e. the wavelength dependence of the absorption) can be used as a simple probe of the *electronic structure* (in particular of the electronic excited state). Finally, it is worth noting that electronic absorption is the first step for several more complex processes such as fluorescence (see later).

- Infrared (IR) absorption refers to the excitation of the molecule to a higher vibrational/rotational level within a given electronic state (usually the ground state). The energy of the absorbed photon is much smaller than for electronic absorption, and lies in the infrared or far infrared ($\lambda \sim$ 3–100 μm). This is then the realm of *infrared (or IR) spectroscopy*, which is accordingly used as a probe of the *vibrational structure* of the molecule.

In typical experiments, a monochromatic light source (for example a laser) is used for excitation, and the number of absorbed photons is directly proportional to the number of incident photons. The absorption rate, i.e. number of photons absorbed per unit time, depends on the molecule, the wavelength, and the incident power density. This will be discussed later in more detail.

2.2.4. Emission and luminescence

Stimulated and spontaneous emission

The opposite of absorption is called *emission*: a molecule relaxes from a level of energy E_b to a lower level of energy E_a while emitting a photon of energy $E = E_b - E_a$. Note that the molecule *must be in an excited state* for this to happen, which we implicitly assume in the following discussion. Radiative emission may occur between vibrational states (the opposite of IR absorption), but in most cases of interest here, it refers to transitions between two electronic states (usually from the first excited to the ground state). We therefore focus on this type of radiative emission. A quantum mechanical treatment [71,80] shows that there are, in fact, two possible processes:

- In the presence of *incident photons*, the interaction of the excited molecule with the incident photons can stimulate the emission of a photon that is identical to the incident ones (i.e. with same energy, direction, and phase). The number of emitted photons is then directly proportional to the number of incident photons. This process, called *stimulated emission*, is the exact opposite to the absorption process and therefore occurs at the same rate for a given incident power density. Note that the incident photons must be at the same energy as an available transition in the molecule for stimulated emission to occur.

- In the absence of any incident photon, one may wonder how a transition is possible since there is no photon for the molecule to interact with. This is where a quantum mechanical treatment of the interaction is necessary. The vacuum state of the electromagnetic field is not empty from a quantum mechanical viewpoint, and exhibits fluctuations. The molecule can therefore interact with this vacuum state, and this then results in the emission a of photon, which in this case may be emitted in any direction (although not isotropically in general). This is called *spontaneous emission*. Spontaneous emission (SE) is intrinsically a *stochastic process* and must be described by the probability that SE occurs per unit time. For a given transition, this probability is a constant (it does not depend on the previous history of the molecule) and is called the *radiative decay rate* Γ_0 [s^{-1}]. Γ_0^{-1} [s] then characterizes the *radiative lifetime* of the excited state, i.e. the typical time for SE to occur, and is of the order of 1–100 ns for typical dye molecules.

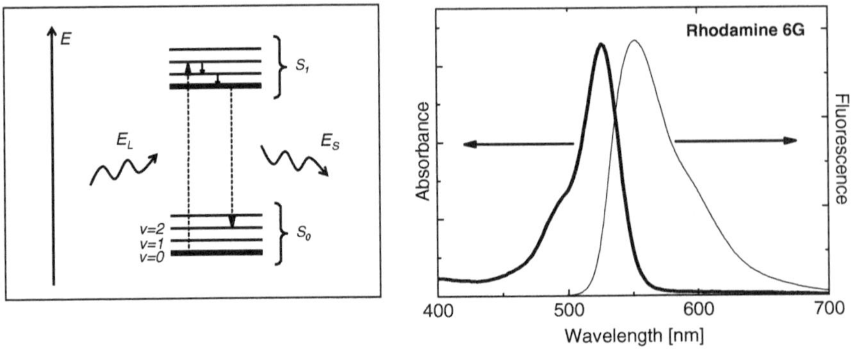

Figure 2.2. Simplified Jablonski diagram (left) illustrating schematically the fluorescence process in a molecule. An electron is excited by an incident photon (of energy E_L) from a sub-level (energy E_a) of the electronic-ground-state manifold S_0 (with its vibrational sub-levels $v = 0, 1, 2$, etc.) into a sub-level (energy $E_b = E_a + E_L$) of the excited (singlet) state (S_1) through a dipole-allowed transition. The electron then undergoes a fast non-radiative decay in the vibrational sub-structure of S_1 (on a picosecond timescale) followed by a radiative transition (spontaneous emission) to S_0, with emission of a photon at energy $E_S < E_L$. The plot (right) shows, as an example, the fluorescence (excited at 514 nm ($E_L = 2.41$ eV)) and UV/Vis (absorption) spectra of rhodamine 6G (a typical SERS dye) in water.

In general, in the presence of an incident beam, both stimulated and spontaneous emission may occur. However, the rate of stimulated emission is in most situations of interest to us much smaller ($<10^9$ s^{-1}) than other energy relaxation rates such as IVR ($\sim 10^{12}$ s^{-1}). Therefore internal energy relaxation occurs much faster than stimulated emission. Because stimulated emission can only occur at the incident photon energy, it can no longer occur after energy relaxation, and is therefore negligible. Stimulated emission will therefore not be relevant to most of the content of this book, except for the fact that it is what makes lasers possible! On the contrary, spontaneous emission may occur at any energy (provided there are allowed transitions), and will accordingly play an important role.

Fluorescence

In order to observe the emission of a photon it is first necessary to excite the molecule to an excited state. The simplest way to do so optically is through optical absorption, i.e. by exciting it with an incident beam. This process of absorption followed by emission is called *luminescence*. The most common (by far) type of luminescence in molecules is called *fluorescence*, which is illustrated in Fig. 2.2 and can be described as follows:

- An incident (laser) beam at energy E_L excites an electron from S_0 to S_1, i.e. this is an absorption (dipole-allowed) process.

- The electron then relaxes down the vibrational sub-structure of S_1 (through IVR, solvent interactions, etc.) on a timescale of the order of 0.1–10 ps [79]. It therefore relaxes down to the lowest energy level of S_1 (up to thermal fluctuations of the order of $k_B T$).

- From there, several transitions are possible. Spontaneous emission may occur with an average lifetime Γ_0^{-1} of the order of 1–100 ns. This is associated with the electron relaxing down to a vibrational level in the sub-structure of the electronic ground state S_0. The emitted photon corresponds to the fluorescence process and has an energy $E_S < E_L$ corresponding to this transition energy. Other transitions are possible (and compete with fluorescence) and will be discussed in Section 2.3.2.

Fluorescence is therefore a *two-step process*, occurring on the same timescale as spontaneous emission. In practice, the energies of the fluorescent photons vary and the wavelength dependence of the fluorescence intensity forms the *fluorescence spectrum*. The spectrum reflects the underlying electronic and vibrational structure of the molecules, in a similar fashion to the absorption spectrum. Because $E_S < E_L$, the fluorescence spectrum peaks at a longer wavelength (lower energy) than the absorption spectrum. The difference in energy between the two maxima is called the *Stokes shift* .

Fluorescence plays a significant role around SERS phenomena; most of the best SERS probes are fluorescent dyes and there are related phenomena associated with the presence of SERS signals like *fluorescence quenching*. Surface-enhanced fluorescence (SEF) is also an important application of plasmonics. We shall come back to specific aspects of the fluorescence process in Section 2.3, and when required to study particular aspects of SERS or plasmonics in the forthcoming chapters.

Other luminescence processes

Many other luminescence processes can be envisioned. We only mention here as an example the case of *phosphorescence*. This occurs as a result of inter-system crossing (mentioned before), in molecules with a large probability of transition from S_1 to the triplet state T_1. The electron may then relax from T_1 to S_0 and this process may be radiative (i.e. through spontaneous emission) in some molecules. Because the transition is normally 'spin-forbidden', the spontaneous emission rate is typically much slower and the lifetime in T_1 accordingly much longer, from $\sim$1 μs to many seconds. This is the phenomenon of phosphorescence, which appears as a strongly delayed luminescence. A transition from T_1 to S_0 normally implies the existence of an additional mechanism to compensate for the required spin change.

2.2.5. Scattering processes

Elastic and inelastic scattering

The optical processes discussed so far all involved the absorption or emission of a single photon (two photons are involved in fluorescence but in two separate steps). Another important family of processes involves the *simultaneous* (instantaneous) absorption of an incident photon and emission of another photon. Such processes are called *scattering processes*, and the emitted photon is called the *scattered photon.*

Scattering processes can further be classified into two main groups:

- *Elastic scattering*, where the incident and scattered photons have the *same energy* (but typically a different direction and/or polarization). For molecules, this process is often referred to as *Rayleigh scattering.* Such a process leaves the molecule in the same energy level after the scattering has occurred (there is no transfer of energy between the molecule and the photon), as illustrated in Fig. 2.3(a), and does not therefore reveal much of the internal structure. Elastic scattering is also common for larger objects, and in particular for nano-particles (metallic or not); it is then usually called Mie scattering.
- *Inelastic scattering*, where the scattered photon is at a different energy E_S from that of the incident photon, E_L. The energy difference corresponds to an accompanying transition between two states in the molecule. One of the most important forms of inelastic scattering in molecules is *Raman scattering*, which involves transitions between the vibrational/rotational levels. The corresponding process is illustrated in Fig. 2.3(b).

Scattering processes, such as Raman scattering, may appear at first to be similar to fluorescence: an incident photon is in both cases 'replaced' by a scattered (or fluorescence) photon. An important difference is that scattering processes are *instantaneous* while fluorescence involves an intermediate step (electronic excitation followed by emission with a finite lifetime). This means that Raman scattering can happen without a direct absorption of the photon (as required for fluorescence), and also therefore even when no electronic transitions exist in the molecule at the incident wavelength. This allows the use of incident light in the transparency region, where there is no molecular absorption and therefore no fluorescence. Scattering processes, however, are intrinsically *weak* phenomena under typical conditions when compared to other optical processes like absorption or fluorescence.

An alternative way to think about scattering processes in quantum mechanical terms is depicted in Fig. 2.3(c–d). In this picture, the scattering process is broken down into two steps and viewed as the combination of absorption of the incident photon and spontaneous emission of the scattered

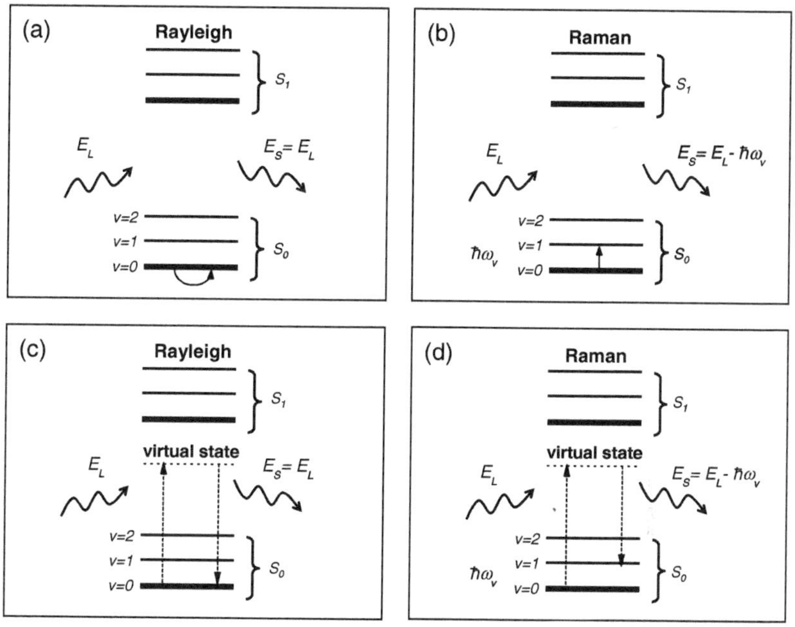

Figure 2.3. Left: Simplified Jablonski diagrams illustrating schematically the Rayleigh (a) and Raman (b) scattering processes. Unlike fluorescence, the incident photon energy E_L does not need to be tuned to a specific transition in the electronic structure of the molecule. In fact, the incident photon might have an energy below the first possible transition (producing absorption). An alternative way of visualizing these scattering processes from a quantum mechanical point of view is shown in (c) and (d). The scattering is then viewed as two simultaneous processes: absorption of a photon through a transition to a *virtual state*, from where a recombination to S_0 follows. If the virtual state coincides with a real electronic state of the molecule (for example in the sub-structure of S_1), the scattering process is said to be *resonant*. Resonant Raman scattering has important implications for both the magnitude and the selection rules of the effect in both normal Raman and SERS conditions.

photon. For absorption to occur, the molecule should in principle be excited to a higher-energy level. Since this energy level may not exist, it is represented as an intermediate *virtual state*. This picture is more 'visual' and is also a good representation of how the effect is modeled using quantum mechanical perturbation theory. The virtual state, as its name suggests, has in most cases no physical reality and should be considered to some degree as a 'mathematical construction of perturbation theory'. However, if the energy of the intermediate virtual state coincides with one of the real electronic (vibronic) levels in the molecule, then we are in the presence of *resonant scattering*. Such resonant effects, such as resonance Raman scattering (RRS), can increase by several orders of magnitude the scattering efficiency. This opens up a whole new variety of scattering conditions, to the extent that many authors treat resonance Raman scattering almost as a different spectroscopic technique. RRS also plays an important role in the context of SERS, and both are sometimes combined under the general denomination of surface-enhanced resonant Raman scattering (SERRS).

More on Raman scattering

Another interesting aspect of inelastic scattering is that the scattered photon may have an energy either lower or higher than the incident photon:

- If the scattered photon has *less energy* than the incident photon ($E_S < E_L$), then the molecule is excited to a higher-energy level by $E_L - E_S$. This is called a *Stokes process* and typically corresponds to the excitation of the molecule from the vibrational ground state $v = 0$ to the first excited state $v = 1$ of a vibrational mode with energy $\hbar\omega_v = E_L - E_S$, where $\hbar\omega_v$ is the energy of the vibration. This is illustrated in Fig. 2.4.
- If, on the contrary, the scattered photon has *more energy* than the incident photon ($E_S > E_L$), then the molecule has relaxed from an excited vibrational state ($v = 1$) to its ground state ($v = 0$). The energy of the vibration is then given by $\hbar\omega_v = E_S - E_L$; this is called an *anti-Stokes process*, and is also illustrated in Fig. 2.4. In order for an anti-Stokes process to take place, the molecule must be in an excited vibrational state in the first place. In typical conditions, this may only occur through thermal excitation and this implies that the anti-Stokes signal depends at equilibrium on temperature (T) through a Boltzmann factor: $\exp(-\hbar\omega_v/(k_B T))$. The anti-Stokes side of a Raman spectrum is accordingly much weaker than the Stokes side and the intensities of the peaks become exponentially weaker the higher the energy of the vibration $\hbar\omega_v$. This can be explicitly seen in the spectrum shown in Fig. 2.4.

Raman spectrum and Raman shift

The energy *lost* by the photons in the scattering event is called the *Raman shift* and is defined in energy as $\Delta E_R = E_L - E_S$. It is therefore positive for a Stokes and negative for an anti-Stokes process. Raman shifts are commonly expressed in wave-numbers and will then be denoted $\Delta\bar{\nu}_R$ (usually in units of [cm^{-1}]). The modulus of the Raman shift corresponds to the wave-number of the vibrational mode ($\bar{\nu}_v = \hbar\omega_v/hc$) that was involved in the scattering event: $\bar{\nu}_v = |\Delta\bar{\nu}_R|$.

The *Raman spectrum* corresponds to the wavelength- (or energy-) dependence of the Raman scattered intensity at a given incident wavelength.

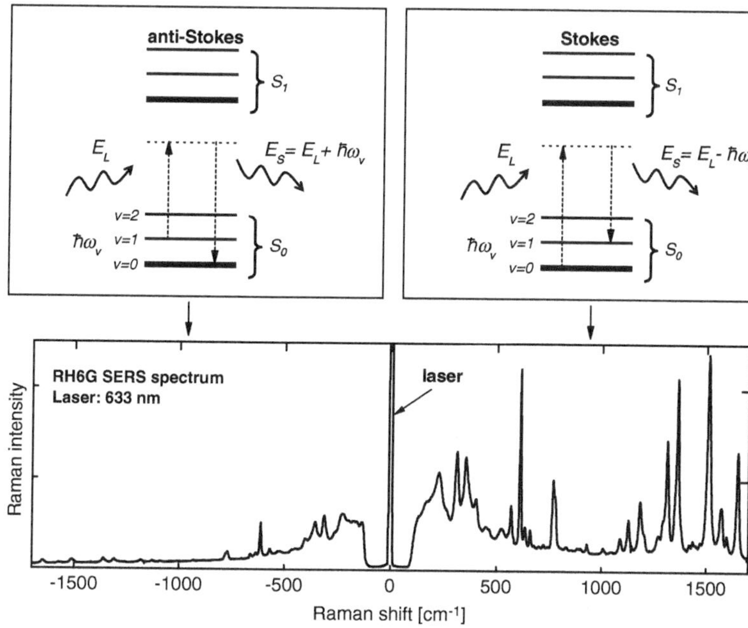

Figure 2.4. Simplified Jablonski diagrams of the anti-Stokes (top left) and Stokes (top right) Raman processes. The Stokes process is the same as depicted in Fig. 2.3(b,d) while anti-Stokes scattering starts with the molecule already in the $v = 1$ vibrational excited state of S_0 and finishes in $v = 0$, thus producing a scattered photon with an energy larger than the incoming one. A typical Raman (in fact SERS here) spectrum (Raman intensity vs Raman shift $\Delta\bar{\nu}_R$) is shown at the bottom (for rhodamine 6G) with several Raman peaks on the Stokes side ($E_S < E_L$, $\Delta\bar{\nu}_R > 0$) and their (weaker) anti-Stokes counterparts ($E_S > E_L$, $\Delta\bar{\nu}_R < 0$).

It is commonly shown as *Raman intensity as a function of Raman shift* (rather than wavelength), as shown in Fig. 2.4 for an example SERS spectrum. Peaks in the Raman spectrum correspond to vibrational modes of the molecule, in a similar fashion as for an infrared absorption spectrum. The Raman shift of a peak is equal to the vibrational energy of the corresponding mode. There is some broadening (homogeneous and inhomogeneous) of the peaks, but Raman peaks for molecular compounds are typically 'narrow' (5–20 cm^{-1}). The total intensity (power) scattered by a given vibrational mode (and detected in our setup) is therefore the *integrated intensity* of the corresponding Raman peak.

Not every vibrational mode, however, leads to strong (or in fact any) Raman scattering, and the Raman intensity varies from mode to mode. This is part of the *selection rules* of the Raman effect which will be treated in more detail later. Despite this, the Raman spectrum as a whole is to a large extent a unique fingerprint of the molecule. The same applies to the IR absorption spectrum, and because the modes appearing in the IR and Raman spectra may be different, the two techniques are in fact complementary probes of the vibrational structure of molecules.

2.2.6. The concept of cross-section

We have so far described the most important optical processes in mostly qualitative terms. We will now focus on how these effects may be rigorously quantified, i.e. for a given molecule, how many incident photons are absorbed, how many are scattered (through Rayleigh or Raman processes), how many produce fluorescence, etc.

Power and power density

Let us consider an incident beam (which we assume to be monochromatic, for example a laser beam) incident on a single molecule. From the experimenter's point of view, the beam is characterized by its power P_{Inc} [W] (proportional to the number of incident photons per unit time). This beam is usually focused to a spot containing the molecule. From the molecule point of view, what matters is the number of photons that can interact with the molecule in a certain way (for example producing absorption), and this is proportional to the *power density* S_{Inc} [W m^{-2}] at the molecule's position[5].

The power density can usually be related to the incident power P_{Inc} if the 'spot size' (beam diameter at focus) and beam profile are known. For example, for a Gaussian beam (a common approximation for focused laser beams) with a waist w_0 [m], the power density at the center of the beam is [8]:

$$S_{\text{Inc}} = \frac{2P_{\text{Inc}}}{\pi w_0^2}. \qquad (2.4)$$

In many theoretical treatments, the incident beam is approximated by a plane wave. The wave is then formally of infinite extent and P_{Inc} cannot be defined. The power density can however be related to the electric field amplitude E_{Inc} [V m^{-1}] as (see Appendix F):

$$S_{\text{Inc}} = \frac{\epsilon_0 c}{2} n_M |E_{\text{Inc}}|^2, \qquad (2.5)$$

where $n_M = \sqrt{\epsilon_M}$ [a.d.] is the index of refraction of the medium in which the incident wave propagates ($n_M = 1$ for air).

---

[5] Unless otherwise stated, we try to reserve as a general rule the notation: P or I for power [W], and S (by analogy with the Poynting vector) for power density [W m^{-2}].

Definition of the cross-section

By definition, the *cross-section* σ [m^2] of an object (here a molecule) for a given linear optical process relates the signal produced by this process, characterized by its intensity or power P [W] (proportional to the number of photons per unit time involved in the process) to the incident power density S_{Inc} [W m^{-2}] at the molecule position as:

$$P = \sigma S_{\text{Inc}}. \tag{2.6}$$

Note that this definition implicitly relies on the fact that the response P is directly proportional to the incident power density. It therefore only applies to linear processes. Also implicit is the fact that σ typically depends on the excitation wavelength. This general definition will be made more precise in specific contexts soon.

Let us first discuss it further, by considering for example optical absorption. From this definition, the absorption cross-section σ_{Abs} [m^2] of a molecule corresponds to the *effective area* of a homogeneous incoming beam from which the molecule will absorb every photon. The absorption cross-section can then be viewed as the 'area absorbed from the incoming beam'[6]. This simple picture is illustrated in Fig. 2.5(a). Of course, the idea of an effective geometrical area being cut out from the beam is not really valid at a molecular level where geometrical optics does not hold. Still, the cross-section is a useful concept with a rigorous definition, given above. Cross-section of optical processes are typically given in units of [cm^2].

It is important to remember that the cross-section applies to *a single molecule*, even if in most experiments, many molecules are involved; it is therefore not straightforward (but not impossible either) to measure it. It does provide a measure of the efficiency or magnitude of the different optical mechanisms. We can compare for example the *fluorescence* cross-section against the *Raman* cross-section. This comparison will yield an immediate relation between how many photons are going into one type of process with respect to another. Once put in terms of a cross-section, we can rigorously compare the efficiency of different optical processes.

Note that cross-sections are not exclusive of optical phenomena and can be defined for any type of scattering process, including scattering and collision phenomena amongst quantum mechanical and/or relativistic particles [84]. In the quantum theory of particle collisions the cross-section measures the 'spatial' extent of the interaction between a target particle and a 'projectile' [84]. A rule of thumb for any cross-section is of course that the larger the cross-section, the stronger the interaction, and the more likely the 'collision'.

---

[6] A highly absorbing molecule will typically have an absorption cross-section of the order of its geometrical area (sometimes called its geometrical cross-section).

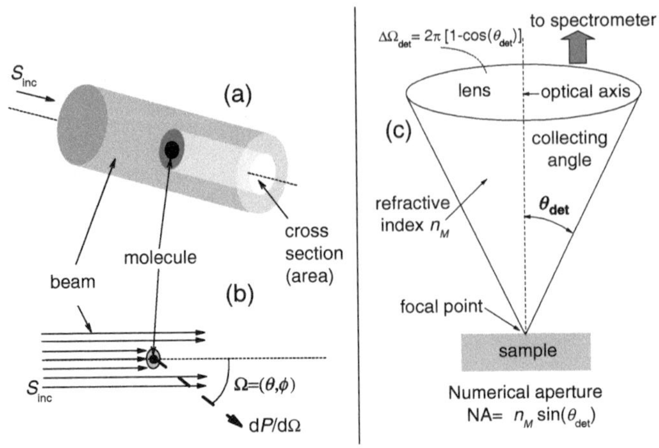

Figure 2.5. (a) Schematic representation of the concept of cross-section, as an area cut out from the incident beam. (b) Similar schematic illustrating the differential scattered power. (c) Schematic representation of the collection optics in a typical Raman scattering experiment. The laser is normally focused to a small spot (typically ~ 1–10 μm diameter depending on the focusing optics). Note that in many implementations the collecting optics is also used for excitation (in a back-scattering configuration).

Finally, note that processes like spontaneous emission cannot be described in terms of a cross-section because they do not involve excitation by an incident beam. Spontaneous emission, which is a stochastic process, is instead characterized by a decay rate (probability of decay per unit time) or lifetime as described earlier.

Differential cross-section

For optical absorption, the cross-section σ_{Abs} simply relates the incident power density to the power absorbed by the molecule P_{Abs}. Similarly, for optical scattering, the cross-section σ_{Sca} relates the incident power density to the power scattered by the molecule P_{Sca}. In this latter case however, this does not fully describe the scattering process, and in particular the *radiation profile*, i.e. how many photons are scattered in *a specific direction* (instead of integrating the signal over all emission directions). In fact, knowing how many photons are scattered in a specific direction is a lot closer to real experimental situations, where a detector is indeed placed along a given direction. We very rarely observe the whole integrated scattered radiation unless specialized experimental arrangements (like integrating spheres) are used. Accordingly, it is interesting to quantify the scattering efficiency in a way that retains information on the radiation profile.

The radiation profile can be characterized by the angular dependence of the scattered power. In 3D, two angles are necessary to specify a direction. They are in this context taken with respect to the incident beam direction and can be grouped into a single variable $\Omega \equiv (\theta, \phi)$ (see Fig. 2.5(b)). The angular dependence of the scattered power can then be formally defined as $\mathrm{d}P_{\mathrm{Sca}}/\mathrm{d}\Omega$ [W sr^{-1}], which is a function of Ω, and is called the *differential scattered power*. By analogy with the previous definition, one can therefore define a *differential scattering cross-section*, $\mathrm{d}\sigma_{\mathrm{Sca}}/\mathrm{d}\Omega$ [m^2 sr^{-1}], also a function of Ω, as:

$$\frac{\mathrm{d}P_{\mathrm{Sca}}}{\mathrm{d}\Omega}(\Omega) = \frac{\mathrm{d}\sigma_{\mathrm{Sca}}}{\mathrm{d}\Omega}(\Omega) S_{\mathrm{Inc}}. \tag{2.7}$$

The function of Ω, $\mathrm{d}\sigma_{\mathrm{Sca}}/\mathrm{d}\Omega$, characterizes the radiation profile of the scattering process. Its value for a given Ω_d characterizes the scattering efficiency in this specific direction. As mentioned earlier, unless a very specialized experimental layout is used (with an integrating sphere) this is what is normally measured in most standard Raman (or scattering) experiments.

Finally, since the total scattered power should be the sum of the scattered power in all directions, we have a relation between the two types of cross-sections in the form of:

$$\sigma_{\mathrm{Sca}} = \int \frac{\mathrm{d}\sigma_{\mathrm{Sca}}}{\mathrm{d}\Omega}(\Omega)\mathrm{d}\Omega. \tag{2.8}$$

These definitions are very general and will soon be specialized to specific cases.

Practical detection and numerical aperture

Note that even if detection is carried out in a single direction specified by Ω_d, the signal is collected from a small cone around this direction, which can be specified by a solid angle $\Delta\Omega_{\mathrm{Det}}$ (as illustrated in Fig. 2.5(c)). The scattered power in the detector [W] is then (for a small $\Delta\Omega_{\mathrm{Det}}$):

$$P_{\mathrm{Sca}}^{\mathrm{Det}} = \frac{\mathrm{d}P_{\mathrm{Sca}}}{\mathrm{d}\Omega}(\Omega_d)\Delta\Omega_{\mathrm{Det}}. \tag{2.9}$$

Moreover, denoting θ_{Det} the half-angle of the maximum cone of light that can enter or exit the lens (see Fig. 2.5(c)), the solid angle for detection is then

$$\Delta\Omega_{\mathrm{Det}} = 2\pi(1 - \cos(\theta_{\mathrm{Det}})). \tag{2.10}$$

In typical setups, this solid angle is determined by the *numerical aperture* (NA) of the collecting optics. Denoting n_M the index of refraction of the medium in which the lens is working (1.0 for air, 1.33 for pure water, and up to ~1.56 for oils), the numerical aperture (NA) is defined as:

$$\boxed{\mathrm{NA} = n_M \sin(\theta_{\mathrm{Det}}).} \tag{2.11}$$

The NA depends on the focal length and diameter of the collecting lens and can be very different depending on the specific experiment being carried out, even for the same spectrometer. The NA is related in classical optics to the *resolving power* of the lens; the larger the NA, the better the resolving power and the more the light collected from the focal point (the better also the quality of an image produced by the lens).

The numerical aperture of many Raman microscopes can be rather large, which means that the detected signal accounts for an average of different scattering directions of the differential cross-section. Typical values of NA can be of the order of ~1 for short working distance high-magnification immersion objectives (water or oil), and as low as ~0.1 for some lower-magnification objectives. A water immersion objective with NA = 1 collects signals spanning a total angle of $2\theta_{\mathrm{Det}} \sim 98°$, equivalent to a solid angle for detection of $\Delta\Omega_{\mathrm{Det}} \approx 2.1$ sr or 17% of the total solid angle (4π).

2.2.7. The Raman cross-sections

Rigorous definition

A definition of the Raman cross-section can be built upon the general definitions given above. It nevertheless requires a number of additional considerations to make it more precise. These are rigorously formulated for example in a review by the International Union of Pure and Applied Chemistry (IUPAC) [85], and we highlight only the most relevant aspects here:

- Firstly, to fully describe the Raman spectrum (and therefore the Raman active vibrational modes) of a given molecule, one must in principle define a cross-section for each vibrational mode.
- Another important aspect is that the scattering process (intensity and radiation profile) depends on the orientation of the molecule with respect to the incident field polarization. In most practical cases, a large number of molecules are measured, and their orientation is random and averaged in the measured signal. The definition of the Raman cross-section therefore refers, by convention, to *a molecule with a randomly-averaged orientation*.
- Note that, rather counterintuitively, the radiation profile may remain non-isotropic even after orientation averaging (because the incident field polarization still breaks the spherical symmetry). For a rigorous

definition of the differential cross-section, one must therefore specify the direction of observation with respect to the incident excitation, i.e. the so-called *scattering configuration*.

- Finally, the Raman scattering process (and therefore the cross-section) depends on a number of additional parameters, which must therefore be specified in the definition, in particular, the energy (or wavelength) of the excitation, and the refractive index of the environment (for example of the solvent for dissolved molecules in liquids).

Synthesizing all these considerations, it is possible to define the *absolute differential Raman cross-section* for a given vibrational mode of energy $\hbar\omega_v$ of a given molecule as follows: consider an incident field that is linearly polarized (always perpendicularly to the incident beam), with a power density of S_{Inc} [W m^{-2}], and an angular frequency ω_L. We denote by $\mathrm{d}P_R/\mathrm{d}\Omega(90°)$ [W sr^{-1}] the molecular-orientation-averaged differential Stokes Raman scattered power (at frequency $\omega_R = \omega_L - \omega_v$), observed in a direction perpendicular to both the incident beam and the incident polarization (the so-called 90°-scattering configuration). The *absolute differential Raman cross-section* $\mathrm{d}\sigma_R/\mathrm{d}\Omega$ [m^2 sr^{-1}] is then *by definition* derived from the expression:

$$\boxed{\frac{\mathrm{d}P_R}{\mathrm{d}\Omega}(90°) = \frac{\mathrm{d}\sigma_R}{\mathrm{d}\Omega} S_{\text{Inc}}.} \tag{2.12}$$

$\mathrm{d}\sigma_R/\mathrm{d}\Omega$ is sometimes also called differential Raman cross-section, or even Raman cross-section, but such omissions may be the source of confusion.

The following remarks can be made:

- Although the definition refers to a 90°-scattering configuration, the same result is obtained for forward-scattering (detection along the direction of the incident beam) or for back-scattering (detection in the direction from where the incident beam originated). For other detection directions however, this is no longer true and one must then consider the radiation profile. More details on this aspect and other scattering configurations can be found for example in Ref. [70].

- The term absolute differential Raman cross-section refers by definition to Stokes scattering. It is however possible to define in the same way the anti-Stokes absolute differential Raman cross-section.

According to this definition, $\mathrm{d}\sigma_R/\mathrm{d}\Omega$ therefore depends on:

- a given vibrational mode for a given molecule (and therefore its energy $\hbar\omega_v$ or the Stokes angular frequency $\omega_R = \omega_L - \omega_v$),

- the excitation wavelength (or angular frequency ω_L),

- and the medium (environment), in which the molecule is dissolved.

Polarized detection and depolarization ratio

The absolute differential Raman cross-section characterizes the magnitude of Raman scattering for a given Raman mode, and is therefore, arguably, its most important characteristic. It is nevertheless possible to extract more information about the Raman mode using a more elaborate detection scheme, the simplest of which being polarized detection. Using the 90°-scattering configuration for example, one can separate the differential scattered power into the sum of two contributions, each associated with the two possible electric field polarizations of the scattered radiation[7]: $\mathrm{d}P_R^{\parallel}/\mathrm{d}\Omega$ for the field component polarized *parallel* to the incident polarization, and $\mathrm{d}P_R^{\perp}/\mathrm{d}\Omega$ for the field component polarized *perpendicular* to it. These correspond to differential Raman cross-sections for parallel, $\mathrm{d}\sigma_R^{\parallel}/\mathrm{d}\Omega$, or perpendicular, $\mathrm{d}\sigma_R^{\perp}/\mathrm{d}\Omega$, configurations, and can be defined as in Eq. (2.12) (for unpolarized detection). The *Raman depolarization ratio*, ρ_R [a.d] of the Raman mode is then defined as:

$$\rho_R = \frac{\mathrm{d}P_R^{\perp}/\mathrm{d}\Omega(90^\circ)}{\mathrm{d}P_R^{\parallel}/\mathrm{d}\Omega(90^\circ)} = \frac{\mathrm{d}\sigma_R^{\perp}/\mathrm{d}\Omega}{\mathrm{d}\sigma_R^{\parallel}/\mathrm{d}\Omega}. \tag{2.13}$$

ρ_R contains information about the *symmetry of the vibrational mode*, and characterizes fully the radiation profile for an orientation-averaged molecule. We will come back to the importance of ρ_R later in this chapter.

The total Raman cross-section

The definition of the Raman cross-section given above is the standard one, and should be preferred whenever possible. It may nevertheless be useful in some situations to derive from it the *total Raman cross-section*, σ_R [m^2], which accounts for scattering integrated over all possible directions. It therefore derives directly from Eq. (2.8). Moreover, since the depolarization ratio fully determines the radiation profile, it is possible to relate σ_R to $\mathrm{d}\sigma_R/\mathrm{d}\Omega$ and ρ_R, and one can show that [8,70]:

$$\sigma_R = \frac{8\pi}{3}\frac{1+2\rho_R}{1+\rho_R}\frac{\mathrm{d}\sigma_R}{\mathrm{d}\Omega}. \tag{2.14}$$

[7] The electric field polarization of a radiation field is always perpendicular to the scattering direction and can therefore be described as the linear combination of two orthogonal linear polarizations.

It is paramount to distinguish clearly between one definition of the cross-section and the other (differential vs total) when making reference to Raman cross-sections in general. The lack of use of the proper definition is sometimes a problem in the scientific (research) literature, and in the SERS literature in particular.

2.2.8. Examples of Raman cross-sections

Measuring the Raman cross-section

The accurate measurement of the absolute differential Raman cross-section is a complex issue that requires careful characterization of the experimental conditions, see for example the comprehensive review in Ref. [86] and references therein. Such studies have been carried out for only a few elementary substances, such as nitrogen (gas), benzene, or cyclohexane, which then serve as standards. The *relative* cross-sections of other compounds are then much more easily measured, and compared to the reference from which the absolute cross-section is derived. As long as the exact same experimental conditions are used, the absolute differential Raman cross-section of a sample can be determined from the relative concentrations and peak intensities of the reference and the sample:

$$\left(\frac{\mathrm{d}\sigma_{\mathrm{R}}}{\mathrm{d}\Omega}\right)_{\mathrm{Sample}} = \left(\frac{\mathrm{d}\sigma_{\mathrm{R}}}{\mathrm{d}\Omega}\right)_{\mathrm{Ref}} \frac{I_{\mathrm{Sample}}}{I_{\mathrm{Ref}}} \frac{c_{\mathrm{Ref}}}{c_{\mathrm{Sample}}}, \tag{2.15}$$

where c is the concentration and I the measured integrated intensities of the Raman peak under study (unpolarized detection is assumed). Ref. [86] provides an extensive list of such measurements for common compounds (gases and liquids).

The measurement of the absolute differential Raman cross-section is essential if SERS enhancement factors (i.e. the scaling factor between SERS and Raman cross-sections) are to be measured and quantified. An extensive study of this aspect can be found in Ref. [8] and we will come back to this issue in Chapter 4.

Examples: Nitrogen gas

Diatomic molecules like N_2, O_2, and H_2 provide some of the simplest (and most abundant) substances that can be used as a reference for Raman cross-sections. They have only *one* vibrational mode, and it is Raman active. Moreover, being gases, their quantities are also easily controllable by both temperature and pressure. Table 2.1 summarizes the absolute differential Raman cross-section of N_2 measured at room temperature at different wavelengths. Note that the cross-section changes with the excitation

Table 2.1 Differential cross-sections of the 2331 cm^{-1} mode of N_2 at different laser excitation wavelengths, from Ref. [86]. The values in the table correspond to averages over different experimental determinations by different authors [86] (measured) and are compared to the result obtained from the empirical expression in Eq. (2.16). On theoretical grounds, the values obtained from Eq. (2.16) for 633 and 488 nm should be more accurate than the experimental values (necessarily subject to uncertainities).

Laser excitation [nm]	$\left(\frac{d\sigma}{d\Omega}\right)$ (Measured) $[10^{-32}\ cm^2\ sr^{-1}]$	$\left(\frac{d\sigma}{d\Omega}\right)$ (from Eq. (2.16)) $[10^{-32}\ cm^2\ sr^{-1}]$
633	21	16.4
515	43	43.0
488	46	54.4
458	75	72.3
436	92	90.1
364	204	200
351	243	234
300	970	462

wavelength, a very general phenomenon to be discussed later in this chapter. In fact, these experimental measurements for N_2 can be condensed in a single formula (which can be justified by theoretical arguments). If ν_L is the absolute wave-number of the incident laser (expressed in cm^{-1}), then the differential cross-section for the 2331 cm^{-1} mode of N_2 is given by [86]:

$$\left(\frac{d\sigma}{d\Omega}\right) = 5\times10^{-48}\left(\bar{\nu}_L - 2331\,cm^{-1}\right)^4\ cm^6\ sr^{-1}, \tag{2.16}$$

where the final units for $(d\sigma/d\Omega)$ are $cm^2\ sr^{-1}$, once $\bar{\nu}_L$ is put in cm^{-1}. This expression can be further verified using DFT (see Appendix A), which predicts a Raman activity of 19.7 $Å^4$/amu for this mode. This expression represents fairly accurately the differential Raman cross-section of N_2 at room temperature for laser excitations in the visible and near UV (up to about $\sim$ 350 nm), as seen in Table 2.1.

Examples: Liquid standards

Simple gases like N_2 provide good textbook examples of absolute differential cross-sections and can be used in many cases as reference standards for Raman measurements in air or gas. There are other instances, however, where it is more convenient to have a 'liquid reference'; particularly important if experiments with immersion objectives are carried out. Table 2.2 (adapted from Ref. [8]) summarizes some examples of absolute differential Raman cross-sections for some common liquid standards. These are relatively small molecules and, therefore, similar results can be predicted and double-checked

Table 2.2 Experimental values (from Ref. [86]) and density functional theory (DFT) predictions for important Raman-active modes of some liquid compounds that can serve as reference standards. Absolute Raman cross-sections refer to 633 nm laser excitation. The values for 2-bromo-2-methylpropane (2B2MP) are obtained from the gas phase data corrected by the local field correction factor $L_M = 3.3$ (see Section 2.4.5). The 516 cm^{-1} mode of 2B2MP can be used as a reference standard in liquid with an absolute differential Raman cross-section of 5.4×10^{-30} cm^2/sr. Values in parenthesis were obtained using this standard [8]. DFT calculations are further discussed in Appendix A.

	Experimental			DFT predictions		
	$\Delta\bar{\nu}_R$ [cm^{-1}]	$\left(\frac{d\sigma}{d\Omega}\right)$ [10^{-32} cm^2/sr]	ρ_R [-]	$\Delta\bar{\nu}_R$ [cm^{-1}]	$\left(\frac{d\sigma}{d\Omega}\right)$ [10^{-32} cm^2/sr]	ρ_R [-]
	302	475	0.26	293	429	0.26
2B2MP	516	558	0.16	509	525	0.18
	806	145	0.60	176	53.4	0.63
Benzene	992	790 (830)				
Toluene	1002	350 (290)				
Dichloro-	282	190 (170)				
methane	713	310 (290)				

by density functional theory (DFT), as described in Appendix A. This is illustrated in Table 2.2 for the case of 2-bromo-2-methylpropane (2B2MP), which constitutes a simple (non-toxic) and relatively inexpensive reference substance that can be used as a standard. The 516 cm^{-1} mode of 2B2MP, for example, can be used as a reference standard in liquid with absolute differential Raman cross-section of 5.4×10^{-30} cm^2/sr as proposed in Ref. [8]. The depolarization ratios of the peaks, introduced in Eq. (2.13), are also shown in Table 2.2 for illustration and will be further discussed later in this chapter.

Examples: SERS dyes and SERS probes

Within the context of SERS, one must characterize and quantify the cross-sections of molecules that are used as SERS probes. Dyes, which strongly absorb in the visible and are therefore in *resonant* conditions, are an important family of SERS probes. However, because of the typically strong fluorescence signal, it is not possible to measure Raman scattering from dyes with conventional approaches (i.e. resonant Raman scattering). One then has to resort to complex time-dependent DFT calculations [60] or difficult experimental setups [87] ; these suggest differential Raman cross-sections of the order of 10^{-23}–10^{-24} cm^2/sr for dyes at resonance (i.e. for laser excitation close to the maximum optical absorption). These are obviously *resonant Raman scattering* conditions.

Other examples of measurements for common SERS dyes and SERS probes at 633 nm excitation are provided in Table 2.3 (from Ref. [8]). A dye like

Table 2.3 Main Raman-active modes of crystal violet (CV), rhodamine 6G (RH6G), benzotriazole dye 2 (BTZ) and benzotriazole (BTA), with their experimentally determined differential Raman cross-sections, obtained by direct comparison, under the same experimental conditions, of integrated intensities against the 516 cm^{-1} mode of 2B2MP with a reference cross-section of 5.4×10^{-30} cm^2/sr. † indicates peaks that are part of a doublet. The table is reproduced from Ref. [8]. Copyright 2007 American Chemical Society.

	$\Delta\bar{\nu}_R$ [cm^{-1}]	$\frac{d\sigma}{d\Omega}$ [cm^2 sr^{-1}]		$\Delta\bar{\nu}_R$ [cm^{-1}]	$\frac{d\sigma}{d\Omega}$ [cm^2 sr^{-1}]
CV	808	3.6×10^{-26}	**RH6G**	612	0.67×10^{-27}
	917	1.1		774	0.76
	1177 †	6.1		1185	0.60
	1200 †	1.5		1311	1.0
	1621	3.6		1364	1.8
				1510	2.4
				1652	1.0
BTZ	1108	1.0×10^{-28}	**BTA**	783	3.6×10^{-30}
	1412	2.5		1019	4.7
	1617	0.87		1376 †	2.8
				1390 †	2.8
				1599	1.3

crystal violet (CV) can be measured in near-resonant conditions here, only because its fluorescence efficiency is very poor. For rhodamine 6G, 633 nm excitation is below its main absorption peak and can therefore be considered as pre-resonant.

Comparison of the magnitude of Raman cross-sections

These examples allow us to summarize the main factors influencing the magnitude of the absolute differential Raman cross-section:

- Firstly, it varies with excitation wavelength or frequency, as ($\propto \omega^4$) for reasons to be explained later. This is evident for example for N_2 in Table 2.1.
- Secondly, a comparison of Tables 2.1 and 2.2 reveals that N_2 has Raman cross-sections in the visible that are typically $\sim$ 10–30 times smaller than typical organic liquids (like benzene or toluene). Part of this difference (a factor $\approx$ 2–4) can be accounted for by the change in refractive index (more precisely, the local field correction factor discussed in Section 2.4.5). The rest is mostly a result of the molecule size. In simple terms, larger molecules typically have larger Raman cross-sections because they are more polarizable.
- Finally, resonant effects, and even pre-resonant effects can have a large influence on the Raman cross-section, as evident in Table 2.3. These

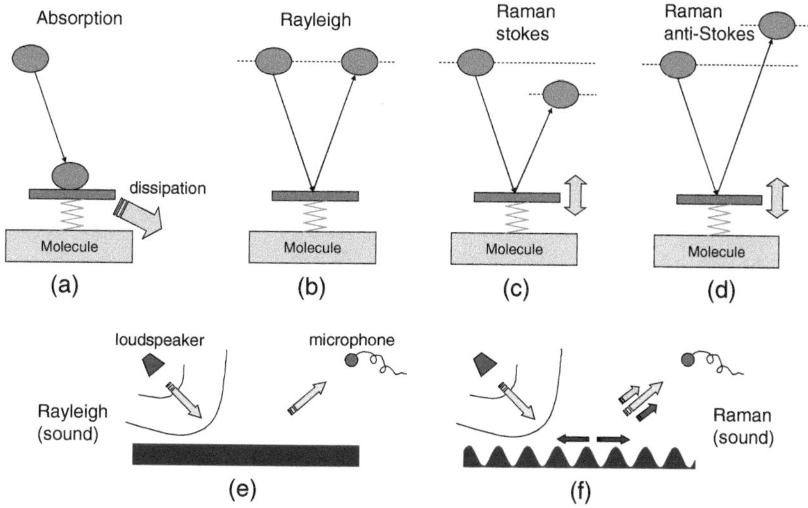

Figure 2.6. An overview of possible simple mechanical analogs to optical processes. At the top (a–d), a moving body representing the 'photon' falls onto a fixed object representing the 'molecule'. (a) The mechanical energy is fully absorbed (and then dissipated), if the particle 'sticks' to the object. (b) Elastic collision, there is no exchange of energy and the full initial mechanical (and therefore potential) energy of the body is recovered, albeit with a change of direction produced by the collision. This is the analog of Rayleigh scattering. (c) Analog of the Stokes Raman process: part of the initial potential energy of the 'photon' is transferred into an internal degree of freedom in the 'molecule'. The bouncing object reaches a smaller height after the collision, which is equivalent to the Raman shift in energy. (d) Analog of the Raman anti-Stokes process. A vibration already existing in the 'molecule' boosts the impinging particle to a height larger than the initial one. At the bottom (e–f), a different type of analogy based on sound waves is shown. (e) A sound wave reflecting on the still liquid surface is reflected and detected with the same frequency of the incoming wave, the analog of Rayleigh scattering for optical waves. (f) The Stokes and anti-Stokes scattering processes appear in this mechanical (sound) model as Doppler-shifted sound waves produced by the interaction of the incoming wave with the traveling waves in the liquid.

can result in variations by *at least four orders of magnitude* in the differential Raman cross-sections of typical SERS probes.

All these effects will be discussed with further details later in this chapter.

2.2.9. Mechanical analogs

We conclude this overview of optical processes in molecules by discussing possible 'mechanical analogs'. These mechanical analogs have to be taken with care, for they can have serious limitations. On the positive side, however, they can be useful as an aid to understanding the main ideas; and particularly useful to the complete newcomer, in order to understand qualitatively the origin of the scattering process.

Particle collision

Possible mechanical analogs of some of the optical processes discussed before are schematically shown in Fig. 2.6. In the first type of model (a–d), the collision of a particle (playing the role of the 'light' or photon) with another fixed object (playing the role of the 'molecule') is considered. The colliding particle is assumed to be at rest at the beginning. Figure 2.6(a) illustrates optical absorption. All the energy of the incident particle is transferred to the molecule (imagine some strong glue on the object!) in the form of an electronic excitation. In Fig. 2.6(b), the incoming particle undergoes an elastic collision, thus regaining its full potential energy. This is the analog of Rayleigh scattering (elastic scattering) in optics. A different situation occurs if the object has an *internal degree of freedom* that can absorb or provide mechanical energy during the collision, as depicted schematically in Fig. 2.6(c–d). If the depicted spring and platform are initially at rest (the mechanical equivalent in this model of the ground vibrational state in a molecule), the incoming particle may transfer part of its initial energy into an internal oscillation in the 'molecule', thus achieving a smaller final height after the collision. The difference in potential energy at the end has information on the characteristics of the internal degree of freedom that was excited in the molecule. This is the mechanical analog of a Raman Stokes scattering process. If the spring and platform system was already oscillating before the collision (the mechanical equivalent of thermal population of an excited vibrational level of a molecule), part of the internal energy of the platform can be transferred to the colliding body under the appropriate conditions, and the latter can reach a higher height (and potential energy) with respect to the initial state. This is the analog of anti-Stokes Raman scattering[8].

Sound waves

Another example of mechanical analogs is the 'sound' analogy of optical scattering, depicted schematically in Figs 2.6(e–f). In this case, the loudspeaker plays the role of the 'laser', the sound wave the role of light, the reflecting surface the role of the molecule, and the microphone the role of the optical detector. If a sound wave bounces back from the 'still' surface of a liquid, it is detected at the same frequency by the microphone. This is the equivalent of Rayleigh scattering. Now if a wave is traveling across the surface, the interaction of the incoming sound with the waves can Doppler shift the frequency to a lower pitch (Stokes scattering) or higher pitch (anti-Stokes scattering) depending on the relative directions of the surface waves with respect to the incoming one. Here the Doppler shift is the mechanical analog of the Stokes and anti-Stokes Raman scattering processes.

---

[8] This mechanical model and the corresponding figures, Figs 2.6(b–d), are courtesy of Matthias Meyer, Victoria University of Wellington, New Zealand.

Limitations

Most mechanical analogs of optical phenomena have limitations. Moreover, for analogs of Raman scattering, it is typically the anti-Stokes process that suffers from the most serious limitations. In the case of the mechanical model in Fig. 2.6(d), for example, the height of the incoming particle after collision depends on the exact time at which the collision happens (with the platform at rest, or moving either upwards or downwards). Similarly, in the sound wave analogy, the Stokes and anti-Stokes scattering processes are distinguished from each other by a relative direction of propagation of the wave producing the scattering with respect to the incoming wave. These aspects do not translate directly to optical Raman scattering.

Despite all the imperfections, the mechanical analogs can be useful in helping to visualize the basic origin of the effect and to put more sophisticated explanations in the right frame of mind. One fundamental concept in all of the mechanical analogs remains true in more sophisticated treatments, namely: the fact that inelastic scattering is produced by a 'modulation' of the intrinsic properties of the medium due to the presence of an internal degree of freedom (vibration in the case of Raman scattering). The same concept of modulation of the intrinsic properties by the presence of an internal excitation will appear in the more detailed treatments of Raman scattering in the following sections.

2.3. ABSORPTION AND FLUORESCENCE SPECTROSCOPY

Having presented the main concepts of molecular spectroscopy, we now return in more detail to some of the processes that are most relevant to us, namely UV/Vis (absorption) and fluorescence spectroscopy in this section, before moving on to Raman scattering itself in the rest of this chapter.

2.3.1. Optical absorption and UV/Vis spectroscopy

The study and analysis of the absorption and/or transmission of light as a function of incident photon energy constitutes the basis of *absorption/reflection spectroscopy*, which probes different excitations in the medium depending on the energy scale (electronic excitations for UV and visible light, and vibrational modes in the far infrared). In addition to changes in light intensity, the study of possible changes in the polarization of the reflected and/or transmitted beams (with respect to the incident one) constitute the realm of *ellipsometry* [88] (either transmission or reflection ellipsometry). We will focus here on UV/Vis spectroscopy of solutions (liquids), which can be used to characterize the optical absorption of molecules, and also (in the context of SERS) the optical properties of metallic substrates.

Absorption, scattering, and extinction

It is difficult to measure absorption directly; one can however measure 'what is not absorbed' and try to infer from it what has been absorbed. A typical implementation of UV/Vis spectroscopy therefore consists in measuring the wavelength dependence of the power *transmitted* through the sample.

The most common optical processes for molecules illuminated by an incident beam are absorption (which may or may not result in fluorescence) and scattering (elastic or inelastic). Photons involved in these processes 'disappear' from the incident beam and their energy is either transformed into heat in the sample or is re-emitted as photons in a different direction. This reduces the power of the beam as it travels through the sample and results in a transmitted power P_{Tra}, smaller than the incident power P_{Inc}. The difference $P_{\text{Inc}} - P_{\text{Tra}}$ is called the extinguished power P_{Ext}. If, as in most situations, only optical absorption (absorbed power is P_{Abs}) and scattering (scattered power is P_{Sca}) contribute to this *extinction*, we must have for energy conservation:

$$P_{\text{Ext}} = P_{\text{Abs}} + P_{\text{Sca}}. \tag{2.17}$$

An *extinction cross-section*, σ_{Ext}, can therefore be defined, and we then have:

$$\boxed{\sigma_{\text{Ext}} = \sigma_{\text{Abs}} + \sigma_{\text{Sca}}.} \tag{2.18}$$

From a measurement of the transmitted power, we can therefore deduce the amount that has been extinguished, which is simply the sum of the absorbed and scattered power.

For molecules, scattering is negligible compared to absorption, and extinction is then equal to absorption, i.e. $\sigma_{\text{Ext}} \approx \sigma_{\text{Abs}}$. This is why UV/Vis spectroscopy is often referred to as absorption spectroscopy, even if it should (strictly speaking) be called *extinction spectroscopy*. This is no longer true however for larger objects, and in particular for nano-particles, where the extinction spectrum (and not the absorption spectrum) is measured by UV/Vis spectroscopy. This is a particularly important distinction for metallic colloids, for example, which constitute an important class of SERS substrates used in applications. UV/Vis spectroscopy for the characterization of SERS substrate will be further discussed in Chapter 7.

The Beer–Lambert law

We have so far discussed optical absorption and scattering only for a single molecule. For an incident power density S_{Inc} [W m^{-2}], the power extinguished by a single molecule is by definition of the extinction cross-section $P_{\text{Ext}} = \sigma_{\text{Ext}} S_{\text{Inc}}$. We now want to relate this single molecule expression to the more practical case of an ensemble of molecules.

Let us, therefore, consider a solution of these molecules of concentration c_m [M] and an elementary volume of length dL [m] along the beam and surface area A [m^2] across (A is chosen small enough to assume that the power density S_{Inc} is uniform across this surface). The power entering this box is therefore $P = S_{\mathrm{Inc}}A$ [W]. Using Avogadro's number $\mathcal{N}$ [mol^{-1}], the volume contains $dN = \mathcal{N}c_m A dL$ molecules, each of which contributes to extinction with a cross-section σ_{Ext}, resulting in an extinguished power $\mathrm{d}P_{\mathrm{Ext}} = \mathcal{N}\sigma_{\mathrm{Ext}}S_{\mathrm{Inc}}c_m A\mathrm{d}L$. The power of the incident beam exiting the box is therefore $P+dP$ with $\mathrm{d}P < 0$ and $\mathrm{d}P = -\mathrm{d}P_{\mathrm{Ext}}$, i.e. we have the differential equation:

$$\frac{\mathrm{d}P}{P} = -\mathcal{N}\sigma_{\mathrm{Ext}}c_m\mathrm{d}L. \tag{2.19}$$

$\mathcal{N}\sigma_{\mathrm{Ext}}c_m$ [m^{-1}] therefore represents the proportion of extinguished power per unit length, sometimes called extinction (or absorption) coefficient of the solution.

The power P_{Tra} transmitted through the solution over a path length L [m] (which is commonly 1 cm in typical experimental implementations) is then related to the incident power by integration of the previous differential equation, i.e.

$$P_{\mathrm{Tra}} = P_{\mathrm{Inc}}\exp(-\mathcal{N}\sigma_{\mathrm{Ext}}c_m L). \tag{2.20}$$

The *transmittance*, T, which is usually defined as the ratio of transmitted over incident power [79], is then given by $T = \exp(-\mathcal{N}\sigma_{\mathrm{Ext}}c_m L)$. Usually, the most relevant quantity, and what is given as output in many UV/Vis spectrometers is the *absorbance* defined as [79]:

$$A = -\log_{10}(T) = \log_{10}\left(\frac{P_{\mathrm{Inc}}}{P_{\mathrm{Tra}}}\right) = \frac{\mathcal{N}\sigma_{\mathrm{Ext}}c_m L}{\ln(10)}. \tag{2.21}$$

Note that some definitions use a natural logarithm (ln), which avoids the factor $\ln(10) \approx 2.3$ in the expression; this depends on the particular instrument under consideration.

The *decadic molar extinction coefficient* , $\bar{e}$ [$\mathrm{m}^2\ \mathrm{mol}^{-1}$ in S.I., usually expressed in $\mathrm{cm}^{-1}\ \mathrm{M}^{-1}$] of the molecule is defined as[9]:

[9] $\bar{e}$ is sometimes quoted in units of M^{-1}. It is then implicit that the length scale is 1 cm and should be understood as $\mathrm{cm}^{-1}\ \mathrm{M}^{-1}$. To confuse the matter further, it is also sometimes referred to as the *molar extinction coefficient* (the adjective 'decadic' is then omitted). This, however, clashes with an alternative definition as $e = \bar{e}\ln(10)$, which is easier to use when the absorbance is defined with a natural logarithm. Care must therefore be taken when absorbance and molar extinction coefficient are handled without specifying the convention used.

$$\bar{e} = \frac{\mathcal{N}\sigma_{\mathrm{Ext}}}{\ln(10)}. \tag{2.22}$$

Like the extinction cross-section, it depends on the excitation wavelength. It is essentially an alternative measure of the extinction cross-section of a molecule, but from an ensemble point of view rather than the single molecule approach. The absorbance then takes the form:

$$\boxed{A = -\log_{10}(T) = c_m \bar{e} L.} \tag{2.23}$$

The absorbance is therefore proportional to both the concentration and the molar extinction coefficient (or extinction cross-section), and should in some respect be called the extinctance! In fact, for molecules, since extinction and absorption are almost equal, the molar extinction coefficient is also sometimes called the *molar absorption coefficient* or *molar absorptivity* . This last expression is called Beer's law or Beer–Lambert law.

UV/Vis spectroscopy consists in measuring the wavelength dependence of the absorbance. For a solution of known concentration, it is then possible to derive the decadic molar extinction coefficient and therefore the extinction cross-section (equal to the absorption cross-section for molecules). A useful expression to this end, which is easily derived from Eq. (2.22), is:

$$\sigma_{\mathrm{Ext}}\,[\mathrm{cm}^2] = 3.82 \times 10^{-21}\,(\bar{e}\,[\mathrm{cm}^{-1}\,\mathrm{M}^{-1}]), \tag{2.24}$$

which gives σ in cm^2 for $\bar{e}$ in $\mathrm{cm}^{-1}\ \mathrm{M}^{-1}$. Alternatively, UV/Vis spectroscopy can be used to determine the concentration of a solution of molecules (or nano-particles) of known molar extinction coefficient. This is, in fact, one of the most popular uses of UV/Vis spectroscopy.

Examples

Examples of UV–Vis absorption for a few dyes dissolved in water are shown in Fig. 2.7. The maximum of the absorption typically corresponds to the main electronic transition from S_0 to S_1 (broadened by the vibrational sub-structure). Typical values for the decadic molar extinction coefficient at the peak of absorption are in the range 5000–2×10^5 $\mathrm{cm}^{-1}\ \mathrm{M}^{-1}$ for dye molecules, and in fact close to or above 10^5 for 'good' dyes. Actually, $\sim 10^5$ is the approximate value, for example, for dyes like rhodamine 6G (RH6G), crystal violet (CV), or fluorescein, at their absorption maxima (528, 590, and 492 nm, respectively, in water). This corresponds to a maximum absorption cross-section of around $\sim 4 \times 10^{-16}\ \mathrm{cm}^2$.

From the UV/Vis spectra, extinction and therefore absorption cross-sections can also be estimated at different wavelengths. For example, at

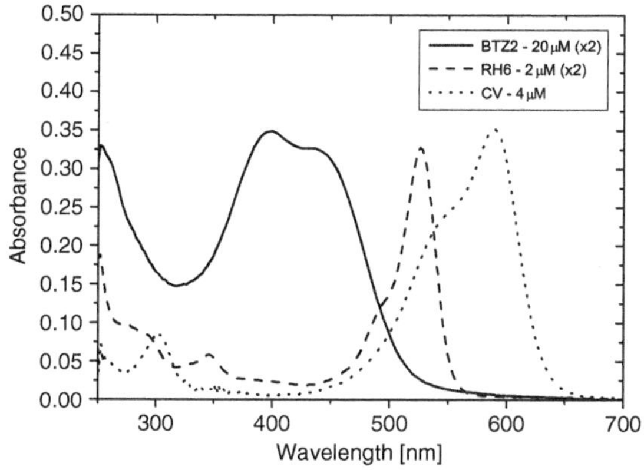

Figure 2.7. Example of UV–Vis extinction spectra for a few dyes of interest to SERS, dissolved in water: rhodamine 6G (RH6G), crystal violet (CV), and a benzotriazole dye (BTZ2 is dye #2 in Ref. [89]). From these, one can extract the decadic molar extinction coefficient, $\bar{e}$. It is around $\sim 10^5$ cm^{-1} M^{-1} at the peak of absorption for RH6G (at 528 nm) and CV (at 590 nm), but 10-times smaller for BTZ2 (at 394 nm).

the Ar^+-ion laser line of 514 nm for rhodamine 6G, we can estimate from the absorbance in Fig. 2.7 that $\sigma_{Abs}(514\,\text{nm}) \approx 2.5 \times 10^{-16}$ cm^2. This is in agreement with experimental observations [81,90] of $\approx 2 \times 10^{-16}$ cm^2.

Finally, note that the absorption spectrum (and cross-section) of molecules can vary (usually slightly) depending on the solvent in which they are dissolved.

2.3.2. Fluorescence spectroscopy

Fluorescence is a somewhat more complex process than absorption, in both its concepts and numerous implementations or applications. In fact, many books are entirely dedicated to the subject [79,91]. Accordingly, we will only provide here a brief overview of the concepts that will be needed to understand the modifications that arise at metallic surfaces, namely fluorescence quenching and surface-enhanced fluorescence (SEF). We focus first here on the basics, and will discuss another important related aspect, photo-bleaching, in the following section.

Radiative and non-radiative decay

As we have seen earlier, fluorescence is a two-step process consisting in the absorption of a photon (through an electronic excitation to the first excited state), followed by spontaneous emission (SE) (through relaxation to the electronic ground state). In general, the electron is excited to a vibrational state of S_1, from which it relaxes very quickly (within a few picoseconds) to the lowest energy state of S_1. This is why fluorescence emission is at a lower energy than excitation (see Section 2.2.4); i.e. the reason why there is a *Stokes shift*. We ignore this aspect in this simple model. The first step, absorption, has been studied in the previous section, and is generally characterized by the absorption cross-section σ_{Abs}. We therefore focus on the second step: spontaneous emission (SE).

As mentioned earlier, this process is independent of the excitation, and characterized by a decay probability per unit time, or decay rate Γ_{Rad} [s^{-1}] called the *radiative decay rate*. However, SE is in general not the only possible mechanism for electron relaxation from S_1 to S_0, which may also occur through non-radiative transitions. The non-radiative transitions from S_1 to S_0 may have several physical origins, but their overall effect can be grouped and described by a single probability per unit time of undergoing such a transition, called the *non-radiative decay rate*, Γ_{NR} [s^{-1}]. The probability (per unit time) of relaxing from S_1 to S_0 is the sum of the probabilities for each decay channel, and we can therefore define a *total decay rate* $\Gamma_{\mathrm{Tot}} = \Gamma_{\mathrm{Rad}} + \Gamma_{\mathrm{NR}}$ [s^{-1}]. $\Gamma_{\mathrm{Tot}}^{-1}$ represents the excited state *lifetime*, i.e. the average time an electron spends in S_1 before relaxation to S_0. Typical lifetimes are in the range 100 ps–10 ns. This is what would be measured in a time-resolved experiment for example. By analogy, $\Gamma_{\mathrm{Rad}}^{-1}$ is called the *radiative lifetime* but it cannot be measured directly if non-radiative processes are present (in fact only the radiative decay rate Γ_{Rad}, not the radiative lifetime, has a direct physical meaning). A given excited electron in S_1 will relax either radiatively (by SE) or non-radiatively, but obviously not both, and this is a stochastic (random) process. There is therefore *competition between the two decay channels*, and their respective importance is determined by their relative probabilities or decay rates.

Fluorescence quantum yield

The *fluorescence quantum yield*, Q [a.d.], is by definition the proportion of excited electrons that decay radiatively to the ground state, hence producing a detectable photon. It characterizes the competition between radiative and non-radiative decay, and is given by:

$$\boxed{Q = \frac{\Gamma_{\mathrm{Rad}}}{\Gamma_{\mathrm{Tot}}} = \frac{\Gamma_{\mathrm{Rad}}}{\Gamma_{\mathrm{Rad}} + \Gamma_{\mathrm{NR}}}.} \tag{2.25}$$

Note that $0 \leq Q \leq 1$.

Fluorescence cross-section

From these definitions, the power radiated as fluorescence photons, P_{Fluo} [W], is simply the proportion Q of the absorbed power P_{Abs} [W] that is re-emitted radiatively (we neglect here the change in fluorescence photon energy compared to the one of the absorbed photon), i.e.

$$P_{\mathrm{Fluo}} = QP_{\mathrm{Abs}} = Q\sigma_{\mathrm{Abs}}S_{\mathrm{Inc}}. \tag{2.26}$$

A fluorescence cross-section, σ_{Fluo} [m^2] can then be defined from this latest expression as:

$$\boxed{\sigma_{\mathrm{Fluo}} = Q\sigma_{\mathrm{Abs}}.} \tag{2.27}$$

This corresponds more precisely to the *total fluorescence cross-section*, since it accounts for every fluorescence photon, at any energy, and emitted in any direction.

It is also possible to define a differential fluorescence cross-section to account for the radiation profile. We do not describe this aspect here since a similar exercise will be carried out in detail for the case of Raman scattering.

Moreover, the fluorescence photons have a distribution of energies (or wavelengths) and this spectral distribution is called the *fluorescence spectrum*; an example was shown in Fig. 2.2. The maximum of fluorescence typically occurs at a lower energy than the absorption maximum (this is called the fluorescence Stokes shift, as mentioned before). The actual spectral shape of fluorescence is determined primarily by the vibrational sub-structure of S_0. This aspect is not included in our model here, which is based on a simple two-level (S_0 and S_1) system. More details and a refinement of this description can be found for example in Ref. [9], and references therein.

Examples

Many common dyes have a good quantum yield, either close to 1, or at least not much smaller than 1. The total fluorescence cross-section is then close to the absorption cross-section. One simple way to estimate it is therefore to measure the absorbance by UV/Vis spectroscopy, as described earlier, and derive the absorption cross-section. From the figures given earlier, the maximum fluorescence cross-section for typical dyes for excitation at their maximum of absorption is therefore of the order of 4×10^{-16} cm^2.

Once the absorption cross-section has been measured, the fluorescence cross-section can be immediately estimated if the quantum yield is known. For example, dyes such as RH6G or fluorescein have a good quantum yield (around $\sim$0.9 [92]). From the result of Section 2.3.1, the total fluorescence cross-section of rhodamine 6G at 514 nm is then $\sigma_{\mathrm{Fluo}} \approx 2 \times 10^{-16}$ cm^2. By contrast, the quantum yield for crystal violet, and therefore its fluorescence cross-section, is much lower because of a fast non-radiative decay involving a rotation of the three phenyl moieties in its molecular structure.

Finally, if the quantum yield is not known, then the total fluorescence cross-section can in general be measured by comparing the integrated fluorescence intensity to that of a known fluorescence standard. The quantum yield can then be inferred from the relative values of the absorption and fluorescence cross-sections.

A note on stimulated emission and saturation effects

It is interesting at this stage to return to a remark made earlier: stimulated emission plays no role (or a negligible role) in molecular fluorescence. We can now assert the validity of this statement more carefully. Let us first recall that stimulated emission is the exact inverse process of absorption. For a given incident excitation, the rate of absorption for a molecule in S_0, is therefore the same as the rate of stimulated emission for a molecule in S_1. The absorption rate (number of photons absorbed per unit time) is:

$$\Gamma_{\text{Abs}} = \frac{P_{\text{Abs}}}{\hbar\omega_L} = \frac{\sigma_{\text{Abs}} S_{\text{Inc}}}{\hbar\omega_L}, \tag{2.28}$$

where $\hbar\omega_L$ is photon energy in the incident beam.

For the sake of argument, let us consider a molecule with absorption cross-section of $\sigma_{\text{Abs}} = 4 \times 10^{-16}$ cm^2 (i.e. a typical value at the maximum of absorption for a dye), and an excitation density of $S_{\text{Inc}} = 10^{10}$ W m^{-2} (which is close to the maximum achievable in a typical Raman microscope, and corresponds to a laser power of $\sim$10 mW uniformly focused on a $\sim$1 μm^2 spot) at an excitation wavelength of 514 nm. The absorption rate (for a molecule in S_0) is in this case $\Gamma_{\text{Abs}} \approx 10^9$ s^{-1} and it is therefore also the stimulated emission rate for an excited molecule in S_1. From our choice of parameters, it is clear that this rate is close to the fastest achievable for typical conditions relevant to us.

It is now important to re-emphasize the fact that stimulated emission (as opposed to spontaneous emission) can only occur at the incident laser energy. In a typical absorption process, the electron is excited to a vibrational sub-level of S_1 from which it is relaxing down to the lowest energy levels of S_1 within 1–10 ps. After this relaxation process, it is no longer possible for stimulated emission to occur, because there is no longer any available electronic transition from S_1 to S_0 at the laser energy. Moreover, before relaxation, the stimulated emission rate ($\sim$10^9 s^{-1} at most) is much smaller than the energy relaxation rate ($\sim$10^{11}–10^{12} s^{-1}) and the chances of it happening are therefore negligible.

This simple argument explains why stimulated emission can be ignored in most situations for molecular fluorescence. However we have shown in the process that the absorption rate in high-power-density conditions can be as large as $\sim$10^9 s^{-1}, and therefore be comparable to the total decay rate from S_1 to S_0. This corresponds to the regime of saturation. We have assumed implicitly so far that the power is small enough to avoid saturation effects in the excited state, i.e. the excited electron has plenty of time to relax to the ground state between each excitation event, and the molecule is accordingly most of the time in state S_0. If, however, the absorption rate becomes comparable to the total decay rate, then the molecule is often in S_1, from where no further absorption may occur (in a two-level system); one then expects a saturation of the fluorescence intensity. More importantly perhaps, we have neglected in this simple model the presence of higher excited states. In the saturated regime, where the first excited state is often populated, excitation to these higher states could become non-negligible and should, in principle, be taken into account. Because of the complexity, we shall not further discuss these aspects here and assume in the rest of the book that the power densities are small enough to avoid any saturation or higher excited state effects.

2.3.3. Photo-bleaching

Photo-bleaching is a well-documented subject in dye spectroscopy, in particular because of its technical relevance in dye lasers and other fluorescent-probe applications, but is still arguably not well understood [82]. Most dyes, under relatively strong resonant excitation, tend to photo-bleach, i.e. stop fluorescing. In typical experimental conditions, where a large number of dyes is studied, this results in a decrease of the fluorescence signal with time (because of the decrease in the number of fluorescing dyes). More recently, with the advent of single molecule fluorescence, photo-bleaching has been observed for single or few molecules, and is then characterized by a step-like decrease in the fluorescence as molecules 'disappear' one at a time [83,93].

Inter-system crossing

The conventional explanation for photo-bleaching (at low or moderate laser power densities) invokes the phenomenon of *inter-system crossing*, i.e. transitions between singlet and triplet states, and is illustrated in Fig. 2.8. In simple terms, for an electron in the singlet excited state (S_1), there is a small probability of transition to the triplet state (T_1). This is what is called inter-system crossing (ISC), and the probability per unit time or decay rate Γ_{ISC} through this pathway is usually very small compared to the total decay rate Γ_{Tot} to the ground state S_0. For a molecule excited in S_1, the chances of such an ISC transition occurring, rather than relaxation to S_0 (radiative or non-radiative), are therefore very small. In fact, the probability of such an event is simply:

$$p_{\text{ISC}} = \frac{\Gamma_{\text{ISC}}}{\Gamma_{\text{Tot}}}. \tag{2.29}$$

However, provided there are sufficiently many excitation–relaxation cycles, the electron will eventually 'cross' to the triplet state T_1.

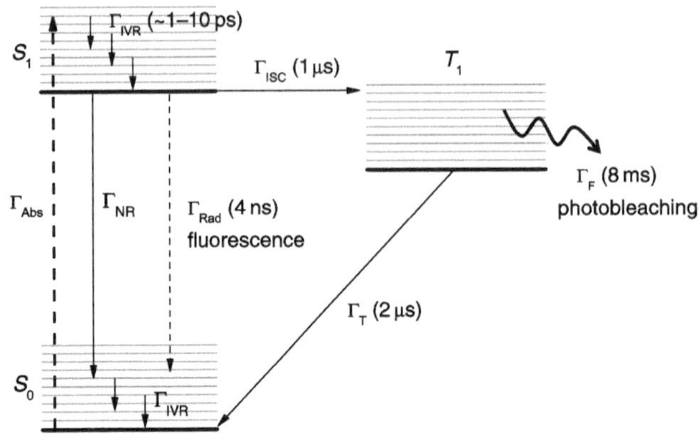

Figure 2.8. Jablonski diagram representing the fluorescence and photo-bleaching processes. Transitions are indicated by arrows, either solid (non-radiative) or dashed (radiative). As an example, the transition rates Γ are shown together with the corresponding lifetime for the special case of rhodamine 6G in water.

The lifetime in the triplet state is much longer than in the singlet state. An electron in T_1 may relax to the ground state S_0 with a decay rate Γ_T, either radiatively and the corresponding emission is called *phosphorescence*,[10] or non-radiatively. Another characteristic of the triplet state is its high chemical reactivity, especially in an oxygen environment [90,94], which can lead to the destruction of the molecule with a fragmentation rate Γ_F, and therefore to photo-bleaching. The probability of this happening for a molecule in its triplet state T_1 is therefore

$$p_F = \frac{\Gamma_F}{\Gamma_T}. \tag{2.30}$$

The photo-bleaching rate

These general considerations can now be used to obtain a simple expression for the photo-bleaching rate. Note that in real life, the situation is much more complicated, and more sophisticated scenarios involving higher excited states and several relaxation and destruction pathways have been proposed and studied [81–83]. The description given here is only aimed at covering the basic aspects.

Because the excitation/relaxation dynamics is much faster than typical timescales involving the triplet state, the result of Eq. (2.26) remains valid, but

[10] In many dyes relevant to SERS, phosphorescence (if at all present), is completely overshadowed by fluorescence and not observed in standard conditions.

will change *slowly* with time due to the decrease in the number of molecules (N). Ignoring saturation effects, the average number of excitation/relaxation cycles per unit time for one molecule is given by Γ_{Abs} (Eq. (2.28)). For each of these cycles, there is a probability p_{ISC} that the molecule crosses to the triplet state, and from there, there is a probability p_F that it photo-bleaches. Overall the probability of photo-bleaching at each cycle is therefore:

$$\phi_B = p_{\text{ISC}}\, p_F = \frac{\Gamma_{\text{ISC}}}{\Gamma_{\text{Tot}}} \frac{\Gamma_F}{\Gamma_T}. \tag{2.31}$$

This probability, ϕ_B, is called the *photo-bleaching quantum yield,* and is the parameter most often quoted to characterize the photo-stability of dyes. The total number of molecules N (and therefore the fluorescence intensity) then decreases exponentially with a *photo-bleaching rate* Γ_B [s^{-1}] as:

$$N(t) = N(t=0)\text{e}^{-\Gamma_B t} \quad \text{with} \quad \Gamma_B = \phi_B \Gamma_{\text{Abs}} = \phi_B \sigma_{\text{Abs}} \frac{S_{\text{Inc}}}{\hbar\omega_L}. \tag{2.32}$$

Hence, in the absence of saturation, *the photo-bleaching rate increases linearly with power.*

In practice, photo-bleaching remains negligible until $(\Gamma_B)^{-1}$ is of the order of the experiment timescale, and this obviously depends on the type of experiment, and the power density. Note also that in practice, the excitation is *not uniform* for typical laser beams, which can result in different Γ_B for different molecules and may lead to a non-exponential decay (i.e. the apparent decay rate decreases with time because the molecules surviving longer are those subject to a smaller power density).

Finally, at higher powers, the photo-bleaching rate should in principle level off in the saturation regime, but other processes involving excitation to higher excited states can then become important [81]. In fact, the photo-bleaching probability ϕ_B itself may sometimes depend on the power density (and the photo-bleaching rate is then no longer linear with power). These aspects reflect the fact that a simple two-level system with a triplet state is no longer a good description at higher powers.

Example

Rhodamine 6G can again be used to illustrate the photo-bleaching effect. Its lifetime is ${\Gamma_{\text{Tot}}}^{-1} \approx 4$ ns and it is also its radiative lifetime since the quantum yield is close to one [92]. Its photo-bleaching quantum yield is of the order of $\phi_B \approx 10^{-6}$ [81], which means that it can on average emit 10^6 photons before being destroyed. The inter-system crossing and radiative triplet lifetime have also been measured [81,90]: $\Gamma_{\text{ISC}}^{-1} \approx 1$ μs and $\Gamma_T^{-1} \approx 2$ μs. From these values, one can deduce a fragmentation lifetime in the triplet state of around

$\Gamma_F^{-1} \approx 8$ ms. This is however likely to depend on the environment, and in particular its oxygen content. These numbers are only yardstick estimates, but give a good idea of the different timescales in the system and are summarized in Fig. 2.8.

From these figures, at a moderate excitation power[11] of $S_{\mathrm{Inc}} \approx 2 \times 10^3$ W cm^{-2}, a photo-bleaching lifetime of 0.3 s is predicted at 528 nm (the absorption maximum). One can also make such estimates for other relevant excitation wavelengths. At 514 nm for example, with $S_{\mathrm{Inc}} \approx 2 \times 10^3$ W cm^{-2}, we obtain $\Gamma_B^{-1} \approx 0.5$ s.

A note on photo-bleaching in liquids

Under photo-bleaching conditions in solutions, another characteristic time has to be taken into account in the problem, namely: the molecule *diffusion time*. The latter may play a substantial role depending on the characteristics of the experiment (integration time, scattering volume, etc.).

As a way of example, the characteristic scattering volume V that is both excited and collected by a water immersion objective with $\times 100$ magnification (NA $= 1$) is $V \sim 1$ µm $\times$ 1 µm $\times$ 10 µm $\sim 10^{-11}$ cm^3. These small collection volumes[12] in optical microscopes allow us to see a few molecules only at small (but measurable) analyte concentrations. For example, a 10 nM solution of an analyte contains typically $\sim$60 molecules in a characteristic scattering volume like this one.

Simple diffusion theory implies that a given molecule will diffuse by a distance $L = 1$ µm over a timescale of the order of $\tau_D \sim L^2/D$ where D [m^2 s^{-1}] is the diffusion coefficient. τ_D is the characteristic timescale over which molecules can diffuse in and out of observation. For a typical value of $D \sim 10^{-5}$ cm^2/sec, we have $\tau_D \sim 1$ ms. Larger values would be obtained for a larger scattering volume. The basic dynamics, therefore, will be dominated either by photo-bleaching or diffusion depending on whether $(\Gamma_B)^{-1} \ll \tau_D$ or $(\Gamma_B)^{-1} \gg \tau_D$. In the latter case, photo-bleaching should not be observable.

2.4. PHENOMENOLOGICAL APPROACH TO RAMAN SCATTERING

Like many other physical processes involving light or atoms (or both), optical scattering should in principle be described by a quantum theory. For Raman scattering, one must first apply quantum theory to describe the molecular energy levels (electronic and vibrational). The interaction with

---

[11] This corresponds for example to the power density at the focal point of a Gaussian beam of 3.5 mW, focused with a beam waist of 10 µm.

[12] This value depends on the exact *confocality* of the collection optics. A value of $V \sim 10^{-11}$ cm^3, however, serves as a yardstick estimation.

photons is then described within quantum electrodynamics by the appropriate interaction Hamiltonian (for example approximated by the electric dipole interaction Hamiltonian). This approach is however far from straightforward, and moreover difficult to generalize to the SERS case. For our purpose here, the main merit of the quantum approach is that it can be used to justify and/or define the range of validity of simpler classical models. With this in mind, we will restrict ourselves in this section to a classical and phenomenological description of Raman scattering, which is not only preferable as a pedagogical tool, but also crucial for the generalization to the case of Raman scattering at metallic surfaces, i.e. to SERS. A more complete classical and semi-classical[13] treatment is then discussed in subsequent sections.

2.4.1. Dipolar emission in vacuum

The oscillating dipole

Within classical electromagnetic (EM) theory, emission (radiation) of light from a localized source can be described in terms of its multipolar components [95,96]. The simplest, and in many cases (especially for small sources) dominant contribution is that of the *electric dipole*. Many electromagnetic radiation processes involving molecules can therefore be described classically by the radiation emitted by an oscillating electric dipole.

Detailed studies of electric dipole radiation can be found in many textbooks [95,96] and we highlight here some relevant results only. An oscillating dipole (at angular frequency ω) can be viewed as two charges $+q$ and $-q$ of opposite sign, harmonically oscillating back and forth with a very small amplitude $x = a\cos(\omega t)$ along a given direction $\mathbf{e}_p$. The dipole is therefore highly localized, in fact point-like in the mathematical treatment, and always neutral in terms of its total charge. It is represented by a vector $\mathbf{p}(t) = qx(t)\mathbf{e}_p$ and its amplitude qa [C m] is called the dipole moment. Charges and currents (moving charges) are sources of electromagnetic fields, and such an oscillating electric dipole behaves as a localized source, i.e. it is radiating energy in the form of an electromagnetic field oscillating at angular frequency ω or, equivalently, photons at energy $\hbar\omega$.

Complex notation for harmonic oscillations

Physical quantities that exhibit *harmonic oscillations* are often represented using *complex notation*, which is defined for example for $\mathbf{p}(t)$ as:

$$\mathbf{p}(t) = \mathrm{Re}\left(\underline{\mathbf{p}} \times \mathrm{e}^{-\mathrm{i}\omega t}\right), \tag{2.33}$$

---

[13] In a semi-classical description, the atomic energy levels are described by quantum mechanics, but the interaction with light is described classically.

where $\underline{\mathbf{p}} = qa\mathbf{e}_p$ is the complex notation for the dipole moment. The latter does not depend on time, and the harmonic time dependence of the physical dipole moment is implicitly contained in the definition of the complex notation as given in the expression above. Complex notations will be used extensively in the following and are discussed in more detail in Section C.1.2. For simplicity in the notation, the under-bar in $\underline{\mathbf{p}}$ will often be omitted, and we will simply write $\mathbf{p}$ (or $\mathbf{p}(\omega)$) for the complex notation, but explicitly $\mathbf{p}(t)$ when referring to the real physical quantity. We will keep the under-bar in this section only for clarity.

Dipolar radiation

The electromagnetic field radiated by an oscillating dipole can be derived from Maxwell's equations [95,96]. The most important aspect to us here is the radiation or far-field properties, in particular the differential radiated power in the far field (per unit solid angle), $\mathrm{d}P_{\mathrm{Rad}}/\mathrm{d}\Omega$ [W sr^{-1}], and the total radiated power, P_{Rad} [W]. The latter represents the total power that can be collected with a detector spanning all directions in space (such as an integrating sphere). The former represents the radiation profile, i.e. how this power is distributed across all directions. These directions are represented as before by $\Omega = (\theta, \phi)$ using spherical coordinates, where $0 \leq \theta \leq \pi$ is the colatitude and $0 \leq \phi \leq 2\pi$ is the longitude (see Section H.3.2). For a dipole $\underline{\mathbf{p}} = \underline{p}\mathbf{e}_z$ (aligned along $\mathbf{e}_z$ for simplicity) located in a dielectric medium of refractive index $n_M = 1$ (i.e. vacuum, or a good approximation for most gases), we have [95]:

$$\frac{\mathrm{d}P_{\mathrm{Rad}}}{\mathrm{d}\Omega}(\Omega) = \frac{\omega^4}{32\pi^2\epsilon_0 c^3}|\underline{\mathbf{p}}|^2 \sin^2\theta. \tag{2.34}$$

The $\sin^2\theta$ dependence leads to a two-lobe radiation pattern characteristic of dipole emission (see for example Fig. 4.3). Note that a dipole (in an infinite dielectric or free-space) does not radiate energy along its axis ($\theta = 0$ or π in this case).

The total radiated power can be obtained by integration of Eq. (2.34) over the full (4π) solid angle:

$$P_{\mathrm{Rad}} = \frac{\omega^4|\underline{\mathbf{p}}|^2}{12\pi\epsilon_0 c^3}. \tag{2.35}$$

One may wonder where this radiated electromagnetic energy comes from. In fact, we have here assumed that the dipole oscillation amplitude is fixed, and something must therefore drive and maintain this oscillation. The energy is therefore radiated at the expense of a source driving and maintaining the dipole oscillation. If nothing drove the oscillations, their amplitude would eventually decrease to zero. This is indeed the case in many electromagnetic problems including moving charges and it is called *radiation damping* [96].

Polarization of the dipole radiation

Denoting $\mathbf{e}_\Omega$ the unit vector for the detection direction, the electric field *polarization* of the dipole radiation is along $(\mathbf{e}_\Omega \times \mathbf{e}_p) \times \mathbf{e}_\Omega$ [96] ; i.e. it is perpendicular to $\mathbf{e}_\Omega$ and in the plane defined by both $\mathbf{e}_\Omega$ and $\mathbf{e}_p$. Therefore, if the polarization of the radiated electric field is analyzed by detecting with a polarizer aligned along a direction defined by a unit vector $\mathbf{e}_P \perp \mathbf{e}_\Omega$ (*perpendicular*[14] to the direction of detection), the polarized differential radiated power is simply given by:

$$\frac{dP_{\mathrm{Rad}}^{\mathrm{Pol}}}{d\Omega}(\Omega, \mathbf{e}_P) = \frac{\omega^4}{32\pi^2\epsilon_0 c^3} |\underline{\mathbf{p}} \cdot \mathbf{e}_P|^2 \sin^2\theta. \tag{2.36}$$

2.4.2. The concepts of polarizability and induced dipole

A scattering process consists of an incident beam of photons giving rise to scattered radiation. We have so far described how the scattered radiation can be classically described as emission from a dipole. In order to complete the picture, it now remains to describe how the incident beam gives rise to this oscillating dipole.

This is described classically by invoking the notions of *induced dipole and polarizability.* These are introduced here largely from a phenomenological point of view, knowing that these concepts could be justified by a more detailed (quantum) treatment.

The static polarizability

It is useful at this stage to consider first the case of a molecule in a *static* (DC) electric field. The structure of a molecule is characterized by the positions of its constituents (atoms and electrons) that are compatible with the multiple interactions among them. It can be defined as the positions that the atoms occupy in space and the corresponding electronic wave-functions (orbitals), in the absence of perturbations, such as external fields, i.e. *at equilibrium.*

In the presence of an external perturbation these positions and probability densities change, which may result in a modification of the molecule properties. For example, in the presence of an external constant electric field,

[14] Most experimental implementations use polarizers perpendicular to the direction of detection. If not, the expression is more complicated than that given in Eq. (2.36).

charge redistribution in general results in a change of the dipole moment of the molecule. This *change with respect to equilibrium* is called the static *induced dipole moment* $\mathbf{p}_{\mathrm{DC}}$ [C m]. It obviously depends on the direction and magnitude of the applied electric field $\mathbf{E}$. In most cases, these changes are small with respect to equilibrium and can therefore be described within the *linear approximation*, i.e.:

$$\mathbf{p}_{\mathrm{DC}} = \hat{\alpha}_{\mathrm{DC}} \cdot \mathbf{E}, \tag{2.37}$$

where $\hat{\alpha}_{\mathrm{DC}}$ [ϵ_0 m^3] is called the *static polarizability tensor* (for a constant applied field). We will come back to the tensorial nature of the polarizability at a later stage. For the moment, $\hat{\alpha}_{\mathrm{DC}}$ can be viewed as the proportionality factor between $\mathbf{p}_{\mathrm{DC}}$ and $\mathbf{E}$. It is a second-rank tensor in the most general case because this provides the most general linear relation between two vectors. This accounts for the fact that $\mathbf{p}_{\mathrm{DC}}$ and $\mathbf{E}$ may not be aligned along the same direction, in general.

It should also be noted that some molecules have an intrinsic or permanent static dipole moment, even in the absence of an applied field, which comes directly from the symmetry of the electronic wave-functions in the ground state at equilibrium; water (H_2O) being a notable example. The dipole $\mathbf{p}_{\mathrm{DC}}$ in Eq. (2.37) is not related to this *permanent* dipole, but describes the *change* in dipole moment with respect to equilibrium, it is an *induced* dipole. In particular, $\mathbf{p}_{\mathrm{DC}}$ can represent an induced dipole in a molecule that has no permanent dipole moment at all by itself.

The polarizability provides a measure of the 'responsiveness' of the electrons in a molecule to the presence of an external electric field. As a rule of thumb, the molecules with a large number of 'free' electrons are therefore the most polarizable ones. These include, for example, molecules containing atoms with lone pairs or delocalized electrons (in π-orbitals), in particular those with one or more 'ring-like' structures. Organic molecules such as dyes are typically among this group of 'highly polarizable molecules'. These considerations will remain true for the optical polarizabilities introduced later.

A note on polarizability units

The S.I. unit for the polarizability tensor components α is [ϵ_0 m^3]. However, they are sometimes (if not often) expressed in Gaussian units, therefore adding a pre-factor $4\pi\epsilon_0$. This leaves α in units of [m^3] or, as it is more often found, in units of [Å^3]. Even worse, Heavyside–Lorentz units are sometimes used, leading to the same units as for the Gaussian system, up to a factor of 4π. This is the perfect recipe for inconsistencies: the factor of 4π leaves no trace since it has no unit! In the absence of clear definitions, it is impossible to know if it is there or not, and this is the source of numerous confusions, if not plain errors, regarding the magnitude of polarizabilities. We will therefore stick to S.I. units as much as possible; α being in units of $[\epsilon_0 \text{ m}^3]$. It may be viewed as the product of ϵ_0 times a volume, which is in fact of the order of the molecular volume.

2.4.3. The linear optical polarizability

Let us now focus on optical processes, starting with the simpler case of elastic (Rayleigh) scattering, whereby scattering occurs at the same frequency as the incident radiation. The response to the incident radiation can then be described, by analogy with the static case, within linear response function theory, which we sketch only briefly below.

Definition of the linear optical polarizability

The incident electric field $\mathbf{E}(t)$ is oscillating at angular frequency ω and described in complex notation by $\underline{\mathbf{E}}(\omega)$. The induced dipole is similarly described by $\mathbf{p}_L(t)$ (real) or $\underline{\mathbf{p}}_L(\omega)$ (complex). One could, by analogy with the static polarizability, assume that $\mathbf{p}_L(t) = \hat{\alpha}_{\text{DC}} \cdot \mathbf{E}(t)$. This means however that the molecular response ($\mathbf{p}_L(t)$) occurs instantaneously with the excitation ($\mathbf{E}(t)$). Such an assumption (called temporal locality) is unfortunately in many cases too restrictive to be sufficiently general for applications. Whilst retaining the assumption of linearity, we can make the less stringent assumption that the response at time t depends linearly on the excitation at all times in the past $t' \leq t$. This can be written as:

$$\mathbf{p}_L(t) = \int_0^{\infty} \hat{\alpha}(\tau) \cdot \mathbf{E}(t-\tau)\mathrm{d}\tau. \tag{2.38}$$

The function $\hat{\alpha}(\tau)$ is no longer directly related to $\hat{\alpha}_{\text{DC}}$ and should be viewed as a linear response function (in fact, this function is never explicitly used as we shall see). Its dependence on $\tau = t - t'$ (rather than t') is a consequence of time invariance. The above expression is too complicated to be useful in practice, but does simplify by carrying out a Fourier transform (see Section C.1.2). We do not necessarily need to explicitly do the full Fourier transform here, since it simplifies greatly for monochromatic excitation when considering complex notations. We then obtain the fundamental definition of the *linear polarizability tensor* $\hat{\alpha}_L(\omega)$ as:

$$\boxed{\underline{\mathbf{p}}_L(\omega) = \hat{\alpha}_L(\omega) \cdot \underline{\mathbf{E}}(\omega).} \tag{2.39}$$

$\underline{\mathbf{p}}_L(\omega)$ is the induced dipole (in complex notation), oscillating at the *same frequency* ω as the applied field $\underline{\mathbf{E}}(\omega)$ (also in complex notation). $\hat{\alpha}_L(\omega)$ is a *linear response function* depending on ω, and related to $\hat{\alpha}(\tau)$ through:

$$\hat{\alpha}_L(\omega) = \int_0^{\infty} \hat{\alpha}(\tau)e^{i\omega\tau}d\tau. \tag{2.40}$$

It fully characterizes the linear optical response of the molecule to an incident electric field and is called the *linear optical polarizability tensor* or *Rayleigh polarizability tensor*. We will exclusively use it for monochromatic harmonic excitation, where Eq. (2.39) applies directly, but it can also describe the response to more complex excitations by applying Eq. (2.39) to the Fourier components.

Note that $\hat{\alpha}_L(\omega)$ was introduced here *phenomenologically*. It is viewed here as a parameter of the problem, characterizing the optical response of the molecule. It is however possible to try and calculate it from a microscopic description of the molecule. One of the simplest models of this is the Lorentz model, briefly discussed in Appendix D.

Using the definition of the Rayleigh polarizability, Rayleigh scattering by a molecule can now be fully described classically: the incident field $\underline{\mathbf{E}}$ (at ω_L) creates an induced dipole (according to Eq. (2.39)) oscillating at the same frequency. This oscillating dipole then radiates (at ω_L) as described in Section 2.4.1. This is the classical description of the Rayleigh-scattered radiation.

Remarks on linear response function theory

Linear response function theory plays an important role in many areas of physics, and here we have only sketched its main features. A similar treatment applies to other forms of matter like crystals. In the case of crystals, the concept of induced polarization can be treated using macroscopic variables in classical electrodynamics [96,97], and the equivalent of the induced molecular dipole becomes the macroscopic polarization of the medium. Except for the so-called *local field correction* [96–98] the total induced macroscopic polarization can be thought of as coming from the sum of all microscopic polarizations (at an atomic level) [75]. The microscopic dipole $\mathbf{p}$ is replaced at the macroscopic level by the concept of dipole moment per unit volume, called electric polarization $\mathbf{P}$ [C/m^2]. This aspect, along with other aspects of response function theory, is further discussed in Appendices C and D.

Non-linear effects

Equation (2.39) is valid within the *linear approximation*, i.e. as long as the applied electric field is a small perturbation from the typical electric fields felt by electrons and atoms in molecules (which is of the order of $\sim$3 10^{10} V/m

[99]); the breakdown of this condition gives rise to additional *non-linear* effects and their associated non-linear spectroscopies. We shall only mention some non-linear optical effects (like hyper-Raman) in the context of SERS later, but non-linear optical effects lie in general beyond the scope of the book and will not be treated otherwise.

We also note that Raman spectroscopy may be considered as a form of non-linear optical process, and it is sometimes grouped in a family of non-linear optical effects in scientific reports. This distinction is to a large degree only semantic, and adds more confusion than anything else for our purpose. We will return to this later.

2.4.4. The Raman polarizability

It is possible to phenomenologically generalize the above argument to the case of inelastic scattering, and in particular to Raman scattering. Let us consider Stokes Raman scattering from a specific vibrational mode of energy $\hbar\omega_v$. The scattered radiation is at a frequency $\omega_R = \omega_L - \omega_v$, different from the incident field frequency ω_L. The induced dipole, called the *Raman dipole*, must therefore oscillate at the radiated frequency ω_R and is denoted $\underline{\mathbf{p}}_R(\omega_R)$ in complex notation. We can then, by analogy with Rayleigh scattering, define the *Raman polarizability tensor* $\hat{\alpha}_R(\omega_L, \omega_v)$ for a specific vibrational mode by the relation:

$$\boxed{\underline{\mathbf{p}}_R(\omega_R) = \hat{\alpha}_R(\omega_L, \omega_v) \cdot \underline{\mathbf{E}}(\omega_L).} \tag{2.41}$$

Note that $\mathbf{p}_R(t) = \mathrm{Re}(\underline{\mathbf{p}}_R(\omega_R)\exp(-\mathrm{i}\omega_R t))$ oscillates at ω_R, while $\mathbf{E}(t) = \mathrm{Re}(\underline{\mathbf{E}}(\omega_L)\exp(-\mathrm{i}\omega_L t))$ oscillates at ω_L. However, $\hat{\alpha}_R(\omega_L, \omega_v)$ is a proportionality factor (tensor) that does not depend on time (it should not be confused with the complex notation of a real physical quantity). Moreover, because of the change in oscillating frequency, $\hat{\alpha}_R(\omega_L, \omega_v)$ cannot be directly considered as a linear response function, as was the case for $\hat{\alpha}_L(\omega_L)$.

Here the factor $\hat{\alpha}_R(\omega_L, \omega_v)$ is fully phenomenological. We have not even justified its validity (as we did for the Rayleigh polarizability by invoking the linear approximation). We have not, in addition, linked it either to the actual properties of the molecule or the nature of the vibrational mode. These tasks will be tackled in the following sections, but are not essential to a basic understanding of the Raman (and SERS) effect.

As for Rayleigh scattering, if $\hat{\alpha}_R(\omega_L, \omega_v)$ is specified for a given vibrational mode and excitation frequency ω_L, then the Raman scattering process can be fully described classically as the induction of a Raman dipole whose radiation produces the Raman scattered light. This description is, in fact, the starting point for most generalizations of the effect to more complex situations, including SERS.

2.4.5. The local field correction

In the previous sub-sections, we have assumed that the molecule was in vacuum or in a gas state (with refractive index $n_M = 1$). Before making the connection between the phenomenological description in terms of a Raman polarizability and the experimental characterization in terms of cross-section, it is necessary for completeness to consider the case of Raman scattering from molecules in liquids. The case of solids (crystals) requires a specific treatment and will be ignored here.

Raman scattering in liquids

If the sample under consideration is in a condensed state (the case of a liquid), at least two effects may affect the Raman scattering process. The first is the possible interaction amongst molecules (either with the solvent in a diluted solution or with themselves at high concentrations). This may affect the equilibrium structure of the molecule and its internal vibrations, and therefore, the Raman polarizability of some modes. Unfortunately, these effects are difficult to predict and can typically be detected only by careful experimental investigations. One should nevertheless bear in mind that the Raman polarizability might depend on the solvent the molecules are dissolved in.

The second factor that can modify the result as presented up to here, is the effect of the optical properties of the liquid. Liquid solutions are typically dielectric (with refractive index $n_M > 1$) and therefore affect the electromagnetic field. This is already evident in relating the incident power to the electric field amplitude as in Eq. (2.5). For our purpose here, it results in at least two additional effects: a modification of the electromagnetic field felt by the molecule, and a modification of the power radiated by the Raman dipole.

Microscopic and macroscopic fields

As explained in Appendix C, Maxwell's equations for media, on which most of our classical EM treatment is based, relate to the *macroscopic electromagnetic fields*, i.e. average fields over relatively small volumes. At the molecular (or atomic) level, the *microscopic fields* are different (and usually much more complicated), than these average fields. These microscopic fields are those that are 'felt' by the molecule and should therefore be used, for example, to calculate the Raman dipole in Eq. (2.41).

The difference between macroscopic and microscopic fields is often referred to as the *local field correction* (local field referring in this context to the microscopic field). It is a well-understood effect in the framework of the optical properties of dielectric media; a correction factor for local fields must be applied to the Raman and every other optical polarizability. This is discussed further in Section C.3.1. For our purpose here, the microscopic electric field is related to the macroscopic one, in the case of a dielectric liquid of refractive index $n_M = \sqrt{\epsilon_M}$ (implicitly at the laser wavelength), by[15]:

$$\mathbf{E}_{\text{Micro}} = (L_M)^{1/4}\mathbf{E}_{\text{Macro}}, \tag{2.42}$$

with

$$\boxed{(L_M)^{1/4} = \frac{n_M^2 + 2}{3} = \frac{\epsilon_M + 2}{3}.} \tag{2.43}$$

As a result, Eq. (2.41) for the Raman dipole can be expressed in terms of the macroscopic laser electric field as:

$$\boxed{\underline{\mathbf{p}}_R(\omega_R) = (L_M)^{1/4}\hat{\alpha}_R(\omega_L, \omega_v) \cdot \underline{\mathbf{E}}(\omega_L).} \tag{2.44}$$

Note that in vacuum or for gases, for which $n_M = 1$, microscopic and macroscopic fields are identical ($L_M = 1$) and this expression simply reduces to the one given previously. Note also that the same correction should be applied to the expressions for the linear optical polarizability and the static polarizability (but ϵ_M is then the static dielectric constant).

Modified dipolar emission

Another (arguably subtler) effect of this local field correction is the modification of the power radiated by a dipole (Eqs (2.34) – (2.35)) when placed in a liquid of refractive index n_M.

The first (simplest) correction derives directly from the macroscopic theory of dipolar emission. One can show that if $n_M \neq 1$, in a similar fashion as for Eq. (2.5) for the power density, an additional factor n_M must be added to Eqs (2.34) – (2.35) for dipole radiation. This is not a local field correction, but simply a consequence of Maxwell's equations for media.

In addition, the field emitted by the dipole is 'reflected' by the dielectric environment onto the dipole and affects the emission process. This phenomenon is not simple, and is in some way related to the SERS enhancement effect itself. It is described in more detail in Section 4.4, and we provide here the final result without further justification: the power radiated by a dipole in a

[15] The 'strange' exponent in the definition of L_M will become clearer later: the fourth power of this factor appears in the expression of the Raman cross-section. It moreover ensures consistency of notations with Ref. [86].

liquid of refractive index n_M is modified by a factor $\sqrt{L_M}$ (where L_M was defined above in Eq. (2.43)), as a result of the local field correction.

Combining these corrections, the differential and total power radiated by a dipole in a liquid of refractive index n_M (at the emission wavelength) are the following generalization of Eqs (2.34) and (2.35):

$$\boxed{\frac{\mathrm{d}P_{\mathrm{Rad}}}{\mathrm{d}\Omega}(\Omega) = \frac{\omega^4 n_M \sqrt{L_M}}{32\pi^2\epsilon_0 c^3}|\underline{\mathbf{p}}|^2 \sin^2\theta,} \tag{2.45}$$

and

$$\boxed{P_{\mathrm{Rad}} = \frac{\omega^4 n_M \sqrt{L_M}|\underline{\mathbf{p}}|^2}{12\pi\epsilon_0 c^3}.} \tag{2.46}$$

Importance of the local field corrections

The factor L_M can have measurable consequences in many standard liquids: for example, $L_M = 2.5$ in water ($n_M = 1.33$) and $L_M = 4.0$ in a typical organic solvent ($n_M = 1.5$). Its importance is rarely emphasized in Raman spectroscopy, mostly because it can be 'absorbed' into the definition of the Raman cross-section (see later), which is what is ultimately measured experimentally. Nevertheless, it does justify the fact that the Raman cross-section for the same molecule is typically larger in liquid phase than in gas phase [86]. It is also of paramount importance when comparing experimental cross-section with theoretical values obtained from quantum chemistry calculations (e.g. DFT): such calculations are usually based on isolated molecules, and the local field correction factor must be applied to model the molecule in a liquid phase (see Appendix A for further details).

2.4.6. Polarizabilities and scattering cross-sections

In the preceding sub-sections, we have laid out the basic ingredients to describe optical scattering in classical terms: the creation of an induced dipole, whose radiation forms the scattering signal. We now put these together to relate the properties of the scattered signal to the polarizability of the molecule. For generality, we consider the case of a molecule in a liquid with refractive index n_M, which we assume to be the same at the excitation and Raman frequency. We will specifically consider the case of Raman scattering, but that of Rayleigh scattering is deduced simply by taking $\omega_R = \omega_L$ and replacing $\hat{\alpha}_R(\omega_L, \omega_v)$ by $\hat{\alpha}_L(\omega_L)$. In fact, we will denote in short $\hat{\alpha} = \hat{\alpha}_R(\omega_L, \omega_v)$ in this section for simplicity, and the induced dipole is therefore written as: $\mathbf{p} = (L_M)^{1/4}\hat{\alpha} \cdot \mathbf{E}$ (we drop the under-bar for complex notations, which is from now on implicitly assumed).

The tensorial nature of the polarizability

As mentioned briefly earlier, $\hat{\alpha}$ is a second-rank tensor, in order to provide the most general linear relation between two vectors, here **p** and **E**. In Cartesian coordinates, $\hat{\alpha}$ can therefore be represented by a 3×3 matrix and we have:

$$\begin{pmatrix} p_x \\ p_y \\ p_z \end{pmatrix} = (L_M)^{1/4} \begin{pmatrix} \alpha_{xx} & \alpha_{xy} & \alpha_{xz} \\ \alpha_{yx} & \alpha_{yy} & \alpha_{yz} \\ \alpha_{zx} & \alpha_{zy} & \alpha_{zz} \end{pmatrix} \begin{pmatrix} E_x \\ E_y \\ E_z \end{pmatrix}. \tag{2.47}$$

Note that except in special circumstances[16], the Raman polarizability tensor is real and symmetric; i.e. $\alpha_{ij} = \alpha_{ji}$.

The tensorial nature of $\hat{\alpha}$ accounts for the fact that **p** may not be, in general, along the same direction as **E**. This has a direct connection with the geometrical structure of the molecule. If **E** is applied along directions of high symmetry, **p** will in general be along the same direction and the polarizability reduces effectively to a scalar magnitude. It is in general possible to find a new system of axes (x', y', z') where $\hat{\alpha}$ is diagonal[17], i.e.

$$\begin{pmatrix} p_{x'} \\ p_{y'} \\ p_{z'} \end{pmatrix} = (L_M)^{1/4} \begin{pmatrix} \alpha_{x'x'} & 0 & 0 \\ 0 & \alpha_{y'y'} & 0 \\ 0 & 0 & \alpha_{z'z'} \end{pmatrix} \begin{pmatrix} E_{x'} \\ E_{y'} \\ E_{z'} \end{pmatrix}. \tag{2.48}$$

The system of axes where $\hat{\alpha}$ is diagonal is called the system of *principal axes* . These axes are *attached to the molecule*, i.e. they are fixed with respect to the equilibrium molecular structure. Polarizability tensors are always given (or calculated) in a coordinate frame attached to the molecule, ideally the principal axes, or other natural axes of the molecule. The induced dipole therefore depends not only on the intrinsic polarizability tensor of the molecule, but also on the *orientation of the molecule* with respect to the incident field. In order to make calculations in practice, either the electric field must be 'rotated' and expressed within the molecular coordinate system, or the polarizability tensor must be 'rotated' and expressed in the fixed laboratory coordinate frame[18]. We will focus in the following on three

[16] This may not be the case for resonant Raman scattering. Moreover, the off-diagonal coefficients in the tensor (2.47) may be complex conjugate magnitudes, i.e. $\alpha_{ij} = \alpha^*_{ji}$, in the presence of optical activity or magnetic fields.

[17] This is always possible for a real symmetric tensor.

[18] A rank-2 tensor is an object (a *physical* object), characterizing the transformation of a vector into another vector. A matrix is only a *mathematical* entity that may be used to do calculations of the result of such a transformation. More precisely, for a given coordinate frame (F), the tensor can be expressed in the form of a *matrix* M. Accordingly, a tensor is characterized by both: a matrix M, *and* the frame F in which the tensor is expressed.

important special cases: a molecule with an isotropic tensor, a fixed molecule, and a randomly-oriented molecule.

Isotropic polarizability tensor

A tensor is said to be *isotropic* if it does not depend on the orientation of the molecule, or equivalently if $\hat{\alpha}$ is proportional to the identity tensor: $\hat{\alpha} = \alpha\hat{1}$, where α is a scalar. The case of a molecule with an isotropic polarizability tensor may not be that common (except for a few molecules with tetrahedral symmetry like CCl_4) but it is a good pedagogical example.

Let us consider an incident beam characterized by its electric field $\mathbf{E}_{\text{Inc}}$ impinging on a molecule with an isotropic Raman polarizability tensor. The excitation, at frequency ω_L, induces a Raman dipole $\mathbf{p}_R = (L_M)^{1/4}\alpha\mathbf{E}_{\text{Inc}}$, at frequency ω_R (which is in this case parallel to $\mathbf{E}_{\text{Inc}}$ and independent of the molecular orientation).

The radiation profile and radiated power of this induced dipole are given by the formulas of Section 2.4.1 and their generalization to liquids in Section 2.4.5. In particular, the differential scattered power in a 90°- or back- or forward-scattering configuration is, from Eq. (2.45):

$$\frac{\mathrm{d}P_{\text{Rad}}}{\mathrm{d}\Omega}(90°) = \frac{\omega_R^4 n_M}{32\pi^2\epsilon_0 c^3} L_M |\alpha|^2 |\mathbf{E}_{\text{Inc}}|^2. \tag{2.49}$$

Moreover, if the scattered light is analyzed into parallel and perpendicular polarizations, as described in Section 2.2.7, then the signal for perpendicular is zero (this is derived from Eq. (2.36) and the fact that here $\mathbf{p}$ is along $\mathbf{E}_{\text{Inc}}$). The Raman depolarization ratio is therefore $\rho_R = 0$.

Moreover, using Eq. (2.5) to relate $\mathbf{E}_{\text{Inc}}$ to the incident power density S_{Inc}, the differential Raman cross-section (defined in Eq. (2.12)) can therefore be written as:

$$\frac{\mathrm{d}\sigma_R}{\mathrm{d}\Omega} = \left(\frac{\mathrm{d}P_{\text{Rad}}}{\mathrm{d}\Omega}(90°)\right)/S_{\text{Inc}} = \frac{\omega_R^4}{16\pi^2(\epsilon_0)^2 c^4} L_M |\alpha|^2. \tag{2.50}$$

This expression relates the phenomenological description of Raman scattering at the molecular level (Raman polarizability) to the observable macroscopic quantity (differential Raman cross-section) in the special case of an isotropic tensor.

General case: Fixed orientation

The previous argument can be generalized to any form of the polarizability tensor. If we consider a molecule with a fixed orientation, the induced Raman dipole is $\mathbf{p} = (L_M)^{1/4}\hat{\alpha} \cdot \mathbf{E}_{\text{Inc}}$. Let us denote $\mathbf{e}_{\text{Inc}}$ the unit vector along the incident polarization $\mathbf{E}_{\text{Inc}}$. The differential scattered power for polarized detection for 90°- or back- or forward-scattering detection, with a polarizer along $\mathbf{e}_P$ (i.e. perpendicular to the detection direction), is therefore from Eq. (2.36) (generalized to $n_M > 1$):

$$\frac{\mathrm{d}P_{\text{Rad}}}{\mathrm{d}\Omega}(90^\circ) = \frac{\omega_R^4 n_M}{32\pi^2\epsilon_0 c^3} L_M |\mathbf{e}_P \cdot \hat{\alpha} \cdot \mathbf{e}_{\text{Inc}}|^2 |\mathbf{E}_{\text{Inc}}|^2. \tag{2.51}$$

If we were able to measure the scattered radiation of this *fixed* single molecule, we would then be able to measure individual components of the tensor in Eq. (2.47) (expressed in the laboratory coordinates). For example, with incident light coming from z and polarized along x, the signal scattered along y and measured with a polarizer along x (i.e. 'parallel' with respect to the incident polarization) will be proportional to $|\alpha_{xx}|^2$. Likewise, the scattered signal along y and polarized along z (i.e. 'perpendicular' to the incident polarization) is proportional to $|\alpha_{xz}|^2$, and so on.

This type of measurement is in fact possible in many forms of scattering (including Raman) in *solids* [100], where the sample preserves order and orientation over macroscopic distances due to the fact that it is a crystal. This is typically no longer possible for molecules because (i) the molecules are usually free to move, and (ii) the scattered signal of a single molecule is by far too small to be detected. These two constraints may be removed under SERS conditions, but other effects (in particular the local field polarization) must then be taken into account instead. We shall come back to the single molecule SERS problem in Chapter 8.

General case: Orientational averaging

The most common situation in *molecular* spectroscopy is the measurement of an ensemble of *many randomly-oriented molecules*. This implies that we normally measure an average signal over random orientations of the molecule (and therefore of the polarizability tensor). *This averaging was in fact an essential part of the definition of the absolute differential Raman cross-section* (see Section 2.2.7). The orientational averaging is therefore a necessary step to relate the cross-section (which is a scalar) to the Raman polarizability tensor.

Let us consider an incident excitation along $\mathbf{e}_x$, and polarized along $\mathbf{e}_{\text{Inc}} = \mathbf{e}_z$. The differential scattered power per molecule for 90° or back- or forward-scattering detection, with polarized detection along $\mathbf{e}_z$ (i.e. parallel to incident polarization), for randomly-oriented molecules is (from Eq. (2.51)):

$$\frac{\mathrm{d}P^{\parallel}_{\text{Rad}}}{\mathrm{d}\Omega}(90^\circ) = \frac{\omega_R^4 n_M}{32\pi^2\epsilon_0 c^3} L_M \langle|\mathbf{e}_z \cdot \hat{\alpha} \cdot \mathbf{e}_z|^2\rangle |\mathbf{E}_{\text{Inc}}|^2, \tag{2.52}$$

where $\langle \dots \rangle$ denotes the averaging over random orientation of the molecules. As for the isotropic tensor case, we can use this expression to deduce the absolute differential Raman cross-section for parallel polarized detection:

$$\frac{\mathrm{d}\sigma_R^{\parallel}}{\mathrm{d}\Omega} = \frac{\omega_R^4}{16\pi^2(\epsilon_0)^2c^4} L_M \langle |\mathbf{e}_z \cdot \hat{\alpha} \cdot \mathbf{e}_z|^2 \rangle. \tag{2.53}$$

A similar treatment gives the absolute differential Raman cross-section for perpendicular polarized detection:

$$\frac{\mathrm{d}\sigma_R^{\perp}}{\mathrm{d}\Omega} = \frac{\omega_R^4}{16\pi^2(\epsilon_0)^2c^4} L_M \langle |\mathbf{e}_y \cdot \hat{\alpha} \cdot \mathbf{e}_z|^2 \rangle. \tag{2.54}$$

It may seem *a priori* that all the information on the polarizability tensor is completely washed out by the averaging process over random orientations. However, this is not the case: $\langle |\mathbf{e}_z \cdot \hat{\alpha} \cdot \mathbf{e}_z|^2 \rangle$ and $\langle |\mathbf{e}_y \cdot \hat{\alpha} \cdot \mathbf{e}_z|^2 \rangle$ do not depend on the choice of $\mathbf{e}_z$ or $\mathbf{e}_y$ thanks to the averaging process, but do remain dependent on intrinsic properties, called *invariants* or *symmetry invariants* [101], of the polarizability tensor $\hat{\alpha}$. This can be proved by carrying out the averaging explicitly, or can be derived from concepts of group theory [102,103]. We shall not reproduce the full calculation here [70,101], and simply provide the final result. For an isotropic distribution of scatterers, i.e. a fully random molecular orientation, we have:

$$\begin{cases} \langle |\mathbf{e}_z \cdot \hat{\alpha} \cdot \mathbf{e}_z|^2 \rangle &= (45\bar{\alpha}^2 + 4\bar{\gamma}^2)/45, \\ \langle |\mathbf{e}_y \cdot \hat{\alpha} \cdot \mathbf{e}_z|^2 \rangle &= 3\bar{\gamma}^2/45, \end{cases} \tag{2.55}$$

where $\bar{\alpha}$ and $\bar{\gamma}$ $[\epsilon_0\ \mathrm{m}^3]$ are two scalar invariants of the tensor $\hat{\alpha}$, defined by using its matrix representation (Eq. (2.47)) as:

$$\bar{\alpha} = \frac{1}{3}\left[\alpha_{xx} + \alpha_{yy} + \alpha_{zz}\right], \tag{2.56}$$

also called the 'reduced trace of the matrix', and:

$$\bar{\gamma}^2 = \frac{1}{2}\left[(\alpha_{xx} - \alpha_{yy})^2 + (\alpha_{yy} - \alpha_{zz})^2 + (\alpha_{zz} - \alpha_{xx})^2\right] \\ + 3\left[\alpha_{xy}^2 + \alpha_{xz}^2 + \alpha_{yz}^2\right], \tag{2.57}$$

also known as the 'anisotropy' parameter of the matrix. These invariants are independent [101] of the coordinate frame in which the matrix representation is given (i.e. of the orientation of the molecule). We mention in passing that there is in addition a third invariant, but it does not play a role in this problem.

The *absolute differential Raman cross-section for parallel and perpendicular polarized detection* can now be expressed as a function of these tensor invariants as:

$$\frac{d\sigma_R^{\parallel}}{d\Omega} = \frac{L_M \omega_R^4}{16\pi^2(\epsilon_0)^2 c^4} \frac{45\bar{\alpha}^2 + 4\bar{\gamma}^2}{45}, \tag{2.58}$$

and

$$\frac{d\sigma_R^{\perp}}{d\Omega} = \frac{L_M \omega_R^4}{16\pi^2(\epsilon_0)^2 c^4} \frac{3\bar{\gamma}^2}{45}. \tag{2.59}$$

In practice, Raman modes are more often characterized by both (i) their *absolute differential Raman cross-section* (which is simply the sum of the parallel and perpendicular contributions), and (ii) their *depolarization ratio* (ratio of perpendicular over parallel contributions). For these, we have:

$$\boxed{\frac{d\sigma_R}{d\Omega} = \frac{L_M \omega_R^4}{16\pi^2(\epsilon_0)^2 c^4} \frac{45\bar{\alpha}^2 + 7\bar{\gamma}^2}{45},} \tag{2.60}$$

and (from Eq. (2.13)):

$$\boxed{\rho_R = \frac{3\bar{\gamma}^2}{(45\bar{\alpha}^2 + 4\bar{\gamma}^2)}.} \tag{2.61}$$

Finally, one may also want in some occasions to calculate the *total Raman cross-section* σ_R [m^2], which within this treatment is given by:

$$\sigma_R = \frac{L_M \omega_R^4}{6\pi(\epsilon_0)^2 c^4} \frac{45\bar{\alpha}^2 + 10\bar{\gamma}^2}{45}. \tag{2.62}$$

In fact, this latter expression (Eq. (2.62)) together with Eqs (2.60) and (2.61)) provides a justification for Eq. (2.14) given earlier.

Depolarization ratio and symmetry of the tensor

The arguments presented so far imply that, for a given vibrational mode, only two independent quantities, usually the differential Raman cross-section and the depolarization ratio, are observable in the Raman scattering signal from an isotropic distribution of molecules. Whilst the cross-section mostly characterizes the 'strength' of the effect, the depolarization ratio is related to the *symmetry* of the Raman polarizability tensor, through the *trace* and *anisotropy* of the matrix representing the polarizability tensor $\hat{\alpha}$. This is a standard result from the theory of Raman scattering in liquids, and has been used many times since the pioneering work of Porto in the 60's [101] as an additional analytical tool to characterize the symmetry of modes of standard molecules and solvents.

We note that the depolarization ratio is always contained within the limits $0 \leq \rho_R \leq 3/4$. $\rho_R = 0$ occurs when $\bar{\gamma}^2 = 0$ ($\bar{\alpha} \neq 0$), which corresponds to a fully *isotropic tensor*. In this case, as mentioned earlier, the scattered light preserves the polarization of the incident beam. For $\rho_R \neq 0$ there is a 'transfer of intensity' from one polarization to the other and the light is therefore partly depolarized by the scattering-plus-averaging process. $\rho_R = 3/4$ occurs when $\bar{\alpha} = 0$ ($\bar{\gamma}^2 \neq 0$), which corresponds to a zero-trace tensor.

Magnitude of the polarizability tensor

By comparison of the general expression in Eq. (2.60) with the special case of an isotropic tensor in Eq. (2.50), it is possible to define the *magnitude* of the Raman polarizability tensor [8], $\tilde{\alpha}$ [ϵ_0 m^3] as the (invariant) scalar given by:

$$\tilde{\alpha}^2 = \frac{45\bar{\alpha}^2 + 7\bar{\gamma}^2}{45}. \tag{2.63}$$

The *absolute differential Raman cross-section* can then be expressed in a simple form, similarly to the isotropic case as:

$$\boxed{\frac{d\sigma_R}{d\Omega} = \frac{L_M \omega_R^4}{16\pi^2(\epsilon_0)^2 c^4}\tilde{\alpha}^2.} \tag{2.64}$$

This expression is equivalent to Eq. (2.60), but is more adequate to definitions and electromagnetic calculations of SERS enhancements, as we shall see in the next chapter[19].

One can also define the *normalized Raman polarizability tensor* as $\hat{\alpha}_N = \hat{\alpha}/\tilde{\alpha}$. This *adimensional* tensor characterizes the type of Raman polarizability tensor (i.e. its symmetry), independently of its strength characterized by the magnitude $\tilde{\alpha}$. Again, these definitions will be most useful in the context of SERS enhancement factors [8].

---

[19] This magnitude is related to the concept of *Raman activity*, as we shall see later. The Raman activity of a mode can be expressed as $R = 45\tilde{R}^2$ where $\tilde{R}$ is the magnitude of its Raman tensor (also defined later).

2.4.7. Final remarks on the phenomenological description

Rayleigh vs Raman scattering of an ensemble of molecules

From a purely 'classical electromagnetic theory' point of view, there is an additional remark we can make regarding the difference between Rayleigh and Raman scattering, and it is related to the *coherence* of the scattering process from an ensemble of molecules. It is only important at a conceptual level, for it does not play a major role in practice in the context of Raman or SERS.

Rayleigh scattering is an inherently *coherent* process, in the sense that the induced dipole (and therefore its scattered electromagnetic radiation) is driven by a constant *phase* relation with respect to the incoming wave. There can be a phase difference between the two (fixed by the polarizability, which is in general a complex magnitude), but this phase difference is the same for all the molecules if we are looking at a scattering volume much smaller than the wavelength. Coherent effects may therefore arise in Rayleigh scattering under appropriate experimental conditions.

However, the situation in Raman scattering is different. The fact that it is an inelastic process implies the participation of another player (the vibration here), the phase of which is in general different for two different molecules. The scattered radiation therefore has an arbitrary phase with respect to the incoming wave; even if the excitation is in phase for two molecules. Raman scattering from an ensemble of molecules is therefore a truly *incoherent* process at a microscopic level.

Resonant Raman scattering

Last, but not least, let us briefly discuss how the concept of *resonance* comes into play in the purely phenomenological (classical) description of the Raman effect. It is worth stressing again here that there are two basic frequency dependencies that arise in the Raman cross-section: one is the natural ω_R^4-factor that appears, in fact, in all optical scattering cross-sections (and is related to the electromagnetic theory of dipole emission). The latter can be seen, for example, in Eq. (2.60) for the differential Raman cross-section, and is largely responsible for the frequency dependence observed, for example, for N_2 (see Table 2.1).

In addition to this standard frequency dependence, there may be an additional contribution from resonance (or pre-resonance) effects (see for example Table 2.3). These can be included phenomenologically in the ω-dependence of the Raman polarizability tensor. The physical origin of the latter is in the electronic structure of the molecule, and in the microscopic origin of the Raman polarizability and usually requires a full quantum description. A comprehensive account of different resonance conditions in molecules and some of their consequences are given in Long's book on the Raman effect [70].

A brief summary of where we are

We have so far described Raman scattering using classical electromagnetism (i.e. induced dipole and dipolar emission) and given a *phenomenological* description of the induced Raman dipole through the Raman polarizability tensor. This approach is somewhat different from many other descriptions, often starting with the theory of molecular vibrations (which, let us not forget, are ultimately the physical origin of the scattering process). This choice to present the subject in this way was a deliberate one and (we hope) it is based on a good reason: this is all that is needed to move on to the next step of describing Raman scattering in the vicinity of metallic surfaces, i.e. SERS. From the point of view of the fundamentals of Raman scattering needed for SERS, we could therefore stop here.

This would however not be very satisfying for the reader (and neither for the authors!). Understanding the connection between the Raman polarizability and its physical origin (molecular vibration) is necessary in many contexts, and fundamental to a complete description of Raman scattering. We therefore dedicate the rest of the chapter to this and other more advanced aspects of Raman scattering.

2.5. VIBRATIONS AND THE RAMAN TENSOR

Vibrations in molecules are ultimately the origin of the Raman effect; it is therefore important to understand how they affect their optical properties. This will justify the introduction of a phenomenological Raman polarizability tensor and link it to the microscopic properties of the molecule.

As before, this could be obtained from a full quantum theory, and would justify in the process the description in terms of a radiating induced Raman dipole. We choose again here to follow a simpler classical approach. This is sufficient to introduce most of the important concepts: vibrational modes, Raman tensor, and their symmetry, etc. The main limitation of this approach, however, is that it is no longer valid in resonant Raman scattering (RRS) conditions. In this case, more elaborate descriptions are necessary (see for example Ref. [70]) and these will be omitted here. This does not prevent us from studying RRS or SERRS using the phenomenological description of the previous section, which remains valid.

2.5.1. General considerations

Implicit in the phenomenological description of the linear optical polarizability ($\hat{\alpha}_L(\omega)$) presented earlier is the fact that the molecule is in its ground-state geometry; i.e. the minimum of its global electronic energy, and that the atoms are static in their positions. This is a good approximation as far as the calculation of $\hat{\alpha}_L(\omega)$ is concerned. Nonetheless, it is quite clear that atoms can vibrate around their equilibrium positions, with restoring forces provided by their interactions with the rest of the atoms in the molecule. Vibrations with small amplitudes will act as a perturbation to the electronic structure and, accordingly, as a small perturbation to $\hat{\alpha}_L(\omega)$. A given vibrational mode will hence induce a 'modulation' of the linear optical polarizability with a frequency equal to that of the internal vibration. This small perturbation gives rise to Raman scattering. It should therefore be possible to derive the Raman polarizability from the study of the small perturbations of $\hat{\alpha}_L(\omega)$ by the internal vibrations. This is the essence of the classical (and semi-classical) approach, as we will now describe.

The first step (Section 2.5.2) is to describe the molecular vibrations themselves, i.e. *vibrational analysis*. Their effect on the linear optical polarizability is then considered in Section 2.5.3, which leads to the definition of the Raman tensor. The connection with the phenomenological approach (the Raman polarizability tensor) can then be made in Section 2.5.4. This completes the fully classical approach to Raman scattering. The rest of this section is then devoted to discussing additional effects that can be understood classically, in particular Raman selection rules.

2.5.2. A primer on vibrational analysis

Here vibrational analysis is described in simple terms, with a minimum of mathematics, only as a preliminary for the introduction of the Raman tensor. A more rigorous and more detailed description is given in Section 2.7.1.

Translations, rotations, and vibrations

A molecule consisting of N atoms ($N \geq 2$) has $3N$ internal degrees of freedom (three coordinates (x_i, y_i, z_i) [m] for each atom i), as far as atomic motion is concerned. Three of these degrees of freedom can be assigned to translations (of the whole rigid structure) along three (independent) directions of space. Three more (or two in linear molecules[20]) degrees of freedom correspond to rotations (of the whole rigid structure) along three (independent) axes of space. The rest, $3N - 6$ ($3N - 5$ in linear molecules) degrees of freedom, correspond to internal deformations of the molecule [104]. These deformations are constrained in a free molecule by the various interactions (chemical bonds) between the constituent atoms (and their electrons), i.e. there is a restoring force preventing them from moving too far away from their equilibrium positions. This results in small oscillations (i.e. vibrations) of the atoms around their equilibrium positions.

[20] There is one degree of freedom less in a linear molecule (like CO_2) because the rotation about the axis has no physical meaning if we consider the atoms as 'point particles'. In other words, there is no *moment of inertia* for a rotation along the axis with point-like masses.

Normal modes

The simplest description of these vibrational degrees of freedom is in terms of $3N - 6$ ($3N - 5$ in linear molecules) *normal vibrational modes* or *vibrational modes* for short. *Vibrational analysis* is concerned with the study of these deformations, i.e. the study of the normal vibrational modes. We shall come back to the details of vibrational analysis in Section 2.7.1. It suffices to say here that a given normal mode k ($k = 1 \ldots 3N - 6$) corresponds to a specific vibrational pattern (displacements from equilibrium) in the molecule, for which all atoms *oscillate at the same frequency* ω_k. In more mathematical terms, it is an eigenvector (with its corresponding eigenvalue ω_k) of the dynamic matrix of the molecule around equilibrium, defined by its geometry and the inter-atomic interactions (coming from the electronic structure). Quantum mechanically, such a normal mode is then associated with a (vibrational) energy $\hbar\omega_k$. Let us also note that the vibrational frequencies ω_k and the pattern of movement of the atoms (relative directions and amplitudes) strongly depend on: (i) the geometry of the molecule, and (ii) the nature of the electronic interactions (chemical bonds).

Finally, we note that the translational and rotational degrees of freedom can also be decomposed into 6 additional normal modes. Because there is no restoring force in this case, these modes have zero frequency[21].

Normal mode coordinates

The positions of the atoms in a molecule can be described by the $3N$ atomic coordinates, or equivalently, by $3N$ *atomic displacements*, ξ_i [m] corresponding to the atomic coordinates ((x_i, y_i, z_i) for each of the N atoms) relative to their equilibrium positions (i.e. all $\xi_i = 0$ at equilibrium). However, it is common in vibrational analysis to use $3N$ *reduced-mass coordinates* or *mass-weighted coordinates*; q_i [kg$^{1/2}$ m] defined as $q_i = \sqrt{m_i}\xi_i$, where m_i is the mass of the atom whose displacement is described by ξ_i.

As explained in more detail in Section 2.7.1, it is also possible to define mass-weighted *normal mode coordinates* or *normal coordinates*, Q_k [kg$^{1/2}$ m], which provide an equivalent description of the atomic displacements that is particularly suited to vibrational analysis. Normal coordinates will be defined rigorously in Section 2.7.1. In short, if a molecule vibrates freely with a pattern corresponding exactly to a given normal mode k (and therefore at a given frequency ω_k), then the normal coordinates are $Q'_k = 0$ for all $k' \neq k$, and the movement of all atoms can then be described in terms of a single scalar $Q_k(t)$ oscillating at ω_k. The normal coordinates Q_k therefore measure the amplitude of the deformation according to the pattern of a given normal mode k.

---

[21] For rotations this is only true in the classical treatment.

More details on vibrational analysis are given in Section 2.7.1. For those readers who would like to acquire a deeper understanding of vibrational analysis it may be better, in fact, to read Section 2.7.1 before continuing further. For those who do not, the 'plain-English' description given above should be sufficient to understand what follows.

2.5.3. The Raman tensor

We now study the small perturbation to the linear optical polarizability introduced by the presence of vibrations; this will lead us to the concept of the *Raman tensor* (not to be confused with the Raman polarizability tensor $\hat{\alpha}_R$ defined earlier).

Definition

Let us consider a molecule excited by a laser at angular frequency ω_L. The linear optical polarizability $\hat{\alpha}_L(\omega_L)$ characterizes the optical response (induced dipole) of the molecule at equilibrium. We will omit the ω_L-dependence in the following for simplicity. If the atoms move from their equilibrium position (which automatically affects the electronic wave-functions/density), the linear optical polarizability changes, i.e. $\hat{\alpha}_L$ depends on the atomic coordinates (x_i, y_i, z_i). One can alternatively describe this dependence as a function of the normal mode coordinates Q_k [$\text{kg}^{1/2}$ m]: $\hat{\alpha}_L(Q_1, Q_2, \ldots)$.

Let us now focus on a specific normal mode k characterized by Q_k (therefore assuming all other $Q_{k'}$ to be zero), and study its influence on $\hat{\alpha}_L$. Because internal vibrations are typically small perturbations of the molecular and electronic structure, we expect the polarizability to be also weakly perturbed by their presence. We can therefore approximate it by a Taylor expansion:

$$\hat{\alpha}_L(Q_k) = \hat{\alpha}_L(0) + \left(\frac{\partial \hat{\alpha}_L}{\partial Q_k}\right)_{Q_k=0} Q_k + \frac{1}{2}\left(\frac{\partial^2 \hat{\alpha}_L}{\partial Q_k^2}\right)_{Q_k=0} Q_k^2 + \ldots, \qquad (2.65)$$

where $\hat{\alpha}_L(0)$ is the unperturbed (at equilibrium) linear optical polarizability and the following terms (which are also second-rank tensors) are the first- and second-order perturbations produced by atomic movements following the normal mode pattern (which is characterized by the scalar amplitude Q_k). Note that the expression is formally written for the tensor, but this is equivalent to writing it for each of the tensor components (in a given fixed coordinate frame).

The *Raman tensor*, $\hat{R}_k(\omega_L)$ [ϵ_0 m^2 kg$^{-1/2}$], of the normal mode k, is defined as:

$$\boxed{\hat{R}_k(\omega_L) = \left(\frac{\partial \hat{\alpha}_L(\omega_L)}{\partial Q_k}\right)_{Q_k=0}.} \qquad (2.66)$$

We will soon show that the Raman tensor, as its name rightly suggests, is intricately linked to the Raman polarizability tensor (defined before) of the corresponding normal mode.

Some properties of the Raman tensor

Before establishing formally the connection with the Raman polarizability tensor, let us list a few characteristics of the Raman tensor:

- In this standard definition (Eq. (2.66)), the Raman tensor corresponds to the derivatives of the linear polarizability $\hat{\alpha}_L(\omega_L)$ with respect to the reduced-mass normal mode coordinate (Q_k) of a particular eigenvector representing a vibration; thus resulting in the units of [ϵ_0 m^2 kg$^{-1/2}$].
- $\hat{R}_k(\omega_L)$ is a measure (to the lowest order) of the *change in linear polarizability* when the molecule is deformed according to the deformation pattern of normal mode k.
- More quantitatively, if the molecular configuration is specified in normal mode coordinates with $Q_{k'\neq k} = 0$ except for one Q_k (assumed small), then the change in linear polarizability is, to lowest order, from Eq. (2.65), given by $\hat{\alpha}_L(Q_k) - \hat{\alpha}_L(0) = \hat{R}_k(\omega_L)Q_k$.
- The single normal mode coordinate Q_k 'hides' the fact that many atoms participate in the vibrational pattern of the corresponding normal mode. As shown in Section 2.7.1, Q_k is a linear combination of the $3N$ reduced-mass Cartesian coordinates q_i for atomic displacements: $Q_k = \mathbf{q} \cdot \mathbf{A}_k$, where $\mathbf{q} = (q_i)_{i=1..3N}$; and the inverse is also true: $\mathbf{q} = \sum_k Q_k \mathbf{A}_k$. Here $\mathbf{A}_k = (A_k^i)_{i=1..3N}$ is the eigenvector describing the vibrational pattern of mode k (see Section 2.7.1). It is therefore possible to recover the individual contribution of each atom to the change in polarizability by expanding the derivative with respect to Q_k in Eq. (2.66) as:

$$\hat{R}_k(\omega_L) = \left(\frac{\partial \hat{\alpha}_L(\omega_L)}{\partial Q_k}\right)_{Q_k=0} = \sum_{i=1}^{3N} A_k^i \left(\frac{\partial \hat{\alpha}_L(\omega_L)}{\partial q_i}\right)_{q_i=0}. \tag{2.67}$$

Such an expansion is more suited to develop an understanding of how symmetries in the atomic motion (say two identical atoms moving in opposite directions with respect to each other) may add up or lead to cancellations in the sum, and possibly result in a zero Raman tensor. This is intimately related also to an approach known as the *bond-polarizability model*, which is further developed in Appendix B for the interested reader.

2.5.4. Link to the Raman polarizability

Classical derivation

Let us now consider a molecule vibrating in a given normal mode k (with vibrational frequency ω_k). This motion can be simply described in normal coordinates by the single scalar coordinate $Q_k(t) = Q_k^0 \cos(\omega_k t + \Phi)$, where Q_k^0 characterizes the amplitude of the oscillations (assumed small) and Φ is an arbitrary phase. Let us denote $\hat{\alpha}_k(\omega_L)$ the Raman polarizability tensor (defined in Eq. (2.41)) for normal mode k (with vibrational frequency ω_k) for excitation at frequency ω_L.

Let us recall at this stage the meaning of the linear and Raman polarizabilities and return to writing explicitly complex notations with an under-bar for clarity:

- The linear polarizability $\hat{\alpha}_L(\omega_L)$ is a *response function*. It relates the *complex amplitude* of the excitation $\underline{\mathbf{E}}(\omega_L)$ (electric field) to that of the response $\underline{\mathbf{p}}(\omega_L)$ (induced dipole). The actual time dependence of the excitation and response is not described directly by $\hat{\alpha}_L(\omega_L)$, but can be recovered from the definition of complex notations, for example: $\mathbf{p}(t) = \mathrm{Re}(\underline{\mathbf{p}}(\omega_L)\exp(-\mathrm{i}\omega_L t))$.
- The Raman polarizability $\hat{\alpha}_k(\omega_L)$ is *not a response function*, but a phenomenological parameter defined (in complex notation again) as $\underline{\mathbf{p}}_R(\omega_R) = \hat{\alpha}_k(\omega_L)\underline{\mathbf{E}}(\omega_L)$, where $\mathbf{p}_R(t) = \mathrm{Re}(\underline{\mathbf{p}}_R(\omega_R)\exp(-\mathrm{i}\omega_R t))$ is the induced Raman dipole oscillating at $\omega_R = \omega_L - \omega_k$ for Stokes or $\omega_R = \omega_L + \omega_k$ for anti-Stokes processes. We ignore here the local field correction factor $(L_M)^{1/4}$, since it does not affect any of the results and discussions that follow.

For small oscillations, the effect of the vibration on $\hat{\alpha}_L$ can be described by a Taylor expansion as in Eq. (2.65), which, using the definition of the Raman tensor $\hat{R}_k(\omega_L)$ for this mode, can be written as:

$$\hat{\alpha}_L(\omega_L, Q_k) = \hat{\alpha}_L(\omega_L, 0) + Q_k \hat{R}_k(\omega_L). \tag{2.68}$$

It is tempting at this stage to write explicitly the time dependence of Q_k and consider a time-dependent linear polarizability. Such an approach however clashes with the basic definition of $\hat{\alpha}_L(\omega_L)$ as a response function. In order to study the time dependence of the system, one must therefore go back a stage and consider the time dependence of the real physical response (not the response function), in this case the induced dipole. This follows simply from the previous expressions:

$$\begin{aligned}\mathbf{p}(t) &= \mathrm{Re}\left[\hat{\alpha}_L(\omega_L, Q_k)\underline{\mathbf{E}}(\omega_L)\mathrm{e}^{-\mathrm{i}\omega_L t}\right] \\ &= \mathrm{Re}\left[\hat{\alpha}_L(\omega_L, 0)\underline{\mathbf{E}}(\omega_L)\mathrm{e}^{-\mathrm{i}\omega_L t}\right] + Q_k(t)\mathrm{Re}\left[\hat{R}_k(\omega_L)\underline{\mathbf{E}}(\omega_L)\mathrm{e}^{-\mathrm{i}\omega_L t}\right] \\ &= \mathbf{p}_L(t) + \mathbf{p}_S(t) + \mathbf{p}_{aS}(t),\end{aligned} \tag{2.69}$$

where

$$\mathbf{p}_L(t) = \mathrm{Re}\left[\hat{\alpha}_L(\omega_L, 0)\underline{\mathbf{E}}(\omega_L)\mathrm{e}^{-\mathrm{i}\omega_L t}\right], \tag{2.70}$$

oscillates at ω_L and can be identified with the induced dipole for Rayleigh scattering. In addition, we have:

$$\mathbf{p}_S(t) = \frac{Q_k^0}{2}\mathrm{Re}\left[\hat{R}_k(\omega_L)\underline{\mathbf{E}}(\omega_L)\mathrm{e}^{-\mathrm{i}(\omega_L-\omega_k)t+\mathrm{i}\Phi}\right], \tag{2.71}$$

which oscillates at $\omega_S = \omega_L - \omega_k$ and can be identified with the induced Raman dipole for Stokes scattering; while:

$$\mathbf{p}_{aS}(t) = \frac{Q_k^0}{2}\mathrm{Re}\left[\hat{R}_k(\omega_L)\underline{\mathbf{E}}(\omega_L)\mathrm{e}^{-\mathrm{i}(\omega_L+\omega_k)t-\mathrm{i}\Phi}\right], \tag{2.72}$$

oscillates at $\omega_{aS} = \omega_L + \omega_k$ and can be identified with the induced Raman dipole for anti-Stokes scattering.

From the last two expressions and the phenomenological definition of the Raman polarizability tensor (Eq. (2.41)), it follows that $\hat{\alpha}_k(\omega_L)$ is the same (up to a phase) for Stokes and anti-Stokes scattering and given by :

$$\boxed{\hat{\alpha}_k(\omega_L) = \frac{Q_k^0}{2}\hat{R}_k(\omega_L).} \tag{2.73}$$

The phenomenological Raman polarizability tensor for mode k ($\hat{\alpha}_k(\omega_L)$) is, therefore, proportional to the Raman tensor $\hat{R}_k(\omega_L)$, which appeared by itself from linear response theory and the microscopic classical description of vibrations. This expression therefore provides the link between the fully phenomenological description in terms of a Raman polarizability tensor $\hat{\alpha}_R$, and the microscopic interpretation in terms of scattering by vibrational modes.

Discussion

The coefficient of proportionality $Q_k^0/2$ simply reflects the amplitude of the vibration that caused the Raman scattering effect. In fact, the scattered intensity will be proportional to $(Q_k^0)^2$ and this is proportional to the classical energy of the vibration. If we transfer this classical result to the quantum case, then $(Q_k^0)^2$ must be replaced by a quantity proportional to the average number n of quanta of vibrations per molecule (this will be made more precise in Section 2.6). Hence, the classical theory predicts the following:

- the same Raman polarizability for Stokes and anti-Stokes scattering, the square of its magnitude being proportional to $(Q_k^0)^2$ ($\propto n$ in the quantum theory),
- and that no Raman scattering should occur when there are no vibrations in the molecule ($Q_k^0 = 0$ or $n = 0$). This is obviously in contradiction with the simplest form of Stokes scattering, where the molecule in its ground vibrational state is excited to a higher level. This deficiency can be tracked down to the existence of a *zero-point energy* for a quantum system, such as a vibrational mode represented by a harmonic oscillator. In reality (see Section 2.6), the Stokes polarizability should actually be proportional to $1 + n$, where the factor 1 arises from interaction with the vibrational ground-state oscillations (vacuum fluctuations) which do not exist classically (but can be fully justified with the addition of quantum mechanics).

This deficiency of the classical treatment is actually not surprising. The situation is similar to the case of absorption/emission of light by an atom or molecule. Absorption and stimulated emission are symmetrical and can be described by a classical electromagnetic field, but spontaneous emission is related to the vacuum state of the quantum electromagnetic field and therefore requires a quantum description. Similarly, the classical approach to Raman scattering predicts absorption (anti-Stokes, proportional to n) or stimulated emission (i.e. creation) of a vibrational quantum (Stokes, proportional to n), but not spontaneous emission (corresponding to the 1 in the total factor $(1+n)$ for the Stokes emission). This deficiency is one of the major drawbacks of this approach, and will be addressed properly in the semi-classical treatment.

This final step essentially completes the classical description of the Raman effect by linking the phenomenological Raman polarizability to the microscopic properties of the molecule. The comparisons with the mechanical analogs of 'modulated scattering' at the beginning of this chapter (Section 2.2.9) should be evident at this stage. The perturbation caused by the vibration produces a 'modulation' of the linear polarizability resulting in a beating of frequencies between the frequency of the excitation ω_L and the frequency of the vibration ω_k; this beating phenomenon results in inelastic scattering; i.e. in Raman scattering.

Further properties of the Raman tensor

In light of this proportionality between Raman tensor and Raman polarizability tensor, we can discuss a few other properties of the Raman tensor and their consequences for the Raman intensities of the mode:

- If $\hat{\alpha}_L(\omega_L)$ does not depend on Q_k (a common situation in molecules with a high symmetry), then $\hat{R}_k(\omega_L) = 0$ and $\hat{\alpha}_k(\omega_L) = 0$; the corresponding mode does not produce Raman scattering and is called a *Raman-inactive* mode. Such occurrences depend on the symmetry of both the molecule and the vibrational pattern (normal mode) and their study constitutes the topic of *Raman selection rules*,[22] a subject further discussed in Section 2.7.2. In the case of small molecular moieties attached to larger molecular structures, the symmetry may be approximate for a sub-section of the molecule. Most selection rules will, nevertheless, be approximately valid in this second case too.

- For Raman-active modes, the symmetry of the vibrational mode (if any) is partly reflected in the *symmetry of the Raman tensor*, which imposes its symmetry on the Raman polarizability tensor. It is impossible in general to measure all the components of the Raman tensor, but its symmetry is partly reflected in the Raman depolarization ratio ρ_R as discussed earlier (see Section 2.4.6). Special cases with a high symmetry include, for example, isotropic ($\rho_R = 0$) or highly uni-axial ($\rho_R = 1/3$) Raman tensors.

- Each normal mode has its *own* Raman tensor, i.e. there are in principle $3N - 6$ Raman tensors for a molecule with N atoms (or $3N - 5$ for a linear molecule). The vast majority of them will be either zero because of symmetry (Raman selection rules) or will have small magnitudes. This leaves only a handful of vibrational modes, like those seen for RH6G in Fig. 2.4, which are the *Raman modes* or *Raman-active modes*[23].

- $\hat{R}_k(\omega_L)$ depends on how large or small the derivative $(\partial\hat{\alpha}_L(\omega_L)/\partial Q_k)_{Q_k=0}$ is at the excitation frequency ω_L. $\hat{R}_k(\omega_L)$ is therefore expected to be larger (in most cases) when $\hat{\alpha}_L(\omega_L)$ is larger. Any electronic resonance of the molecule that results in a resonant increase in $\hat{\alpha}_L(\omega_L)$ (resonant conditions) should also result qualitatively in a resonance for $\hat{R}_k(\omega_L)$, i.e. resonant Raman scattering. In simple terms, the magnitude of the Raman tensor (and Raman polarizability) varies qualitatively with that of the linear polarizability, except where symmetries cancels it out.

---

[22] These symmetries and their consequences are often studied within the framework of group theory [102,103]. A given molecular structure (or its sub-structure) has a certain point-group symmetry, and the different normal modes belong to different irreducible representations of that group. The irreducible representation of a mode then determines whether the mode is Raman active or not.

[23] The denomination of Raman-active modes may sometimes be reserved only for modes that are allowed by symmetry consideration, irrespective of the magnitude of their Raman tensor.

- Apart for this link in magnitude, the symmetry of the Raman tensor $\hat{R}_k(\omega_L)$ is not directly related to that of the linear polarizability $\hat{\alpha}_L(\omega_L)$, but more to that of the normal mode k. For example, we can have a molecule with a diagonal $\hat{\alpha}_L(\omega_L)$ that has a non-diagonal $\hat{R}_k(\omega_L)$, or precisely the opposite. The bond-polarizability model, discussed in Appendix B for its didactic merits, should help in visualizing the origins of such situations.

2.5.5. Limitations of the classical approach

As highlighted in Section 2.4, the classical description of scattering using the *classical theory of electromagnetism* is sufficient in most cases; a quantum description of the electromagnetic field is possible but does not add much new information.

In this section, we have extended the classical treatment to the molecule itself, to link the Raman polarizability to its microscopic properties, i.e. vibrations. This extension is successful to some degree: it correctly predicts the existence of Stokes and anti-Stokes scattering, for example. On the other hand, it does present a number of limitations:

- It does not predict the correct magnitude of the Stokes Raman polarizability (because it ignores the spontaneous creation of a vibration).

- It is based on the knowledge of the linear optical polarizability $\hat{\alpha}_L(\omega_L)$, and its dependence on the molecular structure. Linking quantitatively $\hat{\alpha}_L(\omega_L)$ to the microscopic properties of the molecule (to its electronic structure in particular) requires another step.

- Even if $\hat{\alpha}_L(\omega_L)$ is known, its Taylor expansion in Eq. (2.65) may not always be valid. This is especially the case in resonant Raman scattering conditions where the semi-classical description is then necessary [70].

- The Taylor expansion in Eq. (2.65) is with respect to the normal mode coordinate Q_k, and hence assumes the previous knowledge of the eigenvectors of the different vibrations. This information might be available for some very simple small molecules. But this is not known (in general) and, therefore, our ability to perform a Taylor expansion of that sort remains most of the time a fairly 'formal' construction with limited use.

Despite these limitations, the classical approach is very useful if applied carefully, and can in fact explain many aspects of the problem.

2.5.6. A brief overview of related Raman scattering processes

Overtone and combination bands

The treatment we have applied to the lowest-order term $\propto Q_k$ in Eq. (2.65) can be continued to the next highest-order term in this equation, which is proportional to Q_k^2. This term generates scattering processes at frequencies $\omega_L \pm 2\omega_k$. These should therefore appear on a spectrum at twice the Raman shift of the energy of the corresponding normal mode; and they are called *overtones* . If, in addition, the cross derivatives of the polarizability with respect to two different coordinates Q_k and Q_p are different from zero, then terms proportional to $Q_k Q_p$ appear in the expansion. These generate scattering at $\omega_L \pm (\omega_k \pm \omega_p)$, where ω_k and ω_p are the corresponding frequencies of the k and p normal modes, respectively. These are called *combination bands* (higher-order scattering). In condensed matter theory these processes are called two-phonon scattering.

Note, in addition, that the validity of the Taylor expansion in Eq. (2.65) holds only on the assumption that the normal mode vibrations are a small perturbation to $\hat{\alpha}_L(\omega_L)$, implying that higher-order terms in the expansion become increasingly smaller. Overtones and combination bands are therefore normally much weaker than the normal Raman scattering process, and they are in many cases difficult or impossible to observe (unless certain special resonance conditions are met.[24]). Overtone and combination bands can also be observed under SERS conditions with dyes [44].

Hyper-Raman scattering (HRS)

The classical description of Raman scattering given so far relies on the validity of the linear approximation. That is to say, the starting point is Eq. (2.39), in which we assume a linear dependence of the induced dipole upon the exciting field.

However, when employing high-power lasers (in particular pulsed lasers), this linear approximation may no longer be valid and higher-order (non-linear) terms must be taken into account. The definition of the optical polarizability (Eq. (2.39)) producing the dipole $\mathbf{p}$ must then be generalized to:

$$\mathbf{p} = \hat{\alpha}_L \cdot \mathbf{E} + \hat{\beta} \cdot \mathbf{EE} + \hat{\gamma} \cdot \mathbf{EEE} + \ldots, \tag{2.74}$$

where $\hat{\beta}$ and $\hat{\gamma}$ are tensors (of higher rank) and are called the *hyper-polarizability* and the *second-order hyper-polarizability* tensors, respectively.

[24] If the incident photon plus the energy of two vibrations matches a real electronic state in the molecule, then the condition of *Fermi resonance* is achieved [70]. Under these circumstances overtones and combination bands can be greatly enhanced.

The hyper-polarizability $\hat{\beta}$ results in scattering at frequency $2\omega_L$, called *hyper-Rayleigh* scattering .

The hyper-polarizabilities may be modulated by internal vibrations in a similar way as the linear polarizability. In particular, the modulation of $\hat{\beta}$ by the amplitude of a normal mode Q_k (as done before for Raman scattering) gives rise to scattering components at frequencies $2\omega_L \pm \omega_k$, called *hyper-Raman* scattering [105–107] . An alternative way of thinking about the hyper-Raman process is like a normal Raman process but produced simultaneously (and coherently) by two photons. The magnitude of the hyper-polarizability $\hat{\beta}$ is typically 10^{10}–10^{12} times smaller than that of $\hat{\alpha}_L$ under normal (non-SERS) conditions, which means that hyper-Raman scattering can only be observed with high-power lasers. There have been a few recent reports of hyper-Raman scattering under SERS conditions [108], a technique which is normally abbreviated as SEHRS.

Rotational bands and low-frequency Raman scattering

From the standpoint of the internal vibrations of a molecule, the 6 degrees of freedom representing the rotations and translations of the molecule as a whole play no role whatsoever in the internal dynamics and are discarded in the counting of $3N - 6$ internal modes. A brief comment on these additional degrees of freedom is made here also for the sake of completeness.

The rotational degrees of freedom of a molecule can contribute to inelastic Raman scattering processes through the quantization of the angular momentum [84,109] (which results in a non-zero normal mode frequency in quantum mechanics). Due to the typical size of the moment of inertia of molecules [104] (which fixes the energy scale), rotational bands appear only at relatively low frequencies ($<$150 cm^{-1}) compared to internal vibrations ($>$150 cm^{-1}). Rotational bands in molecules are mainly observable in (low molecular weight) gases and in experimental systems with very good light rejection at low frequencies.

From a SERS standpoint, even in liquids that might be in contact with metals, rotational bands are typically of no concern, not only because the rotations are hindered by the interaction of the molecules on the surface of metals (which are the ones that always contribute the most to the SERS signals) but also because low-frequency inelastic scattering in liquids have additional contributions from the liquid itself. We shall therefore ignore rotational bands in the rest of the book too.

2.6. QUANTUM (OR SEMI-CLASSICAL) APPROACH TO RAMAN SCATTERING

The fully classical approach to Raman scattering presented so far has many advantages. Firstly, the phenomenological description in terms of the Raman polarizability is all that is needed to further develop the argument to the case of SERS. Moreover, the classical link between polarizability and vibrations is a good starting point for the understanding of the microscopic origin of the effect, and already captures many important features, such as symmetries and selection rules (as further illustrated with the bond-polarizability model discussed in Appendix B). However, there is no denying that the molecular electronic and atomic structure is ultimately governed by the laws of quantum mechanics. A quantum treatment is therefore necessary, firstly to justify the classical approach, and secondly to remedy the few issues where the classical approach fails.

We will restrict ourselves here to a brief overview of the quantum approach, with particular emphasis on the quantization of vibrations. This particular aspect will enable us to fill the gaps left in the classical description: the description of spontaneous Stokes Raman scattering, and the zero-point amplitudes of the oscillations. This is then used to complete the link between the Raman cross-section (or Raman polarizability) and the microscopic properties of the normal modes; a link that is necessary, for example, to interpret the results of Raman DFT calculations (presented in Appendix A).

2.6.1. Justification of the classical approach

The foundations of the quantum approach to Raman scattering can be traced back to the polarizability theory developed by Placzek [110]. Note that this approach treats only the molecule quantum mechanically, but not the electromagnetic field. This approach is therefore sometimes called *semi-classical*, and it can be used to rigorously justify the classical approach itself [70,75,110,111]. Note that a 'full' quantum treatment (including the electromagnetic field) is also possible, but it is rarely useful in the context of SERS and certainly beyond the scope of this book. We shall not expand here into the full details of the semi-classical approach, which (like the previous few topics) is mentioned only for completeness and is of relatively minor importance for SERS.

From the viewpoint of quantum mechanics the scattering processes depicted in Fig. 2.4 are described by transition probabilities between an initial state $|i\rangle$ (of energy $\hbar\omega_i$) and a final state $|f\rangle$ (of energy $\hbar\omega_f$), which are typically computed using perturbation theory [84]. A given component of the Raman polarizability can then be shown to be of the form [70]:

$$\alpha_{kl} = \frac{1}{\hbar} \sum_{r \neq i,f} \left\{ \frac{\langle f|p_k|r\rangle\langle r|p_l|i\rangle}{\omega_r - \omega_i - \omega_L - i\Gamma_r} + \frac{\langle f|p_l|r\rangle\langle r|p_k|i\rangle}{\omega_r - \omega_f + \omega_L + i\Gamma_r} \right\}, \tag{2.75}$$

where the sum is in principle over all possible states $|r\rangle$ (with energy $\hbar\omega_r$ and broadening $\sim \hbar\Gamma_r$) of the molecule. p_k and p_l are dipole moment operators (in a quantum mechanical sense). Note also that states $|i\rangle$, $|f\rangle$, and $|r\rangle$, are typically mixed electronic/vibrational (vibronic) states of the molecule.

This expression may not appear particularly useful in itself, and a lot of effort is necessary to recast it into the classical formalism. The transition from the quantum mechanical expression to the classical one with the required approximations and assumptions is treated for example in Ref. [70]. However, this quantum mechanical expression for the Raman polarizability does show (qualitatively) why the effect can be viewed as a two-step transition: ($|i\rangle \rightarrow |r\rangle$ and $|r\rangle \rightarrow |f\rangle$), going through an intermediate state $|r\rangle$ (equivalent to the virtual state in Fig. 2.3). In a typical Raman scattering situation, a direct optical transition from $|i\rangle$ to $|f\rangle$ is not possible and only occurs through second-order perturbation theory as an instantaneous two-step process.

Another feature of the quantum expression is the energy denominators. When these are small, which typically occur when ω_L is close to a *real* electronic transition frequency, $\omega_r - \omega_i$, then a *resonant* response is expected. This provides a qualitative explanation for resonant Raman scattering (RRS). In fact, one major contribution of the semi-classical approach is the correct description of RRS, which cannot be carried out classically (except for a phenomenological description of the Raman polarizability). This aspect is again quite involved and beyond the scope of this book, see for example Ref. [70] for more details.

2.6.2. The quantization of vibrations

Zero-point amplitude

As shown later in Section 2.7.1, the normal vibrational modes can be approximated as independent *quantum harmonic oscillators*. For a Stokes Raman transition (for a normal mode k with frequency ω_k or wave-number $\bar{\nu}_k$), the initial state corresponds to the molecule in a vibrational state $|\nu\rangle$ (usually its ground state $|0_k\rangle$), and the final state to the molecule in the next vibrational excited state $|\nu+1\rangle$ (usually $|1_k\rangle$). The opposite stands for anti-Stokes. The classical treatment of Section 2.5.4 can then be generalized to quantized vibrations, only replacing the classical amplitude of the oscillations $Q_k^0/2$ by its quantum counterpart [70]:

$$\frac{Q_k^0}{2} \rightarrow Q_k^{\mathrm{QM}} = \begin{cases} \langle \nu+1|Q_k|\nu\rangle = (\nu+1)^{1/2}\, b_k & \text{(Stokes)} \\ \langle \nu|Q_k|\nu+1\rangle = (\nu)^{1/2}\, b_k & \text{(anti-Stokes).} \end{cases} \tag{2.76}$$

Here Q_k [kg$^{1/2}$ m] denotes the quantum operator for the harmonic oscillator position, expressed in reduced-mass coordinates, and b_k [kg$^{1/2}$ m] is called the *zero-point amplitude* of the normal vibrational mode:

$$b_k = \left(\frac{\hbar}{2\omega_k}\right)^{1/2} = \left(\frac{h}{8\pi^2 c\bar{\nu}_k}\right)^{1/2}. \tag{2.77}$$

The symmetry between Stokes and anti-Stokes that was predicted classically is no longer present here in the quantum mechanical description. In particular, for a molecule in its ground vibrational state $|0_k\rangle$, there is a non-zero probability for a Stokes Raman transition (but not for anti-Stokes).

Note that, as for the classical treatment, the above results are valid up to a phase (phase between the incident field and the vibrational mode oscillations). Because in most cases of interest, the phases of the vibrations in an ensemble of molecule are random and independent of each other, their various contributions will add up incoherently and what matters is therefore $|Q_k^{\mathrm{QM}}|^2$, which is independent of the phase.

Effect of temperature

For a molecule *in thermal equilibrium* at a temperature T, the probability of occupation of the vibrational levels is given by the standard laws of statistical physics. Vibrational modes, as quanta of a harmonic oscillator, are bosons and follow the Bose–Einstein statistics. After averaging, Eq. (2.76) then reduces to:

$$|Q_k^{\mathrm{QM}}|^2 = \begin{cases} b_k^2\,(1 + n_k^B(T)) & \text{(Stokes)} \\ b_k^2\, n_k^B(T) & \text{(anti-Stokes)} \end{cases}, \tag{2.78}$$

where

$$n_k^B(T) = \left[\exp\left(\frac{\hbar\omega_k}{k_B T}\right) - 1\right]^{-1} \tag{2.79}$$

is the *Bose factor* [112–114] for the vibration under consideration. This expression is implicitly an average over all possible states of a molecule in thermal equilibrium at temperature T.

2.6.3. The full expressions for the Raman cross-section

The above expressions for the zero-point amplitude constitute the final missing link to make the connection between the phenomenological and microscopic descriptions, which we now complete.

Raman polarizability and Raman tensor

We start first by deriving the Raman polarizability tensor $\hat{\alpha}_k(\omega_L)$ of a given mode from the Raman tensor $\hat{R}_k(\omega_L)$, which can in principle be computed from first principle through the calculation of the effects of normal mode oscillations on the linear polarizability.

Within the classical approach, we found that (see Eq. (2.73)):

$$\hat{\alpha}_k(\omega_L) = \frac{Q_k^0}{2}\hat{R}_k(\omega_L), \tag{2.80}$$

but the oscillation amplitude Q_k^0 was *a priori* non-specified.

The quantum approach tells us that we should replace $Q_k^0/2$ by Q_k^{QM}, and we therefore have for the Stokes Raman polarizability (up to a phase):

$$\boxed{\hat{\alpha}_k^{\mathrm{ST}}(\omega_L) = b_k\sqrt{1+n_k^B(T)}\hat{R}_k(\omega_L),} \tag{2.81}$$

while the anti-Stokes counterpart is given by:

$$\boxed{\hat{\alpha}_k^{\mathrm{aS}}(\omega_L) = b_k\sqrt{n_k^B(T)}\hat{R}_k(\omega_L).} \tag{2.82}$$

The Raman activity

The phenomenological expression for the cross-section in terms of the Raman polarizability, obtained in Eq. (2.60), can be recast in terms of the Raman tensor.

To this end, it is customary to introduce the *Raman Activity* R_k $[(\epsilon_0)^2\ \mathrm{m}^4\ \mathrm{kg}^{-1}]$ of the normal mode, derived from the Raman tensor as:

$$\boxed{R_k = 45\bar{\alpha}_k'^2 + 7\bar{\gamma}_k'^2,} \tag{2.83}$$

where $\bar{\alpha}_k'$ and $\bar{\gamma}_k'$ are the isotropic and anisotropic invariants of the Raman tensor $\hat{R}_k(\omega_L)$, defined in Eqs (2.56) and (2.57). R_k is a *scalar* characterizing an average property of the Raman tensor upon arbitrary rotations. In fact, using the definition of the tensor magnitude of Section 2.4.6, we also have $R_k = 45\tilde{R}_k^2$, where $\tilde{R}_k$ is the magnitude of the Raman tensor $\hat{R}_k$ (see Eq. (2.63)).

We note also that the Raman depolarization ratio for the same mode, ρ_k, already given in Section 2.4.6 in terms of the Raman polarizability tensor invariants, can also be expressed in terms of the Raman tensor invariants as:

$$\rho_k = \frac{3\bar{\gamma}_k'^2}{45\bar{\alpha}_k'^2 + 4\bar{\gamma}_k'^2}. \tag{2.84}$$

Note that this is basically the same expression as Eq. (2.61), but written for the Raman tensor invariants. This is simply a consequence of the proportionality between the Raman polarizability tensor and the Raman tensor.

The Raman cross-section

Using the definition of the Raman activity and the previous results, we can now write the *absolute differential Raman cross-section* (implicitly for a Stokes process), for normal mode k with energy $\hbar\omega_k$ (or wave-number $\bar{\nu}_k$), and for excitation at frequency ω_L (or wave-number $\bar{\nu}_L$), at a temperature T, in a medium of refractive index n_M, as (from Eq. (2.60)):

$$\frac{\mathrm{d}\sigma_k^{\mathrm{ST}}}{\mathrm{d}\Omega} = \frac{\hbar\omega_R^4}{1440\pi^2(\epsilon_0)^2c^4\omega_k}(1+n_k^B(T))L_M R_k, \tag{2.85}$$

where $\omega_R = \omega_L - \omega_k$ denotes the frequency of the Stokes Raman photons.

An alternative expression in terms of wave-numbers, where the different contributions are separated is [86]:

$$\boxed{\frac{\mathrm{d}\sigma_k^{\mathrm{ST}}}{\mathrm{d}\Omega} = C\, b_k^2\, L_M\, R_k\, \bar{\nu}_R^4\, (1+n_k^B(T)),} \tag{2.86}$$

where

$$C = \frac{\pi^2}{45\epsilon_0^2}\ [(\epsilon_0)^{-2}] \tag{2.87}$$

is a constant, $\bar{\nu}_R = \bar{\nu}_L - \bar{\nu}_k$ [m^{-1}] is the absolute wave-number of the Stokes-shifted Raman signal, b_k^2 [m^2 kg] is the square of the zero-point amplitude of the normal mode in reduced-mass coordinates, given in Eq. (2.77), L_M [a.d.] is the local field correction factor defined in Eq. (2.43), and R_k [$(\epsilon_0)^2$ m^4 kg^{-1}] is the Raman activity (defined in Eq. (2.83)).

The factor proportional to $\bar{\nu}_R^4$ accounts for the ω^4-dependence of the cross-section (discussed earlier in the classical aspects), while the factor:

$$1 + n_k^B(T) = \left[1 - \exp\left(\frac{-hc\bar{\nu}_k}{k_BT}\right)\right]^{-1} \tag{2.88}$$

accounts for thermal population of the vibrational state. Since at room temperature, $k_B T \sim 200\ \mathrm{cm}^{-1}$, this factor is usually close to 1, except for the modes of lowest energy, for example, it is 1.29 for a mode at $300\ \mathrm{cm}^{-1}$, 1.05 at $600\ \mathrm{cm}^{-1}$, and 1.007 at $1000\ \mathrm{cm}^{-1}$.

The anti-Stokes Raman cross-section

Equivalent expressions can be obtained for an anti-Stokes process. The anti-Stokes differential Raman cross-section for mode k is then given by:

$$\frac{\mathrm{d}\sigma_k^{\mathrm{aS}}}{\mathrm{d}\Omega} = C\, b_k^2\, L_M\, R_k\, \bar{\nu}_{aS}^4\, n_k^B(T), \tag{2.89}$$

where the notations are the same as before, and $\bar{\nu}_{aS} = \bar{\nu}_L + \bar{\nu}_k$ is now the absolute wave-number of the anti-Stokes photons. The main difference in this expression is the factor $n_k^B(T)$ instead of $1 + n_k^B(T)$, and this will be discussed further in the next section.

Note on the Raman activity

In many cases, three quantities are sufficient to obtain most Raman scattering related properties of a normal mode: its frequency ω_k (or energy $\hbar\omega_k$ or wave-number $\bar{\nu}_k$), its Raman activity R_k, and its depolarization ratio ρ_k.

The frequency gives the Raman peak position, the Raman activity characterizes *qualitatively* (but not exactly) its intensity, and the depolarization ratio its polarization properties (symmetry).

These three quantities are typically given in the output of DFT Raman computations in many programs, and we shall review this specifically in Appendix A. An important point to remember (often the source of confusion) is that the *Raman activity does not exactly represent the relative intensities of the Raman peaks.* It can indeed be used to predict them, using Eq. (2.86), but the other factors in this expression, in particular b_k^2, also affect the relative intensities between Raman peaks. In addition, it is worth noting that R_k is commonly expressed in units of [Å^4/amu]. The conversion into S.I. units and the link to the Raman cross-section are further discussed in the context of DFT calculations in Appendix A.

2.6.4. The anti-Stokes to Stokes ratio

Equations (2.86) and (2.89) deserve a brief additional comment. The reason for $\mathrm{d}\sigma_k^{aS}/\mathrm{d}\Omega$ to be proportional to $n_k^B(T)$ is fairly intuitive: anti-Stokes scattering as depicted in Fig. 2.4 does not happen unless the molecule is already in a vibrational excited state, the probability of which is precisely $n_k^B(T)$. The reason for the factor $(1+n_k^B(T))$ for Stokes Raman scattering in Eq. (2.86) is perhaps less obvious. As explained earlier, the factor $n_k^B(T)$ accounts for stimulated creation of a vibration (the exact opposite of the anti-Stokes process), while the factor 1 corresponds to spontaneous creation, which may therefore occur for a molecule in its vibrational ground state, as depicted in Fig. 2.4.

It is common in this context to define the *anti-Stokes to Stokes ratio*:

$$\rho_k^{aS/S} = \frac{\mathrm{d}\sigma_k^{aS}/\mathrm{d}\Omega}{\mathrm{d}\sigma_k^{ST}/\mathrm{d}\Omega} = \left(\frac{\nu_{aS}}{\nu_S}\right)^4 \frac{n_k^B(T)}{(1+n_k^B(T))} = A\,\mathrm{e}^{-\frac{\hbar\omega_k}{k_B T}}, \tag{2.90}$$

where $A = \nu_{aS}^4/\nu_S^4$.

It is important to realize that the *Boltzmann factor* appearing in the last equality is neither an approximation nor a consequence of treating the problem as 'classical', but rather an exact result from the Bose statistics of molecular vibrations. The presence of this factor in Eq. (2.90) provides a rigorous justification for the much smaller intensity observed in a normal anti-Stokes spectrum with respect to its Stokes counterpart, as well as for the experimentally observed temperature dependence.

The anti-Stokes to Stokes ratio can be measured relatively easily and it is to a large extent self-normalizing. By measuring it for several peaks (of known Raman shifts) of a given molecular species, one can easily deduce the temperature (T) of the sample (assumed at thermal equilibrium), without any knowledge of additional parameters (cross-sections, etc.). This technique can be used as a contact-less temperature sensor.

Finally, it is worth mentioning here that these arguments can be partly generalized to the SERS case. In particular, the temperature dependence of the Raman cross-sections and therefore of the anti-Stokes to Stokes ratio is in many cases retained. The Stokes and anti-Stokes cross-sections may however be affected by a different enhancement factor (because the processes occur at different wavelengths) and this will therefore affect the factor A. The factor A then accounts (in addition to the ν^4-dependence of the scattering process) for any additional difference introduced by the SERS enhancement factor and is called the *asymmetry factor* [59,115].

2.7. ADVANCED ASPECTS OF VIBRATIONS IN MOLECULES

The issue surrounding the 'nature' of the vibrations has been mentioned several times already. If we want to deepen our understanding of this concept, there is no option but to review the origin of vibrations in molecules in more detail. This section is therefore presented, for the sake of completeness, as a complement to the brief description of normal modes given earlier in Section 2.5.2. It will provide the reader with a more detailed, but still general, overview of vibrations in molecules.

2.7.1. More on vibrational analysis

Comprehensive treatment of vibrational analysis can be found in classical textbooks in the field [70,74,116,117]. We only highlight here a selection of the most relevant aspects.

General approximations

The starting point for the quantum treatment of vibrations in molecules is the so-called *Born–Oppenheimer approximation* [109,118] in which electronic wave-functions follow 'adiabatically' the movement of the nuclei due to their large difference in mass. In other words, in this approximation the electronic cloud of a molecule can adapt instantaneously to the configuration of the nuclei. The atomic coordinates, denoted as a whole by $\mathbf{R}$, can then be viewed as parameters (and not dynamic variables) in the determination of the electronic wave-function $\Psi^e_{\mathbf{R}}(\mathbf{r})$, where $\mathbf{r}$ represents all electronic coordinates.

The total Hamiltonian of the molecule can then be decomposed as:

$$H_{\text{tot}} = H^n + H^{en}, \tag{2.91}$$

where H^n does not apply to wave-functions depending only on electronic coordinates $\mathbf{r}$; in fact H^n is typically dominated by the kinetic energy T^n of the nuclei. H^{en} may apply to both electronic and nuclei coordinates (typically electronic kinetic energies and Coulomb interactions between nuclei and electrons).

The solution of Schrödinger's equation for the total system can be obtained in two successive steps. First, for a given nuclei configuration $\mathbf{R}$, the electronic wave-functions can be obtained from the solution of the (time-independent) eigenvalue problem:

$$H^{en}\Psi^e_{\mathbf{R}}(\mathbf{r}) = E^e_{\mathbf{R}}\Psi^e_{\mathbf{R}}(\mathbf{r}), \tag{2.92}$$

where $\mathbf{r}$ is the variable, and the coordinates of the nuclei $\mathbf{R}$ enter as external parameters to determine the electronic energies $E^e_{\mathbf{R}}$ and wave-functions $\Psi^e_{\mathbf{R}}(\mathbf{r})$. This step must in principle be repeated for many possible $\mathbf{R}$ to determine the full dependence $E^e_{\mathbf{R}}$ of electronic energies on the nuclei position $\mathbf{R}$.

From there, a solution to the problem with the total Hamiltonian is sought with a wave-function of the form[25]:

$$\Psi^{\text{tot}}(\mathbf{r},\mathbf{R}) = \Psi^e_{\mathbf{R}}(\mathbf{r})\Psi^n(\mathbf{R}), \tag{2.93}$$

[25] In practice several energy solutions of Eq. (2.92) are found and linear combinations are used in this second step.

where $\Psi^n(\mathbf{R})$ is the wave-function for the nuclei (depending on $\mathbf{R}$ only) while $\Psi^e_{\mathbf{R}}(\mathbf{r})$ is that of the electrons (depending on $\mathbf{r}$, and parametrically on $\mathbf{R}$). Using the last three equations, this second step is then equivalent to solving:

$$(H^n + E^e_{\mathbf{R}})\Psi^n(\mathbf{R}) = E_{\text{tot}}\Psi^n(\mathbf{R}), \tag{2.94}$$

where the electronic energy $E^e_{\mathbf{R}}$ now enters as an additional interaction potential affecting the nuclei.

In practice, it is impossible to solve Eq. (2.92) for all possible nuclei configurations $\mathbf{R}$. An initial guess for the equilibrium structure (typically with experimental information or semi-empirical models) must be used for $\mathbf{R}$ and then 'tweaked' according to the result of Eq. (2.94), trying to converge toward a solution. Equations (2.92) and (2.94) must in fact be solved self-consistently (each of them uses the solution of the other as a parameter). This is in general a very difficult undertaking and can be approached with different degrees of approximation in analytic or numerical form [118]. In Appendix A, we come back to this problem in the light of modern computational tools like density functional theory (DFT).

Classical Hamiltonian

To carry out the analysis further, it is useful to focus only on the nuclear motion (Eq. (2.94)) and transfer the problem to its classical equivalent. The classical approach is sufficient to define the concept of normal modes, including most of its important features such as mode symmetries. Its only limitation from a practical point of view is that several parameters must then be introduced empirically. The empirical parameters can however be fitted to the experimental observations either by Raman or infrared spectroscopy. Even better, it may also be possible to calculate these parameters directly from the full solution of the electronic structure, and then apply the classical treatment.

Let us consider, as before, a molecule with N atoms (indexed by $i = 1 \dots N$), each with mass m_i. The coordinates of the nuclei can be expressed in terms of the *displacements* $\mathbf{u}_i$ [m], with $i = 1 \dots N$, i.e. a three-dimensional vector representing the relative departures of the atoms with respect to their equilibrium positions (which must be given empirically). The kinetic energy of the classical Hamiltonian T^n [J] for the nuclei is a simple function of the velocities $\dot{\mathbf{u}}_i = \mathrm{d}\mathbf{u}_i/\mathrm{d}t$ in this representation [104]:

$$T^n = \frac{1}{2}\sum_{i=1}^{N} m_i \left(\dot{\mathbf{u}}_i\right)^2 . \tag{2.95}$$

Instead of (2.95) it is convenient to define $3N$ scalar variables ξ_i (also called displacements) accounting for all the Cartesian components of the N

vectors $\mathbf{u}_i$. For example, $\xi_1 = u_{1x}$, $\xi_2 = u_{1y}$, $\xi_3 = u_{1z}$, $\xi_4 = u_{2x}$, $\xi_5 = u_{2y}$, etc. It is common in this context to use *reduced coordinates* [104] (also known as *reduced-mass coordinates*) , q_i [kg$^{1/2}$ m], instead of Cartesian coordinates. These are defined as:

$$q_i = \sqrt{m_i}\xi_i, \tag{2.96}$$

where m_i $(i = 1 \ldots 3N)$ has been redefined in this expression to refer to the mass of the atom that ξ_i refers to. We then have the simplified expression:

$$T^n = \frac{1}{2}\sum_{i=1}^{3N} \dot{q}_i^2. \tag{2.97}$$

The potential energy, V^n [J], however, is more difficult to define without solving the molecular electronic structure. Nevertheless, for small displacements, it is possible to expand it in powers of q_i's as:

$$V^n = V_0^n + \sum_{i=1}^{3N} \left(\frac{\partial V^n}{\partial q_i}\right)_{q_i=0} q_i + \frac{1}{2}\sum_{i,j=1}^{3N} \left(\frac{\partial^2 V^n}{\partial q_i \partial q_j}\right)_{q_i,q_j=0} q_i q_j$$
$$+ \frac{1}{6}\sum_{i,j,k} \left(\frac{\partial^3 V^n}{\partial q_i \partial q_j \partial q_k}\right)_{q_i,q_j,q_k=0} q_i q_j q_k + \ldots. \tag{2.98}$$

The constant V_0^n is irrelevant as far as the dynamics of the nuclei is concerned and the second term cancels exactly in view of the fact that $(\partial V^n/\partial q_i)_{q_i=0}$ represents the forces on the atoms at their equilibrium positions (in all possible directions), and this must add to zero; we would not be at the equilibrium position otherwise. The classical treatment does not require the solution of the quantum mechanical problem here, for it takes the derivatives of the potential as empirical parameters.

Hence, to lowest order in the small displacements q_i, the potential energy is described by the third term and is given by:

$$V^n = \frac{1}{2}\sum_{i,j=1}^{3N} f_{i,j}\, q_i q_j, \quad \text{where} \quad f_{i,j} = \left(\frac{\partial^2 V^n}{\partial q_i \partial q_j}\right)_{q_i,q_j=0}. \tag{2.99}$$

The scalars $f_{i,j}$ [s^{-2}] are called the *force constants* and are sometimes grouped in a $3N \times 3N$ real symmetric matrix $\hat{F} = (f_{i,j})$ called the Hessian matrix (expressed here with respect to the reduced-mass Cartesian coordinates system, as evident in Eq. (2.99)). These force constants are *empirical parameters* in the classical treatment.

Note that the fourth term in Eq. (2.98), involving third-order derivatives of V^n is responsible for *anharmonic interactions*, to which we shall briefly return later.

Derivation of the normal modes

Once T^n and V^n are known, standard procedures of classical mechanics can be used to analyze the dynamics of the system [104]. One can for example use the *Lagrangian* of the system ($L = T^n - V^n$), and derive the dynamics from the Euler–Lagrange equation [104]:

$$\frac{\mathrm{d}}{\mathrm{d}t}\left(\frac{\partial L}{\partial \dot{q}_i}\right) - \frac{\partial L}{\partial q_i} = 0, \quad (i = 1 \ldots 3N), \tag{2.100}$$

which in this case is:

$$\frac{\mathrm{d}}{\mathrm{d}t}\left(\frac{\partial T^n}{\partial \dot{q}_i}\right) + \frac{\partial V^n}{\partial q_i}, \quad (i = 1 \ldots 3N). \tag{2.101}$$

We therefore obtain a linear system of $3N$ coupled linear differential equations for the reduced coordinates:

$$\ddot{q}_i + \sum_{j=1}^{3N} f_{i,j} q_j = 0, \quad (i = 1 \ldots 3N). \tag{2.102}$$

It is a standard result of classical mechanics that such a system of equations admits $3N$ harmonic independent solutions, each oscillating at a specific frequency ω_k. These solutions are called the *normal modes* of the linear system.

To view this more explicitly, we adopt the complex notation for the time dependence of the q_i's, and look for a harmonic solution of the form $q_i(t) = \mathrm{Re}(A_i \exp(-\mathrm{i}\omega t))$ (where A_i is *a priori* a complex number). We then have the $3N$ linear equations:

$$\sum_{j=1}^{3N} \left(f_{i,j} - \omega^2 \delta_{i,j}\right) A_j = 0 \quad (i = 1 \ldots 3N), \tag{2.103}$$

with $\delta_{i,j} = 1$ if $i = j$, and zero otherwise. This can be written in matrix form as $\hat{F} \cdot \mathbf{A} = \omega^2 \mathbf{A}$, where $\mathbf{A} = (A_i)_{i=1\ldots3N}$ is a $3N$-dimensional vector. The problem is therefore equivalent to finding the eigenvalues (ω^2) and eigenvectors ($\mathbf{A}$) of the Hessian matrix $\hat{F}$. Because $\hat{F}$ is by construction a real symmetric matrix, it must admit exactly $3N$ *real positive* eigenvalues (possibly degenerate), denoted ω_k^2 [s^{-2}], and a corresponding set of $3N$ *mutually orthonormal* eigenvectors $\mathbf{A}_k$ [a.d]. Moreover, $\mathbf{A}_k$ can be taken as real vectors (i.e. with all real components) without any loss of generality.

Some properties of the normal modes

The frequency ω_k [s^{-1}] and the corresponding real orthonormal $3N$-dimensional eigenvector $\mathbf{A}_k$ entirely define the dynamics of normal mode k. For such a normal mode, all atoms oscillate harmonically at the same frequency ω_k, with the $3N$ relative amplitudes for each nucleus coordinates given by the relative absolute values of the corresponding components of $\mathbf{A}_k$. Moreover, because $\mathbf{A}_k$ is real, all atoms oscillate *in phase* (or opposition of phase), i.e. they all reach their maximum amplitude or go through their equilibrium position at the same time. Note that all $\mathbf{A}_k$'s have been chosen, by convention, as unit vectors (they form an orthonormal basis), but any vector proportional to it represents the same normal mode. Therefore, a normal mode pattern is defined up to a proportionality constant for all the amplitudes of oscillations. The frequency of oscillations remains the same independent of the amplitude in the harmonic oscillations[26].

The $3N$ normal modes form a *complete* system (or *basis*, in the vocabulary of linear algebra) for the molecule's *free dynamics*: any solution of Eq. (2.103) can be decomposed as a linear combination of normal modes. An arbitrary pattern of displacement of the atoms left to oscillate freely will therefore in general be composed of more than one frequency. Only the specific patterns corresponding to the normal modes oscillate at a single frequency ω_k.

Finally, from physical considerations, six (five for linear molecules) of the eigenvalues must be zero and their corresponding eigenvectors represent the six (five) degrees of freedom of rigid motion: translation or rotation of the molecule as a whole along or around three independent directions. The other $3N - 6$ (($3N - 5$) in a linear molecule) eigenvalues (indexed by convention from $k = 1$ to $3N - 6$) correspond to internal deformations of the molecule and are called the *vibrational modes* or *normal vibrational modes.*

Normal coordinates

The $3N$ eigenvectors $\mathbf{A}_k$ form an *orthonormal basis* for the $3N$-dimensional vector space of atomic positions. A given molecular configuration, characterized by the atomic displacements q_i ($i = 1 \ldots 3N$) (in reduced-mass coordinates), can also be described by a new type of coordinate Q_k, relative to the orthonormal basis formed by the eigenvectors $\mathbf{A}_k$. Defining the vector $\mathbf{q} = (q_i)_{i=1\ldots3N}$, the new coordinates Q_k ($k = 1 \ldots 3N$) are implicitly defined

[26] The same happens for all harmonic potentials, for example for a pendulum in the small angle approximation; i.e. it has a frequency that is independent of the amplitude of the oscillation, as long as the potential is approximately harmonic.

by the linear transformation:

$$\mathbf{q} = \sum_{k=1}^{3N} Q_k \mathbf{A}_k, \quad \text{from which we deduce} \quad Q_k = \mathbf{A}_k \cdot \mathbf{q}. \tag{2.104}$$

These Q_k ($k = 1 \dots 3N$) are called the *normal coordinates* or *normal mode coordinates*, and are used extensively in the description of Raman scattering.

Note that the Q_k's can be used to characterize *any* molecular configuration, not only movements or internal vibrations. They form an orthonormal coordinate system in the same manner as the original reduced-mass coordinates q_i's. Their main advantage (and the real reason for introducing them) is that they are much more suited to the study of the free dynamics (in particular the vibrations) of the molecule. The normal coordinates simply express how much each normal mode is involved in a given deformation of the molecule. They are particularly useful when studying the effect of a single vibrational mode k, for example, on Raman scattering properties. Then one only need to consider *a single scalar* $Q_k(t)$ to describe the atomic displacements (rather than the $3N$ q_i's).

Finally, it is possible to show [104] that both the kinetic and potential energies do not have any cross-terms when expressed with normal mode coordinates, i.e.:

$$T^n = \frac{1}{2}\sum_{k=1}^{3N} \dot{Q}_k^2, \quad \text{and} \quad V^n = \frac{1}{2}\sum_{k=1}^{3N} \omega_k^2 Q_k^2. \tag{2.105}$$

This simple (uncoupled) form of T^n and V^n makes it relatively easy to transfer the classical results of harmonic oscillators to their quantum equivalents.

Quantization of vibrations

One of the beautiful (and arguably most useful) aspects of harmonic oscillations is that the classical treatment is transferable almost immediately to a quantum picture, with only a minimum of changes. The Hamiltonian for the atomic wave-function of the nuclei is obtained from the quantum version of the classical Hamiltonian $T^n + V^n$ by replacing the momenta and coordinates via the *principle of correspondence* [84]. Using the expressions given above in terms of normal coordinates (Q_k's), we ignore the six degrees of freedom associated with rigid motions ($k > 3N - 6$), and focus on the vibrational Hamiltonian for the quantum problem, which then reads:

$$H_v = \sum_{k=1}^{3N-6} \left[-\frac{\hbar^2}{2}\frac{\partial^2}{\partial Q_k^2} + \frac{1}{2}\omega_k^2 Q_k^2 \right]. \tag{2.106}$$

The vibrational wave-function Ψ_v for the nuclei is a solution of the time-independent Schrödinger equation: $H_v \Psi_v = E_v \Psi_v$, where E_v is the corresponding vibrational energy.

As a result of the decoupling between normal coordinates in Eq. (2.106), H_v is simply the Hamiltonian of $3N - 6$ *independent* quantum harmonic oscillators. The total wave-function can be written as a product of wave-functions of individual Q_k's, and the total energy E_v can then be written as the sum of the contributions of each normal mode, i.e.:

$$E_v = \sum_{k=1}^{3N-6} E_k, \tag{2.107}$$

where the vibrational energy of individual oscillators is:

$$E_k = \hbar\omega_k \left(v_k + \frac{1}{2} \right), \quad v_k = 0, 1, 2, \ldots. \tag{2.108}$$

The corresponding individual wave-functions are the modified Hermite polynomial functions of the standard harmonic oscillator [84,109].

We note at this stage that in most studies, we will focus on a single normal mode k and ignore the energy contribution (and dynamics) of the others. This is justified thanks to the independence (decoupling) of the normal modes.

Anharmonicities

In practice, the harmonic approximation of considering only the lowest-order term in the expansion of Eq. (2.98) is only valid for the lowest vibrational levels $v_k = 0, 1, 2, \ldots$, and even then only approximately. The higher-order terms in this expansion (fourth and beyond) are called anharmonic terms, and their effect is usually grouped under the general term of *anharmonicities.* In a first approximation, anharmonicities can be treated as a small perturbation of the harmonic behavior. This results in additional interactions among normal modes; i.e. the 'independent' harmonic modes become (slightly) coupled.

Anharmonicities play an important role in several aspects of Raman scattering. For example, the coupling among modes is responsible for intra-molecular vibrational redistribution (IVR), i.e. the rapid relaxation of vibrational energy through transfer to lower-energy vibrational (or rotational) modes. This relaxation affects the lifetime of the vibrational mode (which is in principle infinite for an ideal uncoupled mode), and therefore the homogeneous broadening of the corresponding Raman peak [119]. Similarly, anharmonicities are mostly responsible for the (small) temperature dependence of vibrational energies and peak broadenings [120].

A detailed quantitative description of anharmonicities is a difficult task, even for the simplest of molecules. Fortunately, it is rarely necessary for most typical Raman (and SERS) experiments, and we shall therefore not expand this discussion further.

2.7.2. More on symmetries and Raman selection rules

Symmetry of the normal modes and selection rules

For molecules with certain symmetries, the pattern of some of their normal vibrational modes will also necessarily have a certain symmetry, which in some cases automatically result in a zero Raman tensor (i.e. the contribution of some groups of atoms is exactly canceled out by an identical group with a symmetric pattern). A simple illustration of this concept is provided in the framework of the bond-polarizability model in Appendix B. More generally, such properties can in principle be predicted by a symmetry analysis of the molecule structure using concepts from group theory. Of all the subjects we are not going to treat in detail in this introduction to Raman spectroscopy (as a prelude to the SERS problem), the topic of *group theory* is perhaps the most involved. But there are very good reasons for this choice, as we shall explain in what follows. The study of symmetry (group theory) and its applications to spectroscopy can be found for example in Ref. [103]. Detailed accounts of selection rules and symmetry properties for Raman scattering are also given for example in Refs. [100,101,121] where tables of allowed Raman tensors for different symmetries are provided. In many cases, these symmetry considerations are enough to decide what the depolarization ratio of a specific vibration will be (if we can assign the structure of the molecule to a given point group). These can then be compared to experimental determinations, thus reinforcing the assignment of the bands.

It is important to have the right perspective over the understanding provided by symmetry analysis. It is *not* necessary to have a fluent understanding of group theory in order to understand the basic phenomenology of SERS, and Raman scattering by complex molecules in general. In fact, group theory displays its full power and beauty only in highly symmetric (and therefore typically small) molecules, like the ones shown in Ref. [103]. Molecules of moderate size and complexity like many real SERS probes (rhodamine 6G, for example), have typically structures that cannot be assigned to a specific point group. Accordingly, any symmetry analysis cannot be be done unless the eigenvectors are known in advance, which obviously limits its interest. If a mode is 'localized' around a region of high local symmetry (like a ring, for example), then a partial symmetry analysis can still be done and will give approximate results that can compare fairly well with experiments. But this is not the case for many vibrations in molecules (that are of interest for Raman and SERS), which involve collective movements of many atoms with a relatively low symmetry pattern. Finally, in resonant Raman scattering conditions (or even pre-resonant), the symmetry of the Raman tensor and their associated Raman selection rules are typically strongly modified (because they are affected by the strong coupling to an electronic transition in the molecule).

For these reasons, symmetry analysis, although conceptually interesting, is not a critical aspect of Raman scattering, especially in the context of SERS. In fact, for small and medium sized molecules (for which symmetry analysis might be practical), the full results (including the eigenvectors, the Raman tensors, the depolarization ratios, and the Raman cross-sections) can nowadays be obtained directly from a DFT calculation (as explained in Appendix A).

Parity symmetry

There is only one important general rule we would like to rescue from the symmetry analysis of molecules, and that is related to the presence/absence of a parity symmetry. Molecules (or sub-units of larger molecules) with an *inversion center* as a symmetry operation (i.e. inverting all the coordinates by their negatives $\mathbf{r} \rightarrow -\mathbf{r}$ with respect to a fixed point leaves the molecule unchanged) exhibit a mutual exclusion between modes that are Raman or IR active. In this case, both spectroscopic techniques are truly complementary, and the modes observed in one technique will not be active in the other. Classical examples are molecules like CO_2 (which we also consider in Appendix B) or benzene. On the contrary, molecules that do not have a center of inversion can have modes that are simultaneously Raman and IR active. A typical example is water (H_2O). That does not mean that the modes will be observable with the same relative strength in both types of spectroscopies, but they will be (at least in principle) observable with both.

The vibrational density of states of typical SERS probes

Finally, for a Raman-active mode, the Raman intensity of a mode is further dependent on the actual magnitude of the Raman tensor (through the link between Raman polarizability and cross-sections described in Section 2.4.6). Even when a mode has the right symmetry to be Raman active, its intensity (cross-section) may be negligible because all the components of the Raman tensor are very small. This may be a result for example of the nature of the chemical bonds participating in that specific normal mode. Consequently, the Raman spectrum only reveals a relatively small fraction of the total number of normal modes of a molecule.

Besides these visible Raman active modes, molecules with many atoms (such as dyes) have a quasi-continuous density of vibrational modes. This

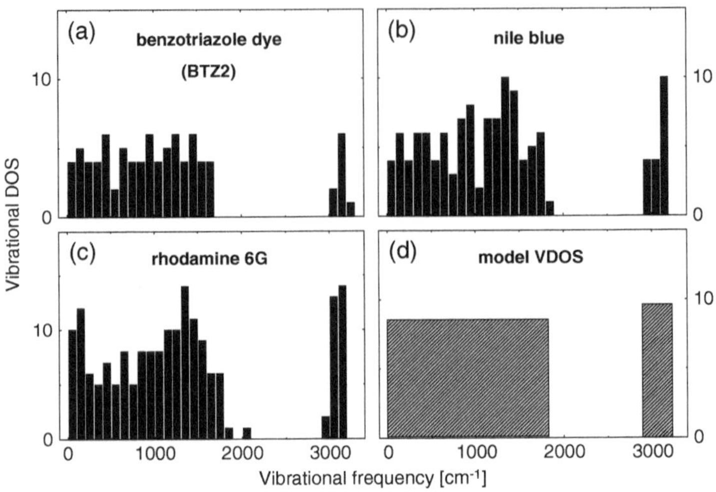

Figure 2.9. Vibrational density of states (VDOS) of three typical SERS probes in different approximations: (a) VDOS from a small probe (benzotriazole dye BTZ2 used in Ref. [34]) obtained by density functional theory (DFT), (b) VDOS of nile blue (NB) calculated with the MM+ molecular mechanics force-field model [122,123], and (c) VDOS of rhodamine 6G (RH6G) obtained by MM+. The main characteristics of most VDOS of interest are summarized in a schematic (d); i.e. a relatively flat and continuous VDOS up to $\sim$ 1700 cm^{-1} followed by isolated H-stretching modes at high frequencies ($\sim$3000 cm^{-1}). Only a small fraction of these modes are actually Raman active.

vibrational density of states (VDOS) is not directly relevant in most Raman experiments, but plays a role in related aspects such as vibrational relaxation, and is also related to the fluorescence spectrum. Several empirical formulas (like the Whitten–Rabinovich approximation [99,124]) have been proposed to calculate the density of vibrational states of a polyatomic molecule. These representations obviously have limitations, but they present a good starting point for analytic evaluations, estimations of orders of magnitude, and qualitative understanding. With the advent of efficient programs and available computational power to perform vibrational calculations in molecules, it is possible to calculate directly the density of states in many cases, including the use of some electronic structure plus vibrational calculations (like density functional theory; treated in Appendix A). The VDOS can be obtained with different levels of accuracy depending on the method being used, but it will not vary drastically, and certainly not enough for the point we want to raise here.

Figure 2.9 shows the VDOS of three different SERS probes, ignoring the rotational degrees of freedom, and calculated with different methods. The probes are three typical examples of SERS analytes (a benzotriazole dye (BTZ2) [34], nile blue (NB), and rhodamine 6G (RH6G)). Regardless of

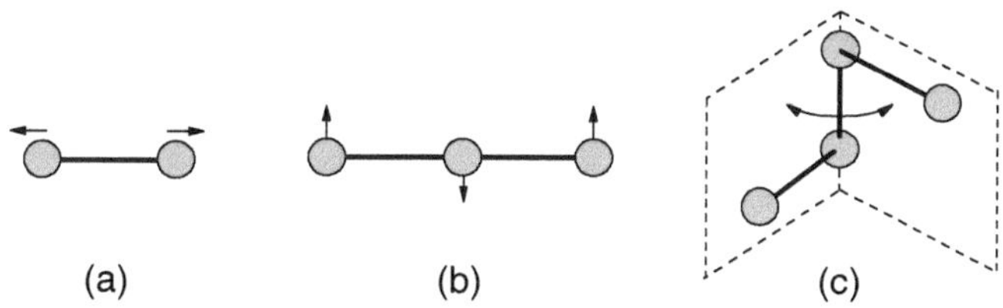

Figure 2.10. Typical local movements considered for *internal coordinates* for localized modes: (a) bond stretching, (b) bond bending, and (c) torsion. The latter is formally a form of bond bending mode, where the bond is being twisted along its main axis.

the approximation being used, the VDOS consists mainly of a relatively flat density of modes in the range ~0–1700 cm^{-1}, followed by a gap (Raman silent region), and an isolated hydrogen stretching region around ~3000 cm^{-1}. Only compounds with triple bonds (C≡C or C≡N) may have vibrational modes in the silent region. The main characteristics of the VDOS are summarized in the schematic plot in Fig. 2.9(d).

2.7.3. Modeling of molecular structure and vibrations

Raman peak assignments

Normal mode coordinates, used for the description of molecular vibrations, have the advantage of producing very compact results, which can be interpreted as the combined effect of many single (independent) oscillators. This advantage is particularly obvious when we are dealing with problems like the Taylor expansion of the polarizability (e.g. Eq. (2.65)). However, we lose the physical meaning of what the vibration actually means in terms of atom (or group of atoms) displacements. It is interesting to note that the experimental situation tends to be the opposite: with the experimental knowledge on the systematic appearance of certain modes in certain energy ranges, one would like to infer something about the nature of the atoms or bonds involved in these vibrations, and assign each Raman peak to a specific dominant vibrational pattern, such as a particular bond stretching or bond bending.

It is an experimental fact that molecules with the same or similar structural units tend to have Raman bands in more or less the same frequency range. This suggests that an analysis of the vibrational patterns producing these bands in terms of *localized movements* should be possible. Such an analysis may be empirical. The Raman band assignments to specific patterns (like a bond stretching) can be made by comparison with simpler compounds possessing similar chemical groups.

Let us mention briefly the types of movements that are typically considered for Raman peak assignments:

- *Bond stretching*, which corresponds to the elongation and contraction of a bond between two neighboring atoms (Fig. 2.10(a)). There are known empirical parameters to describe the most important typical bonds (C–C, C–H, C=C, C–N, etc.).
- *Bond bending* (Fig. 2.10(b)), involving movements in the direction perpendicular to that of the chemical bond between two neighboring atoms.
- *Torsion* (Fig. 2.10(c)), involving the twisting of a bond along its main axis.

In more complex structures such as rings, additional modes, such as *ring-breathing modes*, typically appear.

Force-field models

This can also be approached from a more theoretical point of view and this is called the method of *internal coordinates*. It starts by defining a complete basis of localized movements for certain groups of atoms that appear very often as building blocks of more complex molecules. Figure 2.10 provides examples of some of the simplest localized movements, involving only one or two atomic bonds. Unlike normal mode coordinates, internal coordinates can be defined in the same way for many different molecules, as long as the structural unit in question is present in the structure. A force constant is assigned to these movements based on available experimental data and this is the basis of the *force-field models* [122,123] . The groups can then be made to interact to represent collective movements of more complex structures. Ultimately, when the molecule is considered as a whole, we should get the same results we would have obtained with normal modes, for the difference between the two descriptions (in mathematical terms) is a simple change of basis for the representation of the movements. The advantage of the internal coordinates though is that it gives a description based on *localized movements of groups of atoms*, which are easier to identify spectroscopically in practical situations, either by Raman or IR spectroscopy. This way of analysis by splitting the complex dynamics of a molecule in terms of simpler (more localized) internal modes associated with known groups of atoms is also the basis for the fully empirical approach used by spectroscopists for assigning Raman or IR bands in molecules. It is based on the similarity and transferability of parameters from one molecule with the same local structural motifs into another. This 'transferability' is ultimately justified by the similarities in the nature of covalent chemical bonds.

Force-field models remain to a large extent empirical, even though some DFT calculations on small molecules can also be used to estimate parameters which are then applied in larger molecules. For many years, empirical force-field models have been the only reliable link between the huge wealth of spectroscopic experimental evidence and theory [125]. This situation is now changing slowly with the possibility to do DFT calculations on molecules of small and/or moderate size with modest computational resources (that evolve all the time at an extremely fast pace). Still, a phenomenological description based on local force fields is extremely useful to gain some insight into a problem and also because geometry optimizations of large molecules is an extremely time consuming exercise with DFT. Many DFT calculations start, in fact, by using the geometry optimization of a molecule based on empirical force fields in order to minimize (considerably) the computational time.

Other approaches

We will discuss two other approaches in this book, but both are given as self-contained appendices. The first one, density functional theory (DFT), is presented in Appendix A. Its main interest is practical: it provides with a minimum of effort (at least for small molecules) a wealth of details on the Raman properties: Raman cross-sections, vibrational pattern, Raman tensors, etc. The presentation in Appendix A in fact emphasizes in particular the practical use of DFT for Raman applications.

The second one, the bond-polarizability model, is presented in Appendix B. Its interest is mainly didactic: it provides a hands-on insight into the physical origin of the Raman polarizability and Raman selection rules.

These will be, we hope, interesting 'bonuses' to this chapter, but neither of these is really necessary to understand Raman or its extension to the SERS case.

2.8. SUMMARY

The basic aspects of the theory of Raman scattering by molecules have now been laid out, with a strong emphasis on those most relevant to SERS. It is timely to summarize the main points at this stage, in order not to lose sight of the big picture of the concepts that are specifically useful for SERS.

- Raman scattering can be introduced as a generalization of the (more familiar) linear optical properties of a molecule that produce elastic scattering of light. Central to this description is the concept of *polarizability*. Raman scattering appears thus as a 'modulation' of the linear optical polarizability due to the presence of internal vibrations.
- The phenomenological description given in Section 2.4 is arguably the most important aspect, in particular for the generalization of the effect to SERS (treated in Chapter 4).
- The description of other optical processes like fluorescence, within the same framework, is relevant for many SERS probes (which are fluorescent), and for related effects like surface-enhanced fluorescence.

- The classical link of the phenomenological description of Raman scattering with the microscopic properties (i.e. with molecular vibrations) arises from the expansion of the linear polarizability (Eq. (2.65)) and is extremely successful in explaining the basic phenomenology of the effect. However, it has its limitations. Nonetheless, except for specialized topics far beyond the objectives of this book, the full quantum mechanical description of the Raman effect is normally not needed for SERS. We have attempted to summarize the basic aspects of the latter however, and its links with the missing aspects of the classical description. An important result is the semi-classical expression for the Raman cross-section given in Eq. (2.86).
- Finally, a few elements of DFT calculations are further provided in Appendix A for those readers interested in the practical aspects of predicting the Raman spectrum of a molecule with standard codes and current computational resources. Appendix A should provide a hands-on introduction with further illustration of some of the concepts highlighted in this chapter. Appendix B, on the other hand, should provide an additional (didactic) approach to the understanding of Raman selection rules.

To conclude, we have tried to present in this chapter an approach to Raman scattering that is 'different' in some ways from other standard introductions to the field, and hopefully more adapted to SERS. The chosen topics should cover most of the important aspects of Raman spectroscopy for a potential researcher in the field of SERS. In fact, even only the contents of Sections 2.2 and 2.4 should be sufficient in most cases. We hope that the selection of topics appeals to an audience in between chemistry and physics, and provides a solid background for further studies in either molecular Raman scattering itself or SERS.

Chapter 3

Introduction to plasmons and plasmonics

The aim of this chapter is to give an overview of the physics of plasmons and their applications, so-called plasmonics. The emphasis will be on concepts rather than methods, since the goal is primarily to understand what plasmons are, which types of plasmons can exist under different conditions, and what their relevance to SERS is. Reading this chapter is necessary neither to understand SERS, nor to follow the other chapters of this book. Readers with a more immediate interest in SERS may read only Section 3.2 and then jump directly to Chapter 4.

There are however many reasons, why understanding the physics of plasmons is important to SERS. Firstly, plasmon resonances (in fact a certain type of plasmon resonance) are 'what makes SERS possible', and are mentioned and debated as the origin of the enhancement in almost all SERS discussions. Secondly, *plasmonics* is currently an expanding and very active area, from which SERS can gain further insight and where SERS can play an important role. In fact, this chapter covers many of the 'related plasmonic effects' mentioned in the subtitle of this book.

We discuss first in Section 3.2 the optical properties of noble metals, and in particular of gold and silver, which are most used for SERS. This section is the most directly relevant one to SERS. In Section 3.3, we then give a detailed account of the various types of plasmons, and emphasize their relevance to SERS and other applications[1]. The two most important types of plasmons are then discussed in detail: firstly, Section 3.4 focuses on

[1] As explained in the introductory chapter, SERS is a technique that exists at the boundary between physics, chemistry, biology, engineering, etc. As a result, many terms related to the physics of plasmons have 'evolved' into a terminology that is in many cases ambiguous or inaccurate. The detailed classification of different types of plasmons given in Section 3.3 is 'unnecessary' from a SERS standpoint. Having said this, the present book is aiming at a description of the fundamentals of SERS and, therefore, we felt that it was necessary to

the surface plasmon–polaritons at planar interfaces. These are the basis for most applications of plasmonics related to wave propagation and guiding, and some sensing applications. Section 3.5 then discusses the localized surface plasmon–polaritons, which play a major role in all applications using electromagnetic field enhancements, including SERS and surface-enhanced fluorescence (SEF). We then conclude by giving in Section 3.6 a few examples of plasmonic effects and possible applications (other than SERS).

3.1. PLASMONICS AND SERS

Plasmonics is a relatively new term, encompassing all areas of research and technology concerned with the study, fabrication, and applications of plasmon-supporting structures. This is not exactly new research since, in essence, it is simply the study of the optical properties of noble metals, in particular gold and silver. But the recent advances in nano-technologies and nano-science mentioned in the introductory chapter have opened up new possibilities in the design and fabrication of metallic structures with features in the nanometer size scale. This is sometimes referred to as *nano-plasmonics*. As a consequence, the general interest in plasmonics has dramatically increased over the last few years, and has spread to various other areas of physics, chemistry, and biology. This is reflected in the increase in the (normalized) number of plasmon-related publications as illustrated in Fig. 3.1. Furthermore, this is also evident in the number of recent reviews of the topic published in a wide range of scientific publications [11,62,126–133].

Plasmonics and SERS are two areas of research with a strong overlap and can benefit mutually from each other. SERS can be used as a tool for the study of plasmonic substrates and, reciprocally, some plasmonic substrates can turn out to be very good SERS substrates. Indeed, plasmons are at the core of SERS electromagnetic effects and/or enhancements. Most articles, talks, or even informal discussions about SERS are bound to mention plasmons or plasmon resonances at one point or another. Many different expressions appear in this context, among them: plasmon, plasmon resonance, radiative plasmon resonance, surface plasmon, surface plasmon–polariton, localized surface plasmon–polariton, propagating surface plasmon, and in many cases arbitrary combinations of the aforementioned! It is easy sometimes to lose sight of the main concepts and understand what is really meant from a specific standpoint, especially considering that they are not always used consistently or even correctly across the literature. With this in mind, we will attempt in the following to give a brief overview – at an introductory level – of the various concepts related to plasmons.

attempt to provide a rigorous classification of plasmons for future reference; an undertaking that would never be pursued in conventional papers in the field. This chapter is in that sense a 'bonus' to the content of the book and a prerequisite if a deeper understanding is sought. It is our hope that it will contribute in the future to a more precise vocabulary and denomination of plasmon-related phenomena.

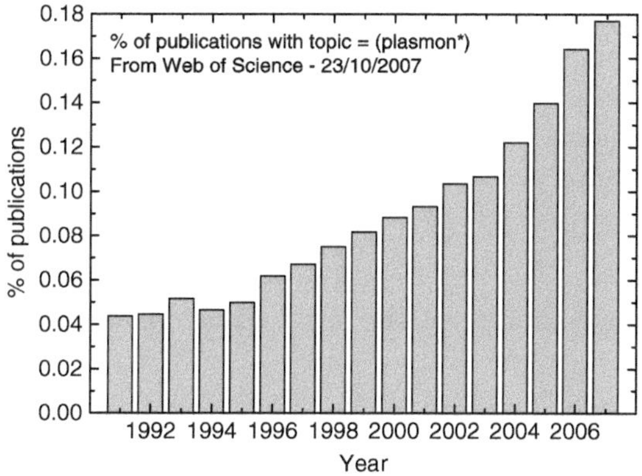

Figure 3.1. Evolution of the fraction of publications related to plasmons in the Web of Science database as of 23 October 2007.

3.2. THE OPTICAL PROPERTIES OF NOBLE METALS

Metals such as gold (Au), silver (Ag), copper (Cu), or aluminum (Al), have long been known to have different optical properties from standard dielectrics. They, for example, reflect light very efficiently in the visible, making them good materials for mirrors of various types (except possibly for their cost, etc.). These particular optical properties, along with many other physical properties (such as heat or electrical conductivity) all have the same physical origin: *the presence of free conduction electrons*. It seems fairly intuitive that these free electrons result in large heat or electrical conductivity[2]. Their connection with the optical response of metals may however appear less obvious at first sight.

The free electrons of a metal move in a background of fixed positive ions (the vibrations of ions, or phonons, are ignored here in a first approximation), which ensures overall neutrality. This forms, by definition, a *plasma* and can be called a free-electron plasma, or solid-state plasma [135]. The study of a solid-state plasma is different from that of a gaseous plasma, since the emphasis is on equilibrium phenomena in the first case, while it is more on instabilities or steady state properties in the latter [135]. The optical response of this free-electron plasma will govern all the optical properties of metals, at least in the visible part of the spectrum where its characteristic resonant energies reside.

To model the optical response of a free-electron plasma, one needs to determine the constitutive equations relating the currents and charges in the plasma to the electromagnetic fields. This is a very difficult undertaking in general

[2] The presence of free electrons in metals is linked to their ability to conduct both electricity and heat. However, it must be kept in mind that heat conduction in solids is in general a much more complex problem that does not only relate to the presence of free electrons. In fact, the best known heat conductor is diamond, which is also one of the best insulators.

because of many possible complications, including: the interaction of electrons with the underlying periodic structure of ions, the electron–electron correlations and the fermionic nature of electrons, the interaction of electrons with impurities and phonons, and the possible presence of surfaces. This response can be described with various degrees of refinements [135,136]. We will simply give here a brief account of the simplest model, namely: the Drude model.

3.2.1. The Drude model of the optical response

The Drude model is described in detail in many textbooks [96,136]. It is sufficient in many instances and for our purpose here and provides, in addition, a simple framework to understand electrical conductivity. It actually leads to the same result as a more elaborate approach (the random phase approximation) for the local dielectric function. We review here the main ingredients, in particular those relating to the optical response.

One simple way to introduce the Drude model is by using the Lorentz model for the atomic polarizability, presented in Appendix D. This model describes the optical response of an electron in an atom or molecule, *bound* with a restoring force characterized by a resonant frequency ω_0. The conduction electrons in a metal are not bound and can therefore, in a first approximation, be described by the Lorentz model , *without restoring force* (i.e. $\omega_0 \approx 0$). Moreover, because the free electrons are distributed uniformly and randomly throughout the metal, their contributions to the total optical susceptibility are simply the sum of their individual polarizabilities, without any local field correction. The Drude relative dielectric function of a metal can then be obtained by taking $\omega_0 = 0$ in the Lorentz model (see Appendix D for more details), i.e.:

$$\epsilon(\omega) = 1 - \frac{ne^2}{m\epsilon_0} \frac{1}{\omega^2 + i\gamma_0\omega}, \tag{3.1}$$

where n [m^{-3}] is the number of free electrons per unit volume and m [kg] their mass[3]. The damping term, γ_0 [rad s^{-1}], here corresponds to the collision rate of free electrons with the crystal or impurities (which also leads to the electrical resistivity in this simple model [96,136]). It is usually small compared to ω in the regions of interest here. The optical response of the positive ions in the crystal has so far been ignored. In a first approximation (which is at least correct at long wavelengths), these contribute to a constant background real dielectric function $\epsilon_\infty \geq 1$. This affects the optical response of the crystal and the dynamics of the free electrons. This can easily be incorporated in the

[3] Strictly speaking, m is the effective mass to partially account for the effect of the surrounding crystal structure and interactions beyond the free-electron approximation.

Drude model and leads to a slightly modified expression for $\epsilon(\omega)$, namely:

$$\boxed{\epsilon(\omega) = \epsilon_\infty \left(1 - \frac{\omega_p^2}{\omega^2 + \mathrm{i}\gamma_0\omega}\right),} \tag{3.2}$$

where we have defined ω_p [rad s^{-1}] as

$$\omega_p = \sqrt{\frac{ne^2}{m\epsilon_0\epsilon_\infty}}. \tag{3.3}$$

In the absence of an external perturbation, the charge density of a plasma is uniform and zero. It can be shown that ω_p is the natural oscillation frequency of the free-electron-plasma charge density and it is therefore called the *plasma frequency*. One can also define the corresponding wavelength $\lambda_p = 2\pi c/\omega_p$.

Taking the real and imaginary parts of the previous expression, we have:

$$\mathrm{Re}(\epsilon(\omega)) = \epsilon_\infty \left(1 - \frac{\omega_p^2}{\omega^2 + \gamma_0^2}\right), \tag{3.4}$$

and

$$\mathrm{Im}(\epsilon(\omega)) = \frac{\epsilon_\infty\omega_p^2\gamma_0}{\omega(\omega^2 + \gamma_0^2)}. \tag{3.5}$$

Keeping in mind that γ_0 is small compared to ω, we see that for a plasma described by a Drude model, the plasma frequency can be obtained from the condition $\mathrm{Re}(\epsilon(\omega_p)) \approx 0$. We also see that in the region where $\omega < \omega_p$ (wavelength longer than λ_p), we have $\mathrm{Re}(\epsilon(\omega)) < 0$. Moreover, if ω is not too small, the absorption, characterized by $\mathrm{Im}(\epsilon(\omega))$, is also small in this region. It is these two conditions, $\mathrm{Re}(\epsilon(\omega)) < 0$ and small $\mathrm{Im}(\epsilon(\omega))$, that make possible a whole range of interesting optical effects, including plasmon resonances. These conditions are never fulfilled in 'standard' dielectrics where $\mathrm{Re}(\epsilon)$ is typically between 1 and ~10. For many metals, the plasma frequency is in the UV part of the electromagnetic spectrum, and the region of interest is therefore in the visible (and close UV, or near infrared, depending on the metal).

3.2.2. The optical properties of real metals

Inter-band and intra-band transitions

The Drude model describes in a relatively simple way the optical response of a plasma, and in our case of the free conduction electrons of a metal.

This corresponds to *intra-band optical transitions*, since the excited electrons remain in the same electronic bands (the conduction band). We have already seen that the optical response of the fixed ions can also be included simply when their contributions are a constant background dielectric function ϵ_∞.

In a real metal, additional optical processes are likely to occur and contribute to the optical response, the most common being *inter-band transitions*, i.e. bound electrons optically excited to a higher energy band. Such transitions can for example be described – in a first approximation – as a collection of Lorentz oscillators, or by more complex expressions that take into account the *band-like* nature of these transitions, as illustrated in Appendix E. Their contribution, $\epsilon_b(\omega)$, adds to the free-electron contribution (from the Drude model). The dielectric function then takes the more general form:

$$\epsilon(\omega) = \epsilon_b(\omega) - \epsilon_\infty \frac{\omega_p^2}{\omega^2 + i\gamma_0\omega}, \tag{3.6}$$

where the term $\epsilon_b(\omega)$ represents the *inter-band transitions* that coexist with the free-electron contribution modeled by the second term. In many cases, inter-band transitions occur at energies in the UV, much higher than the plasma frequency. In this case, their contribution to ϵ in the visible is simply constant and real (no absorption off resonance), i.e. $\epsilon_b(\omega) = \epsilon_\infty$ and the expression reduces to the Drude model of the previous section. This is for example approximately the case for silver (see Appendix E).

If, however, there are inter-band transitions close to or below the plasma frequency, then their optical response needs to be added to the free-electron response as above. This is for example the case for gold (see Appendix E). Note that in this case, $\epsilon_b(\omega_p) \neq \epsilon_\infty$, and the condition $\mathrm{Re}(\epsilon(\omega)) = 0$ then no longer yields the free-electron-plasma frequency ω_p, as for the simple Drude model. This means physically that the natural oscillations of the free-electron plasma are 'affected' by the presence of the inter-band transitions.

Plasma frequency for Ag and Au

This description, although over-simplified, is nevertheless successful in explaining semi-quantitatively the optical response of many real metals. For example, at wavelengths longer than all inter-band transitions, Eq. (3.2) should be valid. The ω-dependence should then be described by the Drude term (and vary as ω^{-2}). This has been verified experimentally [137] for Ag and Au, from which the same value of $\sqrt{\epsilon_\infty}\omega_p \approx 1.4 \times 10^{16}$ rad s^{-1} is derived. This is in remarkable agreement with the expression for ω_p in Eq. (3.3) derived from the Drude model. Taking an effective mass equal to the electron mass and a density of conduction electrons of the order of $n \approx 6 \times 10^{28}$ m^{-3} (corresponding to a full d band with one free s electron per atom) give precisely

$\sqrt{\epsilon_\infty}\omega_p \approx 1.4 \times 10^{16}$ rad s^{-1}, or $\hbar\omega_p \approx 9.1/\sqrt{\epsilon_\infty}$ eV, or a corresponding wavelength of $\lambda_p \approx 136\sqrt{\epsilon_\infty}$ nm. The Drude model therefore gives an excellent description of the long-wavelength optical response of Ag and Au. Only the value of ϵ_∞ is missing. This is understandable though, for ϵ_∞ comes from high energy contributions in the deep UV-range, which are not included in this simple treatment of the problem.

The frequency-dependent dielectric functions of Ag and Au are of crucial importance for many plasmonics problems. They are further discussed in Appendix E, where analytical models are also provided. These analytical models will be used extensively in this book for the modeling of plasmonics and SERS effects.

3.2.3. Non-local optical properties

Many optical properties of metals can be understood using a *local* dielectric function, as described above. This implicitly assumes that the dielectric function of the metal is *only* frequency-dependent. There are, however, a few cases where this approximation is unsuccessful in providing an explanation of the experimental observations. A non-local description, where the dielectric function ϵ of the material depends both on frequency (ω) and wave-vector ($\mathbf{k}$) (see Appendix C for more details), is then required. One such example is the anomalous skin effect [136]. Another situation where non-local effects can be important is that of an emitter very close to a metal surface, which is precisely the case of SERS. However, the inclusion of non-local effects in SERS complicates tremendously the electromagnetic problem, which is already not trivial in the local approximation. The easiest solution is to ignore them altogether, which in many cases does not affect the qualitative and even quantitative conclusions. It is however useful to bear in mind that these effects do exist, and can be occasionally the source of either problems or unexplained phenomena. Their influence can be studied separately for very simple cases, such as emitters close to plane or spherical surfaces. A discussion of non-local effects would take us too far from our main purpose here and can be found (to some degree) in the literature [138].

3.2.4. What makes the metal–light interaction so special?

A brief survey of various metals

A comparison of the optical properties (real and imaginary parts of the dielectric function) of various metals (from Ref. [139]) is given in Fig. 3.2. The monotonous decay of Re(ϵ) from small values in the UV to negative values in the visible, and very negative values in the infrared, is common to all metals and is in fact predicted by the Drude model (for typical plasma frequencies in the UV). *This is one of the most important characteristics of metals*, as far as optical properties are concerned, and it is a consequence of

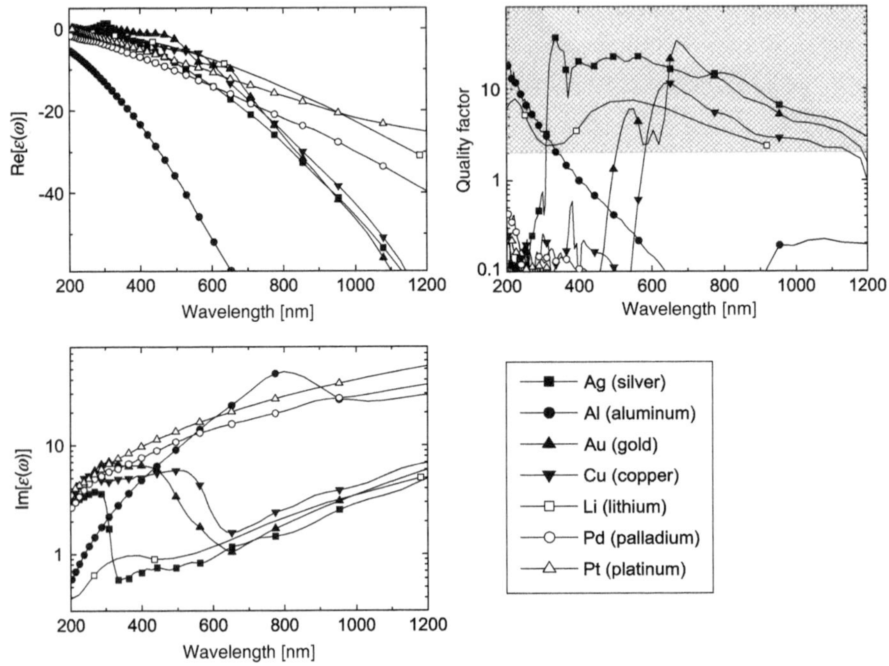

Figure 3.2. Overview of the optical properties of a selection of metals in the (extended) visible range. The real (top) and imaginary (bottom) parts of ϵ are plotted against wavelength on the left (from Ref. [139]). Also shown (top-right) is the predicted approximate quality factor Q of localized surface plasmon resonances for a metal/air nano-particle, as defined in Eq. (3.8). The shaded area is the area of interest to many plasmonics applications.

the optical response of the free (conduction) electrons, as explained simply by the Drude model.

The negativity of the real part of $\epsilon(\omega)$ at visible wavelengths is also the origin of many of the known optical properties of metals, including plasmon-related effects. For example, it implies that the refractive index is smaller than one and even close to zero, while most 'common' materials have a refractive index of $\sim$1 or more. The reflection coefficient at a dielectric/metal interface is derived simply from the relative refractive indices (see for example Appendix F), and the small refractive index of metals directly leads to a reflection coefficient close to one (almost perfect reflector), which is arguably one of the best known and most 'visible' properties of metals.

Plasmon resonances and plasmonics

There are two important types of plasmons, which will be discussed extensively in the rest of this chapter: localized surface plasmon–polaritons

(LSPP or LSP for short) and propagating surface plasmon–polaritons (PSPP). Here we have to anticipate some of these results to carry on the general discussion of the optical properties of metals. As we shall see, the negative real part of $\epsilon(\omega)$ is also linked to the existence of plasmons and plasmon resonances.

To understand this in simple terms, let us consider briefly the problem of a small metallic sphere interacting with an electromagnetic field (in a laser beam). The simplest treatment of this problem is to consider a sphere much smaller than the wavelength of the incident beam and solve the problem within the electrostatic approximation (to be treated in Section 5.1.4). It is then equivalent to the electrostatic problem of a sphere, in a medium of relative dielectric constant ϵ_M, and in a uniform external electric field (to be treated in Section 6.2.1). The (complex) electric field inside the sphere can then be shown to be constant and proportional to the incident field E_0 [96]:

$$E_{\text{In}} = \frac{3\epsilon_M}{\epsilon(\omega) + 2\epsilon_M} E_0. \tag{3.7}$$

The important part in this expression is the denominator. If it were close to zero (a condition which can be achieved if $\epsilon(\omega) \approx -2\epsilon_M$) then the fraction would be *very* large (infinite for perfect equality). This is not possible for standard dielectrics, for which ϵ is typically between 1 and $\sim$10. But for metals, this condition can be approximately met if the absorption is small ($\text{Im}(\epsilon(\omega)) \approx 0$) at a wavelength where $\text{Re}(\epsilon(\omega)) \approx -2\epsilon_M$. The optical response (absorption and scattering) at this particular frequency (or wavelength) is then very large, i.e. this is the signature of a *resonance*. This is explicitly illustrated in Fig. 3.3 where the absorption coefficient (proportional to $|E_{\text{In}}|^2$) is shown as a function of wavelength for a small silver sphere in air ($\epsilon_M = 1$). There is a clear resonant response at $\approx$345 nm, which corresponds to the condition $\text{Re}(\epsilon(\omega)) \approx -2$. It is interesting to note at this stage that while the resonance condition is determined primarily by the real part of $\epsilon(\omega)$, it is actually its imaginary part that limits how large the resonance can be.

This is a simple example of a localized surface plasmon (LSP) resonance. Similar effects will exist for other geometries and configurations, but with different denominators and therefore different resonance conditions. This is an interesting concept that needs to be highlighted: optical resonance conditions in small metallic objects are not purely intrinsic of the material properties but they are strongly linked to the *geometry*. Two objects made with the same metal but with different geometries will have different resonance conditions. These conditions however always correspond to a negative real part of ϵ. Similarly, the existence of propagating surface plasmon–polariton (PSPP) waves at metal/dielectric interfaces also requires a negative real part of ϵ.

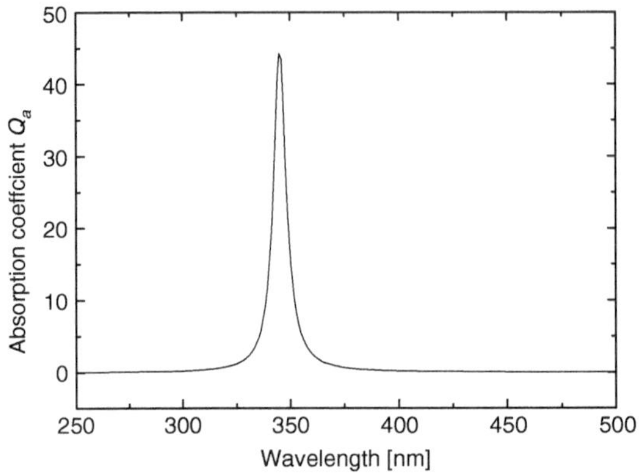

Figure 3.3. Absorption coefficient, $Q_{\rm Abs}^{\rm NP}$ (see Section 5.1.3) as a function of wavelength for a silver sphere of radius 25 nm in air in the electrostatic approximation. $Q_{\rm Abs}^{\rm NP}$ can be expressed as (see Section 6.2.1) $Q_{\rm Abs}^{\rm NP} = 12(\omega a {\rm Im}(\epsilon(\omega)))/(c|\epsilon(\omega)+2|^2)$.

Moreover, both types of plasmons (LSP and PSPP) are strongly affected by optical absorption, i.e. the larger the $\mathrm{Im}(\epsilon)$, the more 'lossy' they are (the meaning of this will be made clearer later). For LSP resonances, this damping can further be characterized by a *quality factor* defined as [140]:

$$Q = \frac{\omega(d\epsilon'/d\omega)}{2(\epsilon''(\omega))^2}, \tag{3.8}$$

where $\epsilon' = \mathrm{Re}(\epsilon)$ and $\epsilon'' = \mathrm{Im}(\epsilon)$. In simple terms, Q is large when $\mathrm{Im}(\epsilon)$ is small and, therefore, characterizes the strength (and width) of the resonance. Q is also plotted in Fig. 3.2 for various metals.

Which metals are good for plasmonics and SERS?

From these qualitative arguments (which will be further justified in the rest of this chapter), one can consider that a metal is good for plasmonics if:

- $\mathrm{Re}(\epsilon)$ is negative in the wavelength range of interest (typically the visible and near infrared). For LSP applications, one in fact needs (as a rule of thumb) $-20 \le \mathrm{Re}(\epsilon) \le -1$.
- $\mathrm{Im}(\epsilon)$ is small (or equivalently Q is large) in the range of interest. Typically, Q must be larger than $\sim$2, preferably larger than $\sim$10.

From the examples of Fig. 3.2, this rules out a number of metals, such as aluminum, palladium, platinum, etc. The latter two present too much absorption, while aluminum would only be suitable for applications in the UV.

Note that these metals do show a metallic behavior (including in their optical properties) but are not expected to exhibit any plasmonic properties[4].

Among the other metals studied in Fig. 3.2, silver is the most promising one (and is in fact widely used for SERS and plasmonics). Gold and copper are also suitable, but only at longer wavelengths (typically more than ~600 nm). At such wavelengths, the optical absorption of gold in fact becomes comparable to that of silver. Finally, lithium also exhibits suitable properties across the whole visible range, as silver, but has not been used much. Lithium reacts easily in water and does not occur freely in nature due to its chemical activity; it is therefore not very easy to be used as a plasmon-supporting material.

To these theoretical considerations, one should add the (very important) practical issues: availability, ease of manipulation, especially for the fabrication of nano-structures, toxicity, durability, cost, etc. Gold is certainly the most promising in these categories, and should therefore be the material of choice for applications beyond ~600 nm (in the red and near IR). Silver, whose absorption is the smallest especially below ~600 nm, can be used for large field enhancement applications (such as single-molecule detection). These two metals are by far the most widely used metals in plasmonics applications, including SERS.

Finally, it is interesting to remark that the conditions listed above also appear in other materials, but at different wavelengths. Similar effects are therefore expected, but are usually not considered as being part of the field of plasmonics. This is the case for example of doped semiconductors, whose conduction electrons result in a similar optical response as metals but with a plasma frequency in the far infrared (due to the smaller density of carriers). Another example is the optical response of phonons in ionic crystals, which also leads to negative $\mathrm{Re}(\epsilon)$, but again at much longer wavelengths. In this final case, plasmons do not even play a role, since the optical response is not dictated by free electrons but by phonons; the related effects may be referred to as 'phononics'.

3.3. WHAT ARE PLASMONS?

In the rest of this chapter, we will focus on plasmons and their relation to SERS and plasmonics.

3.3.1. The plasmon confusion

In the modern SERS literature (and in many other areas), many effects are attributed to plasmons or plasmon resonances, without further details about

[4] For example, SERS has been observed on these, but the enhancements are either 'chemical' (see Chapter 4) or small electromagnetic enhancements not arising from plasmon resonances.

what plasmons actually are. One can sometimes find a sentence or two about the origin of plasmons; the most common assertion being: 'plasmon resonances are due to collective charge oscillations of the free electrons of the metal'. However, there is hardly ever a real concern about these 'electron oscillations' in typical SERS studies. SERS enhancements are usually explained as an electromagnetic effect, not an electron oscillation effect. There seems to be a missing link between the electron oscillations and the large electromagnetic field enhancements. To add to this, the term plasmon is also used in contexts other than SERS; for example as a way to guide light (in a plasmonic waveguide) or in electron energy loss (EELS) experiments. Finally, adjectives are used in many cases to qualify the plasmons, in an attempt to clarify the situation. Examples of these are: surface plasmon, plasmon–polariton, radiative, non-radiative, propagating, and localized plasmons. Unfortunately, these terms are not always used consistently by authors or across disciplines (chemistry and physics in the first instance), and have even evolved over time to designate something different from what they were 40 years ago.

This general vagueness leads to numerous confusions about what plasmons are, which types of plasmons can be encountered, and what their respective importance is for applications such as SERS. This can be the source of frustration for people new to the field. The aim of the following discussion is therefore three-fold:

- Firstly, we will attempt to define clearly what plasmons are, and describe the different types of plasmons. Because of the confusion discussed above, some of the definitions and descriptions may however be slightly biased and not correspond exactly to the choices of other researchers.
- Secondly, we will highlight some of the most important applications of plasmons, including the main topic of this book which is SERS. We will emphasize in particular the different nature of plasmons used for different applications.
- Finally, one important message we wish to convey is that in most cases of interest to SERS and plasmonics, all plasmon-related effects can be understood as *electromagnetic effects*. The relation to the free electrons of a metal is only secondary (although important from a fundamental point of view). All information on plasmons, and plasmon resonances, is fully contained in the dielectric function and the geometry of a specific problem.

3.3.2. Definition and history

Plasmons

The use of the term plasmon has evolved somewhat over the years, mainly because the types of experiments carried out on metals have changed. It is however useful to look at the original definition of the term, which is still valid although it may now be used in a wider sense and in other frameworks.

The term plasmon was introduced by Pines in 1956 in the introduction of a review article [141] about collective energy losses. In Pines' work we find the following definition:

> The valence electron collective oscillations resemble closely the electronic plasma oscillations observed in gaseous discharges. We introduce the term 'plasmon' to describe the quantum of elementary excitation associated with this high-frequency collective motion.

A plasmon is therefore a quantum quasi-particle representing the elementary excitations, or modes, of the charge density oscillations in a plasma. Note that the study of these oscillations started earlier, even if they were not known or identified as plasmons [141]. We will come back to the notion of elementary excitations or modes of a system in the next section.

Although the term 'plasmon' is sometimes used in a broader context, the formal definition given above is the definition of reference. It draws its origin from quantum mechanics, even though we will see that quantum mechanics is, in fact, not necessary to study plasmons. A useful analogy to understand the meaning of this definition is to recall the formal definition of a photon: it is the quantum particle representing the elementary excitations, or modes, of the free electromagnetic field oscillations. *A plasmon is therefore simply to the plasma charge density what photons are to the electromagnetic field.* Many properties of photons can be studied within a classical framework, using Maxwell's equations. Similarly, many properties of plasmons can be studied within a classical description of the plasma and its interactions. There is, may be, a small difference in the vocabulary between plasmons and photons, but it is only artificial: people typically only use the term 'photon' when dealing with quantum aspects of the electromagnetic fields (such as absorption or emission by an atom). In classical situations, the term electromagnetic wave, or electromagnetic mode, is usually preferred. For reasons that are more historical than scientific, the term 'plasmon' tends to be used in all situations, quantum or classical, instead of equivalent (classical) denominations such as *charge density oscillations* . A more important and fundamental difference is that a photon is a real quantum particle while a plasmon is a *quasi-particle* because it is always *'lossy'* and highly interacting. A charge density oscillation, if not maintained by an external source of energy, will always decay because of various loss mechanisms (collisions, etc.).

Plasmon–polaritons

Another important type of elementary excitations, related to plasmons, is that of an electromagnetic wave propagating in a medium. By medium, we mean here an optically responsive medium (with a relative dielectric function $\epsilon \neq 1$ or relative magnetic permeability $\mu \neq 1$). From classical electromagnetism, it is known that such a situation can be described by Maxwell's equations for media, which introduce the notion of internal polarization, **P**, or magnetization **M** (see Appendix C). These account for the fact that the electromagnetic wave excites the internal degrees of freedom of the medium (typically bound electrons in a dielectric). The energy of such a wave is therefore shared between the electromagnetic field oscillations (of **E** and **B**), and the internal excitations of the medium (typically represented by oscillations of **P** and **M**). The corresponding quantum particle is no longer a photon, but a photon coupled to the internal degrees of freedom of the medium. Such modes are usually called *polaritons*[5]. If the internal excitations of the medium are identified, then the polariton can sometimes be further qualified. For example, the optical response of an ionic crystal in the mid-infrared is dominated by the interaction of light with crystal vibrations (phonons). The electromagnetic waves in such a medium are then called phonon–polaritons (mixed photon–phonon modes). Similarly, the optical response of a metal in the visible and infrared is dominated by the interaction of light with the free-electron plasma. The electromagnetic waves in a metal are then called *plasmon–polaritons* (mixed photon–plasmon modes). As we shall see, SERS and plasmonics are mostly concerned with plasmon–polaritons, rather than 'pure' plasmon modes.

Surface Plasmon–polaritons

Finally, in 1957, shortly after the introduction of the term plasmon, Ritchie [143] predicted the existence of another family of plasma modes in thin films, corresponding to longitudinal charge density waves propagating at a metal/dielectric interface. This prediction was confirmed experimentally in 1959 [144] and these modes (once quantized) were called *surface plasmons* a year later in 1960 [145]. In fact, these surface plasmon modes were originally introduced [143] within the electrostatic approximation (to be treated in Section 5.1.4). If however retardation effects are not negligible, then these charge density waves cannot exist without being associated with a transverse electromagnetic wave (a photon). This then corresponds to a mixed mode where the energy is shared between the charge density wave (plasmon) and the electromagnetic wave (photon), and they should therefore be called *surface plasmon–polaritons.*

[5] The exact definition of a polariton remains a matter of choice. Some people reserve this name for media with a strongly resonant optical response (with phonons or plasmons for example), and keep the term photon for 'standard' dielectrics (with a constant relative dielectric function $\epsilon \neq 1$). However, strictly speaking, a photon corresponds only to an electromagnetic wave in vacuum, and it becomes a polariton in any media. See for example Ref. [142] for a detailed discussion on the nature of polaritons.

There is therefore a fundamental difference between plasmon modes and the surface plasmon modes as introduced by Ritchie [143]. Plasmons can exist either by themselves without mixing with a photon, or as a mixed plasmon–photon mode (plasmon–polariton). Surface plasmons however are always strictly speaking surface plasmon–polaritons (mixed modes). The 'pure' surface plasmon modes are only an approximation of a surface plasmon–polariton for which the photon contribution is small or negligible (for example in the electrostatic approximation). We will come back to this distinction later.

3.3.3. The relation between plasmons and the dielectric function

The previous discussion in terms of elementary excitations and their quantization is useful to understand the origin of the term plasmon and the definition of the various types of plasmons. We will indeed come back to it shortly in more detail. However, one should bear in mind that plasmons and plasmon–polaritons are rarely viewed as quasi-particles, and are in fact mostly described as charge density oscillations (for plasmons), or electromagnetic waves in a medium (for plasmon–polaritons).

The body of early work on plasmons in the late 1950s, both theoretical and experimental, was concerned primarily with electron energy loss (EELS) in metals, and not directly with the optical properties of metals themselves. The former relates more to the dynamics of the free-electron plasma by itself rather than to its interaction with an external electromagnetic field. However, this dynamics is partially governed by electromagnetic interactions within the electron gas and with its environment. Charges, static or moving, are sources of electromagnetic field and a charge density wave cannot exist without an associated electromagnetic (or at least electric) wave. Reciprocally, the optical response of the free electrons is determined by their dynamical properties. *The dynamics of the plasma is therefore intricately linked to its optical properties and both can be entirely described using the dielectric function of the metal.*

Depending on the context, it may therefore be useful to emphasize one aspect (charge density and free-electron-gas dynamics) or the other (optical response). In the case of plasmonics, and even more so for SERS, the 'optical response approach' is usually the most relevant. In this sense, *the only thing needed to study SERS and plasmonics effects with a given metal is a knowledge of its optical response described by a relative dielectric function* $\epsilon(\omega)$ (and possibly $\epsilon(\mathbf{k}, \omega)$ if non-local effects are considered). One could then 'forget' about the fact that this optical response is the result of the free-electron dynamics, about the presence or not of charge and/or surface charge oscillations, or collective charge oscillations. Within this 'optical response approach', the plasmons and plasmon–polaritons can then simply be viewed as *electromagnetic modes* of the system under consideration. A detailed general discussion of electromagnetic modes will therefore be given in the rest of

Section 3.3. Its purpose is two-fold: firstly to show that plasmon-related effects are simply a specific class of a wider family of electromagnetic effects. Secondly, to introduce the terminology that is commonly used to qualify plasmon and plasmon–polariton modes. Note that this is somewhat a long theoretical digression in the fundamentals of plasmonics and it is possible to jump directly to the more practical aspects of plasmonics discussed in the rest of this chapter (Section 3.4 and beyond). In fact, this digression may be more digestible (and more useful) as a second read once the rest of the chapter has been understood.

3.3.4. Electromagnetic modes in infinite systems

The concept of elementary excitations or modes in infinite systems

The study of the elementary excitations or modes of a system is common in many areas of physics as part of linear response theory or Fourier analysis. In a nutshell, it consists in finding specific solutions (eigenvectors) of the physical system under study. In infinite systems with translational invariance, these solutions are propagating plane waves, i.e. solutions where all quantities (usually scalars like charge density, or vectors like electric field) have an oscillatory dependence (in space and time) of the form: $\cos(\mathbf{k} \cdot \mathbf{r} - \omega t + \phi)$, or in complex notation $\exp(\mathrm{i}\mathbf{k} \cdot \mathbf{r} - \mathrm{i}\omega t)$ (see Appendix C). Such solutions in general exist only for specific values (eigenvalues) of ω and $\mathbf{k}$. These solutions can usually be described by one or more *dispersion relations* $\omega(\mathbf{k})$. Each of the allowed solution with a given ω and $\mathbf{k}$ is then called an *elementary excitation or mode* of the system, and corresponds to a propagating plane wave. The reason why the modes are indexed here by their $\mathbf{k}$ vector is that the system is assumed to be invariant by translation in all directions.

When such a system is described in quantum mechanics, the elementary excitations are quantized, and they can be viewed either as plane waves, with frequency ω and wave-vector $\mathbf{k}$, or as particles with energy $\hbar\omega$ and momentum $\hbar\mathbf{k}$, linked by the dispersion relation $\omega(\mathbf{k})$. This step is, however, usually not necessary to discuss the classical properties of a system. Despite this, it is common to use the name of the quantum particle to designate the modes or elementary excitations, even when studied within a classical approach. This is in particular the case for plasmons.

Particles and quasi-particles

Many physical systems are affected by damping or losses in one form or another. In this case, a wave cannot propagate unchanged forever. Its amplitude must therefore decay in time and/or space. Such an excitation is called a *quasi-particle* (since it cannot exist 'forever' by itself). Note that plasmons are always quasi-particles (except in ideal non-absorbing metals).

This damping translates mathematically by the fact that ω and $\mathbf{k}$, which are related by the dispersion relation, cannot be both real for a quasi-particle. There are two possible points of view in this case:

- The first and most common approach is to consider ω real, which leads to $\mathbf{k}$ being complex. This implies an exponential decrease of the field amplitudes as $\exp(-(\mathrm{Im}(\mathbf{k}) \cdot \mathbf{r}))$. Such waves are called *evanescent waves or modes*, since they only propagate over a limited distance characterized by $1/|\mathrm{Im}(\mathbf{k})|$. The wave-vector for propagation is then $\mathrm{Re}(\mathbf{k})$.
- The second approach is to take $\mathbf{k}$ real, which leads to a complex frequency $\omega = \omega' - \mathrm{i}\omega''$ (with $\omega'' > 0$). This implies that the field amplitudes decay in time as $\exp(-\omega'' t)$ and such modes are then called *virtual modes* (this is the classical denomination, equivalent to a quasi-particle in the quantum point of view). The theory of virtual modes is common in nuclear and particle physics. These modes cannot exist as such (which is why they are called virtual), but they appear as resonances in the response of the system when the (real frequency) is equal to ω'. ω'' then characterizes the width of the resonance, or the lifetime of the virtual excitation ($\tau = 1/(2\omega'')$).

Both approaches are simply an attempt to represent *damped electromagnetic modes* or, from a quantum perspective, *quasi-particles.*

These two points of view are equivalent and which one is used depends on the exact physical situation and, in particular, on how the mode is excited:

- If an elementary excitation is maintained over time by an external source (acting in a given region of space, and driven at a given frequency ω), then it is logical to take ω real and $\mathbf{k}$ complex, and view the excitation as a wave decaying in space, i.e. an evanescent wave.
- If an elementary excitation is created at a given time by an external source (which is then switched off), then it is more logical to take ω complex and $\mathbf{k}$ real, and view the excitation as propagating in space but decaying over time, i.e. as a virtual mode. This point of view is also the only possible one for problems where $\mathbf{k}$ is not well defined (no translational invariance).

Finally, in infinite systems without losses and damping, elementary excitation can propagate 'forever' without decay in space and time. They are then called *propagating waves,* or from the quantum perspective, simply *particles.*

Longitudinal and transverse modes

One important concept for plasmon modes is that of longitudinal and transverse modes. A mode is described primarily by its frequency ω and wave-vector $\mathbf{k}$, linked by the dispersion relation $\omega(\mathbf{k})$. If the oscillating quantity is a vector, as is the case for the electric field $\mathbf{E}$, then one can distinguish two situations in isotropic and homogeneous media:

- $\mathbf{E}$ // $\mathbf{k}$ everywhere ($\mathbf{k} \times \mathbf{E} = 0$). This is then called a *longitudinal mode or wave*.
- $\mathbf{E} \perp \mathbf{k}$ everywhere ($\mathbf{k} \cdot \mathbf{E} = 0$), which then corresponds to a *transverse mode or wave*.

The origin of the name longitudinal and transverse is then clear; it refers to the orientation of $\mathbf{E}$ with respect to the direction of propagation $\mathbf{k}$. These definitions can also be extended mathematically to a general vector field $\mathbf{E}$, without the need for a wave-vector $\mathbf{k}$ (and therefore also valid in the absence of translational invariance):

- $\nabla \times \mathbf{E} = 0$ for a longitudinal field, and
- $\nabla \cdot \mathbf{E} = 0$ for a transverse field.

For a propagating mode, with a dependence on complex notation of the type $\exp(i\mathbf{k} \cdot \mathbf{r})$, this simply reduces to the previous simple definition in terms of $\mathbf{k}$. Finally, there is a theorem from vector analysis stating that any vector field can be decomposed (uniquely) into the sum of a transverse and a longitudinal field.

Electromagnetic modes in infinite (3D) vacuum – photons

One simple example of an infinite physical system is the electromagnetic field in vacuum. The electromagnetic modes are then derived easily from Maxwell's equations (see Section F.1). In particular, the equations $\nabla \cdot \mathbf{E} = 0$ and $\nabla \cdot \mathbf{B} = 0$ imply that $\mathbf{E}$ and $\mathbf{B}$ are transverse fields. The modes are then *transverse propagating plane waves* characterized by a wave-vector $\mathbf{k}$ and frequency ω, related by the dispersion relation: $\omega = c|\mathbf{k}|$. Each pair of real ω and $\mathbf{k}$ satisfying the dispersion relation corresponds to a *propagating electromagnetic mode*. After quantization, these modes can also be viewed as particles and are then called photons, but most of their properties can also be described classically using Maxwell's equations.

Electromagnetic modes in an infinite (3D) medium – polaritons

A similar situation occurs for transverse electromagnetic waves in a medium[6] with (local) relative dielectric function $\epsilon(\omega)$. The dispersion relation

[6] The media are always assumed to be non-magnetic (with relative magnetic permeability $\mu = 1$), unless otherwise stated.

is then modified to give (Eq. (F.2) in Section F.1):

$$\epsilon(\omega)\omega^2 = c^2\mathbf{k}\cdot\mathbf{k}. \tag{3.9}$$

Such an electromagnetic wave creates in the medium an internal polarization wave: $\mathbf{P} = \epsilon_0(\epsilon(\omega) - 1)\mathbf{E}$. These modes are then called polaritons because they couple transverse electromagnetic excitations (photons) with an internal polarization $\mathbf{P}$, which originates physically from internal excitations of the medium, such as excited bound electrons, phonons or plasmons. In non-absorbing dielectrics, such as glass, the polaritons are quantum particles similar in many ways to photons. For metals, in the region where the optical response is dominated by the free-electron plasma, these modes are usually called *bulk plasmon–polaritons* and are quasi-particles. Note that for these modes, $\mathbf{E}$ and $\mathbf{P}$ are transverse, and the internal charge density is therefore $\rho_{\text{int}} = 0$ everywhere. *There are no macroscopic charge density oscillations.* The denomination plasmon–polariton can be misleading in this respect since there is no net charge density wave, but only a polarization wave.

Longitudinal electric wave in an infinite (3D) medium

Finally, in an infinite medium, there is another family of electromagnetic modes that do not exist in vacuum. Maxwell's equations (C.36) – (C.39) state that the electric displacement $\mathbf{D} = \epsilon_0\epsilon\mathbf{E}$ is transverse: $\nabla\cdot\mathbf{D} = 0$. This condition was previously fulfilled by assuming that the electric field $\mathbf{E}$ was also transverse, but an alternative possibility is that $\epsilon(\omega) = 0$. Such a condition can be fulfilled in some media (and in particular in metals) at one or more specific frequencies ω. If this is the case, then $\mathbf{D} = 0$, and Eq. (C.39) implies that $\mathbf{H}$ is a longitudinal field. Because $\mathbf{H}$ is also a transverse field from Eq. (C.36), it must therefore be zero: $\mathbf{H} = 0$. The remaining equation (C.37) then implies that $\mathbf{E}$ is a longitudinal field. Such a solution therefore corresponds to a *longitudinal electric wave* (with zero magnetic field). This wave is associated with an internal polarization wave $\mathbf{P} = -\epsilon_0\mathbf{E}$ (since $\mathbf{D} = 0$). Moreover, because $\mathbf{P}$ is longitudinal, it also corresponds to an internal charge density wave $\rho_{\text{int}} = -\nabla\cdot\mathbf{P} \neq 0$. These modes therefore correspond to a *real charge density wave*, with an associated longitudinal electric wave. Because of the longitudinal nature of these modes (and of the structure of Maxwell's equations), these modes *cannot couple to light* or photons, which are transverse electromagnetic excitations. For metals, these modes are the 'pure' plasmon modes, as defined previously, and are usually called *bulk plasmons.* In the local approximation, they exist only at specific frequencies for which $\epsilon(\omega) = 0$, and $\mathbf{k}$ can take any values (small enough for the local approximation to remain valid). For a more detailed non-local treatment, the dispersion relation of these modes is given by $\epsilon(\omega, \mathbf{k}) = 0$. Bulk plasmons are longitudinal modes and therefore do not couple to light.

Electromagnetic modes in an infinite (3D) metal

We have introduced in the previous two subsections the two types of electromagnetic modes that exist in an infinite 3D medium. Here we discuss briefly further their properties for the simple case of an ideal metal whose optical properties are described by a simple Drude model. $\epsilon(\omega)$ is then given by Eq. (3.6), where we neglect the damping term ($\gamma_0 \approx 0$), i.e.:

$$\epsilon(\omega) = \epsilon_\infty \left(1 - \frac{\omega_p^2}{\omega^2}\right). \tag{3.10}$$

- The first types of modes are the *bulk plasmon–polaritons.* Using the above expression for $\epsilon(\omega)$, their dispersion relation, Eq. (3.9), can then be simplified as:

  $$\omega^2 = \omega_p^2 + \frac{c^2}{\epsilon_\infty} k^2. \tag{3.11}$$

  These are transverse modes and can essentially be viewed as light or photons, whose properties are modified by the interaction with the electrons of the metal. For a real metal, these modes are damped (i.e. they are quasi-particles) and the corresponding waves are evanescent. Note that it has been pointed out [146,147] that these modes arise from the interaction of the photons with single-electron excitations and have therefore no direct relation to collective excitations as often assumed.

- The second types of modes are the *bulk plasmons,* which are longitudinal modes corresponding to internal charge density oscillations and an associated electric wave (no magnetic field). These modes *do not couple to photons* and are therefore mostly irrelevant to SERS and many plasmonics effects. The condition $\epsilon(\omega) = 0$ reduces simply in the Drude case to $\omega = \omega_p$. These modes are therefore simply collective oscillations of the plasma charge density, and occur only at the plasma frequency ω_p. This is not a surprise since they are the natural modes of oscillation of the free-electron plasma, in the absence of interaction with light or any other electromagnetic sources[7].
  The dispersion relation for these modes could be determined more accurately using a non-local dielectric function. For example, including

[7] Note that for a real metal, the bulk plasmon frequency (determined by $\epsilon(\omega) = 0$) may differ from the plasma frequency ω_p because of the presence of inter-band transitions. This is for example the case for gold.

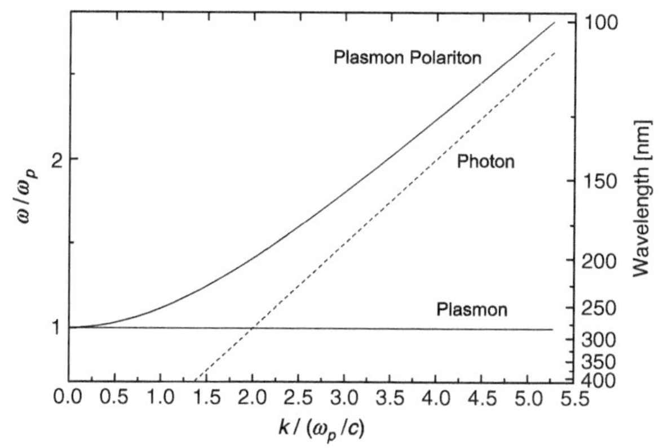

Figure 3.4. Dispersion relations of electromagnetic modes in an infinite 3D metal (silver here), showing the bulk plasmon–polaritons and the bulk plasmons. The dashed line is the dispersion of hypothetic photons non-interacting with the free-electron plasma.

hydrodynamic interactions in the electron plasma, the relation $\epsilon(\omega, \mathbf{k}) = 0$ leads to [146]:

$$\omega^2 = \omega_p^2 + \beta^2 k^2, \tag{3.12}$$

where β is a constant that depends on the chosen hydrodynamic model. In a first approximation, one can assume $\beta \approx \sqrt{0.6}v_F$, where v_F is the Fermi velocity of the metal [146]. β is therefore much smaller than c, which means that the local approximation $\omega \approx \omega_p$ is very good in the optical range.

The dispersion relations of the bulk plasmon–polaritons and bulk plasmons are illustrated in Fig. 3.4 for $\hbar\omega_p \approx 4.4$ eV and $\epsilon_\infty = 4$ (these are the values for silver). One sees that bulk plasmon–polariton modes occur only for energies *larger* than the plasmon frequency. For most metals, these modes are therefore in the UV, or even deep UV. Bulk plasmon–polaritons (like bulk plasmon) are therefore in general not relevant for optical applications, including SERS. They are discussed here as an introduction to the surface modes (of interest to SERS and plasmonics), on which we now focus.

3.3.5. Electromagnetic modes of a system of material bodies

Electromagnetic modes

We have so far discussed the concept of modes or elementary excitation focusing only on infinite systems with translational invariance. To understand the various types of plasmons and plasmon–polaritons, it is now useful to consider in more detail the general problem of the electromagnetic modes of a system of material bodies, i.e. in the presence of interfaces. These considerations can then be applied to metallic structures (and therefore to plasmons and plasmon–polaritons), but the same modes can also exist in other kinds of systems.

The system considered here is restricted to one or more entities of different materials described by a local dielectric function and separated by ideal boundaries. Any such electromagnetic problem can therefore be formally studied by solving Maxwell's equations with the appropriate boundary conditions at the various interfaces (see Appendix C). The modes, or eigenvectors, of the system consist of specific solutions from which any general solution can be inferred. These are no longer necessarily plane waves because there is no translational invariance. Here we consider only electromagnetic fields oscillating at a frequency ω, and the modes then correspond to specific values of ω (continuous or discrete). All physical quantities can be described by a complex amplitude, where a $\exp(-\mathrm{i}\omega t)$ dependence is assumed.

2D and 1D systems with translational invariance

In systems where the translational invariance only exists in 2 dimensions (layered structure), or even 1D (for example for a cylinder) then the modes can again be indexed by a **k** vector parallel to the directions of translational invariance. This leads to a dispersion relation for the modes of the type $\omega(k_x)$, corresponding to propagating modes along (Ox), which can as before be fully propagating, evanescent, or virtual modes.

Studying such systems with a perfect translational invariance may appear as a purely academic exercise, since infinite plane or cylinders never exist in real life. However, it is sufficient that the invariance holds over a typical length scale of the problem, which in many cases is simply the wavelength λ of the electromagnetic radiation, for the approximation to be valid and meaningful. In many experimental situations, all interfaces can indeed be approximated by plane surfaces at length scales of the order of $\sim\lambda$.

Systems without translational invariance

Only when the objects exhibit features on length scales of the order of the wavelength does the translational invariance really fail. In such systems without translational invariance, and in particular for particles of dimensions comparable with λ, the description in terms of a **k** vector becomes irrelevant and inadequate. The electromagnetic modes of the system then correspond to discrete values of ω. If $\omega = \omega' - \mathrm{i}\omega''$ is complex, they again correspond to *virtual modes*. As discussed previously, these are not real modes of the system but resonances with a width ω''. In problems where the oscillation of the fields is not imposed externally, these virtual modes can also be viewed as elementary excitation with a lifetime $\tau = 1/(2\omega'')$.

3.3.6. Classification of electromagnetic modes

We now discuss in more detail a possible classification of some common electromagnetic modes in systems of material bodies as defined before. Here we follow Ref. [148], which gives a detailed description of this classification.

Purely longitudinal modes

Similarly to the case of infinite media, one class of solution regroups those for which $\epsilon(\omega) = 0$ inside the material bodies. They have similar characteristics to those obtained in infinite media, i.e. they are longitudinal electric waves with $\mathbf{H} = 0$ (no magnetic field) and $\nabla \times \mathbf{E} = 0$. Since $\nabla \cdot \mathbf{E} \neq 0$, they are also associated with a bulk internal charge density (ρ_{int}) wave (bulk plasmon in metals). In addition, the boundary conditions at interfaces imply the presence of surface charge density (ρ_s) waves. The condition $\epsilon(\omega) = 0$ is the same as for infinite media. These modes are therefore simply *modified bulk modes*, which are confined inside the material bodies by the interfaces. As before, because these modes are longitudinal, they do not couple to light.

Incident wave modes

We now focus on the situations where $\epsilon(\omega) \neq 0$, and therefore $\nabla \cdot \mathbf{E} = 0$ in the materials. The condition $\nabla \cdot \mathbf{E} = 0$ is, however, not satisfied at the interfaces where the boundary conditions may imply the presence of surface charges. These solutions, although transverse inside the materials, are in fact a *mixture of longitudinal and transverse waves.*

Sufficiently far from all interfaces, 3D electromagnetic modes can exist (plane waves described by $\mathbf{k}$ and ω). A typical experiment will involve sending an *incident wave* of amplitude E_{Inc} toward the interfaces under study, which in general leads to an *outgoing (or scattered) wave* of amplitude E_{Sca}. Due to the linearity of Maxwell's equations, the scattered wave amplitude is proportional to the incident wave amplitude: $E_{\text{Sca}} = fE_{\text{Inc}}$, where f is a factor, possibly complex, depending on the geometry, the optical properties of materials, ω, and $\mathbf{k}$. f characterizes the *optical response* of the system. Note that we have neglected for simplicity here the vectorial nature of the field (i.e. the field polarization). In reality, f should be a matrix and is related to the amplitude scattering matrix [149].

Some electromagnetic modes only exist in the presence of such an incident wave and are therefore called *incident wave modes.* These modes would still exist (as a simple incident wave) if the interface was removed. They can therefore be viewed as a modification of the incident wave due to the interface. The standard reflection or refraction of a plane wave at a dielectric/dielectric interface is an example of an incident wave mode.

Bound modes or surface modes

There can also be modes that exist even in the absence of an incident wave (with only a scattered wave), and they are called *bound modes*. The bound modes owe their existence to the interface and, in contrast to incident wave modes, have no equivalent in infinite media if the interface is removed. They are therefore also called *surface modes*. Note that this denomination of incident wave vs bound mode is similar to the classification of the solutions in quantum mechanics.

The conditions $E_{\text{Sca}} \neq 0$ and $E_{\text{Inc}} = 0$ imply that the optical response, f, is infinite for bound modes. This appears to violate energy conservation, but in reality, the condition $f = \infty$ only occurs for complex values of ω or k. This means that bound modes are damped , and should therefore be viewed either as evanescent waves (ω real, k complex) or virtual modes (k real, ω complex). Because of this damping, it is necessary in practice (but not in theory) to have an incident wave to excite and maintain these modes.

If both ω and k are real, then f presents a sharp peak, instead of a real infinity, for values approaching those of the damped bound modes. This results in strong optical resonances, instead of unrealistic infinite optical response. For example, it can be shown that for a virtual bound mode with imaginary frequency $\omega = \omega' - \mathrm{i}\omega''$, the resonant response occurs for real frequencies equal to ω' and with a half-width of ω''. The smaller the ω'', the sharper the resonance.

Surface plasmon–polaritons at a metal/dielectric interface (discussed in Sections 3.4 and 3.5) are examples of bound or surface modes. In most cases, the *plasmon resonances* mentioned in the context of SERS or plasmonics are resonances associated with these surface plasmon–polariton modes. The resonant optical response can manifest itself differently depending on the type of surface plasmon–polariton creating it, i.e. large optical absorption, large far-field scattering, or large local field. For this reason, it can be useful to further classify the various types of surface modes.

3.3.7. Other properties of electromagnetic modes

Radiative and non-radiative modes

Bound or surface modes and incident wave modes can be associated with an outgoing wave or scattered wave. If this outgoing wave is propagating, it will correspond to a standard 3D mode (photon) when far from the interface, and such a bound mode is then called *radiative*. However, if the outgoing wave is evanescent, then the field decays exponentially away from the surface and this is called a *non-radiative mode*.

For a *bound radiative mode*, the scattered wave radiates energy in the far field but there is no incident wave to provide this energy. Therefore, these modes cannot be real and are always *virtual modes*. If an incident wave excites such a mode at resonance, the strong optical responses will translate in a large far-field radiation, i.e. there is a resonance in scattered intensity. Note that the damping of the mode is in general due not only to radiation, but also to optical absorption in the materials. The resonance in scattering can therefore be accompanied by a resonance in absorption. Radiative modes can be called more or less absorptive depending on the proportion of radiated to absorbed power.

The limiting case of a very absorptive mode is, in fact, a non-radiative mode. When exciting a bound non-radiative mode at resonance, the incident energy is transferred to the mode, but is not re-emitted in the far field because it is non-radiative. There is a resonance in optical absorption, not in scattering.

Localized modes

We have seen that a **k** vector could be defined in the directions of translational invariance. In each of these directions, the mode can either be propagating if **k** is real, or evanescent if not. If the mode is not propagating (evanescent) in all of these directions, then we can call it a *localized mode.* One particular case is that of systems without translational invariance, for which **k** is not relevant. All surface modes are then *localized modes.* This is for example the case of particles small or comparable to the wavelength (i.e. most nano-particles), where all surface modes are necessarily localized modes.

Excitation of electromagnetic modes

The description of a system in terms of electromagnetic modes may appear at first sight to be purely academic in nature. However, once the physical nature of the modes is understood, it is actually a powerful framework to understand the response of a complex system or the interactions between different subsystems. The reason is that a complex system can be divided into subsystems. Furthermore, when the modes of the individual subsystems are known, the response of the whole system can be studied as an interaction or coupling between subsystem modes. For weak coupling, the modes of the subsystems are essentially unchanged, while for stronger coupling the interaction can lead to new types of modes. For two modes to couple, they need to have the same frequency ω (which corresponds to energy conservation), and in the case of translational invariance the same wave-vector **k** (for momentum conservation) is also required. We can give some simple examples in terms of the types of modes discussed previously:

- An incident wave mode is by construction automatically coupled to an incident photon with the same ω and **k**.
- Similarly, a radiative mode is coupled to an outgoing photon with appropriate ω and **k**.

- An incident photon can also couple to a bound mode provided ω (and if relevant $\mathbf{k}$) conservation is fulfilled. For virtual modes ($\omega = \omega' - i\omega''$ complex), the ω conservation applies to real parts ($\omega = \omega'$) and is broadened with a width ω''. Similarly, for evanescent modes ($\text{Im}(k) > 0$), the $\mathbf{k}$ conservation is broadened.

There are specific configurations where one electromagnetic mode may be excited, for example at a given incident angle, incident polarization, or wavelength. One then expects a resonant optical response (for example in reflectivity or absorption) when the parameters match the excitation condition of such a mode. When the electromagnetic modes are plasmons, or plasmon–polaritons, this is called a *plasmon resonance*. The nature and characteristics of such a resonance depends on the nature of the electromagnetic mode giving rise to it. The term *plasmon resonance* can, therefore, have different meanings depending on the context. It can, for example, correspond to enhanced energy loss at a specific electron beam energy, or to a decreased reflectivity at a specific incident angle on a surface, or to an increased field intensity at the surface (which can give rise to a SERS signal) at a specific wavelength, etc. In the context of SERS and plasmonics, plasmon resonances *refer in most cases to bound modes called surface plasmon–polaritons*, discussed in detail in Sections 3.4 and 3.5.

3.3.8. Summary and discussion

The somewhat formal discussion given in the last few subsections should hopefully become clearer when discussing specific examples in the following sections.

The various types of plasmon excitations are tentatively summarized in Fig. 3.5 along with their main properties. The *'pure' plasmon and surface plasmon modes*, as originally defined and studied by Pines [141], Ritchie [143], and others are not directly relevant to SERS and plasmonics because they do not interact with light[8]. This is because they correspond to *longitudinal* excitation of the electric field (electric wave) and cannot therefore couple to photons. These plasmons are also in fact those associated with true collective charge density oscillations.

The electromagnetic modes that interact with light are the *plasmon–polaritons*, which mix photons with internal excitations of the metal. *Bulk plasmon–polaritons* are those modes that exist in an infinite metal and can be viewed simply as photons propagating in a metal. These modes are again not directly relevant to SERS and plasmonics.

[8] They however play an important role in many other properties of metals and metal surfaces [129], in particular for other types of spectroscopies like Electron Energy Loss Spectroscopy (EELS).

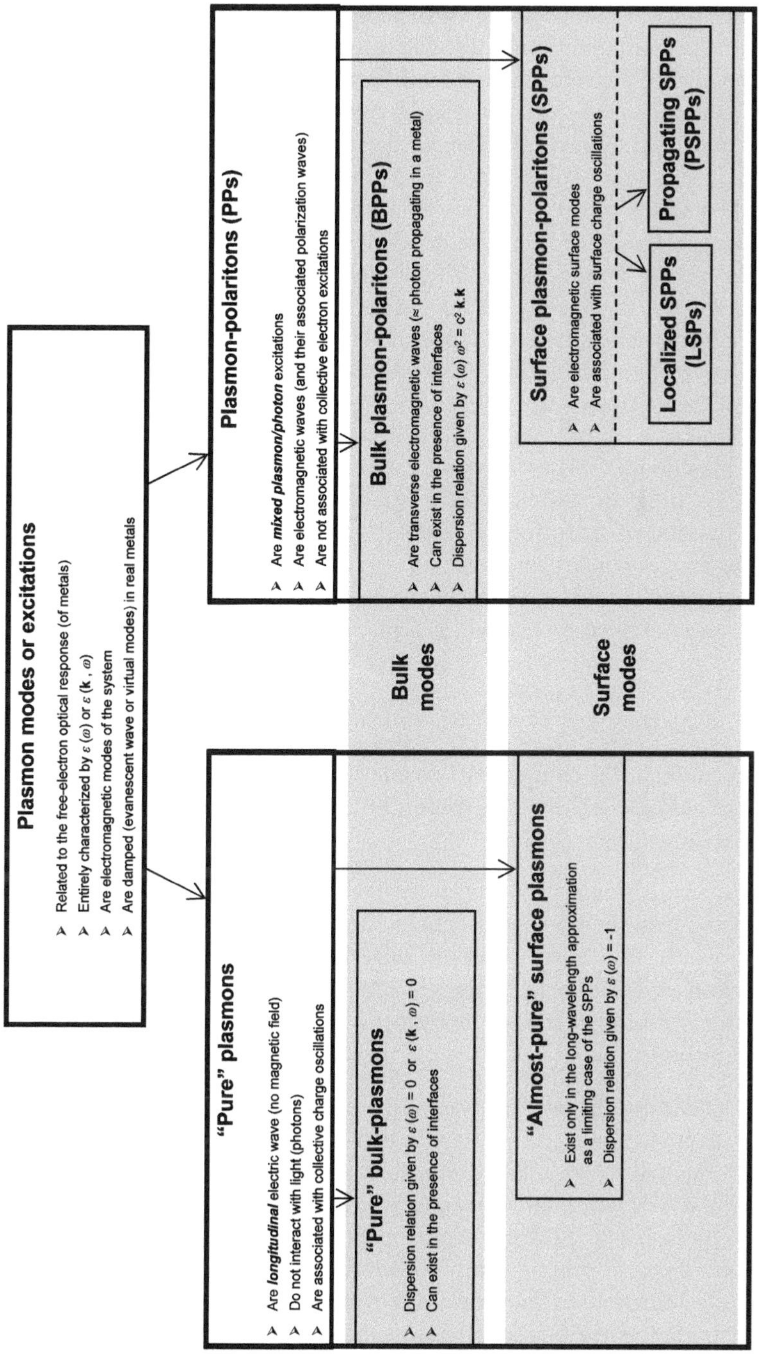

Figure 3.5. Schematic representation of the various plasmon types and their main properties.

The presence of an interface (typically metal/dielectric) gives rise to a new family of electromagnetic modes, which happens to be much more interesting for SERS and plasmonics. These additional modes are bound modes or *surface modes*, as described in the previous section in a more general context. In the case of metals, they are called *surface plasmon–polaritons* (SPPs) .

These electromagnetic surface modes are transverse inside and outside the metal (because $\epsilon \neq 0$). Note however that they do also have a longitudinal component because of the discontinuities at the interface, and are in this sense also partly longitudinal surface plasmon waves. A non-local description of ϵ, where the interface is no longer treated as a singularity, is necessary to identify clearly this mixed longitudinal–transverse nature, see for example Ref. [146]. The 'pure' longitudinal surface plasmons introduced by Ritchie [143] are simply a limiting case where the transverse contribution is negligible (in the electrostatic approximation). The strict decoupling between longitudinal ('pure' plasmons) and transverse (plasmon–polaritons) modes, which was natural for bulk modes in infinite media, is no longer possible for surface modes. Instead of introducing another terminology, these modes are simply called *surface plasmon–polaritons*.

Finally, let us note that there are typically three reasons (two good ones, and the last one partly flawed) that are put forward to justify the denomination of *surface* plasmon–polariton or *surface* mode:

- Firstly, they would not exist without the interface.
- Secondly, the characteristics of these modes depend not only on the optical properties of the metal, but also on that of the dielectric forming the interface.
- Finally, in many cases, these surface modes are localized at the interface (for example the electric field decays exponentially from it). This is however not really a good criterion, since as we will see later, for small metallic objects the electric field of such a surface mode can be almost uniform inside the object and extend infinitely outside (i.e. it is a radiative mode).

After this long digression, these SPP modes are finally the ones that are relevant to SERS and plasmonics. Whenever plasmons or surface plasmons as such are mentioned in a SERS context, it is arguably an abuse of language (although a very common one), and it is strictly speaking referring to surface plasmon–polaritons. As for general electromagnetic modes, these can come in several flavors: propagating, localized, radiating, non-radiating, bound, virtual, or evanescent. They can give rise to various resonance effects, which may be used for various applications. All these different cases will be the subject of the following sections, which focus on a more practical description of SPP modes in metals.

3.4. SURFACE PLASMON–POLARITONS ON PLANAR INTERFACES

The previous discussion was aimed at introducing plasmons from an historical point of view, and emphasized the intricate relationship between optical properties and the free-electron-plasma dynamics. This helped in understanding the origin of the term plasmon and its relation with 'charge oscillations', as often loosely stated. We also highlighted the fact that surface plasmon–polaritons (SPPs) can simply be viewed as *electromagnetic surface modes* of the system under consideration. The study of SPPs then simply reduces to an electromagnetic problem, where the actual role of the electrons can be ignored and is simply contained in the dielectric function describing the metal. This more pragmatic approach is particularly suited to study SPPs in various geometries. One important case is that of a planar metal/dielectric interface, firstly because it is reasonably easy to solve, secondly because many interfaces can be considered as planar over a characteristic length scale of the problem (typically the wavelength λ), and finally because it supports the types of SPPs (propagating SPPs) most useful for several plasmonics applications.

3.4.1. Electromagnetic modes for a planar dielectric/metal interface

Description of the electromagnetic problem

To illustrate this, we will restrict ourselves to a local relative dielectric function $\epsilon(\omega)$ and consider first the textbook example of a plane metal/dielectric interface [150]. The metal (region 2), described by $\epsilon(\omega)$ occupies the half-plane $z > 0$, and a non-absorbing dielectric (region 1), with relative dielectric function $\epsilon_M \geq 1$, real and constant, forms the outside medium in $z < 0$, as shown schematically in Fig. 3.6. Bulk electromagnetic modes exist far away from the interface both in the dielectric (photon modes) and in the metal (bulk plasmon and bulk plasmon–polaritons). Bulk plasmon modes for which $\epsilon(\omega) = 0$ may also exist in the presence of the interface, but we ignore these longitudinal modes here since they do not interact with light. Because of the translational invariance along directions in the plane $z = 0$, the electromagnetic modes should be characterized by their frequency ω, and tangential wave-vector, k_x, in the plane (we assume $k_y = 0$ without loss of generality). Various technical aspects of this problem are treated in Appendix F, which may therefore be read in conjunction with this. As discussed before, the presence of the surface can introduce several types of modes:

- Incident wave modes are those where an incident wave (and possibly a scattered/reflected wave) is present. From the standpoint of a metal/dielectric interface, these modes correspond simply to the classic problem of reflection/refraction of a plane wave at the interface. This problem is treated in many electromagnetic textbooks [96,151], and the most important features are summarized in Appendix F. The refracted/transmitted wave in the metal is always evanescent because of optical absorption in the metal ($\epsilon(\omega)$ complex or real negative). If the incident wave is propagating, and if there is a reflected wave, then it is also propagating, and the mode can be called radiative, and more or less absorptive depending on the reflection coefficient. These incident wave modes exist for all ω and k_x compatible with the dispersion relation in the dielectric, i.e. $k_x^2 + k_z^2 = \epsilon_M \omega^2/c^2$. Note that for TM polarization[9] they are associated with a propagating surface charge density (ρ_s) wave at the interface, but they are not surface modes (or bound modes) as defined previously.

- A particular case of incident wave modes is that for which there is no reflected wave (only a refracted wave). This happens only (see Appendix F) for a specific angle of incidence (called the Brewster angle) . The corresponding Brewster modes are an example of non-radiative incident wave modes.

- The other family of modes, the bound or surface modes, is the one of interest here. They correspond to solutions where no incident wave is present, only the scattered wave. These are derived and discussed in detail in Section F.2.5 of Appendix F and we now discuss in more detail their properties for a metal/dielectric interface.

Surface modes of a metal/dielectric interface

It is shown in Appendix F that there are no surface modes with TE (s-) polarization; only with TM (p-) polarization and we therefore focus on this latter case only. Such a surface mode consists of a scattered wave (no incident wave) in the dielectric with wave-vector $\mathbf{k}_1' = k_x \mathbf{e}_x + k_{1z}' \mathbf{e}_z$ and a transmitted wave in the metal with wave-vector $\mathbf{k}_2 = k_x \mathbf{e}_x + k_{2z} \mathbf{e}_z$, as shown in Fig. 3.6; see Appendix F for more details. The electromagnetic fields of these modes are given by (see Appendix F):

$$\begin{cases} \mathbf{H}_2 = (H_y \mathbf{e}_y) \exp(\mathrm{i}k_x x + \mathrm{i}k_{2z} z) \\ \mathbf{E}_2 = \dfrac{H_y}{\omega \epsilon_0 \epsilon_2} (k_{2z} \mathbf{e}_x - k_x \mathbf{e}_z) \exp(\mathrm{i}k_x x + \mathrm{i}k_{2z} z), \end{cases} \tag{3.13}$$

and similar expressions for $\mathbf{H}_1'$ and $\mathbf{E}_1'$ (the fields have been expressed in terms of a single amplitude: $H_y = H_{1y}' = H_{2y}$).

[9] TM and TE polarizations are defined in Section F.2.1 of Appendix F. For TM polarization, the magnetic field is perpendicular to the plane of incidence, meaning along y here. An example is shown in Fig. 3.6.

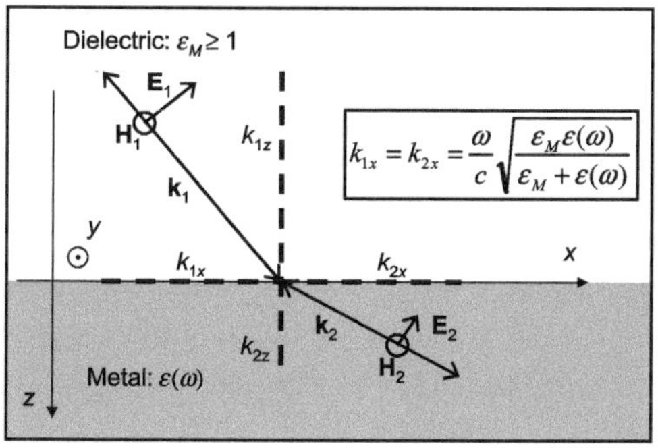

Figure 3.6. Schematic illustration of the electromagnetic problem of two TM waves, on either side of a metal (region 2)/dielectric (region 1) interface. The dispersion relations in each medium, together with the boundary conditions lead to the dispersion relation for SPPs (framed equation). The arrows representing the directions of the k vectors are shown for both directions to show that no assumption is made on the nature of the waves (for example incident or scattered wave in Region 1).

Moreover, the condition for the existence of such a surface mode is (see Appendix F):

$$k_x^2 = k_0^2 \frac{\epsilon \epsilon_M}{\epsilon + \epsilon_M}. \tag{3.14}$$

There are two possible solutions for k_x (of opposite sign). Since they are physically equivalent, we will only consider, by convention, the one given by[10]:

$$\boxed{k_x = \frac{\omega}{c}\sqrt{\frac{\epsilon(\omega)\epsilon_M}{\epsilon(\omega) + \epsilon_M}}.} \tag{3.15}$$

Since $\mathrm{Im}(\epsilon) \geq 0$, we have $\mathrm{Re}(k_x) \geq 0$ and $\mathrm{Im}(k_x) \geq 0$.

One can moreover deduce the corresponding expressions for k'_{1z} and k_{2z}. This is not as trivial as may seem, mostly because of sign issues, a problem often swept under the carpet. This is discussed extensively in Section F.2.5.

This set of expressions entirely defines the TM electromagnetic surface modes for each frequency ω. Each mode is composed of two electromagnetic waves, one on either side of the interface, with $\mathbf{k}$ vectors given by the above

[10] In all these expressions, there are in principle two choices for the square root of a complex number. We use the 'standard' complex square root convention, i.e. the one with a positive real part (or if it is zero, the one with the positive imaginary part). More explicitly: for $-\pi < \phi \leq \pi$, $\sqrt{r\exp(i\phi)} = \sqrt{r}\exp(i\phi/2)$. We then have $\mathrm{Re}(\sqrt{z}) \geq 0$, and $\mathrm{Im}(\sqrt{z})$ has the same sign as $\mathrm{Im}(z)$.

relations. The dispersion relation for the surface modes is Eq. (3.15), which links the frequency ω to its tangential wave-vector k_x.

Finally we note that these modes are associated with a surface charge density ρ_{Surf} created by the discontinuity of the normal component of the electric field. They, therefore, correspond to a surface charge density wave along the interface with wave-vector $k_x \mathbf{e}_x$, propagating, pseudo-propagating, or evanescent depending on k_x (see the next subsection). Note however that this is not a defining characteristic of the surface modes since a similar surface charge density wave exists for most TM incident wave modes (for example for reflection at the metal/dielectric interface).

Classification of surface modes

The exact nature of each mode will depend on whether the wave-vector components k_x, k_{1z}, and k_{2z}, are real or complex. A mode will correspond to a *propagating* wave along a given direction if its wave-vector along this direction is real. Contrarily, if it is imaginary, it is *evanescent* (with an exponential decay of the intensity). If this decay occurs over long distances compared to the wavelength, the wave can be called *pseudo-propagating*. In our example here, if k'_{1z} is real, then we have a propagating scattered wave in medium 1. Contrarily, if k'_{1z} has a non-zero imaginary part, then we have an evanescent wave along the z direction. The field amplitudes then decay as $\exp(-|\text{Im}(k'_{1z})z|)$ and are therefore negligible in the far field ($z \to -\infty$). This corresponds to a *non-radiative* mode, or trapped surface wave. A detailed discussion of the various cases is given in Section F.2.5. Here we only highlight the main results, which are summarized schematically in Fig. 3.7.

It can be useful in this context to consider first the ideal case where the metal (medium 2) is non-absorbing, i.e. its dielectric function is real. Such a metal can in principle support infinitely propagating waves (with k_x and k_{2z} real). In reality, this cannot happen since a real metal always presents a small amount of absorption and all waves are strictly speaking evanescent. However, if $|\text{Im}(k)| \ll |\text{Re}(k)|$ in one direction, then the wave will propagate over long distances (many spatial wavelengths), without substantial decay in amplitude (it is a pseudo-propagating wave).

We can then distinguish three cases for an ideal non-absorbing metal, and extend this classification to a real (absorbing) metal. These situations are represented schematically in Fig. 3.7 and discussed below:

- $\epsilon(\omega) > 0$ (Fig. 3.7(a)) corresponds to an ideal metal at high frequencies, which then behaves as a standard dielectric with no absorption. All wave-vector components, k_x, k'_{1z}, and k_{2z} are then real. k'_{1z} and k_{2z} must also both be positive. These solutions are discussed further in Section F.2.5. They are called *Brewster's modes* and are not strictly speaking surface modes for an ideal metal since k'_{1z} then represents

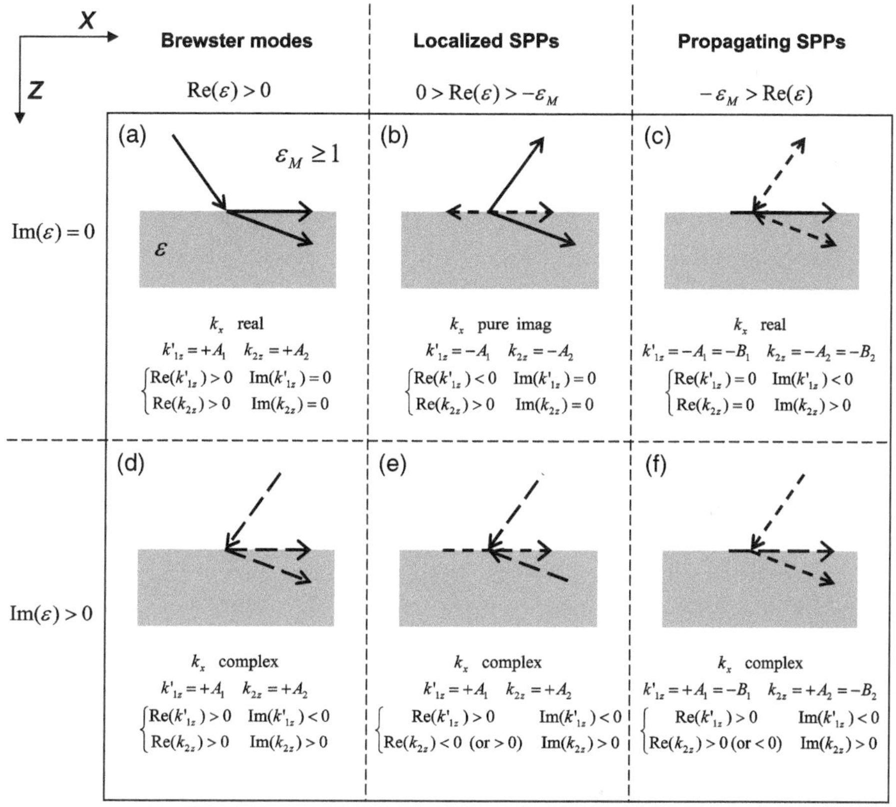

$$k_x = k_0\sqrt{\varepsilon\varepsilon_M/(\varepsilon+\varepsilon_M)} \qquad \begin{cases} A_1 = k_0\varepsilon_M\sqrt{1/(\varepsilon+\varepsilon_M)} \\ A_2 = k_0\varepsilon\sqrt{1/(\varepsilon+\varepsilon_M)} \end{cases} \qquad \begin{cases} B_1 = ik_0\varepsilon_M\sqrt{-1/(\varepsilon+\varepsilon_M)} \\ B_2 = ik_0\varepsilon\sqrt{-1/(\varepsilon+\varepsilon_M)} \end{cases}$$

Figure 3.7. Schematic illustrating various types of electromagnetic modes existing at a planar metal/dielectric interface. Solid (dashed) lines indicate propagating (evanescent) waves. For all evanescent waves, the field amplitudes decay exponentially when moving away from the interface. The direction of propagation (given by $\mathrm{Re}(k_z)$) is indicated by arrows. Note that for SPPs, this may depend on the exact value of $\mathrm{Im}(\epsilon)$ (see Section F.2.5). Long-dashed lines are used for pseudo-propagating waves (only when $\mathrm{Im}(\epsilon)$ remains small). The Brewster modes for $\mathrm{Im}(\epsilon) = 0$ are strictly speaking incident wave modes, but the other situations correspond to surface modes. The appropriate expressions for k'_{1z} and k_{2z} are given for each case, along with the nature (real or imaginary) of the wave-vector components. See Section F.2.5 for more details.

an incident wave. They do become surface modes however, for a non-ideal metal with $\mathrm{Im}(\epsilon) > 0$ (Fig. 3.7(d)), since the wave in region 1 is then evanescent. In this sense Brewster's modes may be viewed as surface modes (then this becomes, ultimately, only a question of vocabulary). Note that these modes are associated with a propagating surface charge wave created by the discontinuity of the electric field. They are a mixture of propagating photons and surface charge waves, and are therefore surface polaritons. They, however, are usually not considered as surface plasmon–polaritons, a denomination reserved for the case where $\mathrm{Re}(\epsilon) < 0$. Their existence is pointed out here to show that surface charge density waves are not a defining characteristic of SPPs.

- $-\epsilon_M < \epsilon(\omega) < 0$ corresponds to the case of an ideal metal at frequencies just below the plasma frequency (Fig. 3.7(b)). We then have $k'_{1z} < 0$ and $k_{2z} > 0$ real of opposite sign, and $k_x = \mathrm{i}\kappa_x$ is pure imaginary. The corresponding surface wave does not propagate at all. These modes therefore correspond to localized modes, and can be called *localized SPPs*[11]. The energy of the evanescent surface wave is dissipated into the two waves propagating away from the surface. These modes are therefore *radiative*. They can in principle be excited by a wave with $k_x \approx 0$, i.e. at normal incidence, but the resonance condition is largely broadened by the strong evanescent nature of the modes. For a real metal with $\mathrm{Im}(\epsilon) > 0$ (Fig. 3.7(e)), both waves in region 1 and 2 become evanescent (pseudo-propagating if the absorption is small), and the corresponding modes are then non-radiative. If the absorption is large, the theory then predicts that these surface modes may become pseudo-propagating but this situation has not been studied in detail.
- Arguably, the most interesting case for plasmonics is when $\epsilon(\omega) < -\epsilon_M$ (Fig. 3.7(c)). We then have for an ideal metal k_x real and therefore a truly *propagating surface wave*. Moreover, k'_{1z} and k_{2z} are both pure imaginary, and therefore correspond to *evanescent waves perpendicular to the surface*. These are *non-radiative surface modes* that are fully trapped at the surface, and that propagate along the interface. There is again a propagating surface charge density wave at the surface. A non-local treatment [146] can show that this corresponds to a longitudinal surface plasmon wave. These modes are the *propagating surface plasmon–polariton* modes of a metal/dielectric planar interface. The energy is trapped at the surface and shared between photon and surface plasmon oscillations. For a real metal (Fig. 3.7(f)), k_x is no longer real, and the surface wave is then a *pseudo-propagating* wave (if the absorption is not too large).

In summary, surface modes exist for (almost) all values of ϵ (and therefore ω). All of them are surface polaritons (surface charge oscillations coupled to electromagnetic fields). When $\mathrm{Re}(\epsilon) < 0$, they are called surface

[11] The term SPP is sometimes reserved for *propagating* or *pseudo-propagating* surface waves and what we call here localized SPPs would then be excluded. This is again only a matter of vocabulary.

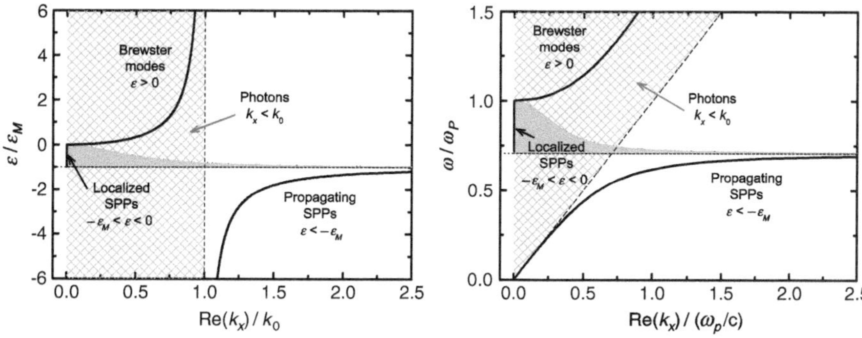

Figure 3.8. Two possible representations of the dispersion relation $\omega(k_x)$ for the surface modes of an ideal metal/dielectric interface ($\mathrm{Im}(\epsilon) = 0$, i.e. no absorption). On the left, ϵ/ϵ_M is plotted against $\mathrm{Re}(k_x)/k_0$. The advantage here is that this plot applies to any (ideal) metal. However, the dispersion relation can only be visualized indirectly since it requires the additional dependence of $\epsilon(\omega)$ (and k_0 also depends on ω). On the right, we show a more direct plot of the dispersion relation for an interface of air ($\epsilon_M = 1$) and an ideal metal following the Drude model (Eq. (3.2)); with $\epsilon_\infty = 1$, a plasma frequency ω_p and no absorption. The plot is shown in terms of adimensional quantities: ω/ω_p and $\mathrm{Re}(k_x)/(\omega_p/c)$. In both plots, the area covered by the dispersion relations of incident photons (with any possible incident angle) is shown as a hatched area. It corresponds to $k_x < k_0\sqrt{\epsilon_M}$ (since $k_x = k_0\sqrt{\epsilon_M}\sin\theta$). The condition $\epsilon(\omega) = -\epsilon_M$, which delimits the region of propagating SPPs is also shown as a straight dashed line. Finally, the broadening of the modes is indicated by a gray shaded area corresponding to the region between $\mathrm{Re}(k_x) - \mathrm{Im}(k_x)/2$ and $\mathrm{Re}(k_x) + \mathrm{Im}(k_x)/2$. Only localized SPP modes are broadened for an ideal metal.

plasmon–polaritons (SPPs) and may either be localized ($-\epsilon_M < \mathrm{Re}(\epsilon) < 0$, if absorption is low) or (pseudo-)propagating ($\mathrm{Re}(\epsilon) < -\epsilon_M$, if absorption is low). If $\mathrm{Re}(\epsilon) > 0$, they correspond to Brewster's modes, but do not play a role in the context of plasmonics. These considerations are summarized in Fig. 3.7 and further discussed in Section F.2.5. In the context of planar metal/dielectric interfaces, the propagating (or pseudo-propagating) SPPs (PSPPs), occurring for $\mathrm{Re}(\epsilon) < -\epsilon_M$, are the most important ones for most plasmonics applications. The rest of Section 3.4 is primarily aimed at discussing some of their properties.

3.4.2. Properties of the SPP modes at planar metal/dielectric interfaces

Dispersion relations for ideal metals

The dispersion relations for the surface modes, which relate the frequency ω, with the wave-vector for propagation k_x are given by Eq. (3.15), provided that the frequency dependence of $\epsilon(\omega)$ is known. The dispersion relations of surface modes for an ideal metal/dielectric interface are shown in Fig. 3.8

where $\text{Re}(k_x)/k_0$ is plotted against ϵ/ϵ_M (which in practice depends on ω). Also shown is the broadening of the mode, characterized here by adding $\pm(1/2)\text{Im}(k_x)/k_0$ to $\text{Re}(k_x)/k_0$. Such plots are more commonly shown as ω as a function of k_x, but for this, one needs to relate ϵ to ω. This is, for example, illustrated in Fig. 3.8 for an ideal Drude metal with plasma frequency ω_p.

The three types of modes that have just been discussed appear clearly in these plots:

- The Brewster modes for $\epsilon > 0$, with no broadening in the ideal case. They appear above the plasma frequency.
- The localized SPPs for $-1 < \epsilon/\epsilon_M < 0$, with $\text{Re}(k_x) = 0$ and a large broadening.
- The propagating SPPs for $\epsilon/\epsilon_M < -1$, with no broadening. The frequency of the SPPs modes approaches for large k_x the limiting frequency $\omega_{\text{SP}} < \omega_p$ corresponding to the condition $\epsilon(\omega_{\text{SP}}) = -\epsilon_M$. Large k_x correspond to the electrostatic (long spatial wavelength) approximation, and these modes therefore resemble the 'pure surface plasmons' introduced by Ritchie [143] and discussed previously. For a perfect Drude model with $\epsilon_\infty = 1$, $\omega_{\text{SP}} = \omega_p/\sqrt{2}$, which is often referred to as the surface plasmon frequency. For a real metal, ω_{SP} can be different due to ϵ_∞ or inter-band transitions. At lower frequencies (longer wavelength), the SPP mode wave-vector approaches that of a photon with grazing angle of incidence and the propagating SPPs are more 'photon-like'. These SPP modes are those most useful for a number of applications in plasmonics.

For a photon incident from the dielectric (region 1) at an angle of incidence θ (angle with respect to the normal of the interface), the tangential component of k is $k_x = k_0\sqrt{\epsilon_M}\sin\theta$. Such a photon would appear on these plots as a straight line of equation $\omega = ck_x/\sin\theta/\sqrt{\epsilon_M}$. The extreme case of a photon with grazing incidence ($\theta = \pi/2$), i.e. $\omega = ck_x/\sqrt{\epsilon_M}$, is shown as a dashed line on the plots in Fig. 3.8. Every point on the graph with $\omega \geq ck_x/\sqrt{\epsilon_M}$ also corresponds to a photon with a given incident angle θ. This region is hatched on the graph and corresponds to modes that can couple (with ω and k_x conservation) to an incident photon from the dielectric. Modes in the other region cannot be directly excited with photons, because of energy and momentum conservation. This is the case of propagating SPP modes and the next section will be entirely dedicated to the important question of how to circumvent this problem and couple photons to PSPPs.

Dispersion relations for real metals

In real metals, ϵ has inevitably a small imaginary part across the visible range. In some cases, the absorption can even be quite large at some

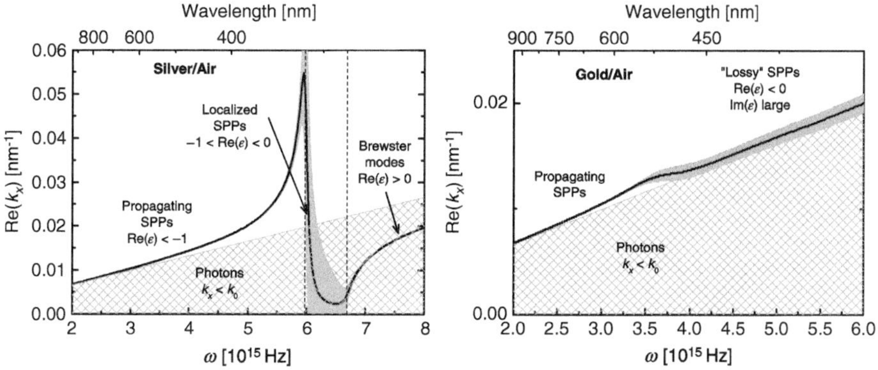

Figure 3.9. Dispersion relations $\omega(k_x)$ for the surface modes of a silver/air (left) or gold/air (right) interface. The legend is the same as that for Fig. 3.8, but note that the axes have been inverted (purely for technical reasons: it is easier to produce the plots).

wavelengths, for example in the case of gold due to inter-band transitions. The dispersion relation of SPP modes can still be obtained from Eq. (3.15) and is plotted in Fig. 3.9 for the model examples of silver and gold in air.

For silver, first, the plot is very similar to the ideal metal case, owing to the relatively small $\mathrm{Im}(\epsilon)$. The Brewster and propagating SPP modes now present a broadening, but it remains very small. One interesting difference is that the dispersion relation no longer diverges for $\epsilon = -\epsilon_M$, and it is actually back-bending in the region of the localized SPPs, i.e. $\mathrm{Re}(k_x)$ decreases with ω, a situation that would never occur for ideal metals. This anomalous dispersion has been observed in experiments [152,153] and bears some similarity with the concept of negative refraction.

For gold, the dispersion plot is quite different from that of an ideal metal, mostly because of the large absorption at wavelengths smaller than 600 nm (due to inter-band transitions). The propagating SPPs still exist for $\lambda > 600$ nm, with similar properties as already discussed. For $\lambda < 600$ nm (for which ϵ lies approximately between -3 and 0), the large absorption significantly affects and broadens the dispersion relation. The previous distinction between propagating and localized SPPs loses its strict meaning. The corresponding modes present a strong damping and are therefore referred to as 'lossy' SPPs, and they are quite localized in nature.

Propagation lengths for propagating SPPs

We now focus more specifically on the properties of propagating SPPs (PSPPs). When absorption is neglected, the surface wave associated with PSPP modes propagates forever along the x direction. Once losses are included ($\mathrm{Im}(\epsilon) > 0$), however, the surface wave is damped by absorption in the

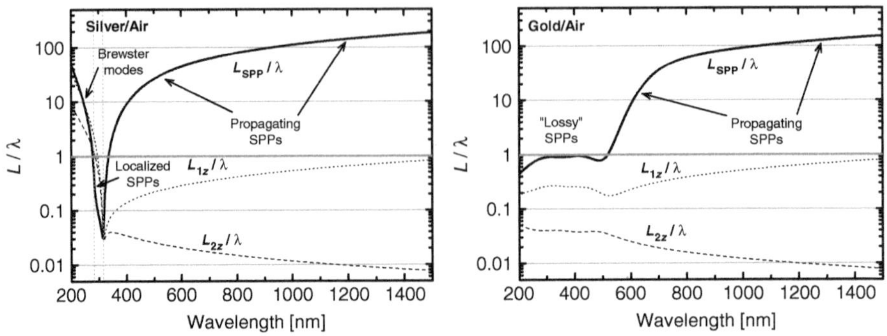

Figure 3.10. Propagation/decay lengths (normalized to the wavelength λ) for surface modes for a silver/air (left) and a gold/air (right) planar interface. L_{SPP} is the propagation length of the surface wave (along x). L_{1z} and L_{2z} are the decay lengths (along z) away from the interface in the dielectric and in the metal, respectively.

metal and the field intensity then decays as $\exp(-2\text{Im}(k_x)x)$. This defines a *propagation length* for the PSPP wave as:

$$\boxed{L_{\text{SPP}} = \frac{1}{2\text{Im}(k_x)}.} \tag{3.16}$$

Using Eq. (3.15), this can be expressed as a function of ϵ in a number of forms, none of which are particularly simple. Writing $\epsilon = \epsilon' + \text{i}\epsilon''$, a much simpler (and arguably more useful) expression can be obtained [11] in the case where ϵ'' remains small compared to ϵ' and $\epsilon' < -\epsilon_M$:

$$L_{\text{SPP}} \approx \frac{\lambda}{2\pi}\left(\frac{\epsilon'(\omega)+\epsilon_M}{\epsilon'(\omega)\epsilon_M}\right)^{\frac{3}{2}}\frac{\epsilon'(\omega)^2}{\epsilon''(\omega)}. \tag{3.17}$$

This propagation length can be much larger than the wavelength if ϵ'' is sufficiently small. It is also larger when $|\epsilon'(\omega)|$ is large, i.e. at longer wavelengths for metals.

To illustrate this, the propagation length (normalized to the wavelength) is plotted in Fig. 3.10 for silver and gold interfaces with air (using the exact expressions in Eq. (3.16)). The definition of L_{SPP} is extended to the region of localized SPP and Brewster modes, although when $L_{\text{SPP}} < \lambda$, it should be viewed as a decay length rather than a propagation length. We first note that the results are similar for silver and gold in the long-wavelength region $\lambda > 600$ nm, where the inter-band transitions no longer play a role. L_{SPP} for the PSPPs is then in the range ~10–100 μm for both silver and gold and can be as large as 0.3 mm in the near infrared ($\lambda \approx 1.5$ μm).

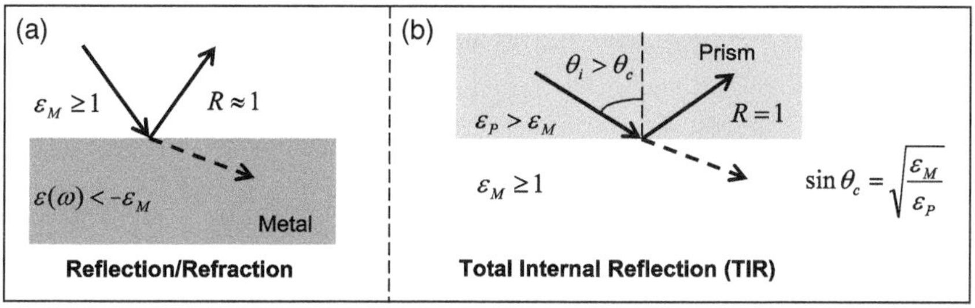

Figure 3.11. (a) Schematic illustrating the standard problem of reflection/refraction at a dielectric/metal interface. No PSPP modes can be excited here because of k_x conservation, and most of the energy is reflected ($R \approx 1$). (b) Schematic illustrating the phenomenon of *total internal reflection* (TIR) for a wave impinging from a high-refractive index dielectric (such as a prism, here with ϵ_P) onto a low-refractive index dielectric (such as air or water, here with ϵ_M). TIR occurs only for incident angles θ_i larger than the critical angle θ_c. In this case, k_x conservation cannot be met for a propagating transmitted wave. The transmitted wave is therefore evanescent and all power is reflected ($R = 1$ exactly for non-absorbing dielectrics).

It is interesting to compare this propagation length along the interface to the confinement of the fields along the z direction. The fields intensities decay as $\exp(-2|\mathrm{Im}(k'_{1z})z|)$ in the dielectric and as $\exp(-2|\mathrm{Im}(k_{2z})z|)$ in the metal. The corresponding decay lengths $L_{1z} = 1/(2|\mathrm{Im}(k'_{1z})|)$ and $L_{2z} = 1/(2|\mathrm{Im}(k_{2z})|)$ are also shown in Fig. 3.10. For PSPPs, it is clear that this decay length or penetration depth in the metal, L_{2z}, is very small ($\approx$10–15 nm). The confinement on the dielectric side is also quite good with a decay length smaller than one wavelength and more than two orders of magnitude shorter than the propagation length along the surface. These PSPP modes are therefore truly trapped electromagnetic surface waves. Provided light can be coupled in and out of them, they can therefore be used as high confinement light wave-guides, called *plasmonic wave-guides*.

3.4.3. Coupling of PSPP modes with light

In the previous description of propagating SPP modes, we have already hinted at the important issue of coupling them to light, a desirable step for many applications, which we now discuss in detail.

Let us first note that PSPP modes are TM (or p-polarized) electromagnetic waves. Because the nature of polarization is conserved at planar interfaces (see Appendix F), *only TM waves can excite PSPP modes.* In what follows, we therefore implicitly assume that all incident waves are TM polarized.

Moreover, one of the peculiarities of PSPP modes is that their wave-vector k_x (or $\mathrm{Re}(k_x)$) is larger than the wave-vector of a photon in the dielectric (which is equal to $k_0\sqrt{\epsilon_M}$). This is evident from the dispersion relation in Eq. (3.15), recalling that $\epsilon(\omega) < -\epsilon_M$ for PSPP modes, or more visually

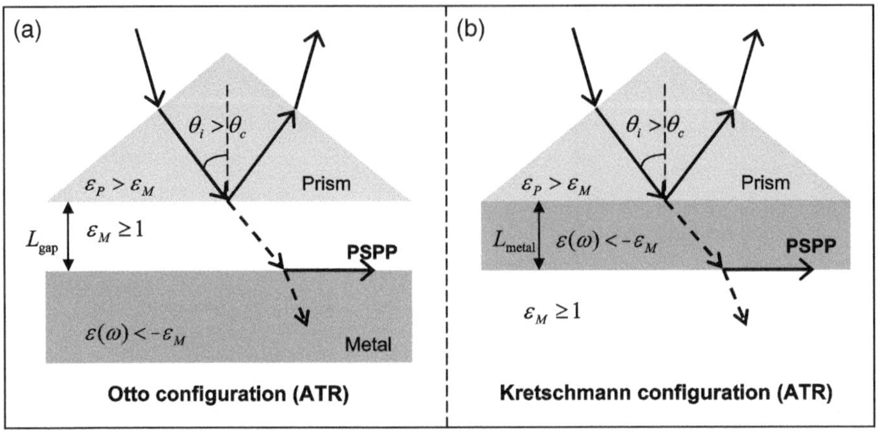

Figure 3.12. Schematic illustrating the two common configurations based on attenuated total reflection (ATR) for exciting PSPP modes with photons: (a) Otto configuration, (b) Prism coated with a thin metal film (Kretschmann configuration). In both situations, the PSPP is excited thanks to the larger momentum of the photon in the prism, which is transferred by the evanescent field, using total internal reflection either through a dielectric spacer (a) or directly through the metal (b).

from the plots of the dispersion relation in Figs 3.8 and 3.9. Let us also recall that $\hbar k$ represents the momentum of the particle or mode (photon or PSPP). The momentum along the x direction for a photon in the dielectric (which is the projection of its total momentum) is therefore always smaller than that of the PSPP modes. Note that the mismatch is usually small in the wave-guiding region (at longer wavelengths for longer propagation, see Fig. 3.10). For example, in the red for silver, we have $k_x \approx 1.03\, k_0$ (in air). However, because momentum must be conserved here (due to the translational invariance), it means that *a photon cannot directly excite the PSPP modes*. In fact, in the situation of an incident photon (electromagnetic wave) impinging on the surface, it is an incident wave mode corresponding to usual reflection/refraction that will be excited as illustrated in Fig. 3.11(a). Moreover for metals in the visible, the reflection coefficient (R^p, see Appendix F) is usually close to 1, i.e. almost no energy is transferred to the metal or to a surface wave.

Several techniques have been devised to impart the missing momentum and excite the PSPP modes with a photon, and we will briefly describe here some of them.

Coupling by total internal reflection

The first technique relies on the phenomenon of total internal reflection (also discussed in Section F.3.5) at a dielectric interface, which is illustrated in

Fig. 3.11(b). This occurs when an incident electromagnetic wave in a medium with refractive index n_1 impinges with a large angle of incidence θ onto a planar interface with a second dielectric of smaller refractive index $n_2 < n_1$. The classical theory of optics (through Snell's law) tells us that there is a critical incident angle defined by $\sin\theta_c = n_2/n_1$, beyond which there is no transmitted wave propagating in medium 2. A more detailed electromagnetic treatment shows that the wave in medium 2 is in this case evanescent (see Section F.3.5). Note that the momentum along the x direction must again be conserved. In medium 1, it is $\hbar k_x = n_1\hbar k_0 \sin\theta$, and is therefore larger for $\theta \geq \theta_c$ than that allowed for a propagating photon in medium 2 (which is always smaller than $n_2\hbar k_0$). This is precisely why the transmitted wave must be evanescent.

This technique therefore allows one to create an evanescent excitation in medium 2 with a parallel momentum $\hbar k_x$ larger than that normally allowed for a propagating photon. Such an excitation could therefore be used to excite the PSPP modes at a (dielectric M)/metal interface. This can be realized as a (dielectric P)/(dielectric M)/metal interface with $n_P > n_M$ (P stands for prism, see also Fig. 3.12(a)). The parallel momentum $\hbar k_x = n_P\hbar k_0 \sin\theta$ in the prism could be sufficiently large to excite PSPP modes of the (dielectric M)/metal interface. However, because the excitation in dielectric M is evanescent in the z direction after going through the P/M interface (because of TIR), the field decays exponentially and the (dielectric M)/metal interface needs to be close enough to the P/M interface to be excited efficiently by this evanescent field. Note however that if the two surfaces are too close to each other (typically smaller than the decay length L_{1z} of the PSPPs modes), then the PSPPs of the metal/(dielectric M) interface are strongly modified and different electromagnetic modes of the (dielectric P)/(dielectric M)/metal system arise, with different properties. There is therefore an optimum separation, of the order of L_{1z} to obtain maximum coupling without affecting the nature of the PSPP modes, and it is typically of the order of $\sim$1 μm.

The simplest way to realize this experimentally is to use a prism on top of a metal surface with a small air gap, as depicted schematically in Fig. 3.12(a). This was first proposed and demonstrated by Otto in 1968 [154] and is now called the *Otto configuration*. This work showed for the first time that PSPP modes could be easily and efficiently coupled to light and triggered a renewed interest in PSPPs and, in some respect, marked the beginning of what is now known as 'plasmonics'.

Thin metal films

One of the problems of the Otto configuration is that it is not straightforward to create an air gap or dielectric spacing with controlled thickness of the order of a micron. It was soon after realized [155,156] that

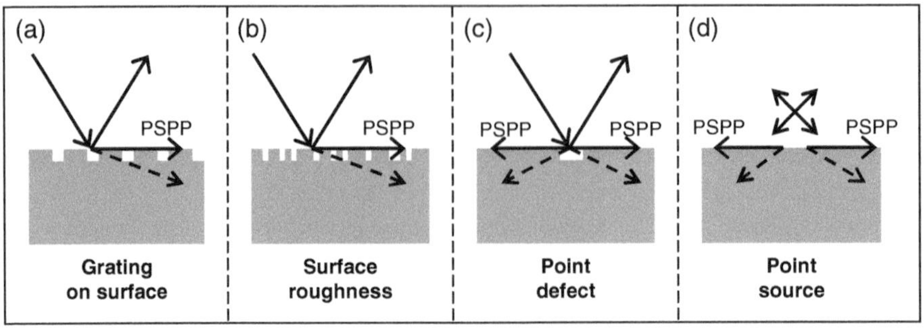

Figure 3.13. Schematic illustrating other techniques of exciting PSPP modes with photons, all based on breaking the translational invariance (and therefore the momentum conservation) by the presence of either a grating (a), surface roughness (b), a point defect (c), or a point source (d).

the metal itself could be used to deliver the photon (through an evanescent field again) using a configuration of the type (dielectric P)/metal/(dielectric M) with $n_P > n_M$ again. This configuration, sometimes called *Kretschmann configuration,* is depicted in Fig. 3.12(b). It only requires to deposit a thin metal film on the prism surface (dielectric P), which is reasonably easy experimentally. The film thickness is much smaller, typically 50 nm, than the dielectric gap in the Otto configuration, due to the stronger decay of the evanescent field in the metal, but must remain larger than the decay length L_{2z} for the metal/(dielectric M) PSPPs to avoid strong coupling between the two interfaces. The parallel momentum $\hbar k_x = n_P \hbar k_0 \sin\theta$ in the prism is then transferred by the evanescent field through the metal and can excite (on the other side) PSPP modes of the metal/(dielectric M) interface.

Both Otto and Kretschmann configurations are based on the same principle of using an evanescent wave to excite the PSPP modes, and will both be referred to as the *ATR configuration, after attenuated total reflection.*

Surface gratings and surface roughness

Another somewhat different approach to coupling PSPPs to light is to relax the conservation of momentum restriction, i.e. break the translational invariance. This can for example be achieved by engraving a periodic structure on the surface along the x direction, i.e. by forming a *surface grating* as illustrated in Fig. 3.13(a). As before, the perturbation to the surface needs to be small enough not to alter substantially the nature of the SPP modes. This is possible for example for a grating whose depth remains small compared to its spatial period, λ_g, and to L_{1z}. Such a periodicity does not fully break translational invariance, but modifies the law of conservation of k_x, allowing addition or subtraction of integer multiples of $k_g = 2\pi/\lambda_g$, and therefore making the coupling of light to PSPPs possible for some specific frequencies and/or angles. The coupling condition remains very sharp due to the discrete nature of the modified conservation law.

Such a conservation law can be removed completely in the case of a *random surface roughness* as illustrated in Fig. 3.13(b). Such a surface behaves like a combination of gratings with any arbitrary λ_g or k_g; k_x conservation is then always possible. Photons then always couple to PSPP modes, but this coupling is not as strong and does not show any sharp resonant effects, which can be a problem for some applications.

Point defects and point sources

A third approach to break the translational invariance is to introduce a spatially localized point defect, such as a bump or hole in the surface (Fig. 3.13(c)). This defect can also be created experimentally by a metal-coated tip of a scanning probe microscope placed in close proximity to the metallic surface. Such a defect breaks locally the law of momentum conservation and enables coupling of light to PSPP modes that will then propagate away from the point. It therefore acts as a *point source for PSPP modes*.

A similar situation happens when a light source (such as a dipolar emitter) is sufficiently close to the metal surface (Fig. 3.13(d)). The translational symmetry of the problem, and k_x conservation, is again broken locally, which enables the light emitted from the source to excite PSPP modes at the dielectric/metal interface. This again acts as a point source for PSPP modes. This can be realized experimentally for example using the optical probe of a scanning near-field optical microscope as the localized source [157].

Strictly speaking the PSPP modes created in this way are slightly different (although similar in many aspects) from those studied so far because they have a symmetry of revolution, instead of translational invariance. A separate mathematical treatment in cylindrical coordinates would be needed to describe them accurately.

A final remark about coupling light to PSPPs

Note that in many of the cases discussed above, the geometry of the problem is changed, for example, by addition of one or more interfaces, or by modification of the surface itself. In this respect, the electromagnetic modes of the system are also changed and are strictly speaking no longer the PSPP modes of a single planar dielectric/metal interface. However, if these modes are weakly coupled to other parts of the system, they then retain essentially their nature. The main change is that they are no longer strictly non-radiative modes because they couple (at least weakly) to radiation. This coupling can affect the propagation length of these modes, since there are now radiative losses in addition to the intrinsic decay through absorption in the metal. If these additional losses are too large, then the modes may lose their propagating nature (when the propagation length is of the order of the wavelength) and should then be considered as localized SPP modes.

3.4.4. PSPP resonances at planar interfaces

Origin of PSPP resonances

One of the important aspects of coupling light with PSPPs is that the coupling condition (conservation of momentum and energy) sets very stringent requirements on the parameters for optimal coupling. When these requirements are met, then the energy of the incident light is efficiently transferred to the PSPP modes, and this can have a dramatic effect on the optical response of the system. This is usually referred to as a *surface plasmon resonance* (SPR). Note that the denomination SPR may sometimes be used in a much more general context, i.e. for resonances arising from coupling to any SPP modes (propagating or localized SPPs for any types of geometries). We reserve it here for resonances arising from coupling to propagating SPP modes at planar interfaces.

In this particular case of PSPPs at planar interfaces, the simultaneous conservation of both momentum and energy (frequency) together with the very small broadening of the modes makes any resonances arising from them particularly sharp. Moreover, the main characteristic of these modes is that they are *non-radiative*. This means that the energy that is transferred to them cannot be re-radiated. It propagates at the surface and unless another coupling mechanism is used to collect it, it will simply dissipate by optical absorption in the metal (typically due to resistive losses related to the imaginary part of the dielectric function). The SPRs for PSPPs at planar interface therefore usually appear as *resonances in absorption*, or equivalently as a *resonant decrease in reflectivity*.

The main parameters that can be varied in a typical experiment are the wavelength λ, the angle of incidence θ_i, the dielectric constant of the outside medium ϵ_M, or the thicknesses of the dielectric layers in a multi-layer configuration (like the ATR configurations). SPRs can appear as functions of any of these parameters depending on the setup. Moreover, in regions of interest (for example visible or near infrared for Ag), the broadening of the PSPP modes, characterized by $\text{Im}(k_x)$ (see Eq. (3.15)), is typically very small. This means that these resonances can be extremely sharp and therefore *very sensitive to the value of the parameters* under study. This makes such SPRs very attractive in applications such as sensors.

The resonance condition

The most common types of optical resonances for PSPPs at planar interfaces are found in the measurement of the reflectivity (i.e. of the reflection coefficient R^p for TM waves, see Appendix F). For a typical metallic plane (with ϵ negative), the reflection coefficient is normally close to 1 (see Appendix F) and does not vary much with λ or θ_i. This is why metals are known as good reflectors. Light does not couple to PSPP modes in this case, and is mostly reflected as illustrated in Fig. 3.11(a).

We now consider a reflectivity experiment in an ATR setup (Otto or Kretschmann configurations, as in Fig. 3.12(a–b)) and look at the resonance condition for coupling to PSPP modes. k_x in the prism is given by $\sqrt{\epsilon_P} k_0 \sin(\theta_i)$, and this is transferred through the dielectric gap or the metal by an evanescent field. The PSPPs excited in the ATR configuration are those of the interface of the metal with a dielectric of lower refractive index than the prism ($\epsilon_M < \epsilon_P$).

The resonance condition resulting from k_x conservation is then obtained from the PSPP dispersion relation for such an interface, given in Eq. (3.15), and takes the form:

$$\sqrt{\epsilon_P} \sin\theta_i = \mathrm{Re}\left(\sqrt{\frac{\epsilon(\omega)\epsilon_M}{\epsilon(\omega)+\epsilon_M}}\right). \tag{3.18}$$

If the absorption is small, i.e. $\epsilon = \epsilon' + \mathrm{i}\epsilon''$ with ϵ'' small compared to $|\epsilon'|$, then this simplifies to:

$$\epsilon_P \sin^2\theta_i \approx \frac{\epsilon'(\omega)\epsilon_M}{\epsilon'(\omega)+\epsilon_M}. \tag{3.19}$$

Moreover, for long wavelengths, where $\epsilon'(\omega)$ is negative and large, the resonance condition can further be approximated by:

$$\sin\theta_i \approx \sqrt{\frac{\epsilon_M}{\epsilon_P}} = \frac{n_M}{n_P}, \tag{3.20}$$

i.e. θ_i at resonance is close to (but still above) the critical angle for total internal reflection at the prism/dielectric interface.

Because the broadening of k_x for the PSPP modes is very small, the resonance condition is extremely sharp, i.e. it is very sensitive to the parameters: angle of incidence, wavelength, and refractive index of the dielectric medium. If this condition is met, then one expects the incident light to couple and transfer its energy to the PSPP modes, resulting in a drop in reflectivity. This drop in reflectivity can be dramatic (and easily detectable), when the configuration is optimized for maximum coupling of the incident light to the PSPP modes. In the Otto configuration, this optimization is achieved by adjusting the size of the dielectric gap L_{Gap}. If L_{Gap} is too large, then the coupling will be poor, even at resonance, because of the evanescent

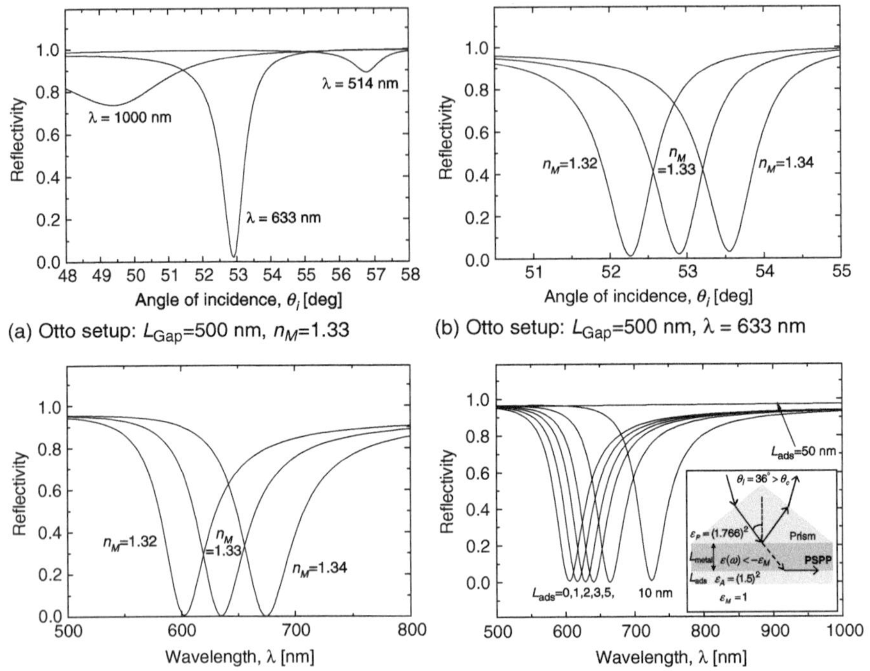

(a) Otto setup: L_{Gap}=500 nm, n_M=1.33

(b) Otto setup: L_{Gap}=500 nm, λ = 633 nm

(c) Kretschmann setup: θ_i=53, L_{Ag}=50 nm

(d) Kretschmann setup: θ_i=36, L_{Ag}=50 nm, n_M=1

Figure 3.14. Examples of SPR experiments involving the excitation of propagating SPP modes at a planar silver/dielectric interface. The ATR configuration is used for coupling light to PSPPs, either in the Otto (a and b) or Kretschmann (c and d) setup (see Fig. 3.12). Angle-modulation (AM) is illustrated in (a) and (b) and wavelength-modulation (WM) in (c) and (d). (b) and (c) illustrate the possibility of refractive-index sensing using SPRs. In (d) an adsorbed layer on the metal changes the effective refractive index of the dielectric medium, thereby resulting in a shift of the SPR condition. Such a configuration can be used for adsorbate detection and/or thickness analysis.

nature of the exciting wave in the gap. If L_{Gap} is too small, then the proximity of the two interfaces may modify the nature of the SPP modes and broaden the resonance. In between the two limits, there is an optimum coupling condition for a given set of experimental variables.

Examples of surface plasmon resonances with PSPPs

The reflectivity in the ATR setup (Otto or Kretschmann configurations) can in fact be calculated by solving Maxwell's equations for the corresponding three-layer system, the details of which are given in Appendix F along with a possible Matlab implementation. Such calculations are shown as an example in Fig. 3.14 for a sapphire prism ($n_P = 1.766$, $\epsilon_P = n_P^2$), using silver as the metal.

In a typical setup, this reflectivity will be measured as a function of angle of incidence, sometimes called angle-modulation (AM), as in Fig. 3.14(a–b), or wavelength, called wavelength-modulation (WM), as in Fig. 3.14(c–d). The position of the resonance can then be used to deduce other parameters, such as the refractive index of the dielectric material, n_M. The angle-dependent reflectivity (R^p) is for example shown for the Otto configuration (see Fig. 3.12(a)) in Fig. 3.14(a). A clear sharp resonance, where the reflectivity drops from 1 to virtually 0 is observed for excitation at 633 nm. This combination of parameters therefore offers optimum coupling to the silver/water PSPP modes at 633 nm. Weaker resonances are also observed (at a different angle of incidence) for 514 nm and 1000 nm excitation, for which the parameters are not optimized. The refractive-index sensitivity of the angle-dependent resonance is illustrated in Fig. 3.14(b) for different values of n_M. The shifts in the resonance are clearly resolved and such a setup could therefore be used to measure the refractive index n_M with high accuracy. A similar example of refractive-index sensing is shown in Fig. 3.14(c) for wavelength-dependent reflectivity in the Kretschmann configuration.

Finally, it is worth pointing out that such a high- refractive-index sensitivity can in fact be used to measure the thickness of an adsorbed layer (of known refractive index n_A), down to only one monolayer in optimized conditions. The adsorbed layer is embedded in the dielectric medium with refractive index n_M and modifies slightly the properties of the metal/(dielectric M) PSPP modes, thereby affecting the resonance condition. A simple way of studying this effect is to add a fourth layer (the adsorbate layer of thickness L_{Ads} and refractive index n_A) to the EM problem and calculate the reflectivity[12]. This is illustrated in Fig. 3.14(d) for WM reflectivity in the Kretschmann configuration. The shift in resonance is clearly observable as a function of adsorbate layer thickness and can therefore be used as a sensor for the presence of adsorbed molecules. In practice, one usually wants to detect one specific type of adsorbed molecule, and it is therefore necessary to functionalize the metallic surface, so that only the desired molecules bind to it. The surface functionalization is in fact a necessary and important step in many sensing applications of plasmonics (including SERS).

3.4.5. Local field enhancements and SPPs at planar interfaces

We have so far concentrated on far-field properties of the SPPs at planar interfaces, i.e. how they couple to incident light, and what their effect is on scattering properties like the reflectivity. For some applications, the near-field properties are also important, since they rely on large local field enhancements at the surface (this is in particular the case of SERS).

[12] This approach has limitations, especially for the smallest thicknesses, but nevertheless illustrates the point.

Due to the strong confinement of the field at the surface (which decays exponentially from it), one could intuitively expect that the field intensity close to the surface is large. In a sense, the electromagnetic energy is confined at the surface and should result in a large local field.

In order to quantify this assertion more accurately, we need to be able to compare the local field at the surface to an exciting field. To do so, one could choose one of the setups previously discussed for coupling an incident wave with amplitude E_{Inc} to the SPP modes, calculate the local field at the surface E_{Loc}, and deduce the local field intensity enhancement factor (LFIEF) $M_{\text{Loc}} = |E_{\text{Loc}}/E_{\text{Inc}}|^2$. We will do that soon, but shall first start with a more general and physical approach.

Local fields for SPP modes

Let us first analyze the local field at a planar metal/dielectric interface for a SPP mode (propagating or localized). The field expressions are given in Eq. (3.13), from which we deduce the electric fields $\mathbf{E}_1$ (in the dielectric) and $\mathbf{E}_2$ (in the metal) at the surface ($z = 0$). They can be decomposed as tangential and perpendicular components, and using the expressions of k_x, k_{1z}, and k_{2z}, we have:

$$\frac{|\mathbf{E}_1^{\perp}(0)|^2}{|\mathbf{E}_1^{\parallel}(0)|^2} = \frac{|k_x|^2}{|k_{1z}|^2} = \frac{|\epsilon|}{\epsilon_M}, \tag{3.21}$$

and

$$\frac{|\mathbf{E}_2^{\perp}(0)|^2}{|\mathbf{E}_2^{\parallel}(0)|^2} = \frac{|k_x|^2}{|k_{2z}|^2} = \frac{\epsilon_M}{|\epsilon|}. \tag{3.22}$$

Moreover, we deduce from these and from the fact that $\mathbf{E}_2^{\parallel}(0) = \mathbf{E}_1^{\parallel}(0)$ (imposed by the boundary condition at the interface) that:

$$\frac{|\mathbf{E}_1(0)|^2}{|\mathbf{E}_2(0)|^2} = \frac{|\mathbf{E}_1^{\perp}(0)|^2 + |\mathbf{E}_1^{\parallel}(0)|^2}{|\mathbf{E}_2^{\perp}(0)|^2 + |\mathbf{E}_2^{\parallel}(0)|^2} = \frac{|\epsilon|}{\epsilon_M}. \tag{3.23}$$

These equalities are useful to bear in mind when considering the local fields for SPP modes at planar interfaces. In particular, there are many common situations where $|\epsilon| \gg \epsilon_M$, for example for PSPP modes at long wavelengths, where $\text{Re}(\epsilon)$ is negative and large, or when losses are important ($\text{Im}(\epsilon)$ large). In these cases, we deduce that:

- The surface electric field inside the metal, $\mathbf{E}_2$, is mostly *parallel* to the interface.

- The surface electric field outside the metal (in the dielectric), $\mathbf{E}_1$, is mostly *perpendicular* to the interface.
- The intensity of the field outside is much larger than that inside the metal. This is the most important effect and can be loosely viewed as an expulsion of the field from the metal and a concomitant concentration of energy at the surface just outside (within a distance spanned by the evanescent field).

We will now attempt to place this final statement on a more rigorous footing, using simple energy considerations.

Some general energy considerations

We consider an experiment where an incident wave in a dielectric (ϵ_{Inc}) is coupled (using any of the methods described before) to *propagating SPP* modes at a dielectric(ϵ_M)/metal(ϵ) interface. The coupling efficiency, i.e. power injected into the PSPP mode over incident power, denoted by η, can in principle be close to 1 at resonance (under optimal coupling conditions). Let us analyze the energy balance in a volume covering a small surface area S on the interface. Because the PSPP modes are non-radiative, no energy is lost to the far field in the dielectric or the metal. Moreover, because of invariance by translation, there is no net flow of energy laterally. The only source of energy loss for the PSPP modes is therefore due to optical absorption in the metal. In the volume delimited by the surface S, the power absorbed is:

$$P_{\mathrm{Abs}} = \int_0^\infty \frac{1}{2}\omega\epsilon_0\epsilon''|\mathbf{E}_2(z)|^2 S dz = S\frac{\epsilon_0 c}{2}\epsilon'' k_0 L_{2z}|\mathbf{E}_2(0)|^2, \tag{3.24}$$

where $k_0 = \omega/c$ is the free-space wave-vector, $L_{2z} = 1/(2\mathrm{Im}(k_{2z}))$ is the penetration depth in the metal defined previously, and $\mathbf{E}_2(0)$ is the electric field at the interface, just inside the metal, (see Eq. (3.13)).

This absorbed power must, by energy conservation, be balanced by the power coupled into the SPP mode, i.e. $P_{\mathrm{Abs}} = \eta P_{\mathrm{Inc}}$. For an incident wave incoming from a dielectric (ϵ_{Inc}) with incident angle θ_{Inc}, the incident power on surface S is:

$$P_{\mathrm{Inc}} = S\frac{\epsilon_0 c}{2}\sqrt{\epsilon_{\mathrm{Inc}}}|E_{\mathrm{Inc}}|^2\cos\theta_{\mathrm{Inc}}, \tag{3.25}$$

where $|E_{\mathrm{Inc}}|$ is the magnitude of the incident electric field. Expanding $P_{\mathrm{Abs}} = \eta P_{\mathrm{Inc}}$, we therefore deduce:

$$\frac{|\mathbf{E}_2(0)|^2}{|E_{\mathrm{Inc}}|^2} = \frac{\eta\sqrt{\epsilon_{\mathrm{Inc}}}\cos\theta_{\mathrm{Inc}}}{\epsilon'' k_0 L_{2z}}. \tag{3.26}$$

Because ϵ'' and $k_0 L_{2z}$ are in typical situations both small, this simple argument naturally predicts a field enhancement at the surface (with respect to the incident field E_{Inc}. But here $\mathbf{E}_2(0)$ is the surface field inside the metal. We are more interested in the surface field $\mathbf{E}_1(0)$ just outside, in the dielectric. These two are related by Eq. (3.23) derived above, which is then the source of an additional enhancement by a factor $|\epsilon|/\epsilon_M$. For PSPPs for example, $\epsilon' < -\epsilon_M$ and usually $|\epsilon'| \gg \epsilon_M$ and this factor is the source of a further field enhancement of the surface field outside compared to the surface field inside. Using the exact expression for $k_0 L_{2z}$, we can finally write the local field intensity enhancement factor (LFIEF) as:

$$M_{\text{Loc}} = \frac{|\mathbf{E}_1(0)|^2}{|E_{\text{Inc}}|^2} = \frac{2|\epsilon|}{\epsilon_M \epsilon''} \text{Im}\left(\epsilon \sqrt{\frac{1}{\epsilon + \epsilon_M}}\right) \eta \sqrt{\epsilon_{\text{Inc}}} \cos\theta_{\text{Inc}}. \tag{3.27}$$

In cases where $|\epsilon'| \gg \epsilon_M$, $\mathbf{E}_1$ is almost perpendicular to the interface, and the LFIEF can be approximated by:

$$M_{\text{Loc}} \approx M_{\text{Loc}}^{\perp} \approx \frac{2|\epsilon'|^{3/2}}{\epsilon_M \epsilon''} \eta \sqrt{\epsilon_{\text{Inc}}} \cos\theta_{\text{Inc}}. \tag{3.28}$$

For a silver/air interface at 633 nm, excited from air, this leads to values of the order of $M_{\text{Loc}} \approx 180$ (from Eq. (3.27)) at normal incidence $\theta_{\text{Inc}} = 0$ for optimum coupling $\eta \approx 1$. Note that this is the maximum LFIEF that can be expected in such a situation. This increases to $M_{\text{Loc}} \approx 200$ at 800 nm. For the same interface in the ATR configuration with a sapphire prism ($\epsilon_P = (1.766)^2$), the incident wave is coming from the prism, i.e. $\epsilon_{\text{Inc}} = \epsilon_P$, and the incident angle must be chosen as $\theta_{\text{Inc}} \approx 36°$ for optimum coupling ($\eta \approx 1$). We then obtain $M \approx 255$ at 633 nm from Eq. (3.27). Note however that the LFIEF is in this case calculated with respect to $E_{\text{Inc}} = E_P$, i.e. the electric field amplitude in the prism. In practice, this needs to be related to the true incident wave that excited the one in the prism from outside, which will be discussed briefly later.

The advantage of the previous argument is that it is very general and gives an upper limit on the field enhancement for SPPs on planar interfaces. In practice however, it can sometimes be difficult to estimate the coupling efficiency η to SPP modes. One must then model the electromagnetic problem to predict the field enhancement. We give here a couple of examples.

Example in the Otto configuration

We first consider the Otto configuration with a sapphire prism ($\epsilon_P = n_P^2 = (1.766)^2$), and an air ($\epsilon_M = 1$)/silver ($\epsilon$) interface. Here the air gap is optimized for PSPP coupling at 633 nm: $L_{\text{gap}} = 850$ nm. We calculate the

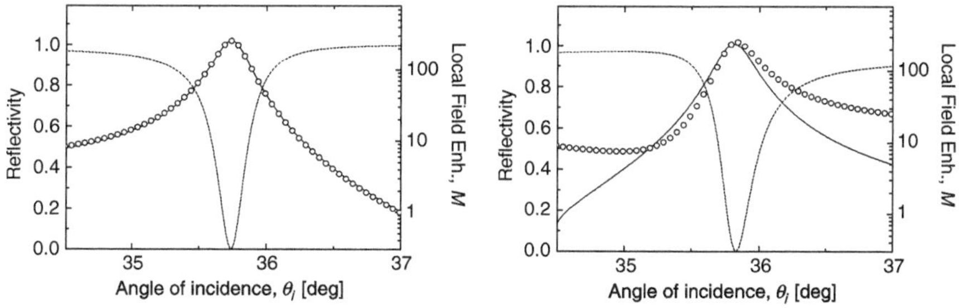

(a) Otto setup: L_{Gap}=850 nm, n_M=1.0, λ=633 nm (b) Kretsch. setup: L_{Ag}=50 nm, n_M=1.0, λ=633 nm

Figure 3.15. Calculated reflection coefficient, R, (dashed line) and local field intensity enhancement factor, M_{Loc}, in the dielectric at the metal/dielectric interface (solid line) when exciting PSPP modes in either the Otto (a) or Kretschmann (b) configurations. The symbols show the predicted LFIEF using Eq. (3.27) and $\eta = 1 - R$.

LFIEF at the metal surface (on the air side), using the tools of Appendix F. This is shown as a function of incident angle θ_i along with the reflection coefficient in Fig. 3.15(a). It is clear that the LFIEF, M_{Loc}, exhibits a sharp resonance, like the reflectivity, R. At the reflectivity minimum, we have $R \approx 0$ and $\theta_i \approx 35.7°$. Assuming all the incident power is coupled into the SPP modes, ($\eta \approx 1$), we can estimate from the simple energy conservation argument (Eq. (3.27)) that $M_{\mathrm{Loc}} \approx 256$, which is precisely what is predicted by the calculation.

Here we can expand further the energy arguments of the previous section. If we assume that even outside resonance the power that is not reflected is coupled into the SPP modes, then we have $\eta = 1 - R$, and the LFIEF can therefore be estimated from Eq. (3.27). This estimation, also shown in Fig. 3.15(a), fits the calculated M_{Loc} extremely well.

Example in the Kretschmann configuration

Finally, to illustrate the limits of this approach, we now study a similar experiment in the Kretschmann configuration. The calculated LFIEF and reflectivity are shown in Fig. 3.15(b) along with the prediction using Eq. (3.27) and $\eta = 1 - R$. The agreement is quite good at resonance, but not that good for other angles. The reason for this discrepancy can be understood, at least qualitatively. In the Otto configuration, the reflection/refraction at the first interface (prism/air) can only excite non-dissipative modes (with no optical absorption), i.e. all power that is not reflected is transmitted to the air/metal interface. In the Kretschmann configuration, however, the excitation must go through the thin metal layer before reaching the metal/air interface supporting the PSPP modes. Dissipation may occur in this layer by optical absorption in the metal. In fact, optical modes other than PSPPs are excited in this configuration. Because these additional modes are dissipative, $\eta = 1 - R$ is only an upper estimate of the coupling to the SPP modes (part of this energy may couple to the other modes and dissipate in the form of heat). Moreover, these modes may contribute to the local electric field at the metal surface, and therefore increase or decrease the LFIEF. This is the reason why, at resonance where SPP coupling dominates, our simple estimate based on Eq. (3.27) is correct, but away from resonance, the contribution of other types of modes becomes important and our simple energy argument is no longer useful.

These two examples illustrate both the power of simple energy arguments in connecting local field enhancements and resonant mode coupling, and its limitations in more complex systems. We will come back to these arguments when discussing localized SPPs in Section 3.5.

These examples also further confirm that: (i) PSPP modes are associated with relatively large electric field at the interface, (ii) This can result in a large local field enhancement *provided the incoming power is efficiently coupled into the PSPP modes.*

The air/prism interface

Finally, we make here a brief digression to discuss the air/prism interface in relation to the energy conservation argument and local field enhancements. In many practical situations, the ATR configuration is used to couple to the PSPP modes, and we must therefore have a prism with $\epsilon_{\text{Inc}} = \epsilon_P > \epsilon_M$. The angle of incidence θ_{Inc} is then not free to vary since it must be chosen appropriately to couple efficiently to the PSPP modes (otherwise η would be very small). In fact, k_x conservation requires:

$$\sqrt{\epsilon_P}\sin\theta_i = \text{Re}\left(\sqrt{\frac{\epsilon\epsilon_M}{\epsilon + \epsilon_M}}\right). \tag{3.29}$$

The expression for the LFIEF can then again be simplified if $|\epsilon'| \gg \epsilon_M, \epsilon''$:

$$M_{\text{Loc}} \approx M_{\text{Loc}}^{\perp} \approx \frac{|\mathbf{E}_1(0)|^2}{|E_P|^2} = \frac{2|\epsilon'|^{3/2}}{\epsilon_M\epsilon''}\eta\sqrt{\epsilon_P - \epsilon_M}. \tag{3.30}$$

Moreover, in practice, the real source of the incident wave is rarely embedded in the high-dielectric constant material (i.e. the prism), but typically comes from air (with $\epsilon_{\text{Source}} = 1$) and is injected into the high-dielectric constant material (ϵ_P) using a prism configuration. To study the true field enhancement in this situation, one must also study the coupling of the light from air into the prism itself, i.e. relate the field amplitude $|E_{\text{source}}|$ of the incident light, to the field amplitude $|E_P|$ refracted in the prism, which then serves as excitation for the PSPP modes. This should be assessed on a case-by-case basis, and should be calculated independently of the rest of the problem because the coupling to the prism breaks the translational invariance and cannot therefore be included as an additional layer. The coupling at the air/prism interface is a simple reflection/refraction problem between two dielectrics and can be modeled using the tools of Appendix F.

We only provide here a model example (in relation to the two previous examples discussed above) for ATR coupling to the silver/air interface at 633 nm. We consider a triangular sapphire prism ($n_P = 1.766$), with (base) angle of 67°, and an incident beam *perpendicular* to the metal/dielectric interface (the base of the prism), i.e. $\theta_{\text{Source}} = 0$. The angle of incidence of this beam with respect to the prism side surface is 67°, and from Snell's law the refracted angle (again with respect to the prism side) is 31.4°, which results in an incident angle for the bottom prism/dielectric interface of $\theta_P = 35.6°$, i.e. the angle required for PSPP coupling at 633 nm (we in fact chose the base angle of the prism for this to be the case).

Assuming optimal coupling from the prism to the PSPP modes, there are two ways of viewing the energy conservation argument with regard to coupling from the air to the prism:

- One may consider the wave from outside the prism as the incident wave. In this case $\epsilon_{\text{Inc}} = \epsilon_{\text{Source}} = 1$ (air) and $\theta_{\text{Inc}} = \theta_{\text{Source}} = 0$. However, the reflection/refraction at the air/prism interface on the side of the prism changes the incident flux of energy (along z) by a factor $(1/n_P)|t^p|^2 \cos\theta_P \approx 0.368$ (this is derived from the expressions of Section F.3.3). This results in a decrease in the eventual coupling to the PSPP modes to $\eta \approx 36.8\%$. Equation (3.27) then gives $M_{\text{Loc}} \approx 66$ for the LFIEF.
- One may consider alternatively the wave in the prism as the incident wave, as was done in the discussion of the examples above. In this case $\epsilon_{\text{Inc}} = \epsilon_P$ and $\theta_{\text{Inc}} = \theta_P$. Equation (3.27) can then be used to estimate the LFIEF and we obtain $M_P \approx 256$, but here this is with respect to the electric field inside the prism E_P, not E_{Source}. These are related (see Section F.3.3) by $|E_P|^2/|E_{\text{Source}}|^2 \approx 0.256$ from which we deduce again the true LFIEF $M_{\text{Loc}} \approx 66$.

Both views are, fortunately, consistent, but they highlight the necessity to take special care when handling this energy conservation argument, and in the definitions of the LFIEF.

3.4.6. SPP modes on planar interfaces: A brief summary

In this section we have discussed mostly the propagating SPP modes at a planar metal/dielectric interface. These modes are important for two reasons:

- They can propagate over relatively long distances on the surface, whilst remaining strongly confined at the surface. This opens up the possibility to use these modes to manipulate and guide light and design plasmonic wave-guides and even plasmonic chips.

- These modes exhibit an extremely narrow broadening and their coupling with light is therefore very sensitive to external parameters. This leads to very sharp surface plasmon resonances, which can be exploited for example as sensors.

Although a planar interface may appear as a very particular case of the possible geometries that may be encountered in real experiments, its reach is in fact much wider: as discussed earlier, small perturbations to the surface (like a shallow grating, roughness, or point defects) do not affect significantly the nature of the PSPP modes. Moreover, there are many situations where a metallic surface can be approximated by a plane over distances of the order of the propagation length of the SPPs (say around 10 μm). In all these cases, the PSPP modes will play an important role. We will now focus on situations where this is no longer the case.

3.5. LOCALIZED SURFACE PLASMON–POLARITONS

3.5.1. Introduction to localized SPPs

The planar approximation is no longer true for small metallic objects, and in particular for nano-particles, where the size becomes comparable or smaller than the wavelength. The nature of the electromagnetic modes of the system is then completely modified. In particular, the description in terms of k vector (k_x for a plane) becomes irrelevant, since the translational invariance is lost. The electromagnetic modes then exist for discrete values of ω (instead of having continuous modes described by the dispersion relation $\omega(k_x)$). These modes are then called *localized surface plasmon–polaritons* (LSPs)[13].

In fact, this is not a property of metals or plasmon–polaritons only. The same happens for photons when the environment exhibits features of the order of the wavelength. Photons correspond to free-space modes of the electromagnetic field (plane waves with well defined ω and $\mathbf{k}$). When boundaries have features much larger than the wavelength, one can apply the 'standard' description giving rise to reflection and refraction at interfaces (Snell's law). All boundaries are approximated by locally planar interfaces, and this 'ray optics' approach is perfectly legitimate. However, when the dimensions of the system become comparable to the wavelength, say in a cavity or a wave-guide, this approach fails. The concept of photon is replaced by that of electromagnetic modes of the cavity (characterized by discrete

[13] In the acronym LSP, the P should be understood as standing for plasmon–polariton, not plasmon. The use (abuse, in fact) of the denomination localized surface plasmon instead of localized SPP is however very common, so common that we have adopted it in this book.

values of ω, or $\omega(k)$ where k is irrelevant). These modes are highly localized inside the cavity. They are sometimes called cavity polaritons to emphasize their mixed nature of a photon with its optical environment, and are the photon analogs of localized SPPs.

Note that the localized SPPs meet the two criteria introduced previously to characterize SPPs (see Section 3.3.8): they would not exist without the presence of the interfaces, and their properties depend on the optical properties of the outside medium. The third criterion (having a field localized at the interface) can lead to misleading interpretations. Finally, LSP modes, and their effects on SERS and other optical properties, will be extensively discussed in Chapter 6 on practical examples, and we therefore restrict ourselves here to general considerations.

3.5.2. LSP on planar structures

A particular case of localized SPP has already been encountered in the discussion of the surface modes of a planar interface. These modes arise when $-\epsilon_M < \mathrm{Re}(\epsilon) < 0$ and are characterized by a large broadening ($\mathrm{Im}(k_x)$), which essentially means that the description in terms of k_x becomes irrelevant. These particular types of localized SPPs have rarely appeared in applications and we will therefore focus on the more important LSPs arising in metallic nano-particles.

3.5.3. LSP modes of a metallic sphere

To understand better the nature of LSP, it is useful to consider the canonical example of a metallic sphere. A full analytical treatment is then possible using Mie theory [149,158]; its results will be discussed in Chapter 6 while the technical details are given in Appendix H. We only discuss qualitatively these results in terms of the electromagnetic modes of the sphere. This discussion is therefore rather abstract.

Thanks to the analytical solution from Mie theory, it is possible to define and study all the electromagnetic modes of the sphere. Many of these modes are irrelevant to SERS or plasmonics but they provide a nice illustration of the general discussion about electromagnetic modes given in Section 3.3. A similar study was carried out for an ionic crystal sphere in Ref. [159], where more details can be found. The only difference here is the form of the dielectric function $\epsilon(\omega)$ and we therefore adapt the discussion to the case of a metal. The k vector is irrelevant here and the modes correspond to a discrete set of frequencies ω. Because of the spherical symmetry, it is convenient to index these modes with an integer $l \geq 1$ corresponding to the total angular momentum.

The longitudinal modes, as is the case for bulk metals must satisfy $\mathrm{Re}(\epsilon(\omega_{\mathrm{BP}})) = 0$ and therefore all occur at the same frequency ω_{BP} as bulk plasmons. For a Drude model, this is simply the plasma frequency. These are 'pure' spherical plasmon modes and behave exactly like bulk plasmon modes. They correspond to longitudinal spherical electric waves (no magnetic field) inside the metal, associated with collective charge oscillations. The field outside the sphere is zero everywhere. These modes are not relevant to the optical properties because of their longitudinal character.

All the other electromagnetic modes are transverse, inside and outside the sphere, but may have a longitudinal character at the interface because of the boundary conditions (and may therefore be associated with surface charge oscillations). Incident wave modes exist for any ω and simply correspond to the solution of the EM problem (using Mie theory). Here we focus on the surface modes (bound modes), which are solutions for which there are no incident waves (eigen-solutions of the linear problem). For a given l, the frequencies of the surface modes are solutions of a complex equation given in Section H.3.5. Because of intrinsic absorption in the metal, all the surface modes must be virtual modes (with $\omega = \omega' - \mathrm{i}\omega''$ complex). The finite lifetime $1/(2\omega'')$, or broadening $2\omega''$ accounts for the absorptive losses in the metal and possible radiative nature of the mode. Coupling to these modes will occur when the incident frequency (real) matches ω', with a broadening of the order of $2\omega''$. This will be discussed further in Chapter 6 and also partly in Appendix H.

The most important surface mode in most cases is the lowest frequency one, which corresponds to $l = 1$ and whose scattered field is that of an electric dipole and it is, therefore, a radiative mode. It should be called the *dipolar localized surface plasmon–polariton* mode of the sphere, but is often loosely called (at best) 'localized surface plasmon' or simply 'surface plasmon', and even sometimes 'plasmon'. Other surface modes exist for $l > 1$, at increasing frequencies, with quadrupolar, octupolar, etc. nature, but are usually less relevant. They are also radiative modes, but with a larger absorptive character than the dipolar modes (i.e. the proportion of losses through absorption, as opposed to radiation, is larger).

The frequency of the dipolar LSP mode of the sphere depends on several parameters:

- Obviously, the metal (through its frequency-dependent optical properties characterized by $\epsilon(\omega)$).
- The environment, through its dielectric constant ϵ_M.
- The size of the sphere (i.e. its radius a).

For the smallest spheres (typically $a < 10$ nm), the dipolar LSP mode frequency can be obtained approximately from:

$$\mathrm{Re}(\epsilon(\omega_{\mathrm{LSP}})) = -2\epsilon_M, \tag{3.31}$$

and therefore lies at a wavelength longer than the longitudinal plasmon–polaritons ($\mathrm{Re}(\epsilon(\omega_{\mathrm{BP}})) = 0$) and the localized SPPs of the plane interface ($-\epsilon_M < \mathrm{Re}(\epsilon(\omega_{\mathrm{SP}})) < 0$). For the smallest spheres, for which the electrostatic approximation applies (see Section 5.1.4), the LSP mode is almost a "pure surface plasmon" in nature, with only a small "photon component" [160]. As the size increases, the dipolar LSP frequency red-shifts (to longer wavelengths) and this is associated with a larger broadening (ω''). These properties will be reviewed in more detail in Chapter 6.

3.5.4. LSP modes of nano-particles

The description of the LSP modes of non-spherical nano-particles in terms of complex frequencies (eigenvalues) is an extremely difficult problem because of the lack of an analytical solution to the EM problem. The standard approach is then to study the EM problem of excitation of the nano-particle by incident waves. The LSP modes of the nano-particle then appear as *resonances* in the optical response and their nature can be inferred from the field solution at their resonance frequency (if there is not too much overlap with other resonances).

The LSP modes of non-spherical nano-particles have the same qualitative features as that of the sphere, but their frequency depends in addition on the geometry (shape) of the particle. This shape may in addition introduce anisotropies, i.e. their coupling to an external field becomes polarization-dependent. This will also be further discussed in Chapter 6.

3.5.5. LSP resonances

The LSP modes of a nano-particle can be excited by an incident wave with the appropriate polarization and frequency. Efficient coupling to LSP modes will then result in a resonant optical response at the LSP frequency. As opposed to PSPP on planar interfaces, LSPs are radiative modes (with an absorptive component because of optical absorption in the metal). The resonant response therefore, not only appears in absorption (which is analogous to the reflectivity experiments for a plane interface), but also in scattering (or similarly extinction) measurements. These resonances, sometimes called LSP resonances (LSPR) to differentiate them from SPR (based on PSPPs), are sensitive to the environment and, like SPR, can be used for applications in refractive-index and chemical sensing.

The LSP resonances also manifest themselves, as for PSPP modes, as large local field enhancements inside the metal, and more importantly on the surface outside. This effect is the basis for most surface-enhanced spectroscopies, including SERS.

It is interesting to highlight the main differences between SPR and LSPR:

- The SPR condition requires conservation of both k_x and ω. This is more difficult to fulfill than only ω conservation for LSPR. In particular, k_x

conservation typically requires a more complex setup, such as the ATR configuration.

- SPRs offer more liberty in the implementation, either in terms of angle-modulation or wavelength-modulation, whereas only wavelength-modulation can be used for LSPRs.
- SPRs are typically much sharper resonances compared to LSPRs. This can be an advantage or a disadvantage depending on the application. It should for example in principle result in a larger sensitivity but only on a more limited range of parameters. For SERS, resonances must be broad enough to encompass both the exciting laser and the Stokes frequencies, and SPRs are typically too sharp to fulfill that condition.
- The active surface for SPRs is a single planar interface, while for LSPRs it is the nano-particle surface (which can therefore be spread in a 3D volume, for example by dispersing the particles in water).
- There are more degrees of freedom to tailor or engineer the LSPRs (shape, size, etc.) as opposed to the SPRs, which may open more possibilities, but also more problems (such as poly-dispersity).

In summary, the use of SPR vs LSPR will depend on the exact application. LSPRs are more versatile (easier to implement) but the resonances are not as well defined as for SPRs.

3.5.6. Local field enhancements and LSP

The local field enhancements arising from LSP excitation in nano-particles (NPs) and more complex structures will be discussed in detail in Chapter 6. Their link to SERS and surface-enhanced fluorescence is moreover the subject of Chapter 4.

Therefore we will only discuss here the local field intensity enhancement factor (LFIEF) in the context of the simple energy conservation argument discussed earlier for PSPPs. There are several differences that make this approach more difficult for LSP modes. Firstly, the full EM solutions is known in the case of PSPPs, therefore enabling the calculation of, for example, the optical absorption accurately. Moreover, the invariance by translation of PSPPs means that the LFIEF is the same everywhere on the surface. This is no longer the case for NPs, where the LFIEF is expected to be non-uniform on the surface. In addition, LSPs are usually radiative modes, which therefore introduces an additional mechanism of energy loss, to be included in the energy balance. Finally, and perhaps most importantly, it is more difficult to define a coupling efficiency η for a LSP mode of a nano-particle excited by an incident beam. The reason is that the extent of the beam is usually much larger than the NP (because of the diffraction limit) and one must therefore reason in terms of incident power density (and cross-section).

These additional features make it difficult to extend simply the energy argument to the case of LSPs [133], but one can nevertheless rescue the qualitative conclusions. We had concluded, for example, that large local field enhancements can be expected when (i) the incoming power is efficiently coupled into one or more electromagnetic modes with the following additional characteristics: (ii) these modes are confined to a small volume (smaller than the metallic object), and (iii) they are not too lossy, i.e. there is little dissipation by optical absorption or loss by radiation. Moreover, (iv) at metal surfaces, the local field enhancement outside the metal is even larger when $|\epsilon| \gg \epsilon_M$.

For example, for the PSPP modes discussed in Section 3.4, the coupling of incoming power can be extremely good (condition (i)). There is a small degree of confinement, at least in one dimension (condition (ii)). The losses are relatively small, since these modes are non-radiative and dissipation is small for good metals like silver or gold (condition (iii)). Finally, the condition of a large local field enhancement outside the metal is also met for long-wavelength PSPP modes (condition (iv)).

These considerations can be qualitatively applied to LSP also. The optimum coupling condition (i) implies that the largest LFIEF will be obtained at resonance with the LSP mode, i.e. at the LSP frequency. Condition (ii) may be ambiguous for LSP modes since the fields inside the NP are not necessarily confined at the surface. However, the fact that it is at least confined to the NP dimensions (which are small) contributes to a larger LFIEF. Condition (iii) implies that the LFIEFs will be larger for low-loss metals like silver in the visible, or gold beyond 600 nm, and for smaller objects (for which radiation losses are smaller). Finally, condition (iv) of a large local field enhancement outside the metal should also result in additional enhancements for LSP modes resonant at longer wavelengths. These qualitative conclusions will in fact be confirmed in Chapter 6 in the study of specific examples. These examples will also highlight additional important considerations regarding the LFIEF arising from coupling to LSP modes: (i) the LFIEF is typically larger at tips, corners, or edges, as opposed to flat surfaces (the so-called *lightning rod effect*), (ii) the LFIEF can also be magnified by LSP mode interactions, i.e. at small gaps between nano-particles. These conditions, together with the factors influencing the resonance frequency of the LSP modes, can be used as a guide to 'engineer' and devise structures with large local field enhancements.

3.5.7. Interaction of SPPs – gap SPPs

Gap SPPs and local field enhancements

Finally, we cannot conclude this chapter on plasmons without mentioning an additional type of plasmonic effect that has become increasingly important in recent research, and which will be dubbed *gap SPPs*. We have just hinted at its importance by mentioning that the LFIEF can be greatly enhanced at the gap between two metallic objects. This effect arises when two metallic objects (typically nano-particles) are brought very close to each other. The LSP modes of each object then interact with each other and form (for the closest distances) hybridized modes – in a similar fashion to atomic orbitals, for example. In fact, most of the work to date on topics like *single-molecule SERS* is based on the use and exploitation of gap SPPs; they provide some of the largest known local field enhancements at surfaces. This will be further justified in Chapter 6.

Let us note that the pair of objects could in fact be considered as a single entity with its own electromagnetic modes, and therefore its own LSP modes. This problem could be studied independently of the study of the LSP modes of each individual objects. It may however be simpler (and more intuitive) to try and deduce the properties of the LSP modes of the pair from those of the components. They then arise as a result of the interaction (coupling) of two modes, and many general results can then be invoked from a general mode-interaction theory.

A simple analog of gap SPPs

An analogy with the case of atomic orbitals (of widespread use in chemistry) can be invoked at this stage.

Let us exemplify the concept of interaction theory with the standard example of the hydrogen molecule. If we start from two isolated hydrogen atoms and we push them together until their electrons start interacting, it is possible to build a *symmetric* (bonding) and *anti-symmetric* (anti-bonding) wave-function by taking linear combinations of the atomic *1s* states in both atoms. Through the interaction between the two atoms, the bonding wave-function has an energy lower than the original degenerate energies, while the anti-bonding state is pushed upwards in energy. The bonding state accumulates electronic charge density in the middle of the molecule while the anti-bonding state does the opposite. By accumulating electronic density in the middle of the molecule, the bonding state achieves a better screening of the Coulomb repulsion of the nuclei, thus resulting in a lower overall energy. This is the basic phenomenology of the general theory of chemical bonding.

Similar concepts apply qualitatively to LSP resonances [134,160,207]. Let us take two metallic nano-particles far away from each other. Under these conditions the two NPs couple independently to an external electromagnetic excitation, for example through their respective dipolar LSP resonance. As the NPs approach each other, their responses start to interact to define a coupled object, and its associated coupled resonances. Even in the simplest minded approximation of a dipolar picture, it is not difficult to imagine that we can take linear combinations of the responses that add them *in phase* or *out of phase* to define the equivalent of the bonding and anti-bonding electronic interactions mentioned above. The 'bonding' resonance is the most important in this framework, for it concentrates its electric field distribution at the gap in between the particles. The resulting resonance is red-shifted. The picture of coupled LSP resonances is a lot more complicated than simple orbital interaction theory of electrons, because of the vectorial nature of electromagnetic fields, retardation effects, and the presence of higher order resonances (quadrupolar, etc.), which play a significant role in particular at short distances. Nevertheless, the most red-shifted resonance, resulting from the coupling between the dipolar LSP resonances of each particle, is usually easy to identify.

Taken in the appropriate context, the concept of interaction theory can be a useful shorthand for the discussion of qualitative effects. More detailed results usually require a solution of the EM problem, as discussed in Section 6.4.

3.6. BRIEF SURVEY OF PLASMONICS APPLICATIONS

Finally, we conclude this chapter by discussing briefly the main possible applications of SPP modes, i.e. of plasmonics [133]. They can be separated into three groups:

- Applications based on surface plasmon resonances or localized surface plasmon resonances, such as chemical sensors.
- Application based on surface wave propagation and guiding.
- Applications based on local field enhancements.

Let us now consider these groups separately.

3.6.1. Applications of surface plasmon resonances

PSPP-based resonances

The sharpness and large sensitivity to parameters of resonances arising from PSPPs on planar interfaces make then well suited for applications as sensors, usually called surface plasmon resonance (SPR) sensors.

The simplest and most direct application is probably to use this sensitivity to measure the optical constants of metals. This is obviously limited to the wavelength range where PSPPs can be excited, but can nevertheless be a valuable approach to complement more conventional measurements, such as ellipsometry. This is particularly important since other optical measurements may be affected by PSPP excitations, and therefore be inaccurate in this region. Examples of such measurements are given in Ref. [150], and references therein.

Many of the other applications of PSPPs as sensors are based on the high-refractive-index sensitivity illustrated and discussed in Section 3.4.4. This means that PSPP-based sensors can (in principle) detect very small changes in the local environment at the metal surface, in particular, the adsorption of molecules. Typical implementations are based on a Kretschmann ATR configuration with a prism and may use angle and/or wavelength-modulation. The main obstacle in the practical implementation is the lack of *specificity*, i.e. one cannot distinguish which type of molecule adsorbs on the metal. As a remedy to this problem, a complementary aspect of SPR sensor research is the study of the *surface functionalization* of metallic surfaces (the application of which extends well beyond the field of plasmonics). Using appropriate chemistry, it is possible to coat the metallic surface with molecules that will bind only to one type of chemical group or even one type of molecule. Examples of such specific binding are antibody–antigen, ligand–receptor, or nucleic acid binding. Implementations based on SPR sensors with surface functionalization are therefore numerous and only limited by the functionalization step. We will not review specific applications here, and instead refer the reader to (for example) Ref. [127] and references therein, where an extensive description of this particular type of application of plasmon resonances is provided.

LSP-based resonances

More recently, there has been an increased interest in using LSP resonances – rather than PSPP resonances – for some sensing applications. The respective merits of these two approaches have been discussed in Section 3.5.5 and we shall not come back to it here. More details on applications of LSP-based resonances can be found in recent reviews, for example in Ref. [126].

3.6.2. SPP propagation and SPP optics

Another active area of research in plasmonics is concerned with the use of PSPPs as optical devices, i.e. to propagate, guide, and manipulate light. The driving force behind this effort is the hope that, thanks to the confined nature of the PSPP modes, large miniaturization of optical devices may be possible, even beyond the diffraction limit, which is the fundamental limit of 'conventional' optics. A related subject is also the use of SPPs for the design of negative refractive index materials, another very active field of research. These aspects of plasmonics, although interesting by themselves, are not directly related to SERS, and will therefore not be discussed further. Recent reviews of pure plasmonics topics are Refs. [11,130,131,133] which constitute a good starting point for further information on these specific aspects.

3.6.3. Local field enhancements

The most relevant types of plasmonics applications to us are those based on the large local field enhancements arising from coupling to SPP modes (primarily localized SPPs). SERS belongs to this group, along with many related techniques, including most surface-enhanced spectroscopies and, in particular, surface-enhanced fluorescence (SEF). Based on similar principles there are numerous applications where plasmonic structures are used to engineer and enhance the optical properties of light emitters; for example to enhance the directionality of emission or the quantum yield of solid-state emitters and absorbers such as semiconductor quantum dots, quantum wells, light-emitting diodes and solar cells, see for example [11,130,131] for further details.

The fundamental principles of these techniques and their relation to local field enhancements and SPPs will be discussed extensively in the rest of the book.

Chapter 4

SERS enhancement factors and related topics

Having discussed in detail in the last two chapters the two basic ingredients of SERS: *Raman spectroscopy* and *plasmon resonances*, we are now in a position to mix them together to understand the fundamentals of the SERS effect.

In short, *surface-enhanced Raman scattering* (SERS) consists in using the large local field enhancements that can exist at metallic surfaces (under the right conditions, typically by profiting from localized surface plasmon resonances) to boost the Raman scattering signal of molecules at (or close to) the surface. A similar approach can be used to boost the fluorescence signal and it is called *surface-enhanced fluorescence* (SEF)[1]. This technique will also be discussed here in the context of SERS.

From both a fundamental and an applied point of view, one of the most important aspects of the technique is to know by how much the signal can be boosted, i.e. to know what the *enhancement factor* (EF) is and what its physical origins are. The purpose of this chapter is to begin addressing these issues, first by defining rigorously the SERS enhancement factors, and then by discussing their origin, with a strong emphasis on the electromagnetic (EM) mechanisms. Here we only focus on the methodology and the physics of the phenomena, not on the actual solution of the EM problem for a particular geometry. We therefore always assume that we have a solution of the EM problems we may encounter. The many techniques that can be used to obtain such solutions will be discussed in the next two chapters, which are entirely dedicated to this other aspect of EM enhancements and their relation to plasmon resonances. In particular, the theoretical results of this chapter will be illustrated with concrete examples obtained from analytical solutions of EM problems in Chapter 6.

[1] More recently, it has also been called *metal-enhanced fluorescence.*

This chapter is organized as follows:

- Before studying the physical origin of the SERS enhancement, we first review some possible definitions of SERS enhancement factors in Section 4.1. These definitions are important for a rigorous classification of the electromagnetic enhancements contributing to SERS.
- In Section 4.2, we discuss some practical aspects of their experimental measurements, providing a few examples to illustrate the theoretical treatment to follow.
- The SERS EM problem is then discussed in simple terms in Section 4.3. We highlight there the main features, and lay out the notations for the rest of the chapter.

These three introductory sections, in fact, contain most of the important aspects of SERS enhancements. *They are sufficient for a basic understanding of the SERS effect.* The rest of the chapter is then dedicated to a more formal study and more rigorous (and therefore more complicated) justifications of these concepts, as well as other more advanced aspects of the problem:

- In Section 4.4, we review the most important aspects of modified spontaneous emission.
- The concepts and tools introduced there are then applied to the formal description of the EM-SERS enhancements in Section 4.5.
- The formalism is then adapted to the description of surface-enhanced fluorescence in Section 4.6.
- These are then used in Section 4.7 to discuss a number of related EM mechanisms, which may not have a direct contribution to the SERS enhancement itself, but are still relevant in a number of situations.
- We then conclude the chapter in Section 4.8 by discussing the possible contribution of a *chemical enhancement* to SERS.

4.1. DEFINITION OF THE SERS ENHANCEMENT FACTORS

SERS enhancement factors (EFs) are central to SERS, both for research and applications. There are however a number of difficulties when one attempts to either predict, measure, or compare EFs. This is reflected in the wide variability of values quoted in the literature, probably from $\sim$10 to $\sim 10^{15}$! This spread is partly a result of real differences between various SERS substrates. But for identical substrates, quoted values can still be many orders of magnitude apart, depending on how the EF is calculated or measured. This makes it very difficult to compare the relative merits of two SERS substrates for a given application. This also poses problems when comparing theoretical predictions with experiments. It is therefore crucial to define clearly what is meant by SERS enhancement factor. In this respect, there are several possible definitions or points of view, which are discussed below. We follow largely in this section the arguments given in Ref. [8] (and the accompanying supplementary information), which provides a detailed study of many aspects of the SERS EFs, including their definition and measurement.

4.1.1. General considerations

Electromagnetic and chemical enhancements

By default, we will refer in the following to Stokes Raman scattering by a given vibrational mode, unless otherwise stated. In normal Raman spectroscopy (see Chapter 2), the average Raman intensity of a molecule is directly proportional to the laser power density and to the Raman cross-section of the molecule. This simple fact can be generalized to SERS. In simple terms, the SERS intensity for a given vibrational mode of a given analyte should also be proportional to the laser intensity and to the normal Raman cross-section but *affected by an enhancement factor* (EF). This enhancement is the essence of SERS; without it SERS would not be a promising tool, and there would be little point in studying it either!

One important aspect of SERS EFs is that they should be *'real'* enhancement factors, i.e. they characterize the enhancement obtained with respect to *what would be obtained under non-SERS conditions for the same molecule.* This statement may seem trivial, but ignoring it has been one of the major reasons why erroneously large (up to $\sim 10^{15}$!) SERS enhancements have been repeatedly reported in the literature (see for example Ref. [8] for more details).

The SERS EFs are traditionally separated into two main *multiplicative* contributions:

- The *electromagnetic (EM) enhancement factor*, F_{EM}, is thought to be, by far, *the main contribution.* It is due to the coupling of the incident and Raman electromagnetic fields with the SERS substrate and it can usually be separated into two multiplicative EFs, one for the incident field, and one for the re-emitted (Raman) field, as will be justified later. The EM enhancement relies on the large local field enhancements that occur close to metallic surfaces when localized surface plasmon (LSP) resonances are excited (see Chapter 3). To profit from these, the molecule must then be in close proximity to the surface (typically within ~ 10 nm from the surface, at most). In most implementations, the molecule is actually directly *adsorbed* on the surface, either through physisorption or chemisorption (i.e. through the formation of a chemical bond with the metal).

- Another (multiplicative) contribution to the EF is the so-called *chemical enhancement (CE) factor*, F_{Chem}. Its existence – and even its very definition – are still subject to controversy [7] and its contribution is, in any case, believed to be much smaller than the EM effect. The CE factor is sometimes viewed as a modification of the electronic polarizability of the probe, which can induce resonant-Raman scattering (and therefore enhanced signals) at wavelengths where the non-adsorbed molecules would not be resonant [7]. The most widely accepted explanation for this is the so-called charge-transfer (CT) mechanism [161]. This requires the molecule to be *chemically adsorbed* on the surface (hence the name chemical enhancement). These aspects are discussed in more detail in Section 4.8, and will be mostly ignored until then.

Note that knowing the exact origin of the SERS EFs is in fact not necessary when defining them or measuring them experimentally. In the following, we will mostly ignore this origin, therefore encompassing all possible types of enhancements, but still bearing in mind that the EM effect is the dominant one, as shown by an overwhelming majority of experimental evidence.

Factors influencing the SERS enhancements

The SERS process (and therefore the enhancement factor) depends on a long list of parameters, including:

- *Characteristics of the laser excitation*, in particular: wavelength, polarization, angle of incidence (for a planar substrate), etc.
- *Detection setup*, in particular: scattering configuration (e.g. back-scattering geometry), solid angle for collection, polarized and/or unpolarized detection, etc.
- *SERS substrate*, in particular: material (usually silver or gold), geometry, orientation with respect to incident beam direction and polarization, and refractive index of the environment n_M. The dimensionality of the substrate (e.g. 2D planar substrate or 3D particles in solution) is also an important parameter since it requires different sample preparation procedures.
- *Intrinsic properties of the analyte*, in particular: Raman polarizability tensors of the modes (or intrinsic Raman cross-sections).
- *Analyte adsorption properties*, in particular: adsorption efficiency and analyte concentration (surface coverage), distance from the surface, adsorption orientation (fixed or random), and the possible modification of the intrinsic Raman polarizability induced by adsorption. This latter aspect is essentially the chemical contribution to SERS while, the previous one provides the background and origin of surface selection rules [52].

Many of these aspects will be discussed in more detail in the following sections. Note that the nature of the substrate environment (typically air, water, or another solvent) and in particular its refractive index n_M affects not only the SERS signal, but also the non-SERS cross-sections. All enhancement factors will therefore be defined *with respect to the non-SERS properties of the same molecule in the same environment as used for the SERS experiment.*

It is difficult to account for all of these parameters and many of them may, in fact, be unknown. In defining a SERS EF, the aim is to find a definition that:

- is independent of as many parameters as possible (or at least gives a good estimate for a wide range of parameters),
- can easily be either measured experimentally, or predicted theoretically (ideally both), and
- allows for a direct comparison of the merits of different SERS substrates.

It is impossible to meet all these criteria, since an accurate SERS EF taking into accounts all possible parameters is necessarily complex and suited only to very specific and limited conditions. Contrariwise, a simple definition of the SERS EF that would apply to most situations can only be approximate, since it does not take into account, say, the symmetries of the Raman tensors or chemical properties of the probe.

Two types of enhancement factors

Before introducing any definition, it is probably useful to emphasize an important characteristic of most SERS substrates. The local field enhancements depend strongly on the exact position of the molecule on the surface. As a result, the SERS EFs on the surface of a typical SERS substrate are *highly non-uniform*, even at a molecular scale of nanometers. This will be justified in detail in Chapter 6. Points of very large enhancements, the so-called 'hot-spots', are generally highly localized, and can be within tens of nanometers of points with little or negligible enhancement.

For some applications, such as single-molecule (SM) SERS detection, it is desirable to profit from these points of highest enhancements. On the other hand, because this is not always straightforward and usually not easily reproducible, many applications (for example in analytical chemistry) are interested instead in the average enhancement factor, i.e. for molecules with random positions on the substrate. Most SERS EF definitions will therefore fall into one of two categories:

- *Average EFs*, which represent an average property of the substrate.
- *Single-Molecule EFs* (SMEFs), which represent the enhancement only at specific positions on the substrate.

Single-molecule EFs can be calculated theoretically at all points on a given substrate; provided the solution of the electromagnetic problem is known. The full *distribution of single-molecule EF* on the entire SERS substrate surface is the ultimate characterization of SERS EFs, since most other properties (including any type of average SERS EFs) can be deduced from it. However, measuring this experimentally is rather challenging. One would therefore typically focus on the *maximum single-molecule EF* on the substrate. Because the maximum SMEF only applies to one or a few localized regions of the surface, the average EFs are typically much smaller, by several orders of magnitude (see for example Ref. [162]). *There is, therefore, no such thing as a unique SERS EF for a given substrate.* Which type of EF definition is relevant ultimately depends on the application/experiment that is carried out.

With these considerations in mind, we will consider in the following several useful definitions of the SERS EF and discuss their merits and problems. The definitions will start accumulating and, for clarity, they have been summarized in Table 4.1. We will also ignore in the following the possible complications associated with photo-bleaching or photo-desorption of the analytes. These effects can strongly affect any experimental measurements but can (and should) be avoided by a careful choice of the probe and/or power density for excitation.

4.1.2. The analytical point of view

Definition

This first approach is arguably the simplest and most intuitive. For a given concentration of analyte c_{RS}, the Raman signal is I_{RS}. Under identical experimental conditions (laser wavelength, laser power, microscope or lenses, spectrometer, etc.), and for the same preparation conditions, the same analyte on a SERS substrate, possibly with different concentration c_{SERS}, now gives a SERS signal I_{SERS}. The *Analytical Enhancement Factor* (AEF) can then be defined as:

$$\boxed{\mathrm{AEF} = \frac{I_{\mathrm{SERS}}/c_{\mathrm{SERS}}}{I_{\mathrm{RS}}/c_{\mathrm{RS}}}.} \tag{4.1}$$

Implicit in this definition are the assumptions that:

- I_{RS} scales linearly with incident power density and molecule concentration c_{RS}. This is reasonable in most situations of interest. For dyes, one should however ensure that effects such as photo-beaching, molecular aggregation, molecular absorption, or even stimulated Raman scattering are avoided. Reducing the analyte concentration will in general prevent the latter three effects, if necessary.

- I_{SERS} scales linearly with incident power density and molecule concentration c_{SERS}. This latter condition is certainly less general and should be carefully assessed when using this definition, as discussed below.

Discussion

The AEF can be very useful for specific practical applications but tends to depend strongly on many factors and in particular on the adsorption properties (e.g. efficiency) of the probe and the analyte concentration c_{SERS}; or more precisely on the surface coverage (sub-monolayer vs multi-layer). It is also strongly dependent on the sample preparation procedure, especially for planar SERS substrates (e.g. spin-coating, dipping, or drying). In fact, c_{SERS} does not characterize well the number of adsorbed molecules. For example, the AEF would be zero for molecular species that do not adsorb at all on the SERS substrate. This does not mean that the same SERS substrate could not be very good for other analytes. For this reason, it is not a good characterization of the SERS substrate itself, and cannot be used to easily compare the performances of different substrates (except possibly for some analytes where the adsorption properties have been carefully assessed in a separate study).

In spite of these shortcomings, and provided all experimental procedures are clearly stated, the AEF represents a simple yardstick figure for the average SERS EF, whose measurement is easily implemented and reproducible. From its definition in terms of volumic molecular concentrations, it is also clear that the AEF is particularly suited to the case of SERS active liquids, e.g. colloidal solutions, as opposed to planar SERS substrates.

4.1.3. The SERS substrate enhancement factor – Experimental approach

One important drawback of the previous definition is that it ignores the fact that SERS is a type of *surface spectroscopy*. This means that only the adsorbed molecules contribute to the signal (and even only the first few monolayers for very large coverage). To remedy this problem, the following definition for the *SERS substrate enhancement factor* (SSEF) can be used:

$$\boxed{\mathrm{SSEF} = \frac{I_{\mathrm{SERS}}/N_{\mathrm{Surf}}}{I_{\mathrm{RS}}/N_{\mathrm{Vol}}},} \tag{4.2}$$

where $N_{\mathrm{Vol}} = c_{\mathrm{RS}}V$ is the average number of molecules in the scattering volume, V, for the Raman (non-SERS) measurement, and N_{Surf} is the average number of *adsorbed* molecules in the same scattering volume for the SERS experiments. This expression has been used several times in the past (see for example Ref. [163]) and is usually considered as the best estimate of the *average SERS EF* for a given SERS substrate.

It is, however, necessary to include in this definition an additional constraint to make it more accurate: *the surface coverage must remain smaller than one monolayer* when using Eq. (4.2). This is because the SERS effect is distance-dependent and the SERS signals from molecules on the second monolayer and beyond are therefore typically reduced (see Section 6.2.4). This additional constraint ensures that the average SERS intensity is proportional to the average number of adsorbed molecules, which is necessary for Eq. (4.2) to remain independent of N_{Surf}. This would no longer be the case beyond monolayer coverage. This is not a limitation since the interest for many practical applications is in the low (or even ultra-low) concentration limit. Even with this additional constraint, this definition cannot be considered to be strictly rigorous, and we will come back to this point in Sections 4.1.5 and 4.2.3.

From an experimental point of view, we note that $I_{\mathrm{RS}}/N_{\mathrm{Vol}}$ is simply the average (over molecular orientations) Raman intensity per free molecule, denoted $\langle I_{\mathrm{RS}}^{\mathrm{SM}} \rangle$ [W]. It is directly related to the differential Raman cross-section $\mathrm{d}\sigma_{\mathrm{RS}}/\mathrm{d}\Omega$ [$\mathrm{m}^2\ \mathrm{sr}^{-1}$], defined in Section 2.2.7, by:

$$\langle I_{\mathrm{RS}}^{\mathrm{SM}} \rangle = \frac{\mathrm{d}\sigma_{\mathrm{RS}}}{\mathrm{d}\Omega} S_{\mathrm{Inc}} \Delta\Omega_{\mathrm{Det}}, \tag{4.3}$$

where S_{Inc} [$\mathrm{W\ m}^{-2}$] is the incident power density at the molecule position and $\Delta\Omega_{\mathrm{Det}}$ [sr] is the solid angle for detection (see Section 2.2.6). The difficulty in the practical estimation of the SSEF lies in the determination of N_{Surf}, and to a lesser extent of N_{Vol} (or equivalently of the scattering volume V). A rigorous definition of the SSEF from a theoretical point of view will be given in Section 4.1.5, since it requires the definition of single-molecule EF first. The link between this rigorous definition and the experimental approach described above (Eq. (4.2)) is discussed in Section 4.2.3.

4.1.4. The SERS cross-section and single-molecule EF

Objectives

As mentioned already, SERS enhancements on most SERS substrates are *highly non-uniform*, even at a molecular scale of nanometers. *The AEF and SSEF are spatially-averaged properties* that ignore this non-uniformity.

To account for this, it is therefore useful to define a *single-molecule enhancement factor* (SMEF). This is the SERS enhancement felt by a given molecule at a specific point. It is in general dependent on the symmetry of the Raman tensor of the mode and on the probe orientation on the SERS substrate. It is also dependent on the orientation of the SERS substrate with respect to the exciting polarization. It therefore requires the exact knowledge of the SERS substrate geometry, and the exact position and orientation of the probe on it. Because of these constraints, this definition is typically more suited to theoretical estimations.

The problem of molecular orientation

The SMEF could be defined simply as the ratio of the differential SERS intensity of a single molecule $dP_{SERS}/d\Omega$ to the differential Raman intensity $dP_{RS}/d\Omega$ of the same molecule under the exact same conditions, but in the absence of the metallic SERS substrate. The definition as stated above, however, presents a major shortcoming because the SERS intensity is compared to that of the free molecule in a *fixed* position and orientation. To illustrate why, one can consider the case of a highly uni-axial Raman tensor with axis perpendicular to the exciting polarization. The Raman intensity of the fixed molecule would then be zero, while the SERS intensity could be finite (possibly small) since the SERS substrate can modify the local field polarization. This would result in an infinite EF, which would certainly not reflect the overall magnitude of the SERS signal.

Note that a similar issue arose when defining the absolute differential Raman cross-section. This was simply solved by averaging over all orientations of the molecule; an acceptable approach for molecules in gases or liquids. Such a trick cannot be directly generalized to SERS, because the adsorbed molecule may, in some situations, have a fixed orientation (or a limited movement) with respect to the metallic surface (for example, the case of a covalently-bound molecule to the surface).

The problem of the scattering configuration

Another necessary assumption when we defined the absolute differential Raman cross-section concerned the scattering configuration. As discussed in Section 2.2.7, for linearly polarized excitation and upon molecular-orientation averaging, the differential Raman scattered power $dP_{RS}/d\Omega$ [W sr^{-1}] for 90° scattering, back-scattering (BS) and forward-scattering (FS) configurations are the same and independent of the incident polarization. This is in fact used for the definition of the differential Raman cross-section, $d\sigma_{RS}/d\Omega$ [m^2 sr^{-1}], as given in Eq. (2.12) and repeated here:

$$\left\langle \frac{dP_{RS}}{d\Omega}(90^\circ, \text{or BS or FS}) \right\rangle = \frac{d\sigma_{RS}}{d\Omega} S_{Inc}, \tag{4.4}$$

where S_{Inc} [W m^{-2}] is the incident power density at the molecule position, and $\langle \ldots \rangle$ denotes the molecular-orientation averaging.

Again, this cannot be generalized to the case of SERS. In particular, under SERS conditions, the differential scattered SERS power $\mathrm{d}P_{\mathrm{SERS}}/\mathrm{d}\Omega$ depends in general both on the incident polarization (with respect to the SERS substrate) and on the detection direction (BS and 90° configurations are no longer equivalent for example).

Definitions

For these reasons, it is not possible to define a general differential SERS cross-section as was the case for normal Raman. However, for a given incident polarization (IP), scattering configuration (SC), molecule position ($\mathbf{r}$) and molecule orientation (MO), one can still define the *absolute differential SERS cross-section* (or SERS cross-section for short) $\mathrm{d}\sigma_{\mathrm{SERS}}/\mathrm{d}\Omega$ [$\mathrm{m}^2\ \mathrm{sr}^{-1}$] as:

$$\begin{aligned}\frac{\mathrm{d}P_{\mathrm{SERS}}}{\mathrm{d}\Omega}&(\mathrm{IP},\mathrm{SC},\mathbf{r},\mathrm{MO},\omega_L,n_M)\\ &=\frac{\mathrm{d}\sigma_{\mathrm{SERS}}}{\mathrm{d}\Omega}(\mathrm{IP},\mathrm{SC},\mathbf{r},\mathrm{MO},\omega_L,n_M)S_{\mathrm{Inc}}.\end{aligned} \tag{4.5}$$

Like the Raman cross-section, the SERS cross-section depends on excitation wavelength or frequency ω_L, the refractive index of the environment n_M, and obviously the vibrational mode under consideration. It depends in addition on a collection of additional parameters: IP, SC, $\mathbf{r}$, and MO. Its utility is therefore limited unless these are clearly stated or further assumptions are made. In a similar fashion, the *single-molecule enhancement factor* (SMEF) can also be defined as:

$$\mathrm{SMEF}=\frac{\mathrm{d}\sigma_{\mathrm{SERS}}/\mathrm{d}\Omega}{\mathrm{d}\sigma_{\mathrm{RS}}/\mathrm{d}\Omega}=\frac{\frac{\mathrm{d}P_{\mathrm{SERS}}}{\mathrm{d}\Omega}(\mathrm{IP},\mathrm{SC},\mathbf{r},\mathrm{MO},\omega_L,n_M)}{\langle\frac{\mathrm{d}P_{\mathrm{RS}}}{\mathrm{d}\Omega}(90^\circ,\mathrm{or\ BS\ or\ FS})\rangle}. \tag{4.6}$$

Note that the normal Raman cross-section in the denominator should be that measured in the *same dielectric environment* (typically air, water, or another solvent) as that used for the SERS experiment[2].

The SMEF obviously depends on the same parameters as the SERS cross-section. Although it appears here as a simple ratio between cross-sections, it is important to realize the differences in the meaning of the SERS and non-SERS cross-sections. The former depends on many external parameters, while the latter is an intrinsic property of the analyte (for a given environment and excitation wavelength).

[2] If this is not the case, then the local field correction discussed in Section 2.4.5 could introduce artificial factors that have nothing to do with SERS. For example, an enhancement of ≈ 2.5 would be measured for a molecule in water compared to air, even in the absence of any plasmon resonances.

SMEF for polarized detection

For some experimental situations, and also for the formal description of EM-SERS EFs given in Section 4.5, it is also useful to consider the case of polarized detection of the SERS signal. One can then define the corresponding polarized differential SERS cross-section $d\sigma^P_{SERS}/d\Omega$ and polarized SMEF ($SMEF^P$), where the superscript P characterizes the polarizer direction $\mathbf{e}_P$ used for detection:

$$\frac{dP^P_{SERS}}{d\Omega}(IP, SC, \mathbf{e}_P, \mathbf{r}, MO, \omega_L, n_M) = \frac{d\sigma^P_{SERS}}{d\Omega}(IP, SC, \mathbf{e}_P, \mathbf{r}, MO, \omega_L, n_M) S_{Inc}, \tag{4.7}$$

and

$$SMEF^P = \frac{d\sigma^P_{SERS}/d\Omega}{d\sigma_{RS}/d\Omega}. \tag{4.8}$$

Orientation-averaged SMEF

For a given substrate configuration (IP and SC) and a given molecule position/orientation, the definition of the SMEF reflects directly the SERS intensity. However, the molecule orientation is not necessarily known or even fixed during the time of a measurement. It can therefore also be useful to define the *orientation-averaged SMEF* (OASMEF) at a given position as:

$$\boxed{OASMEF = [SMEF] = \frac{[\frac{dP_{SERS}}{d\Omega}(IP, SC, \mathbf{r}, MO, \omega_L, n_M)]}{\langle \frac{dP_{RS}}{d\Omega}(90^\circ, \text{or BS or FS})\rangle},} \tag{4.9}$$

where the average denoted by [...] is taken over all *allowed orientations* of the molecule under SERS conditions. There are situations where this average is not necessary. For example, for an isotropic Raman tensor, OASMEF and SMEF are identical since the SMEF depends only on the position of the probe, not on its orientation. However, for a uni-axial tensor, the SMEF then depends on $\mathbf{e}_m$, orientation of the main molecular axis. If the molecule is known to adsorb with its axis perpendicular to the metallic surface, then $\mathbf{e}_m$ is well defined and equal to the local normal unit vector $\mathbf{n}$, and again no averaging is needed. Now, if the molecule adsorbs with its axis tangential to the metallic surface, then there remains one degree of rotational freedom, the axis is not well defined, and averaging is needed over one angle (rotation in the tangent plane). Similarly, if adsorption can occur in any possible orientation, then averaging is needed over two angles representing all possible orientations in space. In this final case only, the average over *allowed* orientations [...] has the same meaning as the full average over *random* orientations ⟨...⟩.

Discussion

The SMEF or OASMEF provides a measure of the SERS EF for a given SERS substrate at a given position, for example a 'hot-spot'. Using these definitions, SERS enhancements can then be easily compared for different positions on a given SERS substrate, or between two different SERS substrates, at least at a theoretical level. Moreover, when no chemical enhancements are involved, the SMEF should be the same for two modes of a given analyte with the same Raman tensor symmetry, or even for two modes of two analytes with the same Raman tensor symmetry and same adsorption properties (orientation). This definition of the SMEF is therefore the most unbiased estimation of the SERS enhancement for a single molecule. For SERS substrate comparison, it should ideally be calculated or measured with probes showing no chemical enhancements, and with a representative Raman tensor/adsorption orientation.

In fact, in an attempt to recover the generality of the normal Raman cross-section, it is possible to define a *Standardized SMEF* (StdSMEF) as the SMEF of a molecule:

- with an *isotropic Raman tensor*, and
- subject only to the *EM enhancement mechanism* (i.e. no chemical enhancement, which would be probe specific).

The StdSMEF (called IsoEMSMEF in Ref. [8]) is no longer dependent on the analyte properties and therefore reflects directly the *EM-SERS properties of the substrate.* Note that it still depends on the incident polarization and scattering configuration (these should arguably be part of the SERS substrate specifications), and most importantly, on the molecule position.

Finally we note again that in most cases, what is important is the maximum SMEF (or StdSMEF) on a SERS substrate. This is particularly true for experiments, where single-molecule signals are only detectable from the points of highest enhancements (hot-spots). When possible, knowing the full spatial distribution of the SMEF (or StdSMEF) on a substrate obviously provides a wealth of additional information, such as average properties or degree of spatial localization of the hot-spot (see for example Section 6.4 or Ref. [162]).

Other SERS cross-sections

Although it is not directly related to SERS EFs it is worth digressing here to discuss other possible types of SERS cross-sections.

By analogy with normal Raman scattering, it is possible to define the *integrated radiative SERS cross-section* σ_{SERS}, obtained by integrating the differential SERS cross-section over all detection directions. As for the non-SERS case, this quantity is difficult to measure (it requires an integrating sphere, for example). Moreover, because the radiation profile for SERS is not as 'regular' as for Raman, σ_{SERS} cannot be expressed in simple terms (as was done in Eq. (2.14)) as a function of the differential SERS cross-section and the depolarization ratio (which is in fact also changed under SERS conditions).

Moreover, there is an additional peculiarity to the SERS case: because the metallic SERS substrate is optically absorbing, *some of the SERS-scattered photons are absorbed and cannot be detected in the far-field.* They correspond to *non-radiative SERS processes,* where the Raman scattering event does occur in the molecule (excitation of a vibration), but the Stokes-scattered photon is re-absorbed in the metal (and therefore not detectable). The radiative SERS cross-section σ_{SERS} does not include such processes and is therefore not an exact representation of what is actually felt by the molecule (in terms of the number of vibrations created). We can therefore define a *total SERS cross-section* σ_{Tot}, which includes both radiative events (in all directions) and non-radiative ones. One could argue that this subtlety is irrelevant, since it is usually non-observable. There is however at least one situation where this cross-section is measured: in SERS vibrational pumping experiments [59] where the total number of vibrations produced in the molecule is proportional to the *total* SERS cross-section.

4.1.5. The SERS substrate enhancement factor – Formal definition

Having defined rigorously the differential SERS cross-section and the SMEF (and OASMEF), we can now return to the average SERS EF and provide a definition of the *SERS substrate enhancement factor* (SSEF) from a more formal (and rigorous) perspective.

Rigorous definition of the SSEF

To this end, let us consider a uniform coverage of a probe molecule on the surface of a SERS substrate. The SERS EF of each molecule can be characterized by the OASMEF at its position **r**. The SSEF can therefore be rigorously defined as the spatially-averaged OASMEF on the surface, i.e.

$$\boxed{\mathrm{SSEF} = \frac{1}{A_M}\int_{A_M} \mathrm{OASMEF}(\mathbf{r})\,\mathrm{d}S,} \tag{4.10}$$

where A_M represents the surface area of the metallic substrate. This last expression can be rewritten as:

$$\boxed{\mathrm{SSEF} = \{\mathrm{OASMEF}\} = \{[\mathrm{SMEF}]\},} \tag{4.11}$$

where $\{\ldots\}$ denotes the spatial averaging (on the substrate surface) and $\{[\text{SMEF}]\}$ is therefore the spatial-and-allowed-orientation-averaged single-molecule EF.

Discussion

The above definition of the SSEF provides a strictly rigorous definition of the *average* SERS EF and is well suited to theoretical predictions, but an additional step is necessary to relate it to experimental measurements and to the commonly-used expression given earlier in Eq. (4.2). This will be discussed in detail in Section 4.2.3.

As mentioned earlier, the SSEF characterizes the SERS intensity of only the *first monolayer* (directly on the surface). Multi-layer (distance dependence) effects are not accounted for in this definition, although one could define a similar SSEF for a fixed distance d from the surface.

Finally, the SSEF, being an average property, no longer depends on position. The same dependence over other parameters as the SMEF however remains. A less specific definition can be given in terms of the spatial average of the standardized SMEF: the resulting *standardized SSEF* (StdSSEF) corresponds to the SSEF for an isotropic mode subject only to EM enhancements. Then it only depends on the incident polarization (IP) and scattering configuration (SC), two properties that should arguably be part of a SERS substrate's specifications. *The StdSSEF is therefore the best figure to compare the average SERS EF of different SERS substrates.*

4.1.6. Discussion and merits of the various definitions

Let us now compare the merits of the previous definitions of the SERS EFs.

AEF vs SSEF/SMEF

The AEF is the easiest to measure experimentally, and is directly relevant to many SERS applications. It provides a meaningful and easy-to-interpret figure for the EFs *when conditions are clearly defined.* This figure may however change easily with the exact experimental conditions (even with analyte concentration for example). It is also difficult to link it directly to theoretical predictions of the SERS EF. For these reasons, it is not an appropriate definition to characterize and compare SERS substrates.

SSEF/SMEF for specific predictions

The definitions of the SSEF and SMEF (maximum SMEF on the substrate being in general sufficient) are more suited to this end. They characterize two important aspects of a SERS substrate: its average and maximum SERS

EF, respectively. The SSEF and SMEF definitions can be used in two ways: the first is to choose a specific analyte/metal combination, along with a set of parameters (polarization, wavelengths, etc.). The SSEF and SMEF are then specific to this situation, but can nevertheless be used to compare some characteristics of SERS substrates with, say, different geometries. They can also be used to probe some advanced effects, such as the surface selection rules or the presence of any additional chemical enhancement. If an analyte exhibits a large chemical enhancement, for example, the SSEF or SMEF would be larger for this analyte than for some others on the same substrate. Because the results are specific to the metal/analyte combination, the SSEF and SMEF are not intrinsic properties of the substrate in general.

SERS substrate characterization

The second approach seeks to find such an intrinsic EF, and this can be achieved by taking the following steps:

- To properly characterize a SERS substrate, the SMEF (and SSEF) should be independent of the chosen analyte. The first pre-requisite to this is to measure it using a molecule for which there is no chemical enhancement (CE). The absence of CE can sometimes be guessed but is not always easy to prove experimentally. The best approach to resolve this (experimentally) is therefore to measure the SMEF (SSEF) with different analytes. If similar values are obtained, it is a strong indication that the SMEF (SSEF) is meaningful, and that the chosen analytes do not experience any appreciable CE (or it would be a CE of comparable magnitude for all of them, an unlikely coincidence for different molecules). The obtained value is then the *electromagnetic SMEF* (EM-SMEF), or *electromagnetic SSEF* (EM-SSEF), mostly independent of the analyte.
- The symmetry of the Raman tensor of the mode and molecule orientation can still affect the value of the EM-SMEF because of surface selection rules [52], as we shall see in Section 4.5. One simple convention is then to consider the EM-SMEF for fully isotropic modes, which we have called *standardized SMEF* (StdSMEF), and the corresponding *standardized SSEF* (StdSSEF). This makes theoretical calculations easier but is not always practical experimentally. However, if the EM-SMEF is measured for a mode where the Raman tensor and molecular orientation are known, it is then in principle possible to deduce the StdSMEF from theoretical arguments (see Section 4.5 for details).
- Finally, the StdSSEF and StdSMEF remain dependent on a (smaller) number of parameters, which therefore needs to be clearly stated when quoting or comparing values. They are dependent on the excitation wavelength λ_L and, to a lesser degree, on the vibrational mode

energy $\hbar\omega_k$. They are also dependent on the incident polarization and scattering configuration, but these may be viewed as part of the specifications of the SERS substrate itself (in general, one will always choose these parameters so as to optimize the StdSSEF and/or StdSMEF). With this in mind, these two SERS EFs are currently the most general and rigorous definitions to compare the merits of various SERS substrates in terms of average and single-molecule enhancements.

Final comments

Table 4.1 provides a summary of most of the definitions introduced in this section. These definitions clearly highlight the need to characterize first the non-SERS properties of the molecule under consideration, an obvious step when speaking of an 'enhancement', but a step that has been very often overlooked.

This classification is a first step toward removing one significant artificial source of variability in the SERS EFs quoted in the literature, which is simply the lack of consistent definitions. Using the rigorous definitions of this section, any theoretical calculations should in principle agree on the SERS EF for a given substrate. From the experimental point of view, there remain however many problems associated with the assumptions that are made when estimating SERS EFs from a given SERS measurement; for example, how to estimate surface coverage, how to ensure the single-molecule nature of a SERS signal, etc. These assumptions must be clearly stated and explained to enable others to analyze critically the validity of the measured SERS EFs. Some of these experimental aspects of SERS EF measurements will be discussed shortly in Section 4.2.

Finally, it is also important to realize that the standardization and definitions of the SERS EFs are still, surprisingly, a recent development; the first comprehensive attempt being reported in Ref. [8]. These definitions, and their relation to experimental measurements, are therefore still likely to evolve. They will, however, represent for us a starting point to quantify rigorously the magnitude of the SERS EFs and understand further their physical origins.

4.2. EXPERIMENTAL MEASUREMENT OF SERS ENHANCEMENT FACTORS

Some of the theoretical concepts highlighted in the previous sections can be easily demonstrated through practical examples of SERS EFs measurements. We give here a few representative examples that will enhance the meaning and clarify the details of some of the definitions provided earlier and, by the same token, demonstrate a few practical aspects of the problem. The first overriding consideration for any serious estimation of SERS EFs is the knowledge of the *non-SERS cross-section*, a concept extensively discussed in Chapter 2, which we briefly recall here in the context of SERS EFs.

Table 4.1 Summary of acronyms, notations, and definitions pertaining to SERS EFs. The dependence of some of these enhancement factors upon the following parameters is also specified: molecule position **r** (for non-averaged EFs), excitation wavelength (λ_L), mode vibrational energy or wave-number $\bar{\nu}_k$, Raman polarizability tensor $\hat{\alpha}_k$ or its symmetry represented by the normalized Raman polarizability tensor $\hat{\alpha}_k^N$ (see Section 2.4.6), molecular adsorption or orientation (MO), substrate orientation with respect to incident polarization (IP), and scattering configuration (SC).

Name	Acronym	Definition	Depends on
Enhancement Factor	EF	–	–
Electromagnetic EF	F_{EM}	–	–
Chemical EF	F_{Chem}	–	–
Analytical EF	AEF	$(I_{\text{SERS}}/c_{\text{SERS}})\,/\,(I_{\text{RS}}/c_{\text{RS}})$ (Eq. (4.1))	λ_L, $\bar{\nu}_k$, $\hat{\alpha}_k^N$, MO, IP, SC
SERS Substrate EF	SSEF	(Eqs (4.2), (4.10))	λ_L, $\bar{\nu}_k$, $\hat{\alpha}_k^N$, MO, IP, SC
Single-Molecule EF	SMEF	$I_{\text{SERS}}^{\text{SM}}/\langle I_{\text{RS}}^{\text{SM}}\rangle$ (Eq. (4.6))	λ_L, $\bar{\nu}_k$, $\hat{\alpha}_k^N$, MO, IP, SC
Orientation-Averaged SMEF	OASMEF	$[I_{\text{SERS}}^{\text{SM}}]\,/\langle I_{\text{RS}}^{\text{SM}}\rangle$ (Eq. (4.9))	λ_L, $\bar{\nu}_k$, $\hat{\alpha}_k^N$, IP, SC
Electromagnetic SSEF	EM-SSEF	SSEF for purely EM enhancement	λ_L, $\bar{\nu}_k$, $\hat{\alpha}_k^N$, MO, IP, SC
Electromagnetic SMEF	EM-SMEF	SMEF for purely EM enhancement	λ_L, $\bar{\nu}_k$, $\hat{\alpha}_k^N$, MO, IP, SC
Standardized SSEF	StdSSEF	EM-SSEF for an isotropic Raman mode	λ_L, $\bar{\nu}_k$, IP, SC
Standardized SMEF	StdSMEF	EM-SMEF for an isotropic Raman mode	λ_L, $\bar{\nu}_k$, IP, SC
Total SERS Substrate EF	TSSEF	see Section 4.2.3	λ_L, $\bar{\nu}_k$, $\hat{\alpha}_k^N$, MO, IP, SC
Polarization-Averaged SSEF	PASSEF	SSEF averaged over IP (see Section 4.2.2)	λ_L, $\bar{\nu}_k$, $\hat{\alpha}_k^N$, MO
Differential Raman cross-section	$\text{d}\sigma_{\text{RS}}/\text{d}\Omega$	see Section 2.2.7	λ_L, $\bar{\nu}_k$, $\hat{\alpha}_k$
Differential SERS cross-section	$\text{d}\sigma_{\text{SERS}}/\text{d}\Omega$	see Eq. (4.5)	λ_L, $\bar{\nu}_k$, $\hat{\alpha}_k$, MO, IP, SC
Integrated radiative SERS cross-section	σ_{SERS}	see Section 4.1.4	λ_L, $\bar{\nu}_k$, $\hat{\alpha}_k$, MO, IP, SC
Total SERS cross-section	σ_{Tot}	see Section 4.1.4	λ_L, $\bar{\nu}_k$, $\hat{\alpha}_k$, MO, IP, SC

4.2.1. The importance of the non-SERS cross-section

General considerations

The non-SERS cross-section of any probe used in SERS provides the proper normalization condition for any consideration regarding the SERS enhancement factor. Any EF should correspond to a ratio of the SERS cross-section over the non-SERS cross-section *of the same molecule under identical conditions.* This may seem as stating the obvious, but failing to account properly for the non-SERS cross-section has been a recurrent problem in the SERS literature. Moreover, the large differences between the non-SERS cross-sections of various SERS probes have been generally underestimated. As discussed in Section 2.2.8, the Raman cross-section can span 6–7 orders of magnitude going from small non-resonant molecules to larger dyes at resonance. Examples of non-SERS cross-sections of representative SERS probes were given in Table 2.3, where variations by a factor of $\sim 10^4$ are evident. This means that – unless careful normalization with respect to the non-SERS cross-section of the appropriate probe is done – large errors can arise in the estimations, by as much as 10^4! The 'myth' of 'huge' single-molecule SERS EFs, as large as 10^{14}–10^{15}, largely comes from this error, and it is therefore worth discussing further its origin.

The 'myth' of 'huge' single-molecule EFs

The first estimates of single-molecule EF came from the two original reports of single-molecule detection with SERS [30,31]. In both cases, the SERS probes were dyes (RH6G [30] or CV [31]) under resonant or pre-resonant conditions, but surprisingly, the SMEF was estimated by comparison with the non-SERS cross-section of small non-resonant molecules.

Let us focus for the sake of argument on the case of Ref. [31], where the SMEF of crystal violet (CV) excited at 830 nm was estimated by comparison with the non-SERS cross-section of methanol (a clearly non-resonant molecule). The only way to justify this would be to assume that CV at 830 nm (arguably under pre-resonant conditions) would have a comparable cross-section to that of methanol (which is not only non-resonant, but also a small molecule). The discussion in Section 2.2.8 is sufficient to guess that this is unlikely, and this can indeed be shown experimentally. For example, at 785 nm laser excitation (close to 830 nm used in Ref. [31]) the following differential Raman cross-sections for methanol (for the ~ 1030 cm^{-1} mode) and CV (for the ~ 1170 cm^{-1}mode) are obtained[3]: $(d\sigma/d\Omega)_{\text{Meth.}} = 2.1 \times 10^{-31}$ cm^2/sr, and $(d\sigma/d\Omega)_{\text{CV}} = 2 \times 10^{-28}$ cm^2/sr, i.e. a factor of $\sim 10^3$ difference. This

[3] These values are obtained in solution by direct comparison with the standard reference of 2-bromo-2-methylpropane (whose reference cross-section is scaled to 785 nm from 633 nm by the ω^4-dependence) as explained in Section 2.2.8.

is simply a manifestation of the pre-resonance effect for CV (and probably its larger size). Even further away from resonance, at 1064 nm excitation for example (Nd–YAG laser), the difference in cross-sections is still a factor of $\sim$350. As a consequence, the SMEF estimated in Ref. [31] is an overestimate by a factor $\sim 10^3$, possibly more if other uncertainties in the scattering volume characterization are accounted for.

This is in fact a general rule: all reports of 'huge' SMEF of the order of 10^{14}–10^{15} to date are a consequence of an erroneous normalization with respect to the non-SERS cross-sections. More careful estimations [8,32] bring it back to more reasonable values of $\sim 10^{10}$ (which could still be considered as 'huge' by many standards...).

The problem has been in reality more serious than a mere disagreement on a factor of $\sim 10^3$–10^4. Since enhancement factors of the order of 10^{14}–10^{15} are really difficult to justify theoretically, these overestimated values have led to a lot of speculation on the existence of an additional contribution to the enhancement, like the *chemical enhancement*. It is obvious from here that a real debate on the existence/absence of additional contributions to the enhancement can only arise if the EFs are measured as accurately as possible; and this can only be achieved through proper normalizations. This highlights further the importance of rigorous and accurate experimental estimations of SERS EFs, which play an important part in the understanding of the physical origin of the effect.

4.2.2. Example of AEF measurements

Experimental results

As mentioned before, the simplest possible experimental example of EF determination is the *analytical enhancement* factor (AEF), albeit being the most difficult in general to relate to EM properties of the substrate. Table 4.2 shows a list of experimental measurements of AEF for different common SERS probes in colloidal Ag solutions (taken from Ref. [8]).

Link of the AEF with theoretical estimates of the SSEF

It is possible, at least in theory, to relate these measured AEF to the more rigorous SSEF, but this step is quite case specific. One needs to take into account the possible effects of adsorption efficiency of the analyte, number of layers of analytes on the substrate if there is more than one monolayer (distance dependence), orientation of the substrate with respect to incident polarization, poly-dispersity of the SERS substrate (e.g. for colloids), preparation conditions, etc.

Let us consider for example the case of a solution of colloidal aggregates as in the experimental results of Table 4.2, and let us assume sub-monolayer coverage and total adsorption of the analyte for simplicity (a case relevant to

Table 4.2 Main Raman active modes of rhodamine 6G (RH6G), crystal violet (CV), benzotriazole dye 2 (BTZ) [89], and benzotriazole (BTA), with their experimentally-determined non-SERS and SERS frequencies († indicates peaks that are part of a doublet, and the concentration for the SERS experiments is given in the first column). For each mode of each molecule, the analytical enhancement factor (AEF) corresponds to that measured for the same Ag colloidal solution (Lee & Meisel prepared as detailed in Chapter 7 with the addition of 10 mM KCl). Reproduced from Ref. [8].

	$\bar{\nu}_i$ (Raman) [cm^{-1}]	ν_i (SERS) [cm^{-1}]	AEF –
RH6G	612	612	5.0×10^5
5 nM	774	768–778 †	4.2
	1185	1181 (1198) †	4.5
	1311	1312 (1292) †	4.4
	1364	1363 (1349) †	3.4
	1510	1511	3.4
	1652	1651	3.1
CV	808	804	1.0×10^5
5 nM	917	914	1.8
	1177 (1200) †	1176	0.53
	1621	1622	0.86
BTZ	1108	1106 (1125) †	6.4×10^5
100 nM	1412	1390 (1413) †	6.2
	1617	1616	8.2
BTA	783	787	3.5×10^5
1 μM	1019	1035 (1021) †	4.5
	1376–1390 †	1369–1394 †	5.4
	1599	1579	7.8

these experiments). Even if the SSEF of a particular aggregate is known, one first needs to average over possible orientations of the aggregate in solution, resulting in a polarization-averaged SSEF (PASSEF), where the averaging is over all possible orientations of the substrate with respect to the incident polarization. This would correspond to the actual AEF only if all aggregates were exactly identical. This is unrealistic in practice, where aggregates differ in particular in their resonance condition with the laser (and therefore in the magnitude of their SSEF or PASSEF). An averaging of the SSEF over aggregate orientation and over all types of aggregates is then required to obtain the AEF, a difficult undertaking in general. See Refs. [8,48] for a more detailed discussion.

Another example of interest is the case of planar SERS substrates. Well-defined structures can now be fabricated using nano-lithography techniques [38,41]. The SSEF can therefore be in principle calculated. The AEF could then be predicted for a specific probe, but for this, one needs to link c_{SERS} (analyte concentration for the AEF) with N_{Surf} (number of molecules

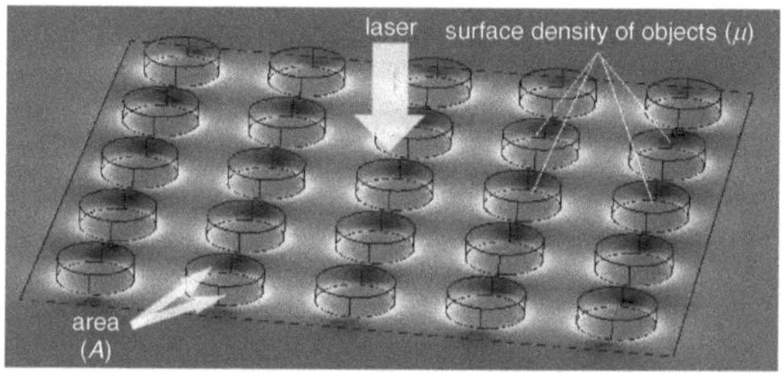

Figure 4.1. A planar SERS substrate formed by a repeating pattern of individual 3D sub-wavelength metallic objects on an underlying planar substrate. The surface density of individual nano-structures is μ_M [m^{-2}] and their metallic surface area is A_M [m^2] (each) (which includes both the top and side areas of each object). The array is illuminated from above with a laser.

adsorbed on the surface in the scattering volume). This will depend on the preparation conditions, i.e. how the analyte is transferred from solution to the substrate (dipping, drying, spin-coating, etc.). This is a much more complex situation than what it would appear at first sight, and we therefore dedicate the next sub-section to this specific issue.

4.2.3. Link between SSEF definition and experiments

The formal definition of the SSEF, as a spatial average of the OASMEF, was given in Eq. (4.10). Although rigorous, it is somewhat distant from the more commonly-used expression given in Eq. (4.2). We will make the connection here between these two, focusing on the most relevant case of planar SERS substrates. This is also the most difficult case, because of the problems of comparing signals originating from a 2D structure to that of a 3D volume. This connection will give a rigorous meaning to N_{Vol} and N_{Surf} in Eq. (4.2), therefore allowing for a meaningful experimental estimation of the SSEF. We follow largely Ref. [8] in the presentation.

Notations

Let us consider a SERS experiment on a typical planar SERS substrate consisting of a repeating pattern of individual 3D sub-wavelength metallic structures as illustrated in Fig. 4.1. Let μ_M [m^{-2}] be the surface density of the individual nano-structures with respect to the main plane forming the substrate, and A_M [m^2] be the metallic surface area in each structure (usually larger than the projected surface area or shadow of the structure on the main plane). The excitation is usually non-uniform and characterized in the focal plane by an intensity profile $S_{\text{Inc}}(\rho)$ [W m^{-2}] and a corresponding power P_{Inc} [W], which can be obtained by integration of $S_{\text{Inc}}(\rho)$. We assume that the excitation area is much larger than the individual structures forming the substrate and that the exciting intensity is approximately uniform for a given single structure.

Let us now work out the SERS and non-SERS intensities: I_{SERS} [W] and I_{RS} [W] used in Eq. (4.2).

SERS intensity

From the definitions of the previous section, one molecule at a given position $\mathbf{r}'$ on the surface of an individual metallic object at a position ρ within the beam, emits a SERS signal equal to:

$$I(\mathbf{r}') = \mathrm{OASMEF}(\mathbf{r}')\frac{\mathrm{d}\sigma_{\mathrm{RS}}}{\mathrm{d}\Omega}S_{\mathrm{Inc}}(\rho)\Delta\Omega_{\mathrm{Det}}. \tag{4.12}$$

Assuming a surface density of molecules μ_S [m^{-2}] on the metal, the SERS signal from this single object is then:

$$I_{\mathrm{SERS}}^{\mathrm{NP}} = \mu_S A_M\{\mathrm{OASMEF}\}\frac{\mathrm{d}\sigma_{\mathrm{RS}}}{\mathrm{d}\Omega}S_{\mathrm{Inc}}(\rho)\Delta\Omega_{\mathrm{Det}}, \tag{4.13}$$

where $\{\ldots\}$ denotes spatial averaging as before.

Adding the contribution of all the metallic objects in the beam (by integrating over ρ), we obtain the total SERS signal as:

$$I_{\mathrm{SERS}} = \mu_M\mu_S A_M\{\mathrm{OASMEF}\}\frac{\mathrm{d}\sigma_{\mathrm{RS}}}{\mathrm{d}\Omega}P_{\mathrm{Inc}}\Delta\Omega_{\mathrm{Det}}. \tag{4.14}$$

Non-SERS intensity

We now compare this SERS signal to that obtained with the same setup under non-SERS conditions (I_{RS}) from a solution of concentration c_{RS} [molecules m^{-3}] (note that c_{RS} is taken in [molecules m^{-3}] and not in [mol m^{-3}] here for simplicity). We must relate I_{RS} to $\mathrm{d}\sigma_{\mathrm{RS}}/\mathrm{d}\Omega$ and P_{Inc}. For a small element of volume $\rho\mathrm{d}\rho\,\mathrm{d}z$, we have $\mathrm{d}I_{\mathrm{RS}} = c_{\mathrm{RS}}\,(\mathrm{d}\sigma_{\mathrm{RS}}/\mathrm{d}\Omega)S_{\mathrm{Inc}}(\rho,z)\rho\mathrm{d}\rho\,\mathrm{d}z\Delta\Omega_{\mathrm{Det}}$. Integration over ρ simply gives P_{Inc}, while integration over $\mathrm{d}z$ gives the height H [m] of the scattering volume, resulting in: $I_{\mathrm{RS}} = c_{\mathrm{RS}}\,(\mathrm{d}\sigma_{\mathrm{RS}}/\mathrm{d}\Omega)HP_{\mathrm{Inc}}\,\Delta\Omega_{\mathrm{Det}}$.

In practice, one must define (and measure) the height of the scattering volume. In fact, the collection efficiency is not uniform along the beam axis and depends on the experimental setup (confocal depth). This together with the possible non-uniform excitation along the beam axis and the 3D nature of the scattering volume can cause additional complications. These issues are discussed in full detail in Section S.IV of the supplementary information of Ref. [8], from where we borrow here the final result:

$$I_{\mathrm{RS}} = c_{\mathrm{RS}} H_{\mathrm{Eff}} \frac{\mathrm{d}\sigma_{\mathrm{RS}}}{\mathrm{d}\Omega} P_{\mathrm{Inc}} \Delta\Omega_{\mathrm{Det}}, \tag{4.15}$$

where H_{Eff} [m] is the *effective height* of the scattering volume (with a precise definition given in Section S.IV of Ref. [8]).

Rigorous connection

Recall now that the SSEF is formally defined as {OASMEF} (see Eq. (4.10)). Combining the last two equations, this can then be rewritten in terms of the experimental observables I_{SERS} and I_{RS} as:

$$\mathrm{SSEF} = \frac{I_{\mathrm{SERS}}/(\mu_M \mu_S A_M)}{I_{\mathrm{RS}}/(c_{\mathrm{RS}} H_{\mathrm{Eff}})}. \tag{4.16}$$

This expression reduces to the 'non-rigorous' definition of the SSEF (Eq. (4.2)) provided that:

$$\boxed{N_{\mathrm{Surf}} = \mu_M \mu_S A_M A_{\mathrm{Eff}}, \quad \text{and} \quad N_{\mathrm{Vol}} = c_{\mathrm{RS}} V = c_{\mathrm{RS}} H_{\mathrm{Eff}} A_{\mathrm{Eff}},} \tag{4.17}$$

where A_{Eff} [m^2] is the effective surface area of the scattering volume (the equivalent of a scattering volume but for 2D). The exact definition of A_{Eff} is not important since it cancels out in any case in the expression for the SSEF. It is only introduced here in order to interpret the meaning of Eq. (4.16) in terms of number of molecules (on the surface or in the volume).

Discussion

This rigorous derivation highlights a number of important aspects for the experimental determination of the SSEF:

- As mentioned earlier, the parameter A_{Eff} is irrelevant to the final result. The only relevant aspect of the scattering volume here is therefore the effective height H_{Eff}, which needs to be characterized carefully (see Section S.IV of Ref. [8]). It is precisely this length scale that allows *the connection between 2D and 3D measurements.*
- In this derivation, we have considered only molecules directly adsorbed on the metal surface. In experiments, it is therefore important that the surface coverage remains smaller than (or equal to) one monolayer when

using Eq. (4.2) or (4.16). This is because the SERS effect is distance-dependent and the SERS signal from molecules on the second monolayer is usually reduced. This additional constraint ensures that the average SERS intensity is proportional to the average number of adsorbed molecules, which is necessary for Eqs (4.2) and (4.16) to represent a physically meaningful quantity.

Additional considerations

Finally, in the previous definitions, N_{Surf} corresponds to the number of molecules adsorbed *on the metallic surface*. This choice emphasizes the EF of each individual structure (for periodic patterns), as opposed to that of the entire substrate as a whole. If, alternatively, the total number of molecules (i.e. those adsorbed on the metal plus those adsorbed elsewhere on the surface in between the metallic structures) is chosen, then a smaller EF is obtained and we call it the *Total SERS Substrate EF* (TSSEF). The TSSEF is then simply the SSEF times the percentage surface coverage of the metal surfaces on a planar SERS substrate. The distinction between SSEF and TSSEF has already been emphasized [38] and can be important when comparing EFs across substrates. The proportion of the total surface that is occupied by the metal can be as low as 7% (as for example in Ref. [41]), which results in a factor of $\approx$15 between SSEF and TSSEF. The SSEF is more relevant in situations where the analyte selectively attaches on the metal only, but the TSSEF is more important if adsorption is not selective (in this case many molecules adsorb in between the metallic objects and do not contribute to SERS).

Experimental problems

Once a precise definition such as those above is agreed, comparison between SERS substrates is in principle possible, but there can still be many discrepancies in the measured SSEF or TSSEF because of experimental problems, in particular:

- It is not always easy to ensure that one (or less) monolayer is adsorbed. In practice, it is often assumed that exactly one monolayer is adsorbed based on chemical arguments (such as covalent bonding of the first layer, followed by rinsing of subsequent layers), but no independent confirmation is sought. Alternatively, it is sometimes argued that only the first layer gives rise to SERS, thereby eliminating the problem associated with subsequent layers. However, simple theoretical considerations (see Section 6.2.4) and recent experimental results [42] show that the distance dependence of the SERS effect is not in general as dramatic as sometimes believed, and the first layer argument should be considered as wrong, unless justified by other means.

- Even if exactly one monolayer is adsorbed, an accurate estimation of the molecule surface density μ_S is in most cases difficult, which results in uncertainties in the derived SSEFs.
- Other uncertainties can be associated with the value of H_{Eff} if the scattering volume is not carefully characterized.

Overall, the rigorous definition and careful derivation of the SSEF as described in this section are only a first step toward reducing uncertainties and 'standardizing' SSEF measurements. However, until new tools are developed to address some of the issues mentioned above, the comparison of SSEFs among substrates remains problematic.

4.3. OVERVIEW OF THE MAIN EM EFFECTS IN SERS

As mentioned already, the SERS enhancement is often separated into two contributions, the first electromagnetic, the second chemical. Within this approach, *electromagnetic enhancements apply to all analytes*, while *chemical enhancements are probe-dependent* and in general require chemical bonding of some sort to the metal substrate. If both mechanisms coexist, then their effect are in general assumed to be cumulative (i.e. the enhancement factors are multiplicative). The relative contribution, or even the existence of the chemical enhancement, has been the subject of much debate over the years, and is still arguably not fully resolved (we come back to this issue in Section 4.8). It is however widely accepted that the *main origin of the SERS enhancement is electromagnetic*. Understanding the electromagnetic contributions to the SERS effect is therefore paramount. In the rest of this chapter, we describe in detail the origin of the EM-enhancement contribution to the SERS effect and derive most of the necessary expressions describing the various enhancement mechanisms.

We start in this section by a general overview of the main EM effects that are relevant to SERS. In particular, we introduce *in simple terms* the local field and radiation enhancement factors and the common $|E|^4$-approximation to the single-molecule EF (SMEF). This section attempts to be a plain-English introduction to these important concepts. In fact, it may be sufficient for most applications of SERS. Further justifications and more rigorous definitions will be given and discussed in subsequent sections.

4.3.1. Analysis of the EM problem of SERS

The study of the EM-SERS enhancement factor is equivalent to the EM problem of Raman scattering by a molecule in the vicinity of a metallic surface[4]. Many approaches can be used, for example: a full ab initio description

---

[4] Note that the same approach can be used for any other type of surface, in particular dielectrics, but metals provide the largest (and most useful) enhancements.

of the molecule, or a quantum-mechanical treatment of the molecule and/or the EM field. The simplest approach, and arguably the one that gives most insight into the SERS effect is the *phenomenological description of Raman scattering within classical EM theory* (fully explained in Chapter 2). Most, if not all, aspects of SERS can be understood within this framework, which can in addition be justified by more complex quantum treatment in simple cases.

Hence, in this book we will almost exclusively focus on this approach. The classical and phenomenological approach to the Raman effect was described in detail in Section 2.4. Let us now recall the main ingredients before moving on to SERS.

A reminder on the phenomenological approach to Raman scattering

Within this approach, the Raman response of a given Raman mode of a molecule is fully characterized by its *Raman polarizability tensor* $\hat{\alpha}_R$. Under monochromatic light excitation at a frequency ω_L, the electric field $\mathbf{E}_{\text{Inc}}$ at the molecule position induces a Raman dipole $\mathbf{p}_R = \hat{\alpha}_R \cdot \mathbf{E}_{\text{Inc}}$, oscillating at the Raman frequency ω_R. In this introductory section, we will ignore for simplicity the tensorial character of $\hat{\alpha}_R$ and simply write $\mathbf{p}_R = \alpha_R \mathbf{E}_{\text{Inc}}$, equivalent to considering an isotropic tensor (a rigorous general treatment is given in the following sections). This oscillating Raman dipole radiates a power proportional to $|\mathbf{p}_R|^2$ at frequency ω_R, and it is this radiation that we detect in the far-field as the Raman signal. The differential cross-section of this Raman scattering process can be calculated, as was done in Section 2.4. We stress again here that although the Raman process can be described formally as a two-step process (induced dipole + re-emission), it is in reality *instantaneous*, and these two steps cannot be decoupled.

Application to SERS

The same phenomenological description can be applied to a Raman scatterer under SERS conditions. However, the presence of the metal will affect these processes in a number of ways, which we list here briefly and discuss in more detail afterward.

- Firstly, the electromagnetic field at the molecule position is modified. In cases of interest, its amplitude can be dramatically enhanced. This results in a possible *local field enhancement.*
- Secondly, the radiation properties of the Raman dipole $\mathbf{p}_R$ are modified. This results in a possible *radiation enhancement.*
- Thirdly, the Raman polarizability tensor $\hat{\alpha}_R$ may be modified. Such a modification would typically be classified as the *chemical enhancement* (CE), although it may be electromagnetic in origin. This aspect is

secondary in most cases and we will neglect it here until the discussion of the chemical enhancement in Section 4.8.

The SERS EM problem therefore consists mainly in understanding the first two effects: modification of the EM field and of the radiation properties of a dipole. This can further be separated into two largely independent steps:

- Firstly, solving the EM problem itself, for a given geometry. This aspect, along with the connection of the EM properties with plasmon resonances will be mostly dealt with in the next two chapters. We assume therefore in this chapter that we have a solution available when needed.
- Secondly, relating the EM field solution to the actual SERS properties (intensity, cross-section, etc.). This aspect, which is essentially the understanding of the fundamental EM mechanisms giving rise to the SERS enhancement and of the tools needed to study them, is the main subject of the rest of this chapter. As will be shown, it can also be readily applied to other optical processes and, in particular, to surface-enhanced fluorescence (SEF).

Important parameters in the EM problem

The majority of SERS experiments consists in studying the Raman signal emitted by one or more molecules, in close proximity of a metallic substrate with sub-wavelength features. As for the definitions of the SERS EFs, a number of parameters need to be given to study this problem:

- Metallic substrate: exact geometry and optical properties (depending on the material used).
- Probe molecule: exact orientation and position (including distance to surface). Also optical properties of the probe, such as Raman polarizability tensor.
- Optical setup: excitation wavelength, polarization, angle of incidence, power density, and beam profile (plane-wave approximation or Gaussian beam, for example). Also important is the scattering configuration, which defines the direction from which the emitted light is collected with respect to excitation. Standard configurations include back-scattering (when the delivering optics is also used for collection), forward-scattering, or 90°-scattering. Moreover, polarizers can also be used in the detection, for example to study depolarization ratios.

It is clear that in a real-life experiments, it is difficult to actually know accurately all of these parameters. Some of them can be estimated or measured, such as the beam waist of a Gaussian laser beam using beam profiling techniques. Others, like the adsorption geometry of a probe or its distance from the surface, are more difficult to measure, and in general some 'reasonable' assumptions have to be made. These are necessary to study the problem theoretically, and may sometimes be justified through comparison with experiments.

4.3.2. Local field enhancement

The local field intensity enhancement factor

As mentioned, the electromagnetic field is strongly modified in the vicinity of metallic objects. This is particularly true when the excitation wavelength λ_L is close to the electromagnetic resonances of the system (in particular localized surface plasmon resonances for metals, see Chapter 3). This means that the electric field $\mathbf{E}_{\mathrm{Loc}}$ at the molecule position can be very different, *both in magnitude and orientation*, to the incident field $\mathbf{E}_{\mathrm{Inc}}$. $\mathbf{E}_{\mathrm{Loc}}$ is usually called the *local field*[5] and is the *macroscopic* electric field felt by the molecule. It depends on excitation wavelength and polarization, and can also vary dramatically with position.

In cases of interest to SERS, the magnitude $|\mathbf{E}_{\mathrm{Loc}}|$ on the metallic surface can be much larger than $|\mathbf{E}_{\mathrm{Inc}}|$. This increase can sometimes be particularly spectacular at some specific (localized) positions on the surface, the so-called 'hot-spots'. The local field induces a Raman dipole $\mathbf{p}_R = \alpha_R \mathbf{E}_{\mathrm{Loc}}(\omega_L)$, whose magnitude is therefore enhanced by a factor $|\mathbf{E}_{\mathrm{Loc}}(\omega_L)|/|\mathbf{E}_{\mathrm{Inc}}|$. If such a dipole radiates in free-space (no metal), then the energy radiated, which is proportional to $|\mathbf{p}_R|^2$, would then be enhanced by a factor:

$$\boxed{M_{\mathrm{Loc}}(\omega_L) = \frac{|\mathbf{E}_{\mathrm{Loc}}(\omega_L)|^2}{|\mathbf{E}_{\mathrm{Inc}}|^2}.} \tag{4.18}$$

This factor, which we will call the *local field intensity enhancement factor* is associated with the *excitation* of the Raman dipole. For an absorbing molecule, it also characterizes the enhancement of the absorption cross-section (see Section 4.6.2). Note that it characterizes the enhancement of the *electric field intensity*, but ignores any modification of the electric field *polarization*.

[5] The denomination of 'local field' here should not be confused with that used for the *microscopic* field (often called also the local field). The difference between macroscopic and microscopic field is discussed in Section C.3.1 and is the origin for example of the local field correction factor discussed in Section 2.4.5. In a SERS context, the local field is the *macroscopic electric field* at the molecule position, which is different from the incident field due to the presence of the metal substrate.

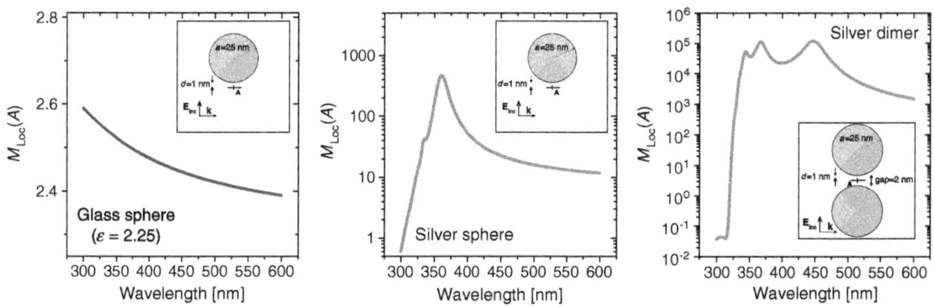

Figure 4.2. Examples of modification of the local field intensity, characterized here by the wavelength dependence of the LFIEF $M_{\rm Loc}$ defined in Eq. (4.18). This is calculated for a molecule in air at a distance $d = 1$ nm from the surface at point A (shown in the insets) using Mie theory. The cases of a 50 nm-diameter dielectric glass or metallic silver spheres are shown along with that of a molecule at the gap between two closely-spaced silver spheres (a dimer). For a glass sphere, there is a small enhancement, hardly dependent on wavelength. For a silver sphere, $M_{\rm Loc}(A)$ varies widely with wavelength, showing resonances associated with the *localized surface plasmon resonances* of the silver sphere. Large enhancements ($\sim$470) are predicted at the main resonance (λ = 360 nm). Such effects are even more striking in the case of a molecule in between two silver spheres separated by a 2 nm gap (note the log-scales for the silver sphere and dimer). This case is shown since it is believed to be a good representation of a SERS hot-spot: the LFIEF can reach values in excess of $\sim 10^5$, and it is consistently large across most of the visible range.

Illustration

The local field enhancement effect is illustrated in Fig. 4.2 by considering the wavelength dependence of $M_{\rm Loc}$ at a specific point at a distance $d = 1$ nm from the surface of a sphere. We compare the results for two materials: a non-absorbing dielectric such as glass ($\epsilon = 2.25$), and a metal like silver (whose optical properties are described in Appendix E). The case of a dipole in the gap of a silver dimer, which is more relevant to SERS, is also shown. These simple examples illustrate a number of important features:

- Large local field intensity enhancement factors are possible close to the surface of a metallic object. These are associated with a coupling to *localized surface plasmon resonances*, as evidenced by the resonant response as a function of wavelength. This was already highlighted in the general discussion of plasmons in Chapter 3. The presence of the metal in this case acts as 'funnel' to concentrate the electromagnetic fields on the surface.

- Under the right conditions, as in the gap between two metallic objects, huge local field enhancements of the order of $\sim 10^5$ can be obtained. This is the type of enhancement that we are looking to exploit in SERS.

- Additional aspects of the local field, such as its spatial distribution and its polarization (orientation) are not apparent in Fig. 4.2, but will be further illustrated in Chapter 6.

4.3.3. Radiation enhancement

The radiation enhancement factor

Under SERS conditions, the Raman dipole radiates, not in free-space, but in close proximity to the metal. Dipole radiation is strongly affected by the metal, in a similar fashion as the exciting field is modified. This phenomenon is well known in the case of atomic spontaneous emission, and will be referred to as *modified spontaneous emission* (MSE). The environment affects the dipole emission in two ways:

- Firstly the *radiation pattern* $\mathrm{d}P_{\mathrm{Rad}}/\mathrm{d}\Omega$ (power radiated in the far-field per unit solid angle in a given direction) may be modified. This is for example obvious for a dipole emitting close to a perfectly reflecting plane: the field is entirely reflected by the plane and no power is radiated in this half-space.
- Secondly, the *total power radiated* by the dipole, P_{Rad}, can also be enhanced or quenched compared to that in free-space, P_0. We can therefore define a *radiation enhancement factor* as $M_{\mathrm{Rad}} = P_{\mathrm{Rad}}/P_0$. This second aspect is sometimes the source of confusion and deserves particular attention. It is wrong to consider that the dipole first emits radiation, as if in free-space, and this emitted field is then modified by its environment. *The emission process itself is directly affected by the environment.* This is somewhat less natural to understand and we will come back to it in detail in the next section. It means that for a given fixed oscillating dipole amplitude, more or less energy will be extracted from this dipole, depending on its electromagnetic environment. The power extracted from a dipole oscillating close to a metal surface can, for example, be many orders of magnitude larger than that radiated in free-space *by the same dipole with the same amplitude.* This does not contradict energy conservation, since an external source of energy is always formally required to maintain the amplitude of the dipole oscillations, which would otherwise decrease to zero as energy is radiated.

These effects will depend on several factors including the substrate geometry and optical properties, the dipole position, orientation, and its emission frequency ω_R. Overall, this results in a modification by a factor $M^d_{\mathrm{Rad}}(\omega_R)$ of the collected Raman signal compared to what would be collected if the Raman dipole radiated in free-space (a more rigorous definition of this factor is given later). This *radiative enhancement factor* is associated with the *re-emission* step of the Raman process.

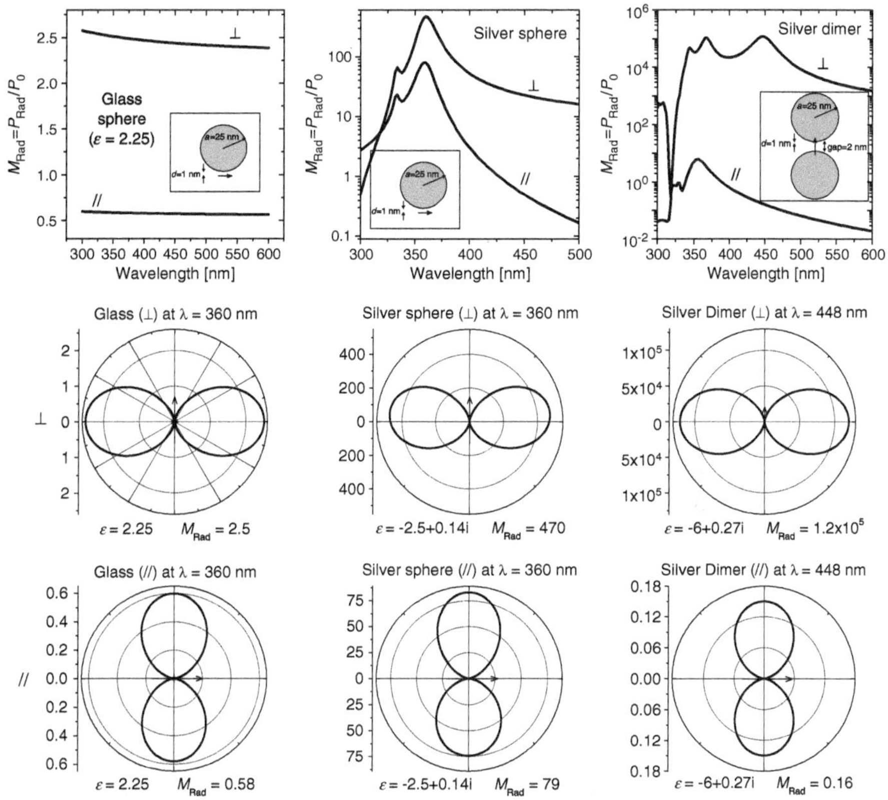

Figure 4.3. Examples of modification of dipolar emission in air at a distance $d = 1$ nm from a glass sphere (first column), a silver sphere (second column), or in the gap of a silver dimer (third column). The first row shows the wavelength dependence of the radiation enhancement, $M_{\text{Rad}} = P_{\text{Rad}}/P_0$ for a dipole perpendicular ($\perp$) or parallel (//) to the surface. This characterizes how much more (or less) energy is radiated in the far-field compared to the same dipole in free-space. For a silver sphere or dimer, M_{Rad} varies widely with wavelength, showing resonances associated with the *localized surface plasmon resonances* of the system (note the log-scales for the silver sphere and dimer). Large enhancements ($\sim$470) are predicted at the main resonance ($\lambda = 360$ nm) of the sphere, and even more spectacular figures are obtained for the dimer, $\approx 1.2 \times 10^5$ at $\lambda = 448$ nm. Second and third rows show examples of radiation patterns at chosen wavelengths. They represent the angular dependence of $\text{d}P_{\text{Rad}}/\text{d}\Omega$, in units of $(3P_0)/(8\pi)$.

Illustration

The modifications of the radiation intensity and pattern of a dipole are illustrated in Fig. 4.3 for a few simple cases of a dipole in air close to the surface of a sphere (at a distance $d = 1$ nm). We again compare the results for two materials: a dielectric such as glass (absorbing or not), and a metal like silver (whose optical properties are described in Appendix E). The case of a dipole in the gap of a silver dimer, which is relevant to SERS, is also shown. These simple examples illustrate a number of important features:

- The total power radiated by the dipole can be either quenched or enhanced, depending on the relative dielectric function $\epsilon(\omega)$ of the objects, their geometry, and the dipole position/orientation. A small quenching is for example observed for a dipole parallel to the surface of a non-absorbing dielectric sphere with $\epsilon(\omega) > 1$ like glass. Since there is no absorption in this case, this re-emphasizes the fact that the object directly modifies the power emitted by the dipole.
- When $\mathrm{Re}(\epsilon(\omega)) < 0$, as is the case of metals in the plasmonics region, large enhancements in the radiated power can be obtained. These are associated with a coupling to radiative *localized surface plasmon resonances* of the metallic objects. The presence of the metal in this case 'forces' the dipole to radiate more energy.
- In the latter case, the dipole radiates paradoxically more energy in the direction of the absorbing metallic sphere. One could have naively expected that part of the radiation in this direction would be absorbed or reflected, and would result in a deficit in radiation in this direction.
- Under the right conditions (like in the gap of a dimer), huge radiation enhancements of the order of $\sim 10^5$ can be obtained. This is again the type of enhancement that we are looking to exploit in SERS.
- Finally, as seen by comparing the emission from dipoles parallel and perpendicular to the surface, these effects depend dramatically on the orientation of the dipole (and also obviously on its position).

It is clear from these simple examples that this radiation enhancement is not a trivial effect. We will come back to it in more detail in this chapter and in Chapter 6.

4.3.4. Other EM effects

Other electromagnetic effects can influence the SERS signal or are relevant to SERS and related techniques like SEF. For example non-radiative processes (emission of a photon that is subsequently absorbed in the metal and is therefore undetectable in the far-field) can be important in some situations, in particular for SEF. Also, the Raman and optical polarizabilities of a probe can be modified when in close proximity to a metal because of the reflected field. These aspects are believed to contribute only marginally, if at all, to the main SERS enhancement mechanisms. They are difficult to describe in simple terms and we will come back to them later in this chapter once the full theory of EM SERS has been developed.

4.3.5. The common $|E|^4$-approximation to SERS enhancements

SMEF in the $|E|^4$-approximation

When considering only the two main mechanisms of enhancement, namely local field enhancement (excitation) and radiation enhancement (re-emission), the SERS EM enhancement for a single molecule can be simply expressed as:

$$\text{SMEF} \approx M_{\text{Loc}}(\omega_L) M^d_{\text{Rad}}(\omega_R). \tag{4.19}$$

Note that what is important here is the radiation enhancement factor *in the direction of detection of the SERS signal*, which we have denoted M^d_{Rad} and will call the *directional radiation enhancement factor*. This is closely related to M_{Rad}, the total radiation enhancement factor discussed earlier, but also to the radiation pattern. For example, if radiation is enhanced but also redirected away from our detector, we might observe a quenching. M^d_{Rad} is therefore the relevant factor for SERS, not M_{Rad}, although the two are in many instances related. These will be defined more precisely in the next section.

M_{Loc} can be found by solving the electromagnetic problem under specific external excitation conditions with an incident field $\mathbf{E}_{\text{Inc}}$, which yields the local field $\mathbf{E}_{\text{Loc}}$ everywhere. Estimating M_{Rad} is *a priori* a more difficult task. One would need to solve the electromagnetic problem of dipolar emission, instead of external excitation. This means working with dipolar singularities, which can cause problems both numerically and theoretically. It also means solving the problem for each possible position (and for at least three perpendicular orientations) of the dipole, which becomes in most cases intractable. To avoid these complications, it is often assumed that $M^d_{\text{Rad}}(\omega) \approx M_{\text{Loc}}(\omega)$, which means that the SERS enhancement can then be expressed simply as:

$$\boxed{\text{SMEF}(\omega_L, \omega_R) \approx M_{\text{Loc}}(\omega_L) M_{\text{Loc}}(\omega_R) \approx \frac{|\mathbf{E}_{\text{Loc}}(\omega_L)|^2}{|E_{\text{Inc}}|^2} \frac{|\mathbf{E}_{\text{Loc}}(\omega_R)|^2}{|E_{\text{Inc}}|^2}.} \tag{4.20}$$

This expression has been used extensively in the literature, and we will refer to it as the $|E|^4$-*approximation*. It provides a *fairly simple way of estimating the single-molecule enhancement factor from a calculation of the local field* at the excitation and Raman frequencies. Moreover, in many cases, the Raman shift is also small and one can make the additional approximation that $\omega_R \approx \omega_L$. This leads to the even more famous expression of the SERS enhancement for zero-Stokes shift in the $|E|^4$-approximation as:

$$\boxed{\text{SMEF}(\omega_L) \approx \frac{|\mathbf{E}_{\text{Loc}}(\omega_L)|^4}{|E_{\text{Inc}}|^4}.} \tag{4.21}$$

One could argue that *this is the most important expression in SERS.* In many instances, this approximation is sufficient to obtain the right order of magnitude of the SMEF. Average SERS EFs (like the SSEF) can also be derived within this approximation by surface averaging.

However, despite its uncontested simplicity and usefulness, this expression hides the underlying physics of the SERS process. The radiation enhancement is swept under the carpet in favor of the local field enhancement, much simpler to understand and to study, and it is easy to 'forget' about the real origin of this $|E|^4$ factor. This can be a problem, because *this factor remains an approximation* and it may be necessary for some studies to go back to the original description of local field + radiation enhancement; one example of this situation is the study of *polarization effects* or *surface selection rules* in SERS, discussed in more detail later.

Qualitative justification

A rigorous justification of this expression will be given in the following sections and a discussion of its range of validity is provided in particular in Section 4.5.3. It is however interesting to note at this stage that the assumption $M^d_{\mathrm{Rad}}(\omega) \approx M_{\mathrm{Loc}}(\omega)$ is not *a priori* obvious, but a few hand-waving arguments can nevertheless be used to justify it:

- Firstly, the local field enhancement and radiation enhancement have the same physical origin: the electromagnetic field (of the incident beam in one case or emitted by the dipole in the other) couples to electromagnetic resonances of the metallic substrate, namely localized surface plasmon resonances. Although the coupling may not be identical, the resonances should at least be qualitatively similar for both enhancement mechanisms. This is already evident in the similarities of some of the plots in Figs 4.2 and 4.3.
- Secondly, in simple cases where analytical calculations are possible, the total EM enhancement was found to reduce to the $|E|^4$ factor [164,165].
- Finally, arguments involving optical reciprocity have been quoted as a justification of this approximation. Indeed, a rigorous approach requires the use of the optical reciprocity theorem and will be described later.

From this brief overview, one aspect of electromagnetic theory stands out as being highly relevant to SERS, namely the radiation enhancement, which has been the subject of many studies in the context of modified spontaneous emission of atoms. Because of its relevance to the EM-SERS problem (and to related effects like SEF), Section 4.4 will be dedicated to this aspect, before a rigorous discussion of the EM-SERS EF can be given in Section 4.5. These two sections, which are more technical, are aimed at understanding the details of the EM-SERS enhancements and their relations to the SERS EF definitions.

4.4. MODIFIED SPONTANEOUS EMISSION

We focus first in this section on a general treatment of *modified spontaneous emission* (MSE). This phenomenon has traditionally been studied in the context of atomic spontaneous emission, but applies equally to molecules and is directly related to the radiation enhancement contributing to SERS (and SEF), as will be shown in the following sections. This section is somewhat more technical than the rest of this chapter and may be browsed quickly at first if it is not the main interest of the reader. The most relevant aspects to SERS are discussed in Sections 4.4.3 (in particular Table 4.3) and 4.4.6.

4.4.1. Introduction

The emission properties of a dipole, or any emitter, are affected by its environment. This fact was first pointed out by Purcell in 1946 [166] and has since then been demonstrated experimentally in a wide range of situations; for example for excited Eu^{3+} ions in front of a plane metallic surface, as in the original observation by Drexhage in 1968 [167], or more spectacularly for Rydberg atoms in an optical cavity [168]. This effect can be particularly strong in close proximity of metallic structures [138,169–171]. It is for example evident in the modification (quenching or enhancement) of spontaneous emission (SE) rates [172].

Quantum and classical descriptions

The optical process of *spontaneous emission* (SE) has been introduced and discussed in simple terms in Section 2.2.4, in the context of other photon emission processes. The role of SE in fluorescence processes was further developed in Section 2.3.2. These arguments, on which we will rely here, relate only to emitters in free-space (or possibly in an infinite non-absorbing dielectric of refractive index n_M). We now focus on extending them to the case of an emitter in the presence of (and usually close to) optically active objects (or surfaces), in short to *modified spontaneous emission* (MSE).

Spontaneous emission is an intrinsically *quantum effect*, where a dipolar (atomic or electronic) transition couples to the *vacuum state of the electromagnetic field*. This coupling can only occur with the quantized electromagnetic field, where the vacuum state has a zero-point energy and exhibits quantum fluctuations. However, because of the difficulties in the quantization of the field in complex structures, the modification of SE rates is most often studied within a classical model of an oscillating electromagnetic dipole [169]. This leads (in most cases) to results in agreement with experiments and with the full quantum-mechanical treatment.

Several complementary classical techniques can be used; the first of which considers the self-reaction field and is closest to the quantum treatment (and can be shown to be formally equivalent in most cases [12,173]). It enables the derivation of a total decay rate but does not distinguish between radiative and non-radiative processes. To avoid this problem, one can use a second technique based on the Poynting vector, which emphasizes energy conservation. We review here in Sections 4.4.4 and 4.4.5 the derivation and main results of these two methods. We will then describe in Section 4.4.6 a third approach based on the *optical reciprocity theorem*, which has not been commonly used but is important in connection with the SERS enhancement factors.

All these classical approaches are based on the study of the emission of a dipolar point source (in short, a dipole). It is therefore necessary to first understand the link between dipolar emission and spontaneous emission.

Emitters in a non-absorbing dielectric

The modification of spontaneous emission has been mostly studied for emitters embedded in vacuum or air, i.e. in optically inactive media. There are many situations, in particular for SERS or SEF, where the emitters are located in a non-absorbing dielectric medium, such as water or organic solvents. The effect of the embedding medium is typically secondary compared to modifications induced by other objects like metallic particles. It is nevertheless there, and does affect the emission property even in the absence of any other object or surfaces (what we would consider as non-modified conditions). Extending the conventional treatment of MSE to the case of a non-absorbing dielectric of refractive index $n_M \neq 1$ is not conceptually difficult, but it does introduce a number of 'cumbersome' factors. Because of their relevance to SERS, and because these factors are not easily tracked in the literature, we will include them explicitly. In the following, a free-space emitter should therefore be understood as an emitter embedded in an infinite dielectric of refractive index n_M, and we will study the modifications arising from the presence of nearby objects (the emitter remaining locally embedded in the same dielectric at all times). The subscript M will be used to refer to this free-space situation. The case of an emitter in vacuum can easily be recovered by taking $n_M = 1$, and the subscript 0 will be used when referring specifically to vacuum or air.

4.4.2. The link between spontaneous emission and dipolar emission

A reminder on dipolar emission

Here we take elements of the dipole emission problem sketched in Section 2.4.1 and we apply them to the problem of spontaneous emission. We repeat here some of the ingredients of Section 2.4.1 so that the presentation is made as self-contained as possible.

A dipolar emitter is one of the simplest (and most common) sources of radiation in classical electromagnetic theory. It is defined by its position $\mathbf{r}_0$, the angular frequency of the oscillation ω [rad s^{-1}], and its dipole moment, a vector denoted $\mathbf{p}$ [C m] in complex notations, where an $e^{-i\omega t}$ dependence is assumed (see Sections 2.4.1 and C.1.2 for details). Charges and currents (moving charges) are sources of the electromagnetic field, and such a dipole behaves as a localized current source with current density:

$$\mathbf{j}(\mathbf{r}) = -i\omega\delta(\mathbf{r} - \mathbf{r}_0)\mathbf{p}. \tag{4.22}$$

In principle, by solving Maxwell's equations with the above source term and appropriate boundary conditions, dipolar emission in any environment can be modeled. This is, unfortunately, usually not an easy task as we shall see in the next chapter. Even for a dipole in vacuum with no boundaries, the treatment is somewhat mathematical [96]. The most important aspect to us is the radiation or far-field properties, in particular:

- The *polarized differential power* radiated in the far-field per unit solid angle, $dP_{\text{Rad}-P}/d\Omega$ [W sr^{-1}], where P here indicates detection with a polarizer along unit vector $\mathbf{e}_P$ (implicitly *perpendicular* to the detection direction).
- The *differential radiated power* (i.e. for unpolarized detection), $dP_{\text{Rad}}/d\Omega$ [W sr^{-1}]; which is simply the sum of $dP_{\text{Rad}-P1}/d\Omega$ and $dP_{\text{Rad}-P2}/d\Omega$ where $P1$ and $P2$ refer to two mutually orthogonal polarization detections.
- The total radiated power, P_{Rad} [W].

We consider first a dipole $\mathbf{p} = p\mathbf{e}_p$ in free-space and denote by Ω the direction of observation (along unit vector $\mathbf{e}_r$). The polarized differential radiated power in this direction is then [12,96]:

$$\boxed{\frac{dP^M_{\text{Rad}-P}}{d\Omega}(\Omega) = \frac{n_M\omega^4|\mathbf{p}|^2}{32\pi^2\epsilon_0 c^3}|\mathbf{e}_p \cdot \mathbf{e}_P|^2.} \tag{4.23}$$

Summing this expression over two orthogonal detection polarizations, $P1$ and $P2$, results in the unpolarized differential radiated power (independent of the choice of $P1$ and $P2$):

$$\boxed{\frac{dP^M_{\text{Rad}}}{d\Omega}(\Omega) = \frac{n_M\omega^4|\mathbf{p}|^2}{32\pi^2\epsilon_0 c^3}\left[1 - |\mathbf{e}_p \cdot \mathbf{e}_r|^2\right].} \tag{4.24}$$

Note that the term in brackets [...] can be expressed simply as $\sin^2\theta$, with θ the angle between the dipole ($\mathbf{e}_p$) and the detection direction ($\mathbf{e}_r$).

Integrating over all directions (over the full solid angle with $\mathrm{d}\Omega = \sin\theta \mathrm{d}\theta \mathrm{d}\phi$, $0 \leq \theta \leq \pi$, $0 \leq \phi \leq 2\pi$), we obtain the total radiated power as:

$$\boxed{P_{\mathrm{Rad}}^{M} = \frac{n_M \omega^4 |\mathbf{p}|^2}{12\pi\epsilon_0 c^3}.} \tag{4.25}$$

These formulas were further discussed in Section 2.4.1.

One may wonder where the radiated electromagnetic energy is actually coming from. In fact, we have assumed here that the dipole oscillation amplitude is fixed, and something must therefore drive and maintain this oscillation. The energy is therefore radiated at the expense of whichever source is driving and maintaining the dipole oscillation. Energy conservation is therefore preserved.

Dipole emission and spontaneous emission in vacuum

Spontaneous emission (SE) from an atom or a molecule is the result of the decay of an electron from an excited state to a lower energy state (usually the ground state), whereby the energy lost in the transition is emitted in the form of a photon[6]. Spontaneous emission may look at first very distant from the phenomenon of dipole emission described above. However, it can be shown formally using quantum mechanics that the two processes are equivalent [12,173]. This is because the quantum transition is a result of a dipolar interaction between the atomic or molecular electronic energy states and the electromagnetic field.

Let us briefly go through the argument is simple terms. The decay of the electron from an excited state is a random process (with a given probability per unit time of happening), and spontaneous emission is accordingly a *stochastic process*. The relevant physical quantity is therefore the rate at which the decay takes place. This is the *spontaneous emission rate*, and it arises naturally in the quantum-mechanical picture by applying the Fermi golden rule. For an atom in vacuum (the only 'easy' case for a quantum-mechanical treatment), the SE rate is then obtained as:

$$\Gamma_{\mathrm{Rad}}^{0} = \frac{\omega^3 p_{\mathrm{qu}}^2}{3\pi\epsilon_0 \hbar c^3}, \tag{4.26}$$

where p_{qu} is the dipole moment of the transition between the upper ($|b\rangle$) and lower ($|a\rangle$) electronic states: $p_{\mathrm{qu}}^2 = |\langle b| \mathbf{p} |a\rangle|^2$.

[6] We ignore here other radiative processes like phosphorescence, discussed briefly in Section 2.2.4.

Moreover, the energy is emitted as quanta of energy (photons), each with an energy $\hbar\omega$ [J], equal to the difference in energy between the two electronic states involved in the transition. A given atom in an excited state can only emit one such photon, unless there is a process which re-excites it back to the excited state. If a source of energy is constantly available to re-excite the atom, then the (average) total power radiated for an atom in vacuum is simply obtained from:

$$P^{0\ \mathrm{qu}}_{\mathrm{Rad}} = \Gamma^{0}_{\mathrm{Rad}}\hbar\omega = \frac{\omega^4 p_{\mathrm{qu}}^2}{3\pi\epsilon_0 c^3}. \tag{4.27}$$

This expression does not correspond exactly to that obtained within the classical treatment, which (taking Eq. (4.25) with $n_M = 1$) differs by a factor of 4. To reconciliate both approaches, one would need to use a classical dipole amplitude of $p_{\mathrm{cl}} = 2p_{\mathrm{qu}}$. The reasons for this factor of 4 are not straightforward [174], but ultimately irrelevant to our purpose here. It suffices to know that the classical approach leads to the same results as the full quantum treatment up to a factor of 4. Moreover, this factor is independent of the environment of the emitter, and therefore disappears when studying modification to the spontaneous emission rates within a given approach. *Modified spontaneous emission can therefore be studied within the framework of classical dipole emission, as long as we do not seek exact values for the absolute SE rate.* In summary, the absolute vacuum spontaneous emission rate Γ^0_{Rad} should be calculated within the quantum-mechanical framework if required (but it is often in practice an experimentally-determined parameter). Modifications to this rate (and to the radiation profile) can then be estimated within the classical EM theory of dipole emission.

Generalization to a dielectric environment

A direct consequence of this conclusion, is that we can in principle easily derive the spontaneous emission rate of an emitter in a dielectric, Γ^M_{Rad} from that of the same emitter in vacuum Γ^0_{Rad}. Γ^0_{Rad} is known from the quantum treatment and given in Eq. (4.26). Modification to this rate can be estimated in the classical EM theory as the modification of the total power radiated by the dipole, i.e. $P^M_{\mathrm{Rad}}/P^0_{\mathrm{Rad}} = n_M$ (from Eq. (4.25)). We therefore deduce:

$$\Gamma^M_{\mathrm{Rad}} = n_M \Gamma^0_{\mathrm{Rad}}. \tag{4.28}$$

Local field correction

There is however an additional subtlety for a dipole embedded in a dielectric. All our calculations so far have related to the *macroscopic* electric field (i.e. the one appearing in Maxwell's equations). What is really felt by a molecule is the *microscopic* electric field, which is the same in vacuum, but differs in the dielectric from the macroscopic field by the *local field correction factor*, given here by $(L_M)^{1/4}$, where L_M was defined and discussed already in Section 2.4.5. This factor affects the electric field created by a dipole at its own position, i.e. the self-reaction field, and therefore its rate of emission (this will be justified shortly). As a result, the spontaneous emission rate for the emitter embedded in the dielectric should, in fact, be expressed as:

$$\Gamma_{\text{Rad}}^{M} = (L_M)^{1/2} n_M \Gamma_{\text{Rad}}^{0}. \tag{4.29}$$

The same local field correction factor applies for an emitter close to any object as long as the emitter is still embedded in the same dielectric. This will always be the case when estimating enhancement factors in the following and *the local field correction therefore cancels out in any enhancement factor expressions*. We will therefore largely ignore it in the following. This correction would only be relevant when relating the spontaneous emission rate to the microscopic properties of the molecule (i.e. to the dipole moment p_{qu} of the quantum transition). For our purpose, Γ_{Rad}^{M} is always taken as a given parameter of the problem (usually obtained from experimental measurements).

Intrinsic non-radiative decay

Note that non-radiative decay (without photon emission) may also occur, especially in molecules, with a decay rate Γ_{NR}^{M} [s^{-1}]. This decay channel is usually an intrinsic property of the molecule (and possibly its environment, for example: the solvent) and can therefore be studied independently of the radiative decay (spontaneous emission). Note that the *total decay rate* Γ_{Tot}^{M} for an emitter in free-space is therefore the sum of the radiative and intrinsic non-radiative decay rates: $\Gamma_{\text{Tot}}^{M} = \Gamma_{\text{Rad}}^{M} + \Gamma_{\text{NR}}^{M}$. The lifetime of the excited state is then given by $(\Gamma_{\text{Tot}}^{M})^{-1}$.

4.4.3. Modification of dipole emission: definitions of enhancement factors

General considerations

The fact that the radiative properties of a source are modified by its environment is somehow intuitive. Take a hypothetical isotropic source (which emits in a first approximation the same power in all directions), and place a mirror in front of it. It is obvious that no light will be detected in the direction behind the mirror, while more light will be detected in the opposite direction because of reflection. In this example, the radiation pattern is modified, but the total power emitted is the same (the source does not feel the presence of the mirror). When the mirror becomes much closer (within a wavelength or less) to the source, there can however be additional effects, which are much less intuitive than what we get from our daily experience with ray optics.

The reason for this is that the environment can affect the electromagnetic field around the source and in particular at the source position itself. For an emitting dipole, this field, created by the dipole at its own position through interaction with the environment, is called the *self-reaction field.* In the classical oscillating dipole picture, the self-reaction field can either oppose or add to the dipole oscillations. Because the dipole amplitude is fixed, this results in more or less power needed to maintain the oscillations, and therefore, by conservation of energy, more or less power being extracted from the dipole to the electromagnetic field. *This effect plays a major role in the radiation enhancements of SERS.*

One simple situation of this modification of extracted power is, in fact, the case of a dipole emitting from inside an infinite non-absorbing dielectric material, as discussed in Section 4.4.2. For glass, for example, the refractive index is $n_M = 1.5$, and Eq. (4.25) shows that the very same dipole inside glass radiates 50% more power than it would in vacuum[7]. Other examples have also already been discussed in Fig. 4.3.

Absorption and non-radiative emission

In addition, many materials, metals in particular, absorb electromagnetic energy (which is then dissipated in the form of heat). For dipole emission in the presence of such media, a proportion of the total power extracted from the dipole (P_{Tot}) is absorbed and is therefore missing from the modified radiated power (P_{Rad}) in the far-field. This is usually referred to as *non-radiative emission*, and it cannot be detected in the far-field. Denoting P_{NR} the emitted power lost through non-radiative emission (absorption in the metal), energy conservation leads to:

$$P_{\mathrm{Tot}} = P_{\mathrm{Rad}} + P_{\mathrm{NR}}. \tag{4.30}$$

Obviously, for emission in a non-absorbing environment, $P_{\mathrm{NR}} = 0$ and therefore $P_{\mathrm{Tot}} = P_{\mathrm{Rad}}$.

Note that, as for emission in free-space, the modified radiative power P_{Rad} can be expressed as an integral over the full solid angle of the modified differential radiative power $\mathrm{d}P_{\mathrm{Rad}}/\mathrm{d}\Omega$ in a given direction $\Omega = (\theta, \phi)$, i.e.:

$$P_{\mathrm{Rad}} = \int\int \frac{\mathrm{d}P_{\mathrm{Rad}}}{\mathrm{d}\Omega}(\Omega)\mathrm{d}\Omega. \tag{4.31}$$

[7] Taking into account the local field correction, the emitted power is even three times larger in glass than it is in vacuum.

Finally, by analogy with the free-space case, one can also define the modified polarized differential radiative power $\mathrm{d}P_{\mathrm{Rad}-P}/\mathrm{d}\Omega$ for detection in a given direction with a polarization along $\mathbf{e}_P$.

Definition of enhancement factors

The different types of emitted power, radiative, non-radiative, and total, that can be considered for modified dipolar emission can be conveniently characterized by a number of enhancement factors[8], defined with respect to the free-space (in a dielectric of refractive index n_M) radiated power P^M_{Rad} given in Eq. (4.25) as:

$$\boxed{\begin{aligned} M_{\mathrm{Rad}} &= P_{\mathrm{Rad}}/P^M_{\mathrm{Rad}}, \\ M_{\mathrm{Tot}} &= P_{\mathrm{Tot}}/P^M_{\mathrm{Rad}}, \\ M_{\mathrm{NR}} &= P_{\mathrm{NR}}/P^M_{\mathrm{Rad}}. \end{aligned}} \tag{4.32}$$

In some situations, it is also convenient to define the *EM radiative efficiency* as:

$$\boxed{\eta^{\mathrm{EM}}_{\mathrm{Rad}} = \frac{P_{\mathrm{Rad}}}{P_{\mathrm{Tot}}} = \frac{M_{\mathrm{Rad}}}{M_{\mathrm{Tot}}}.} \tag{4.33}$$

Energy conservation moreover ensures that:

$$\boxed{M_{\mathrm{Tot}} = M_{\mathrm{Rad}} + M_{\mathrm{NR}}.} \tag{4.34}$$

These definitions, along with many of the notations used in this section, are summarized in Table 4.3 for convenience.

Let us discuss them in more detail:

- The *radiative enhancement factor*, M_{Rad}, characterizes the enhancement of the signal detected in the far-field, *when integrating over all directions*. From the point of view of modified spontaneous emission, it also characterizes the enhancement factor for the *modified radiative decay rate*: $\Gamma_{\mathrm{Rad}} = M_{\mathrm{Rad}}\Gamma^M_{\mathrm{Rad}}$.
- The *total EM enhancement factor*, M_{Tot}, characterizes the enhancement in the total energy extracted from the dipole. This does not translate directly in a detectable signal in the far-field, because part of this

[8] Strictly speaking, these factors should be called 'modification factors' since they can represent both *quenching* and *enhancement* depending on whether they are smaller or larger than one, respectively. However, most applications are interested in situations of enhancement, not quenching.

Table 4.3 Summary of notations and definitions for modified spontaneous emission.

Notation	Name	Definition/ expression
Free-space dipole in a dielectric (index of refraction: n_M)		
P^M_{Rad}	Integrated radiated power	Eq. (4.25)
$\mathrm{d}P^M_{\mathrm{Rad}}/\mathrm{d}\Omega$	Differential radiated power	Eq. (4.24)
Γ^M_{Rad}	Radiative (SE) decay rate	Section 4.4.2
Γ^M_{NR}	Intrinsic non-radiative decay rate	Section 4.4.2
Γ^M_{Tot}	Total decay rate	$\Gamma^M_{\mathrm{Rad}} + \Gamma^M_{\mathrm{NR}}$
Modified spontaneous emission		
Notation	**Name**	**Definition/ expression**
P_{Rad}	Integrated radiated power	
$\mathrm{d}P_{\mathrm{Rad}}/\mathrm{d}\Omega$	Differential radiated power	
$\mathrm{d}P_{\mathrm{Rad}-P}/\mathrm{d}\Omega$	Same with polarized detection along $\mathbf{e}_P$	
P_{Tot}	Total power extracted from dipole	
P_{NR}	Power emitted in non-radiative modes	$P_{\mathrm{Tot}} - P_{\mathrm{Rad}}$
M_{Rad}	Radiative enhancement factor	$P_{\mathrm{Rad}}/P^M_{\mathrm{Rad}}$
$M^d_{\mathrm{Rad}}(\Omega)$	Directional radiative enhancement factor	Eq. (4.36)
$M^d_{\mathrm{Rad-P}}(\Omega)$	Same with polarized detection along $\mathbf{e}_P$	Eq. (4.35)
M_{Tot}	Total EM enhancement factor	$P_{\mathrm{Tot}}/P^M_{\mathrm{Rad}}$
M_{NR}	Non-radiative EM enhancement factor	$P_{\mathrm{NR}}/P^M_{\mathrm{Rad}}$
$\eta^{\mathrm{EM}}_{\mathrm{Rad}}$	EM radiative efficiency	$M_{\mathrm{Rad}}/M_{\mathrm{Tot}}$
Γ_{Rad}	Modified radiative (SE) decay rate	$M_{\mathrm{Rad}}\Gamma^M_{\mathrm{Rad}}$
$\Gamma^{\mathrm{EM}}_{\mathrm{Tot}}$	Total EM decay rate	$M_{\mathrm{Tot}}\Gamma^M_{\mathrm{Rad}}$
$\Gamma^{\mathrm{EM}}_{\mathrm{NR}}$	EM non-radiative decay rate	$M_{\mathrm{NR}}\Gamma^M_{\mathrm{Rad}}$
Γ_{Tot}	Modified total decay rate	$\Gamma^{\mathrm{EM}}_{\mathrm{Tot}} + \Gamma^M_{\mathrm{NR}}$

energy corresponds to non-radiative emission (it is absorbed by the environment). From the point of view of modified spontaneous emission, M_{Tot} is also related to the *total EM decay rate* $\Gamma^{\mathrm{EM}}_{\mathrm{Tot}} = M_{\mathrm{Tot}}\Gamma^M_{\mathrm{Rad}}$. This rate, added to the intrinsic non-radiative decay rate Γ^M_{NR} (assumed to be non-modified here), gives the *modified total decay rate* of the excited state $\Gamma_{\mathrm{Tot}} = \Gamma^{\mathrm{EM}}_{\mathrm{Tot}} + \Gamma^M_{\mathrm{NR}}$ and therefore its modified lifetime, $(\Gamma_{\mathrm{Tot}})^{-1}$. This modified lifetime can be measured experimentally, for example in a time-resolved experiment, as in Ref. [172].

- The *EM radiative efficiency*, $0 \le \eta^{\mathrm{EM}}_{\mathrm{Rad}} \le 1$, may be used to characterize the competition between radiative and non-radiative emission ($\eta^{\mathrm{EM}}_{\mathrm{Rad}} = 1$ in a non-absorbing environment). Note that a large radiative enhancement can coexist with a small radiative efficiency, and vice versa, depending on the weight of non-radiative processes. In particular, a large value of M_{Tot} does not necessarily imply a large radiative enhancement, M_{Rad}, especially in the presence of absorbing media such as metals [172,173,175].

- The *non-radiative EM enhancement factor*, M_{NR}, is in a way a redundant definition, since it can usually only be inferred from the knowledge of M_{Rad} and M_{Tot} using Eq. (4.34).

The advantage of working with enhancement factors M's instead of powers P's (or rates Γ's) is that exact results can be obtained from the classical EM theory for the M's, while the exact absolute values of the P's (or Γ's) would require the quantum treatment, as already mentioned.

Directional radiative enhancement factor

Moreover, in many cases of interest, radiation is detected in only one direction (over a small solid angle $\Delta\Omega_{\text{Det}}$), and possibly analyzed with a polarizer (along $\mathbf{e}_P$, implicitly perpendicular to the detection direction). It is therefore best characterized by the polarized differential power, $dP_{\text{Rad}-P}/d\Omega$, radiated in the detection direction defined by $\Omega = (\theta, \phi)$. It is therefore convenient to define also the *polarized directional radiative enhancement factor*, $M^d_{\text{Rad}-P}(\Omega)$, and the corresponding *directional radiative enhancement factor* for unpolarized detection, $M^d_{\text{Rad}}(\Omega)$.

There is one technical problem here, because even in free-space, $\mathrm{d}P^M_{\text{Rad}-P}/\mathrm{d}\Omega$ is not constant and can be zero in some directions (along the dipole axis for example). If we use this as our denominator, then $M^d_{\text{Rad}-P}(\Omega)$ could potentially be indefinite or infinite. We will therefore use the following definition for polarized detection:

$$\boxed{\begin{aligned} M^d_{\text{Rad}-P}(\Omega) &= \left(\frac{\mathrm{d}P_{\text{Rad}-P}}{\mathrm{d}\Omega}(\Omega)\right) \Big/ \left(\frac{\mathrm{d}P^M_{\text{Rad}}}{\mathrm{d}\Omega}\right)_{\text{Max}} \\ &= \left(\frac{\mathrm{d}P_{\text{Rad}-P}}{\mathrm{d}\Omega}(\Omega)\right) \Big/ \left[\frac{3}{8\pi}P^M_{\text{Rad}}\right], \end{aligned}} \tag{4.35}$$

and the corresponding expression for unpolarized detection:

$$\boxed{M^d_{\text{Rad}}(\Omega) = \left(\frac{\mathrm{d}P_{\text{Rad}}}{\mathrm{d}\Omega}(\Omega)\right) \Big/ \left[\frac{3}{8\pi}P^M_{\text{Rad}}\right].} \tag{4.36}$$

This definition has been chosen because it satisfies several convenient criteria:

- The denominator is a constant value (with respect to Ω). This ensures that $M^d_{\mathrm{Rad}-P}(\Omega)$ gives a direct representation of the *modified radiation pattern*.
- The denominator is chosen equal to the maximum value of $\mathrm{d}P^M_{\mathrm{Rad}}/\mathrm{d}\Omega$ for a free-space dipole (i.e. that obtained in the directions perpendicular to its axis). The maximum value of $M^d_{\mathrm{Rad}-P}$ for a dipole in free-space is therefore 1. This choice will, in addition, avoid the presence of additional constants when applying the optical reciprocity theorem (see later). This, however, implies that there is a constant appearing when relating the radiative and directional radiative enhancement factors:

$$M_{\mathrm{Rad}} = \frac{3}{8\pi}\int M^d_{\mathrm{Rad}}(\Omega)\mathrm{d}\Omega. \tag{4.37}$$

Finally, as for the unpolarized differential radiated power, M^d_{Rad} can be derived from the sum of the polarized directional radiative enhancement factor for two perpendicular detection polarizations ($P1, P2$):

$$M^d_{\mathrm{Rad}}(\Omega) = M^d_{\mathrm{Rad}-P1}(\Omega) + M^d_{\mathrm{Rad}-P2}(\Omega). \tag{4.38}$$

With these definitions in mind, we will now briefly review how these enhancements (and related rates) can be calculated. We assume again that (whenever necessary) we have the solution of Maxwell's equations with appropriate boundary conditions and sources.

4.4.4. Spontaneous emission and self-reaction

Principle

We consider a dipole $\mathbf{p} = p\mathbf{e}_p$, oscillating at a frequency ω, possibly in the presence of other optically active objects. Within the classical treatment [169], SE can be seen as an effect of the self-reaction field, i.e. the electromagnetic field $\mathbf{E}_{\mathrm{SR}}$ created at the dipole position by itself, either directly or through its interaction with the environment (reflected field). Because of the linearity of Maxwell's equations, the self-reaction field can be written as $\mathbf{E}_{\mathrm{SR}} = \hat{\mathbf{G}}(\omega)\cdot\mathbf{p}$, where $\hat{\mathbf{G}}(\omega)$ is a second-rank tensor[9]. It can be viewed as the *classical self-reaction Green's tensor*, obtained by solving Maxwell's equation for the dipolar source with the appropriate boundary conditions [169].

The total EM decay rate can then be calculated from:

$$\Gamma^{\mathrm{EM}}_{\mathrm{Tot}} = \frac{2}{\hbar}p^2\mathrm{Im}(\mathbf{e}^*_p\cdot\hat{\mathbf{G}}(\omega)\cdot\mathbf{e}_p). \tag{4.39}$$

[9] A second-rank tensor provides the most general linear relation between $\mathbf{E}_{\mathrm{SR}}$ and $\mathbf{p}$.

This expression can be rigorously justified using *quantum* arguments [12,173]. The reason why the *classical* self-reaction Green's tensor $\hat{\mathbf{G}}(\omega)$ can be used to determine the quantum property $\Gamma_{\text{Tot}}^{\text{EM}}$ is that it is intricately linked to the vacuum fluctuations through the *quantum fluctuation-dissipation theorem* [173], or the local density of EM states [12]. Without going into more details, a convenient feature here is that once the validity of this expression is known, we can ignore completely the quantum aspects and use it to calculate $\Gamma_{\text{Tot}}^{\text{EM}}$ from the *classical* calculation of the Green's tensor [173].

This is even true for an atom in free-space, although the calculation of $\hat{\mathbf{G}}(\omega)$ is not so easy. It can be shown [12,176] that for an atom in vacuum, $\hat{\mathbf{G}}_0(\omega) = G_0(\omega)\hat{\mathbf{1}}$, with:

$$\text{Im}(G_0(\omega)) = \frac{\omega^3}{6\pi\epsilon_0 c^3}, \tag{4.40}$$

which leads to the correct spontaneous emission lifetime in vacuum, given in Eq. (4.26).

Similarly, for an emitter in a non-absorbing dielectric[10], we have:

$$\boxed{\text{Im}(G_M(\omega)) = \frac{n_M\omega^3}{6\pi\epsilon_0 c^3} = \frac{k_M^3}{6\pi\epsilon_0\epsilon_M},} \tag{4.41}$$

where we have used the wave-vector $k_M = n_M\omega/c$ in the second equality.

Total EM enhancement factor

In the presence of boundaries, in particular close to metallic surfaces, the self-reaction field can be strongly modified. For a dipole embedded in a dielectric, one can then write $\hat{\mathbf{G}} = \hat{\mathbf{G}}_M + \hat{\mathbf{G}}_r$, where $\hat{\mathbf{G}}_r$ corresponds to the self-reaction due to the field reflected by the boundaries at the dipole position. *This can be calculated by solving Maxwell's equation for the dipole emission in the presence of boundaries.* The total EM enhancement factor is then derived from Eq. (4.39) and reads:

$$\boxed{M_{\text{Tot}}(\omega) = \frac{\Gamma_{\text{Tot}}^{\text{EM}}}{\Gamma_{\text{Rad}}^{M}} = 1 + \frac{\text{Im}(\mathbf{e}_p^* \cdot \hat{\mathbf{G}}_r(\omega) \cdot \mathbf{e}_p)}{\text{Im}(G_M(\omega))},} \tag{4.42}$$

where $\text{Im}(G_M(\omega))$ is given by Eq. (4.41).

---

[10] Note that the local field correction is ignored in this treatment. This is not important here since we focus on rate enhancements, not absolute rates.

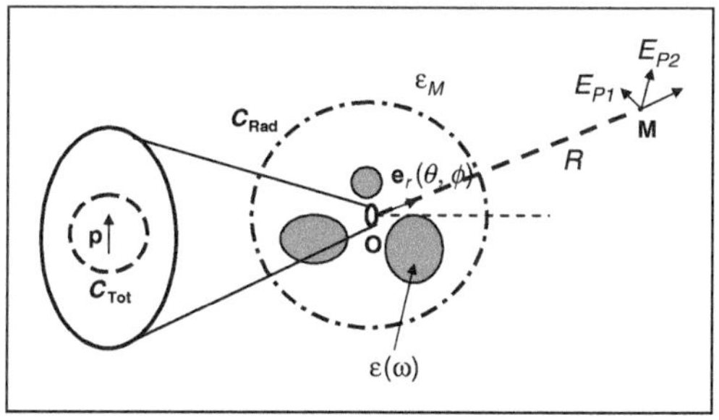

Figure 4.4. Schematic representation of the problem of an emitter (dipole) in close proximity to a metallic environment. The surfaces C_{Tot} and C_{Rad} are used to evaluate P_{Tot} and P_{Rad} within the Poynting vector approach. In the top-right corner is shown the radiation field, in relation with the application of the optical reciprocity theorem (Section 4.4.6).

Calculating the reflected field (or $\hat{\mathbf{G}}_r$) is usually not straightforward, but it is possible in simple cases. This approach has for example been used to study dipole emission close to planar surfaces [138,169], or to a sphere [177]. The main advantage of this approach is its direct connection with quantum mechanics. It is however important to stress here that M_{Tot} corresponds to the modification of the *total EM decay rate* of the emitter (radiative + non-radiative). Therefore the main inconvenience of this approach is that it does not yield any information about the radiative properties, except in the case of a non-absorbing environment (where $M_{Tot} = M_{Rad}$).

4.4.5. The Poynting vector approach

To study radiative properties (EM radiative efficiency and radiation pattern), it is much easier to return to the fully-classical EM treatment.

Total power extracted

Let us first consider again the total EM decay rate enhancement factor M_{Tot}. The total power P_{Tot} extracted from a classical dipole $\mathbf{p}$ can be estimated classically by solving Maxwell's equations for the dipole in the presence of the boundaries and then integrating the flux of the Poynting vector, $\mathbf{S}$, through a closed surface outside the dipolar source [169,175,177]. To understand how this works, let us consider a dipolar emitter located close to (but not inside) one or several finite-size objects. The situation is depicted schematically in Fig. 4.4. We can draw a surface around the emitter, C_{Tot}, small enough to enclose no absorbing media, and calculate the flux of the Poynting vector, P_{Tot}, through this surface. Because of energy conservation (no absorption inside this surface), P_{Tot} must be equal to the power extracted from the dipole by the EM field, as required.

The equivalence of this classical approach with the (quantum) self-reaction approach can now be made clearer by considering the following alternative (fully-classical) derivation of Eq. (4.42). From standard EM theory [96], the power extracted from a current source is directly related to the electric field at the source position, i.e. the self-reaction field in the case of dipole emission:

$$P_{\mathrm{Tot}} = -\mathrm{Re}\left(\frac{1}{2}\int \mathbf{j}^* \cdot \mathbf{E}_{\mathrm{SR}}\right) = \frac{\omega}{2}\mathrm{Im}\left(\mathbf{p}^* \cdot \hat{\mathbf{G}}(\omega) \cdot \mathbf{p}\right), \tag{4.43}$$

where we used Eq. (4.22) and $\mathbf{E}_{\mathrm{SR}} = \hat{\mathbf{G}}(\omega) \cdot \mathbf{p}$ to obtain the second equality. Combining this expression with Eq. (4.25) for P^M_{Rad}, we obtain for $M_{\mathrm{Tot}} = P_{\mathrm{Tot}}/P^M_{\mathrm{Rad}}$ the same expression as obtained earlier in Eq. (4.42). This therefore establishes formally *the equivalence between the quantum and classical calculations of the total EM decay rate enhancement factors.* The interested reader can also consult Chapter 8 of Ref. [12] for a more detailed argument.

The two approaches are therefore fully consistent owing to energy conservation (at least when the dipole is assumed to be located in a non-absorbing medium, the only case considered here). Note, however, that the self-reaction approach, thanks to its direct justification from the quantum treatment, can be used to calculate the absolute decay rates, as in Eq. (4.39). The Poynting vector approach only yields the decay rate enhancement factors, for example $M_{\mathrm{Tot}} = \Gamma^{\mathrm{EM}}_{\mathrm{Tot}}/\Gamma^M_{\mathrm{Rad}}$. This is not a limitation here, since we are interested precisely in this decay rate enhancement.

Radiative properties

Besides, the Poynting vector approach also yields additional information, namely: the radiative enhancement factor M_{Rad}, its angular dependence and polarization dependence (radiation pattern) and therefore the polarized directional radiative enhancement factors $M^d_{\mathrm{Rad}-P}$, and the non-radiative EM enhancement factor M_{NR}. This can for example be obtained as follows [169,175]:

- In the situation of Fig. 4.4, where the objects in the environment are bounded, one can enclose the emitter and all absorbing media in a large sphere, C_{Rad}, and calculate the power outflow on this surface, which corresponds to the power radiated by the source in the far-field, P_{Rad}. From there, $M_{\mathrm{Rad}} = P_{\mathrm{Rad}}/P^M_{\mathrm{Rad}}$ can be deduced.
- Moreover, if this surface is sufficiently far from any objects and sources, such that the lines of the Poynting vector are approximately radial (radiation field), then one can also obtain $(\mathrm{d}P_{\mathrm{Rad}}/\mathrm{d}\Omega)(\Omega) = r^2\mathrm{Re}(\mathbf{S} \cdot \mathbf{e}_r)$, i.e. the power radiated per unit solid angle, and deduce the angular radiation pattern and $M^d_{\mathrm{Rad}}(\Omega)$. A similar approach can be used for

$M^d_{\text{Rad}-P}(\Omega)$. Alternatively, the properties of the radiation field can also be deduced from the EM solution on the object boundaries; using for example the Stratton–Chu formula [95].

- Finally, from the solution of the EM field *inside* the absorbing objects, it is also possible to calculate by volume integration the total absorbed power, P_{NR}, and therefore $M_{\text{NR}} = P_{\text{NR}}/P^M_{\text{Rad}}$, which characterizes the non-radiative losses. Alternatively, one can also use energy conservation (Eqs (4.30) and (4.34)), to obtain directly the non-radiative power from the total and radiated powers.

In summary, as long as a solution of the EM problem of dipolar emission with appropriate boundaries can be obtained, all the necessary enhancement factors can then be deduced from it.

4.4.6. Spontaneous emission and the optical reciprocity theorem

The two approaches to modified spontaneous emission presented so far require solving Maxwell's equations *for the dipolar singularity* in the presence of boundaries, either numerically or analytically. This can present some problems. Analytically, the dipolar singularity can significantly complicate the problem (for example for a dipole near a sphere [177]). Numerically, singularities are not always straightforward to introduce. An additional major inconvenience is that the problem needs to be solved for *each dipole position* that one wants to study (and in principle for three orthogonal dipole orientations). We discuss in this section a third approach, based on the optical reciprocity theorem (ORT), which can circumvent some of these problems. It is also particularly suited to the calculation of SERS enhancements, as we shall see.

Statement of the theorem

The *optical reciprocity theorem* (ORT) states (see for example p. 308 of Ref. [178]) that the field $\mathbf{E}$ created at a given point M by a dipole $\mathbf{p}$ (at point O) is related to the field $\mathbf{E}_2$ at O created by a dipole $\mathbf{p}_2$ at M according to:

$$\boxed{\mathbf{p} \cdot \mathbf{E}_2 = \mathbf{p}_2 \cdot \mathbf{E}.} \tag{4.44}$$

This theorem is also sometimes referred to as the *Lorentz reciprocity theorem.* It has appeared (surprisingly) rarely in the scientific literature and has mostly been used in the study of antennas. A formal demonstration can be found in Ref. [178], but it remains fairly vague about its conditions of validity, especially about boundary conditions in unbounded geometries.

For our purpose here, it suffices to know that the validity is very general, and in particular in the presence of boundaries and absorbing media such as metals (assuming a local dielectric function). One situation where it may fail is when the boundaries are not bounded (i.e. extend to infinity), but this can always be avoided for our purpose here.

Application to modified dipolar emission

We now show that the ORT enables us to derive the far-field properties of an emitter *in a given direction* from the solution of two plane-wave-excitation (PWE) problems (one for each possible detection polarization), without any source singularities. Although the following is somewhat technical, the end result given in Eq. (4.51) will be of great importance in the formal description of EM-SERS EFs.

We consider a dipole $\mathbf{p} = p\mathbf{e}_p$ at O and focus on its far-field emission at a point M at a distance R in direction $\mathbf{e}_r$ defined by angles (θ, ϕ) (see Fig. 4.4). The radiation field is transverse in the far-field (no radial component) and its electric field at point M, denoted $\mathbf{E}$, can be decomposed into two perpendicular polarizations along unit vectors $\mathbf{e}_{P1}$ and $\mathbf{e}_{P2}$: $\mathbf{E} = E_{P1}\mathbf{e}_{P1} + E_{P2}\mathbf{e}_{P2}$.

To apply the ORT, we then consider the (separate) problem of a dipole $\mathbf{p}_2 = p_2\mathbf{e}_2$ situated at point M and with $\mathbf{e}_2 \perp \mathbf{e}_r$. The ORT then yields:

$$p_2\mathbf{e}_2 \cdot \mathbf{E} = p\mathbf{e}_p \cdot \mathbf{E}_2(\mathrm{O}), \tag{4.45}$$

where $\mathbf{E}_2(\mathrm{O})$ is the field created by $\mathbf{p}_2$ at O. To determine $\mathbf{E}_2(\mathrm{O})$, let us first consider the field $\mathbf{E}_2(\mathbf{r})$ created by $\mathbf{p}_2$ at a point $\mathbf{r}$ in the absence of any objects. For sufficiently large R (and we can take R as large as we like), the field of this dipole in the region of interest (i.e. the region where the objects are, with $|\mathbf{r}| \ll R$) can be approximated by expanding the free-space dipole field:

$$\mathbf{E}_2(\mathbf{r}) \approx E_p\mathbf{e}_2\, \mathrm{e}^{-\mathrm{i}k_M\mathbf{e}_r\cdot\mathbf{r}}, \quad \text{with } E_p = \frac{k_M^2 p_2 \mathrm{e}^{\mathrm{i}k_M R}}{4\pi\epsilon_0\epsilon_M R}. \tag{4.46}$$

This is the field of a plane wave with wave-vector k_M, propagating along $-\mathbf{e}_r$, polarized along $\mathbf{e}_2$, and with complex amplitude E_p. The problem of finding $\mathbf{E}_2(\mathrm{O})$, created by the dipole $\mathbf{p}_2$ at O in the presence of the objects is therefore equivalent to finding the local field at O, $\mathbf{E}_{\mathrm{Loc}}(\mathrm{O})$ for plane-wave excitation with the characteristics given above. Moreover, using the ORT expression in Eq. (4.45), we deduce:

$$\mathbf{e}_2 \cdot \mathbf{E} = \frac{E_p}{p_2} p \frac{\mathbf{e}_p \cdot \mathbf{E}_{\mathrm{Loc}}(\mathrm{O})}{E_p}. \tag{4.47}$$

By choosing $\mathbf{e}_2 = \mathbf{e}_{P1}$ (or $\mathbf{e}_2 = \mathbf{e}_{P2}$), we can therefore, using Eq. (4.47), determine the polarization component E_{P1} (or E_{P2}), of the radiation field of the dipole $\mathbf{p}$ in the direction defined by $\mathbf{e}_r$, by computing the local field $\mathbf{E}_{\mathrm{Loc}}^{\mathrm{PW}-P1}$ (or $\mathbf{E}_{\mathrm{Loc}}^{\mathrm{PW}-P2}$) created at O by a *plane wave* of amplitude E_p incident along $-\mathbf{e}_r$ and polarized along $\mathbf{e}_{P1}$ (or $\mathbf{e}_{P2}$). The EM problem of *dipolar radiation in a given direction* can therefore be replaced by *two* EM problems of *plane-wave excitation* (PWE) from this same direction.

Note that for the PWE problems, by linearity, $\tilde{\mathbf{E}}_{\mathrm{Loc}}^{\mathrm{PW}-P1} = \mathbf{E}_{\mathrm{Loc}}^{\mathrm{PW}-P1}/E_p$ is the relevant physical quantity (and is independent of E_p). It is a unit-less vector characterizing the direction of the local electric field, and its magnitude enhancement (or quenching) due to the presence of the objects.

Using Eq. (4.47) and the expression for E_p in Eq. (4.46), we therefore have:

$$E_{P1} = \frac{k_M^2 e^{ik_M R}}{4\pi\epsilon_0\epsilon_M R}\mathbf{p}\cdot\tilde{\mathbf{E}}_{\mathrm{Loc}}^{\mathrm{PW}-P1}. \tag{4.48}$$

A similar expression is obtained for E_{P2}, but note that it requires the solution of a *different PWE problem* (with a different polarization). From there, we can deduce the complex Poynting vector for this radiation field at M: $\mathbf{S} = (n_M\epsilon_0 c/2)(|E_{P1}|^2 + |E_{P2}|^2)\mathbf{e}_r$, where the contributions for each detection polarization are clearly distinguished. The time-averaged differential power radiated per unit solid angle in this direction is then:

$$\begin{aligned}\frac{dP_{\mathrm{Rad}}}{d\Omega}(\Omega) &= R^2\mathrm{Re}(\mathbf{S}\cdot\mathbf{e}_r)\\ &= \frac{n_M\omega^4|\mathbf{p}|^2}{32\pi^2\epsilon_0 c^3}\left[|\mathbf{e}_p\cdot\tilde{\mathbf{E}}_{\mathrm{Loc}}^{\mathrm{PW}-P1}|^2 + |\mathbf{e}_p\cdot\tilde{\mathbf{E}}_{\mathrm{Loc}}^{\mathrm{PW}-P2}|^2\right].\end{aligned} \tag{4.49}$$

Similarly, the polarized (along $\mathbf{e}_P$) differential radiated power is:

$$\frac{dP_{\mathrm{Rad}-P}}{d\Omega}(\Omega) = \frac{n_M\omega^4|\mathbf{p}|^2}{32\pi^2\epsilon_0 c^3}|\mathbf{e}_p\cdot\tilde{\mathbf{E}}_{\mathrm{Loc}}^{\mathrm{PW}-P}|^2. \tag{4.50}$$

It is easy to verify (by taking $\tilde{\mathbf{E}}_{\mathrm{Loc}}^{\mathrm{PW}-P} = \mathbf{e}_P$) that the above expression is fully consistent with that of an isolated dipole in free-space, yielding the same expression as the standard EM approach (Eq. (4.23)).

Radiation enhancement factors within the ORT approach

From its definition in Eq. (4.35), the *polarized directional radiative enhancement factor* is then given simply within the ORT approach by:

$$\boxed{M_{\mathrm{Rad}-P}^d(\Omega) = |\mathbf{e}_p\cdot\tilde{\mathbf{E}}_{\mathrm{Loc}}^{\mathrm{PW}-P}|^2,} \tag{4.51}$$

where $\tilde{\mathbf{E}}_{\text{Loc}}^{\text{PW}-P}$ can be obtained from the solution of a PWE problem. This is the main result of this section as far as SERS is concerned. Note that M_{Rad}^{d} follows naturally from Eq. (4.38) but requires the solution of *two* different PWE problems (for $P1$ and $P2$).

We repeat the important conclusion of this derivation: *the radiative property in a given direction of a dipole at any position in a complex environment can be obtained by modeling two (one for each polarization detection) PWE problems, without a dipolar singularity.* This is a substantial improvement when modeling a typical experimental situation where detection is carried out in a given fixed direction. The solution of these two PWE problems is sufficient to calculate the radiation properties in this particular direction for any dipoles *at any possible positions and with any orientations.* Hence, this essentially replaces an infinity of EM problems with a singular source by only two PWE problems; a much easier endeavor from a mathematical (and numerical) standpoint.

Equation (4.51) also shows that the far-field emission of a dipole in a given direction is in some way related to the local field intensity enhancement factor: $M_{\text{Loc}}^{\text{PW}-P} = |\tilde{\mathbf{E}}_{\text{Loc}}^{\text{PW}-P}|^2$ for PWE *from this direction.* This will be the basis for the generalization of the $|E|^4$-approximation of SERS enhancements in the next section.

Remarks

We conclude this presentation of the ORT with a few remarks:

- In principle, it is also possible within this approach to obtain the radiative enhancement M_{Rad} by integrating $M_{\text{Rad}}^{d}(\Omega)$ according to Eq. (4.37). However, it then requires the solution of a large (in principle infinite) number of PWE problems from all possible incoming directions. This is unpractical, except in the simplest situations (for example for a dipole near a plane as shown in Appendix F, or for a sphere thanks to the symmetry of the problem). The Poynting vector approach is therefore usually better suited to the calculation of M_{Rad}.
- The ORT approach, contrary to the previous two, yields no information on the total EM enhancement factor M_{Tot} (i.e. on non-radiative emission).
- This approach has hardly been used until now [45,179,180], but is actually well suited to many experimental situations where only radiative properties are of interest, and where detection is performed in only one direction. This is particularly true for SERS (and other spectroscopies in general).
- A simple example showing the power of this approach is the study of the emission of a dipole close to a sphere. This problem can easily be

solved using Mie theory for plane waves and the ORT. This is equivalent – and much simpler – than including the dipolar singularity in the Mie scattering formalism (see Appendix H for more details). In a similar fashion, the ORT can simplify greatly the study of dipolar emission close to a plane, as discussed in Appendix F.

- Finally, it is worth stressing that the ORT (as stated here) does not have, to our knowledge, any direct physical meaning related to the interaction energies between two dipoles, as its expression, $\mathbf{p}\cdot\mathbf{E}_2 = \mathbf{p}_2\cdot\mathbf{E}$, may at first suggest. One would in fact need the complex conjugates of the fields (or the dipoles) to be able to translate this in a statement about the mutual interaction energies of each dipole with the field of the other. The demonstration of the ORT [178] does not involve, in fact, the interaction energy. Moreover, the ORT considers the solution of two independent problems with a single dipole in each and not the problem of the two dipoles at the same time, thus avoiding explicitly mutual interaction. For these reasons, the ORT has to be viewed as a *mathematical symmetry relation* embedded in Maxwell's equations, which relates the field solution of *two independent electromagnetic problems*.

4.5. FORMAL DERIVATION OF SERS EM ENHANCEMENTS

We now have all the tools necessary to tackle one of the main subjects of this chapter: the rigorous derivation of the EM-SERS enhancements.

4.5.1. Definitions, notations, and assumptions

Intrinsic analyte properties

We have already provided rigorous definitions of the SERS enhancement factors (EFs) and cross-sections in Section 4.1. We will now describe how the electromagnetic contribution to these EFs can be calculated from standard EM theory. We will mainly focus here on the *SERS cross-section* and *single-molecule enhancement factor* (SMEF), as defined in Section 4.1.4.

Note that these definitions refer to *one given vibrational mode* of the molecule under study. As discussed extensively in Chapter 2, the Raman properties of this mode are characterized by its vibrational energy $\hbar\omega_v$ and its *Raman polarizability tensor*, denoted $\hat{\alpha}$ hereafter. The latter depends on the excitation frequency ω_L, but we will omit this dependence in the notation for simplicity. The Raman photons are emitted at the Raman frequency ω_R ($\omega_R = \omega_L - \omega_v$ for Stokes processes).

As mentioned in Chapter 2, the phenomenological description of Raman scattering is sufficient (and better suited) to the case of SERS, and we will focus exclusively on this approach. We therefore assume the Raman polarizability tensor $\hat{\alpha}$ as known. This can either be measured, guessed empirically, deduced from the Raman tensor (if known or calculated), or obtained from more complex quantum treatments for resonant-Raman-scattering conditions. A major assumption here is that *the Raman polarizability tensor is the same under normal and SERS conditions.* The validity of this should be assessed on a case-by-case basis; any modification to it would act in addition to the EM enhancement mechanisms described here (and would be therefore classified as a chemical enhancement, see Section 4.8).

Let us also recall that the SERS EFs were defined with respect to the non-SERS properties of *the same molecule embedded in the same environment.* Hence, any local field correction factor (L_M or similar, as discussed in Section 2.4.5) cancels out in all the expressions of any enhancement factors (including the SERS EFs). Consequently, we will omit for simplicity in the following any local field correction.

Finally, as for Raman scattering, the symmetry of $\hat{\alpha}$ plays an important role in the description of the SERS process. It is, accordingly, useful to recall here some of the most important definitions from Chapter 2, in particular:

- The *magnitude of the Raman polarizability tensor*, $\tilde{\alpha}$ was defined in Eq. (2.63) in terms of the two tensor invariants. It is a *scalar* and enables us to express the absolute differential Raman cross-section as Eq. (2.64), which we rewrite here:

$$\boxed{\frac{\mathrm{d}\sigma_R}{\mathrm{d}\Omega} = \frac{\omega_R^4}{16\pi^2(\epsilon_0)^2c^4}\tilde{\alpha}^2.} \tag{4.52}$$

- The *Raman depolarization ratio*, ρ, another *scalar* that directly depends on the two tensor invariants (see Eq. (2.61)), characterizes (partially) the symmetry of the Raman polarizability tensor. It can be readily measured with polarized Raman experiments.

- The *normalized Raman polarizability tensor* is defined as:

$$\boxed{\hat{\alpha}_N = \frac{\hat{\alpha}}{\tilde{\alpha}}.} \tag{4.53}$$

  It is a tensor characterizing fully the *symmetry* of $\hat{\alpha}$, *independently of its magnitude.*

These definitions will be useful to account properly for the normalization of SERS EFs with respect to the non-SERS cross-section.

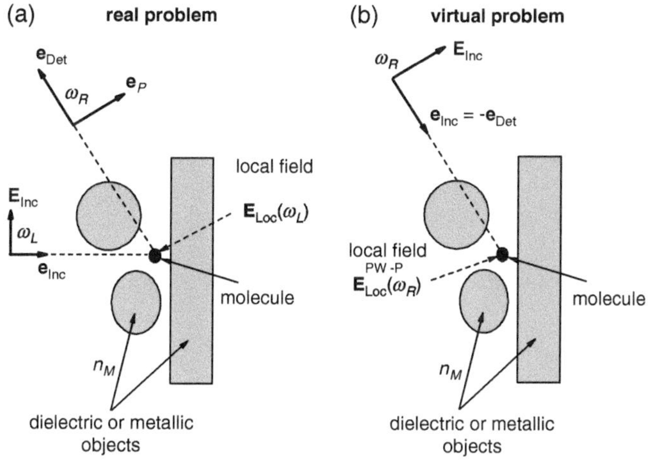

Figure 4.5. Schematic representation of the SERS EM problem (for polarized detection), which is solved by considering both the real excitation problem (a) and the 'virtual' plane-wave-excitation problem for re-emission (b). Using the optical reciprocity theorem, the 'virtual' problem (b) allows the calculation of the SERS EF without solving the EM problem of emission from a localized dipole at the molecule position.

The EM problems

Let us consider a fixed molecule at a given position O, embedded in a dielectric (of refractive index n_M), and in the vicinity of a number of interfaces (dielectric and/or metallic objects). The situation is schematically depicted in Fig. 4.5. The system is excited by an incident laser at frequency ω_L, incoming along a direction $\mathbf{e}_{\text{Inc}}$ and polarized along $\mathbf{e}_E \perp \mathbf{e}_{\text{Inc}}$. This will be modeled as a plane wave with complex electric field $\mathbf{E}_{\text{Inc}} = E_{\text{Inc}} e^{ik_M \mathbf{e}_{\text{Inc}} \cdot \mathbf{r}} \mathbf{e}_E$. The incident power density S_{Inc} is related to the field amplitude E_{Inc} by $S_{\text{Inc}} = n_M \epsilon_0 c |E_{\text{Inc}}|^2/2$ (see Appendix F). This excitation interacts with the objects and creates at the molecule position a local electric field $\mathbf{E}_{\text{Loc}}(\omega_L)$, which by linearity can be normalized against the incident amplitude to give: $\tilde{\mathbf{E}}_{\text{Loc}}(\omega_L) = \mathbf{E}_{\text{Loc}}(\omega_L)/E_{\text{Inc}}$.

We will study the SERS (Raman) signal emitted by the molecule in a given direction Ω_{Det}, along unit vector $\mathbf{e}_{\text{Det}}$. This signal can also be analyzed along two perpendicular polarizations defined by $\mathbf{e}_{P1}$ and $\mathbf{e}_{P2}$, themselves perpendicular to $\mathbf{e}_{\text{Det}}$. The corresponding differential radiated SERS powers will be denoted by $\mathrm{d}P^P_{\text{SERS}}/\mathrm{d}\Omega$, with $P = P1, P2$.

To apply the ORT, we will also need to consider the *'virtual' problem* of plane-wave excitation with frequency ω_R incoming along $-\mathbf{e}_{\text{Det}}$ and polarized along $\mathbf{e}_P$. We denote $\tilde{\mathbf{E}}_{\text{Loc}}^{\text{PW}-P}(\omega_R)$ the local field solution to this problem at the molecule position (normalized against the incident plane-wave amplitude as before). This 'virtual' problem has no relation whatsoever with the real excitation here (which is along $\mathbf{e}_{\mathrm{Inc}}$) but will be used as a mathematical trick with the ORT to study the re-emission problem, as discussed in Section 4.4.6.

Armed with these definitions, we can now describe rigorously the most important mechanisms of electromagnetic enhancement in SERS.

4.5.2. The SERS EM enhancement: general case

Let us first start with the most general case.

SMEF derivation

Following the phenomenological treatment of Raman scattering, the local field at the molecule position induces a Raman dipole:

$$\mathbf{p} = \hat{\alpha} \cdot \mathbf{E}_{\mathrm{Loc}}(\omega_L), \tag{4.54}$$

oscillating at the Raman frequency ω_R.

The differential radiative power emitted at frequency ω_R in direction Ω_{Det} is modified by the environment, according to the directional radiative enhancement factor $M^d_{\mathrm{Rad}}(\Omega_{\mathrm{Det}}, \omega_R, \mathbf{e}_p)$. We write explicitly here the dependence of this factor on the dipole orientation $\mathbf{e}_p$ to emphasize it. From the definition of $M^d_{\mathrm{Rad}-P}$ in Eq. (4.35), we can write:

$$\frac{\mathrm{d}P^P_{\mathrm{SERS}}}{\mathrm{d}\Omega} = \frac{n_M \omega_R^4}{32\pi^2 \epsilon_0 c^3} |\mathbf{p}|^2 \, M^d_{\mathrm{Rad}-P}(\Omega_{\mathrm{Det}}, \omega_R, \mathbf{e}_p). \tag{4.55}$$

Writing the incident power density as $\mathrm{S}_{\mathrm{Inc}} = n_M \epsilon_0 c |E_{\mathrm{Inc}}|^2/2$, the *differential SERS cross-section for P-polarized detection* is then obtained from its definition in Eq. (4.7) as:

$$\frac{\mathrm{d}\sigma^P_{\mathrm{SERS}}}{\mathrm{d}\Omega} = \frac{\omega_R^4}{16\pi^2 \epsilon_0^2 c^4} |\hat{\alpha} \cdot \tilde{\mathbf{E}}_{\mathrm{Loc}}(\omega_L)|^2 \, M^d_{\mathrm{Rad}-P}(\Omega_{\mathrm{Det}}, \omega_R, \mathbf{e}_p). \tag{4.56}$$

Note that in this expression, the factor $M^d_{\mathrm{Rad}-P}$ depends on the orientation of the induced Raman dipole $\mathbf{e}_p$, which itself depends on $\hat{\alpha}$ and $\mathbf{E}_{\mathrm{Loc}}(\omega_L)$ through Eq. (4.54), making it very impractical. This inconvenience can be removed by using Eq. (4.51), derived from the ORT, from which we deduce:

$$\begin{aligned} |\mathbf{p}|^2 \, M^d_{\mathrm{Rad}-P}(\Omega_{\mathrm{Det}}, \omega_R, \mathbf{e}_p) &= |\tilde{\mathbf{E}}^{\mathrm{PW}-P}_{\mathrm{Loc}}(\omega_R) \cdot \mathbf{p}|^2 \\ &= \left|\tilde{\mathbf{E}}^{\mathrm{PW}-P}_{\mathrm{Loc}}(\omega_R) \cdot \hat{\alpha} \cdot \mathbf{E}_{\mathrm{Loc}}(\omega_L)\right|^2, \end{aligned} \tag{4.57}$$

where $\tilde{\mathbf{E}}_{\mathrm{Loc}}^{\mathrm{PW}-P}$ is the normalized local field solution of the 'virtual' problem. We then obtain the more convenient expression:

$$\frac{\mathrm{d}\sigma_{\mathrm{SERS}}^{P}}{\mathrm{d}\Omega} = \frac{\omega_R^4}{16\pi^2\epsilon_0^2 c^4} \left| \tilde{\mathbf{E}}_{\mathrm{Loc}}^{\mathrm{PW}-P}(\omega_R) \cdot \hat{\alpha} \cdot \tilde{\mathbf{E}}_{\mathrm{Loc}}(\omega_L) \right|^2 . \tag{4.58}$$

The *single-molecule enhancement factor for P-polarized detection* then follows simply as the ratio of this to the non-SERS differential cross-section in Eq. (4.52), giving:

$$\boxed{\mathrm{SMEF}^P = \left| \tilde{\mathbf{E}}_{\mathrm{Loc}}^{\mathrm{PW}-P}(\omega_R) \cdot \hat{\alpha}_N \cdot \tilde{\mathbf{E}}_{\mathrm{Loc}}(\omega_L) \right|^2 ,} \tag{4.59}$$

where $\hat{\alpha}_N$ is the *normalized Raman polarizability tensor* defined in Eq. (4.53). Note that this SMEF only depends on the symmetry of the Raman tensor, not on its magnitude, as expected for an enhancement factor.

Similar expressions can be obtained for the differential SERS cross-section and SMEF in the case of unpolarized detection, by summing the results for the two perpendicular polarization detections, $P1$ and $P2$. For the SMEF for example:

$$\boxed{\begin{aligned}\mathrm{SMEF} = &\left| \tilde{\mathbf{E}}_{\mathrm{Loc}}^{\mathrm{PW}-P1}(\omega_R) \cdot \hat{\alpha}_N \cdot \tilde{\mathbf{E}}_{\mathrm{Loc}}(\omega_L) \right|^2 \\ &+ \left| \tilde{\mathbf{E}}_{\mathrm{Loc}}^{\mathrm{PW}-P2}(\omega_R) \cdot \hat{\alpha}_N \cdot \tilde{\mathbf{E}}_{\mathrm{Loc}}(\omega_L) \right|^2 .\end{aligned}} \tag{4.60}$$

Link with the local field intensity enhancement factors

From this expression of the SMEF, it is clear that in the most general case, the local field enhancement (for excitation at ω_L) and the radiative enhancement (for re-emission at ω_R) cannot be entirely decoupled. This is because the radiative enhancement depends on the induced dipole orientation, itself directly related to the local field polarization.

Nevertheless, except for these orientation/polarization effects, the magnitude of the enhancement remains primarily dominated by the magnitude of both the local field and radiative enhancements. To highlight this, we can write the normalized local field as:

$$\tilde{\mathbf{E}}_{\mathrm{Loc}}(\omega) = \tilde{E}_{\mathrm{Loc}}(\omega)\mathbf{e}_{\mathrm{Loc}}(\omega), \tag{4.61}$$

where $\mathbf{e}_{\mathrm{Loc}}(\omega)$ is, by definition, a unit vector (possibly complex), characterizing the *local field polarization.* The local field intensity enhancement factor then takes the form:

$$M_{\mathrm{Loc}}(\omega) = |\tilde{\mathbf{E}}_{\mathrm{Loc}}(\omega)|^2 = |\tilde{E}_{\mathrm{Loc}}(\omega)|^2. \tag{4.62}$$

A similar expression can be written for the 'virtual' PWE problems:

$$\tilde{\mathbf{E}}_{\mathrm{Loc}}^{\mathrm{PW}-P}(\omega) = \tilde{E}_{\mathrm{Loc}}^{\mathrm{PW}-P}(\omega)\mathbf{e}_{\mathrm{Loc}}^{\mathrm{PW}-P}(\omega), \tag{4.63}$$

and for the corresponding $M_{\mathrm{Loc}}^{\mathrm{PW}-P}(\omega)$.

Inserting these expressions into Eq. (4.59), the SMEF for polarized detection can then be rewritten as:

$$\boxed{\mathrm{SMEF}^P = M_{\mathrm{Loc}}(\omega_L)\, M_{\mathrm{Loc}}^{\mathrm{PW}-P}(\omega_R)\, T(\hat{\alpha}_N, \mathbf{e}_{\mathrm{Loc}}(\omega_L), \mathbf{e}_{\mathrm{Loc}}^{\mathrm{PW}-P}(\omega_R)),} \tag{4.64}$$

where:

$$\boxed{T(\hat{\alpha}_N, \mathbf{e}_L, \mathbf{e}_R) = \left|\mathbf{e}_R \cdot \hat{\alpha}_N \cdot \mathbf{e}_L\right|^2.} \tag{4.65}$$

Let us pause and discuss this expression further. It contains three multiplicative factors:

- $M_{\mathrm{Loc}}(\omega_L)$ represents the LFIEF for *excitation* of the Raman dipole. It has essentially the same role as described earlier in simple terms in Section 4.3. It characterizes the enhancement in the intensity of the *'real' local electric field* felt by the molecule.
- $M_{\mathrm{Loc}}^{\mathrm{PW}-P}(\omega_R)$ characterizes the magnitude of the *radiation enhancement* for the Raman dipole. Thanks to the ORT, it can be written here as the LFIEF that would be obtained for a *'virtual'* problem with plane-wave excitation incoming from the direction of detection and polarized along the polarization of detection $\mathbf{e}_P$.
- $T(\hat{\alpha}_N, \mathbf{e}_{\mathrm{Loc}}(\omega_L), \mathbf{e}_{\mathrm{Loc}}^{\mathrm{PW}-P}(\omega_R))$ is an additional factor, which couples the excitation and re-emission problems through the normalized Raman polarizability tensor (characterizing the symmetry of the Raman tensor and the molecular orientation). In a first approximation, T has only a secondary influence on the overall magnitude of the SMEF, except for some specific combinations of Raman tensor symmetry and molecule orientation for which it would cancel out to zero. It can nevertheless account for changes in the *relative intensities* of Raman peaks of different symmetries under SERS conditions. We will call it the *surface selection rules factor*, and now discuss some of its properties in more detail.

Surface selection rules

The surface selection rules factor (SSRF) is admittedly a fairly complicated factor in general, involving the polarization of the local field for both the real and 'virtual' problems. In fact, in most SERS studies, experimental or theoretical, it is largely ignored. The possible additional influence of the *local field polarization* (in contrast to the main effect of the local field intensity enhancement) had, however, been recognized early by Moskovits [52]. This effect was originally studied for molecules adsorbed on *planar metallic surfaces*, for which analytic EM solutions exist. For a fixed molecular adsorption on the surface, changes in the relative intensities of Raman modes with different Raman tensor symmetries were predicted and, this was dubbed *surface selection rules* (SSRs). Only recently [48,49], this study has been re-discussed in a more modern context and extended to more realistic SERS substrates, such as those containing hot-spots. Surface selection rules on planar surfaces possess the big advantage of simplicity (and the possibility of having analytic solutions for the EM problem), but the SERS enhancement factors on them are too small for any real application of the effect. Therefore, it has primarily an 'academic' interest in that sense.

Surface selection rules represent the starting point for any studies of polarization effects in SERS and possible investigations of molecular adsorption or molecular orientation using SERS. Within the formal derivation described above, all these effects are contained in the surface selection rules factor T. Note that T depends on the local field polarization at the molecule position for both the real excitation ($\mathbf{e}_L = \mathbf{e}_{\mathrm{Loc}}(\omega_L)$) and the 'virtual' problem ($\mathbf{e}_R = \mathbf{e}_{\mathrm{Loc}}^{\mathrm{PW}-P}(\omega_R)$)). The orientation of these vectors should ultimately come from the EM solution of these problems. Taking these as parameters we can, nevertheless, study some general properties of T.

Firstly, T must remain positive, and may be zero for an appropriate combination of $\mathbf{e}_L$ and $\mathbf{e}_R$. Moreover, since by definition the vectors in Eq. (4.65) are unit vectors, one can show that for a given tensor $\hat{\alpha}_N$, the maximum possible value of T is $|\mu_{\mathrm{Max}}|^2$ where μ_{Max} is the largest eigenvalue of $\hat{\alpha}_N$. Finally, when considering all possible Raman tensors[11], one can show that:

$$0 \leq T(\hat{\alpha}_N, \mathbf{e}_L, \mathbf{e}_R) \leq 15/4, \tag{4.66}$$

the largest value being obtained for a uni-axial tensor when both $\mathbf{e}_L$ and $\mathbf{e}_R$ are along the tensor axis. This inequality justifies the statement made earlier that T had only, in most cases, a secondary effect on the overall magnitude of the SMEF (this effect might be dramatic, but only in highly specific cases, e.g. when $T \approx 0$).

[11] We restrict ourselves here to real symmetric tensors.

Moreover, these arguments refer to the case of a fixed molecule. We can also consider the case of orientation averaging. We will restrict to the most general case of full orientation averaging, denoted $\langle \dots \rangle$ (as opposed to averaging over a limited number of degrees of freedom, denoted $[\dots]$). It is then possible to show that the orientation-averaged SSRF (denoted as $\langle T(\hat{\alpha}_N, \mathbf{e}_L, \mathbf{e}_R) \rangle$) is:

$$\langle T(\hat{\alpha}_N, \mathbf{e}_L, \mathbf{e}_R) \rangle = \frac{\rho + (1-\rho)|\mathbf{e}_L \cdot \mathbf{e}_R|^2}{1+\rho}, \tag{4.67}$$

where ρ is the non-SERS depolarization ratio. The exact value therefore depends on the relative orientation of $\mathbf{e}_L$ and $\mathbf{e}_R$, but remains in the range:

$$\frac{\rho}{1+\rho} \leq \langle T(\hat{\alpha}_N, \mathbf{e}_L, \mathbf{e}_R) \rangle \leq \frac{1}{1+\rho}, \tag{4.68}$$

the maximum being obtained for $\mathbf{e}_R = \mathbf{e}_L$. A corollary is that *regardless of the specific type of Raman tensor, we always have* $0 \leq \langle T \rangle \leq 1$.

Other SERS EFs and other SERS cross-sections

From the SMEF expressions given in this section, the orientation-averaged SMEF (OASMEF) can be obtained by averaging over the *allowed* molecule orientation. Note that this averaging should only affect the surface selection rules factor, which must then be replaced by $[T]$. This can be case specific, since some molecules may have a fixed orientation, while others may have one (for planar adsorption for example), or even two or three (for fully random orientation) degrees of rotational freedom.

Moreover, the average SERS substrate defined as the SERS substrate EF (SSEF) can in principle be obtained by *spatial averaging* of the OASMEF over the entire metallic surface as in Eq. (4.10). Note that all three factors in Eq. (4.64), $M_{\mathrm{Loc}}(\omega_L)$, $M_{\mathrm{Loc}}^{\mathrm{PW}-P}(\omega_R)$, and $T(\hat{\alpha}_N, \mathbf{e}_{\mathrm{Loc}}(\omega_L), \mathbf{e}_{\mathrm{Loc}}^{\mathrm{PW}-P}(\omega_R))$, depend on position $\mathbf{r}$. They cannot therefore (in general) be decoupled in the spatial averaging, and the SSEF for polarized detection is therefore:

$$\begin{aligned} \mathrm{SSEF}^P = \{ & M_{\mathrm{Loc}}(\mathbf{r}, \omega_L) M_{\mathrm{Loc}}^{\mathrm{PW}-P}(\mathbf{r}, \omega_R) \\ & \times \left[T(\hat{\alpha}_N, \mathbf{e}_{\mathrm{Loc}}(\mathbf{r}, \omega_L), \mathbf{e}_{\mathrm{Loc}}^{\mathrm{PW}-P}(\mathbf{r}, \omega_R)) \right] \}, \end{aligned} \tag{4.69}$$

where $\{\dots\}$ denotes spatial averaging, and $[\dots]$ represents averaging over allowed molecular orientations. The SSEF for unpolarized detection can be obtained simply as before by summing the SSEF^P for two perpendicular polarization detections $P1$ and $P2$.

Finally, all the expressions so far relate to the differential SERS power, i.e. for detection in a single direction. We simply mention here the possibility of extending these to the other types of SERS cross-sections defined in Section 4.1. For example, the integrated radiative SERS cross-section, σ_{SERS}, can in principle be obtained by integrating the differential SERS cross-section over the full solid angle Ω. This is however not practical within the ORT approach. One can instead go back to Eq. (4.56) and write a similar expression for σ_{SERS}:

$$\sigma_{\text{SERS}} = \frac{\omega_R^4}{6\pi\epsilon_0^2 c^4} |\hat{\alpha} \cdot \tilde{\mathbf{E}}_{\text{Loc}}(\omega_L)|^2 M_{\text{Rad}}(\omega_R, \mathbf{e}_p). \tag{4.70}$$

Similarly, the *total* SERS cross-section (which includes non-radiative SERS processes) can be expressed as:

$$\sigma_{\text{Tot}} = \frac{\omega_R^4}{6\pi\epsilon_0^2 c^4} |\hat{\alpha} \cdot \tilde{\mathbf{E}}_{\text{Loc}}(\omega_L)|^2 M_{\text{Tot}}(\omega_R, \mathbf{e}_p). \tag{4.71}$$

Note that in both cases, the enhancement factor for re-emission M_{Rad} or M_{Tot} depends on the orientation of the induced dipole $\mathbf{e}_p$, which itself depends on the local field polarization. These cross-sections are in fact rarely used in a SERS context and are included here only for completeness.

4.5.3. SERS EM enhancements in the back-scattering configuration

The derivation of the previous section is very general and can be applied to most SERS experiment configurations. In order to link these expressions with the commonly-used $|E|^4$-approximation, and to discuss a few examples, we will now restrict ourselves to a specific configuration, namely, the *back-scattering configuration*. This is done for three main reasons: (i) it is one of the most widely used for SERS experiments (in particular in Raman microscopes, where the delivery and collection optics are the same), (ii) it makes the theoretical description much simpler since the 'virtual' (re-emission) problem corresponds to plane-wave excitation along the same direction as the real excitation, (iii) it allows a formal justification of the $|E|^4$-approximation, under an appropriate set of conditions to be defined.

SMEF in the back-scattering configuration

In the back-scattering (BS) configuration, excitation and detection are along the same direction. This corresponds in the real problem of Fig. 4.5(a) to $\mathbf{e}_{\text{Inc}} = -\mathbf{e}_{\text{Det}}$. We choose them along the x-axis in the following. The incident wave is taken as a plane wave polarized along z (necessarily $\perp$ to x). The determination of the local field for the real excitation therefore requires the solution of one PWE problem, incoming along x and polarized along z.

To fully characterize the emitted SERS signal in the back-scattering direction, we have to consider two polarizations for detection (which is also along x): we choose them either parallel to the incident polarization (along z) or perpendicular to it (along y). This re-emission problem then requires the solution of *two* 'virtual' plane-wave-excitation problems, with incoming beam along x and polarization along z *and* y. The first one is in fact the same as the real excitation problem, while only the polarization changes for the second. Let us therefore denote $M^P_{\mathrm{Loc}}(\omega)$ and $\mathbf{e}^P_{\mathrm{Loc}}(\omega)$ the LFIEF and local field polarization for PWE incoming along x and polarized along $P = Z$ or $P = Y$. $P = Z$ applies to the real excitation and 'virtual' problem for parallel detection, while $P = Y$ is for the 'virtual' problem for perpendicular detection.

Following the results obtained earlier in the general case, the SMEF in the BS configuration for parallel and perpendicular detections then take the form:

$$\begin{aligned}\mathrm{SMEF}^{\parallel}_{\mathrm{BS}} &= M^Z_{\mathrm{Loc}}(\omega_L)\; M^Z_{\mathrm{Loc}}(\omega_R) T^{\parallel}_{\mathrm{BS}} \\ &\quad \text{with} \quad T^{\parallel}_{\mathrm{BS}} = |\mathbf{e}^Z_{\mathrm{Loc}}(\omega_R) \cdot \hat{\alpha}_N \cdot \mathbf{e}^Z_{\mathrm{Loc}}(\omega_L)|^2, \end{aligned} \tag{4.72}$$

and

$$\begin{aligned}\mathrm{SMEF}^{\perp}_{\mathrm{BS}} &= M^Z_{\mathrm{Loc}}(\omega_L) M^Y_{\mathrm{Loc}}(\omega_R) T^{\perp}_{\mathrm{BS}} \\ &\quad \text{with} \quad T^{\perp}_{\mathrm{BS}} = |\mathbf{e}^Y_{\mathrm{Loc}}(\omega_R) \cdot \hat{\alpha}_N \cdot \mathbf{e}^Z_{\mathrm{Loc}}(\omega_L)|^2. \end{aligned} \tag{4.73}$$

Note also that for unpolarized detection, the SMEF is simply the sum of the two:

$$\mathrm{SMEF}_{\mathrm{BS}} = \mathrm{SMEF}^{\parallel}_{\mathrm{BS}} + \mathrm{SMEF}^{\perp}_{\mathrm{BS}}. \tag{4.74}$$

Comparison with the $|E|^4$-approximation

The $|E|^4$-approximation to the SERS EF (discussed in Section 4.3) would be written here as:

$$\mathrm{SMEF}_{\mathrm{BS}} \approx M^Z_{\mathrm{Loc}}(\omega_L) M^Z_{\mathrm{Loc}}(\omega_R). \tag{4.75}$$

The exact expressions given above have some similarities with this approximation but differ from it in at least two aspects:

- Firstly, the factor $M^Z_{\mathrm{Loc}}(\omega_R)$ appears only *for parallel polarized detection*. For perpendicular polarized detection, it should be replaced by $M^Y_{\mathrm{Loc}}(\omega_R)$, which may be very different.
- Secondly, the surface selection rules factor, T, does not appear in the $|E|^4$-approximation.

For the $|E|^4$-approximation to be exact, we therefore require that $\mathrm{SMEF}^{\perp}_{\mathrm{BS}} = 0$ (or $\ll \mathrm{SMEF}^{\parallel}_{\mathrm{BS}}$) and $T^{\parallel}_{\mathrm{BS}} = 1$. The only general case where this is true is for *an isotropic Raman tensor*, with the additional assumptions that (i) $\mathbf{e}^{Z}_{\mathrm{Loc}}(\omega_R) \approx \mathbf{e}^{Z}_{\mathrm{Loc}}(\omega_L)$, i.e. the local field polarization does not change much between the laser and Raman frequency; and (ii) $M^{Y}_{\mathrm{Loc}}(\omega_R) = 0$ (or $\ll M^{Z}_{\mathrm{Loc}}(\omega_R)$) or $\mathbf{e}^{Y}_{\mathrm{Loc}}(\omega_R) \perp \mathbf{e}^{Z}_{\mathrm{Loc}}(\omega_L)$.

Only under these conditions and in the back-scattering configuration is the $|E|^4$-approximation an *exact* description of the SMEF. As a matter of fact, this situation is rarely achieved in practice. Fortunately, the discrepancies between the approximation and the exact results are most of the time secondary, at least for the back-scattering configuration. This happens for reasons that we can now pin down more accurately:

- In general, the SMEF for parallel detection $\mathrm{SMEF}^{\parallel}_{\mathrm{BS}}$ is much larger or at least of the same order as $\mathrm{SMEF}^{\perp}_{\mathrm{BS}}$. It is therefore a good estimate of the SMEF, even for unpolarized detection (a very common configuration). The main reason for this is that, in general, $M^{Z}_{\mathrm{Loc}}(\omega)$ is larger (sometimes much larger) than $M^{Y}_{\mathrm{Loc}}(\omega)$; otherwise it would simply make more sense to carry out the SERS experiment with excitation polarized along Y to maximize the signal. Obviously, this argument fails when $T^{\parallel}_{\mathrm{BS}} \approx 0$ for which $\mathrm{SMEF}^{\parallel}_{\mathrm{BS}} \approx 0$. This however requires specific conditions, and cannot be considered as a typical situation.

- In fact, as discussed earlier, in many instances, the SSRF $T^{\parallel}_{\mathrm{BS}}$ (or its orientation-averaged equivalent) is of the order of 1 (within a factor of 2, say, except in exceptional circumstances).

In summary, the $|E|^4$-approximation provides in most situations a good (and 'simple') estimate of the magnitude of the SMEF in the back-scattering configuration, say within a factor $\approx$2. One should, however, be aware of its limitations:

- Since it ignores the SSRF, T, it cannot be used to understand the (typically small) differences between the SMEF of Raman modes of the same molecule but with different symmetries, i.e. to study the *surface selection rules* in SERS (for example changes in the relative intensities of the peaks under SERS conditions).

- It cannot predict any *polarization properties of the SERS signal*, such as the SERS depolarization ratio.

- Finally, it applies to *the back-scattering configuration only.*

A 'historical' note

It is worth highlighting the fact that the general description of the SERS EM enhancement factor as given above is fairly recent, if not new for some of its aspects. Most SERS studies, experimental or theoretical, have almost exclusively used the $|E|^4$-approximation, and in most cases without justification. There are a few theoretical studies where the exact EM descriptions was studied, but only in the simplest cases such as a planar surface [52] or a sphere [164]. The optical reciprocity theorem has, in fact, rarely been used in the context of SERS (an example is Ref. [180] where it was applied to study silver gratings). It is only recently that the ORT was put forward as a means for a general formulation of the SERS EM problem [45], and it has since been applied to (for example) the study of surface selection rules at a SERS hot-spot [48] or the study of polarization effects in SERS [49]. The presentation given above is a synthesis and in some instances an extension of these recently proposed concepts. We believe it is, at the time of writing, the best approach to a rigorous and general treatment of SERS EM enhancements.

4.6. SURFACE-ENHANCED FLUORESCENCE (SEF)

The enhancement mechanisms that are relevant to SERS are also present in several other processes such as absorption, fluorescence, or other types of scattering. In this section, we will briefly show how the formalism developed for SERS can be applied to these processes.

Many applications, including SERS, make use of fluorophores, sometimes in complex environments. The problem of the modification of fluorescence in a SERS-type environment, called *surface-enhanced fluorescence* (SEF)[12], is an active area of research. In addition to its many applications as a method to enhance existing fluorescence-based techniques, it is also relevant to SERS because many SERS probes are, indeed, fluorescent dyes.

4.6.1. Similarities and differences between SEF and SERS

Fluorescence is very similar to scattering; it involves absorption of a photon, followed by emission. The same enhancement mechanisms as that derived for SERS are therefore expected. The absorption should be subject to the local field enhancements, and the emission, which is simply spontaneous emission, should follow the radiative enhancements. *The crucial difference is that scattering (SERS) is instantaneous, while fluorescence is a multi-step process.* This means that for SERS, both enhancements (local field

[12] In recent years, the denomination metal-enhanced fluorescence (MEF) has also appeared in the literature.

and radiative) contribute to the SERS cross-section. For fluorescence, the local field enhancement does also lead to a modification of the absorption cross-section, but the radiative (SE-type) enhancements only lead to a *modification of the decay rates* (radiative and non-radiative), which then compete with each other. To put it differently, once a photon is absorbed (possibly with enhanced absorption) and excites an electron in fluorescence, no enhancement mechanism can lead to more energy being extracted from this single excited electron, but only to energy being extracted *faster*. After all, the quantum yield cannot be larger than 1, and for a dye with a good free-space quantum yield, the only source of enhancement is therefore in the absorption cross-section.

Let us now put these qualitative arguments on a firmer footing. The basic principles of fluorescence were presented in simple terms in Section 2.2.4 and in more detail in Section 2.3. We use here the same notations and extend the treatment to fluorophores in the vicinity of optically active objects. The two main characteristics of a fluorophore, its absorption cross-section and its quantum yield, are both modified by the environment and we now apply the arguments developed in this chapter to the description of these modifications.

4.6.2. Modified (enhanced) absorption

Phenomenological approach to absorption

Optical absorption can be described phenomenologically in a similar fashion as Raman scattering by considering the *linear optical polarizability tensor*, $\hat{\alpha}_L(\omega)$ of the molecule, already discussed in Section 2.4.3.

Excitation by an incident field $\mathbf{E}$ at frequency ω_L results in an induced dipole $\mathbf{p}_L = \hat{\alpha}_L(\omega_L) \cdot \mathbf{E}$ oscillating at the same frequency ω_L (and therefore different from the induced Raman dipole discussed earlier). Within the classical phenomenological approach, the power absorbed P_{Abs} by the molecule is equal to the power absorbed by this induced dipole, i.e.:

$$P_{\text{Abs}} = \frac{\omega_L}{2} \text{Im}\left(\mathbf{E}^* \cdot \mathbf{p}_L\right) = \frac{\omega_L}{2} \text{Im}\left(\mathbf{E}^* \cdot \hat{\alpha}_L(\omega_L) \cdot \mathbf{E}\right). \tag{4.76}$$

The most common type of linear polarizability tensor is the uni-axial tensor along an axis $\mathbf{e}_m$ (attached to the molecule). The reason is that most common molecular absorption processes occur as a result of dipolar interaction with the electromagnetic field. In the rest of this section, for simplicity we will restrict ourselves to this case, which simplifies greatly the surface selection rules factor. We therefore take: $\hat{\alpha}_L = \alpha_L \mathbf{e}_m \otimes \mathbf{e}_m$, where α_L is the (complex) scalar polarizability (implicitly at frequency ω_L). The absorbed power can then be simplified to:

$$P_{\text{Abs}} = \frac{\omega_L}{2} \left|\mathbf{e}_m \cdot \mathbf{E}\right|^2 \text{Im}(\alpha_L). \tag{4.77}$$

Free-space absorption

For a molecule with fully random orientation in free-space (i.e. in a dielectric of refractive index n_M), excited with a plane wave with electric field amplitude E_{Inc}, we can then deduce:

$$P_{\text{Abs}}^M = \frac{L_M^{1/2}\omega_L}{2}|E_{\text{Inc}}|^2\frac{\text{Im}\,(\alpha_L)}{3}, \tag{4.78}$$

where we have included explicitly the local field correction (see Section 2.4.5), and where the factor of 1/3 arises from the orientation averaging ($\langle|\mathbf{e}_m\cdot\mathbf{E}|^2\rangle = 1/3$). Since the power density for excitation is $S_{\text{Inc}} = n_M\epsilon_0 c/2|E_{\text{Inc}}|^2$, we also deduce the free-space absorption cross-section as:

$$\sigma_{\text{Abs}}^M = \frac{P_{\text{Abs}}}{S_{\text{Inc}}} = \frac{(L_M^{1/2})\omega_L}{n_M\epsilon_0 c}\frac{\text{Im}\,(\alpha_L)}{3}. \tag{4.79}$$

Absorption enhancement factor

The same expressions can be written for a fixed molecule close to surfaces. Using the notations of the previous sections, the field felt by the molecule is then $\mathbf{E}_{\text{Loc}} = E_{\text{Loc}}\mathbf{e}_{\text{Loc}}$. We write here only the resulting *absorption enhancement factor* defined as $M_{\text{Abs}}(\omega_L) = P_{\text{Abs}}/P_{\text{Abs}}^M = \sigma_{\text{Abs}}/\sigma_{\text{Abs}}^M$, i.e.

$$\boxed{M_{\text{Abs}}(\omega_L) = M_{\text{Loc}}(\omega_L)\;T_{\text{Abs}}, \quad \text{with } T_{\text{Abs}} = 3|\mathbf{e}_m\cdot\mathbf{e}_{\text{Loc}}(\omega_L)|^2.} \tag{4.80}$$

$M_{\text{Loc}} = |\mathbf{E}_{\text{Loc}}|^2/|E_{\text{Inc}}|^2$ is the local field intensity enhancement factor encountered earlier. The second term, T_{Abs}, is the absorption equivalent of the surface selection rules factor obtained for SERS. Note that for a fully random molecular orientation, $\langle T_{\text{Abs}}\rangle = 1$ and therefore $\langle M_{\text{Abs}}\rangle = M_{\text{Loc}}$.

4.6.3. Modified fluorescence quantum yield

Let us now consider a fluorophore with a free-space radiative decay rate Γ_{Rad}^M, an intrinsic non-radiative decay rate Γ_{NR}^M, and a total free-space decay rate $\Gamma_{\text{Tot}}^M = \Gamma_{\text{Rad}}^M + \Gamma_{\text{NR}}^M$ (see Section 4.4.2 and Table 4.3). As explained in Section 2.3.2, the fluorescence quantum yield is by definition the proportion of excited electrons that decay radiatively to the ground state. In free-space, we therefore have (Eq. (2.25)):

$$Q^M = \frac{\Gamma_{\text{Rad}}^M}{\Gamma_{\text{Tot}}^M} = \frac{\Gamma_{\text{Rad}}^M}{\Gamma_{\text{Rad}}^M + \Gamma_{\text{NR}}^M}. \tag{4.81}$$

Following Section 4.4, close to a surface (a metal in particular), there are two effects that will affect this quantum yield: (i) the radiative decay rate Γ_{Rad} is modified by a factor M_{Rad}: $\Gamma_{\text{Rad}} = M_{\text{Rad}}\Gamma^M_{\text{Rad}}$, and (ii) there is, in addition to the intrinsic non-radiative decay with rate Γ^M_{NR} (assumed to be non-modified here), another non-radiative path corresponding to emission that is subsequently absorbed in the metal (i.e. non-radiative emission), with a non-radiative EM decay rate $\Gamma^{\text{EM}}_{\text{NR}} = M_{\text{NR}}\Gamma^M_{\text{Rad}}$. The modified quantum yield is therefore:

$$Q = \frac{\Gamma_{\text{Rad}}}{\Gamma_{\text{Rad}} + \Gamma^{\text{EM}}_{\text{NR}} + \Gamma^M_{\text{NR}}}. \tag{4.82}$$

Using the definitions of Section 4.4.3 (see Table 4.3), and the expression for Q^M in Eq. (4.81), the modified quantum yield for fluorescence emission at frequency ω_S can be expressed as:

$$\boxed{Q = \frac{M_{\text{Rad}}(\omega_S)}{M_{\text{Tot}}(\omega_S) + (Q^M)^{-1} - 1}.} \tag{4.83}$$

Note that since $0 \leq Q \leq 1$, no quantum yield enhancement can be obtained for a good fluorophore (with Q^M in free-space already of the order of ~ 1).

Moreover, in many cases of interest, M_{Tot} is much larger than $(Q^M)^{-1} - 1$ (this is especially true for all fluorophores with a good free-space quantum yield $Q^M \approx 1$). The modified quantum yield then reduces to the EM radiative efficiency $\eta^{\text{EM}}_{\text{Rad}}$ defined in Section 4.4.3:

$$Q \approx \frac{M_{\text{Rad}}(\omega_S)}{M_{\text{Tot}}(\omega_S)} = \eta^{\text{EM}}_{\text{Rad}}(\omega_S). \tag{4.84}$$

In this approximation, the modified quantum yield then becomes entirely governed by the EM interaction with the metal, and *all fluorophores should therefore exhibit the same modified quantum yield.*

It is important to note that the modified quantum yield, as in the free-space case, is determined by the competition between two decay channels: radiative and non-radiative. This is conceptually very different from scattering processes such as SERS discussed previously. A large radiative enhancement, M_{Rad}, always results in enhanced scattering or SERS but this is not true for the quantum yield in most cases. Even if the radiative decay rate is increased, non-radiative decay rates may also be enhanced, and will compete with radiative processes. It is therefore the relative value of M_{Rad} and M_{Tot} that is relevant, i.e. the modified EM radiative efficiency $\eta^{\text{EM}}_{\text{Rad}}$. This distinction is crucial to understand the difference between SERS and SEF.

4.6.4. Fluorescence quenching and enhancement

Derivation of the fluorescence enhancement factor

We can now put together the last two subsections to study the whole fluorescence process. Using the same notations; we will neglect in the following the shift between excitation (ω_L) and fluorescence (ω_S) frequencies. Ignoring also saturation effects[13], the fluorescence cross-section (characterizing the integrated power radiated by fluorescence) is simply the product of the absorption cross-section and quantum yield (see Section 2.3.2). The *fluorescence enhancement factor*, M_{Fluo}, defined as the ratio of the modified fluorescence cross-section over the free-space one is then given by:

$$M_{\text{Fluo}} = \frac{\sigma_{\text{Abs}} Q}{\sigma_{\text{Abs}}^M Q^M} = M_{\text{Loc}} T_{\text{Abs}} \frac{M_{\text{Rad}}}{Q^M M_{\text{Tot}} + (1 - Q^M)} \approx M_{\text{Loc}}\, \eta_{\text{Rad}}^{\text{EM}}\, (Q^M)^{-1}, \tag{4.85}$$

where the latter approximation has been discussed earlier and also assumes $T_{\text{Abs}} \approx 1$. Despite the name of M_{Fluo} as a fluorescence *enhancement* factor, it obviously also applies to (common) situations of fluorescence *quenching* (when $M_{\text{Fluo}} < 1$). Accordingly, a more appropriate name would be the one of 'fluorescence modification factor', but we will stick to the name of 'enhancement factor' for consistency.

This expression clearly identifies the sources of fluorescence enhancement or quenching. There are in fact three competing mechanisms:

- The first term, the LFIEF M_{Loc}, is in most situations of interest larger or much larger than 1, and contributes to *enhanced absorption* (and therefore enhanced fluorescence).
- The second term, the EM radiative efficiency $\eta_{\text{Rad}}^{\text{EM}} = M_{\text{Rad}}/M_{\text{Tot}}$, is the modified quantum yield for the emitter in EM interaction with the metal. This term is smaller (and sometimes much smaller) than 1 and contributes to *fluorescence quenching.*
- The third term, $(Q^M)^{-1}$ is irrelevant for good fluorophores (with a free-space quantum yield close to 1). It can however contribute significantly to the fluorescence enhancement for fluorophores with a poor quantum yield. This enhancement is then a consequence of the small free-space

[13] The saturation intensity should also be modified by a factor $\sim M_{\text{Tot}}/M_{\text{Loc}}$. In most situations this would increase its value and, hence, make saturation effects less likely. We will not discuss this further here and simply assume that the power is not high enough for saturation to occur.

fluorescence cross-section, and not directly of the EM interaction with the metal.

The various reports of quenching and enhancement of fluorescence are simply different situations where either the first (and possibly the third) term, or the second term in Eq. (4.85) dominates with respect to the other.

Remarks

In addition, let us make a few important remarks on the applicability of Eq. (4.85):

- The re-emission factors M_{Rad} and M_{Tot} depend on the emitting dipole orientation and therefore on the molecule orientation. For the uni-axial case considered here, the emitting dipole is along $\mathbf{e}_m$.
- M_{Fluo} describes the enhancement for the integrated fluorescence cross-section. Following the discussion in Section 4.4.3, the modified radiation pattern in SEF simply follows $M^d_{\text{Rad}}(\Omega, \omega_L)$ (which again depends on $\mathbf{e}_m$). One can, in fact, define a *differential fluorescence enhancement factor* as:

$$M^d_{\text{Fluo}}(\Omega) = M_{\text{Loc}} T_{\text{Abs}} \frac{M^d_{\text{Rad}}(\Omega)}{Q^M M_{\text{Tot}} + 1 - Q^M} \approx M_{\text{Loc}} \frac{M^d_{\text{Rad}}(\Omega)}{Q^M M_{\text{Tot}}}. \quad (4.86)$$

  By analogy with Raman, one can also define the free-space absolute differential fluorescence cross-section $\mathrm{d}\sigma^M_{\text{Fluo}}/\mathrm{d}\Omega$ as that obtained from randomly-oriented molecules in 90° or back-scattering configurations. Denoting by $\mathrm{d}\sigma_{\text{SEF}}/\mathrm{d}\Omega$ the differential SEF cross-section in the detection direction Ω_{Det}, we then conveniently have:

$$\frac{\mathrm{d}\sigma_{\text{SEF}}}{\mathrm{d}\Omega}(\Omega_{\text{Det}}) = M^d_{\text{Fluo}}(\Omega_{\text{Det}}) \frac{\mathrm{d}\sigma^M_{\text{Fluo}}}{\mathrm{d}\Omega}. \quad (4.87)$$

- Finally, M_{Loc} should in principle be evaluated at ω_L (for excitation), while M_{Rad} at ω_S (for emission). However, if one wants to study the spectral profile of the modified fluorescence ($M_{\text{Fluo}}(\omega_S)$), a more complex treatment is necessary to account for the fact that all decay channels into fluorescence photons at every possible frequencies ω_S compete with each other. Such a description has recently been provided in Ref. [9]. We only mention here that the spectral profile of fluorescence is also modified in SEF, sometimes to a point where the original fluorescence spectrum may no longer be recognizable. Moreover, the ω_S-dependence of the fluorescence enhancement factor is predicted to follow that of $M_{\text{Rad}}(\omega_S)$. See Ref. [9] for more details.

4.7. OTHER EM EFFECTS IN SERS

Before concluding this chapter, we will discuss a few secondary electromagnetic effects in SERS in this section, and additional non-electromagnetic enhancement mechanisms in the next section. These are not crucial to a basic understanding of the main source of EM enhancements, but are still relevant in a number of situations.

4.7.1. Fluorescence quenching in SERS

Comparison between SERS and SEF enhancements

We consider here the common case of a fluorescent dye under SERS conditions and carry on the discussion of SEF of Section 4.6 in the context of SERS. It is well known that fluorescence is strongly quenched when large SERS signals are observed. This should in principle be derived naturally within this framework.

Ignoring for simplicity the surface selection rules factor and any polarization effects, and omitting the Stokes shift, the SERS EF (SMEF) can be expressed as:

$$\mathrm{SMEF} = \frac{\mathrm{d}\sigma_{\mathrm{SERS}}/\mathrm{d}\Omega}{\mathrm{d}\sigma_{\mathrm{RS}}^{M}/\mathrm{d}\Omega} \approx M_{\mathrm{Loc}} M_{\mathrm{Rad}}^{d}. \tag{4.88}$$

The fluorescence enhancement for the same molecule is characterized by the differential fluorescence enhancement factor:

$$M_{\mathrm{Fluo}}^{d} = \frac{\mathrm{d}\sigma_{\mathrm{SEF}}/\mathrm{d}\Omega}{\mathrm{d}\sigma_{\mathrm{Fluo}}^{M}/\mathrm{d}\Omega} \approx M_{\mathrm{Loc}} \frac{M_{\mathrm{Rad}}^{d}}{Q^{M} M_{\mathrm{Tot}}}. \tag{4.89}$$

The SERS and SEF enhancement factors are therefore related by:

$$\boxed{\frac{\mathrm{SMEF}}{M_{\mathrm{Fluo}}^{d}} \approx Q^{M} M_{\mathrm{Tot}}.} \tag{4.90}$$

It is important to note that this relation applies to enhancement factors, not to the signals themselves. The ratio of non-modified cross-sections must be taken into account to compare directly the SERS and SEF signals.

Discussion

It is interesting to discuss this result qualitatively using 'order-of-magnitude' estimates only. Note that by definition M_{Tot} is at least of the order of M_{Rad} and possibly larger if non-radiative emission is important. Moreover, in a first approximation, M_{Rad} is of the order of M^d_{Rad}, which owing to the ORT is itself of the order of M_{Loc}, i.e.:

$$M_{\text{Tot}} \geq M_{\text{Rad}} \approx M^d_{\text{Rad}} \approx M_{\text{Loc}}. \tag{4.91}$$

Let us then consider, for the sake of argument, a fluorophore with $Q^M \approx 1$ subject to a large SERS enhancement factor, say SMEF $\approx 10^8 - 10^{10}$, which from Eq. (4.88) is equivalent to $M_{\text{Loc}} \approx 10^4 - 10^5$. The fluorescence is also enhanced by a factor M^d_{Fluo} (Eq. (4.89)) of the order of M_{Loc} (10^4–10^5) or possibly less if non-radiative emission dominates. The SERS EF is therefore much larger than the fluorescence EF, by a factor 10^4–10^5 (Eq. (4.90)) (and possibly more if non-radiative emission dominates). The relative importance of Raman peaks with respect to the fluorescence background should therefore be much larger under SERS conditions. This is what fluorescence quenching under SERS conditions really means: *a fluorescence 'quenching' with respect to Raman peak intensities*, despite the fact that the fluorescence signal itself may also be enhanced (but much less than Raman). This fluorescence (SEF) background has in fact recently been proposed as the main origin of the so-called *SERS background* or *SERS continuum* [9], although the issue is not resolved yet.

An important consequence relates to the special case of resonant-Raman scattering (RRS), i.e. when the excitation wavelength is close to the absorption maximum. The fluorescence cross-section is then so large that it completely overwhelms the Raman signal under normal conditions. This makes it impossible to measure RRS spectra with conventional techniques. Under SERS conditions, the relative intensities of the Raman peaks with respect to fluorescence are magnified by several orders of magnitude, and it becomes possible to resolve them clearly (albeit on top of a fluorescence background in many cases).

Finally, when comparing fluorescence and Raman signals (which in general can both appear on the same spectrum), it is important to highlight the fact that the fluorescence signal is typically spread over a much wider range of energies ($\sim$1000 cm^{-1}) than the signal of a Raman peak (typically $\sim$10 cm^{-1}). If the Raman peaks are of comparable intensity as the fluorescence background underneath, this therefore usually means that the differential fluorescence cross-section is $\approx$100 times larger than that for Raman.

4.7.2. Photo-bleaching under SERS conditions

Introduction

A related aspect is the study of photo-bleaching under SERS conditions, also a common feature in many SERS experiments with dyes. This is not so surprising at first, since the excitation conditions are often comparable to those where photo-bleaching would be observed in any case (under non-SERS conditions). When considering the large local field experienced by the molecules, one could at first expect them to photo-bleach at an even much faster rate than under non-SERS conditions.

The experimental study of photo-bleaching under SERS conditions is a difficult problem, and there have been only a few attempts [181,182]. There are many other possible sources of signal fluctuations in SERS, and it is often difficult to decouple these from true photo-bleaching (especially when the single-molecule limit is approached). Nonetheless, even without quantitative measurements, it is clear that photo-bleaching is present, and it is often ignored for the sake of simplicity.

It is fair to say that the details of photo-bleaching under SERS conditions have not yet been elucidated. We use here the formalism presented so far for a basic discussion of what can be expected for 'conventional' photo-bleaching (through inter-system crossing) under SERS conditions. This does not exclude the possibility that photo-bleaching in SERS might be dominated by other mechanisms, not yet identified.

A simple model

We follow the same model of photo-bleaching through inter-system crossing (ISC) as presented in Section 2.3.3 and extend it to the case of SERS or SEF (the notations, accordingly, are those of Section 2.3.3).

We use as before the superscript M for free-space quantities: the ISC rate Γ_{ISC}^M and the fragmentation probability p_F^M are in a first approximation assumed to be unchanged for an adsorbed molecule[14]. The total decay rate is modified and the ISC probability p_{ISC} is therefore modified accordingly:

$$p_{\text{ISC}} = \frac{\Gamma_{\text{ISC}}^M}{\Gamma_{\text{Tot}}} = \frac{p_{\text{ISC}}^M}{Q^M M_{\text{Tot}} + 1 - Q^M}. \tag{4.92}$$

Similarly, the photo-bleaching quantum yield, defined in Section 2.3.3 is also modified in the same proportion:

$$\phi_B = p_{\text{ISC}}\, p_F^M = \frac{\phi_B^M}{Q^M M_{\text{Tot}} + 1 - Q^M}. \tag{4.93}$$

In most cases of interest, the probability of photo-bleaching – once in the excited state – is therefore reduced (and often by a large amount $\approx M_{\text{Tot}}$)

---

[14] The reason for this assumption is that these processes do not involve emission/absorption of photons. This implicitly assumes, among other things, that the decay path from the triplet to the ground state is non-radiative, an assumption that may not always be valid.

under SERS conditions. The dye should therefore be able to sustain many more excitation/relaxation cycles before photo-bleaching. The number of such cycles per unit time is however also greatly increased because of enhanced absorption. The relevant quantity here is therefore the photo-bleaching rate, Γ_B, which is affected not only by the value of ϕ_B, but also by the absorption cross-section – itself modified under SERS conditions. The modification of the photo-bleaching rate therefore takes the form:

$$\frac{\Gamma_B}{\Gamma_B^M} = \frac{\phi_B}{\phi_B^M}\frac{\sigma_{\mathrm{Abs}}}{\sigma_{\mathrm{Abs}}^M} = \frac{M_{\mathrm{Loc}}}{Q^M M_{\mathrm{Tot}} + 1 - Q^M} \approx \frac{M_{\mathrm{Loc}}}{Q^M M_{\mathrm{Tot}}}. \tag{4.94}$$

This expression shows that the photo-bleaching rate under SERS conditions should, in most situations, be smaller or comparable to that in free-space:

- If M_{Tot} is dominated by radiative emission, $M_{\mathrm{Tot}} \approx M_{\mathrm{Rad}} \approx M_{\mathrm{Loc}}$, then the photo-bleaching rate is unchanged for a good fluorophore, and may increase for bad fluorophores (with a small Q^M). Moreover, the modified photo-bleaching rate is independent of M_{Loc}.
- If M_{Tot} is dominated by non-radiative emission: $M_{\mathrm{Tot}} \approx M_{\mathrm{NR}}$ (and therefore $M_{\mathrm{Loc}} \ll M_{\mathrm{Tot}}$), then Γ_B can be greatly reduced. Photo-bleaching is then reduced. Moreover, Γ_B is then proportional to M_{Loc}, i.e. the photo-bleaching rate is larger for molecules subject to a larger enhancement. In SERS experiments where 'many molecules' are involved producing the signal, this would result in a non-exponential signal decay, the details of which reveal (in principle) information on the distribution of enhancements on the surface; see Ref. [162] for more information.

There have not been many studies of photo-bleaching under SERS or SEF conditions, and it is therefore difficult to elaborate further. This simple model can nevertheless be used as a basis for designing experiments to investigate further photo-bleaching under these conditions; an issue that currently remains largely open and at the boundary of what has been understood in the field so far.

4.7.3. Non-radiative effects in SERS

In fluorescence, non-radiative effects play a major role, and manifest themselves as a reduced quantum yield, because of the competition between radiative and non-radiative decays. In SERS, because scattering is instantaneous, there is not such a competition, and radiative and non-radiative emission can occur independently of each other. The non-radiative component of SERS is therefore absent from most SERS EFs of interest. It only appears when considering the total SERS cross-section given in Eq. (4.71).

This cross-section may be much larger than the observable radiative SERS cross-section of Eq. (4.70). The difference corresponds to SERS processes resulting in emission of a Raman photon in the non-radiative modes of the metal. These Raman photons are obviously not observable in the far-field, but they do correspond to a real Raman event, and therefore to the creation (for Stokes events) or destruction (for anti-Stokes) of a vibration in the molecule. This total SERS cross-section is not relevant to most applications, but it is the one that is relevant when studying phonon population dynamics, in particular in effects such as *vibrational pumping* [59].

4.8. THE CHEMICAL ENHANCEMENT

4.8.1. Introduction

The relative contribution, or even the existence of a 'chemical enhancement', has been the subject of much debate over the years, and is still, to some extent, not resolved. This is partly because many other chemical effects can affect the SERS intensity, such as molecule adsorption or orientation, but should not be strictly speaking considered as a SERS enhancement factor. For example, a negatively charged dye cannot adsorb on a negatively charged silver colloid (because of strong electrostatic repulsion). Its SERS signal is therefore absent, while a positively charged dye in the same silver colloid solution would exhibit a strong SERS signal. This difference could be seen as a 'chemical' effect, but is not related to the SERS process itself, but only to the characteristics of the sample.

The exact definition of what should be called 'chemical enhancement' in SERS in fact has a long history in the field, with some practitioners taking diametrically opposite views. According to Ref. [4] the chemical enhancement is not only difficult to measure experimentally but also (quote):

> ...the magnitude of this effect rarely exceeds a factor of ∼10, and is best thought to arise from the modification of the Raman polarizability tensor of the adsorbate, resulting from the formation of a complex between the adsorbate and the metal. Rather than an enhancement mechanism, the chemical effect is more logically to be regarded as a change in the nature and identity of the adsorbate.

We adhere to this view: the 'chemical enhancement' corresponds to any *modification of the Raman polarizability tensor upon adsorption* of the molecule onto the metal surface.

Note that such a change could in fact result in a *quenching* as much as an *enhancement*, depending on the situation. The most relevant situation, however, arises when the modified polarizability is more resonant with the excitation than the original one (as a result of a charge-transfer mechanism in the metal–adsorbate complex, for example). The Raman

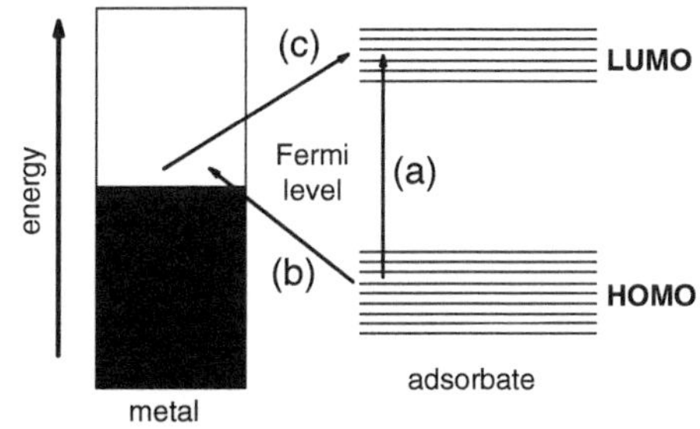

Figure 4.6. Schematic representation of a charge-transfer mechanism in the SERS cross-section, appearing as a 'resonant contribution' to the measured intensity (inspired by Ref. [6]). The laser energy can be directly in resonance with an electronic transition of the molecule–metal complex (case (a)), or can profit from an indirect coupling (charge transfer) through the metal ((b) + (c)).

intensity is then naturally increased as a result of this resonant or pre-resonant condition. Within this definition, the 'chemical enhancement' is not restricted to chemically-bound molecules, but the most common description through charge transfer does require covalent bonding. In all cases, this 'chemical enhancement' is multiplicative and acts in conjunction with the electromagnetic enhancement discussed earlier.

A detailed discussion of the main experimental arguments for the chemical enhancement has been provided in Ref. [4]. In the same review, it is also discussed in detail how most of these reported evidence could in fact be explained using the EM model of SERS. Despite this, most of the evidence for the CE to date is still along the same lines, and its interpretation in terms of CE or EM enhancement remains controversial. Examples of experimental results interpreted in terms of a chemical enhancement mechanism can be found for example in Refs. [183–186].

4.8.2. The charge-transfer mechanism

The most studied mechanism for the chemical enhancement is the so-called *charge-transfer* mechanism. This is a subject treated in many articles [4–6,161,187] and we will not repeat here the detailed account given in the specialized literature.

This mechanism is represented schematically in the diagram of Fig. 4.6. Using this diagram, let us discuss briefly the main scenarios for the chemical enhancement that have been proposed to exist [54,188–191]. We follow closely the discussions in Refs. [5,187].

- Type I is the simplest and may occur when the adsorbate does not bind covalently to the metal. In this case, the presence of the metal acts only as a *perturbation* to the electronic structure of the analyte causing a 'mild' change in its electronic distribution. The latter can result in a corresponding change of polarizability and, ultimately, in a change of the Raman efficiency of the mode. It has been argued, for example, that the change in *relative* Raman efficiency of the 1008 cm^{-1} and 1035 cm^{-1} modes of pyridine is due to this effect [188–191].

- Type II is more in the spirit of what is represented in Fig. 4.6, and it involves the presence of a *surface complex* either by direct (covalent) binding to the metal, or by indirect binding with the assistance of an electrolyte ion (typically chloride). This may produce a substantial change in the intrinsic polarizability of the molecule. The easiest way to understand this is to realize that the magnitude of the polarizability depends explicitly on the available optical transitions. The 'new' indirect transitions provided by the overlap of molecular orbitals provide a channel for the polarizability to be modified. It is possible also that the surface complex creates a new electronic state (similarly to the gap states created at the surfaces of semiconductors) that is explicitly in resonance – or close to resonance – with the laser, hence providing a contribution to the enhancement of a resonant-Raman type.

- Type III is a more sophisticated version of type II (typically more difficult to pin down also), which involves the process of photo-driven charge transfer between the analyte and the metal [163,191]. This situation can occur when the difference between the Fermi level (E_F) of the metal and the HOMO or LUMO energies is matched by the laser. A photo-driven charge-transfer mechanism between the HOMO and unoccupied states above the Fermi level (or between the LUMO and occupied states slightly below E_F) can be triggered. This mechanism was uncovered only through experiments in electrochemical cells (which is how the SERS phenomenon was observed in the first place [1]) where it is possible to change the difference in energy between the adsorbed analyte and the metal through an external potential. The maximum SERS intensity is observed at different potentials [163,191] for different incident laser energies. The interpretation of this is that different 'tuning' conditions are needed depending on the laser energy being used, and this tuning is provided by the external potential. These experiments are, arguably, one of the clearest proofs that charge-transfer mechanisms between the molecules and the substrates are real and can play a non-negligible role in the magnitude of the SERS enhancement.

Overall, the main message is that the understanding of the 'chemical' contributions to the enhancement in SERS is difficult and should ideally be aided by suitable electronic structure calculations of the molecule plus the metal. It is only recently that sophisticated computational methods like density functional theory (DFT) and molecular dynamics [188,192] have been applied to the study of molecular interactions with silver, for some of the simplest SERS probes like pyridine.

4.8.3. Electromagnetic contribution to the chemical enhancement

The 'chemical enhancement', defined as a modification of the polarizability upon adsorption, can in fact also have an electromagnetic origin.

Image dipole enhancements

This arises when considering the effect of a dipole self-reaction field on its polarizability. Since the self-reaction field is often approximated by that created by an image dipole, we call this mechanism the *image dipole enhancement*. This mechanism has in fact been discussed very early after the discovery of SERS [193–195], and was even (at first) thought to be one of the main SERS enhancement mechanisms, but it has since been overlooked. It actually appears naturally when considering SERS in the framework of modified spontaneous emission. As discussed in Section 4.4, for SE, the self-reaction field modifies the ability of a dipole to radiate energy. Although the 're-emission' enhancement in SERS (and other scattering processes) follows the physics of modified SE, we also have to consider an additional ingredient: in SE the dipole has a *fixed amplitude* (determined by the nature of the molecular 'excited' state), whereas for scattering, the dipole amplitude is determined by the interaction with the electromagnetic field. Since the dipole is driven by an external field, a constant source of energy is readily available from the laser. The self-reaction field can therefore for scattering processes also *oppose or amplify the dipole amplitude* (which is normally fixed by the transition dipole moment in SE). Note that the distinction between this effect and the radiation enhancement itself is quite subtle. In the latter case, the self-reaction field can facilitate or prevent extraction of energy from the dipole, but has no effect on its amplitude (because it is fixed). In the former, the self-reaction fields partially adds to, or cancels, the action of the external field; it therefore affects the dipole amplitude when it is free to vary. This operates in addition to the standard radiation enhancement discussed earlier.

Effective polarizability

Let us consider the simplest case of a polarizable dipole $\mathbf{p}$ with isotropic *linear* optical polarizability α_L and assume that the self-reaction tensor $\mathbf{G}_r$

(see Section 4.4.4) is also isotropic. The dipole is driven by both the external, $\mathbf{E}$, and reflected ($\mathbf{E}_{\mathrm{SR}} = G_r\mathbf{p}$) fields:

$$\mathbf{p} = \alpha_L (\mathbf{E} + G_r\mathbf{p}) . \quad (4.95)$$

This can be rewritten as $\mathbf{p} = \alpha_m\mathbf{E}$, with a modified polarizability α_m of the form [193–195]:

$$\alpha_m = \alpha_L (1 - \alpha_L G_r)^{-1} . \quad (4.96)$$

α_m is often called the *effective linear polarizability.*

The image dipole enhancement factor

The same effect applies to SERS through the effective Raman polarizability. The treatment is more complex but it has been shown [193–195] to lead to an additional multiplicative enhancement (or quenching) factor of the form: $M_{\mathrm{Im}}(\omega_L)M_{\mathrm{Im}}(\omega_R)$, where the *image dipole enhancement factor* is defined as:

$$M_{\mathrm{Im}}(\omega) = \frac{1}{|1 - \alpha_L(\omega)G_r(\omega)|^2}. \quad (4.97)$$

Note that it is the *linear optical polarizability* of the molecule $\alpha_L(\omega)$ that appears in this expression (not the Raman one).

For a dipole at a short distance d from a metallic plane (with dielectric function ϵ), the self-reaction field in the electrostatics approximation corresponds to the field created by the image dipole, and we have (see Section 6.1.1):

$$G_r(\omega) \approx \frac{1}{16\pi\epsilon_0\epsilon_M d^3}\frac{\epsilon(\omega) - \epsilon_M}{\epsilon(\omega) + \epsilon_M}, \quad (4.98)$$

for a dipole perpendicular to the surface (half of this result for a parallel dipole). This approximation is also quite good for any type of metallic surface when d is small.

Discussion

The image dipole enhancement factor involves both the real and imaginary parts of the self-reaction tensor G_r, whereas only the imaginary part of G_r contributes to M_{Tot} (Eq. (4.42)). M_{Im} and M_{Tot} are therefore conceptually different, although both are related to the self-reaction field. M_{Tot} characterizes the ability of the environment to extract energy from a dipole with a fixed amplitude, whereas M_{Im} represents the ability of the environment to amplify or decrease the dipole amplitude of a polarizable element. These factors are multiplicative for a polarizable dipole.

The image dipole enhancement factors had been proposed in the past to explain SERS enhancements [193–195], but were then thought to be too small and have not been since then part of the mainstream discussions in the field. Now that the main EM enhancement factor mechanisms are well understood, it remains possible that the image dipole enhancement factors play a small role, introducing additional factors of say ≈ 10. As seen in Eq. (4.98), its effect depends critically on the exact distance between the surface and the probe. Unfortunately, at distances where it may become non-negligible (typically less than 1 nm), one could argue that the local description of the EM problem should fail, and quantitative predictions are therefore very difficult. Such a mechanism would however have many of the characteristics of other CE mechanisms, and cannot therefore be excluded from any discussions or interpretations.

4.8.4. The chemical vs electromagnetic enhancement debate

Conceptually, there is no real controversy on the issue of chemical vs electromagnetic enhancement, despite some claims that still populate the literature in the field. The electromagnetic enhancement would exist by itself, even if the probe were not there. When the probe is present, *the electromagnetic enhancement applies to the intrinsic properties of the adsorbate.* It is a fact that some probes have different intrinsic properties when adsorbed on the metal compared to free-space, as a result of chemical and/or electromagnetic interactions. This effect should not be considered as SERS itself, although it cannot be decoupled from it. The way around it then, is to choose probe/metal pairs that do not exhibit this effect (although this has to be assessed carefully). This would avoid a lot of confusion for many fundamental studies of other aspects of SERS. The study of adsorption can then be carried out separately, possibly using SERS itself as a tool. SERS experiments in electrochemical cells [163,191] provide the best chance to analyze some of the subtleties of this effect due to the additional degree of freedom provided by the electrode potential.

Finally, it is probably fair to emphasize the fact that the presentation given here of an aspect as controversial as the chemical enhancement is necessarily very biased by the authors' own opinions. We therefore strongly recommend consulting, as a complement, reviews on this subject written by authors with a probably very different view [5,6,187].

4.9. SUMMARY

This chapter presents a comprehensive account of the physical origin of the SERS enhancement factor, which is (arguably) the very essence of SERS. As such, this can be considered to be a core chapter of the book with important concepts and formulas. The first three sections were aimed at a non-technical introduction to the most important aspects: EF definition, measurement, and physical origin. These are sufficient for a basic understanding of the effect. The rest of this chapter could be considered as much more advanced. It is our hope that the concepts developed here will serve in the future either as reference, or as a solid step forward from where further developments in the microscopic theory of SERS can be developed.

Nonetheless, what is still missing for a practical understanding of the SERS effect is how the necessary local field enhancements arise in the first place, how they can be modeled and predicted, and eventually engineered for a particular application. These aspects are the subject of the next two chapters.

Chapter 5

Calculations of electromagnetic enhancements

The presence of a metallic object strongly affects the electromagnetic field in its vicinity, and this is the main origin of most plasmon-related effects, including SERS. In the previous chapter, we have described in detail how these local-field modifications can affect the optical properties of a molecule, namely: absorption, emission, and scattering. All these properties can therefore (in principle) be predicted using the tools of the previous chapter, as long as the electromagnetic problem can be solved. This aspect, which was taken for granted in Chapter 4, is in fact a difficult task in all but the simplest of situations. Although it is not *a priori* necessary to understand the EM mechanism of SERS or SEF, it is in fact required for any type of quantitative prediction, and even to understand when an enhancement (rather than a quenching) may be obtained.

This chapter is therefore devoted to the issue of solving the electromagnetic problem. We will first discuss in Section 5.1 what is actually required of the EM solution in the context of SERS and SEF, and the possible approximations that may be used. The solutions to the EM problem can then be separated into two approaches: either analytical or numerical. The analytical approaches, discussed in Section 5.2, although restrictive to specific simple examples, provide an easier way to understand the general physical principles. The numerical approaches, discussed in Section 5.3, have the advantage of being applicable to a wider range of systems. Note that Section 5.2 provides only an overview of the main analytical approaches without going into the technical details, which have been included as appendices rather than in the main text. What matters most are the final results of the analytical predictions, which can then be used for a physical understanding of the origin of the various

EM enhancements relevant to SERS and SEF. This important aspect will be extensively discussed in the next chapter using the tools and techniques developed here.

5.1. DEFINITION OF THE PROBLEM AND APPROXIMATIONS

5.1.1. The EM problem

The EM problem in plain English

Let us consider a SERS metallic substrate. This may be a single metallic nano-particle or a collection of nano-particles, embedded in a dielectric (typically air or water) or lying on a 2D planar substrate (typically glass or metal). It could also be an ordered array of such objects (for example a grating-like structure). From the considerations of the previous chapter, two main types of EM problems are of interest here: either *excitation* by a monochromatic laser beam, or *emission* from a localized emitter (point source). These two examples (and combinations thereof) will embrace most situations which are relevant from the standpoint of understanding SERS and SEF. In fact, thanks to the optical reciprocity theorem, solving the excitation problem is sufficient in many instances as far as SERS is concerned (see Sections 4.4.6 and 4.5).

Solving the EM problem means determining the electric field at every point in space (and, therefore, the magnetic field through Maxwell's equations). This requires solving Maxwell's equations for dielectric media, with appropriate dielectric functions for the optical properties of the objects and surrounding medium, and appropriate boundary conditions describing the geometry of the objects. The source of the field must also be included. For excitation, this usually means having an additional boundary condition 'at infinity', i.e. far from any objects. For emission, a dipolar source term must be added to Maxwell's equations as a source. Both problems deal basically with the same issues of matching appropriate boundary conditions for the fields at interfaces, but the nature of the excitation is of course different (plane wave coming from infinity or localized emitter close to the SERS substrate/object). Maxwell's equations for dielectric media are discussed in further detail in Appendix C.

Some common approximations

To simplify the problem, we will first make a number of approximations, which are commonly made for SERS and plasmonics problems. As a matter of fact, these are so common that they are usually implicit in most treatments of the problem found in the literature. We mention them here for completeness:

- We assume that *macroscopic* Maxwell's equations apply for our systems.

- We only consider *time-harmonic* problems, i.e. all fields have a oscillating harmonic dependence at a single frequency ω. The fields are then defined by the complex fields (see Appendix C) unless otherwise stated, with the convention of a time dependence as $\exp(-\mathrm{i}\omega t)$.
- The materials considered are assumed to be *non-magnetic, isotropic, and homogeneous* media.
- Their optical response is described by a dielectric function ϵ within linear response theory (*linear approximation*).
- We also assume that the dielectric response is local spatially (*local approximation*), i.e. ϵ only depends on frequency ω. The introduction of spatial dispersion in the dielectric function would add a great deal of complication to an already complex problem, and it is normally neglected for simplicity. It could, in principle, be reintroduced in situations where its presence is absolutely essential [138].

The meaning of these approximations is discussed in more detail in Appendix C. Some of these assumptions are in fact often applied even at length scales where justifying them theoretically would be difficult (e.g. ~1 nm). This is partly because a more accurate description (including for example a non-local description of the dielectric function or the importance of the microscopic local field) is, in most cases, intractable. Moreover, most of the phenomenology of SERS and SEF can be understood within these assumptions. Finally, there is no experimental evidence so far that these assumptions may need to be reconsidered. We only need to remember here that, within these approximations, the predictions agree well with experiments in many situations of interest to SERS and plasmonics, and that looking beyond these is extremely difficult except for the simplest of cases [138].

Formal layout of the EM problem

More formally, the problem is defined in an infinite non-absorbing medium with relative dielectric constant ϵ_M (real positive), or equivalently with a refractive index n_M ($\epsilon_M = n_M^2$). In air or vacuum $\epsilon_M = n_M = 1$.

The geometry of the objects under consideration can be described by a number of volumes V_i, separated by interfaces (surfaces) S_j. Each volume V_i is composed of a uniform homogeneous material with optical properties defined by its relative dielectric function $\epsilon_i(\omega)$ (as pointed out before, we ignore magnetic materials here). We must therefore solve Maxwell's equations for harmonic fields (Eqs (C.36) – (C.39)) with the corresponding constitutive relation (Eq. (C.40)) in each volume V_i. The boundary conditions (Eqs (C.41) – (C.42)) must also be fulfilled on all interfaces S_j.

EM problem for excitation

If we assume (without any loss of generality) excitation along (Ox), the following additional boundary condition: $\mathbf{E}(\mathbf{r}) = \mathbf{E}_{\mathrm{Inc}}$ must be fulfilled for $x \to -\infty$. Here $\mathbf{E}_{\mathrm{Inc}}$ is the incident electric field propagating along (Ox), i.e. for a plane wave, it is $\mathbf{E}_{\mathrm{Inc}}(\mathbf{r}) = \mathbf{E}_0 \exp(\mathrm{i}k_M \mathbf{e}_x \cdot \mathbf{r})$, where $k_M = n_M\omega/c$.

This leads to the following set of equations for the *excitation* case:

$$\forall i\, \forall \mathbf{r} \in V_i : \begin{cases} \nabla \cdot \mathbf{H}(\mathbf{r},\omega) &= 0 \\ \nabla \times \mathbf{E}(\mathbf{r},\omega) - \mathrm{i}\omega\mu_0 \mathbf{H}(\mathbf{r},\omega) &= 0 \\ \nabla \cdot \mathbf{D}(\mathbf{r},\omega) &= 0 \\ \nabla \times \mathbf{H}(\mathbf{r},\omega) + \mathrm{i}\omega \mathbf{D}(\mathbf{r},\omega) &= 0 \\ \mathbf{D}(\mathbf{r},\omega) &= \epsilon_0 \epsilon_i(\omega)\mathbf{E}(\mathbf{r},\omega) \end{cases} \tag{5.1}$$

$$\forall j\, \forall \mathbf{r} \in S_j : \begin{cases} (\mathbf{E}_j^+ - \mathbf{E}_j^-) \times \mathbf{n} &= 0 \\ (\mathbf{D}_j^+ - \mathbf{D}_j^-) \cdot \mathbf{n} &= 0 \end{cases} \tag{5.2}$$

$$\mathbf{E}(\mathbf{r}) = \mathbf{E}_{\mathrm{Inc}}(\mathbf{r}) \quad \text{for} \quad x \to -\infty. \tag{5.3}$$

In fact, the two problems of finding $\mathbf{E}$ and $\mathbf{H}$ can be decoupled by combining these equations and using some standard vector calculus formula. The problem then reduces to finding $\mathbf{E}$ that satisfies:

$$\forall i\, \forall \mathbf{r} \in V_i : \begin{cases} \nabla \cdot \mathbf{E}(\mathbf{r},\omega) &= 0 \\ \nabla^2 \mathbf{E}(\mathbf{r},\omega) + k_i^2 \mathbf{E}(\mathbf{r},\omega) &= 0 \end{cases} \quad \text{where} \quad k_i^2 = \frac{\omega^2}{c^2}\epsilon_i(\omega) \tag{5.4}$$

$$\forall j\, \forall \mathbf{r} \in S_j : \begin{cases} (\mathbf{E}_j^+ - \mathbf{E}_j^-) \times \mathbf{n} &= 0 \\ (\epsilon^+(\omega)\mathbf{E}_j^+ - \epsilon^-(\omega)\mathbf{E}_j^-) \cdot \mathbf{n} &= 0 \end{cases} \tag{5.5}$$

$$\mathbf{E}(\mathbf{r}) = \mathbf{E}_{\mathrm{Inc}}(\mathbf{r}) \quad \text{for} \quad x \to -\infty. \tag{5.6}$$

From there, the magnetic field can simply be deduced as:

$$\mathbf{H}(\mathbf{r},\omega) = \frac{1}{\mathrm{i}\omega\mu_0} \nabla \times \mathbf{E}(\mathbf{r},\omega). \tag{5.7}$$

The second equation in Eq. (5.4) is called the *Helmholtz equation* or vector wave equation. Due to its vectorial nature, it is much more complicated to solve than the scalar wave equation, and only reduces to three scalar wave equations when expressed in Cartesian coordinates.

EM problem for emission

In the case of emission by a dipole $\mathbf{p}$, the boundary condition at infinity does not exist, but a localized source term at the dipole position $\mathbf{r}_p$ (assumed to be in the external medium with ϵ_M) must be included instead. This source term corresponds to a point current source: $\mathbf{j} = -\mathrm{i}\omega\delta(\mathbf{r}-\mathbf{r}_p)\mathbf{p}$ (see Eq. (4.22) in Section 4.4.2). The full set of equations then reads:

$$\forall i\,\forall\mathbf{r}\in V_i : \begin{cases} \nabla\cdot\mathbf{H}(\mathbf{r},\omega) &= 0 \\ \nabla\times\mathbf{E}(\mathbf{r},\omega) - \mathrm{i}\omega\mu_0\mathbf{H}(\mathbf{r},\omega) &= 0 \\ \nabla\cdot\mathbf{D}(\mathbf{r},\omega) &= 0 \\ \nabla\times\mathbf{H}(\mathbf{r},\omega) + \mathrm{i}\omega\mathbf{D}(\mathbf{r},\omega) &= -\mathrm{i}\omega\delta(\mathbf{r}-\mathbf{r}_p)\mathbf{p} \\ \mathbf{D}(\mathbf{r},\omega) &= \epsilon_0\epsilon_i(\omega)\mathbf{E}(\mathbf{r},\omega) \end{cases} \tag{5.8}$$

$$\forall j\,\forall\mathbf{r}\in S_j : \begin{cases} (\mathbf{E}_j^+ - \mathbf{E}_j^-)\times\mathbf{n} &= 0 \\ (\mathbf{D}_j^+ - \mathbf{D}_j^-)\cdot\mathbf{n} &= 0. \end{cases} \tag{5.9}$$

In terms of the Helmholtz equation, the dipole emission problem is the same as for plane-wave excitation (Eqs (5.4), (5.5), (5.7)) but with a (non-homogeneous) source term:

$$\nabla^2\mathbf{E}(\mathbf{r},\omega) + k_i^2\mathbf{E}(\mathbf{r},\omega) = -\omega^2\mu_0\delta(\mathbf{r}-\mathbf{r}_p)\mathbf{p}. \tag{5.10}$$

Further considerations

From a full solution of the EM problem, all physical properties can be derived using, for example, the tools developed in the previous chapter. However, it is not always necessary to find the full solution, and partial solutions (such as finding the electric field at a given point, or determining the far field only) can be sufficient in many cases.

For SERS and other plasmonic effects, the most important characteristic is the electric field in the vicinity of the metallic substrate, often referred to as *local field* or *near field* [1]. As was discussed in the previous chapter, determining the local field (amplitude and polarization) at a single point under plane-wave excitation is in many cases sufficient to analyze the SERS properties of an emitter at that point. Experimentally, the most common tool to characterize a substrate is to measure its extinction, which is a far-field property. It is therefore equally important to be able to predict the far-field characteristics.

5.1.2. Far field and local/near field

In fact, this constant 'juggling' between far-field and near-field effects is a recurring issue in SERS and plasmonics. The following statement is, we believe, accurate:

[1] Both denominations, local field and near field, are confusing and may have another meaning in a different context, as discussed later.

The SERS or SEF signal is a *far-field* signal governed by the *near-field* properties.

We shall attempt hereafter to clarify this statement without going into the full details of scattering theory [149,196].

Incident and scattered fields

Let us consider an object or substrate in 3D, excited by a laser beam. The laser beam can be approximated by a well-defined propagating incident electromagnetic field $\mathbf{E}_{\mathrm{Inc}}(\mathbf{r})$ (we focus on the electric field for simplicity), the simplest description being that of a plane wave incident from a fixed direction. The presence of the (optically active) object modifies the electromagnetic field, inside and around the object, because of the boundary conditions on the object surface. This modification is associated with the appearance of surface charges/currents. Another way of viewing this is that the incident field creates a polarization and magnetization inside the object, which automatically results in a surface charge/current at its boundaries. These surface charges/currents are sources of the electromagnetic field and therefore create an electromagnetic field called the *scattered field* $\mathbf{E}_{\mathrm{Sca}}(\mathbf{r})$ outside the object. The solution of the EM problem outside the object is simply the sum of the incident and scattered fields: $\mathbf{E}_{\mathrm{Out}}(\mathbf{r}) = \mathbf{E}_{\mathrm{Inc}}(\mathbf{r})+\mathbf{E}_{\mathrm{Sca}}(\mathbf{r})$. This physical interpretation in terms of induced surface charges/currents that create a scattered field is helpful in understanding conceptually the problem, but is not often used in practice to solve it (i.e. it is usually easier to find directly the field solution rather than determining the surface charges and currents and deducing the scattered field from them).

Properties of a radiation field

At a point M sufficiently distant from the object (i.e. much larger than both the wavelength and the object dimensions), the object can be considered as a point O, and the induced surface charges/currents can then be viewed as point sources. It can be shown that the scattered field must then be a *radiation field*, which has the following characteristics [149]:

- It is a spherical electromagnetic wave propagating along wave-vector $\mathbf{k} = k_M \mathbf{e}_R$, where $\mathbf{OM} = R\mathbf{e}_R$ and $k_M = n_M \omega / c$.
- Denoting an angular direction by $\Omega = (\theta, \phi)$ for short, the electric field at a point defined by (R, Ω) can be written as:

$$\mathbf{E} = \frac{\mathrm{e}^{\mathrm{i}k_M R}}{R} \left[A_{P1}(\Omega)\mathbf{e}_{P1}(\Omega) + A_{P2}(\Omega)\mathbf{e}_{P2}(\Omega) \right], \tag{5.11}$$

where $\mathbf{e}_{P1}(\Omega)$ and $\mathbf{e}_{P2}(\Omega)$ are mutually orthogonal unit vectors and also orthogonal to $\mathbf{e}_R$. These, along with the two complex amplitudes $A_{P1}(\Omega)$ and $A_{P2}(\Omega)$, entirely define the radiation field in this direction, i.e. its (real) amplitude and its state of polarization. The magnetic field is simply obtained from $\mathbf{H} = (n_M \epsilon_0 c)\mathbf{e}_R \times \mathbf{E}$. Note that the $1/R$-dependence of the electric field amplitude is the most important characteristic of a radiation field.

- The (complex) Poynting vector, $\mathbf{S} = \frac{1}{2}\mathbf{E} \times \mathbf{H}^*$, is radial (i.e. along $\mathbf{e}_R$) and given by:

$$\mathbf{S}(\Omega, R) = \frac{n_M \epsilon_0 c}{2R^2} \left[|A_{P1}(\Omega)|^2 + |A_{P2}(\Omega)|^2 \right] \mathbf{e}_R. \tag{5.12}$$

The power radiated per unit solid angle in a given direction is then independent of distance R:

$$\frac{\mathrm{d}P}{\mathrm{d}\Omega}(\Omega) = \frac{n_M \epsilon_0 c}{2} \left[|A_{P1}(\Omega)|^2 + |A_{P2}(\Omega)|^2 \right], \tag{5.13}$$

where the two terms in the sum corresponds to polarized detections along either $\mathbf{e}_{P1}$ or $\mathbf{e}_{P2}$. The angular variation of $\mathrm{d}P/\mathrm{d}\Omega(\Omega)$ is called the *radiation profile* or *radiation pattern* and can also be used to characterize the radiation field. The total radiated power can also be derived by integration over Ω.

Definition of far field and near field

This 'radiation field' is what is commonly called *the far field.* To put it differently, the far field is the region that is sufficiently far from the objects for the scattered field to be well approximated by a radiation field. It can then be characterized by the radiation profile $\mathrm{d}P/\mathrm{d}\Omega(\Omega)$ and polarization state, which are independent of the distance R to the object (provided it is large enough). It contains some information (but not all[2]) about the nature of the electromagnetic sources (electric or magnetic, dipolar or multipolar nature, etc.) that produce the radiation. In most standard optical experiments, it is always the far-field properties that are measured. In many cases the polarization is not measured, and $\mathrm{d}P/\mathrm{d}\Omega(\Omega)$ is actually detected in a single direction[3].

By exclusion, *the near field* refers to everything that is not well described by a radiation field. This therefore corresponds to the field in the vicinity

[2] Here we have ignored any phase information, a property typically difficult to measure.
[3] A real system will always detect signals within a range of directions defined by the numerical aperture of the collecting optics (see Section 2.2.6).

of the objects, and more generally in the region where the radiation-field approximation is not valid[4]. There is no exact boundary between far- and near-field regions, but rather a gradual transition from one to the other.

Fields in near-field and far-field regions

In these definitions, the far field and near field refer to *regions of space* where the electromagnetic field has certain characteristics, but not to the electromagnetic field itself. It is however common to express the scattered electromagnetic field as the sum of two fields: the 'far-field field' $\mathbf{E}_{\mathrm{FF}}$, which is a radiation field as defined above with a $1/R$-dependence, and the 'near-field field' $\mathbf{E}_{\mathrm{NF}}$, which is simply the rest (with an implicit faster-than-$1/R$ distance dependence). The electric field in the far-field region is then simply $\mathbf{E}_{\mathrm{FF}}$ (plus possibly the incident field $\mathbf{E}_{\mathrm{Inc}}$ in the appropriate direction). Unfortunately, the electric field in the near-field region is not $\mathbf{E}_{\mathrm{NF}}$, but $\mathbf{E}_{\mathrm{NF}}+\mathbf{E}_{\mathrm{FF}}+\mathbf{E}_{\mathrm{Inc}}$. This can be the source of confusion and we will therefore attempt to refer to far-field and near-field *regions only*, not electromagnetic fields. Instead, we will refer to the electromagnetic field in the near-field region as the *local field*. Note that this denomination is not without its shortcomings: the local field for example may also refer to the microscopic field (as opposed to the macroscopic field, see Appendix C). This is however rarely the source of confusion in a SERS context where mostly the macroscopic field is studied.

Far field and near field in SERS

At this stage, we can now go back to the statement at the beginning of this section, i.e. 'the SERS or SEF signal is a far-field signal governed by near-field properties'. The SERS and SEF signals (as most spectroscopic and optical signals) are indeed far-field signals, because they are measured in the far-field ('radiation-field') region. However, these signals typically come from molecules located in the near-field region, i.e. close to the objects. It is the interaction of these molecules with the local field (in the near-field region) that produces the SERS or SEF signals. Even when we are measuring these signals in the far field, their characteristics remain determined by what happened in the near field. The molecules can be viewed as local 'sensors' or probes located in the near field, which 'transmit' the information to the far field for us to measure it. SERS and SEF can therefore be considered as near-field effects, even though their signal is measured in the far field. This is to be distinguished from truly far-field properties, such as the extinction spectrum. In this case, there is no sensor to 'capture' the near-field information, which is therefore lost by the time it is measured in the far field.

[4] This region is also called the *convection field* in many books on classical electrodynamics [96].

In order to directly measure the local field (without using a molecule as a 'sensor'), one needs to position the measuring instrument sufficiently close to capture this information (without affecting the field too much). This is possible to some extent and is called near-field optics – an active research area in itself – but typically requires complex experimental setups, such as in scanning optical near-field microscopy [12].

5.1.3. Some key EM indicators

Aims of the EM calculations

From a modeling perspective, accessing near-field information is not as demanding as it may be experimentally. To model a typical SERS or SEF experiment, one must first solve the EM problem of the laser excitation of the substrate/objects. The most relevant quantity is then the electromagnetic field $\mathbf{E}_{\text{Loc}}(\mathbf{r})$ at the molecule(s) position. From this, both the local field intensity enhancement factor M_{Loc} and the local field polarization $\mathbf{e}_{\text{Loc}}$ can be derived. These are sufficient for example to determine the modified absorption cross-section (see Section 4.6.2).

This step is necessary for SERS and SEF, but we need, in addition, to model the emission problem. Modeling the dipolar emission from a localized source is in general more difficult than the excitation problem. For SERS, emission is usually studied in a single direction (the direction of the collecting optics), and one can then derive the directional radiative enhancement factor M^d_{Rad} from the solution of two excitation EM problems using the optical reciprocity theorem, as shown in Sections 4.4.6 and 4.5. For SEF, however, one usually requires the integrated radiative enhancement factor M_{Rad} and the total decay rate enhancement factor M_{Tot} (see Section 4.4.3). It is then usually necessary to model the dipolar emission problem and use the tools developed in Section 4.4.

To model the SERS or SEF situations, one may in addition need to define the optical properties of the molecule itself (Raman polarizability tensor, adsorption orientation, etc.). Once these quantities are determined, the SERS or SEF intensities (and other properties) can then be predicted following the treatment given in Sections 4.5 and 4.6 of the previous chapter. Depending on the problem under consideration, this could be at a single position, or at all points on a surface (for a monolayer or molecule), or even at all points within a given distance from the surface (for multiple layers).

Finally, it is also important to have tools to predict the far-field properties of a given substrate such as extinction or scattering cross-sections. These are much more easily compared to experiments, since they do not involve any adsorbed molecules and their associated problems. They will also reveal some of the most important electromagnetic resonances (as functions of wavelength) that are present in the system.

The need for key EM indicators

We should note here that studying the full details of the EM properties (SERS and SEF) for a given substrate is an immensely complicated task. Parameters such as Raman tensors, adsorption geometry of the molecules, coverage, distance from the surface, laser polarization, etc. can in principle be varied across endless combinations. If all of them are known, it is then possible to predict the SERS and SEF properties accurately. However, the results would then be very specific to the chosen combination of parameters and, most likely, not so useful from the point of view of understanding general experimental data. In fact, in most real situations a big fraction of these parameters is not known at all. It is therefore equally important to define a few key EM characteristics, which provide a good idea of the performances that can be expected from a given substrate independent of any adsorbed molecules, and can further be used for a more elaborate description of specific conditions.

These key indicators can be separated into two groups: those representing *average* EM properties and those representing *punctual* EM properties, i.e. the EM response at a given point on the substrate. This latter group is relevant to single molecule applications, where only a specific position on the substrate provides the required response. In many applications for SERS and plasmonics, the wavelength (or frequency) dependence is one of the most important features and these indicators are therefore functions of excitation frequency (ω_L) and/or emission frequency (ω_R). The polarization of the incident field in excitation, or the dipole orientation in emission, are also important parameters. The key EM indicators can therefore be evaluated for two relevant (orthogonal) incident polarizations and/or two or three relevant (orthogonal) dipole orientations.

Position-dependent key EM indicators

We now list examples of key EM indicators, starting with those characterizing a given position $\mathbf{r}$ on the substrate:

- <u>The local field intensity enhancement factor:</u> $M_{\text{Loc}}(\omega_L, \mathbf{r})$
  This simply characterizes the intensity enhancement at a given position $\mathbf{r}$ for incident *plane-wave excitation* at a frequency ω_L:

$$\boxed{M_{\text{Loc}}(\omega_L, \mathbf{r}) = \frac{|\mathbf{E}_{\text{Loc}}(\omega_L, \mathbf{r})|^2}{|E_{\text{Inc}}|^2} = |\tilde{\mathbf{E}}_{\text{Loc}}(\omega_L, \mathbf{r})|^2,} \tag{5.14}$$

  where $\tilde{\mathbf{E}}_{\text{Loc}}(\omega_L, \mathbf{r})$ is defined (as in Section 4.5) as the local field normalized with respect to the incident electric field amplitude E_{Inc}. By linearity, $\tilde{\mathbf{E}}_{\text{Loc}}(\omega_L, \mathbf{r})$ (and therefore M_{Loc}) does not depend on E_{Inc}. Note that by considering M_{Loc} only (as opposed to $\tilde{\mathbf{E}}_{\text{Loc}}$), any information on the local field polarization is lost. Its effect is typically secondary and can be studied separately if needed. Part of the polarization information can in fact be recovered by expressing the local field as the sum of a perpendicular and parallel contributions (defined with respect to the metallic surface). The corresponding LFIEF, $M_{\text{Loc}}^{\perp}$ and $M_{\text{Loc}}^{\parallel}$, then reflect at least partially the local field polarization with respect to the metallic surface. Moreover, to characterize a substrate, M_{Loc} should be calculated for relevant incident polarizations (typically two perpendicular polarizations), and at relevant positions on the substrate (if possible at every point on the surface).

- The zero-Stokes-shift SERS EF in the $\mid E \mid ^4$-approximation: $F_{\text{E4}}^{0}(\omega_L, \mathbf{r})$
  It is simply expressed as:

$$F_{E4}^{0}(\omega_L, \mathbf{r}) = M_{\text{Loc}}(\omega_L, \mathbf{r})^2 = |\tilde{\mathbf{E}}_{\text{Loc}}(\omega_L, \mathbf{r})|^4. \tag{5.15}$$

  The conditions of validity of this approximation and its relevance to real SERS enhancements have been discussed extensively in Section 4.5. It gives a general figure for the SERS enhancement factor that can be expected at a given position $\mathbf{r}$. It is arguably the most important indicator as far as SERS is concerned, the remaining indicators in this list applying mostly to SEF.

- The radiative decay rate enhancement: $M_{\text{Rad}}(\omega_R, \mathbf{r})$
  This was defined in Section 4.4.3. For completeness, it should be calculated for a least two dipole orientations, for example: perpendicular ($M_{\text{Rad}}^{\perp}(\omega_R, \mathbf{r})$) or parallel ($M_{\text{Rad}}^{\parallel}(\omega_R, \mathbf{r})$) to the surface.

- The total decay rate enhancement: $M_{\text{Tot}}(\omega_R, \mathbf{r})$
  This was defined in Section 4.4.3. As before, it should be calculated for at least two dipole orientations, for example: perpendicular ($M_{\text{Tot}}^{\perp}(\omega_R, \mathbf{r})$) or parallel ($M_{\text{Tot}}^{\parallel}(\omega_R, \mathbf{r})$) to the surface. Its magnitude with respect to $M_{\text{Rad}}(\omega_R, \mathbf{r})$ gives a measure of the importance of non-radiative effects for emitters at this frequency and position.

- The EM radiative efficiency: $\eta_{\text{Rad}}^{\text{EM}}(\omega_R, \mathbf{r})$
  This is defined as (see Section 4.4.3):

$$\eta_{\text{Rad}}^{\text{EM}}(\omega_R, \mathbf{r}) = \frac{M_{\text{Rad}}(\omega_R, \mathbf{r})}{M_{\text{Tot}}(\omega_R, \mathbf{r})}. \tag{5.16}$$

  $\eta_{\text{Rad}}^{\text{EM}}(\omega_R, \mathbf{r})$ characterizes the relative importance of radiative and non-radiative emission for an emitter at this frequency and position. For an emitter with an intrinsic quantum yield of 1, $\eta_{\text{Rad}}^{\text{EM}}$ is exactly the modified quantum yield under SEF conditions. For less efficient emitters, $\eta_{\rm Rad}^{\rm EM}$ is in many cases a good approximation of its modified quantum yield (see Section 4.6).

- The approximate fluorescence EF: $M_{\rm Fluo}$
  This is defined as

$$M_{\rm Fluo}(\omega_L, \mathbf{r}) = M_{\rm Loc}(\omega_L, \mathbf{r})\eta_{\rm Rad}^{\rm EM}(\omega_L, \mathbf{r}). \tag{5.17}$$

  Many approximations are made in writing this expression (good fluorophore with free-space quantum yield $Q^M \approx 1$, negligible Stokes shift of fluorescence, no spectral dependence of fluorescence, only integrated fluorescence signal considered, etc.), but $M_{\rm Fluo}$ still represents a good general indicator of the fluorescence enhancement expected from a good fluorophore excited at frequency ω_L at a given position (see Section 4.6 for details).

- The approximate differential fluorescence EF in the $| E |^4$-approximation: $M_{\rm Fluo}^{d-E4}$
  This is an alternative approximate expression for the fluorescence enhancement factor. It is based on the comparison between the SERS and SEF enhancement factors discussed in Section 4.7.1, combined with the $|E|^4$-approximation of the SERS EF. It results in:

$$M_{\rm Fluo}^{d-E4}(\omega_L, \mathbf{r}) = \frac{F_{E4}^{0}(\omega_L, \mathbf{r})}{M_{\rm Tot}(\omega_L, \mathbf{r})}. \tag{5.18}$$

  Similar approximations as before are made here. $M_{\rm Fluo}^{d-E4}$ should provide a similar estimate as $M_{\rm Fluo}$, but is based on the approximate fluorescence enhancement factor for detection *in a given direction* (as opposed to integrated scattering).

All these indicators can be calculated at relevant points on the surface, for example at the points of highest local field enhancement (so-called hot-spots). Ideally, they could even be calculated at all points on the surface to study the effects of their spatial distribution [162].

Average key EM indicators

We can also define a few average indicators, which characterize the EM response of the substrate as a whole. For this, we note that by calculating the value of a quantity $A(\mathbf{r})$ at every point on a substrate surface, it is possible to derive its surface-average value $\langle A \rangle$. This can be formally written as:

$$\langle A \rangle = \frac{1}{S} \int_S A(\mathbf{r}) \mathrm{d}S. \tag{5.19}$$

$\langle A \rangle$ then represents the average response from adsorbed molecules (either a monolayer, or a smaller number of randomly adsorbed molecules).

We can highlight the following quantities as key average EM indicators:

- The surface-averaged local field intensity EF: $\langle M_{\text{Loc}}(\omega_L) \rangle$
- The surface-averaged zero-Stokes-shift SERS EF in the $|E|^4$-approximation: $\langle F^0_{\text{E4}}(\omega_L) \rangle$
  This gives an estimate of the performance of the substrate for SERS toward analytical applications. It can be viewed as an approximate average SERS enhancement factor.
- The approximate average fluorescence EF: $\langle M^{d-E4}_{\text{Fluo}}(\omega_L) \rangle$
  This is based on the averaging of M^{d-E4}_{Fluo} defined above, since it is usually easier to calculate than its cousin M_{Fluo}:

$$\langle M^{d-E4}_{\text{Fluo}}(\omega_L) \rangle = \left\langle \frac{F^0_{\text{E4}}(\omega_L)}{M_{\text{Tot}}(\omega_L)} \right\rangle. \tag{5.20}$$

  This definition is chosen to provide a reasonable estimate of the performance of a substrate for SEF toward analytical applications (for example with a monolayer or randomly positioned fluorophores). Note in this context that dielectric spacers are often used on top of the metal surface for SEF. The average should then be taken on the dielectric spacer surface (i.e. at a fixed distance from the metal surface). Another possible (and similar) definition for this indicator would be to use $\langle M_{\text{Fluo}}(\omega_L) \rangle$.
- The optical density or extinction coefficient: $Q_{\text{Ext}}(\omega_L)$
  This is a far-field property, which is important because it is in general easily accessible experimentally, and to a lesser extent theoretically. The extinction coefficient is usually obtained from a transmittance measurement (see Section 7.2.2 for details) by simply measuring the transmitted intensity $I(\omega_L)$ and comparing it to the reference transmitted intensity I_0 (in the absence of the metallic substrate). It is then usually defined as:

$$Q_{\text{Ext}}(\omega_L) = \log_{10} \frac{I_0}{I(\omega_L)}. \tag{5.21}$$

  The magnitude of $Q_{\text{Ext}}(\omega_L)$ characterizes the strength of the optical response of the system at frequency ω_L. Its frequency dependence can be used to identify some of the optical resonances of the system. These resonances should also appear in some of the local field properties, but their strength may be very different. The extinction coefficient can be viewed (at least formally) as the sum of two contributions:

$$Q_{\mathrm{Ext}} = Q_{\mathrm{Abs}} + Q_{\mathrm{Sca}}, \tag{5.22}$$

where Q_{Abs} and Q_{Sca} are the absorption and scattering coefficients, respectively. They correspond to the two physical processes by which the energy of the incident beam can be extinguished: either by optical absorption in the metallic objects (producing heat), or by scattering, i.e. re-emission of the energy in other directions. This decomposition may be used to calculate Q_{Ext}, but measuring Q_{Abs} and Q_{Sca} experimentally is usually more difficult: only Q_{Sca} can in general be measured, but typically in only one detection direction (for example, in a reflection measurement or in a dark-field configuration). Q_{Abs} and Q_{Sca} can also be associated qualitatively with the non-radiative and radiative contributions to extinction.

- <u>The extinction cross-section:</u> $\sigma_{\mathrm{Ext}}(\omega_L)$ $[\mathrm{m}^2]$
For individual objects or nano-particles, it is usually more relevant to consider the *extinction cross-section* $\sigma_{\mathrm{Ext}}(\omega_L)$, rather than the extinction coefficient $Q_{\mathrm{Ext}}(\omega_L)$. This is because the power extinguished by a single particle, P_{Ext} [W], depends linearly on the incident power density S_{Inc} $[\mathrm{W/m}^2]$ on the particle, not on the incident power. The extinction cross-section is therefore defined as:

$$\sigma_{\mathrm{Ext}} = \frac{P_{\mathrm{Ext}}}{S_{\mathrm{Inc}}}. \tag{5.23}$$

Note that the extinction coefficient of a single nano-particle, $Q_{\mathrm{Ext}}^{\mathrm{NP}}$, is sometimes defined as the ratio of its extinction cross-section over its geometric cross-section, i.e. $Q_{\mathrm{Ext}}^{\mathrm{NP}} = \sigma_{\mathrm{Ext}}/\sigma_{\mathrm{Geom}}$. This definition has a very different physical meaning from the more general definition of Q_{Ext} given earlier, and will be denoted by $Q_{\mathrm{Ext}}^{\mathrm{NP}}$ to avoid confusion with the 'real' extinction coefficient or optical density Q_{Ext}. The concept of optical density, in fact, recovers its meaning for an ensemble of particles and then depends naturally on the concentration of such particles. For particles in solution, this is discussed further in Section 7.2.2. For particles on a planar substrate with a surface density μ_M $[\mathrm{m}^{-2}]$, the extinction coefficient of the substrate is then given by:

$$Q_{\mathrm{Ext}} = \mu_M \sigma_{\mathrm{Ext}}. \tag{5.24}$$

Finally, similar considerations apply to the absorption and scattering cross-sections of the particle: σ_{Abs} and σ_{Sca}.

5.1.4. The electrostatic approximation (ESA)

We have already discussed some basic approximations necessary to solve the EM problem. These are approximations that are almost always made (and are usually implicit). There are a few other approximations that can be made to simplify further the problem, but these might not always be valid and need to be justified. Among them, one of the most important and most commonly used is the *electrostatic approximation* (ESA), also called the quasi-static approximation, or the long-wavelength approximation, or the Rayleigh approximation[5]. It is in particular quite common in the study of the optical properties of nano-particles. The reason is simple: it works well when the dimensions of the object are much smaller than the wavelength of light, with a typical limit being of the order of $\lambda/20$ [196], possibly $\lambda/10$. In the visible, this means that it is roughly valid for particles of typical dimensions of the order of 20–40 nm (at most). Despite this, it is regularly applied to larger systems, simply because it is much simpler. In this case, one has to be careful about the conclusions, since this approximation may miss out on some important features.

Formulation of the ES approximation

The basic principle of the ES approximation is to solve an equivalent electrostatics problem instead of the full EM equations, but *still using the complex frequency-dependent dielectric functions* to describe the optical response of the materials. This is equivalent to assuming an *instantaneous* response of the electromagnetic fields, a fact often described as neglecting any *retardation effects*, which can also be viewed as letting the speed of light c tend to ∞. The solution then corresponds to the complex fields (with an implicit time-harmonic dependence at the incident frequency ω). One important consequence is that the magnetic field is zero everywhere: $\mathbf{H} = 0$. Also the incident field cannot propagate (there are no propagating waves in electrostatics!) and is therefore typically taken as a constant external field, $\mathbf{E}_{\text{Inc}}$. The series of 8 equations describing the excitation problem (given in Section 5.1.1), for example, is then reduced to:

$$\forall i\,\forall \mathbf{r} \in V_i : \begin{cases} \nabla \times \mathbf{E}(\mathbf{r},\omega) &= 0 \\ \nabla \cdot \mathbf{D}(\mathbf{r},\omega) &= 0 \\ \mathbf{D}(\mathbf{r},\omega) &= \epsilon_0 \epsilon_i(\omega) \mathbf{E}(\mathbf{r},\omega) \end{cases} \tag{5.25}$$

$$\forall j\,\forall \mathbf{r} \in S_j : \begin{cases} (\mathbf{E}_j^+ - \mathbf{E}_j^-) \times \mathbf{n} &= 0 \\ (\mathbf{D}_j^+ - \mathbf{D}_j^-) \cdot \mathbf{n} &= 0 \end{cases} \tag{5.26}$$

[5] The exact meaning of these approximations may differ depending on the authors. For example, the quasi-static approximation sometimes refers to the electrostatics solution combined with the magnetostatics solution [138]. Such a distinction is usually irrelevant in the context of SERS.

$$\mathbf{E}(\mathbf{r}) = \mathbf{E}_{\text{Inc}} \quad \text{for} \quad x \to -\infty. \tag{5.27}$$

Moreover, in electrostatics, the electric field can be derived from a single *scalar potential* V. The (vectorial) Helmholtz equation for the electric field then reduces to the Laplace equation for the potential V: $\nabla^2 V = 0$. The solution is in general much simpler, because the problem has been reduced to a *scalar problem* instead of a vectorial one.

One major characteristic of the ES solution is that it is *scale-invariant*, i.e. it does not depend on the size of the objects, but only on their shape and optical properties. Obviously, for this reason, it cannot be valid for all sizes and we shall see that it only applies to the smallest objects.

Validity of ES approximation

Let us now discuss *qualitatively* the validity of this approximation:

- We can distinguish two types of retardation effects. The first one relates to the propagation of the exciting field, i.e. the fact that it is not uniform, but may vary (with an $\exp(ik_M x)$-dependence, where $k_M = n_M\omega/c$) from one side of the object to the other. This approximation is *a priori* reasonable if $\exp(ik_M x) \approx 1$ within the object, i.e. $k_M x \ll 1$. Denoting D the (largest) dimension of the object, this can be rewritten as $D/\lambda \ll (2\pi n_M)^{-1}$. This is more or less the condition $D \leq \lambda/20$ quoted earlier. If this is not fulfilled, then there is a phase difference between the exciting field on two opposite sides of the object, which could lead to interference effects that would not be captured in the ESA.

- The second aspect is related to retardation effects for the internal field inside the metallic particle, which are characterized by an $\exp(ikr)$-dependence, where $k = \sqrt{\epsilon}\omega/c$. Because $\epsilon(\omega)$ is complex, it is more difficult to understand these. In the context of Rayleigh scattering by small particles [196], it is often assumed that the condition $D/\lambda \ll (2\pi|\sqrt{\epsilon(\omega)}|)^{-1}$ must be fulfilled. This may be more restrictive than the first condition given above, especially for metals. From an alternative point of view, we can write $\sqrt{\epsilon} = n + i\kappa$, where n characterizes wave propagation and κ optical absorption (damping). n is typically smaller than 1 and κ larger than 1 for metals like gold or silver under conditions considered here. Since κ is much larger than n, retardation effects are strongly damped inside the metal, i.e. retardation effects inside the particle should therefore be secondary compared to those outside.

- Finally, following similar arguments, the ES solution is only valid around the object within a region of space of dimension $\approx \lambda/20$. This implies that the far-field properties (in particular the radiation field) are not directly described in the ESA. This is also a natural consequence of the fact that there is no propagating wave in electrostatics (and therefore no radiation field).

In practice, assessing *a priori* the validity of the ES approximation is a difficult undertaking in general. This is especially true for metallic objects because of their strong EM response (in the region of localized surface plasmon resonances). One rigorous approach to this issue is to justify the validity of the ESA by comparing its predictions with those of the full EM solution. This is not always possible and obviously diminishes the interest of solving the ES approximation in the first place (if the full EM solution is obtainable).

Alternatively, it is also in principle possible to justify the ES approximation by calculating rigorously the following terms in the approximation, i.e. terms of order $k_M D$, $(k_M D)^2$, etc. A general and rigorous treatment of this problem has been given by Stevenson [197,198]. This elegant theory is arguably the best formal justification of the ES approximation. It is however only practical in the simplest cases like a sphere [198], where a full EM solution is typically no more difficult. Finally, it is worth highlighting here that the ES approximation (or Rayleigh approximation) is different from another commonly-used approximation for EM scattering by small particles called the Rayleigh–Gans or Rayleigh–Debye–Gans approximation [149,196]. The latter only applies to small particles that are only mildly optically active, i.e. with $\epsilon \approx 1$. This condition automatically excludes any metallic particles.

Overall, the ES approximation remains a very useful tool and is in fact commonly used in situations where its formal justification would be difficult or impossible (particle sizes larger than $\lambda/10$). The reason is that it still captures many important aspects of the EM problem to be solved and, as long as one bears in mind its limitations, an ES solution is better than no solution. In particular, the local field on the surface of small nano-particles is in principle well described in the ES approximation within its range of validity. One of the main shortcomings of the electrostatic approximation, however, is that it is scale-invariant, i.e. it does not predict any size-related effect. Further discussion of the ESA validity using several examples will be given in Chapter 6.

Far-field properties in the ES approximation

Provided the field solution inside the particle is valid within the ESA, the absorbed power, P_{Abs} [W] can be easily computed by simple integration over the volume V_P of the particle. It is given by [96]:

$$P_{\text{Abs}} = \int_{V_P} \frac{1}{2}\omega\epsilon_0 \text{Im}(\epsilon)|\mathbf{E}(\mathbf{r})|^2 \text{d}r. \tag{5.28}$$

From this, one can for example predict the absorption cross-section, σ_{Abs}. Note that since the ES solution is scale-invariant, P_{Abs} and σ_{Abs} scale proportionally with the particle volume V_P. Moreover, as mentioned earlier, there is no propagating wave or EM energy flow in electrostatics (and therefore no radiation field). The radiation field and scattering cross-section are therefore formally zero in electrostatics.

This conclusion could however appear as a serious limitation of the ES approximation in terms of far-field predictions. Even if the scattered power is negligible compared to the absorbed power, it is nevertheless the most relevant in terms of the radiative properties of the object. For a non-absorbing particle, for example, standard electrostatics theory would predict $\sigma_{\text{Ext}} = \sigma_{\text{Abs}} = \sigma_{\text{Sca}} = 0$; not a very useful prediction... It is therefore desirable to compute, if possible, the scattered field and radiative properties from the ES solution and there are in fact several ways to achieve this:

- Firstly, as explained in Section 4.4.6, the radiative properties of a dipole located close to a metallic object can in principle be obtained from a solution of one or more plane-wave excitation (PWE) problems by applying the optical reciprocity theorem (Chapter 4). If the ES approximation provides a valid solution of the local field at the dipole position for the PWE problems, then the radiative properties of this dipole can be deduced.

- In a similar fashion, if the local field at the surface of an object is well approximated by the ES solution, then additional tools can be used to deduce far-field properties, even though electrostatics theory would fail to do so. One can show from standard EM theory that the EM solution of Maxwell's equations is entirely determined by a knowledge of the EM fields on a closed surface. In practice one can therefore use the Stratton–Chu formula [95] – or any similar expressions [198] – to derive the full EM solutions from the ES approximation (provided it gives an accurate description of the local field at the object surface). The radiation field in particular (intensity and radiation profile) can be calculated with such tools. This approach is fairly general, but can be cumbersome to implement, and may therefore be more suited to numerical computations. More details on these far-field/near-field relations can be found, for example, in Ref. [198].

- As far as the radiation field is concerned, a somehow simpler approach can be used, based on the dipolar approximation, which we now discuss in more detail.

The dipolar approximation

The *dipolar approximation* is a common approach in the context of the ES approximation for nano-particles. Some authors may also call it the *Rayleigh approximation*[6], due to its similarities with the treatment of Rayleigh scattering by small particles [149,196]. Its principle is very simple, although its justification and range of validity are more difficult to assess. It consists in modeling the EM response of a small nano-particle by that of an induced dipole. One must first determine the dipolar polarizability of the nano-particle, which is typically derived from the ES solution. The radiative properties of the induced dipole then derive simply from standard EM theory.

To describe these steps in more detail, let us consider a nano-particle (with dielectric function $\epsilon(\omega)$) embedded in a dielectric medium (with dielectric constant ϵ_M) and assume that the ES solution for a given constant external field $\mathbf{E}_{\text{Inc}}$ is known. This external field induces an electric polarization $\mathbf{P}(\mathbf{r})$ inside the particle. $\mathbf{P}$ is discussed (and defined) in Appendix C for example, but because our particle is embedded in a dielectric medium here, it is more convenient (and more physical) to consider the electric polarization $\mathbf{P}_M$ with respect to this medium (rather than vacuum). It is then defined by the relation $\mathbf{D} = \epsilon_0\epsilon_M\mathbf{E} + \mathbf{P}_M$, where the usual constitutive relation $\mathbf{D} = \epsilon_0\epsilon\mathbf{E}$ still holds. We then have inside the particle:

$$\mathbf{P}_M(\mathbf{r}) = \epsilon_0(\epsilon - \epsilon_M)\mathbf{E}(\mathbf{r}). \tag{5.29}$$

Within the ESA, this electric polarization exhibits a time-harmonic dependence at frequency ω, as imposed by the incident field and is, hence, a source of electromagnetic fields and radiation. Viewed from the far field, it can be considered as a punctual source, and its radiated field can be expressed in terms of its multipolar components as described for example in Chapter 9 of Ref. [96]. We only retain here the dipolar component, which is the dominant one in most cases. It can be simply described as the field radiated by a dipole $\mathbf{p}_M$ (embedded in a dielectric with ϵ_M), with:

$$\mathbf{p}_M = \int_{V_P} \mathbf{P}_M(\mathbf{r})\mathrm{d}V = \epsilon_0(\epsilon - \epsilon_M)\int_{V_P} \mathbf{E}(\mathbf{r})\mathrm{d}V. \tag{5.30}$$

Thanks to the linearity of the problem, this can be rewritten as

$$\mathbf{p}_M = \hat{\alpha}_M \cdot \mathbf{E}_{\text{Inc}}, \tag{5.31}$$

where $\hat{\alpha}_M$ $[\epsilon_0\ \text{m}^3]$ is the *dipolar polarizability tensor of the particle* in the ES approximation, defined by the previous two equations. $\hat{\alpha}_M$ depends on the optical properties of the material (through $\epsilon(\omega)$) and its environment (through

---

[6] There is a disparity of choices in terms of what the Rayleigh approximation and Rayleigh scattering really refer to. An interesting discussion and historical perspective are given in Ref. [199] and a detailed theoretical account in Ref. [200].

ϵ_M), and on the nano-particle shape and size (through the volume integration). In fact, because the ES solution is scale-invariant, $\hat{\alpha}_M$ scales proportionally with the particle volume V_P (all other parameters being fixed).

From there, the radiated field (intensity and radiation profile) can simply be obtained from the standard radiative properties of an oscillating dipole. Let us for example derive here the scattering cross-sections in the simple case of a scalar dipolar polarizability α_M (equivalent to an isotropic tensor). The scattered (or radiated) power [W] is (Eq. (4.25)):

$$P_{\text{Sca}} = \frac{n_M \omega^4 |\mathbf{p}_M|^2}{12\pi\epsilon_0 c^3} = \frac{n_M \omega^4 |\alpha_M|^2}{12\pi\epsilon_0 c^3} |\mathbf{E}_{\text{Inc}}|^2. \tag{5.32}$$

Recalling that the incident power density is given by $S_{\text{Inc}} = (n_M \epsilon_0 c)/2 |\mathbf{E}_{\text{Inc}}|^2$ [W/m^2], the scattering cross-section [m^2] in the dipolar approximation is:

$$\boxed{\sigma_{\text{Sca}} = \frac{P_{\text{Sca}}}{S_{\text{Inc}}} = \frac{(k_M)^4}{6\pi} \frac{|\alpha_M|^2}{(\epsilon_0 \epsilon_M)^2},} \tag{5.33}$$

where $k_M = \sqrt{\epsilon_M}\omega_L/c$ [m^{-1}] is the wave-vector in the embedding medium. This can be summed to the absorption cross-section σ_{Abs} (derived from Eq. (5.28)) to obtain the extinction cross-section $\sigma_{\text{Ext}} = \sigma_{\text{Abs}} + \sigma_{\text{Sca}}$.

An alternative expression can be obtained from the power [W] extinguished by the induced dipole, given by [96,149]:

$$P_{\text{Ext}} = \frac{\omega}{2} \text{Re}\left(i\mathbf{p}_M^* \cdot \mathbf{E}_{\text{Inc}}\right) = \frac{\omega_L}{2} \text{Im}(\alpha_M) |\mathbf{E}_{\text{Inc}}|^2, \tag{5.34}$$

and resulting in the extinction cross-section [m^2] in the dipolar approximation:

$$\boxed{\sigma_{\text{Ext}} = \frac{P_{\text{Ext}}}{S_{\text{Inc}}} = k_M \frac{\text{Im}(\alpha_M)}{\epsilon_0 \epsilon_M}.} \tag{5.35}$$

The two approaches do not give the same result for σ_{Ext}, and this is a consequence of the various approximations made in obtaining them. In fact, if they differ too much, it is a strong indication that the ES approximation is no longer valid.

These expressions, together with Eq. (5.28) for absorption, highlight two important characteristics of scattering by small nano-particles and typically define the Rayleigh scattering regime:

- The scattering cross-section scales as ω^4 ($1/\lambda^4$), while extinction and absorption cross-sections scale as ω ($1/\lambda$).

- Moreover, since α_M is proportional to the particle volume V_P, the predicted scattering cross-section scale as $(V_P)^2$, while the extinction and absorption cross-sections scale as V_P. A direct consequence is that the equality $\sigma_{\text{Ext}} = \sigma_{\text{Abs}} + \sigma_{\text{Sca}}$, which relates to energy conservation, is only valid in the limit of vanishing volume V_P. The fact that it breaks down as the size increases simply reflects the inadequacy of the ES approximation, except for the smallest particle sizes.

5.1.5. Other approximations

Two-dimensional (2D) approximations

A somewhat different approach to simplifying the EM problem is to approximate it as a two-dimensional (2D) problem, rather than the (real) 3D one. This is particularly suited to numerical solutions, where the gains in memory and computing time can be dramatic by going from 3D to 2D.

There are two very different situations where 2D models may be used:

- Firstly, there are a small number of situations that are, thanks to translational symmetry, truly 2D in nature. Among these is the typical example of an infinite cylinder. The reduction to a 2D problem is then fully mathematically justified and the results are equivalent to solving the 3D problem. One must however be careful with the inclusion of any aspects that are 3D in nature, such as an emitting dipole. This breaks the translational symmetry and voids any formal justification of the results in a real 3D situation.
- The other class of 2D models includes those that are used to understand some aspects of the EM problem that may be difficult to model in 3D. An example would be the comparison of the excitation of a 2D metallic disk with a 2D metallic triangle or square to study the effect of corners. This may indeed be instructive (and better than nothing), but one should always bear in mind that 2D electromagnetism is different in many respects to its 3D counterpart. The electrostatic field of a point charge and the field created by an oscillating dipole are for example very different; radiation fields decay as $1/\sqrt{r}$ in 2D, not $1/r$ (for more details see e.g. Chapter 12 in Ref. [198]). These differences must be accounted for carefully when interpreting the results of simulation. Despite these, 2D models can be useful for understanding qualitatively many features of the problem. Quantitative predictions (such as enhancement factors) cannot however be trusted, unless justified independently by a full 3D model.

Remarks on the various approximations

It is sometimes possible to use these approximations even when they are not valid, since they may still capture some features of the problem. This may be because the approximated solution is much faster to obtain. In this case, one may consider carrying out a more complete simulation on a few carefully chosen examples. By comparing these results with those obtained from the approximated solution, it is then possible to assess clearly the validity of the approximation over a well-defined parameter space, and use confidently the approximated solutions.

Sometimes, approximations are used because the problem would otherwise be intractable. It is then important to understand clearly the implications of the approximation, in order to identify which features are real physical effects and which are only artifacts of the approximation.

5.2. ANALYTICAL TOOLS AND SOLUTIONS

With the background of the first section, we can now move on to the issue of actually solving the electromagnetic problem. In this respect, analytical solutions have many advantages over numerical ones. Unfortunately, there are not that many problems that are tractable analytically, and among them, a large part are mathematically challenging. In particular, in most cases, numerical tools remain necessary to compute analytical solutions expressed as series or integrals. Nevertheless, analytical solutions are useful for a number of reasons, including:

- to study the validity of approximations (such as the ESA);
- to validate numerical methods by comparing their results with reliable analytical solutions on a few simple examples;
- to carry out extensive parametric studies (for example: wavelength, distance, or polarization dependence), which are often intractable numerically.
- Finally – and most importantly for our purpose here – analytical solutions, especially the simplest ones mathematically, can serve as a didactic tool to pinpoint and understand the underlying physical concepts.

We therefore review in this section some of the most common (and simplest) analytical solutions relevant to SERS and plasmonics electromagnetic calculations. We have deliberately left out the technical details, which can be found in the appendices, and only provide here a general discussion of the available tools. This discussion should hopefully be a sufficient starting point to use the numerical implementations of some of these solutions, many of which can be downloaded from the book website (www.victoria.ac.nz/raman/book). Some of these tools will moreover be extensively illustrated and discussed on specific examples in the next chapter.

5.2.1. Plane surfaces

A semi-infinite planar metallic surface could appear at first as the simplest substrate to study. This is partly true, and for example, reflection/refraction of a plane wave at a planar interface is one of the most basic electromagnetic problems, and is treated in most electromagnetic textbooks and in Appendix F. In its most basic form, it can simply be described by Snell's law (or Descartes' law) of reflection/refraction. The inclusion of absorbing materials like metals already complicates the matter and a more accurate treatment then consists in defining and using the Fresnel coefficients for reflection and transmission. The complexity can be further increased by including more than one interface (i.e. layered structures). This increased complexity (either by using metals, multi-layers, or both) is rewarded by a wealth of additional interesting physical phenomena, many of which are directly related to plasmonics. Understanding these EM calculations and/or being able to implement them numerically is therefore well worth the effort for anyone interested in plasmonics and related effects, such as SERS. Appendix F describes the derivation of the simplest and most important results and their practical use, and additional details can be found in numerous EM textbooks (e.g. Ref. [96,151]).

Another important problem in this class is that of emitters (as opposed to incident plane waves) close to plane surfaces. This problem can also be treated entirely analytically but is much more involved mathematically, and the solutions must be expressed as integrals, which must then be estimated numerically. The most important results are also summarized in Appendix F. Although the full EM solution is obtainable, the ES solution of a dipole emitting in close vicinity of a metallic plane provides a good example of the utility of the ESA. This will be discussed in more detail in Section 6.1.

Finally, Appendix F also discusses briefly the numerical computation of these analytical solutions at planar interfaces. The corresponding ready-to-use Matlab codes can be found on the book website.

5.2.2. The perfect sphere

Mie theory

The problem of scattering by a sphere can be solved analytically using the formalism of *Mie theory* [149,158,196]. It is, actually, one of the simplest solutions for the full EM solution of scattering by a particle. In this respect, it is a very important 'toy model' to understand the validity of approximations, and many important features of the optical properties of metal particles. Unfortunately, 'the simplest' does not mean really 'simple' here, and it should more be understood here as 'less complicated'. The mathematics of Mie theory is indeed (in places) challenging, and we have accordingly left its description as an appendix (Appendix H).

In short, the electromagnetic field solutions can be expressed analytically as series (infinite sums) of vectors called *vector spherical harmonics* (which are functions of position). From these it is possible to express analytically, also as series, most EM properties, such as extinction cross-section, etc. The analytical expressions remain complicated (with Bessel functions and spherical harmonics, for example), and can in practice be evaluated numerically by truncating the series (summing the first N terms only). With modern computers and software like Matlab, sums with $N \approx 100$ take a negligible time, and ensure convergence of the series for most cases of interest here, i.e. for metallic spheres of radii up to at least $a \approx 100$ nm. Smaller N's can in fact be used in many cases, provided one is satisfied that convergence has been reached (for example, by ensuring that the results are not changed for larger N's). Examples of Matlab codes that can be used for Mie theory and are based on the analytical expressions given in Appendix H can be downloaded from the book website.

We have, in addition, attempted in Appendix H to describe first the principles of Mie theory in simple, non-mathematical terms, and have also emphasized the numerical implementation of the theory. Both of these aspects should, we hope, enable the interested reader to use Mie theory as a tool, without having to delve into its mathematical details. With Mie theory in our toolbox, it is then possible to work with much simpler approximations (like the ES approximation for nano-particles described in the following) to understand more clearly the important physical principles. Numerical implementations of Mie theory can then be used to validate, and possibly extend, these conclusions. This approach will be followed in Section 6.2 in the next chapter to discuss many features of the localized surface plasmon resonances of metallic nano-particles. We will also discuss the much simpler treatment of the problem in the ES approximation and both approaches will then be compared to discuss the plasmonics properties of small metallic spheres.

Generalized Mie theory

As briefly mentioned in Chapter 3, coupled-LSP resonances arising from the interaction of two closely-spaced metallic objects ('gap-plasmon resonances') typically exhibit the largest local field enhancements and accordingly play an important role in SERS. An exact analytical solution that would serve as a model example of such a situation is therefore desirable. Such a solution in fact exists for the simplest case of two (or in fact more) spheres. It is based on an extension of Mie theory to the case of an arbitrary number of spherical particles, which, although conceptually simple, is considerably more involved mathematically and numerically (for its implementation). It is in general called *generalized Mie theory* (GMT), although this denomination may also refer to other types of extensions of the Mie theory (for example, for a single multi-layer sphere, or for Gaussian beam excitation). Its mathematical details are described for example in Ref. [201]. Because of its complexity, its use in a SERS context has been limited but it has nevertheless played an important role in recent years in the discussion of SERS from a single molecule at a gap or junction between two metallic nano-particles [32,162,202–204]. Finally, the solution of the problem in the ES approximation is also possible [165], but the simplifications are arguably not sufficient to justify its use over the more general GMT solution. We will provide examples of results from GMT in Section 6.4 in the simplest case of a dimer of two metallic spheres.

5.2.3. Ellipsoids

The sphere is a great simple model system, but many interesting effects may be absent because of the high level of symmetry. It is therefore interesting to study other model systems with a lower symmetry. The next logical step is therefore to study ellipsoids, or for simplicity ellipsoids of revolution or spheroids (i.e. ellipsoids with two axes of the same length). The solution of the full electromagnetic problem of scattering by a spheroid is in fact possible analytically [205], along the lines of Mie theory (but much more challenging, even to implement numerically). We will however restrict ourselves in this book to the study of ellipsoids and spheroids in the ES approximation, which is sufficient to highlight most of the physical phenomena of interest. Moreover, the comparison between the ES approximation and the exact results for the case of the sphere can serve as a guide to the validity of the ES approximation in the spheroid case. The mathematical details of the solution are given in Appendix G. It contains most of the important expressions, in particular in the context of SERS. These expressions also form the basis for the related Matlab codes on the book website.

5.2.4. Other approaches

All the solutions discussed so far are general and would apply to many types of particles, not only metallic ones. A somewhat different analytical approach, based on the so-called 'plasmon hybridization', has been proposed recently to model specifically the localized surface plasmon resonances of complex systems [134,160,206,207]. This method does not yield a solution for the local field, and its use for SERS predictions is rather restricted. However, its main interest is pedagogical. It provides a rather simple and intuitive view to understand LSP resonances in complex systems, where two or more LSP resonances are made to interact with each other. A typical example of this is the qualitative description of LSP resonances in dielectric-core/metallic-shell structures (nano-shells) [206,207], and it will be discussed briefly in this context in Chapter 8.

5.3. NUMERICAL TOOLS

5.3.1. A brief overview of the EM numerical tools

Electromagnetic modeling

In many practical cases, numerical approaches are the only option. With the wider availability and rapidly increasing computing power of desktop PCs, EM modeling should no longer be confined to dedicated groups or facilities. Nowadays, it is possible to run routinely fairly advanced EM simulations on PCs, using either freely available codes that can be adapted to the user's need, or even stand-alone commercial software with full graphical user interface. Examples of the latter include software implementing advanced mathematical tools, such as finite-element methods (FEM), or finite-difference time-domain (FDTD) techniques. However, many full 3D electromagnetic simulations in the most general cases are still beyond the capabilities of desktop PCs, and even supercomputers or PC clusters. This means that it is not yet possible to simply define an EM problem at will, and let the computer do the rest of the work. The user's input (and knowledge) remains very important to set up the simulation: choice of the appropriate approximations and of the various parameters (solver parameters, algorithms, mesh precision, bounding box, and other boundary conditions). A tractable solution, i.e. with acceptable accuracy and manageable computing time, will only be obtained if these choices have been optimized and adapted to the problem at hand. The user must, therefore, be sufficiently 'educated' to understand the minimum required to make these choices. It is the aim of this section to describe briefly a few of the numerical tools available. This should serve as a starting point for understanding the specialized literature, and hopefully guide the reader toward the right numerical method for her/his needs.

General considerations for SERS and plasmonics problems

Most numerical methods rely at some point on a discretization (meshing) of the objects under study into (small) cells. The characteristics of the EM fields inside each cell are then computed. The size and shape of the cell is one of the most important issues for the convergence and accuracy of the numerical solution. In particular, since the EM solution is determined by boundary conditions at interfaces, it is paramount that the interfaces are accurately described by the chosen mesh. The simplest mesh is a cubic lattice of cubic cells, and it is used in many numerical methods. The interfaces are then approximated by staircase-like boundaries. This will only be satisfactory if the cell dimension (or equivalently the lattice constant) is very small. In many instances, decreasing sufficiently the lattice constant will result in prohibitive computer power requirements (CPU speed or memory), especially in 3D where this scales as the cube of the lattice constant (or faster).

There are at least two ways to remedy this problem. The first approach is to change the cell geometry. For example, using tetrahedral cells in 3D, a surface may then be approximated by planar (but not necessarily parallel) triangles. The second approach is to use a position-dependent cell size. In this case, a fine meshing can be applied at interfaces, where it matters the most, together with a coarser mesh elsewhere to keep the number of cells manageable. Note however that in most EM problems, the cell size must remain everywhere much smaller than the wavelength for waves to propagate correctly. Finite-element methods (FEM), for example, use both approaches: tetrahedra of varying dimensions to accurately approximate interfaces while keeping the number of cells to a minimum.

Finally, in most applications for SERS and plasmonics, the boundaries (metallic/dielectric interfaces) play an even more important role than in conventional EM problems. The local fields in their vicinity may be very large and vary widely over small distances. The discretization problems at interfaces are therefore magnified and the use of advanced meshing techniques is necessary to accurately predict the local fields close to metallic surfaces.

Two types of numerical methods

The numerical methods can be further classified into two groups:

- Partial differential equation solvers:
  These are 'simply' numerical methods to solve partial differential equations (and in particular Maxwell's equations) with the appropriate boundary conditions. Because of the importance of such problems in many other areas (physics, engineering, etc.), these tools benefit from years of intense research by applied mathematicians, computer scientists, and physicists. The mathematical methods and solvers are therefore highly optimized, and in most cases, beyond the scope of the users. However, with an adequate software package, and a minimum of effort, these can be used for fairly advanced EM simulation. In this group of methods are, for example, those based on finite-difference time-domain (FDTD) techniques, and finite-element methods (FEM). One advantage of these tools is that they can be used to solve many types of physical problems and not only electromagnetic ones. This is in fact also a shortcoming, since they are not optimized for the specificity of the EM problems (for Maxwell's equations).

- Semi-analytical methods: A number of methods were designed precisely to remedy this problem. They have been developed as Maxwell's equations' solvers only, and can therefore use additional analytical tools to reduce the computing requirements. Examples of these are the discrete dipole approximation (DDA) and its derivatives or the multiple multipole (MMP) method. These methods are very interesting conceptually, but are reserved in practice to dedicated researchers. This is because their implementation can be complex, and although codes are freely available, they still require a large user input and a solid knowledge of the methods to make them work efficiently.

With this in mind, our approach here will be to give a very brief overview of some of the approaches cited above, i.e. only an introduction to the concepts with pointers to the specialized literature for the interested reader.

5.3.2. A semi-analytical approach: the discrete dipole approximation

Introduction

The discrete dipole approximation (DDA) is a conceptually simple numerical method proposed originally by Purcell and Pennypacker [98] to model the far-field properties (absorption and extinction) of dielectric particles of arbitrary shape. This technique was in fact developed for astrophysics and therefore applied to interstellar grains. However, it can be easily applied to most EM problems and has therefore extensively been used in the context of plasmonics, i.e. for the study of (small) metallic particles, including SERS (see, for example, Ref. [208]).

The original method has also been studied in more detail, extended in various ways, and placed on a more rigorous theoretical footing. It was also shown to be equivalent to other, independently derived, methods. The various formulations and approximations of the DDA method and its derivatives have now been clearly laid out and studied. These may appear under various names: discrete dipole approximation (DDA), coupled dipole method (CDM), method of moments (MOM), digitized Green's function method (DGF). Moreover, progress has also been made on the computing side, with improved algorithms. Reference [209] provides a recent review of all the issues mentioned above. They are also discussed for example in Chapter 15 of Ref. [12].

Principle

Here we only discuss the method qualitatively. The actual equations (along with more details) can be found, for example, in Refs. [12,98,208,209].

Like most numerical methods, the first step of the DDA is to discretize the objects under consideration into a (large) number N of (small) cells. In its simplest and most common form, the cells are spherical cells of equal size arranged in a cubic lattice. Each cell i is then considered as a polarizable dipole, i.e. when subjected to an electric field $\mathbf{E}_i$ at the cell position, an electric dipole $\mathbf{p}_i = \alpha_i \mathbf{E}_i$ is induced. This dipole then creates an electromagnetic field that contributes to the field and therefore to the induced dipole at every other cell position. The dipole moment at each cell i, therefore, depends on that at every other cell (and on any additional incident or external fields). Solving the problem therefore requires solving a coupled system of linear equations, with $3N$ equations and $3N$ complex unknowns (3 electric field components for each of the N cells), i.e. inverting a $3N \times 3N$ complex matrix. Once the dipole moments are found, the electric field and any other optical properties can be easily computed.

One of the most crucial steps for such an approach to work is in the choice of the cell polarizabilities, α_i, which should correctly represent the optical response of the material composing the cell. α_i should obviously depend on $\epsilon(\omega)$, the dielectric function of the material at the frequency under study. Many expressions have been proposed for this dependence and each corresponds to a specific DDA formulation. Accordingly, they will have different accuracies and convergence properties, see Ref. [209] for more details.

Two approaches are generally used to solve the linear system: Either a direct exact inversion of the matrix (using for example LU-factorization), or an iterative approach such as the conjugate gradient algorithm. The latter is usually the only option for systems with a large number of dipoles N, since the memory requirements for direct matrix inversion become prohibitive.

A popular suite of Fortran codes implementing the DDA method, called DDSCAT, is freely available for download [210].

Advantages and shortcomings

One nice feature of the DDA is that it is conceptually simple, and its physical interpretation is appealing. The induced dipoles in each cell simply represent a discretized version of the induced electric polarization in the medium. Another important advantage of the DDA is that one only needs to discretize the objects under study, not their environment. For a single 3D object, where a typical bounding box would be at least three times the dimensions of the object, this results in a reduction of the order of $\approx$30 for the number of cells. Such a gain may be even more dramatic in sparse structures consisting of several objects, such as colloidal clusters.

Moreover, the DDA methods result in fairly accurate predictions of the far-field properties (extinction, absorption, and scattering cross-sections). It is, actually, simple to understand why. The far-field properties are sometimes analyzed in terms of their multipolar components: electric dipole, magnetic dipole, electric quadrupole, etc. A small collection of closely-spaced electric dipoles is sufficient to represent the multipolar components of the lowest orders. For example, two closely-spaced opposite dipoles viewed from the far field are equivalent to a quadrupole. A sufficient number of coupled dipoles should therefore represent accurately the far-field properties, provided that their positions reasonably describe the geometry of the object, as seen from far away.

The problems arise when one gets closer to the object. If the distance of a point from the object is no longer much larger than the cell size, the field at this point will be very sensitive to its closest cell, and therefore to numerical artifacts such as the cell shape. The predictions of the DDA for the local field are therefore not valid at short distances from the object, which is precisely one of the most desirable properties in many plasmonics problems.

Let us consider the simplest DDA implementation with identical cells arranged in a cubic lattice of lattice constant b. Let us assume, for the sake of argument, that only predictions at a distance of more than $\approx 10b$ are valid. For typical SERS applications, one would need to predict the local field at a distance of $\approx$1 nm from the metallic surface, which would then require $b < 0.1$ nm. A typical nano-particle of, say, 40 nm would then have to be discretized in $N \approx 400^3 = 64 \times 10^6$ cells. Such large numbers are well beyond the available computing powers of most. The local field predictions of DDA calculations should therefore be considered with great care in most cases. This limitation of the DDA can in principle be overcome through the use of cells of different shapes and/or variable (or adaptive) meshing. Such approaches are however still confined to research labs and no user-friendly codes are available.

As a conclusion, the DDA can be a great tool to study the far-field properties (and indirectly the EM resonances) of complex substrates such as colloidal clusters. It is however inadequate for quantitative local field predictions.

5.3.3. Direct numerical solutions

The DDA is a fairly elaborate approach to solving Maxwell's equations and very specific too. Another type of approach is to solve numerically Maxwell's equations using standard techniques that have been developed to solve general partial differential equations. These techniques are applicable to a wide range of problems; such as mechanical engineering, fluid dynamics, acoustics, etc. They have accordingly benefited from a tremendous effort of research and development, resulting in many well-tested and optimized algorithms, often implemented in free or commercial softwares (some with a full graphical user interface). These tools can be used without a detailed knowledge of the underlying mathematics. If it was not for the computing power (that still limits to a large extent what is solvable), one could even use them without understanding much of the underlying physics...

In the context of electromagnetic simulations (i.e. solving Maxwell's equations), two methods stand out:

- the finite-difference time-domain (FDTD) method;
- the finite-element method (FEM).

Both are extensively used and implemented in a large number of general partial differential equation solvers or dedicated electromagnetic softwares.

There are a number of differences between these two approaches. Some of them are fairly mathematical, such as the way the differential equation is numerically solved. One important physical difference is in the differential equation that is actually solved. In FDTD, it is the time dependence of the fields that is sought, while in FEM one generally solves the (complex) frequency-dependent fields (for a given frequency).

The finite-difference time-domain (FDTD) method

Finite-difference is a simple and intuitive way of solving numerically partial differential equations. It is accessible to all in its simplest form, and can even be implemented in a simple spreadsheet editor [211]. FDTD consists in solving, using a finite-difference approach, the time-dependent Maxwell's equations under a prescribed exciting field. This field can be an oscillating field at a fixed frequency ω, but the full power of FDTD calculations is obtained when the exciting field is a single short pulse. The response to such a short pulse contains information on the Fourier components of the fields at all frequencies. The response of the system at all desired frequencies can therefore be computed from a single time-dependent calculation, followed by appropriate Fourier transforms. The simplicity of the FDTD scheme is often cited as an advantage over, for example, FEM. However, this simplicity is immediately lost as soon as the FDTD method is extended to more elaborate meshes (non-cubic, varying dimensions, etc.), which is a prerequisite for many SERS and plasmonics applications. Entire textbooks are dedicated to applying the FDTD method to EM problems, for example Refs. [212,213], and numerous free and commercial software also implement the method.

The finite-element method (FEM)

The finite-element method is more difficult to comprehend for the non-specialist because it uses more involved mathematics. In a first approximation, however, it suffices to know that the FEM is simply another way to solve Maxwell's equations. Using an appropriate software, the role of the user is essentially the same as for FDTD: setting up the EM problem in a way that it is tractable (and physical). The main difference regarding the physics is that, using FEM, one usually solves Maxwell's equations for *harmonic fields* at a given frequency ω. The simulation must then be repeated for each desired frequency (but each simulation at a given ω should be much faster than solving the full time-dependent problem). Textbooks such as Ref. [214] extensively discuss the use of FEM for electromagnetic problems. FEM techniques for EM problems are also implemented in software such as Femlab (www.comsol.com). Examples of simulation results with this method will be provided in Chapter 6.

Advantages and shortcomings

Let us briefly discuss the merits of these two methods within the context of numerical simulations relevant to plasmonics and SERS, i.e. for nano-scale metallic objects. FDTD methods and FEM have a number of common characteristics:

- Both methods rely on a discretization (meshing) of the volume under study. Contrary to most semi-analytical methods however, both the (dielectric or metallic) objects *and* the environment have to be discretized. This increases substantially the number of mesh elements in most applications.
- Since the volume under study cannot be infinite, one needs to truncate it, by the so-called bounding box. The choice of the geometry and size of the bounding box is crucial to both methods. Moreover, imposing external fields that in principle extend to infinity (like an incident plane wave) on such a problem with a bounding box requires extra care. Similarly, radiated fields extend to infinity, and one must make sure that the presence of the bounding box does not influence these fields. This is usually done by applying special boundary conditions at the bounding box edges (such as absorbing boundary conditions or perfectly matched layers) to make sure that the bounding box does not reflect the scattered fields toward the objects (and becomes part of the problem too).
- Since the bounding box is not infinite, far-field properties are in general not straightforward to extract (compared to local fields close to the objects), but methods exist to deduce them from the local field properties.
- In the context of plasmonics and SERS, elaborate meshing techniques are necessary to model the local fields in general. These techniques alone, independently of the differential equation solver itself, are sufficiently complicated to make such calculations from scratch impossible except for the specialists. FDTD and FEM should therefore be viewed as tools, and the details of their implementation (meshing, solver, and other algorithmic tasks) left to specialists (either through freely available codes or commercial software). What the user needs to learn however is how to best use these tools to his/her own ends; in our case the modeling of plasmonics and SERS substrates. Because of the limited computing power, this means being able to make judicious choices for setting up the EM problem: bounding box size and shape, boundary conditions, finesse of the mesh, etc.

5.3.4. Other approaches

We conclude by briefly mentioning a few other approaches to electromagnetic simulations.

The *Multiple Multipole (MMP) method*, or *Fast Multipole Method (FMM)*, is a semi-analytical approach where the field solutions are expressed in terms of multipole fields in a similar fashion as for Mie theory. The fields are therefore exact solutions of Maxwell's equations and the coefficients of the expansions are sought numerically through application of the boundary conditions at a finite number of points. Several multipole centers can be chosen to improve computations. A more detailed description of the method can, for example, be found in Chapter 15 of Ref. [12].

Another method that has been developed more recently is called the *Boundary Element Method (BEM)*, or *Boundary Integral Equations (BIE)*. It consists in solving an integral equation on the surfaces (rather than the volumes; as done for volume-integral methods such as the formal theory of the DDA). This can be advantageous for problems with a large surface to volume ratio. This method is still confined to specialists, and we discuss it no further here.

It is also worth mentioning a semi-analytical approach called the *T-matrix*, which is particularly suited to, and has been extensively used for, the study of the far-field properties of small particles of arbitrary shape. This method is described in detail, along with many of the other approaches described in this chapter in Ref. [215], which can be downloaded freely from http://www.giss.nasa.gov/~crmim/books.html.

Finally, we note that the classification (and denominations) of the semi-analytical approaches is not rigid and many of these are in fact connected to each other and sometimes equivalent at some level of their formulation.

Chapter 6

EM enhancements and plasmon resonances: examples and discussion

In Chapter 4, we have discussed in detail the physical mechanisms by which SERS or SEF enhancements arise from the modification of the EM fields in the vicinity of metallic objects. Then, in Chapter 5, we have discussed how these modified EM fields could be calculated with various techniques. The results for a few specific geometries of interest, a plane, an ellipsoid in the electrostatics approximation, and a sphere (Mie theory), are described in detail in Appendices F, G and H, respectively. We will now discuss the physical meaning of these results in the context of SERS and related plasmonic effects. The main goals of this chapter are therefore:

- To understand more quantitatively the link between the EM enhancements (in SERS or SEF) and the underlying localized surface plasmon (LSP) resonances of the substrate. This will connect with the general introduction to plasmons and plasmonics given in Chapter 3.
- To understand in more detail the characteristics of the LSP resonances and how they relate to the geometry and other characteristics of the substrate.
- To identify the various factors governing the magnitude of the EM enhancements in SERS and SEF.
- To provide and discuss simple examples of EM predictions for a few model systems.

This chapter is, in fact, organized around these examples. We use for these the dielectric functions of silver and gold as given in Appendix E. We start in Section 6.1 with the case of planar surfaces, which is arguably not very relevant to SERS, but provides a good introduction to non-radiative effects, relevant to SEF. We then move on to the canonical case of a metallic sphere in Section 6.2. Many important aspects of plasmon resonances and EM enhancements are illustrated with this simple example, which accordingly plays a central role in this chapter. The following sections then focus on additional aspects that could not be accounted for in the sphere model. Ellipsoids (in the ES approximation) are used as a model example to highlight the effects of shape and 'corners' in Section 6.3. The dimer of two metallic spheres is briefly discussed in Section 6.4 to illustrate gap effects and interacting particles. A few additional effects are then mentioned in Section 6.5. This chapter concludes in Section 6.6 by a synthetic summary of the various factors affecting the plasmon resonances and EM enhancements.

6.1. QUENCHING AND ENHANCEMENT AT PLANAR SURFACES

Planar surfaces have played an important role in the theoretical understanding of SERS and SEF thanks to the availability of analytical solutions [138,169]. They are, however, not often used in experiments (except plasmonics applications using propagating surface plasmon–polaritons) for reasons that should become clear. We discuss in the following those aspects of EM enhancements at planar surfaces that are most relevant to SERS and SEF. The details of the analytical solutions of these problems can be found in Appendix F.

6.1.1. The image dipole approximation for the self-reaction field

Dipolar emission in the ES approximation

We first discuss an approximate solution of the dipolar emission problem close to a plane, which plays an important role in the understanding of non-radiative effects at metal surfaces. We consider a dipole $\mathbf{p}$ at a *short* distance d from a semi-infinite metal plane and propose to solve the problem within the electrostatic approximation (ESA). The rationale is simple: when d is small, both the EM source (dipole) and boundaries (planar interface) are within a very small volume, where the ES solution should be a very good approximation to the full EM solution. In particular, the self-reaction field (field created by the dipole onto itself as a result of the interface) should be well approximated within the ESA.

This problem is a classical ES problem and is discussed in many textbooks [12]. It is usually solved using the *method of images* [96] . To this end, one replaces the effect of the metallic plane by a fictitious dipole $\mathbf{p}'$ positioned symmetrically to $\mathbf{p}$. More precisely, the metallic substrate occupies the $z < 0$ half-space, $\mathbf{p}$ is at position $\mathbf{r}_p = d\mathbf{e}_z$, and $\mathbf{p}'$ is at $\mathbf{r}'_p = -d\mathbf{e}_z$. The dielectric constants (at a given frequency ω) are ϵ in the metal ($z < 0$) and ϵ_M

(real positive) outside ($z > 0$). The electric fields created by each dipole are, by definition, solutions of the electrostatic problem (Laplace equation) everywhere in a homogeneous infinite space (except at the dipole position), and so are any superposition of them.

We therefore postulate that the field solution outside ($z > 0$) is the superposition of the field created by both the dipole and its (virtual) image:

$$\mathbf{E}_{\text{Out}}(\mathbf{r}) = \mathbf{E}_{\mathbf{p}}(\mathbf{r}) + \mathbf{E}_{\mathbf{p}'}(\mathbf{r}), \tag{6.1}$$

while the field inside the metal ($z < 0$) is proportional to the field created by another fictitious dipole $\mathbf{p}''$ at the same position as the real dipole $\mathbf{p}$, but with possibly a different dipole moment[1]:

$$\mathbf{E}_{\text{In}}(\mathbf{r}) = \mathbf{E}_{\mathbf{p}''}(\mathbf{r}). \tag{6.2}$$

The location of the image dipole (symmetrically to $\mathbf{p}$) ensures that the boundary conditions can be easily matched at all points on the planar interface ($z = 0$). Here these boundary conditions are:

$$\mathbf{E}_{\text{out}}(x, y, 0) \cdot \mathbf{e}_x = \mathbf{E}_{\text{in}}(x, y, 0) \cdot \mathbf{e}_x \tag{6.3}$$

$$\mathbf{E}_{\text{out}}(x, y, 0) \cdot \mathbf{e}_y = \mathbf{E}_{\text{in}}(x, y, 0) \cdot \mathbf{e}_y \tag{6.4}$$

$$\epsilon_M \mathbf{E}_{\text{out}}(x, y, 0) \cdot \mathbf{e}_z = \epsilon \mathbf{E}_{\text{in}}(x, y, 0) \cdot \mathbf{e}_z. \tag{6.5}$$

Using the standard expression for the electrostatic field of a dipole, the dipole moments $\mathbf{p}'$ and $\mathbf{p}''$ of the images can easily be deduced from these three equations (see for example Ref. [12], Section 10.10).

It is convenient here to distinguish two situations: a dipole perpendicular to the surface ($\mathbf{p}_\perp = p_\perp \mathbf{e}_z$) and one parallel to the surface ($\mathbf{p}_\parallel \perp \mathbf{e}_z$). A general dipole can always be considered as the superposition of two such dipoles, $\mathbf{p} = \mathbf{p}_\perp + \mathbf{p}_\parallel$, and the solution is then the superposition of the two solutions. On can show that the continuity of the tangential electric field (Eqs (6.3)–(6.4)) can be ensured by taking:

$$\begin{cases} \mathbf{p}_\perp - \mathbf{p}'_\perp &= \mathbf{p}''_\perp \\ \mathbf{p}_\parallel + \mathbf{p}'_\parallel &= \mathbf{p}''_\parallel. \end{cases} \tag{6.6}$$

---

[1] Note that the electrostatic field created by a dipole embedded in two different dielectrics with ϵ_1 and ϵ_2 are the same up to a proportionality constant. In our approach here, we consider that the virtual image dipoles are embedded in the outside medium (ϵ_M). If one considers that they are embedded in the metal (with ϵ), $|\mathbf{p}'|$ and $|\mathbf{p}''|$ are then different, but the field solutions are (fortunately) the same, i.e. the two approaches are equivalent.

A similar condition for the continuity of the normal electric displacement (Eq. (6.5)) applies:

$$\begin{cases} \epsilon_M(\mathbf{p}_\perp + \mathbf{p}'_\perp) &= \epsilon \mathbf{p}''_\perp \\ \epsilon_M(\mathbf{p}_\parallel - \mathbf{p}'_\parallel) &= \epsilon \mathbf{p}''_\parallel. \end{cases} \tag{6.7}$$

From these, the virtual dipoles amplitudes are easily derived. It is convenient to introduce the non-dimensional parameter

$$\boxed{\beta_P = \frac{\epsilon - \epsilon_M}{\epsilon + \epsilon_M}.} \tag{6.8}$$

By analogy with the problem of the sphere (discussed later), β_P can be viewed as the non-dimensional polarizability of the metal/dielectric interface.

For a perpendicular dipole, we then have:

$$\begin{cases} \mathbf{p}'_\perp &= \beta_P \mathbf{p}_\perp, \\ \mathbf{p}''_\perp &= \dfrac{2\epsilon_M}{\epsilon + \epsilon_M} \mathbf{p}_\perp = (1 - \beta_P)\mathbf{p}_\perp. \end{cases} \tag{6.9}$$

For a parallel dipole, we have:

$$\begin{cases} \mathbf{p}'_\parallel &= -\beta_P \mathbf{p}_\parallel, \\ \mathbf{p}''_\parallel &= \dfrac{2\epsilon_M}{\epsilon + \epsilon_M} \mathbf{p}_\parallel = (1 - \beta_P)\mathbf{p}_\parallel. \end{cases} \tag{6.10}$$

These expressions complete the solution of the electrostatic problem.

Decay rate modification in the ES approximation

As mentioned earlier, for a small distance to the surface, $d \ll \lambda$, the ES solution should be a good approximation of the self-reaction field, on which we now focus. Here this field, which is the field reflected on the dipole by its environment (see Section 4.4.4) is simply the field created by the image dipole $\mathbf{p}'$ at the dipole position, i.e.:

$$\begin{aligned} \mathbf{E}_{\mathrm{SR}} = \mathbf{E}_{\mathbf{p}'}(d\mathbf{e}_z) &= \frac{1}{4\pi\epsilon_0\epsilon_M(2d)^3}\left[2\mathbf{p}'_\perp - \mathbf{p}'_\parallel\right] \\ &= \frac{\beta_P}{4\pi\epsilon_0\epsilon_M(2d)^3}\left[2\mathbf{p}_\perp + \mathbf{p}_\parallel\right], \end{aligned} \tag{6.11}$$

where the second equality is expressed in terms of the real dipole.

We can then deduce using Eq. (4.42) the total EM decay rate enhancement factor for this dipole:

$$M_{\mathrm{Tot}} = 1 + \frac{3}{16(k_M d)^3} \frac{2|\mathbf{p}_\perp|^2 + |\mathbf{p}_\parallel|^2}{|\mathbf{p}_\perp|^2 + |\mathbf{p}_\parallel|^2} \mathrm{Im}(\beta_P), \tag{6.12}$$

where $k_M = (\omega/c)\sqrt{\epsilon_M}$. This expression simplifies for a dipole either perpendicular or parallel to the plane:

$$\boxed{M_{\mathrm{Tot}} = 1 + \frac{3\kappa}{16(k_M d)^3} \mathrm{Im}(\beta_P),} \tag{6.13}$$

where $\kappa_\perp = 2$ (for a $\perp$ dipole) and $\kappa_\parallel = 1$ (for a // dipole).

This is the main result of this section. It provides a fairly simple approximate analytical expression for the total (radiative+non-radiative) decay rate modification for a dipole close to a planar interface. In particular a fairly strong $(1/d)^3$ distance dependence is predicted. This is approximate, because it was obtained by deriving the self-reaction field in the ESA. However, for a dipole very close to the surface (say, within ~10 nm), one expects it to be sufficiently accurate. The validity of this expression and its consequences will now be discussed.

Comparison of the ESA predictions with exact results

We first compare in Fig. 6.1(a and b) the results of the previous expression (Eq. (6.13)) obtained in the ESA, to the exact results (given in Section F.5 of Appendix F). As expected, the ESA results are in almost perfect agreement with the exact predictions for the smallest distance ($d \approx 1$ nm) and remain a good approximation up to 5–10 nm separation. Figure 6.2(a) illustrates further the properties of M_{Tot} for a dipole near a planar metal surface and confirms further this agreement. This shows that Eq. (6.13) provides a simple expression for M_{Tot} close to a plane, which is valid is most cases of interest to SERS and SEF, i.e. at short distances, typically up to ≈10 nm.

6.1.2. Enhancement and quenching at plane metal surfaces

Radiative emission close to a metallic plane

Predictions of the radiative decay rate enhancement factor, M_{Rad} (defined in Chapter 4) for a dipole close to a metallic plane are shown in Fig. 6.1(c and d). These are obtained from the exact analytical results of Section F.5. A quick comparison with the predictions for M_{Tot} in 6.1(a and b) shows that M_{Rad} is much smaller than M_{Tot}, especially for the dipoles closest to the plane.

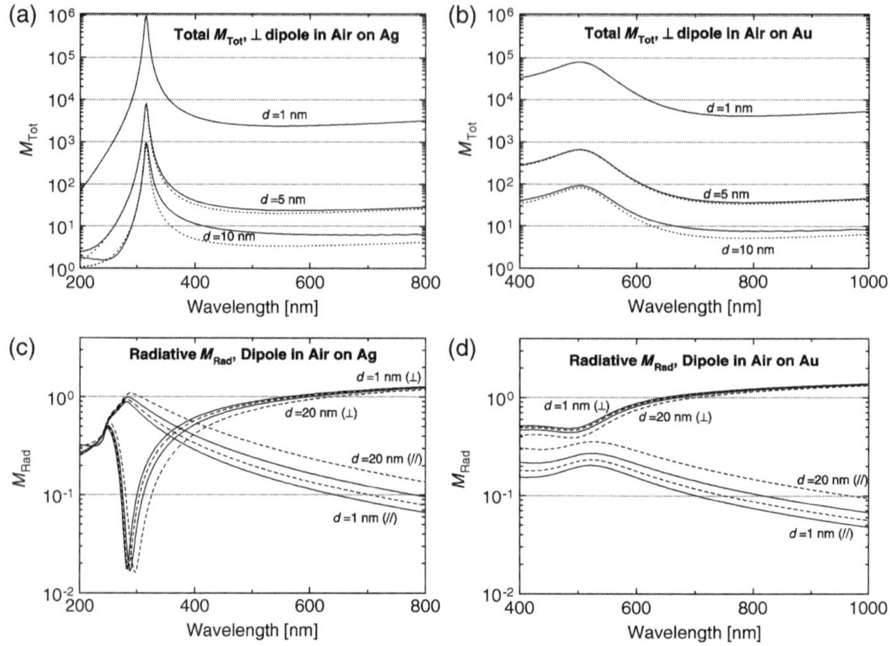

Figure 6.1. (a and b) Wavelength dependence of M_{Tot} for a dipole in air at a distance d from a planar metal surface (silver or gold) and perpendicular to it. The solid lines correspond to the exact result and the dotted line to the ESA (Eq. (6.13)). The two are almost identical for $d = 1$ nm and the ESA remains quite good up to $\approx$5 nm. A comparable agreement is obtained for a parallel dipole (for which M_{Tot} is a factor of $\approx$2 smaller). (c and d) Wavelength dependence of M_{Rad} (exact results) for a dipole in air at a distance d from a planar metal surface (silver or gold). Both perpendicular and parallel dipoles are considered, and four distances for each: $d = 1$ (solid), 5 (dashed), 10 (solid), and 20 nm (dashed).

This is further illustrated in Fig. 6.2(b and c). In particular, the EM radiative efficiency $\eta_{\text{Rad}}^{\text{EM}}$, which characterizes the approximate modified quantum yield of a fluorophore, can be extremely small $\approx 10^{-5}$–10^{-4} at $d = 1$ nm. This shows that dipolar emission very close to a metallic plane is *almost purely non-radiative.*

Despite this fact, as shown in Figs 6.1(c and d) and 6.2(b), the radiative decay rate EF itself, M_{Rad}, is in general not so small, in fact of the order of 1. The non-radiative nature of the total decay rate therefore comes from the large non-radiative component, not from a real radiative quenching. To put differently, the dipole still radiates the same power to the far field, but also emits a much larger power in the form of photons that are subsequently absorbed in the metal (and therefore are not detected in the far-field).

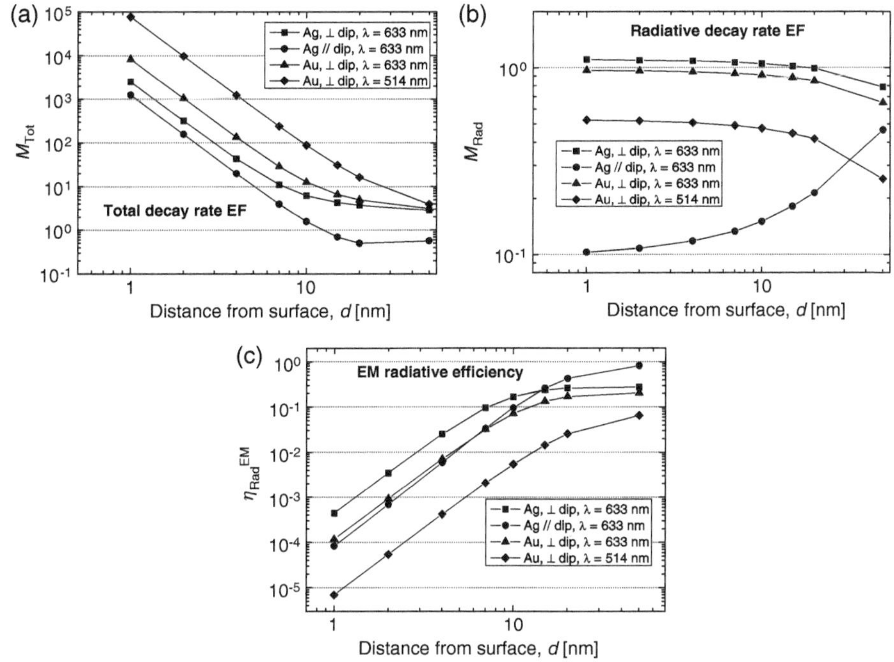

Figure 6.2. (a) Distance dependence of the total EM decay rate EF M_{Tot} for a few selected cases of a dipole at a distance d above a planar metal surface (silver or gold) and perpendicular ($\perp$) or parallel ($\|$) to it. The d^{-3}-dependence at short distances is clearly apparent (all curves approach parallel straight lines). (b) Same for the radiative decay rate EF M_{Rad}. M_{Rad} hardly changes with distance, especially for the smallest distances. (c) Same for the EM radiative efficiency $\eta_{\text{Rad}}^{\text{EM}} = M_{\text{Rad}}/M_{\text{Tot}}$. $\eta_{\text{Rad}}^{\text{EM}}$ can be very small at the shortest distances and decreases as d^3, a direct consequence of the strong dependence of M_{Tot}.

SEF and SERS at plane surfaces

In order to focus now on SERS and SEF signals, we first need to consider the local field intensity enhancement factor M_{Loc}, which plays a role in both cases. For single planar metal/dielectric interfaces, M_{Loc} is typically at most of the order of $\approx$3 at the surface, and this only when excited at an optimal angle of incidence (around 60°). The tools of Appendix F can be used for such predictions.

The approximate SERS EF can therefore be expected to be of the order of $F_{E4}^{0} \approx 10$ in optimum conditions. This is, by most standards, a fairly *weak* SERS enhancement. However, it is worth pointing out that this EF is achieved at *every point on the surface.* The average SERS EF (the SSEF in Chapter 4, for example) is therefore of the same order. It is also worth noting that the large non-radiative component in M_{Tot} does not affect the SERS EF, a fact already emphasized several times in Chapter 4.

This is certainly not the case for fluorescence (SEF in this case). The fluorescence enhancement can be characterized by the approximate fluorescence enhancement factor: $M_{\rm Fluo} = M_{\rm Loc}\eta^{\rm EM}_{\rm Rad}$. The moderate value of $M_{\rm Loc}$ is far from compensating the very small value of $\eta^{\rm EM}_{\rm Rad}$ at the shortest distances from the surface. The overwhelming influence of non-radiative emission over radiative emission therefore results in a *strong fluorescence quenching*. This is the origin of the well-known effect of fluorescence quenching for molecules adsorbed on metals. Most of the emitted photons are absorbed in the metal[2]. This can only be reduced by placing the emitter further away from the surface to increase $\eta^{\rm EM}_{\rm Rad}$ (see Fig. 6.2(c)). But even then, no fluorescence enhancement is predicted because of the small local field intensity EF.

Non-radiative effects at metal surfaces

We have seen that the total EM decay rate EF $M_{\rm Tot}$ at short distances is almost entirely dominated by its non-radiative component. It is moreover well described by the simple expression given in Eq. (6.13) obtained in the ES approximation. In fact, this expression can also be used to account approximately for the non-radiative decay rate EF in the general case of dipolar emission sufficiently close to any metallic objects (independent of their geometry). This fact is loosely based on the fact that at sufficiently short distances most geometries will look locally like a 'plane'; it will be also further justified and discussed in the case of a sphere later.

Equation (6.13), hence, *plays a major role in the description of non-radiative processes (and therefore of SEF) for emitters very close to any metallic object*, not only metallic planes. Using Eq. (6.13) and the examples of Figs 6.1(a and b), we can identify some important features of the non-radiative decay rate EF $M_{\rm NR}$ that will remain valid in most cases for a dipole in close proximity to a metal surface:

- $M_{\rm NR}$ is *strongly distance-dependent*, varying as d^{-3} at short distances. This implies large changes for example between adsorbed molecules in the first and second monolayers. Assuming a typical molecular size of $\sim$1 nm, the first and second monolayers can have differences in $M_{\rm NR}$ by a factor of the order of $\sim$10.
- For silver, $M_{\rm NR}(\omega)$ exhibits a very strong resonance at $\lambda = 316$ nm, as can be clearly seen in Fig. 6.1(a). It is easy to show that this resonance occurs at the wavelength for which $\mathrm{Re}(\epsilon(\omega)) = -\epsilon_M$. This can be seen

---

[2] A more detailed analysis can show that most of these photons are emitted into propagating surface plasmon–polariton (PSPP) modes at the air/metal interface. Even if some of these waves may propagate quite far along the interface (see Chapter 3), they are intrinsically non-radiative for a perfect infinite plane and these photons are eventually absorbed. Some of this emission could however (in principle) be recovered at defects in the plane (roughness or point defects), see Chapter 3.

directly from Eq. (6.13), recalling the definition of β_P given in Eq. (6.8). This resonance corresponds to the localized surface plasmon resonance of the silver/air planar interface in the ES approximation. It can be viewed as a result of the large number of surface plasmon–polariton (SPP) modes existing around this particular wavelength[3]. At this resonance, $M_{\rm NR}$ can reach values as high as 10^6 for a dipole at $d = 1$ nm from the surface. The situation for gold is not as clear because of the large optical absorption in the resonance region. $M_{\rm NR}$ nevertheless reaches values of almost $\sim 10^5$ around 500 nm for a dipole at $d = 1$ nm from the surface.

- At wavelengths longer than the resonance, $M_{\rm NR}$ becomes approximately constant and remains quite large for both silver and gold, of the order of $\sim 2\text{–}6 \times 10^3$ for $d = 1$ nm. This is in fact a consequence of Eqs (6.13) and (6.8), and the fact that $\mathrm{Re}(\epsilon(\omega))$ is negative with a large absolute value.
- Finally, it is important to bear in mind that the predictions for the smallest distances ($d < 0.5 - 1$ nm) should be taken with care. A non-local treatment of the dielectric function of the metal is in principle required for calculations at such short separations [138]. The results then depend on the chosen non-local dielectric function model. A further increase in $M_{\rm NR}$, by a factor as large as ≈ 8 has been predicted at distances $d \approx 0.3$ nm [138].

The case of a planar metallic substrate allows us to discuss these non-radiative effects without any interference from radiative effects and local field enhancements. Its interest is primarily pedagogical: the poor local field enhancements are, in fact, not ideal for any SERS or SEF implementations. We can now focus on more relevant cases, where local field enhancements (and therefore SERS enhancements) are much more important; the simplest, and again most pedagogical examples among these being the case of a metallic sphere.

6.2. A SIMPLE EXAMPLE IN DETAIL: THE METALLIC SPHERE

The metallic sphere presents a major didactic advantage: a reasonably 'easy' analytical solution of the EM problem exists in the form of Mie theory (see Appendix H for details). Using this exact solution, we can discuss in detail many aspects of electromagnetic properties of metallic spherical nano-particles and study the spectral dependence of the various key EM indicators and how they relate to SERS or SEF. This will introduce most of the important physical

[3] Note that the translational invariance is broken here by the dipole, and dipole emission can in principle couple to any of the SPP modes (there is no k-conservation).

concepts, which are then easily generalized to more complex cases. Moreover, the exact solution can be used for a justification of the validity of the much simpler electrostatic approximation (ESA). In fact, as we will show, many concepts can be readily (and more easily) understood, at least qualitatively, within the ESA, which we therefore discuss first.

We consider here a metallic sphere of radius a and dielectric function $\epsilon(\omega)$ embedded in a non-absorbing infinite dielectric medium of dielectric constant ϵ_M. As in the previous cases, the discussion will be illustrated using the two most common metals in plasmonics: silver and gold, whose dielectric functions can be conveniently expressed analytically (Appendix E).

6.2.1. Metallic sphere in the ES approximation

The electrostatics solution

In the ESA (discussed in Section 5.1.4), the incident field is represented by a constant field $\mathbf{E}_{\text{Inc}}$. The solution of the electrostatics problem for a sphere is treated in many textbooks (see for example Ref. [96]) and we only focus here on the final results, which we adapt to our purpose. The electric field solution inside the sphere is given by:

$$\mathbf{E}_{\text{In}} = \frac{3\epsilon_M}{\epsilon + 2\epsilon_M}\mathbf{E}_{\text{Inc}} = (1 - \beta_S)\mathbf{E}_{\text{Inc}}, \tag{6.14}$$

where we have introduced

$$\boxed{\beta_S = \frac{\epsilon(\omega) - \epsilon_M}{\epsilon(\omega) + 2\epsilon_M}.} \tag{6.15}$$

The electric field is therefore constant and uniform inside. Here it is interesting to consider as well the electric polarization, $\mathbf{P}$ inside the sphere. $\mathbf{P}$ is discussed and defined (for example) in Appendix C, but because our sphere is embedded here in a dielectric medium, it is more convenient (and more physical) to consider the electric polarization $\mathbf{P}_M$ with respect to this medium, rather than vacuum (the same approach was used to discussed the dipolar approximation in Section 5.1.4). It is then defined by the relation $\mathbf{D} = \epsilon_0\epsilon_M\mathbf{E}+\mathbf{P}_M$, where the usual constitutive relation $\mathbf{D} = \epsilon_0\epsilon\mathbf{E}$ holds. We then have $\mathbf{P}_M = \epsilon_0(\epsilon - \epsilon_M)\mathbf{E}$. The electric polarization inside the sphere is therefore also uniform and given by (using Eq. (6.14)):

$$\mathbf{P}_M = 3\epsilon_0\epsilon_M\beta_S\mathbf{E}_{\text{Inc}}. \tag{6.16}$$

Hence, the electric polarization is uniform inside and, by integration over the volume of the sphere (denoted V_S), is equivalent to a dipole $\mathbf{p}_M = \alpha_S\mathbf{E}_{\text{Inc}}$

(embedded in an ϵ_M-dielectric) where:

$$\boxed{\alpha_S = 3\epsilon_0\epsilon_M V_S\beta_S = 4\pi\epsilon_0\epsilon_M a^3\frac{\epsilon(\omega)-\epsilon_M}{\epsilon(\omega)+2\epsilon_M}.} \tag{6.17}$$

α_S is called the polarizability (or dipolar polarizability) of the sphere in the ES approximation and characterizes its ES response. It is proportional to β_S, which can therefore be viewed as the 'non-dimensional' polarizability of the sphere.

Moreover, the ES solution outside the sphere can be written as:

$$\mathbf{E}_{\text{Out}} = \mathbf{E}_{\text{Inc}} + \mathbf{E}_{\mathbf{p}_M}, \tag{6.18}$$

where $\mathbf{E}_{\mathbf{p}_M}$ is the (electrostatic) field created by the dipole $\mathbf{p}_M = \alpha_S\mathbf{E}_{\text{Inc}}$ positioned at O, center of the sphere, and embedded in a medium with ϵ_M.

In writing these expressions, we have anticipated the interpretation of this solution. It is indeed convenient to view this EM problem as follows: the incident field $\mathbf{E}_{\text{Inc}}$ induces a polarization in the sphere (or equivalently surface charges at the surface of the sphere). This polarization (or surface charge) is equivalent to a point dipole $\mathbf{p}_M = \alpha_S\mathbf{E}_{\text{Inc}}$ at O, which is sometimes called the *induced dipole*. This induced dipole in turn creates a field (the scattered field) outside the sphere, which adds up to the incident field to yield the full electrostatics solution. This approach is appealing by its simplicity, and gives here the exact electrostatic results, but is by no means general. It would for example fail for a non-spherical object, and even for a sphere if higher-order multipoles are excited (i.e. if the external field is not uniform). This interpretation is also the basis for the *dipolar approximation* discussed earlier in Section 5.1.4, and which we now use to express the far-field properties of the sphere in the ES approximation.

Far-field properties in the ES approximation

As explained in Section 5.1.4, we can calculate directly the optical absorption (for example power absorbed, P_{Abs}, or absorption cross-section σ_{Abs}) from the ESA solution, by integration over the volume of the metallic object. For a sphere, the integral is trivial since the field is uniform inside the metal. Recalling that the incident power density is $S_{\text{Inc}} = (n_M\epsilon_0 c)/2|E_{\text{Inc}}|^2$ [W/m^2], the absorption cross-section [m^2] in the ESA is obtained as:

$$\sigma_{\text{Abs}} = \frac{P_{\text{Abs}}}{S_{\text{Inc}}} = 4\pi k_M a^3\frac{3\epsilon_M \text{Im}(\epsilon)}{|\epsilon+2\epsilon_M|^2}. \tag{6.19}$$

This can also be expressed in terms of the single-particle absorption coefficient for easier comparison with the Mie theory and takes the concise form:

$$Q_{\mathrm{Abs}}^{\mathrm{NP}} = \frac{\sigma_{\mathrm{Abs}}}{\pi a^2} = 4k_M a \mathrm{Im}(\beta_S). \tag{6.20}$$

Moreover, the scattering and extinction cross-sections can be obtained using the dipolar approximation discussed in Section 5.1.4. The dipolar polarizability tensor of the sphere in the ESA is isotropic and is given in Eq. (6.17). We therefore deduce σ_{Sca} and σ_{Ext} from Eqs (5.33) and (5.35). The single-particle scattering and extinction coefficients are therefore:

$$Q_{\mathrm{Sca}}^{\mathrm{NP}} = \frac{\sigma_{\mathrm{Sca}}}{\pi a^2} = \frac{8}{3}(k_M a)^4 |\beta_S|^2, \tag{6.21}$$

and

$$Q_{\mathrm{Ext}}^{\mathrm{NP}} = \frac{\sigma_{\mathrm{Ext}}}{\pi a^2} = 4k_M a \mathrm{Im}(\beta_S). \tag{6.22}$$

Note that within these approximations, the extinction and absorption cross-sections are equal. Energy conservation cannot therefore be fulfilled: $\sigma_{\mathrm{Ext}} \neq \sigma_{\mathrm{Abs}} + \sigma_{\mathrm{Sca}}$. This is simply a reminder that these results are not exact but only an approximation. As a matter of fact, in the limit of small sizes ($k_M a \ll 1$), energy conservation is recovered (as pointed out in Section 5.1.4).

At this stage, we will digress in the next few pages to discuss other more advanced aspects of the ESA solution for a sphere. It is possible to skip these in the first reading and move directly to Section 6.2.2.

The depolarization and radiative corrections

One of the main shortcomings of the electrostatic approximation is that it is scale-invariant, i.e. it does not predict any size-related effects such as the increasing role played by radiation at larger sizes. This is in fact the reason why energy conservation is only approximate. Attempts have been made at applying corrections to the ESA approach, thereby retaining its simplicity whilst improving its range of validity and in particular recovering the energy conservation condition. This can for example be achieved within the framework of the dipolar approximation.

One possible correction is called the *radiative correction or radiation damping correction* [216] and consists in replacing the dipolar polarizability (or its non-dimensional equivalent β_S), by:

$$\beta_S^{\mathrm{RC}} = \frac{\beta_S}{1 - \frac{2}{3}i(k_M a)^3 \beta_S}. \tag{6.23}$$

This correction is suggested primarily by energy considerations [216] (it ensures conservation of energy for scattering by a polarizable dipole). Note that the correction is of order $(k_M a)^3$.

This correction can be compared to that obtained by a direct expansion of the exact result for the dipolar polarizability (from Mie theory, Eq. (H.53)). Although it matches exactly the term of order $(k_M a)^3$, the exact expansion reveals that there should be, in addition, a term of order $(k_M a)^2$ in the expansion. Being of lower order, this term should in fact dominate over the radiative correction, whose utility is then questionable.

This was first pointed out by Meier and Wokaun [217], who in the process studied the physical origin of the $(k_M a)^2$ terms and proposed a scheme to calculate them approximately. They dubbed the effect *dynamic depolarization* and showed that it resulted, for a sphere, in the following correction:

$$\beta_S^{\mathrm{DD-RC}} = \frac{\beta_S}{1 - (k_M a)^2 \beta_S - \frac{2}{3} i (k_M a)^3 \beta_S}, \tag{6.24}$$

where the radiative damping term of order $(k_M a)^3$ has also been included. This approach is quite appealing, since it could be generalized to more complex geometries. However, if one compares it to the exact expansion obtained from Mie theory (Eq. (H.53)):

$$\beta_S^{\mathrm{Mie}} = \frac{\beta_S}{1 - (k_M a)^2 \left[1 - \frac{2\epsilon+1}{5(\epsilon-1)}\right] \beta_S - \frac{2}{3} i (k_M a)^3 \beta_S}, \tag{6.25}$$

the term of order $(k_M a)^2$ remains inexact, which casts doubts on the usefulness of this expression. If any, the best correction to apply is therefore the latter expression, obtained directly from Mie theory. It cannot, unfortunately, be generalized to objects other than the sphere, as could the dynamic depolarization and radiative corrections.

If one of these corrections is applied, the following expressions should then be used to determine extinction and scattering coefficients (shown here for β_S^{Mie}):

$$Q_{\mathrm{Ext}}^{\mathrm{NP}} = 4 k_M a \mathrm{Im}(\beta_S^{\mathrm{Mie}}), \tag{6.26}$$

and

$$Q_{\mathrm{Sca}}^{\mathrm{NP}} = \frac{8}{3} (k_M a)^4 |\beta_S^{\mathrm{Mie}}|^2. \tag{6.27}$$

The absorption coefficient is then obtained from $Q_{\rm Abs}^{\rm NP} = Q_{\rm Ext}^{\rm NP} - Q_{\rm Sca}^{\rm NP}$, and note that it is different (by terms of order $(k_M a)^2$ or more) from that obtained directly in the electrostatics approximation. The expressions given above are then equivalent (up to order $(k_M a)^3$) to the results of Mie theory when the electric dipolar response *only* is considered. However, the correct expansion within Mie theory of $Q_{\rm Ext}^{\rm NP}$ to order $(k_M a)^4$, and $Q_{\rm Sca}^{\rm NP}$ to order $(k_M a)^6$, should take into account the additional effect of the induced magnetic dipole and the induced electric quadrupole [218], whose lowest order is actually comparable to the dynamic depolarization correction (i.e. equivalent to an order $\sim(k_M a)^2$ correction in β_S).

In conclusion, these possible corrections to the ES polarizability (and their generalization to more complex objects [218,219]), although conceptually interesting, are not of much use because they account only for part of the terms in the expansion. The omitted terms, of comparable order, may therefore entirely flaw the results of these approximations. Unless further work is carried out to correct this problem, it is therefore better to either stick to the ES approximation, or use the full apparatus of Mie theory instead.

Higher-order excitations in the ES approximation

We have treated so far only the special case of a sphere in a *uniform* external field. The problem can be generalized to an arbitrary external electrostatic field. The treatment is then similar (although simpler) to Mie theory and will not be described here, see e.g. [96]. We only mention that the solution can be described as a sum of spherical harmonics. For a given order (angular momentum) l of the spherical harmonics, the ES response is characterized by a multipolar polarizability α_l (of order l), which is proportional to [165]:

$$\alpha_l \propto a^{2l+1} \frac{\epsilon - \epsilon_M}{\epsilon + \frac{l+1}{l}\epsilon_M}. \tag{6.28}$$

This expression will be used later for the discussion of higher-order localized surface plasmon resonances.

Dipole emission in the ES approximation

Finally, as for the metal plane, it is possible to study the problem of dipolar emission close to a metal sphere in the ESA. As before, one expects the result to be fairly accurate for dipoles very close to the metal surface, which is the case in most of our applications.

We provide here for illustration the final result for the total decay rate enhancement in the ESA of a dipole perpendicular to the sphere surface, and at a distance d:

$$M_{\text{Tot}}^{\perp} = 1 + \frac{3}{2(k_M d)^3} \times \sum_{l=1}^{\infty} (l+1)^2 \left(\frac{d}{a+d}\right)^3 \left(\frac{a}{a+d}\right)^{2l+1} \text{Im}\left(\frac{\epsilon - \epsilon_M}{\epsilon + \frac{l+1}{l}\epsilon_M}\right). \tag{6.29}$$

Such a series does not converge as fast as most other series we may encounter in this book, in particular when d is small compared to a. A more careful study in fact highlights that the first few terms are almost negligible compared to the rest. Even if a resonance condition in the ϵ-dependent term is fulfilled, for example for $l = 1$, this term will hardly have any influence on the sum except for small values of $\text{Im}(\epsilon)$ (very strong resonance), in fact much smaller than what exists in typical metals. Since relatively large l dominates for $d \ll a$, we can make the approximation $(l+1)/l \approx 1$ and the ϵ-dependent term can be moved outside the sum; it then reduces to $\text{Im}(\beta_P)$, where β_P is the non-dimensional polarizability introduced earlier for *planar surfaces* (Eq. (6.8)). The series can then be evaluated analytically, and for $d \ll a$, we obtain:

$$M_{\text{Tot}}^{\perp} \approx 1 + \frac{3}{8(k_M d)^3} \text{Im}\left(\beta_P\right). \tag{6.30}$$

This is the same expression as obtained earlier for a dipole perpendicular to a planar metal surface. In fact, this can be interpreted simply physically and was already pointed out: when the dipole is very close to the sphere, the sphere surface can be approximated as a plane over a small region of space around the dipole position. The reflected electric field at the dipole position is therefore similar to that obtained for the plane case, hence the approximate equality of the *total* decay rates. This result is important, since it can be generalized to any geometries (as long as the distance d of the dipole to the surface is very small compared to the object's dimension). This is the reason why the expression for M_{Tot} obtained for a dipole near a plane in the ES approximation (Eq. (6.13)) is so important: it applies approximately to most situations for small d. Note that this argument is no longer valid for *radiative* properties, since they are far-field properties and therefore are not determined by what happens only in the close vicinity of the dipole.

Finally, this ES result is in fact also useful for Mie theory. Because of the slow convergence of the series, numerical problems arise when calculating M_{Tot} within Mie theory, and a workaround these problems is to use the ES approximation for the higher-order terms in the series. This is discussed in more detail in Appendix H.

6.2.2. Localized surface plasmon resonances and far-field properties

To understand the localized surface plasmon (LSP) resonances of the sphere, we will first use the electrostatic approximation discussed in the previous section, for which simple expressions have been derived. This is valid only for very small spheres but then we use exact results from Mie theory to ascertain the range of validity of this approximation and discuss the effect of sphere size on these resonances.

Dipolar LSP resonance in the electrostatic approximation

The following discussion is based mostly on the analytical results of Section 6.2.1 for a sphere in a uniform external field in the ES approximation. We showed there that the EM response (in the ESA) of the sphere is characterized by a dipolar polarizability α_S given in Eq. (6.17), or more conveniently by the non-dimensional polarizability β_S, which we recall here:

$$\beta_S(\omega) = \frac{\epsilon(\omega) - \epsilon_M}{\epsilon(\omega) + 2\epsilon_M}. \tag{6.31}$$

β_S is wavelength-dependent, through the wavelength (or ω) dependence[4] of ϵ for the metal. Note also that β_S is in general (and in particular for a metal) complex, since ϵ is itself complex:

$$\epsilon(\lambda) = \epsilon'(\lambda) + i\epsilon''(\lambda). \tag{6.32}$$

Any resonant EM response of the sphere (in the ESA) should therefore translate in a resonance for β_S. This occurs if $|\beta_S|$ is large, so when the denominator goes to zero, i.e.: when $\epsilon(\lambda) = -2\epsilon_M$. Because $\epsilon''(\lambda) \neq 0$ for a metal, this condition cannot be met exactly, but it will be met approximately when $\epsilon'(\lambda) = -2\epsilon_M$. Since the embedding medium is non-absorbing (ϵ_M real positive), this therefore requires $\epsilon'(\lambda) < 0$. Only dispersive media may have this property, and this is in particular the case of metals at wavelengths longer than their bulk plasmon resonance (see Chapter 3). One can show that the resulting resonance corresponds to the *dipolar localized surface plasmon resonance* of the sphere. In the ES approximation, it therefore corresponds to the condition:

$$\boxed{\epsilon'(\lambda_{\text{Res}}) = -2\epsilon_M,} \tag{6.33}$$

[4] We will write this dependence as $\epsilon(\omega)$ or $\epsilon(\lambda)$ depending on the context.

Table 6.1 Summary of the predicted LSP resonance properties of small spheres. These are computed from Eqs (6.33) and (6.35) using the optical properties of the metal found in the indicated reference: Palik [139], J & C [137], or the analytical approximations provided in Appendix E (Anal.). For weak resonances (small $|\beta_{\text{Res}}(\lambda_{\text{Res}})|^2$), these predicted values may differ slightly from the actual maximum of $|\beta_{\text{Res}}(\lambda_{\text{Res}})|^2$. Note the discrepancy between the predictions for silver depending on the source chosen for the optical properties (see Appendix E for more details on this specific point).

Metal interface	Ref.	ϵ_M	λ_{Res} [nm]	$\epsilon''(\lambda_{\text{Res}})$	$\|\beta_S(\lambda_{\text{Res}})\|^2$
Ag/Air	Anal.	1	346	0.12	596
	J & C	1	354	0.28	114
	Palik	1	354	0.6	26
Ag/Water	Anal.	1.77	387	0.17	965
	J & C	1.77	382	0.19	799
	Palik	1.77	394	0.68	62
Ag/Oil	Anal.	2.25	411	0.205	1085
	J & C	2.25	402	0.212	1015
	Palik	2.25	414	0.73	86
Au/Air	Anal.	1	467	4.8	1.39
	J & C	1	482	4.34	1.48
	Palik	1	485	4.0	1.56
Au/Water	Anal.	1.77	512	3.2	3.7
	J & C	1.77	513	2.94	4.3
	Palik	1.77	510	2.8	4.6
Au/Oil	Anal.	2.25	530	2.66	7.5
	J & C	2.25	528	2.46	8.5
	Palik	2.25	520	2.45	8.6
Cu/Air	Palik	1	354	4.94	1.37
Cu/Oil	Palik	2.25	456	5.6	2.45
Al/Oil	Palik	2.25	187	0.46	216
Pt/Air	Palik	1	275	5.6	1.29
Pt/Oil	Palik	2.25	376	8.8	1.6
Pd/Air	Palik	1	215	3.0	2.0
Li/Air	Palik	1	359	0.97	10.6
Li/Oil	Palik	2.25	490	0.92	55

and is usually fulfilled for one wavelength only, which depends on the metal and the embedding medium through ϵ_M. The resonance wavelength is typically red-shifted (i.e. moved to longer wavelengths) for a larger ϵ_M. As an example, Table 6.1 summarizes the resonance wavelengths for some common metal/dielectric pairs. Note that within the ESA, the dipolar LSP resonance wavelength λ_{Res} does not depend on the radius a of the sphere.

Moreover, if the resonance condition (Eq. (6.33)) is met, then:

$$\beta_S(\lambda_{\text{Res}}) = 1 + i\frac{3\epsilon_M}{\epsilon''(\lambda_{\text{Res}})}. \tag{6.34}$$

One therefore concludes that the resonance will be particularly strong if, in addition, $\epsilon''(\lambda_{\text{Res}})$ is small at resonance. In this case, $|\beta_S|$ is large and more precisely:

$$\boxed{|\beta_S(\lambda_{\text{Res}})|^2 \approx \frac{9\epsilon_M^2}{(\epsilon''(\lambda_{\text{Res}}))^2}.} \tag{6.35}$$

For a given metal, the strength of the sphere dipolar LSP resonance is therefore fixed (as is the resonant wavelength) for small particles (for which the ESA is valid), and depends only on the dielectric constant ϵ_M of the environment. This determines λ_{Res} through the condition (6.33) which, in turn, determines the value of $\epsilon''(\lambda_{\text{Res}})$. These values are summarized in Table 6.1 for a selection of metal/dielectric pairs.

It is clear from this table that the strongest resonances are to be expected from silver spheres, thanks to the relatively low absorption (ϵ'') in the resonance region. However, resonances are then in the near UV, which may not be appropriate for many applications. On the other hand, the LSP resonances of small gold nano-particles are expected to lie in the green region of the spectrum, which is much more interesting for applications. However, the large value of ϵ'' in this region damps strongly these resonances, as testified in the small values of $|\beta_S|^2$. We will come back to this shortly.

Far-field properties in the ES approximation

The LSP resonance should be evident in the far-field properties through the relations given in Section 6.2.1. Within the dipolar approximation, we have:

$$Q_{\text{Ext}}^{\text{NP}} = Q_{\text{Abs}}^{\text{NP}} = 4k_M a \text{Im}(\beta_S) \quad \text{and} \quad Q_{\text{Sca}}^{\text{NP}} = \frac{8}{3}(k_M a)^4 |\beta_S|^2, \tag{6.36}$$

where $k_M = n_M \omega / c = 2\pi n_M / \lambda$. Note that within the range of validity of the ESA, we should also have $Q_{\text{Sca}}^{\text{NP}} \ll Q_{\text{Ext}}^{\text{NP}}$ to ensure approximate energy conservation.

The wavelength dependence of these functions (arising mostly from the λ-dependence of ϵ, and therefore of β_S) is illustrated in Fig. 6.3 for silver and gold spheres in water. A radius of $a = 10$ nm is assumed in this figure, but the spectral profiles are independent of a; only their magnitude changes: linearly with a for $Q_{\text{Ext}}^{\text{NP}}$, and as a^4 for $Q_{\text{Sca}}^{\text{NP}}$. We highlight here the much stronger intensity of the LSP resonance for silver compared to gold, which is simply another manifestation of the stronger dipolar LSP resonance for Ag. As mentioned already, the strong damping for gold is the result of a relatively large ϵ'' at resonance. This in fact results in a smaller ratio of $Q_{\text{Sca}}^{\text{NP}}/Q_{\text{Ext}}^{\text{NP}}$ for gold than for silver. The condition of approximate energy conservation within

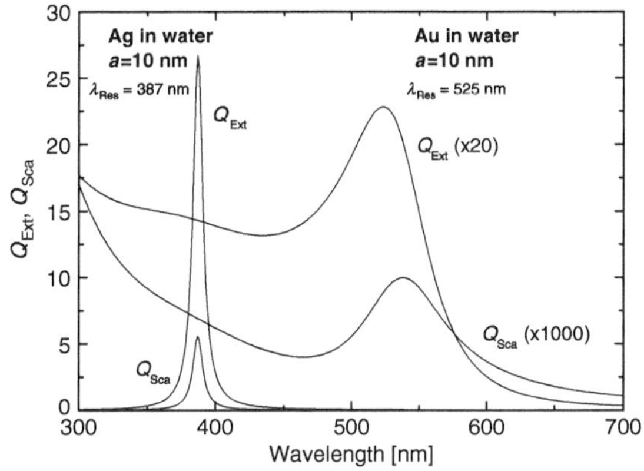

Figure 6.3. Calculated far-field properties of small silver and gold spheres immersed in water in the ES approximation, as a function of incident wavelength. Extinction coefficient, $Q_{\mathrm{Ext}}^{\mathrm{NP}}$, and scattering coefficient, $Q_{\mathrm{Sca}}^{\mathrm{NP}}$ are shown for a sphere radius of $a = 10$ nm.

the ESA, $Q_{\mathrm{Sca}}^{\mathrm{NP}} \ll Q_{\mathrm{Ext}}^{\mathrm{NP}}$, is barely met for silver, even for spheres as small as 10 nm radius, but is easily met for gold (up to $a \approx 25$ nm by extrapolation of the results of Fig. 6.3). These figures will be further confirmed later by direct comparison with Mie theory.

Higher-order LSP resonances in the ES approximation

When treating the case of the sphere in the ES approximation, it is generally assumed that the exciting (external) field is uniform. In this case, as discussed above, only the dipolar LSP resonance of the sphere is excited. For a more general excitation, higher-order multipoles may be excited with a multipolar polarizability given in Eq. (6.28). In analogy with the dipolar case, multipolar LSP resonances arise when the denominator of the polarizability approaches zero, which for a multipole of order l corresponds (in the ESA) to the condition:

$$\epsilon'(\lambda) = -\frac{l+1}{l}\epsilon_M. \tag{6.37}$$

Note that the dipolar LSP resonance condition is recovered for $l = 1$. Moreover, the quadrupolar LSP resonance condition (in the ESA) corresponds for example to $\epsilon'(\lambda) = -(3/2)\epsilon_M$. For a typical metal, the multipolar LSP resonances are therefore blue-shifted (i.e. moved to shorter wavelengths) with respect to the main dipolar LSP resonance. Moreover, as l increases, $(l+1)/l$ quickly approaches 1, and the multipolar LSP resonance wavelengths approach quickly the resonance wavelength of the corresponding *planar* metal/dielectric interface, i.e. $\epsilon'(\lambda) = -\epsilon_M$. As illustrated in Table 6.2, the multipolar

Table 6.2 Summary of the predicted dipolar and quadrupolar LSP resonance wavelengths of small spheres (the intrinsic planar LSP resonance is also shown). These are computed from Eq. (6.37) using the optical properties of Au or Ag from Appendix E.

Metal interface	ϵ_M	λ_{Res} [nm]		
		$l = 1$	$l = 2$	plane
Ag/Air	1	346	331	316
Ag/Water	1.77	387	364	339
Ag/Oil	2.25	411	383	353
Au/Air	1	467	383	342
Au/Water	1.77	512	491	442
Au/Oil	2.25	530	508	478

resonances are in general fairly close (in wavelength) to the dipolar LSP resonance. For silver, the resonances are relatively sharp, and it may therefore be possible to distinguish between them when several resonances are excited. For gold, the resonances are much broader (more damped), and are therefore indistinguishable from each other. We will come back to these higher-order resonances within the exact results of Mie theory.

LSP resonances and far-field properties from Mie theory

The previous considerations, within the ES approximation, provide a useful introduction to the LSP resonances of the sphere because of the simplicity of the analytical expressions. They are, nonetheless, strictly valid for the smallest spheres. The exact results of Mie theory will now be used to assess the validity of the ES approximation and understand additional effects related to larger sizes, in particular the effects of retardation and radiation. The LSP resonances can be formally studied within Mie theory by studying directly the Mie coefficients or susceptibilities (see Appendix H for full details). Here we follow for simplicity the more indirect (but more applied) approach of inferring the LSP resonances from the predicted wavelength dependence of the far-field properties of metallic spheres.

The wavelength dependence of the extinction, absorption, and scattering coefficients is shown in Fig. 6.4 for a silver sphere of varying size in air or water (radii of $a = 10$, 25, and 50 nm). These plots reveal clear resonances in the optical properties of these objects, which we now discuss and compare to the ESA results:

- The longest wavelength resonance (indicated in the figure as λ_{Res}) corresponds to excitation of the lowest-order (electric dipolar) localized surface plasmon (LSP) resonance (it is associated with the coefficient d_1 or susceptibility Δ_1 within the Mie theory framework, see Appendix H). For a small sphere, as seen for example for $a = 10$ nm, the resonance condition corresponds to $\epsilon'(\lambda_{\text{Res}}) \approx -2\epsilon_M$, as shown before in the ESA.

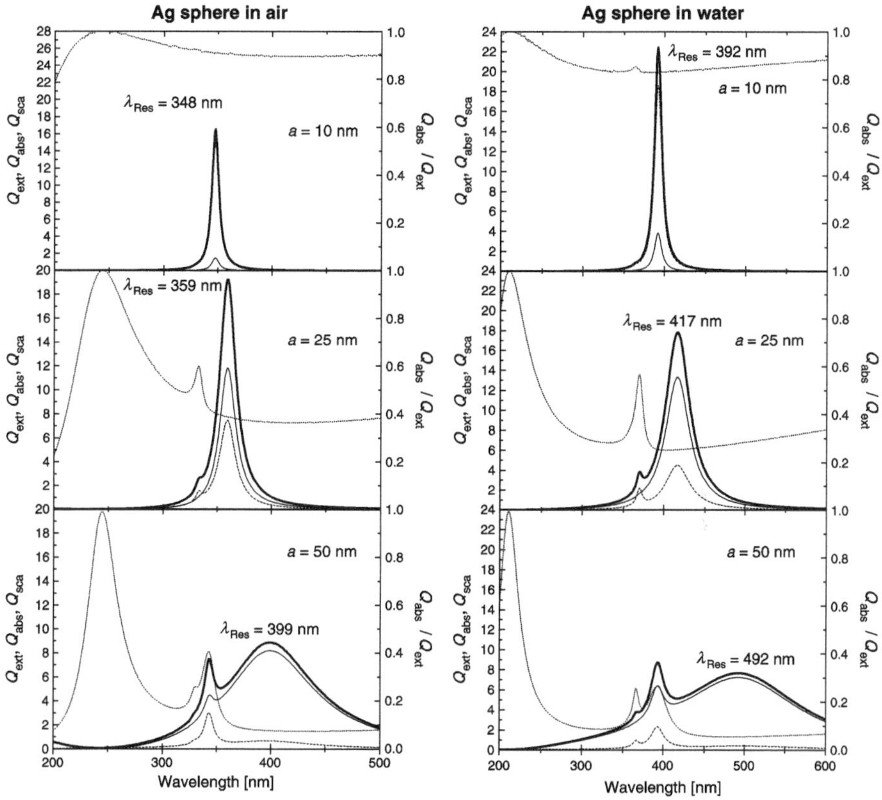

Figure 6.4. Calculated (using Mie theory) far-field properties of silver spheres of varying radii a, in either air ($\epsilon_M = 1$, left) or water ($\epsilon_M = 1.77$, right) as a function of incident wavelength: extinction coefficient, $Q_{\text{Ext}}^{\text{NP}}$ (thick solid lines), absorption coefficient, $Q_{\text{Abs}}^{\text{NP}}$ (dashed lines), scattering coefficient, $Q_{\text{Sca}}^{\text{NP}}$ (thin solid lines). Also shown is the ratio $Q_{\text{Abs}}^{\text{NP}}/Q_{\text{Ext}}^{\text{NP}}$ (dotted lines), with scale on the right-hand-side axis. The scales are kept identical for all plots for easier comparison.

- As the size increases, *this resonance is red-shifted to longer wavelengths, and it broadens.* This is shown more explicitly in Fig. 6.5, where the radius dependence is plotted. The red-shift/broadening is due to radiation/retardation effects (and is therefore not predicted in the ESA). The broadening of the resonance is associated with a decreased quality factor, i.e. the resonance is damped. This damping is mostly due to radiation effects (energy is lost through scattering, i.e. radiation). This reduced quality factor is also associated with reduced local field enhancements as will be shown later. The radiation damping also prevents $Q_{\text{Ext}}^{\text{NP}}$ from increasing as a^4 as predicted in the ESA: it compensates this increase and eventually results in a decreasing $Q_{\text{Ext}}^{\text{NP}}$ above a certain size as shown in Fig. 6.5.

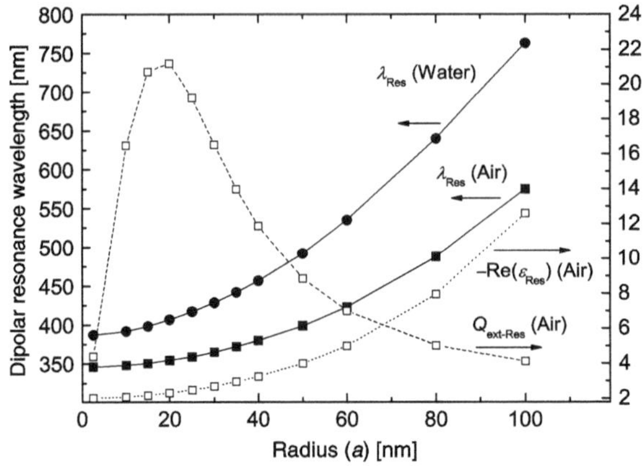

Figure 6.5. Calculated (using Mie theory) dipolar LSP resonance wavelength as a function of sphere radius, a, for a silver sphere in air ($\epsilon_M = 1$) or water ($\epsilon_M = 1.77$). Also shown with scale on the right-hand-side axis are the corresponding extinction coefficient $Q_{\text{Ext}}^{\text{NP}}$ and $-\text{Re}(\epsilon(\lambda_{\text{Res}}))$ at the resonance (in air only). The resonances beyond $a = 50$ nm ($\lambda > 400$ nm in air or 500 nm in water) are weak (and broad), as evidenced by the small $Q_{\text{Ext}}^{\text{NP}}$. This is the result of a strong radiation damping.

- From these exact results, we conclude that the prediction of the ESA is no longer valid for spheres larger than $a \approx 10$ nm radius, in the case of silver spheres. The failure of the ESA in fact coincides approximately with a size where scattering (instead of absorption) starts to dominate the extinction (see for example $a = 25$ nm). This is not so surprising since scattering (radiation) effects are not accounted for in the ESA.
- Higher-order LSP resonances are invisible for the smallest spheres, but start to appear as the size is increased. The main reason is that multipolar polarizabilities increase faster with size than the dipolar one (see the ESA expression in Eq. (6.28)). In addition, the red-shift of the dipolar LSP resonance reduces any overlap with the higher-order LSP resonances. The electric quadrupolar resonance is clearly visible for $a = 25$ nm, and appears at a shorter wavelength than the dipolar resonance. Another resonance (an octupolar resonance) also appears at an even shorter wavelength for $a = 50$ nm. These higher-order LSP resonances also red-shift and broaden as the sphere size is increased, but this occurs at larger sizes than for the main dipolar resonance. Accordingly, they remain quite sharp (with a large quality factor) in the range of sizes investigated here.
- For the smallest spheres, extinction is mainly dominated by absorption, with little scattering. This trend is reversed as the size of the sphere increases. This is a well-known effect in general scattering theory and was already highlighted in the framework of the dipolar approximation in Section 5.1.4. Note that the dipolar resonances are not associated

with a peak in the ratio $Q^{\rm NP}_{\rm Abs}/Q^{\rm NP}_{\rm Ext}$, while the higher-order resonances are. This reflects the fact that the higher-order resonances have a larger non-radiative component than the dipolar resonance.

- Note also that these LSP resonances depend on the environment, and are red-shifted for larger ϵ_M (water compared to air for example), as already predicted in the ES approximation.

- Finally, note that the peak in the ratio $Q^{\rm NP}_{\rm Abs}/Q^{\rm NP}_{\rm Ext}$ around 245 nm in air or 210 nm in water simply corresponds to the condition $\epsilon'(\lambda) = \epsilon_M$. In this case, little scattering is expected and $Q^{\rm NP}_{\rm Sca} \approx 0$.

Influence of optical absorption: gold vs silver

It is now interesting to carry out the same study for gold spheres and this is shown in Fig. 6.6. The results are in sharp contrast with those obtained for silver.

Firstly, the main (dipolar) LSP resonance occurs at a longer wavelength for gold than for silver. This is simply a result of the different optical properties: the condition $\epsilon'(\lambda_{\rm Res}) \approx -2\epsilon_M$, which characterizes the dipolar LSP resonance of a small sphere (in the ESA), occurs at a longer wavelength for gold ($\lambda \approx 480$ nm in air) compared to silver ($\lambda = 346$ nm in air), see Table 6.1.

In addition, we observe that the dipolar LSP resonance is much weaker and much broader for gold. This was already predicted in the ES approximation. Moreover, the radiation/retardation effects (red-shift and broadening of the resonance) are much less evident for gold. The reason for these is that the metal-like optical properties of gold are strongly affected at shorter wavelengths ($\lambda < 600$ nm) by a large optical absorption, i.e. $\mathrm{Im}(\epsilon)$ is large for gold in this region, as shown in Fig. 6.7(a). This strong optical absorption damps significantly the LSP resonances that occur in this region. This explains the large discrepancy in the strengths of the dipolar LSP resonance between gold and silver for the small spheres ($a = 10$ and 25 nm). These occur (for these radii) in the region of large absorption for gold and are therefore strongly damped. Only for $a = 50$ nm in water, the LSP resonance for gold is sufficiently red-shifted (to 572 nm) to become less affected by absorption and thus more prominent. However, this is also the size where radiation damping starts to become important (as for silver), and the resonance is therefore not as pronounced as optical absorption alone would allow.

In summary, the LSP resonances of gold spheres are always strongly damped, either because of the large optical absorption of gold (small sizes) or because of radiation damping (large sizes). Silver is affected by the latter effects, but not by the former because the LSP resonances of small spheres occur in a region of low absorption. It is interesting to note however that if one could tune the resonance of a sufficiently small gold particle further in the red, then both effects would disappear and stronger resonances are then

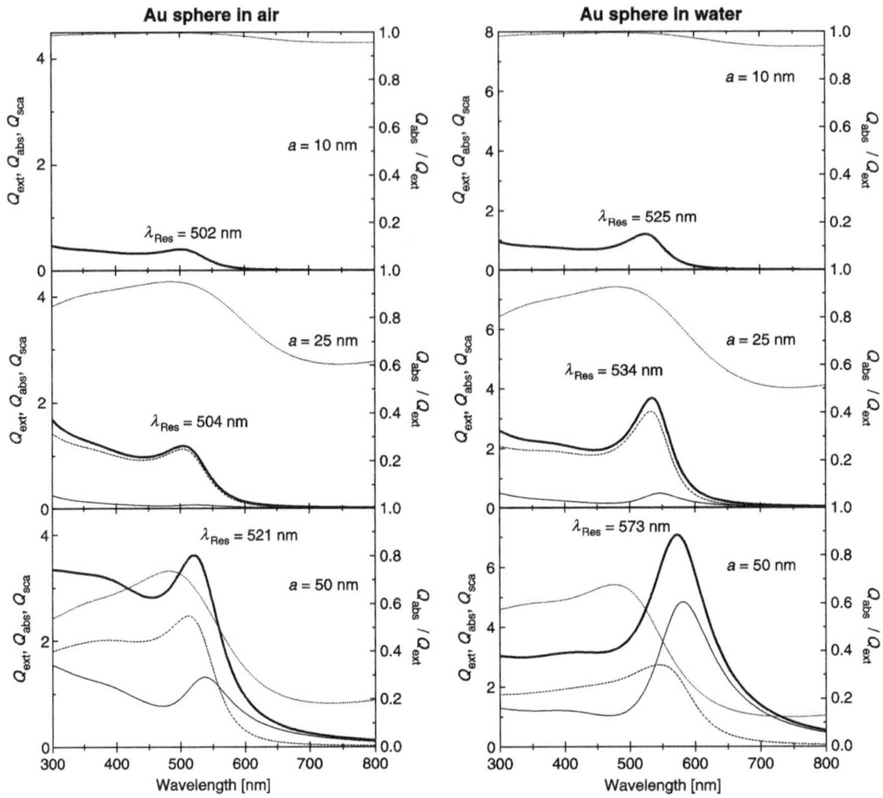

Figure 6.6. Calculated (using Mie theory) far-field properties of gold spheres of varying radii a, in either air (left) or water (right) as functions of incident wavelength: extinction coefficient, $Q^{\mathrm{NP}}_{\mathrm{Ext}}$ (thick solid lines), absorption coefficient, $Q^{\mathrm{NP}}_{\mathrm{Abs}}$ (dashed lines) and scattering coefficient, $Q^{\mathrm{NP}}_{\mathrm{Sca}}$ (thin solid lines). Also shown is the ratio $Q^{\mathrm{NP}}_{\mathrm{Abs}}/Q^{\mathrm{NP}}_{\mathrm{Ext}}$ (dotted lines), with scale on the right-hand-side axis. The scales are kept identical for all plots for easier comparison.

expected. The results so far already point toward one way of achieving this: by embedding the sphere in a high dielectric constant environment (large ϵ_M). This would red-shift the resonance of small gold spheres, possibly beyond the large-absorption region and therefore considerably sharpen the resonances. We will discuss later other approaches to achieve this (for example by changing the shape of the particle).

Quality factor of the resonance

To further characterize this intrinsic influence of the optical absorption, it is possible to define a quality factor for the LSP resonances, at least in

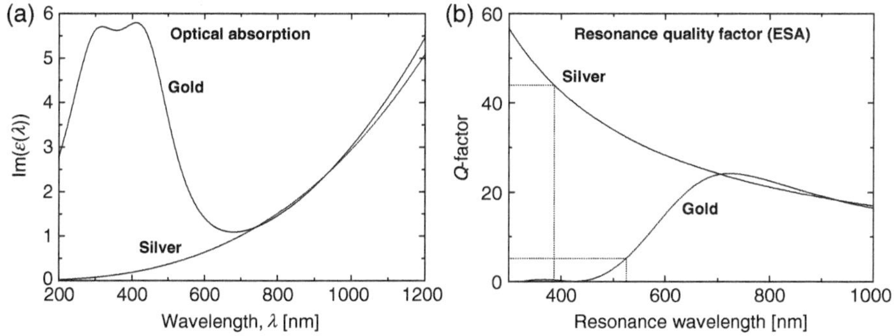

Figure 6.7. (a) Wavelength dependence of the optical absorption (characterized by $\mathrm{Im}(\epsilon)$) of gold and silver (using the analytical expressions of the dielectric functions given in Appendix E). (b) Calculated quality factors of LSP resonances for metallic objects of silver or gold. Q is calculated from Eq. (6.38) in the ESA and is predicted to be independent of geometry within this approximation. The geometry (and dielectric environment) however dictates the position of the resonance wavelength. The dotted lines indicate the resonant wavelength for a small sphere in water. Note also that in the case of silver, one can show analytically that the observed dependence corresponds to a resonance with a wavelength-independent FWHM of $\gamma_0 = 73$ meV, where γ_0 is the loss parameter of the Drude model used for the dielectric function of Ag (see Appendix E).

the ESA [140]. Under a number of reasonable assumptions [140], the quality factor Q in the ESA is *independent of the geometry* and takes the simple general form:

$$Q = \frac{\omega_{\mathrm{Res}} \left(\mathrm{d}\epsilon'/\mathrm{d}\omega\right)_{\mathrm{Res}}}{2\epsilon''_{\mathrm{Res}}}, \tag{6.38}$$

where ω_{Res} is the frequency of the LSP resonance under consideration (which is determined by material, geometry, and dielectric environment). Q characterizes the sharpness of the resonance and can also be expressed as $Q \approx \omega_{\mathrm{Res}}/\Delta\omega$, where $\Delta\omega$ is the full-width at half-maximum (FWHM) of the resonance response (for example $Q^{\mathrm{NP}}_{\mathrm{Ext}}$ or $Q^{\mathrm{NP}}_{\mathrm{Sca}}$ in our case). For example, for a silver sphere in water, the dipolar LSP resonance quality factor is $Q \approx 44$ (at $\lambda_{\mathrm{Res}} = 387$ nm), corresponding to a FWHM of $\approx$9 nm (73 meV). For gold in water, it is $Q \approx 6$ (at $\lambda_{\mathrm{Res}} = 518$ nm), corresponding to a FWHM of $\approx$90 nm. This difference in FWHM is clearly evident in Fig. 6.3.

The wavelength dependence of Q is shown in Fig. 6.7(b) for silver and gold and reflects directly the intrinsic optical absorption shown in Fig. 6.7(a). It further confirms the earlier discussion, and generalizes it to any metallic particle shape (although it applies here only to small particles in the ESA). If the LSP resonances are at wavelengths shorter than $\approx$600 nm, the resonances should be very strong (and very sharp) for silver, but weak and strongly damped for gold. However, *beyond* ≈600 nm (for longer wavelengths), *both metals should behave similarly.* In all cases, one should add to the effect of this intrinsic (ESA) quality factor the possible radiation damping losses that arise at larger particle sizes.

Because gold spheres have LSP resonances in the high-absorption region (which are therefore strongly damped), they do not provide a good model system to further illustrate the LSP resonance characteristics. In view of the fact that LSP resonances are much sharper in silver spheres, we will therefore focus in the following mostly on silver rather than gold to illustrate other aspects of these resonances. We also constrain ourselves mostly to the case of water as the environment, since the resonances are then more separated.

6.2.3. Local field effects

Having identified and discussed the LSP resonances of the sphere from the study of its far-field optical properties, we now focus on their effects on the local field. We consider in particular the local field enhancement effects, i.e. local field intensity enhancement factor (LFIEF), M_{Loc}, and approximate SERS EF, F^0_{E4}. The local field EFs are expected to exhibit similar resonances as the far-field properties (both are governed by the underlying LSP resonances of the substrate), but their magnitude and many other properties (such as their spatial distribution on the surface or local field polarization) cannot be inferred directly from the far-field properties.

Local fields in the ES approximation

As before, let us start with the predictions for a sphere within the ES approximation and focus, in particular, on some of the key EM indicators discussed earlier.

The local field just outside the sphere on its surface (i.e. at $r = a$) is easily obtained from Eq. (6.18). Using the previous notations, and assuming an exciting polarization along z ($\mathbf{E}_{\text{Inc}} = E_{\text{Inc}}\mathbf{e}_z$), it can be written as:

$$\mathbf{E}_{\text{Out}}(r = a) = E_{\text{Inc}}\left[(1 - \beta_S)\mathbf{e}_z + 3\beta_S\left(\mathbf{e}_r \cdot \mathbf{e}_z\right)\mathbf{e}_r\right]. \tag{6.39}$$

The components either normal (along $\mathbf{e}_r$) or parallel (perpendicular to $\mathbf{e}_r$) to the sphere surface can then easily be extracted, from which we derive the corresponding LFIEF, normal ($\perp$), and parallel ($\|$) to the surface:

$$\begin{cases} M^{\perp}_{\text{Loc}}(\mathbf{r},\omega) = \dfrac{|\mathbf{E}_{\text{Out}} \cdot \mathbf{e}_r|^2}{E^2_{\text{Inc}}} = A^{\perp}_S(\omega)\left|\mathbf{e}_r \cdot \mathbf{e}_z\right|^2 = A^{\perp}_S(\omega)\cos^2\theta, \\ \text{where} \quad A^{\perp}_S(\omega) = \left|1 + 2\beta_S(\omega)\right|^2 = \dfrac{9|\epsilon|^2}{|\epsilon + 2\epsilon_M|^2}; \end{cases} \tag{6.40}$$

and

$$\begin{cases} M^{\parallel}_{\text{Loc}}(\mathbf{r},\omega) = A^{\parallel}_{S}(\omega)[1 - |\mathbf{e}_r \cdot \mathbf{e}_z|^2] = A^{\parallel}_{S}(\omega)\sin^2\theta, \\ \text{where} \quad A^{\parallel}_{S}(\omega) = |1 - \beta_S(\omega)|^2. \end{cases} \tag{6.41}$$

The total local field intensity enhancement factor defined in Eq. (5.14) is therefore:

$$M_{\text{Loc}}(\mathbf{r},\omega) = M^{\perp}_{\text{Loc}} + M^{\parallel}_{\text{Loc}} = A^{\parallel}_{S}(\omega) + (A^{\perp}_{S}(\omega) - A^{\parallel}_{S}(\omega))\cos^2\theta, \tag{6.42}$$

and the SERS EF for zero-Stokes shift in the $|E|^4$-approximation can be obtained from its definition:

$$F^{0}_{E4}(\omega_L) = (M_{\text{Loc}}(\omega_L))^2. \tag{6.43}$$

Furthermore, we have the useful expression:

$$\frac{A^{\parallel}_{S}(\omega)}{A^{\perp}_{S}(\omega)} = \frac{(\epsilon_M)^2}{|\epsilon(\omega)|^2}. \tag{6.44}$$

In many cases of interest, this ratio is smaller than 1, and even much smaller than 1 for metals at long wavelengths (for which $\text{Re}(\epsilon(\omega)))$ is large and negative.

Main features of the local field in the ES approximation

Let us note first that in the ES approximation, the predicted local field is independent of the size of the object, here the radius of the sphere.

The LFIEF and the SERS EF are maximum for $\theta = 0$, i.e.: at the two points A and A′ on the axis of the incident field polarization (z-axis here). The local field is normal to the surface at these points, and more precisely:

$$\mathbf{E}_A = (1 + 2\beta_S)E_{\text{Inc}}\mathbf{e}_z. \tag{6.45}$$

$A^{\perp}_{S}(\omega) = |1 + 2\beta_S(\omega)|^2$ therefore represents the maximum LFIEF on the surface (at points A, see Fig. 6.8 and A′ diametrically opposite to A).

The LFIEF and the SERS EF are minimum for $\theta = \pi/2$, i.e. at the four points B, B′, C, C′ on the axes perpendicular to the incident field polarization (x- and y-axes here). The local field is tangential to the surface at these points, and more precisely,

$$\mathbf{E}_B = \mathbf{E}_{B'} = \mathbf{E}_C = \mathbf{E}_{C'} = (1 - \beta_S)E_{\text{Inc}}\mathbf{e}_z, \tag{6.46}$$

$A_S^{\|}(\omega) = |1-\beta_S(\omega)|^2$ represents the minimum LFIEF on the surface (at points B, B′, C, and C′).

Moreover, using Eq. (6.44), we have at all points on the surface ($r = a$):

$$\frac{M_{\text{Loc}}^{\|}(\mathbf{r},\omega)}{M_{\text{Loc}}^{\perp}(\mathbf{r},\omega)} = \frac{(\epsilon_M)^2}{|\epsilon(\omega)|^2}\tan^2\theta. \tag{6.47}$$

Therefore, at most intermediate points (except close to B, B′, C, C′), *the local field polarization is approximately normal to the surface* when $|\text{Re}(\epsilon(\omega))| \gg \epsilon_M$. At the dipolar LSP resonance for example ($\text{Re}(\epsilon(\omega)) = -2\epsilon_M$), we have $M_{\text{Loc}}^{\|}(\mathbf{r})/M_{\text{Loc}}^{\perp}(\mathbf{r}) = (\tan^2\theta)/4$. It is often assumed in a SERS context that the local field polarization is always perpendicular to the metal surface. These simple results for a sphere in the ESA partly support this assumption, but also highlight its limitations: it is not true for all points on the surface and is only approximate unless $|\text{Re}(\epsilon(\omega))| \gg \epsilon_M$.

Local field enhancements at resonance in the ES approximation

Finally, at the dipolar LSP resonance, we can show using Eq. (6.34) that:

$$A_S^{\|}(\lambda_{\text{Res}}) = \frac{9(\epsilon_M)^2}{(\epsilon''(\lambda_{\text{Res}}))^2} \approx |\beta_S(\lambda_{\text{Res}})|^2, \tag{6.48}$$

and using Eq. (6.44):

$$A_S^{\perp}(\lambda_{\text{Res}}) = \frac{9|\epsilon(\lambda_{\text{Res}})|^2}{(\epsilon''(\lambda_{\text{Res}}))^2} \approx 4|\beta_S(\lambda_{\text{Res}})|^2, \tag{6.49}$$

where the approximations are valid when $|\beta_S(\lambda_{\text{Res}})|^2 \gg 1$ (strong resonances). The maximum (approximate) SERS EF on the surface at resonance is thus of the order of:

$$\text{Max}\left(F_{E4}^{0}(\lambda_{\text{Res}})\right) \approx 16|\beta_S(\lambda_{\text{Res}})|^4, \tag{6.50}$$

obtained at the two points on the sphere surface located along the axis of the incident field polarization (A and A′). The minimum SERS EF at resonance is of the order of $|\beta_S(\lambda_{\text{Res}})|^4$, i.e. 16 times less than the maximum SERS EF.

This highlights an important and general aspect of the SERS EF (and to a lesser extent LFIEF) at metal surfaces: *they typically exhibit large variations in magnitude over the surface.* In the case of the sphere in the ESA, those molecules located at the right spot (A or A′) will contribute to the SERS signal ≈16 times more than those at the minima of the SERS EF. Note however that although much smaller at the minima, the SERS EF there may still amount to a reasonable EF. In general, what matters is the convolution of the enhancement magnitude with its spatial localization on the surface. If molecules are uniformly distributed on the surface, there could be larger areas with lower enhancements that contribute with a SERS signal comparable to smaller areas with larger enhancements; and the opposite might also be true for extreme distributions with the widest (by many orders of magnitude) spread of SERS EFs.

An estimate of these maximum EFs for various metal spheres can be inferred from the values of $|\beta_S(\lambda_{\mathrm{Res}})|^2$ given in Table 6.1. Note that these are only valid for the smallest spheres (for which the ESA applies). Large maximum SERS EFs are predicted for silver spheres, in the range $\sim 10^5$–10^7 depending on the optical properties used for silver, but these occur for resonance wavelengths in the near UV (350–400 nm). Much more moderate values, 10–10^3 are obtained for gold spheres depending on their dielectric environment. This is again a result of the strong damping due to large optical absorption in the resonance region (460–530 nm). Among the other metals, it is worth highlighting the case of aluminum, for which SERS EFs of $\approx 10^5$ are predicted, but for a resonance in the deeper UV (≈200 nm). Lithium is also predicted to exhibit relatively strong SERS enhancements in the UV/visible region, but (as pointed out in Chapter 3) is not used much because of its poor chemical stability and high reactivity under typical ambient conditions.

Average enhancement factors in the ES approximation

Of interest in many experiments are the average enhancement factors, rather than their maximum value. Using the following surface averages for a sphere: $\langle \cos^2\theta \rangle = 1/3$ and $\langle \cos^4\theta \rangle = 1/5$, we have:

$$\langle M^{\perp}_{\mathrm{Loc}}(\omega) \rangle = \frac{1}{3} A^{\perp}_S(\omega) \tag{6.51}$$

$$\langle M_{\mathrm{Loc}}(\omega) \rangle = \frac{1}{3} (A^{\perp}_S(\omega) + 2A^{\parallel}_S(\omega)) = 1 + 2|\beta_S(\omega)|^2, \tag{6.52}$$

$$\langle F^0_{E4}(\omega) \rangle = \frac{1}{15} \left[3\left(A^{\perp}_S(\omega)\right)^2 + 4A^{\perp}_S(\omega)A^{\parallel}_S(\omega) + 8(A^{\parallel}_S(\omega))^2 \right]. \tag{6.53}$$

At the dipolar LSP resonance (assuming again $|\beta_S(\lambda_{\text{Res}})| \gg 1$), the average SERS EF is, therefore, of the order of:

$$\langle F^0_{E4}(\lambda_{\text{Res}})\rangle \approx \frac{24}{5}|\beta_S|^4. \tag{6.54}$$

As expected from the previous analysis of the SERS EF maxima and minima, the average SERS EF on the surface is obviously smaller than the maximum EF, here by a factor $\approx$3. In other words, if one were able to place all molecules at the position of highest enhancement, the SERS signal would be $\approx$3 times stronger than for the same number of randomly-adsorbed molecules on the surface. The contrast between the points of highest enhancements and the rest of the surface is not too dramatic for a single sphere, but it can be truly extreme in other situations, as we shall see later.

Local fields from Mie theory

As for the far-field properties treated before, the ES approximation provides a useful analytical insight into the physical aspects of the local field enhancements on a metallic sphere. We will now confirm these results and discuss additional features using the exact solution provided by Mie theory.

We consider a metallic sphere excited by a plane wave with polarization[5] along (Ox) and wave-vector along (Oz) as described in Section H.4 of Appendix H. We first calculate the wavelength dependence of M_{Loc} at five important points inside the sphere or outside on its surface (see the caption of Fig. 6.8 for details). The results are shown in Fig. 6.8 and we now discuss the most important features.

- The local field dependence exhibits strong resonances, and their positions almost match (up to small shifts) those already observed in the far-field properties. This is somewhat expected, since these resonances correspond to a resonant electromagnetic response of the sphere; i.e. they should therefore manifest themselves in most of its electromagnetic properties.
- There are however large differences in how these resonances affect the local field at different points. For example, the field inside the sphere (at the center) is mostly affected by the dipolar LSP resonance (the one at longer wavelength), which also appears clearly for the local field at all points on the surface.
- The influence of the higher-order resonances can be observed for a medium-sized silver sphere. For example, the quadrupolar LSP

[5] Note that, for mathematical convenience, the incident field polarization is along (Ox) and different from that used in the ES approximation.

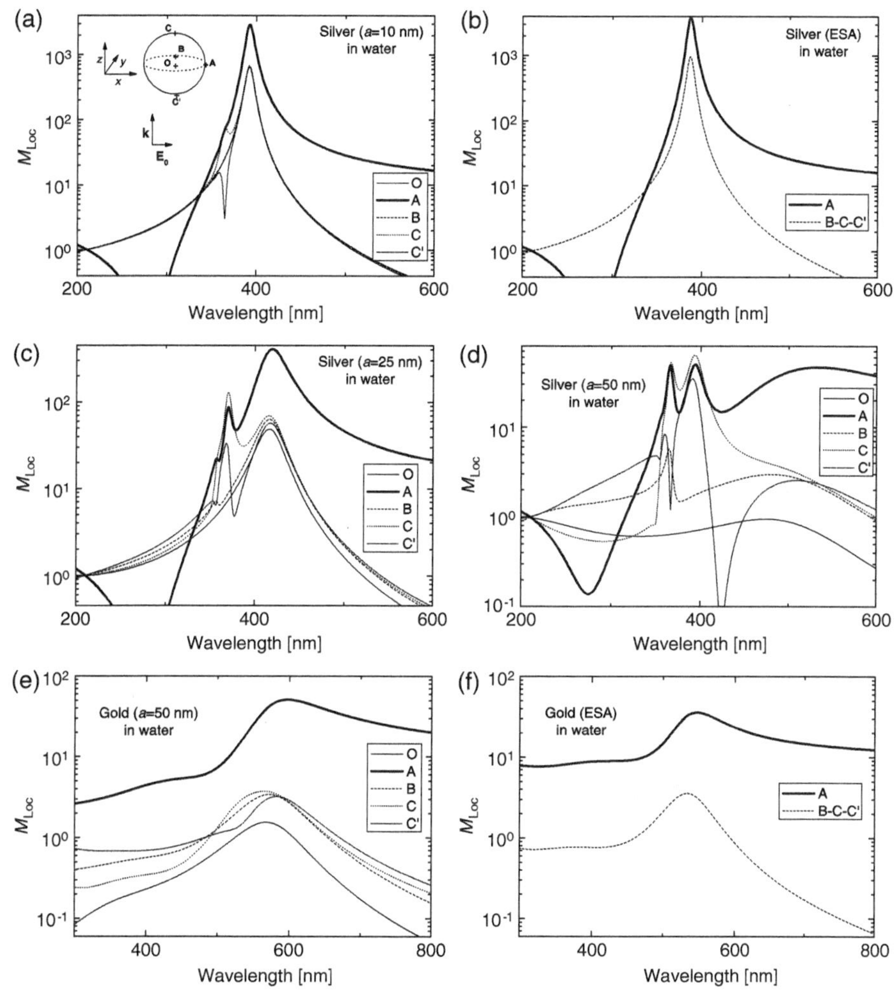

Figure 6.8. Wavelength dependence of the local field intensity enhancement factor, M_{Loc} for six representative cases, all in water ($\epsilon_M = 1.77$): (a) a silver sphere of radius 10 nm (Mie theory), which can be compared to (b) the ES approximation results for a silver sphere (size-independent). The next three plots all show calculations with Mie theory, for a silver sphere of radii (c) 25 nm, (d) 50 nm, and (e) a gold sphere of radius 50 nm. The latter can be compared to (f), which is the ES approximation results for a gold sphere (size-independent). In most cases, the field is calculated for 5 important points, as shown in the top-left of (a). O is the center of the sphere (and therefore inside the metal); this is not directly relevant to SERS, but is shown here for completeness. The four other points are outside the sphere, on the surface. A is on the incident polarization axis, C and C' are opposite on the propagation axis of the incident wave, and B is on the third axis (magnetic field polarization).

resonance appears clearly (blue-shifted with respect to the dipolar LSP resonance) for the local field at points A, C, and C', etc.[6].

- The local field at point A (along the field polarization) always exhibits a dip around 282 nm for silver. This corresponds to the condition $\mathrm{Re}(\epsilon) = 0$, i.e. it is associated with the *bulk plasmon resonance* (see Chapter 3). This effect is irrelevant to SERS in most cases of interest.
- For many applications, including SERS, the dipolar LSP resonance (i.e. the most red-shifted one) is the most important. As we can see here, the LFIEF at this resonance can be quite large, and affects all points on the surface of the sphere. This resonance is particularly prominent at point A (along the axis of incident polarization), and this simply reflects its dipolar nature. We also note that the LFIEF at this point remains fairly large even beyond the resonance (at longer wavelengths). These features were already predicted and discussed in the ES approximation.
- The magnitude of $M_{\mathrm{Loc}}(\mathrm{A})$ is largest for the smallest spheres. The maximum possible value at the dipolar LSP resonance (as can be seen for $a = 10$ nm) is of the order of ~3000 for silver in water. This is comparable, for $a = 10$ nm, with the predictions of the ES approximation. This figure however decreases sharply as the size increases, to reach around 45 for $a = 50$ nm. This decrease is a consequence of the damping of the resonance as a result of radiation, as discussed earlier for the far-field properties. This however highlights the fact that the resonance damping dramatically affects the local field enhancements, much more than the far-field properties. As a corollary, the ES approximation (which does not include radiation effects) fails dramatically for the local field enhancement predictions as soon as radiation becomes non-negligible. This is evident in comparing the results of $a = 25$ nm or $a = 50$ nm with the ESA predictions.
- For gold spheres, for which the resonances are strongly damped even in the absence of radiation (because of intrinsic optical absorption), radiation effects play a much less important role (at least for the sizes considered here). The ES approximation therefore remains approximately valid up to $a \approx 50$ nm. This strong damping however

[6] The behavior of point C' around the quadrupolar resonance is quite peculiar, it increases sharply at a wavelength shorter than this resonance and is then followed by a pronounced decrease on the other side. This dip is even observed for the smallest sphere ($a = 10$ nm), but disappears at even smaller radii ($a \approx 1$ nm) as predicted in the ES approximation. Although an interesting EM effect in its own right, it will not be discussed further since it is mostly irrelevant in the context of SERS. It does highlight an interesting fact though: the ESA is broadly valid up to $a \approx 10$ nm as discussed earlier, but it may still miss out on some very specific features (like the field at C$'$). It is therefore strictly speaking only correct in the limit of vanishing a.

means that the magnitude of the LFIEF at resonance is much smaller, of the order of $M_{\text{Loc}} \approx 50$ at point A. It does not vary much with size since radiation effects remain negligible up to $a \approx 50$ nm. Moreover, it becomes comparable to silver for sizes of the order of $a = 50$ nm.

- Finally, for the largest silver spheres ($a = 50$ nm), the LFIEF at the quadrupolar (and even higher order) resonances can become comparable in magnitude of even larger than that at the dipolar resonance. Point C may then experience a larger field enhancement than point A. In fact, a more careful study (see below) will show that the point of highest enhancement in this case is in between A and C.

Spatial distribution of the local field enhancement

This final remark highlights the importance of not overlooking other points on the surface of the sphere. A full wavelength dependence at all points of the sphere is possible, but the results are difficult to visualize. We only show here representative examples of the field distribution on the surface at specific wavelengths. We first show in Fig. 6.9 two examples of the local field distribution at the dipolar LSP resonance: $\lambda = 392$ nm here for a $a = 10$ nm silver sphere in water, and $\lambda = 494$ nm for a larger silver sphere ($a = 50$ nm) in water. Note the log-scale used for the LFIEF, M_{Loc}. This figure illustrates in addition the spatial variation of the local field polarization. At each point, the local field is decomposed into its normal and tangential components: $\mathbf{E}_{\text{Loc}} = \mathbf{E}^{\perp} + \mathbf{E}^{\parallel}$. The plots on the right-hand side show the spatial distribution of the normal contribution expressed as: $|\mathbf{E}^{\perp}|^2/|\mathbf{E}_{\text{Loc}}|^2$, i.e. a value of one corresponds to a field normal to the surface while a value of zero to a tangential field.

The dipolar nature of this resonance is evident in the field distribution at the surface of the sphere, which resembles that created by a dipole at the center of the sphere, and oriented along the axis of the incident field polarization $\mathbf{E}_{\text{Inc}}$. The field magnitude is maximum at points on this axis, and the local field polarization there is also along this axis (i.e. normal to the surface). The field magnitude is minimum in the plane perpendicular to $\mathbf{E}_{\text{Inc}}$, and the local field polarization is tangential to the surface (along $\mathbf{E}_{\text{Inc}}$) there. These results are basically the same as those predicted in the ES approximation.

The field enhancement remains large everywhere on the surface for the small sphere and its magnitude exhibits a good 'uniformity', with a factor of only ≈ 5 between the largest and smallest M_{Loc} (a factor of ≈ 4 was predicted in the ESA). This is no longer true for the larger sphere ($a = 50$ nm), where this factor increases to around ≈ 20. This translates into a factor ≈ 400 between the maximum (≈ 2000) and minimum (≈ 5) SERS EFs on the surface.

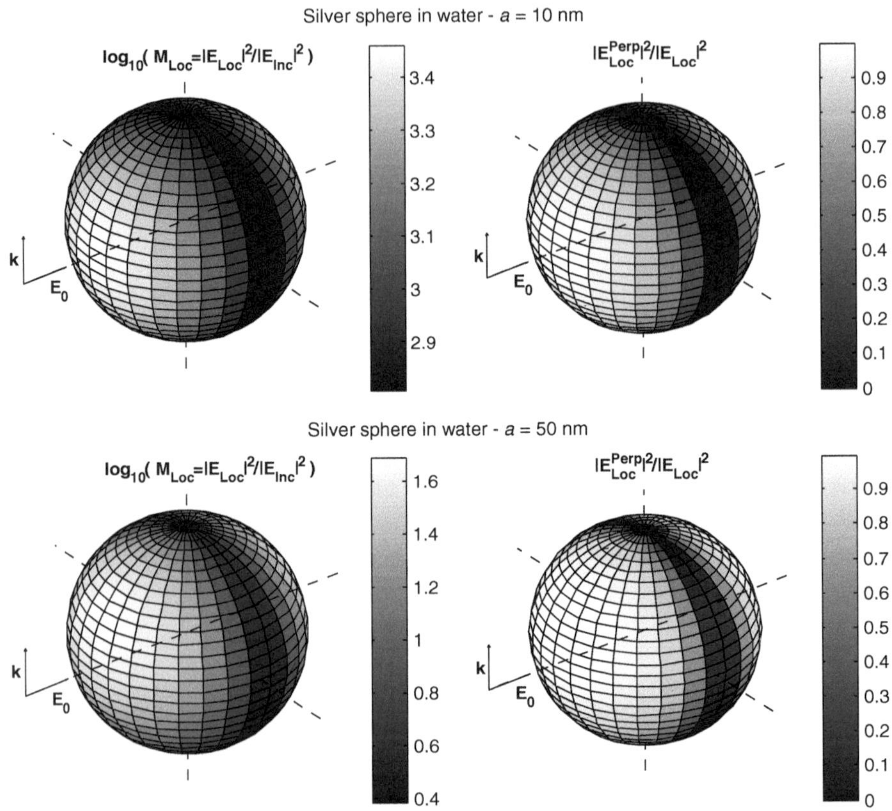

Figure 6.9. Spatial distribution of the local field, characterized by the enhancement factor $\log_{10}(M_{\text{Loc}})$ (left), and the proportion of the normal component $|\mathbf{E}^{\perp}|^2/|\mathbf{E}_{\text{Loc}}|^2$(right), both computed at the main dipolar LSP resonance: (top) for a 10 nm radius silver sphere in water at $\lambda = 392$ nm, and (bottom) for a 50 nm radius silver sphere in water at $\lambda = 492$ nm.

Overall these field distributions only show – in a different way – what could already be inferred from the inspection of the 5 previously selected points in Fig. 6.8.

The situation for the quadrupolar resonance reveals additional effects as shown for example in Fig. 6.10(a). The field distribution resembles, as expected, that created by a quadrupolar source at the center of the sphere. The most important characteristic here is that the maximum field enhancement occurs at a point half-way between points C and A (or C and A′). Note that another local maximum is observed at a point between C′ and A, but its magnitude is not as large. This aspect is shown more clearly in Fig. 6.10(b). The local field polarization is again normal to the surface at these points of highest enhancement.

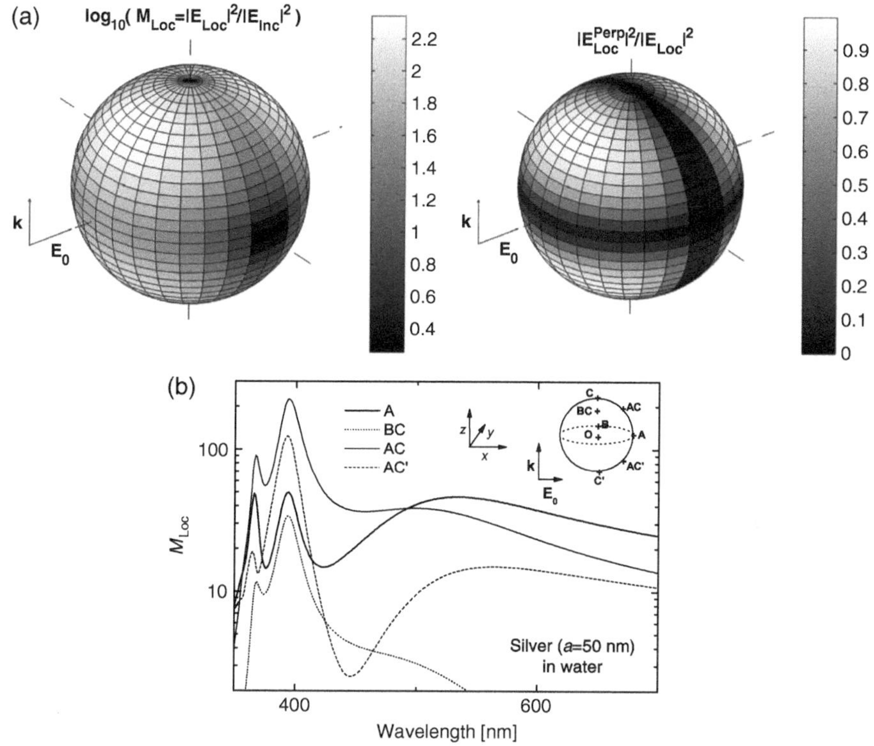

Figure 6.10. (a) Spatial distribution of the local field, characterized by the enhancement factor $\log_{10}(M_{\text{Loc}})$ (left), and the proportion of the normal component $|\mathbf{E}^{\perp}|^2/|\mathbf{E}_{\text{Loc}}|^2$ (right), for a 50 nm radius silver sphere in water at $\lambda = 393$ nm (quadrupolar LSP resonance). (b) Wavelength dependence of the LFIEF, M_{Loc} for a 50 nm silver sphere in water. The field is calculated here for additional points (compared to Fig. 6.8(d)), as shown in the inset. This highlights the important role, especially for larger spheres, of points like AC and AC$'$, which are not conventionally looked at in most simple descriptions.

Spatially-averaged local field enhancement

These latest remarks highlight the importance of considering not only the local field at a few selected points, as often done in the literature, but also spatially-averaged properties. This is particularly true for most SERS applications where the molecules to be analyzed are randomly adsorbed on the metallic surface. We discussed earlier two important indicators for spatially-averaged properties: $\langle M_{\text{Loc}} \rangle$, which characterizes the average local field intensity enhancement factor, and $\langle F_{\text{E4}}^{0} \rangle = \langle (M_{\text{Loc}})^2 \rangle$, which characterizes the approximate average SERS enhancement factor. The wavelength dependence of these two indicators is illustrated in Fig. 6.11 for a few representative cases.

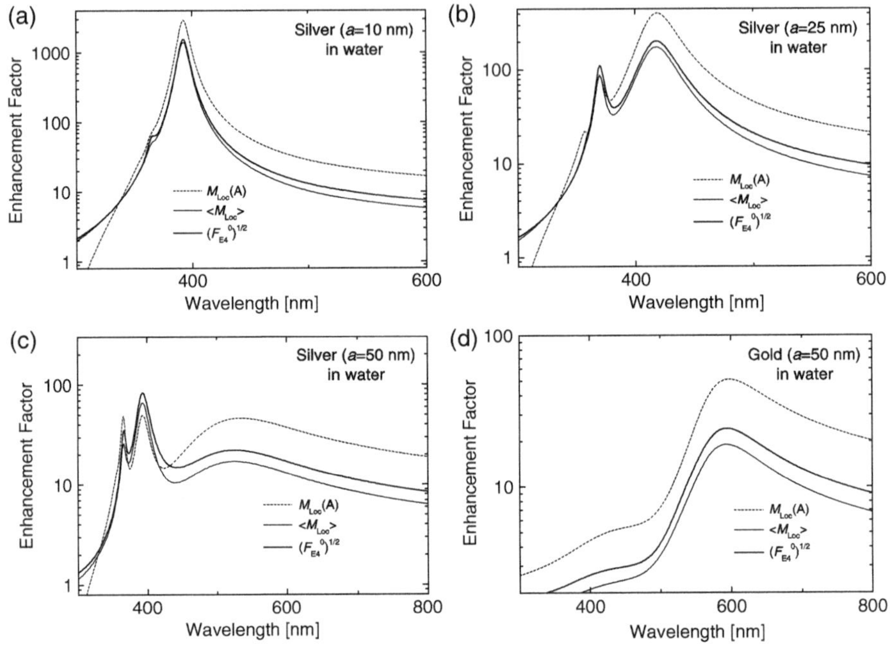

Figure 6.11. Wavelength dependence of the average local field intensity enhancement factor, $\langle M_{\text{Loc}} \rangle$, and average SERS enhancement factor (characterized here by its square root $(\langle F^0_{\text{E4}} \rangle)^{1/2}$ for plotting convenience). These are shown in comparison with the local field intensity enhancement factor at point A (at the surface along the incident polarization axis). The same four cases as in Fig. 6.8 are considered.

The average SERS EF and the average LFIEF follow a similar spectral profile (resonances) as their counterpart predicted at point A. For $a = 10$ and 25 nm for silver, and $a = 50$ nm for gold, the magnitude of the average LFIEF is a factor of $\sim$2–3 smaller than its maximum value at A (point of maximum enhancement in these cases). This translates into a factor 4–9 between the maximum and average SERS EFs, and is similar to what was obtained in the ES approximation (a factor of $\sim$3). Such factors can be considered as 'good uniformity' as far as SERS EFs are concerned. However, we will see later that this is no longer true for non-spherical geometries, and even less for interacting objects, where much wider spatial distributions are typically predicted.

Note finally that for the case of the $a = 50$ nm silver sphere, the average EFs are predicted to be larger than the punctual EFs at point A, in particular at shorter wavelengths where higher-order resonances become important. This simply reflects the fact discussed previously that there the point of maximum enhancement is moved to a different position on the sphere.

6.2.4. Distance dependence

It is common in EM calculations to consider only the points directly on the metallic surface. The variation of the local field enhancement as a function of distance from the surface is however important in several situations, for example: (i) in relation to the discussion of the first-layer effect in SERS, and (ii) in surface-enhanced fluorescence (SEF), where the optimal position of the fluorophore is slightly away from the surface (to avoid excessive non-radiative emission, treated earlier).

Distance dependence in the ES approximation

As before, it is convenient to consider first the analytical formulas of the ES approximation for a qualitative understanding of the main features. We will restrict ourselves to situations of reasonably large enhancements, where the scattered field dominates over the incident field. The electric field outside the sphere in the ESA is then simply that created by an electrostatic dipole at the sphere center. For a molecule at a distance d from the sphere surface, the electric field amplitude should then decay as $1/(a+d)^3$. Punctual and average LFIEFs should therefore decay as $1/(a+d)^6$ and SERS EFs as $1/(a+d)^{12}$. Such a power law with an exponent as large as 12 could be interpreted as a dramatic decay, and it may have contributed to the common belief that the SERS signal is dominated by the first layer of adsorbed molecules. This is not so however, simply because the distance dependence is not $1/d^{12}$, but $1/(a+d)^{12}$, where a is typically larger or much larger than d. In fact, it is easy to see that, with respect to $d = 0$, the SERS EF decreases by a factor of only 2 at $d \approx 0.06a$ and by a factor of 10 at $d \approx 0.2a$. For a typical radius of $a \approx 30$ nm, this corresponds to $d \approx 2$ nm and $d \approx 6$ nm respectively, i.e. much more than the thickness of a molecular monolayer (typically smaller than 1 nm). This is further illustrated in Fig. 6.12(a). Note also that a factor of 10 decrease in the SERS EF is not that large considering that variations of at least the same order of magnitude already occur as functions of position on the surface even at $d = 0$.

In conclusion, although the SERS EF is maximum at $d = 0$ (first layer), the *EM enhancements are long-range effects*, and typically extend to at least $\approx$10 nm away from the metallic surface. This discussion, for a sphere within the ES approximation, can moreover be qualitatively extended to most typical SERS substrates.

Coverage dependence in the ES approximation

Predictions of the coverage dependence (i.e. on the number of molecular monolayers on the surface) of the SERS signal can in principle be deduced directly from the distance dependence of the SERS EF, provided the inter-layer distance d_L is known. An example is shown in Fig. 6.12(a) for a

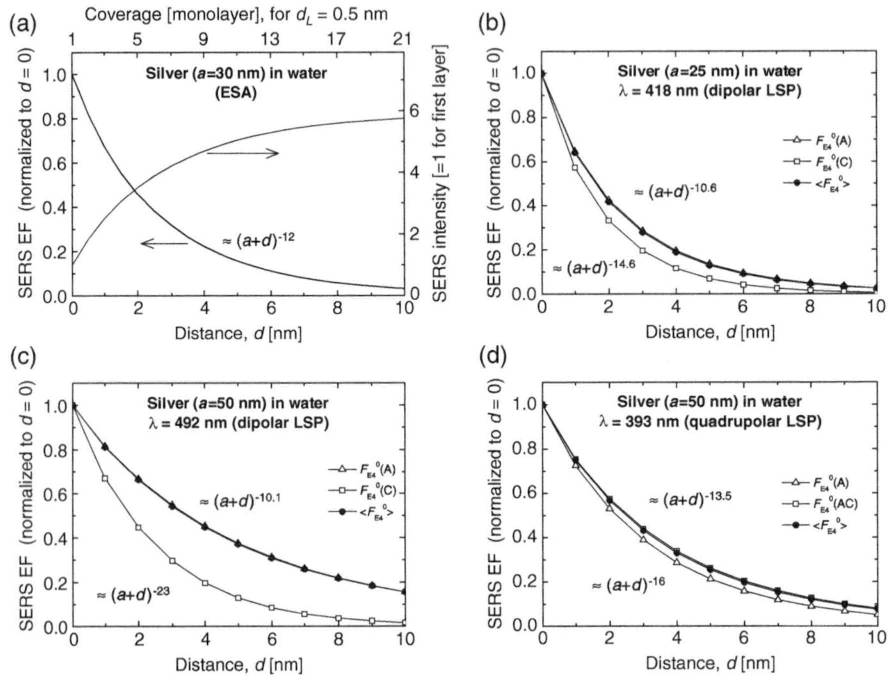

Figure 6.12. Dependence of punctual and average EM SERS EFs with the distance d from the metal surface of a silver sphere in water. Points A, C, and AC have been defined in Fig. 6.10. (a) ES approximation for a sphere of radius $a = 30$ nm. Also shown in this plot is the cumulated SERS intensity (scale on the right axis) as a function of coverage assuming a monolayer thickness of $d_L = 0.5$ nm. (b and c) Mie theory predictions for $a = 25$ and 50 nm radius spheres at their respective dipolar LSP resonance wavelengths. (d) Same for a sphere with $a = 50$ nm at the quadrupolar LSP resonance wavelength.

sphere with $a = 30$ nm in the ES approximation assuming an inter-layer distance of $d_L = 0.5$ nm. Because the SERS EF decreases with distance, the cumulated SERS signal varies sub-linearly with coverage and eventually saturates. Nevertheless, it is clear that the contribution of the subsequent layers can be larger than that of the first layer: the cumulated SERS signal is for example 5 times larger than the first-layer signal at a coverage of 10 monolayers in the example of Fig. 6.12(a).

These figures are provided here only as a starting point for coverage dependence studies. In practice, many additional effects should be considered at coverages larger than a monolayer. Among other things, the molecules on the first layer may be chemically influenced by the metallic surface (this the source of the chemical enhancement discussed in Section 4.8). The adsorption geometry may be different in subsequent layers. The LSP resonances (and therefore the EM properties of the substrate) may also be affected by the presence of the adsorbate at large coverage (in a similar manner as the dielectric constant of the environment ϵ_M affects them).

Distance dependence from Mie theory

We conclude this section by a few examples of the distance dependence predicted for a sphere from the exact EM results of Mie theory. Fig. 6.12(b–d) illustrates this distance dependence of punctual and average enhancement factors for representative cases of a sphere. From these, the following facts are worth highlighting:

- The average and maximum EF follow the same distance dependence. The distance dependence is *not* a property solely of the place with the largest intensity (point A), but rather it is a widespread property of the enhancement on the entire surface.
- At the dipolar LSP resonance, the dependence is similar to the ESA predictions.
- At the quadrupolar LSP resonance, the decay is slightly faster. This would in fact also be predicted in the ESA if a quadrupole was excited (i.e. if the assumption of a constant external field is dropped). The faster decay with distance is also a well-known property in the comparison of quadrupoles vs dipoles, even in the electrostatic approximation.
- Specific points on the surface may experience a slightly different distance dependence depending on the relative importance of the dipolar or quadrupolar components at those points. For example $F^0_{E4}(C)$ decays faster than $F^0_{E4}(A)$. The reason for this is simply that the EF at C is governed by the quadrupolar component of the field. This subtle differences, although conceptually interesting, are mostly irrelevant to most SERS situations.

Overall, the Mie theory results confirm the discussion given earlier within the ESA, except for secondary effects that are typically irrelevant to most SERS experiments.

6.2.5. Non-radiative effects – surface-enhanced fluorescence

The local field intensity and SERS EF, which were studied in the previous sections, correspond to the 'excitation' problem. We now focus on the 'emission' problem, i.e. the properties of an emitter in close vicinity of a metallic sphere. As discussed in Chapter 4, the emission properties in a given direction can in fact be linked to the corresponding excitation problem through the optical reciprocity theorem. This is the reason why the quantity F^0_{E4} provides a reasonable estimate of SERS enhancements in many situations.

For SEF however, the emission properties in a single given direction are not sufficient to make physical predictions. This is because SEF is a two-step process and emission in one direction (that of our detector) must compete with emission in other directions or into non-radiative channels. These other decay channels must therefore also be studied and (as discussed in Section 5.1.3) it can be discussed in terms of a number of key EM indicators:

- the total EM decay rate enhancement factor, M_{Tot},
- the radiative decay rate enhancement factor, M_{Rad},
- and the EM radiative efficiency or approximate modified quantum yield, $\eta_{\mathrm{Rad}}^{\mathrm{EM}} = M_{\mathrm{Rad}}/M_{\mathrm{Tot}}$.

The approximate SEF enhancement factor at a given point can then be characterized by M_{Fluo} (Eq. (5.17)) or M_{Fluo}^{d} (Eq. (5.18)), while the average SEF EF is characterized by $M_{\mathrm{Fluo}}^{\mathrm{Ave}}$ (Eq. (5.20)). All these quantities can be estimated within the framework of Mie theory (or its extension to dipolar emitters as detailed in Appendix H), the predictions of which are now discussed.

Modified decay rates and quantum yield

As shown in Section 6.1, M_{Tot} can be very large for an emitter close to a planar metallic surface; for example of the order of 10^6 at the LSP resonance of the planar interface ($\mathrm{Re}(\epsilon) \approx -\epsilon_M$) at a distance $d = 1$ nm. However, the radiative enhancement is then (at best) of order $\sim$1, sometimes much smaller. Most of the energy is therefore emitted into non-radiative modes, and is eventually dissipated in the metal. These considerations, valid for plane metallic surfaces, are the origin of the common belief that fluorescence is always quenched close to metals. We will show here that for other geometries (illustrated here by the example of spherical nano-particles) enhancements with a large radiative component can exist.

To understand this, we show in Fig. 6.13 the wavelength dependence of M_{Tot}, M_{Rad}, and the corresponding modified quantum yield $\eta_{\mathrm{Rad}}^{\mathrm{EM}}$ for a selection of dipole configurations close to the surface of a silver sphere in water. Results are shown for three sizes of spheres with radii $a = 10$, 25, and 50 nm, respectively; typical for colloidal Ag solutions. Varying the radius allows us to understand the role of the main dipolar LSP resonance of the sphere, which red-shifts from about 390 to 490 nm as the size increases. M_{Rad} clearly exhibits maxima at the LSP resonances (both dipolar and higher order). In contrast, M_{Tot} remains virtually unchanged for all sizes, peaking at 340 nm ($\mathrm{Re}(\epsilon) \approx -\epsilon_M$), corresponding to the intrinsic LSP resonance of a planar Ag/water interface; a conclusion already reached in Section 6.2.1 in the ES approximation.

These results and other features of Fig. 6.13 can be understood simply by the following qualitative arguments:

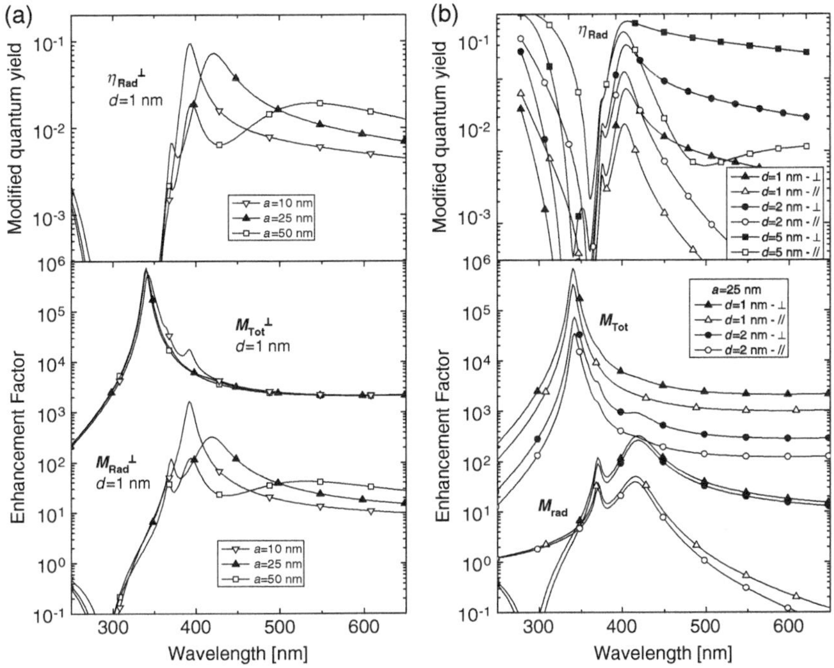

Figure 6.13. (a) EM radiative efficiency, $\eta_{\rm Rad}^{\rm EM}$ (top); and both total ($M_{\rm Tot}$) and radiative ($M_{\rm Rad}$) decay rate enhancement factors (bottom) for a dipole perpendicular (at a distance $d = 1$ nm) to silver spheres of radii $a = 10$, 25 and 50 nm, in water. The variation of a allows us to extract the influence of the main LSP dipolar resonance on these EFs, which appears as a peak in $M_{\rm Rad}$ that red-shifts with increasing a. The peak in $M_{\rm Tot}$, however, does not change with a and remains at ~340 nm. This corresponds to the condition $\text{Re}(\epsilon) \approx -\epsilon_M$ and is the intrinsic LSP resonance of a planar silver/water interface. (b) Same plots for a silver sphere in water of fixed size ($a = 25$ nm), but varying the dipole orientation (perpendicular, $\perp$, or parallel, $\parallel$, to the surface) and its distance, d, from the surface. The total EM decay rate EFs are strongly increased as the dipole gets closer to the surface, while the radiative enhancements do not change much. Consequently the modified quantum yields are small for $d = 1$ nm, even at resonance, and increase substantially when moving away (even to only $d = 2$ nm).

- The sphere LSP resonances are mostly *radiative* (especially the dipolar one). They result in a larger radiative decay rate EF, $M_{\rm Rad}$, when the dipole couples to them efficiently. These radiative resonances, and the main dipolar LSP resonance in particular, are strongly *size-dependent* for spheres and more generally are *geometry-dependent*. Here, they are red-shifted for larger spheres.

- The coupling of the emitter to these resonances is *not strongly distance-dependent*: $M_{\rm Rad}$ does not drastically vary with d. It is however sensitive to the dipole orientation. This is illustrated in Fig. 6.13(b) where perpendicular and parallel dipoles are compared: the perpendicular dipole couples more efficiently to the LSP resonances.

- The intrinsic LSP resonance of the planar Ag/water interface at 340 nm is however strongly *non-radiative*. It is related to the strong reflected field created by the dipole image. This is *independent of geometry* because the surface is approximately a plane when viewed from the dipole at very short distances. The intensity of the reflected field however *strongly depends on the distance* d of the dipole from the surface, and decreases as d^{-3} as discussed in detail in Section 6.1. This non-radiative contribution can in fact be modeled to a good approximation by that obtained for a planar surface (Eq. (6.13)).
- The total EM decay rate EF M_{Tot} simply reflects these two contributions: radiative and non-radiative. For a dipole very close to the surface ($d \leq 1$ nm), the non-radiative component (reflected field) almost entirely dominates, except possibly when the radiative resonance is very strong (for example the small 'bump' in Fig. 6.13(a) for $a = 10$ nm). M_{Tot} is then very similar to that obtained for the planar interface (Eq. (6.13)).

Finally, we can also identify from Fig. 6.13 the situations where the radiative decay rate EF (M_{Rad}) is the largest. Firstly, dipoles perpendicular to the surface present larger enhancements than those parallel to it. Moreover, M_{Rad} is maximum at the LSP resonances of the sphere, which are size-dependent. At these maxima, M_{Rad} is largest for the smallest sphere, with values of $\approx$1100 for $a = 10$ nm, down to $\approx$330 for $a = 25$ nm, and $\approx$40 for $a = 50$ nm. For a given radius, the maximum M_{Rad} varies little with d, from $\approx$330 for $d = 1$ nm, to $\approx$260 for $d = 2$ nm, and $\approx$ 150 for $d = 5$ nm (for $a = 25$ nm). However, because M_{Tot} decreases strongly, the EM radiative efficiency $\eta_{\text{Rad}}^{\text{EM}} = M_{\text{Rad}}/M_{\text{Tot}}$ increases markedly, from $\eta \approx 0.07$ for $d = 1$ nm to $\eta \approx 0.3$ for $d = 2$ nm, and $\approx$0.6 for $d = 5$ nm (for $a = 25$ nm).

Fluorescence enhancement factors for SEF

This final remark is crucial to any applications in surface-enhanced fluorescence (SEF). SEF profits mostly from the local field intensity enhancement factor, M_{Loc}, which results in enhanced absorption. The benefits of this enhancement are only useful if the modified quantum yield is not too small (which would otherwise cancel the enhancement in absorption). From the previous discussion, the distance of the emitter from the surface is therefore the most important parameter in SEF. The emitter must be reasonably close to the surface to profit from a good local field enhancement, but if it is too close, the sharp decrease in quantum yield cancels any benefits. There is therefore an optimum distance, typically in the range $\sim$2–10 nm depending on the structure.

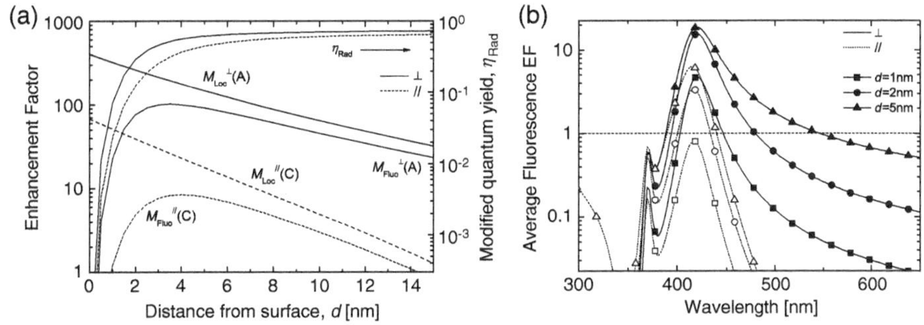

Figure 6.14. SEF close to a silver sphere ($a = 25$ nm) in water: (a) Distance dependence of the local field EF, M_{Loc}, the approximate modified quantum yield, $\eta_{\text{Rad}}^{\text{EM}}$, and the approximate fluorescence EF, $M_{\text{Fluo}} = M_{\text{Loc}}\eta_{\text{Rad}}^{\text{EM}}$, at $\lambda = 418$ nm (dipolar LSP resonance) for two cases of emitters: a dipole perpendicular to the surface at A (along the incident polarization), and a dipole parallel to the surface at C (along the incident wave-vector). (b) Wavelength dependence of the *average* fluorescence EF for a monolayer of fluorophores either perpendicular or parallel, and at a constant distance d from the surface.

This is illustrated in Fig. 6.14(a) for a silver sphere ($a = 25$ nm) in water at the dipolar LSP resonance ($\lambda = 418$ nm). We consider an emitter at two positions (same as defined earlier): point A, where the LFIEF is maximum and the local field perpendicular to the surface (and the emitting dipole is therefore chosen as perpendicular) and point C, where the local field is smaller but parallel to the sphere surface (and the emitting dipole is then chosen to be parallel). The distance dependence of the LFIEF, M_{Loc}, and the approximate modified quantum yield $\eta_{\text{Rad}}^{\text{EM}}$ are shown to have opposite contributions, as discussed earlier. Accordingly, the fluorescence EF, M_{Fluo}, which is simply the product of the two, exhibits in both cases a maximum, of ≈ 100 at $d = 3.5$ nm ($\perp$ at point A) and of ≈ 8.5 at $d = 4$ nm ($\|$ at point C).

Fluorescence EFs are in general much smaller (here up to ≈ 100) than SERS EFs, and in fact more often result in quenching rather than enhancement. This is normal since SERS, being a scattering (instantaneous) process, benefits from both the local field (absorption) and radiative (emission) EF. SEF, on the other hand, only benefits from the local field (absorption) EF. The best possible outcome for emission is to emit all the energy radiatively ($\eta_{\text{Rad}}^{\text{EM}} = 1$); no enhancement occurs if the emitter was already very efficient (with a free-space quantum yield of 1). Only intrinsically poor emitters may benefit from an additional enhancement in the 'emission' part. In this context, it is worth noting the following additional points:

- Perpendicular dipoles exhibit in general larger fluorescence enhancements. This is a result of both a larger LFIEF (absorption) and modified quantum yield (re-emission).
- The distance dependence (for short distances) is mostly dominated by that of $\eta_{\text{Rad}}^{\text{EM}}$ (imposed by M_{Tot}) and therefore drops sharply as d^{-3}

as d approaches zero. M_{Fluo} may then be smaller than 1, i.e. there is a fluorescence quenching, a well-known aspect of fluorescence for emitters adsorbed on metal surfaces.

- However, at (or close to) the radiative resonance of a nano-particle, as is the case in the example of Fig. 6.14(a), the large LFIEF can partially compensate the strong non-radiative quenching, even for small distances (i.e. for adsorbed molecules). The quenching may then not be as dramatic as usually assumed, and fluorescence emission could still, in principle, be observable (albeit with a smaller cross-section). Such residual fluorescence from adsorbed molecules could in fact be in many cases the origin of the so-called SERS continuum, although such an interpretation is still subject to debate [9].
- As the distance increases, the modified quantum yield increases toward a roughly constant value. The distance dependence of the fluorescence EF is then governed by that of the LFIEF and therefore decreases, but much slower than for short distances. In fact, at resonance, a fluorescence EF larger than 10 can still be obtained up to $d = 15$ nm (and even beyond) as shown in Fig. 6.14(a).

Average fluorescence enhancement factor

The above discussion focused on the case of a single emitter at a given position. In many practical cases, an ensemble of fluorophores (a monolayer or even multiple layers) contribute to the SEF signal, which can then be characterized by the average fluorescence EF, defined in Eq. (5.20). A similar distance and wavelength dependence is obtained for the average EF as for the position-dependent fluorescence EF, but the magnitude of the enhancement becomes even smaller because not all emitters experience an optimum enhancement (some may even experience quenching). This is illustrated in Fig. 6.14(b). It is clear that average fluorescence EFs are much smaller than typical average SERS EF on the same substrate, reaching $\approx$20 at most. Moreover, the range of distances and wavelengths where an enhancement (and not a quenching) is predicted is also fairly restricted.

To conclude this discussion of SEF on metal spheres, it is worth highlighting the fact that many aspects of this section can in fact be generalized to more complex structures. We will therefore focus in the rest of this chapter on local field and SERS enhancements. Any predictions for SEF can be obtained by simply 'adapting' the arguments of this section.

6.3. THE EFFECT OF SHAPE ON THE EM ENHANCEMENTS

The simple model of a spherical metallic nano-particle has enabled us to introduce and discuss many important features of the EM enhancements and their connection to plasmon resonances. In the rest of this chapter, we will highlight additional aspects that were absent from the sphere case because of its intrinsic high symmetry. Hence the next logical step is to study *ellipsoids*, for which analytical solutions are still obtainable, at least in the ES approximation. This model system will enable us to discuss many features of the effect of *shape* on plasmon resonances and EM enhancements. We will rely here on the analytical solution for the ellipsoid in the ES approximation discussed in detail in Appendix G. We therefore consider a metallic ellipsoid of semi-axis lengths $a \geq b \geq c > 0$ along its main axes x, y, and z. The optical response is characterized by its dielectric function $\epsilon(\omega)$, and it is embedded in a non-absorbing dielectric with ϵ_M (as in the previous cases for spheres).

6.3.1. Shape effects on localized surface plasmon resonances

Localized surface plasmon resonances of the ellipsoid

The study of the localized surface plasmon (LSP) resonances of the ellipsoid in the ESA can proceed, thanks to the similarities, along the same lines as for the sphere. Hence the optical response is characterized by three (one for each axis) non-dimensional polarizabilities, β_i ($i = 1, 2, 3$ for incident polarization along axes x, y, and z):

$$\boxed{\beta_i(\omega) = \frac{\epsilon(\omega) - \epsilon_M}{3L_i\epsilon(\omega) + (3 - 3L_i)\epsilon_M},} \tag{6.55}$$

where

$$0 < L_1 \leq L_2 \leq L_3 < 1, \quad \text{and} \quad L_1 + L_2 + L_3 = 1. \tag{6.56}$$

The L_i's are called *geometrical factors* or *depolarization factors* (see Appendix G). They essentially characterize the *curvature* of the ellipsoid along the corresponding axis. $L_1 = L_2 = L_3 = 1/3$ corresponds to the special case of a sphere. In simple terms:

- if $L_i > 1/3$, the ellipsoid has less curvature along the corresponding axis, i.e. it is *'flatter' than a sphere*;
- if $L_i < 1/3$, it is the opposite, i.e. it is *'more pointy' than a sphere* along this axis.

Out of the three 'corners' of the ellipsoid, one is necessarily 'pointy' ($L_1 < 1/3$ in our case), and one is necessarily 'flat' ($L_3 > 1/3$ in our case).

The resonance conditions, which in analogy with the sphere correspond to a zero real part in the denominators of β_i, are:

$$\epsilon'(\lambda_i) = -\left(\frac{1}{L_i} - 1\right)\epsilon_M. \tag{6.57}$$

There are therefore three resonance wavelengths, λ_i ($i = 1, 2, 3$), associated with each principal axis, and whose positions depend on the corresponding L_i. Each resonance will only be excited if the incident polarization has a non-zero component along the corresponding axis.

Let us focus on the resonance condition given above and in particular on the (positive) coefficient $1/L_i - 1$. For a sphere, it is equal to 2, and the corresponding resonance wavelength is denoted λ_S. We also recall that for a typical metal, $\epsilon'(\lambda)$ is negative and its absolute value increases with λ in the region of interest to plasmonics. In fact, the condition $\epsilon'(\lambda) = 0$ corresponds to the bulk plasmon resonance of the metal, which therefore occurs at a wavelength $\lambda_B < \lambda_S$. We can therefore distinguish two cases for the resonances of the ellipsoid:

- For axes with 'flat' curvatures ($L_i > 1/3$), then $0 < 1/L_i - 1 < 2$. The resonance wavelength is *blue-shifted* compared to the sphere, but remains bounded by the bulk plasmon resonance wavelength: $\lambda_B < \lambda_i < \lambda_S$.

- For axes with 'pointy' curvatures ($L_i < 1/3$), then $1/L_i - 1 > 2$. The resonance wavelength is *red-shifted* compared to the sphere. The amount of the red-shift will depend on the magnitude of the coefficient $1/L_i - 1$. The smaller the L_i (i.e. the more 'pointy'), the larger the red-shift. Moreover, the red-shift can in principle be as large as possible, since the coefficient tends to ∞ as L_i tends to zero.

This can be summarized as:

$$\begin{array}{lll} L_i > 1/3 \quad \text{(flat)}, & \lambda_B < \lambda_i < \lambda_S & \text{(blue-shift)}, \\ L_i < 1/3 \quad \text{(pointy)}, & \lambda_i > \lambda_S & \text{(red-shift)}. \end{array} \tag{6.58}$$

From the above discussion, the most important parameters characterizing the LSP resonances of the ellipsoid are therefore the L_i's. Thus, it is useful to understand how the L_i's relate to the actual shape of the ellipsoid. To this

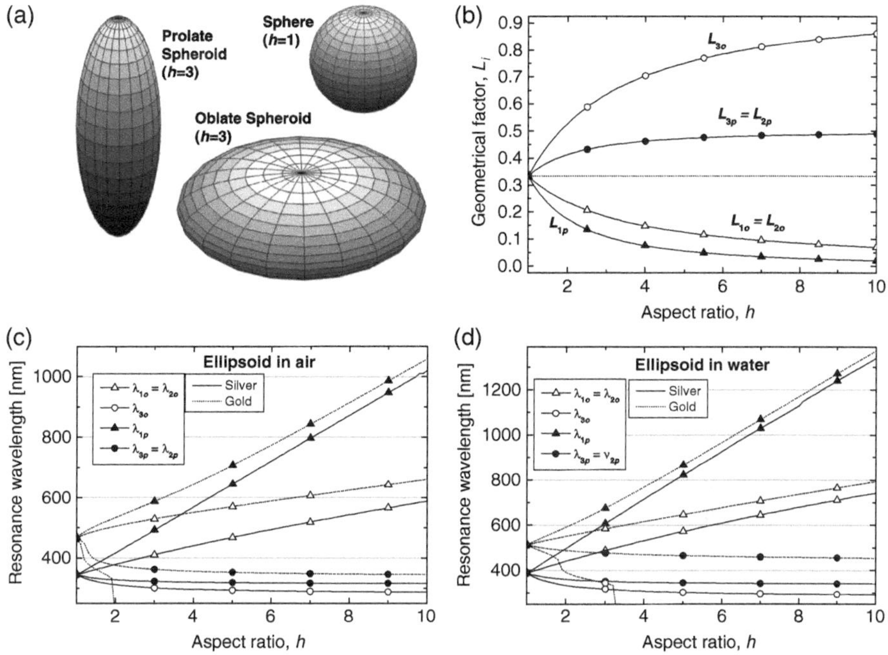

Figure 6.15. (a) Schematic representation of a sphere, and prolate and oblate spheroids with an aspect ratio of $h = 3$. (b) The geometrical factors L_i for oblate (o) and prolate (p) spheroids as functions of aspect ratio. (c-d) The corresponding resonance wavelengths (from Eq. (6.57)) for silver and gold embedded either in air (c) or water (d). For a given spheroid, one resonance is slightly blue-shifted, while the other is red-shifted. The latter is the one most relevant to SERS since it provides the largest enhancements. The red-shift can be significant at a large aspect ratio (very pointy corners), and is more pronounced for the prolate spheroid (but is then associated with only one axis, instead of two for an oblate spheroid).

end, it is convenient to consider the two special cases of ellipsoids of revolution (see the schematic in Fig. 6.15(a)):

- the *oblate* (pumpkin-like) spheroid, for which $a = b > c$ and $L_{1o} = L_{2o} < 1/3 < L_{3o}$, i.e. the two large axes are 'pointy' and the small one is 'flat'.

- and the *prolate* (rugby-ball-like) spheroid, for which $a > b = c$ and $L_{1p} < 1/3 < L_{2p} = L_{3p}$, i.e. the large axis is 'pointy' and the two small ones are 'flat'.

Analytical expressions can then be obtained (see Appendix G for full details), and the results can be conveniently presented for both cases as functions of the *aspect ratio* $h = a/c$ (as shown in Fig. 6.15).

'Strength' of the LSP resonances for ellipsoids

Let us now assume that the resonance condition (Eq. (6.57)) is fulfilled for a given axis i. We then have:

$$\beta_i(\lambda_i) = \frac{1}{3L_i} + i\frac{\epsilon_M}{3L_i^2\epsilon''(\lambda_i)}. \tag{6.59}$$

As for the sphere, the resonance will be particularly strong if $\epsilon''(\lambda_i)$ is small and the magnitude of the resonant response is then characterized by:

$$|\beta_i(\lambda_i)|^2 \approx \frac{9\epsilon_M^2}{(3L_i)^4(\epsilon''(\lambda_i))^2}. \tag{6.60}$$

Comparing this result to that of the sphere (Eq. (6.35), which can be recovered here by taking $L_i = 1/3$), we note the additional $(3L_i)^4$ term in the denominator. This term will have opposite effects whether the resonance corresponds to a 'flat' or 'pointy' axis:

- For excitation along axes with 'flat' curvatures ($L_i > 1/3$), the magnitude of the resonance is *damped* compared to that of the sphere.
- For excitation along axes with 'pointy' curvatures ($L_i < 1/3$), the magnitude of the resonance is *enhanced* compared to that of the sphere.

These general considerations on the strength of the resonance will now be applied to the predictions of local field enhancements at the surface of ellipsoids.

6.3.2. Shape effects on local fields

Local field intensity and polarization

In a similar fashion as for the sphere, we can now consider the local field intensity enhancement factor (LFIEF) at selected positions on the ellipsoid surface, in particular at the 'corners' (along the main axes). We will confine ourselves for simplicity to the two special cases of oblate and prolate spheroids. We will moreover only consider the case of an incident field polarized along one of the axes of the spheroid. The necessary expressions have been given in Appendix G. We only analyze and discuss these predictions here.

The LFIEF at all points on the surface can be expressed in terms of the non-dimensional polarizability $\beta_i(\omega)$ along the incident polarization axis ($i = 1, 2, 3$ or x, y, z, which correspond to incident polarization along each of the principal axes). More precisely, distinguishing between local field perpendicular and parallel to the surface, the LFIEF is characterized by analogy with the sphere in terms of:

$$\boxed{\begin{aligned} A_i^{\perp}(\omega) &= |1 + (3 - 3L_i)\beta_i(\omega)|^2 &= \left|\frac{\epsilon(\omega)}{L_i\epsilon(\omega) + (1 - L_i)\epsilon_M}\right|^2, \\ A_i^{\|}(\omega) &= |1 - 3L_i\beta_i(\omega)|^2 &= \left|\frac{\epsilon_M}{L_i\epsilon(\omega) + (1 - L_i)\epsilon_M}\right|^2. \end{aligned}} \tag{6.61}$$

The LFIEF at a point $\mathbf{r}$ on the surface can then be expressed as ($\mathbf{e}_\xi$ is the unit normal vector at $\mathbf{r}$, see Section G.1.4):

$$\boxed{M_{\text{Loc}}(\mathbf{r}, \omega) = A_i^{\|}(\omega) + (A_i^{\perp}(\omega) - A_i^{\|}(\omega)) \left|\mathbf{e}_\xi \cdot \mathbf{e}_i\right|^2.} \tag{6.62}$$

Interestingly, the same expression as obtained for the sphere also holds:

$$\frac{A_i^{\|}(\omega)}{A_i^{\perp}(\omega)} = \frac{(\epsilon_M)^2}{|\epsilon(\omega)|^2}. \tag{6.63}$$

The physical interpretation of these expressions is very similar to that given for the sphere. $A_i^{\perp}(\omega)$ and $A_i^{\|}(\omega)$ characterize the maximum and minimum LFIEF for incident polarization along $\mathbf{e}_i$. The maximum enhancement is obtained at the two points on this axis (the tips) and the local field is perpendicular to the surface at these two points. The minimum enhancement is obtained at the four points on the two other axes (those perpendicular to the incident polarization) and is then tangential to the surface. For other points, the wavelength dependence of the LFIEF is dictated by $A_1^{\perp}(\omega)$ and $A_1^{\|}(\omega)$, while the position dependence is governed by the factor $|\mathbf{e}_\xi \cdot \mathbf{e}_i|^2$. For oblique incident polarizations (with respect to the main axes), several resonances are excited and the situation is more complicated but can be analyzed using the tools of Appendix G.

Local field enhancements at resonance

Furthermore, for a given incident polarization (along $\mathbf{e}_i$), we have at resonance (i.e. for $\lambda = \lambda_i$), using Eq. (6.59):

$$A_i^{\|}(\lambda_i) = \frac{9(\epsilon_M)^2}{(3L_i)^2\left(\epsilon''(\lambda_i)\right)^2} \approx (3L_i)^2 |\beta_i(\lambda_i)|^2, \tag{6.64}$$

and using Eq. (6.63):

$$A_i^{\perp}(\lambda_i) = \frac{9|\epsilon(\lambda_i)|^2}{(3L_i)^2\left(\epsilon''(\lambda_i)\right)^2}. \tag{6.65}$$

These expressions are very similar to those obtained for the sphere, except for the $(3L_i)^2$ factor in the denominator (which reduces to 1 for a sphere). In addition, in the case of $A_i^{\perp}$, one must bear in mind that $\epsilon(\lambda_i)$ is determined by the resonance condition (Eq. (6.57)), and therefore intricately linked to L_i, in particular:

$$|\epsilon(\lambda_i)|^2 = \left(\frac{1}{L_i} - 1\right)^2 (\epsilon_M)^2 + (\epsilon''(\lambda_i))^2. \tag{6.66}$$

To discuss this further, we need again to distinguish between excitations along a 'pointy' or a 'flat' axis:

- For excitation along axes with 'flat' curvatures ($L_i > 1/3$), the factor $(3L_i)^{-2}$ contributes to a decrease of the LFIEF at resonance compared to the sphere case. This decrease is even larger for $A_i^{\perp}$ because of the dependence of $|\epsilon(\lambda_i)|^2$ on L_i (Eq. (6.66)).
- For excitation along axes with 'pointy' curvatures ($L_i < 1/3$), the factor $(3L_i)^{-2}$ contributes to an increase of the LFIEF at resonance compared to the sphere case. This increase is further magnified for $A_i^{\perp}$ because $|\epsilon(\lambda_i)|^2$ increases as the resonance wavelength red-shifts (which is the case for decreasing L_i). From Eq. (6.66), this gives an additional factor of $\approx(1/L_i - 1)^2/4$ for the enhancement compared to the sphere (the 4 comes from the value for the sphere).

For applications exploiting large LFIEFs, and in particular for SERS, the red-shifted resonances are therefore the most important ones. They can be excited with polarization along the most 'pointy' axes. The SERS EF is then increased at every point on the surface compared to the sphere, by a factor at least $(3L_i)^{-4}$, a result of the 'stronger' resonance. In addition, the SERS EF at the tips on this axis (points of highest enhancement) is increased even more than at other positions on the surface, by an extra factor $\approx ((1/L_i - 1)/2)^4$. This additional enhancement effect at corners or tips, illustrated here for the case of an ellipsoid, is in fact fairly general and sometimes known as the *'lightning rod effect'*, by analogy with a similar phenomenon in standard electrostatics (creation of high fields at sharp corners or points). As a result of this, the distribution of SERS EF on the surface widens.

Because the SERS EFs scale as $(3L_i)^{-4}$ (at least), the additional gain compared to the sphere can be quite large as L_i approaches 0. One must, however, bear in mind that the resonance wavelength λ_i red-shifts as L_i goes to zero and may then lie outside the range of interest. As an example, let us consider that $L_i = 1/6$ (corresponding to an oblate spheroid with an aspect ratio of $h = 3.4$ or a prolate spheroid with $h = 2.1$). At resonance (i.e. for $\epsilon'(\lambda_i) = -5\epsilon_M$), $A_i^{\parallel}$ is then 4 times larger than that for a sphere (we ignore the slight wavelength dependence of ϵ''), while $A_i^{\perp}$ is further increased by a factor $5^2/4$ due to the lightning rod effect, i.e. it is a total of 25 times larger than that for a sphere. The minimum SERS EF is then 16 times larger than that for the sphere, while the maximum SERS EF (at the pointy tips) is $25^2 = 625$ times larger. This clearly demonstrates that shape effects can play an important role in SERS, and can be exploited to obtain both large punctual and average SERS EFs. We also note that the spread of the distribution is then widened by the lightning rod effect, from a factor of 16 between maximum and minimum SERS EF for the sphere, to a factor of $\approx$600 in this example. This will be a characteristic that is preserved in more complicated examples; i.e. the higher the enhancement at a specific point on the substrate, the wider the spread of enhancements.

Finally, it should be noted that all the EF figures quoted in this section are obtained within the ES approximation and are accordingly *only valid for the smallest particles.* As for the sphere, radiation damping is expected to decrease substantially these values as the size increases. They should therefore be viewed as *upper estimates.*

Examples for silver and gold spheroids: wavelength dependence

Figure 6.16 illustrates the wavelength dependence of the LFIEF for the red-shifted (A_1) and blue-shifted (A_3) resonances of oblate and prolate spheroids of aspect ratio $h = 3$. Also shown are the corresponding results for a sphere (denoted A_{Sphere}). All the features discussed so far are evident in these plots. In particular, it is clear from these plots that the red-shifted resonance is much stronger than that of the sphere while the blue-shifted one is weaker. This simply reflects the respective value of L_i, as discussed earlier. For silver, the red-shifted resonance is in addition in the region of interest (the visible range) while the other one is closer to the UV. For these two reasons, the red-shifted resonance is the relevant one for most SERS (and related) effects.

Maximum LFIEFs in excess of $\sim 10^4$, corresponding to a SERS EF of $\approx 10^8$, are predicted for the silver spheroids. For the oblate gold spheroids with $h = 3$ in air, the enhancements are much smaller because the red-shift of the resonance is not quite sufficient to push the resonance toward the lower-absorption region of gold. For prolate gold spheroids with $h = 3$ in air, the red-shift is slightly larger and maximum SERS EFs of almost $\approx 10^8$ are then predicted.

This highlights an additional effect of shape in the particular case of gold. We have already stressed when studying the sphere that the dipolar LSP

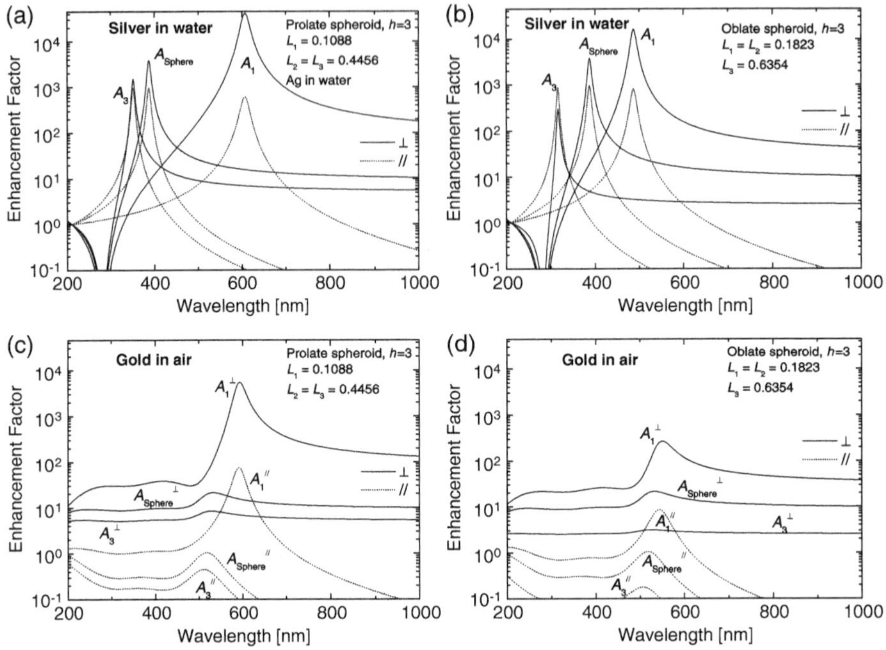

Figure 6.16. Wavelength dependence of the maximum ($A_i^{\perp}$) and minimum ($A_i^{\parallel}$) LFIEFs on the spheroid surface. $i = 1$ ($i = 3$) corresponds to the red-shifted (blue-shifted) resonance of the spheroid, and A_{Sphere} to the corresponding quantity for a sphere, given here for comparison. We consider either prolate (a, c) or oblate (b, d) spheroids, all with an aspect ratio of $h = 3$. They are either silver spheroids in water (top – a, b) or gold spheroids in air (bottom, c, d). The dielectric functions are taken from Appendix E. In all cases, the $\perp$-cases are represented by full lines, while the corresponding $\parallel$-cases are shown with dotted lines.

resonance for gold spheres tends to be strongly damped because it occurs at wavelengths where the optical absorption is large (i.e. large $\epsilon''(\lambda_S)$). For a gold ellipsoid, the red-shifted resonance (if it moves beyond ≈600 nm) is no longer affected by this problem. This resonance should then have comparable properties as that of the silver ellipsoid, a situation very different from the perfect sphere case. Elongated shapes can therefore be used as a means of obtaining gold nano-particles with a resonance beyond 600 nm, where optical absorption no longer damps it significantly.

Examples for silver and gold spheroid: influence of aspect ratio

We now focus on the variation of the maximum LFIEF, $A_1^{\perp}(\omega)$, with aspect ratio and metal (silver or gold). This dependence is illustrated in Fig. 6.17.

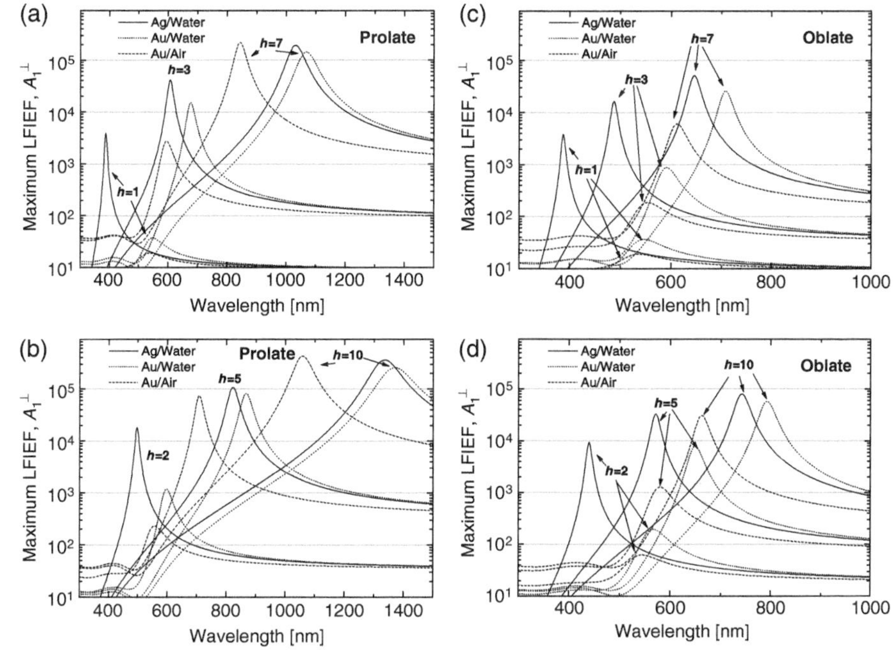

Figure 6.17. Wavelength dependence of the maximum LFIEF, $A_1^{\perp}$, for excitation along the long axis of either prolate (a and b) or oblate (c and d) spheroids of various aspect ratios h.

This figure further illustrates most of the effects discussed previously, in particular:

- For a given type of spheroid, the maximum EF increases with the aspect ratio h.
- For a given aspect ratio, prolate spheroids (rugby balls) exhibit a larger maximum EF than oblate spheroids (pumpkins).

These two facts are a manifestation of the same effect: the more 'pointy' the shape, the larger the EF.

Large LFIEF, of the order of 10^5 (corresponding to SERS EF of the order of 10^{10}) and even beyond are predicted for the largest aspect ratios of prolate spheroids. The largest of these occur at relatively long wavelengths, typically in the near infrared. Moreover, LFIEF of the order of $\sim 10^4$ or larger occur over a relatively large parameter range: for silver prolate spheroids with $h \geq 2$, gold prolate spheroids with $h \geq 3$, silver oblate spheroids with $h \geq 3$, and gold oblate spheroids with $h \geq 5$. Finally, these results further illustrate an important point, often the source of confusion: *gold substrates are as efficient as silver substrates when resonances are sufficiently red-shifted* (typically

$\lambda > 600$ nm). This is illustrated here for situations where the red-shift arises from shape effects, but it is a much more general conclusion.

EF distribution and average EF

It is clear from the previous arguments that large additional enhancements can be achieved at specific points (hot-spots) for large aspect ratio objects, in particular prolate spheroids. A needle-like object is a good model example. There is a chance, however, that the place where the large EF is achievable becomes increasingly localized to a very small area (the tip of the needle) for large aspect ratios. Such a situation may therefore not be ideal for 'average' measurements, where molecules are adsorbed randomly on the surface. To assess this, we will now focus on the average LFIEF and average SERS EFs on the spheroids.

General expressions for these EFs have been given in Section G.1, and further analytical expressions for spheroids are given in Appendix G. Let us consider for the sake of argument:

$$\langle M_{\text{Loc}}^{\perp} \rangle = A_1^{\perp} \langle |\mathbf{e}_{\xi} \cdot \mathbf{e}_x|^2 \rangle. \tag{6.67}$$

It is clear in this expression that a large maximum LFIEF, $A_1^{\perp}$, may be compensated by a small geometrical average $\langle |\mathbf{e}_{\xi} \cdot \mathbf{e}_x|^2 \rangle$. In fact, in the limit of a needle-like object (with longest axis x), the factor $|\mathbf{e}_{\xi} \cdot \mathbf{e}_x|^2$ is close to 0 at every point on the surface, except at the tips. One can show that its average value tends to 0 as $1/h^2$ (see Section G.3). The maximum LFIEF and SERS EF are very large but they only occur in a highly localized region. These are in fact the two defining characteristics of what is generally referred to as a *hot-spot*.

It is possible to be more quantitative using the tools developed in Appendix G, but we will restrict ourselves here to the most important results. These are illustrated, along with a summary of many other aspects of shape effects, in Fig. 6.18. It is convenient here to focus on the case of silver spheroids in water to discuss the results. As concluded earlier for the maximum SERS EF, the average SERS EF is larger for more pointy structures, i.e. it increases with aspect ratio. And for equal aspect ratio, it is larger for prolate than for oblate spheroids. However, the gains associated with this 'pointness' are much less spectacular than those predicted for the maximum SERS EF. This is simply a result of the *localization* discussed earlier. The maximum SERS EF increases but applies to a smaller proportion of the surface area. Nevertheless, as the plots in Fig. 6.18 indicate, upon surface averaging, the former effect dominates and still ensures an increase in the average SERS EF, but this increase is partly compensated by the latter effect (localization). Note that the results for gold can be explained in the same way with the added complication of the

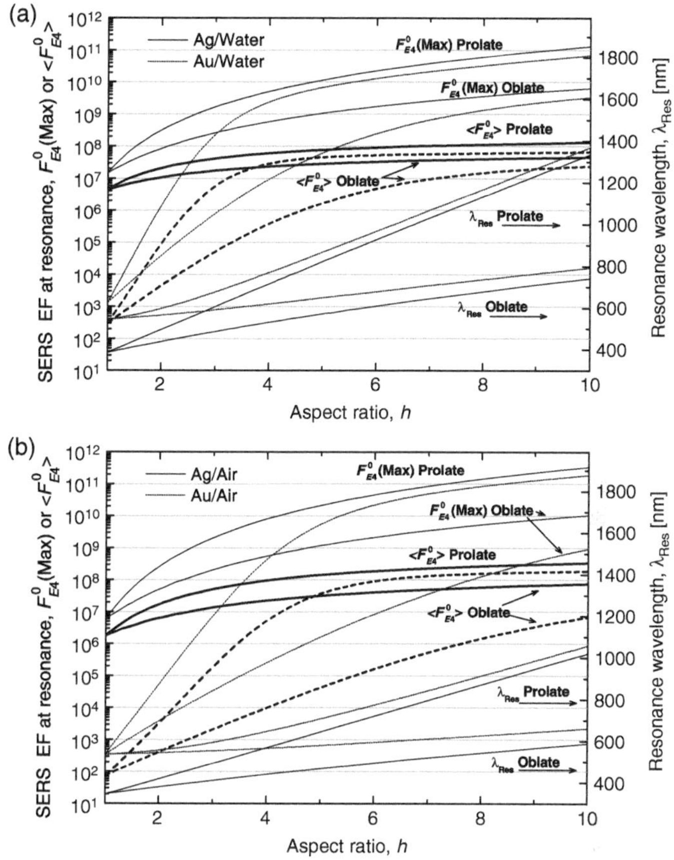

Figure 6.18. Summary of the maximum SERS EF F^0_{E4}(Max) (thin lines) and average SERS EF $\langle F^0_{E4}\rangle$ (thick lines) at the red-shifted resonance (whose wavelength is denoted $\lambda_{\rm Res}$ and also plotted for reference with scale on the right-hand-side axis). These quantities are plotted here as functions of the aspect ratio h for either silver (Ag, solid lines) or gold (Au, dotted lines) prolate or oblate spheroids. The top plots (a) refer to a water environment ($\epsilon_M = 1.77$), while the bottom one (b) is the equivalent for air ($\epsilon_M = 1$).

strong damping of the EFs for $\lambda \leq 600$ nm (an effect discussed extensively in this chapter).

6.3.3. Summary of shape effects

At this stage we could extend the considerations of the previous sections to more complicated shapes; say, polyhedra for example, or any other object using combinations of sharp/round edges and corners. All of them will show, obviously, a much larger diversity of plasmon resonances [208] and, concomitantly, much more complex polarization properties.

However, to a large extent, the basic phenomenology contained in the case of the ellipsoid, actually describes to a good approximation *most* of the general situations that can be found for other more complicated shapes. The main concepts of: (i) shape-dependent resonances; (ii) red- and blue-shifting and/or splitting of resonances with respect to a parent geometry; (iii) enhancement localization at 'pointy' ends (as in the ellipsoid case); (iv) wider spread of the enhancement distribution when hot-spots are formed; (v) polarization-dependent coupling to the resonances; are all captured qualitatively in the ellipsoid model. More complicated shapes can always be studied quantitatively numerically [208] if required.

6.4. GAP EFFECTS – JUNCTIONS BETWEEN PARTICLES

The previous sections dealt with shape and size effects of *single* objects on their plasmon resonances and EM enhancements. There is, nonetheless, one additional effect that was *not* covered in the aforementioned phenomenology: the plasmon resonances and EM enhancements that arise from the *interaction* of two (or more) objects. The 'interaction' becomes particularly important at small separation between objects (small gaps), where large local fields result in some of the largest known values for the SMEF $\sim 10^{11}$. Here we will follow the spirit of the previous sections and choose a limited number of model examples that best represent the basic phenomenology of gap-plasmon resonances. We choose this approach both for its didactic value and its conciseness.

One of the main problems of gap-plasmon resonances is that there are not many examples where analytic solutions are available. For the case of ellipsoids in the study of shape effects in Section 6.3 we already had to resort to the electrostatic approximation to obtain a simple analytic solution of the problem. The situation becomes a lot more intractable in general for multiple objects. There is one 'semi-exception' in the case of spheres, where the results of *generalized Mie theory* (GMT) (See Chapter 5 and Appendix H) may be used; albeit with the proviso that the approach is only 'semi-analytic', in the sense that the analytic expressions (series) must usually be computed numerically. Here we will apply GMT to the simplest possible case: that of a dimer of two identical metallic spheres, as a model example to illustrate the most important characteristics of gap-plasmon resonances.

6.4.1. Coupled localized surface plasmon resonances and SERS

A representative example

We study the case represented schematically in Fig. 6.19(a) (and repeated in Fig. 6.20(a)); i.e. a dimer formed by two identical spheres (with radius $a = 25$ nm here) of dielectric function $\epsilon(\omega)$, separated by a gap g, and

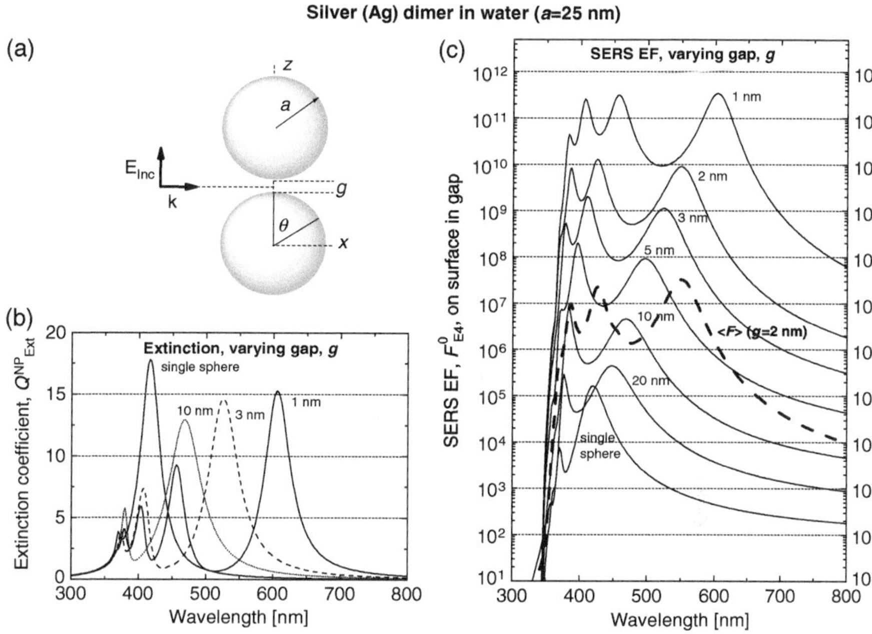

Figure 6.19. (a) Schematic configuration of a dimer (along z) formed by two spheres of radii a and separated by a gap g. The incoming wave is polarized along z (axis of the dimer) and with wave-vector $\mathbf{k}$ along x. (b) Extinction coefficients for varying gaps as functions of incident wavelength. Note the presence of a dominant red-shifted LSP resonance that moves further to the red for smaller gaps. (c) SERS EF, F^0_{E4}, at the point on the surface in the gap (on the dimer axis), i.e. at the 'hot-spot'. Also shown (thick dashed line) is the *average* SERS EF, $\langle F^0_{E4}\rangle$, in the case of a $g = 2$ nm gap. Note that the resonances in the SERS EF have their counterparts in the extinction spectra, although their respective magnitudes are very different [57,204]. Also note the similarities in the spectral profile (but not the intensity) between the average EF $\langle F^0_{E4}\rangle$ and the SERS EF at the hot-spot for the case $g = 2$ nm.

embedded in a non-absorbing dielectric with ϵ_M (in water here, $\epsilon_M = 1.77$). The only difference between Figs 6.19 and 6.20 is the metal; Ag or Au, respectively. The exciting beam is considered to be incident along x (perpendicular to the dimer axis along z) and we consider only excitation with *incident polarization along the dimer axis*, for which maximum coupling to the main LSP resonance is obtained. Note that the incident-direction and incident-polarization dependence may nevertheless be important in some instances, see for example Refs. [48,57,204] for more details.

Nature and position of the coupled-LSP resonances

As before, we can start the discussion of the main characteristics of the resonances by looking first at the extinction spectra shown in Fig. 6.19(b) (for Ag), and 6.20(b) (for Au). The main attributes of the extinction spectra for a dimer can be summarized as follows:

- The interaction between the two objects create *new* LSP resonances which are a result of the coupling between the various individual LSP resonances of the isolated particles. We shall call them *coupled-LSP resonances* or gap-plasmon resonances.
- As for single particles, the most red-shifted resonance is typically the most important one for SERS; *it arises from the dipolar interaction or dipolar coupling between the two single-sphere dipolar LSP resonances.* We shall call it loosely the *main* coupled-LSP resonance of the dimer. This resonance, similar to a bonding state between two orbitals, is red-shifted with respect to the original single-sphere dipolar resonance. The red-shift depends on the strength of the interaction and, therefore, increases for smaller gaps (stronger interactions). For the smallest gap, it is extremely sensitive to the exact gap length; a change of ~1 nm can make a large difference in the resonance wavelength.
- Other resonances appear at shorter wavelengths from the main coupled-LSP resonance (especially for Ag). They result either from higher-order interaction between the single-sphere dipolar LSP resonances, or from dipolar coupling between single-sphere higher-order (e.g. quadrupolar) LSP resonances. Although it is (in general) not of much use to identify the exact nature of these resonances (except as an academic exercise), this information is readily available in the coefficients of the interaction matrix of the GMT-method used to solve the electromagnetic problem. It can always be retrieved from the solution if required for a specific case. The absence of these secondary resonances in Au compared to Ag can be accounted for by the much larger intrinsic absorption of Au below 600 nm (in wavelength).
- It is important to highlight the fact that the magnitude of the extinction coefficient is not much affected by the interaction (only the resonance positions change).

EM enhancements in gaps

The most important aspect of coupled-LSP resonances is not really their position or red-shift but the large (or very large) local field enhancements (and therefore SERS EFs) that arise, in particular in the region of the gap between the two particles, the so-called 'hot-spot'. This is illustrated in Figs 6.19(c) (for Ag), and 6.20(c) (for Au). The approximate SERS EF, F_{E4}^{0}, is calculated immediately above the surface of one of the spheres, in the gap, and along the z-axis; i.e. at the position that would be occupied by a molecule (on the

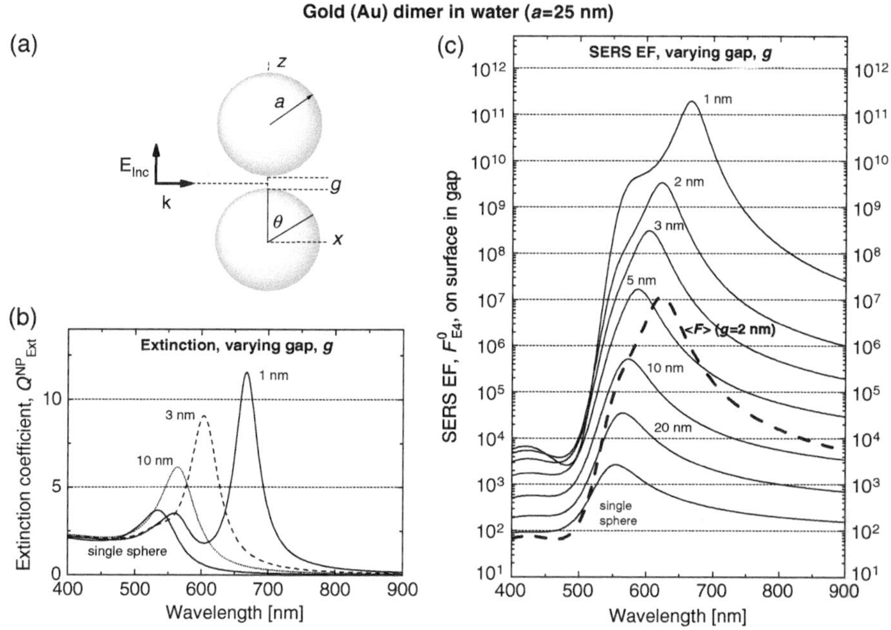

Figure 6.20. Same as Fig. 6.19 but for a gold (Au) dimer in water. Note the differences in the magnitudes of the SERS EFs compared to the case of Ag in Fig. 6.19. The main cause is the larger optical properties of Au below 600 nm. Once the coupled-LSP resonances red-shift below ~600 nm, enhancements comparable to Ag ($\sim 10^{11}$) are recovered.

surface of one of the spheres) in the gap. Let us summarize the main features revealed in Figs 6.19(c) and 6.20(c):

- *Large SERS EFs* are predicted (up to $\sim 5 \times 10^{11}$) at the hot-spot. They exhibit resonances at the various coupled-LSP resonances of the system (that were evident in the extinction spectra), but it is worth highlighting that the SERS EF also remains very high even *outside* these resonances (in fact across most of the visible range).
- The magnitude of the SERS EF at the hot-spot is *highly* sensitive to the gap between the particles for the smallest gaps, with differences of the order of ~1 nm in the separation producing changes of an order of magnitude (or more) in the SERS EF at the hot-spot.
- Note that the extinction spectrum reflects the position of the resonances, but says nothing about the *magnitude* of the SERS EF. It is impossible to infer the SERS EF from an extinction measurement in the presence of interacting particles [204]. This, one more time, highlights the fundamental differences between far-field and local-field properties.

- The relatively smaller SERS EFs obtained for gold (with respect to silver) are a consequence of the larger optical absorption for Au below 600 nm. When the main coupled-LSP resonance is sufficiently red-shifted (to above ~600 nm), the dissipative properties of Au then become comparable to those of Ag (see Appendix E), and the performance of the two metals – as far as enhancements are concerned – becomes comparable. SERS EFs of the order of $\sim 10^{11}$ are obtained at the main coupled-LSP resonance for gold when $g = 1$ nm, comparable to Ag.

- The *average* SERS EF is also large, but not enhanced as spectacularly as the SMEF at the hot-spot. It is shown in Figs 6.19(c) (for Ag) and 6.20(c) (for Au) as a dashed line for a gap of $g = 2$ nm only. This can be compared to the SMEF at the hot-spot for the same gap $g = 2$ nm. The same resonances are observed in both spectra, but with an overall difference of approximately two orders of magnitude between them. We will come back to this shortly, in the context of the spatial localization of the hot-spots.

Summary of the properties of the main coupled-LSP resonance

We concentrate now on the properties of the main coupled-LSP resonance. It is characterized here by its resonance wavelength λ_{Res} and the LFIEF $M_{\text{Loc}}(\lambda_{\text{Res}})$ at the hot-spot (on the metal surface along the dimer axis in the gap). The gap dependence of these two quantities is summarized in Fig. 6.21 for a silver dimer in water ($a = 25$ nm), and in Fig. 6.22 for a gold dimer also in water ($a = 25$ nm). It is worth stressing the fact that the resonance wavelength and the LFIEF at the hot-spot are plotted on semi-log and log–log scales, respectively. A sharp change is seen for both for gap distances g below ~20 nm. Actually, we can use this information as an 'empirical' rule to establish in this case a characteristic distance beyond which the spheres can be considered as 'independent'. Spheres with $a = 25$ nm (of both, Ag and Au) can be effectively considered as electromagnetically independent if they are separated by more than ~20 nm in the gap. For shorter distances, the interaction starts playing a role, and the main LSP resonance red-shifts with a concomitant steep increase in the hot-spot LFIEF at that wavelength. The data in Figs 6.21 and 6.22 only convey, in a different way, what is already visible in the more elaborate plots in Figs 6.19 and 6.20.

The fits to the data in Figs 6.21 and 6.22 are *empirical* and, therefore, not intended to represent any specific physical model. It is useful nevertheless to have these empirical representations of the data for typical parameters (like the ones used here), to produce back-of-the-envelope estimations for real cases.

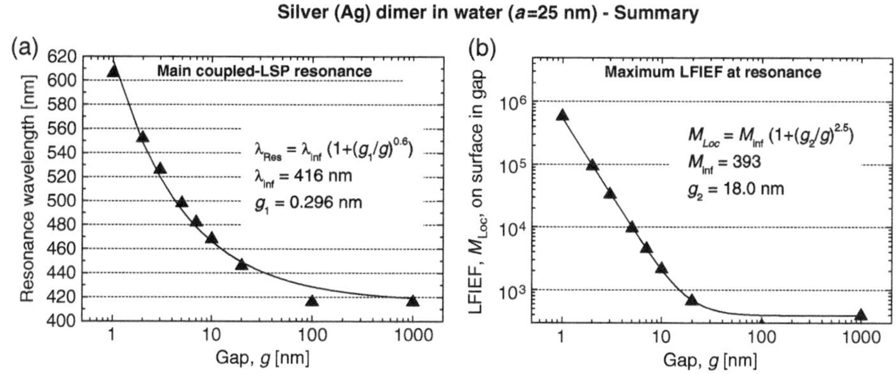

Figure 6.21. Gap dependence of the main coupled-LSP resonance: (a) resonance wavelength and (b) LFIEF, M_{Loc} (Eq. (4.18)), at the 'hot-spot' in the gap. The fits given as solid lines are *empirical*, and not based on any physical model.

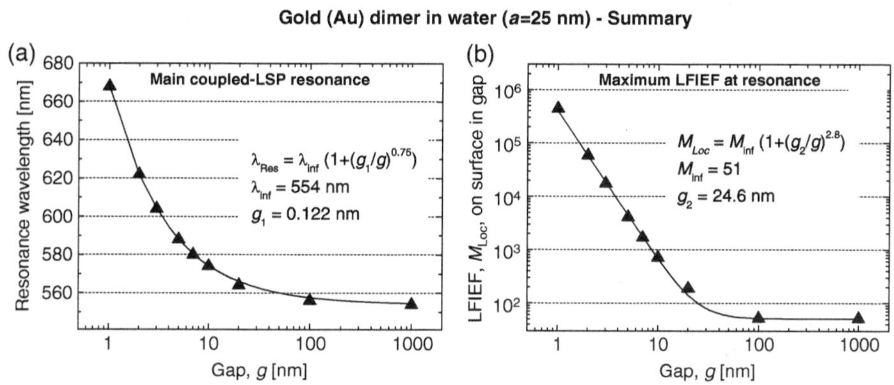

Figure 6.22. Same as Fig. 6.21 but for a gold dimer in water.

6.4.2. EF distribution and hot-spot localization

In the spirit of addressing the most general properties of gap resonances (that would be transferable to other problems), an interesting aspect is the study of the *spatial distribution* of the enhancement on the surface around a hot-spot. We take again the case of the dimer of spheres as a paradigm that actually represents a much wider group of situations. Some of the general characteristics of the spatial distribution of the enhancement have been already hinted at in this chapter, and we reinforce those concepts here with a particular example.

A rule of thumb is that: *the highest the enhancement, the more spatially localized it becomes*, and the wider the difference in enhancement between the 'hottest' place and the rest of the enhancement distribution. The spatial

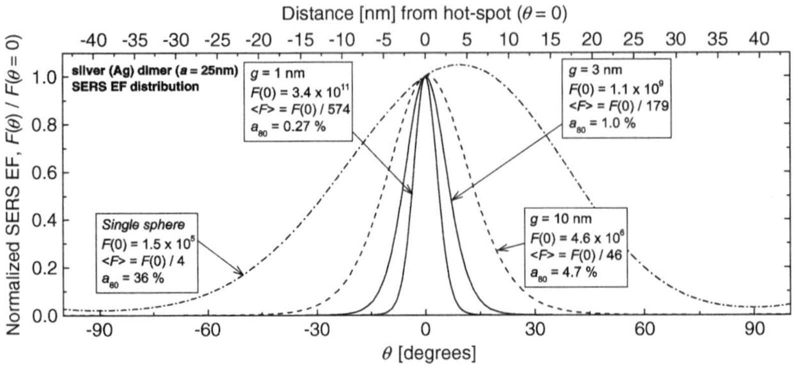

Figure 6.23. Illustration of the *hot-spot localization* effect. The position-dependent normalized SERS EF, $F^0_{E4}(\theta)/F^0_{E4}(\theta=0)$ on the sphere surface is plotted as a function of θ (bottom axis) or, equivalently, distance from the hot-spot (top axis). See Fig. 6.19(a) for a schematic diagram ($\phi = 0$ is assumed here). The results are plotted for dimers with gaps g of 1, 3, and 10 nm, and at the corresponding resonance wavelengths of the main coupled-LSP resonances ($\lambda = 604$ nm, 526 nm, and 468 nm, respectively). The single-sphere case is also shown for comparison ($\lambda = 416$ nm). Indicated in the insets for each case are: the SERS EF at the hot-spot $F(0)$, the average SERS EF $< F >$ (given here as a fraction of the maximum EF to highlight the extreme statistics), and the hot-spot area a_{80}, defined as the percentage surface area on the particle from which 80% of the total SERS signal originates [162]. These quantities also reflect the localization effect. The 'narrowing' of the hot-spot size is clearly evident as the gap decreases. It is directly correlated with the strength of the maximum SERS EF $F(\theta = 0)$: the stronger the hot-spot, the more localized it is. The full-width at half-maximum of the normalized SERS EF for $g = 1$ nm is ~3.3 nm, i.e. comparable to a typical molecular size for dyes.

localization of a typical 'hot-spot', at the highest achievable enhancements, can be over distances comparable to the size of a molecule (~1–2 nm). This results in very interesting characteristics for the statistical distribution of enhancement factors, the most important of which is that it typically follows a *long-tail* probability distribution[7]. Extensive discussions on these topics are given in Ref. [162] ; we shall only provide here a simple illustration based on the example of the Ag dimer ($a = 25$ nm) with different gaps (g).

Figure 6.23 shows very clearly the concept of *hot-spot localization.* The figure shows the position-dependent SERS EF (normalized to its maximum value) on the surface of one of the spheres, here as a function of angle θ (as depicted in Fig. 6.19(a)), and for different gaps. The case of a single sphere

[7] The EF distribution measures the probability for a randomly-adsorbed molecule of having a given EF. Long-tail distributions lead to some extreme statistics. Typical distributions of this kind can have, for example, a standard deviation comparable or even much larger than the average value [162]. Extreme fluctuations become the norm and issues of *sampling* require special care. This is intimately connected to the single-molecule SERS problem, and will be further discussed in Chapter 8.

is also shown for comparison[8]. Note that Fig. 6.23 displays the *normalized* SERS EF (with respect to the value at $\theta = 0$), but the real enhancements differ by several orders of magnitude depending on the gap.

The data in Fig. 6.23 are conclusive in the fact that *the largest enhancements display a strong degree of spatial localization.* This can be considered to be one of the basic characteristics of coupled-LSP gap resonances. Similar spatial localization characteristics are also present in tip-enhanced Raman scattering (TERS) experiments to be described later in Section 8.2. For smallest gaps for which the largest SERS EFs $\sim 10^{11}$ are achieved, the spatial localization of the hot-spot can be within distances that are comparable to typical sizes of molecules. Moreover, the average SERS EF is then ~ 580 times smaller than the SMEF at the hot-spot. In other words, the SERS intensity for a single molecule at the hot-spot is as strong as that of ~ 580 molecules randomly adsorbed on the surface. The high localization however makes it difficult to profit from the large SMEF at the hot-spot. It would imply that a single molecule has to be precisely positioned at the right spot on the surface, with an accuracy comparable to its size (in a gap which is already comparable to its size, ~ 1 nm). Controlling the gap size *and* the molecule position with nanometer precision is required therefore for single-molecule SERS spectroscopy under *controlled conditions*. Needless to say, this is a very difficult (if not impossible) task in general. The *single-molecule SERS problem* will be further discussed in Section 8.1.

The hot-spot localization effect together with the high sensitivity of the SERS EF to the exact gap results in numerous problems in terms of reproducibility and controllability in any SERS experiments with gap-containing substrates. These should therefore only be used when the large EF that they provide is an absolute necessity (typically for single-molecule SERS experiments).

6.5. ADDITIONAL EFFECTS

The most important effects influencing plasmon resonances and their associated EM enhancements have been discussed through the model examples of: the plane, the sphere, the ellipsoid, and the dimer. This completes a basic phenomenological description of the main plasmonics effects that will appear by themselves, or as combinations thereof, in most practical cases. We conclude by briefly discussing a few additional effects that were not captured within these simple 'toy models'.

[8] Note that the maximum EF for the single sphere is shifted to the right (to positive θ's); an effect arising from the 'symmetry breaking' introduced by the wave-vector of light (see the diagram in Fig. 6.19(a)). This effect is negligible for a dimer, in which case the SERS EF distribution is mostly centered around $\theta = 0$ and dominated by the interaction between the spheres.

6.5.1. Nano-particles on a supporting substrate

We have already highlighted in each of the previous examples the importance of the dielectric constant of the environment (ϵ_M), i.e. of the dielectric in which the metallic substrate is embedded; typically air or water. This has a direct effect on the position of the LSP resonances, as readily seen in Eq. (6.33) or Table 6.1, for example.

Nonetheless, there are many common practical situations where this environment may not be as simple as a homogeneous, isotropic, embedding medium; a typical example being that of a metallic nano-particle (NP) lying on a planar supporting substrate, which could be a dielectric like glass, or a silicon substrate, or even a metal. In principle, the EM problem is then entirely different from the bare nano-particle case (there are additional boundaries!), and should be treated as an independent problem. Nevertheless, one would expect that many of the EM properties of the bare nano-particle would remain, but only to be modified by the environment (as is the case when it is fully embedded in a dielectric with ϵ_M). In a naive approximation, the particle is in some way 'half-embedded' in the dielectric with ϵ_M, and half-embedded in the substrate itself. If the substrate is a dielectric (with ϵ_S), one may therefore guess that the situation could be approximated by a nano-particle embedded in a medium with an effective dielectric constant ϵ_{Eff}, lying somewhere between ϵ_M and ϵ_S.

Although it is difficult to justify this formally (from Maxwell's equations), this simple reasoning is intuitively appealing and can be useful in many cases, at least for qualitative or phenomenological analysis. This is exemplified in Fig. 6.24, where we consider the example of a nano-particle, here a gold nano-disk, lying on top of a supporting substrate, which we assume to be dielectric with $\epsilon_M = 4$ (constant throughout the entire wavelength range). The latter is a good representation for an ITO-glass substrate, which is common in gold nano-structures fabricated by nano-lithography such as those shown in Fig. 7.4. The problem is solved numerically using FEM in the electrostatic approximation (ESA). The average zero-Stokes shift SERS EF in the $|E|^4$-approximation, $\langle F^0_{E4}\rangle$, is computed by averaging over the top and side surfaces of the disk; i.e. the surfaces that would be available for a molecule to adsorb. To understand the effect of the substrate, calculations are performed for three different cases: (i) the isolated NP, i.e. no substrate; (ii) NP on top of the dielectric substrate ($\epsilon_M = 4$); and (iii) NP embedded in an homogeneous dielectric medium with $\epsilon_M = 2.5$ (and no substrate). The main characteristics of the results can be summarized as follows:

- The isolated NP presents a clear (shape-related) LSP resonance around $\sim$520 nm with a moderate maximum average enhancement of $\sim 10^2$.
- Upon introducing the substrate, the main resonance red-shifts to $\sim$620 nm and the maximum average enhancement is now almost 10^6

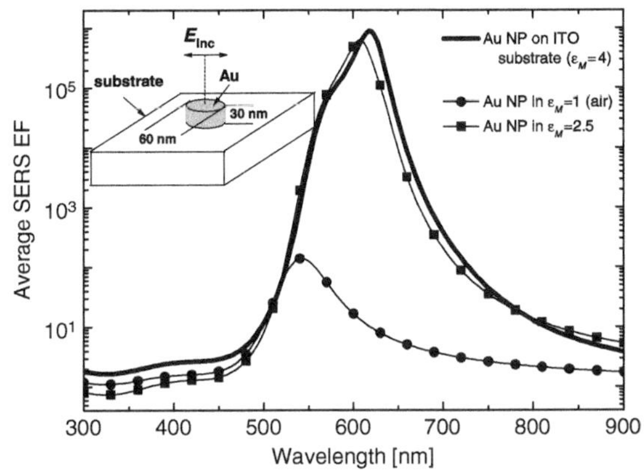

Figure 6.24. Illustration of the effect of the underlying (supporting) substrate for a single nano-particle (NP), a gold nano-disk here (with the dielectric function of gold given in Appendix E), supported by a substrate of constant dielectric function $\epsilon_M = 4$ (modeling the dielectric function of ITO-glass). A schematic representation of the geometry is given on the top-left diagram. Light is assumed to impinge from the top, with polarization along $\mathbf{E}_{\mathrm{Inc}}$. The problem is solved numerically with the finite-element method in the electrostatics approximation. The average SERS EF is evaluated over the top and side surfaces of the disk. Three calculations are shown: (i) single NP with no supporting substrate, (ii) single NP with supporting substrate ($\epsilon_M = 4$), and (iii) single NP embedded in an effective homogeneous medium with ($\epsilon_M = 2.5$).

(this increase is again related to the lower absorption of gold beyond 600 nm).

- A 'similar' behavior of the resonance can be obtained when we immerse the disk in a homogeneous dielectric medium with $\epsilon_M = 2.5$, which is in between the ϵ of air and the ITO-glass. This latter case, represents the approximate equivalent 'effective medium' to the substrate case.

Based on this example, placing a metallic nano-particle on a substrate will in general affect its EM properties and in particular shift its LSP resonance in a similar fashion as if it were placed in a fully embedding medium with an effective dielectric constant. This reasoning can usually only be qualitative and quantitative predictions must ultimately rely on EM modeling.

One important point to extract from these results is the fact that *the supporting substrate can have dramatic consequences for the electromagnetic response of the NPs*; a point that is not always appreciated or explicitly taken into account. Such effects can be spectacular, as shown in Fig. 6.24, and could be even more important on a substrate like silicon, whose dielectric constant is very large ($\epsilon \approx 15$ at 633 nm).

6.5.2. Surface roughness

Another inevitable experimental aspect, which is not included in the general description of plasmon resonances in 'perfect shapes' like planes, spheres, ellipsoids, etc., is the presence of *surface roughness* on metallic surfaces. This is a topic that has a long history in SERS, for the study of SERS began through observations in electrochemically roughened electrodes [1–3]. The effect of roughness on planar surfaces received a considerable amount of attention in the early days because it was thought to be the simplest possible model of SERS enhancement (and one of the few known experimentally at the time). Several of the early reviews in the field in Refs [4,5] discuss some of the EM models, and we shall not repeat that information here. Surface roughness is not easy to handle theoretically (except possibly on planar surfaces). For flat metallic surfaces, one of the consequences of the introduction of roughness is that it *breaks momentum conservation*; thus allowing the coupling to propagating surface plasmon–polaritons at the interface that would be otherwise forbidden (see Chapter 3). The local field enhancements on the surface are therefore increased (see Chapter 3), and this results in a reasonable SERS EF, say of the order of 10^3–10^4. Rough surfaces are therefore a simple example of SERS substrate, although with moderate enhancements. In this particular case, roughness is used as a means to change a 'bad' substrate (the metallic plane which hardly couples to light) into a 'better' one (with increased coupling). Whether roughness can be used to improve the performance (SERS EF) of an already 'good' substrate, like most nanoparticles and their derivatives, is debatable and requires further investigation. It is possible that roughness contributes to a small proportion of the EM SERS EF, through mechanisms similar to the additional enhancements arising from shape effects in NPs. Such an effect remains however secondary except for the 'worst' substrate, like a metallic plane or electrodes.

6.6. FACTORS AFFECTING THE EM ENHANCEMENTS: SUMMARY

This chapter has described all the main factors affecting the properties of plasmon resonances and their associated EM enhancements, with a strong emphasis on SERS EFs. We have chosen to highlight the main effects through the discussion of model examples. The cases treated in this chapter (which illustrate the general phenomenology), and their link to relevant aspects of plasmon resonances and EM enhancements, can tentatively be summarized as follows:

- The case of the *plane* highlights the most important aspects of *non-radiative emission* close to metal surfaces. The results in this context can moreover be generalized to more complex cases. The non-radiative components of the emission, therefore, can be estimated to a good approximation using the results of a plane. Non-radiative emission is mostly dependent on (and strongly sensitive to) the *distance* between the emitter and the surface. The distance dependence is not as marked for radiative properties and local field enhancements.

- The *link between LSP resonances and local field enhancements* has been discussed extensively in the context of the metallic sphere. Large local field enhancements are obtained at resonance. The dipolar LSP resonance (which is the one occurring at the longest wavelength) is typically that of most interest to SERS.

- *Size effects* can be exemplified by the electromagnetic response of metallic spheres. The EM properties of NPs are independent of size for the smallest NPs (for which the electrostatic approximation is valid). As the size increases, (i) LSP resonances tend to *red-shift* toward longer wavelength, and (ii) the resonances broaden and the associated *local field enhancements decrease* dramatically; this is primarily a result of increased damping due to *radiation losses.*

- The basic phenomenology of *shape effects* can be exemplified with ellipsoids. Additional LSP resonances arise, some of which can be strongly red-shifted, and are then associated with *larger local field enhancements.* Moreover, further enhancements can be expected at specific positions on the surface, typically 'pointy' ends or corners (the 'lightning rod effect').

- *Gap-plasmon resonances* (of major importance for SERS), can be exemplified by the case of the dimer. This highlights the phenomena of: (i) *red-shifting* of the main coupled-LSP resonance [204] ; (ii) *large local field enhancement in the gap* (at the hot-spot); (iii) hot-spot *localization,* (iv) *extreme sensitivity* to the separation (gap) between the two objects. These phenomena form part of the general phenomenology of interacting metallic NPs and can be qualitatively adapted to other cases of gap-plasmon resonances when required.

- Additional effects that appear in practical implementations; like the effect of the *supporting substrate* or the existence of (intrinsic) *surface roughness* should be taken into account when necessary.

These examples cover most of the main phenomenological aspects of plasmon resonances and EM enhancements that are relevant to SERS. At this level, we have therefore 'closed the circle' as far as the fundamental aspects of the SERS effect are concerned. We will now explore their practical implementation in real systems in the following two chapters.

Chapter 7

Metallic colloids and other SERS substrates

The SERS literature makes often reference to '*SERS substrates*', which generally mean *any* metallic structure (or nano-structure) that produces the SERS enhancement. As mentioned in Chapter 1, SERS substrates can be tentatively classified into three main classes:

- Metallic particles (usually nano-particles) in solution, such as colloidal solutions.
- 'Planar' metallic structures, such as arrays of metallic nano-particles supported on a planar substrate (glass, silicon, metals, etc).
- Metallic electrodes.

This is a rough classification of no consequence for any subsequent analysis, but it does try to highlight three different types of environments and it is convenient for our purpose here.

Electrodes were briefly discussed in Chapter 1. Because the physical and chemical mechanisms involved in the electrochemical aspects of SERS are somewhat distant from the other topics treated in this book, we shall not discuss them further. We instead refer the reader to the specialized literature [4–6].

Amongst the other two classes of SERS substrate, solutions of metallic colloids, predominantly made of silver (Ag) or gold (Au), represent one of the simplest and most accessible routes to SERS. Because colloids are relatively easy to produce in the laboratory, many researchers get involved in SERS through them, rather than with more sophisticated SERS substrates. Metallic colloids are strongly related in addition to some of the most important and relatively recent developments in SERS; the first claims of single-molecule SERS detection were made, indeed, for metallic colloids in water [31] or deposited on a planar substrate [30].

In this chapter, we will therefore discuss some important aspects of SERS substrates in general, with a strong emphasis on metallic colloids in particular. These will serve as an example to discuss more general aspects of SERS such as the characterization of SERS substrates, or the importance of molecular adsorption. We will in addition discuss some basic aspects of colloid science that are particularly relevant to SERS experiments, in particular the problem of colloid stability.

7.1. METALLIC COLLOIDS FOR SERS

Metallic Ag and Au colloids are typically produced by a reduction reaction in solution through several possible chemical routes. The colloids exist in solution (often in water for SERS applications) only because they are stabilized by Coulombic (or sometimes steric) repulsions among particles. This normally requires the presence of a 'stabilizing agent', which coats the surface of the colloids and prevents them from aggregating. In some cases, a single chemical compound plays both the role of reducing agent and stabilizer; this is the case of sodium-citrate-reduced colloids, one of the most commonly-used types in SERS (also called Lee-&-Meisel colloids). We discuss further the problem of the stability of colloids later in this chapter.

7.1.1. Silver vs gold

Gold colloids are more easily stabilized by chemical means over very long periods of time than silver colloids. This partly explains their more widespread commercial availability, while Ag colloids have to be prepared – as a rule of thumb – in the laboratory and remain stable for a shorter (but still sufficiently long) period of time (~1 year). In addition, Au colloids have been of widespread use in biology for many years, and a wealth of knowledge has been accumulated for their synthesis and properties. Gold colloids are used as 'immunogold' for electron microscopy enhancement of images, for example. The 'chemistry' of molecular binding to gold (using thiol groups for example) is also better understood in general. Last, but not least, gold is much more stable under ambient conditions than silver, being less susceptible to oxidation.

Despite these apparent advantages, Ag colloids have been (and are still) extensively used in a SERS context. This is partly due to historical reasons, but also because Ag particles results in much larger electromagnetic enhancements in the visible (at least up to ≈ 600–650 nm, see for example Chapter 6). We will, therefore, mainly focus on the example of Ag colloids in the following, even though many of the arguments equally apply to Au colloids.

7.1.2. Citrate-reduced colloids

Preparation of Citrate-reduced silver colloids

Let us summarize very briefly the standard 'recipe' to obtain citrate-reduced Ag colloids. The most cited paper describing the preparation of Ag sols is Ref. [220] ; in particular for the citrate-reduction route. Careful investigation and optimization of this approach in a SERS context were carried out in Ref. [221]. The production of Ag colloids is extremely simple, and we summarize here the essential steps following basically Ref. [220]:

- 90 mg of silver nitrate ($AgNO_3$) are dissolved in 500 mL of distilled water (H_2O) and brought to boiling.
- A solution of 10 mL of 1% (by weight) sodium citrate (more precisely trisodium citrate, $Na_3C_6H_5O_7$) is added while boiling. It serves both as a reducing and stabilizing agent.
- The solution is kept on boiling for ~1 h, while refluxing the evaporating water, and then cooled down to room temperature.

The initial concentration of the reagents in solution is therefore: 1.04 mM of $AgNO_3$ and 0.76 mM of trisodium citrate. We shall refer to this specific colloid preparation as the 'Lee-&-Meisel colloids', as they are widely known (and referred to) in the literature.

Properties of Lee-&-Meisel Ag colloids

This procedure results in a grey-yellow solution which has a UV/Vis absorption maximum at ~400–430 nm (depending on the details of the preparation), as shown in Fig. 7.1. This optical absorption/extinction is produced by the average single-particle localized surface plasmon resonance of the particles. The characterization of colloidal solutions and other SERS substrates by UV/Vis extinction spectroscopy is further discussed later in Section 7.2.2.

The colloidal sols produced by this method (sodium-citrate reduced) have a fairly important spread of sizes and shapes, as evidenced for example by SEM imaging[1], as shown in Fig. 7.1. The mean particle size is around ~60 nm in diameter. Assuming that the reduction of $AgNO_3$ is complete and that all the silver is in the form of colloids of 60 nm diameter, the colloid concentration can be estimated[2] to be $\approx 10^{11}$ colloids/cm^3, or equivalently $\approx$0.17 nM. This is

[1] All SEM images are courtesy of David Flynn and Richard Tilley, Victoria University of Wellington, New Zealand.

[2] One 60 nm diameter silver particle contains $\approx 6.6 \times 10^6$ Ag atoms or equivalently $\approx 1.1 \times 10^{-17}$ mol of Ag per colloid. The nominal starting Ag concentration is $\approx$1 mM. The ratio of these two estimates leads to a colloid concentration of $\approx 10^{11}$ colloid per cm^3.

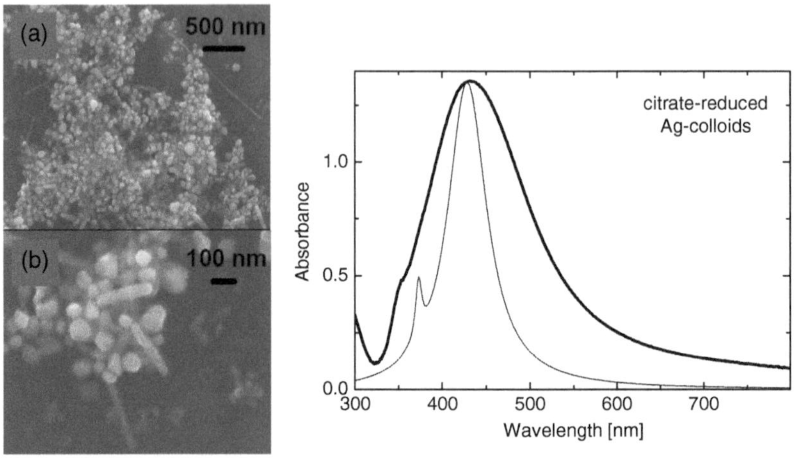

Figure 7.1. (Left) Scanning electron microscope (SEM) images of citrate-reduced Lee-&-Meisel silver colloids at two different magnifications. The colloidal solution was dried on an Al stub for imaging. The average 'diameter' of the colloids is ~60 nm, although there is an important poly-dispersity in both size and shape. Note for example in (b) the presence of 'rod-like' colloids. (Right) Corresponding UV/Vis absorption spectrum (dotted line is the prediction from Mie theory for a radius $a = 30$ nm). The main peak is the contribution of the dipolar localized surface plasmon resonances of the individual colloids. Comparison with Mie-theory predictions clearly indicates that the peak is strongly *inhomogeneously broadened* because of the poly-dispersity in size and shape. The smaller peak at ~380 nm is the quadrupolar resonance, and its presence may be hinted at also in the experiment as a shoulder in the main peak.

not very accurate and is most likely an overestimate of the real concentration since a non-negligible proportion of silver could be in a small number of very large particles (like the rod-like particles seen in Fig. 7.1). Note that these are relatively diluted colloidal solutions, i.e. the particle density is approximately one colloid per 10 μm^3, and they may be diluted further (if required) for SERS experiments.

Lee-&-Meisel silver colloids might display signs of 'aging' after long periods of time, but can be preserved for periods up to ~1 year in a fridge (at ~ 4 °C) without any significant sign of deterioration.

Note finally, that citrate ions $C_6H_5O_7^{3-}$ are triply negative and are a weak (triple) base whose triple acid form is citric acid $C_6H_8O_7$ (with pK_a's of 6.40/3.76/3.13). This may be relevant in pH-dependent SERS experiments.

Citrate-reduced gold colloids

Similar reduction routes can be followed for Au sols. For example, for citrate-reduced Au colloids, the steps are:

- 240 mg $HAuCl_4$ are dissolved in 500 mL of distilled H_2O and brought to boiling.
- A solution of 1% sodium citrate (50 mL) is added while boiling.
- The solution is kept on boiling for ~1 h (with reflux), and then cooled down to room temperature.

This procedure results in a wine-red solution with UV/Vis absorption maximum at approximately ~520–530 nm.

7.1.3. Other types of colloids

Use of other reducing/stabilizing agents

Many other combinations of reducing/stabilizing agents are possible for the fabrication of Ag or Au colloids. These may result in colloids of different sizes and/or shapes and therefore solutions with slightly different colors (and UV/Vis absorption maxima).

One of the most common alternatives to citrate as a (very strong) reducing agent is borohydride, usually in the form of sodium borohydride ($NaBH_4$). Borohydride-reduced colloids prepared as reported in Ref. [224] have also been extensively used for SERS. The starting reagent is again $AgNO_3$, but these particles are in this case not stabilized (except possibly by bonding of borate ions BO_3^{3-} on the surface). They are, therefore, much less stable than the Lee-&-Meisel type unless another stabilizer is added to the reaction.

Other types of borohydride-reduced colloids have been synthesized, for example, by reduction of silver perchlorate, $AgClO_4$, as reported in Ref. [225] (perchlorate may then serve as a stabilizer).

Improved size/shape distribution

A two-step synthesis was also reported in Ref. [222], where very small particles are first fabricated by borohydride reduction and they then serve as seeds for growth of larger particles in a second, slower, and more controlled reduction step (using ascorbic acid as a reducing agent for example [222]). Much better size uniformity and greater control over the final size can be achieved using such a two-step (seed + growth) synthesis [222]. This approach was also applied for the synthesis of very mono-disperse Ag colloids by reaction of silver nitrate with hydrazine hydrate at room temperature and fixed pH [223]. This latter recipe results typically in a very good size/shape distribution of the colloids, as seen in Fig. 7.2. Despite these improvements in colloid synthesis, the Lee-&-Meisel Ag colloids have been by far the most widely used in SERS, for a mixture of historical reasons and reproducibility issues.

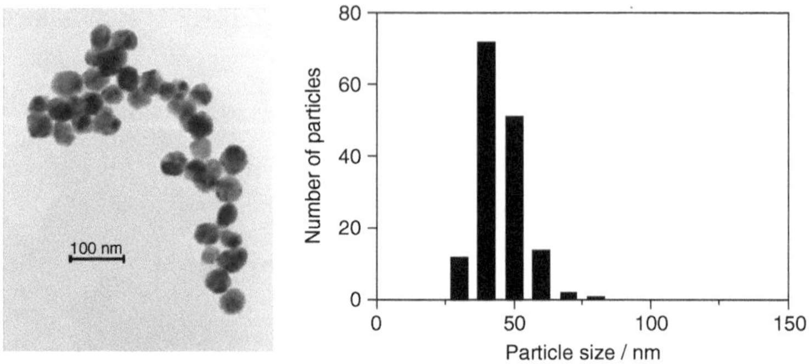

Figure 7.2. A relatively homogeneous size/shape distribution for Ag colloids can be obtained using a two-step synthesis (seed + growth), as described for example in Refs. [222,223]. This is illustrated above in the electron microscopy image and histogram of size distribution (from Ref. [223] reproduced with permission, copyright 2004 Wiley).

'Off-the-shelf' Au colloids

The goal of absolute control of size and shape over very large numbers of particles by chemical synthesis has not been achieved yet, even though very good results are obtained for gold colloids. Highly mono-disperse gold colloidal solutions (of various sizes) are even available commercially. The properties (SEM image and UV/Vis absorption) of commercial 20 nm diameter gold colloids are for example illustrated in Fig. 7.3. These colloids have a very uniform size distribution, as evident in the few colloids seen in the SEM image. These solutions certainly represent an interesting alternative (to in-house chemistry) for SERS substrates. However, the surface chemistry (stabilizing agents) of commercially-available particles is often not disclosed by the manufacturers, and this may sometimes limit their range of applicability to SERS.

Other approaches

Commercially-available nucleating silver salts (silver enhancer solutions) can also be used as an alternative. This allows, among other things, a control of the average size over a wider range, since it depends typically on the amount of enhancer added to the solution. The way the particles grow by these methods is somewhat different from the reduction route, and different options are available for nucleating the growth. For example, nucleation can be started by small (5 nm) gold particles as described in Ref. [226]. Other more 'exotic' routes to form silver nano-particles are possible, including the formation of Ag colloids in non-aqueous environments [227].

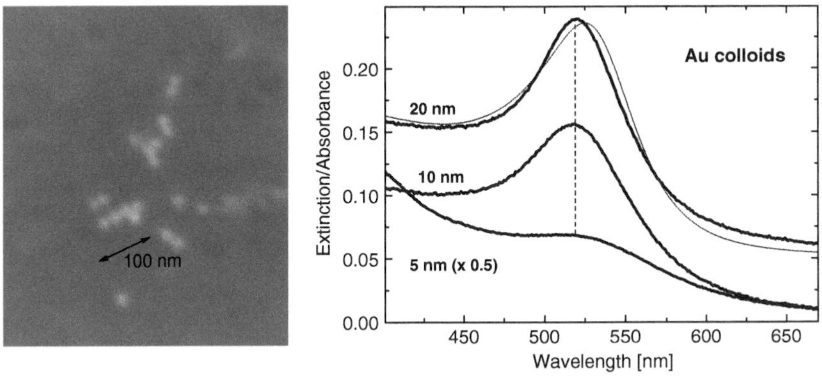

Figure 7.3. (Left) Commercial Au colloids (20 nm in diameter, from Sigma-Aldrich Chemicals) imaged by SEM. The colloids are directly dried from the solution onto an Al stub, which serves as sample holder for SEM. (Right) UV/Vis Absorption spectra for (mono-disperse) commercial Au colloids with different diameters: 5, 10, and 20 nm (dotted line is the prediction from Mie theory for 20 nm diameter). The curves have been vertically displaced for visualization purposes. The main peak corresponds to the dipolar localized surface plasmon resonances of individual particles. As long as the diameter of the colloids is $\ll\lambda$, the resonance is at the same wavelength (i.e. the electrostatic approximation applies). It only scales in intensity proportional to the volume of the particles, and their relative concentrations.

7.1.4. Remarks on colloid fabrication methods

It would be erroneous to transmit the impression to the reader that the formation of colloids through chemical reactions like the ones depicted above is completely understood. Colloid formation starts by a nucleation process, which triggers the growth of a silver nano-particle (the colloid) to a certain size and shape. The reaction continues in principle until the amount of metal or reducing agent is extinguished, but it can also be quenched at different stages by different physical (temperature) of chemical variables. The exact mechanisms of nucleation and growth at an atomic/molecular level are not fully understood, and neither are the fine details of the changes produced by different reducing and stabilizing agents. A fair amount of work has been dedicated in the literature to the understanding of the factors that control the size, shape, and poly-dispersity of the particles in these reactions. The chemical details of the reduction process have also been studied in some detail for citrate-reduced [221,228] or borohydride-reduced [225] colloids.

7.1.5. Dry colloids and other '2D planar' SERS substrates

'2D planar' substrates are also widely used in SERS (as much as colloidal liquids). The main differences with colloidal solutions are (i) the 2D nature of these substrates as opposed to the intrinsic 3D nature of solutions, and

(ii) their *fixed* geometry, as opposed to the complex dynamics that may exist in colloidal solutions. Such 2D substrates can simply be obtained by drying colloidal solutions (of Ag or Au colloids) onto for example a glass slide or a silicon substrate. Other more involved options imply the direct engineering of metallic nano-structures through a variety of methods, some of which are briefly reviewed in Chapter 8. Here we mention a few of their main characteristics.

Planar substrates obtained from colloidal solutions

The simplest way to create a 'planar' SERS substrate is by drying colloidal solutions (of the Lee-&-Meisel type, for example). Typical underlying substrates are silicon or glass, possibly coated with specific chemicals. The drying on the surface of the substrate can be made with a variety of schemes, that go from simple drop-casting to dipping of the substrate at a controlled speed. This results in colloidal clusters of different types. Straightforward drop-casting typically results in fractal-like clusters with very high EM enhancement in gaps between colloids, but with large spatial inhomogeneities and non-uniformity in the coverage of the surface. Dipping of substrates at a controlled speed results on the other hand in more uniform (but typically sparser) distribution of clusters, which is preferred in many cases to study the effects of small isolated clusters (sometimes with the addition of microscopy imaging).

A strategy often used in the literature is that of *grafting*, in which the surface is coated with a substance that will facilitate the adhesion of colloidal clusters or single colloids. Poly-lysine provides a common example of this approach. Its acts as a 'glue' for the negatively-charged colloidal particles by providing a thin (positively charged) layer on the surface. In practice, the surface of the substrate is exposed to a solution of poly-lysine in water [229] for a few minutes and then rinsed and dried. Thereafter, the colloidal solution is deposited on the poly-lysine coated surface and left for a short period of time ($\sim$1 min). The substrate is then rinsed with distilled water and dried. The few clusters/colloids that make contact with the surface during the exposure time attach to it by Coulombic interaction. In this manner, a distribution of isolated, relatively sparse, homogeneous, and firmly attached colloidal clusters or single colloids can be obtained on the surface. These substrates find many applications in fundamental studies. In a way, their disadvantage is their intrinsic 'randomness'; i.e. we cannot decide where the clusters are going to be (and the characteristics of their plasmon resonances). But this can be also be seen as their strength: virtually all possible resonance conditions are achieved in a sufficiently large population of clusters. Such an approach provides a good 'snapshot' of the colloidal aggregates present in solution. One should however bear in mind that the EM properties of these aggregates are typically modified once they are attached to the planar substrate (see Chapter 6).

Roughened surface and metal-island films

There are other strategies for the fabrication of planar SERS substrates that do not start with the synthesis of nano-particles in solution. Two such strategies are the 'roughening' of metallic surfaces by chemical means, or the evaporation of metallic islands on surfaces. There are very many combinations to obtain and generate rough (SERS active) surfaces which include hot/cold metal evaporation and even electrochemical and/or roughening by ion bombardment, and miscellaneous alternatives like HNO_3-roughened Ag foils [230]. Electrochemically roughened electrodes and the so-called 'metal-island film' approach (which is basically a metal evaporation on a substrate, at cold or room temperature) occupied the attention of most of the early studies. In fact, many early results were obtained with pyridine as a probe on such substrates. A detailed account of the early 'models of roughness' and their connection with experimentally observed phenomena is given in Ref. [4].

Self-organized and 'ordered' substrates

The approach to produce planar SERS substrates for typical applications is currently being pursued along many different lines. At this point, the presentation here makes contact with the content of Chapter 8, in which some of the novel approaches like nano-sphere lithography, island lithography, or patterning through nano-lithography are explained in some more detail. Here we only mention one example and refer the reader to Chapter 8 for further details. Figure 7.4 shows a complementary example to those shown in Chapter 8 of a planar SERS substrate of the type being explored at present [47].

Highly ordered SERS substrates like the one shown in Fig. 7.4 are fabricated by modern lithographic techniques and achieve a very high control over the geometry and homogeneity of the patterned features. In the case of Fig. 7.4 this is done for gold nano-structures of different shapes and sizes (thus allowing for a control of their plasmon resonances). As further pointed out in Chapter 8, these planar SERS substrates normally 'sacrifice' the largest achievable enhancement at the expense of uniformity in the signal, which is obviously a *very* desirable property for applications and for some fundamental studies. These types of nano-structures represent a growing trend in SERS at the moment. Other approaches to produce similar patterns are also based on self-organization. Further details will be given in Chapter 8.

7.2. CHARACTERIZATION OF SERS SUBSTRATES

The characterization of SERS substrates (planar substrates or colloidal solutions) is an important first step toward an understanding of the SERS signals in these systems. We briefly discuss here some of the most common techniques employed to this end.

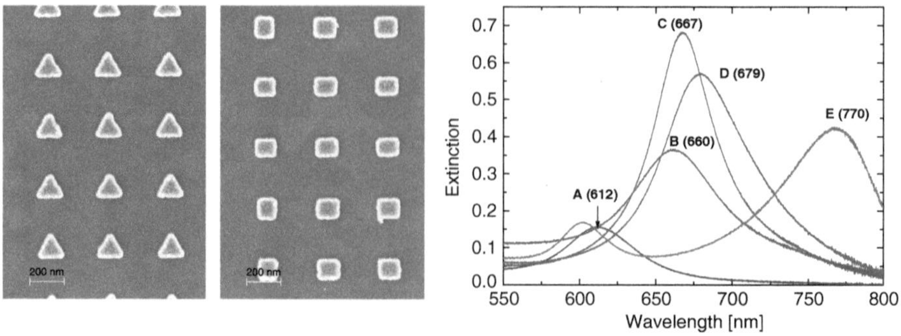

Figure 7.4. Figure adapted from Ref. [47] (reproduced with permission, copyright 2008 Elsevier). (Left) SEM images of arrays of gold nano-particles fabricated by e-beam nano-lithography. Exquisite control of the shape and high uniformity can be achieved. (Right) Corresponding UV/Vis extinction spectra for similar arrays. A: triangles (equilateral) with a side length of $a = 100$ nm; B: triangles (equilateral) with $a = 150$ nm; C: dots (circles) with a diameter of $a = 100$ nm; D: squares with a side length of $a = 120$ nm (shown in the SEM); E: triangles (equilateral) with $a = 200$ (shown in the SEM). The height is $h = 40$ nm in all cases.

7.2.1. Microscopy

The geometrical structure of a SERS substrate, such as size and shape (and indirectly the size distribution or poly-dispersity) of particles, plays a decisive role in the properties of the plasmon resonances and EM enhancements the substrate can sustain. The most direct measure of this structure is by using electron *microscopy*.

The most common technique is scanning electron microscopy (SEM) which, however, requires a conducting substrate or evaporation of a thin metallic layer (usually platinum) on the substrate. Examples of SEM images of a planar substrate fabricated on ITO (which is conducting) are shown in Fig. 7.4. Accurate size and shape information can be extracted from these for the in-plane geometry. Height information can also be obtained by inclining the sample at an angle, but this is more difficult to carry out and interpret and the accuracy is limited. Alternatively, it is possible to use atomic force microscopy (AFM) instead. This yields very accurate height measurements, but the in-plane measurements are usually not as accurate as with SEM (and are also more tedious to obtain through mapping).

Colloidal solutions cannot be imaged directly with SEM (in liquid), but it is possible to do it indirectly by drying them onto an appropriate substrate. Examples have been shown in Figs. 7.1–7.3. There is always a possibility, however, that this drying process may affect the geometry of the colloids. The 'clustering' of colloids observed in the pictures is precisely a result of the process of drying the solution. This approach is usually suitable to characterize the properties of individual particles, but it does not provide any insight into the clustering that may already be present in the solution itself [229]. To avoid this problem, it is necessary to attach or graft individual clusters onto a substrate, thereby taking a 'snapshot' of the clusters (as they are in solution). This can be achieved, as explained in the previous subsection, by dipping a poly-lysine coated glass slide into the colloidal solutions and rinsing it afterward. Individual colloids (or clusters) attach to the positively-charged poly-lysine layer, thus giving a 'snapshot' of the solution. For SEM imaging, it is however necessary to further coat the slide with a thin conducting layer ($\sim$5 nm platinum for example). One can then characterize the geometry of the clusters with some limitations imposed by the imperfections introduced by the additional capping layer.

SEM provides a direct look into the basic shapes and sizes of the produced particles, but it is certainly time consuming and costly. Fortunately, other easier-to-implement experimental approaches (mainly optical; light scattering methods) are available. Even if their interpretations are more indirect, they are often sufficient for routine characterization of new SERS substrates. We describe in the following the standard method of UV/Vis absorption (extinction) due to its overwhelming importance and ease of implementation.

7.2.2. Extinction or UV/Vis spectroscopy of SERS substrates

Arguably, the simplest form of optical characterization of SERS substrates is by direct optical absorption measurements or equivalently UV/Vis spectroscopy, typically measured in a transmission configuration. UV/Vis spectroscopy has been discussed already in Section 2.3.1 for the characterization of dye solutions. The same principles are readily applied to colloidal solutions and can be simply extended to the measurement of transmission (and therefore extinction) on planar SERS substrates; provided the underlying substrate is transparent.

Resonance wavelength in extinction spectra

Examples of UV/Vis extinction spectra have been given in the previous subsection: for Lee-&-Meisel Ag colloids in Fig. 7.1, for Au colloids in Fig. 7.3, and for gold-nano-particle arrays on planar substrates in Fig. 7.4. All these spectra exhibit one, or sometimes more, peaks associated with a resonant optical response at the wavelength corresponding to the *localized surface plasmon resonances* of the substrate, as discussed theoretically in Chapter 6. This peak may be inhomogeneously broadened, because of substrate non-uniformity, (like the poly-dispersity in citrate-reduced Ag colloids). The most important characteristic that is extracted from extinction spectra is precisely this *resonance wavelength*. Because SERS occurs precisely as a result of the interaction with such localized surface plasmon resonances, the largest SERS enhancement is expected when the incident laser or Stokes-shifted SERS photon (or both) is at a wavelength close to the resonance wavelength.

The *wavelength dependence* of the extinction spectrum and the SERS EM enhancement are therefore intricately related. The connection between the extinction spectrum and the *magnitude* of the electromagnetic SERS enhancements is, however, indirect in general. For example, resonances associated with gap-plasmons are not more intense in the extinction spectrum, despite their much larger local field enhancements and therefore SERS enhancements [204] (see Section 6.4) In situations where the SERS signal is dominated by a small number of high-enhancement regions (hot-spots), the extinction spectrum can be a bad indicator of where (in wavelength) the maximum SERS enhancements occur. This is the case for example of aggregated colloidal solutions, which will be discussed further in this chapter. For a more complete discussion of the connection between extinction and SERS enhancements, see for example Ref. [204].

Size and shape determination

The extinction spectrum can, moreover, be compared to theoretical predictions to try to extract the properties of the sample under consideration. This is a difficult exercise in general (especially in the absence of additional structural information from microscopy) since a variety of size/shape/poly-dispersity combinations can lead to very similar spectra. Even if structural information is available, the position of the single-particle resonance tends to be relatively insensitive to the size of the particles (at small sizes). Indeed, as long as the electrostatic approximation is a good description of the optical properties of a metallic particle, the resonance position is independent of size (but not of shape). An example is explicitly given in Fig. 7.3 for Au colloids of diameters 5, 10, and 20 nm, respectively.

Comparison with theory nevertheless provides an additional level of understanding in some cases. For example, in Fig. 7.3 we show the extinction predicted from Mie theory for 20 nm gold spheres in water. The agreement is good and enables one to predict the colloid concentration (see later for details). A similar prediction is shown in Fig. 7.1 for Lee-&-Meisel colloids, assuming 60 nm diameter silver spheres in water. Here the predicted resonance wavelength is in agreement with experiment, but there is a substantial additional broadening of the resonance in the measured spectrum. A significant poly-dispersity of the solution (size and/or shape variations) can therefore be inferred from this comparison. Regarding the shape determination, an interesting series of experiments to correlate the size and shape of individual Ag nano-particles with their plasmon resonance characteristics has been carried out for example in Ref. [226]. In all these cases, comparison with theory is used more as a consistency check, rather than a really predictive tool of the structure. Extinction measurements are therefore good 'quality-control' measurements to compare different batches of samples, but tend to be rather inaccurate for the estimation of geometrical properties.

Colloid concentration from UV/Vis absorption

In addition to the resonance peak position, the maximum extinction at resonance is also of interest even though, as mentioned already, the latter is not a good indicator of the magnitude of the SERS enhancement. For example, planar arrays of nano-particles may exhibit a low extinction because of a small surface density of particles (wide apart, for example). This does not mean that each individual particle is not highly SERS active. If the extinction properties of an individual particle are known, then the extinction value can be used to derive the surface density of particles. A similar argument can be made for a mono-disperse colloidal solution: the colloid concentration can be inferred from the extinction magnitude, as we show now.

For dyes, UV/Vis absorption can be used to deduce the solution concentration from a knowledge of its absorption properties (in particular its decadic molar extinction coefficient), or vice versa. This is also true of colloidal solutions and the principles are exactly the same (see Section 2.3.1). The absorbance , which should in fact be called 'extinctance' here, is related to the concentration by the Beer–Lambert law, discussed in Section 2.3.1, which we recall here:

$$A = -\log_{10}(T) = c_m \bar{e} L = \frac{\mathcal{N} \sigma_{\text{Ext}} c_m L}{\ln(10)}, \tag{7.1}$$

where $\mathcal{N}$ is Avogadro's number and L is the path length (typically 1 cm). When applied to a mono-disperse colloidal solution, c_m [M] is the colloid concentration, and $\bar{e}$ [M^{-1} cm^{-1}] and σ_{Ext} [m^2] are the decadic molar extinction coefficient and extinction cross-section of an individual colloid, respectively. Let us also recall the useful conversion expression:

$$\sigma_{\text{Ext}}[\text{cm}^2] = 3.82 \times 10^{-21} (\bar{e}[\text{cm}^{-1}\ \text{M}^{-1}]). \tag{7.2}$$

In general, colloids are rarely characterized by the knowledge of $\bar{e}$, but more often by σ_{Ext}. Hence, the first equality in Eq. (7.1) is more often used in the context of dye or molecular absorption, while the last one is more suited to colloidal solutions. Finally, σ_{Ext} is not often known, but it can in simple cases be predicted fairly accurately from theory.

For example, for 20 nm spherical gold colloids in water, the single-particle extinction coefficient $Q_{\text{Ext}}^{\text{NP}}$ (see Chapter 5) at resonance (λ = 525 nm) is predicted to be $Q_{\text{Ext}}^{\text{NP}} = 1.1$, equivalent to a cross-section of $\sigma_{\text{Ext}} = 3.8 \times 10^{-12}$ cm^2 or $\bar{e} \approx 10^9$ M^{-1} cm^{-1}. This is much larger than that for the best absorbing molecules, owing to the much larger size of the colloidal particles (compared to molecules). Since the extinction at resonance is measured experimentally (see Fig. 7.3) to be $A \approx 0.19$, we can easily deduce from the Beer–Lambert law that the colloid concentration is $c_m \approx 0.2$ nM or equivalently 1.2×10^{11} colloids per cm^3.

The same exercise can be carried out for Lee-&-Meisel Ag colloids. For 60 nm diameter spherical silver particles in water, the extinction coefficient $Q_{\text{Ext}}^{\text{NP}}$ at resonance ($\lambda = 429$ nm) is predicted to be $Q_{\text{Ext}}^{\text{NP}} = 14.5$, equivalent to a cross-section of $\sigma_{\text{Ext}} \approx 4 \times 10^{-10}$ cm^2 or $\bar{\epsilon} \approx 10^{11}$ M^{-1} cm^{-1}. Since the extinction at resonance is measured experimentally (see Fig. 7.1) to be $A \approx 1.4$, we can easily deduce from the Beer–Lambert law that the colloid concentration is $c_m \approx 14$ pM or equivalently $\approx 10^{10}$ colloids per cm^3. Since the spectrum was taken for a colloidal solution diluted by half in water, the concentration in the original solution is $\approx 2 \times 10^{10}$ colloids per cm^3. Note that this is a factor of 5 smaller than the simple concentration estimate obtained earlier from silver conservation. We noted already that the previous estimate was most likely an overestimate because of the large amount of silver contained in the rod-like particles. A discrepancy is also not so surprising here given the fact that the theoretical spectrum is much narrower than the measured extinction, and therefore does not account for colloids of smaller (at shorter wavelength) or larger (at longer wavelength) sizes (or of different shapes). The poly-dispersity here makes it much more difficult to extract the concentration. The real colloid concentration is therefore between 2×10^{10} and 10^{11} colloids per cm^3, but a more elaborate theoretical model (with various sizes) would have to be used to obtain a better estimate from the extinction spectrum.

Additional considerations for colloidal solutions

Extinction spectra provide a good rough assessment of the colloidal solution quality and possibly information on the colloid size and/or concentration, but the technique does have its limitations and problems; among them are:

- Because the technique relies on a measurement of the transmitted light, one must ensure that a sufficient fraction of the incident light is transmitted to avoid errors. This also minimizes the possible undesirable effects of multiple scattering. Assuming for example that we need a minimum of $\approx 1\%$ transmission for an accurate reading (this depends on the instrumentation), then absorbance values above ≈ 2 cannot be trusted. A rule of thumb is that liquid samples for UV/Vis measurements should have a low turbidity. If it is not the case, then it is necessary to dilute the sample before the measurement, but this may in turn introduce unwanted effects. A concentration-dependence study can be carried out to rule out any problem associated with dilution and/or multiple scattering.

- Extinction spectra on colloidal solutions tend to be rather insensitive to the presence of a small fraction of aggregates (clusters of colloids) inside the solution and their associated gap-plasmon resonances. These play an important role in the SERS enhancement mechanism thanks to the large local field enhancements in gaps. In the presence of colloidal clusters, the SERS enhancement factor is therefore only indirectly related (if related at all) to the extinction spectrum [204]. This is, in fact, a well-known phenomenon in (for example) standard Lee-&-Meisel Ag colloids.

Rayleigh and Mie scattering

There are a number of possible variations of the extinction spectrum measurements, for example by carrying out measurements of optical absorption as functions of scattering angle (typically at a single wavelength). Their denomination refers more to the theoretical tools needed to interpret the result, rather than to different experimental instrumentations. Depending on the relative size of the particles (a) with respect to the wavelength (λ), different regimes of scattering are identified: for $a \ll \lambda$ (Rayleigh scattering), or $a \sim \lambda$ (Mie scattering). Such angle-dependent scattering experiments may provide additional information than simple extinction spectra, in particular on the size of the particles. However, in a SERS context, the extinction spectra in a simple transmission configuration (as described previously) are often sufficient to characterize the substrate. They provide more directly relevant information (position of the optical resonances, for example) than angle-dependent experiments.

7.2.3. Other techniques: dynamic light scattering (DLS) for colloidal solutions

There are numerous other techniques available, specifically designed for colloid characterization (sedimentation, centrifugation, etc.), which are occasionally reported in the SERS literature. We briefly mention one of them here: *dynamic light scattering* (DLS), for its additional insight into colloid dynamics (in particular diffusion). The reader is directed to the more specialized colloid literature in Refs. [231,232] for details on additional techniques.

Principle of DLS

The principles of DLS (also known as quasi-elastic light scattering, or photon correlation spectroscopy) have been known for a very long time [233], but it has only become a 'routine' technique with the advent of powerful electronic auto-correlators and the associated electronics in computers. DLS relies on the measurement of intensity fluctuations (or equivalently spectral broadening) in the light scattered by a monochromatic source[3]. These fluctuations are directly related to the movement (Brownian motion) of the scatterers, in our case the colloids. It is then possible (with some basic assumptions about the scatterers) to estimate their diffusion coefficient from the intensity fluctuations and thus, indirectly, their size. In principle, the different time constants in the intensity fluctuations (or the frequency components of the spectral broadening) can be resolved, and DLS is accordingly capable of measuring the size distribution. This however requires a 'clean' system (with perfectly spherical dielectric particles for example) and the interpretations are not always easy.

Practical implementation

A very schematic representation of a DLS experiment is given in Fig. 7.5(a). The necessary formulas are given here without further explanation. For a more detailed discussion of the principles of DLS and its applications, see for example Ref. [233]. The intensity fluctuations can be measured for different scattering directions θ, which define the scattering vector $\mathbf{q}$, i.e. the change in direction of the wave-vector of the scattered light with respect to the incident beam: $\mathbf{q} = \mathbf{k}_{\text{Inc}} - \mathbf{k}_{\text{Sca}}$. Its modulus [$\text{m}^{-1}$] is therefore [231]:

$$|\mathbf{q}| = \frac{4\pi n_M}{\lambda_L} \sin(\theta/2), \tag{7.3}$$

where n_M [a.d.] is the refractive index of the medium (water here, $n_M = 1.33$), and λ_L [m] the wavelength of the incident light ($\lambda_L = 532$ nm here).

The most convenient way to extract the time constants involved in the intensity fluctuations $I_s(t)$ is to compute the auto-correlation function: $G(\tau) = \langle I_s(t+\tau)I_s(t)\rangle / \langle I_s(t)I_s(t)\rangle$, where $\langle \ldots \rangle$ denotes the time average and τ is a *delay*. For identical scatterers moving by Brownian motion in an otherwise homogeneous liquid, $G(\tau)$ [a.d.] should take the form [233]:

$$G(\tau) = G_\infty \left[1 + b\,\mathrm{e}^{-2\tau/T_D}\right], \tag{7.4}$$

where b [a.d.] is a constant (determined by the geometry of the experiment) and T_D [s] is the characteristic diffusion time for the colloids to come out of phase in the incident monochromatic beam. This depends on the scattering angle, and is the time taken to travel a distance $L_D = 1/|\mathbf{q}|$ (with $\mathbf{q} \equiv$ scattering vector). Moreover, T_D [s] and L_D [m] are related to the diffusion coefficient D [$\text{m}^2\ \text{s}^{-1}$] through the standard expression $L_D = \sqrt{DT_D}$.

[3] Note that techniques like Rayleigh or Mie scattering are referred to by contrast as static light scattering experiments since only the average scattered intensity is measured.

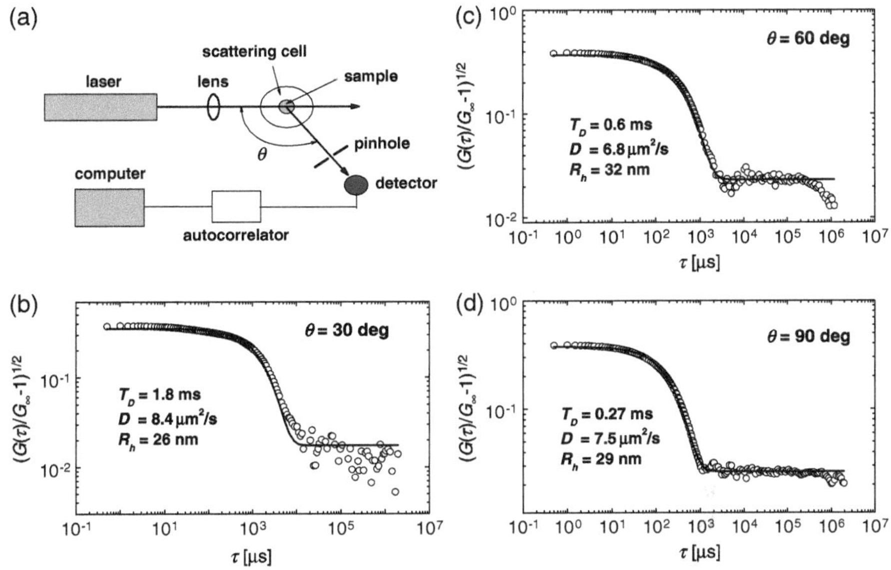

Figure 7.5. (a) Schematic representation of a DLS experiment to characterize metallic colloids. A laser is focused on the middle of a cell with circular windows, and the scattering coming from the center of the cell is collimated through a pinhole (which defines the scattering angle θ) and sent to a photomultiplier. The signal of the photomultiplier is analyzed by an auto-correlator and sent to the computer. Examples of auto-correlation functions (plotted as $\sqrt{G(\tau)/G_\infty - 1}$ on a log–log plot) are shown in (b–d) for Lee-&-Meisel Ag colloids at different scattering angles, along with a single-exponential fits to the data. The diffusion time T_D is deduced from these fits, from where the diffusion coefficient D and hydrodynamic radius R_h follow.

For a mono-disperse solution, T_D can be deduced from the experimental auto-correlation function $G(\tau)$. This can be conveniently obtained from a single-exponential fit to the function $\sqrt{G(\tau)/G_\infty - 1}$.

Example for a metallic colloid solution

Figures 7.5(b–d) show examples of such fits for a DLS experiment at various scattering angles for Lee-&-Meisel Ag colloids (described earlier in this chapter). T_D varies with scattering angle, and this can be used to probe different timescales, or as a consistency check of the results. The physical properties of the system, and in particular the diffusion coefficient D, should not depend on the scattering angle. From T_D, the diffusion coefficient D is obtained by:

$$D = \frac{1}{T_D |\mathbf{q}|^2}, \tag{7.5}$$

where $|\mathbf{q}|$ is given in Eq. (7.3). The results in Figs. 7.5(b–d) indeed show that the experimentally measured diffusion coefficient is (within experimental errors) the same for all scattering angles, as expected. For Lee-&-Meisel Ag colloids, $D \approx 7.5\ \mu\text{m}^2/\text{s}$. Because of the poly-dispersity, this should be viewed as the diffusion coefficient for an average-sized colloid in the distribution.

Note that metallic colloids are *a priori* not the ideal system for DLS, firstly because of the poly-dispersity in size and shape, and secondly because of their (inevitable) optical absorption, which may cause problems in the interpretation. The results in Figs. 7.5(b–d), nevertheless, show that DLS can be used as an additional characterization tool to determine the average diffusion coefficient of the colloids.

Size determination with DLS

Under the assumption that the particles are spherical in shape, the radius R_h [m] of the particle is directly related to the diffusion coefficient D through the Stokes–Einstein equation [233]:

$$D = \frac{k_B T}{6\pi\eta R_h}, \tag{7.6}$$

where η [kg m^{-1} s^{-1}] is the viscosity of the solvent ($\eta = 10^{-3}$ kg m^{-1} s^{-1} for water here). R_h is called the *hydrodynamic radius* of the particle, and may in some cases (especially for non-spherical shapes, or coated particles) differ from the real radius a. In our example, we find $R_h \approx 30$ nm for a standard Lee-&-Meisel Ag colloid. This is in good agreement with previous measurements by SEM or UV/Vis spectroscopy performed on the same samples.

Colloid diffusion and SERS

The value of the diffusion coefficient of $D \approx 7.5\ \mu\text{m}^2/\text{s}$ for a typical Lee-&-Meisel colloid is a useful figure to bear in mind when using them for SERS. Brownian motion of colloids in-and-out of the scattering volume can be a source of SERS intensity fluctuations, especially for experiments at high magnification (like many single-molecule SERS experiments).

Consider for example a $\times 100$ immersion objective. The scattering volume is of the order of 10–20 μm^3 [8], corresponding to one colloid on average in the scattering volume at a time. Moreover, the typical size of the beam is of the order of $L = 1\ \mu\text{m}$ (laterally). This corresponds to a diffusion time of $L^2/D \approx 130$ ms. This is comparable to the integration time in many practical SERS experiments. Hence, the colloid dynamics (Brownian motion) is expected to play a major role in this case in the SERS intensity fluctuations. We shall come back to this point in Section 7.4.5. For colloidal clusters (which will be discussed later), the hydrodynamic radius is expected to increase, and the diffusion coefficient to decrease, accordingly. The effect would therefore be important on even longer timescales.

Remarks on DLS

These results demonstrate that DLS can be used as an additional tool for metallic colloid characterization, at least for their average properties. The technique has its drawbacks though, and we mention very briefly some of them:

- Poly-dispersed solutions (for particles with many different shapes and sizes) can be particularly problematic to analyze. The use of a single *hydrodynamic radius* from a sphere to match the experimental results is clearly an approximation that has to be judged on a case-by-case basis. It is possible to apply a more complex analysis to the data to try to extract the size distribution. The reliability of such approaches is, however, somewhat questionable (in particular for complex systems like the Lee-&-Meisel Ag colloids).
- In line with the previous observation, the scattering cross-section for DLS increases very rapidly with the size of the particle. This tends to produce a biased overestimation of the average size in situations where groups of different particles coexist.
- The treatment presented here is only valid when multiple scattering is negligible. Ideally, one should carry a series of concentration studies to show that this is, indeed, the situation. In addition, several scattering angles (θ) are typically required to ensure consistency among the values deduced for D.

Despite the shortcomings, DLS is a technique that is widely used for a quick and reliable characterization of dielectric colloids and is also applicable to metallic colloids (as shown above). It moreover provides an insight into an unavoidable and important aspect of colloidal solutions for SERS: Brownian motion (diffusion).

7.3. THE STABILITY OF COLLOIDAL SOLUTIONS

7.3.1. Introduction

The importance of colloid stability for SERS

The single-colloid dipolar localized surface plasmon resonances – readily seen in Figs. 7.1–7.3 – provide the simplest possible form of SERS enhancement in liquids. Simple estimates of SERS EFs (maximum and average) have been provided in Chapter 6. As shown in Fig. 6.11(b), for 50 nm diameter Ag spheres in water (typical of Lee-&-Meisel Ag colloids described in this chapter), the predicted average SERS EFs (SSEF) and maximum SMEF per colloid are approximately $\sim 4 \times 10^4$ and $\sim 1.6 \times 10^5$, respectively (at resonance, i.e. 420 nm). This enhancement is already not negligible, but it is not large enough for many applications (like single-molecule SERS spectroscopy). Moreover, the resonance wavelength is not in a convenient range for silver spheres (too close to the UV).

The largest SERS enhancements – the ones that provide in particular enough sensitivity to observe single molecules – in general occur at the junction between two particles. They arise from coupled-LSP resonances (or gap-plasmon resonances) as described in Section 6.4. Colloidal sols therefore need to be *aggregated*, at least partially, to become useful SERS active liquids with high enhancements. Understanding the way colloids aggregate and form clusters is therefore necessary to understand their use as SERS substrates. The problem of colloid stability will be discussed in the following few sections from the general (and simplified) point of view of colloid science. To this end, a brief introduction to the Derjagun–Landau–Verwey–Overbeek (DLVO) theory[4] of colloid stability and the essential forces involved in the process will be given. The discussion will then be specialized to SERS.

Main ingredients of colloid stability

One of the most remarkable properties of colloidal solutions [232] is that they actually exist. The equilibrium achieved in the steady state of a colloidal system is, in fact, *metastable*. It is a fine balance between electrostatic (and steric) *repulsions*, van der Waals *attractions*, and hydrodynamic forces (interaction through movement in the fluid). The different roles, magnitudes, and relative strengths of these contributions produce an amazing variety of phenomena that go from glass formation through crystallization (used in opal photonic crystals for example [234]), to a long list of many-body phenomena, which are extensively covered in the specialized literature [231,232,235].

In many cases of interest to SERS, two basic forces are responsible for colloid stability: (i) the attractive van der Waals interactions, and (ii) the (possibly screened) Coulomb (electrostatic) repulsions. The combination of these two forces – together with certain approximations to treat them analytically – constitutes what is known as the DLVO theory of colloid stability [231,232,236,237]. We describe briefly its physical origin and some of the most important consequences for aggregation and SERS experiments in liquids.

[4] After Derjaguin, Landau, Verwey, and Overbeek who developed several aspects of the theory independently.

7.3.2. The van der Waals interaction between metallic particles

The origin of van der Waals forces

Van der Waals interactions have their origin in the theory of *dispersion forces* developed by London. The explanation requires a few basic concepts of quantum mechanics and electromagnetic theory and full details can be found, for example, in Refs. [238,239]. The concepts can be extended to macroscopic objects afterward, through the so-called *Hamaker theory* [231] (treated briefly in the next section), or through the much more advanced *modern dispersion theory* of Dzyaloshinskiǐ and Pitaevskiǐ [240]. From the point of view of colloid stability, only the final results of Hamaker theory are necessary in most cases. Nonetheless, we describe first the microscopic origin of van der Waals forces to give a taste of its real physical origin.

London dispersion forces can be introduced by studying the quantum mechanical interaction of two neutral molecules. Consider the total charge of a molecule to be $Q = \sum q_i$, and its dipole moment $\mathbf{p} = \sum q_i \mathbf{r}_i$; both given in terms of its constituent charges q_i and their respective coordinates. We consider explicitly the example of the interaction between two neutral and non-polar molecules; i.e. $\langle \Psi_0 | Q | \Psi_0 \rangle = 0$ and $\langle \Psi_0 | \mathbf{p} | \Psi_0 \rangle = 0$, where $|\Psi_0\rangle$ is the ground state wave-function.

The interaction potential $V_{\text{int}}(R)$ between two neutral molecules A and B separated by a distance R is typically dominated by the *dipole–dipole interaction* of the total (electrostatic) dipoles $\mathbf{p}^A$ and $\mathbf{p}^B$, which is of order $1/R^3$. Classically, since the molecules are assumed to be non-polar (with no net dipole), the interaction should be *zero*. Quantum mechanically however, only the *average* dipole moments are zero, but there are small *quantum fluctuations* around this average. The non-zero dipoles resulting from the fluctuations can interact with each other, and it turns out that after averaging this effect, the interaction energy is no longer zero. The full description of this effect requires the use of second-order quantum mechanical perturbation theory, which accounts for quantum fluctuations (while first-order perturbation theory accounts for average quantities and results in a zero interaction energy). Without going into the full details here, this results in an interaction energy of the form [231]:

$$\Delta E \sim -\frac{C_{AB}}{R^6}, \tag{7.7}$$

where C_{AB} is a constant depending on the matrix elements of V_{int} with respect to the electronic wave-functions of the molecules A and B. This therefore results in an *attractive* interaction energy proportional to $1/R^6$ (as a second-order perturbation of $V_{\text{int}} \propto 1/R^3$). The interaction between the fluctuating dipole moments of the molecules is therefore the microscopic origin of the $\propto 1/R^6$ attractive van der Waals interaction between neutral molecules.

Van der Waals interaction between objects – Hamaker theory

The London dispersion forces described in the previous section hold for the interaction between two molecules. For the interaction between two nanometer-sized particles, we can think of each particle as being formed by small infinitesimal elements that interact among themselves in the form described in the previous section. The simplest possible approach is to assume that the interaction energy will be a simple pairwise summation of these microscopic interactions. This is the principle of *Hamaker theory.* It provides the simplest extension of van der Waals interactions to nanoscopic/macroscopic objects.

Consider two different objects (A and B) divided into N and M small cells, respectively. The interaction energy between the two is then:

$$\Delta E \sim \frac{1}{2} \sum_{i}^{N} \sum_{j}^{M} V_{\text{int}}(\mathbf{R}_{i,j}), \tag{7.8}$$

where $\mathbf{R}_{i,j}$ is the distance between cell i in A and cell j in B. Equation (7.8) will be, in general, replaced by suitable integrals over the combined coordinates of the two bodies. The following important aspects should be highlighted:

- The sums in Eq. (7.8) (or corresponding integrals over the volume) affect the distance dependence of the total interaction, which decreases slower than $1/R^6$.
- For the same reason, the total pairwise interaction depends on the *geometry* of both bodies A and B. For homogeneous bodies with simple shapes, colloid-science books provide lists of analytic expressions for the interaction energy. For poly-disperse colloidal solutions like the Lee-&-Meisel Ag colloids, the approximation of 'effective' spheres with an average radius is necessary, albeit with obvious limitations.

Hamaker theory suffers from a number of limitations:

- Equation (7.8) ignores microscopic local-field corrections (similarly to that described in Section C.3.1). In reality, every 'cell' of both bodies is affected by the presence of all other cells.
- Another serious limitation is that this interaction, as described above, is purely electrostatic. This is typically *not* an issue for molecules at short distances, but can become more problematic for larger objects.

A proper theory of dispersion forces that takes into account the local-field corrections and retardation effects has been developed by Dzyaloshinskiĭ and Pitaevskiĭ in the Russian School of Landau and Lifshitz [240], and is usually called the 'modern' dispersion theory. Its complexity, however, makes it difficult to be used as a basis for further studies, including colloid stability. Hamaker theory, despite its conceptual problems, therefore represents the best approach for this task and is a powerful tool to understand the phenomenology of colloid interactions.

Van der Waals attraction for two spherical colloids

For the specific case of interest here (colloids), the total effective van der Waals interaction potential $V_{\text{vdW}}(r)$ [J], which accounts for the attraction between two identical spheres of radii a [m], separated by a distance r [m] (from center to center), can be written within Hamaker theory as [231]:

$$V_{\text{vdW}}(r) = -\frac{A_{\text{HC}}}{6}\left[\frac{2a^2}{r^2-4a^2} + \frac{2a^2}{r^2} + \ln\left(1-\frac{4a^2}{r^2}\right)\right], \tag{7.9}$$

where A_{HC} [J] is the Hamaker constant, which is related to the constant C_{AB} in Eq. (7.7) and depends on both the material of the objects under consideration, and on the medium where they are immersed in.

Example

For two silver colloids, A_{HC} is of the order of 3.1 eV in air and 2.5 eV in water. For gold colloids, A_{HC} is of the order of 2.5 eV in air and 1.9 eV in water. Let us consider spherical silver colloids of radius $a = 30$ nm in water (representative of the Lee-&-Meisel Ag colloids) and at room temperature ($T = 290$ K). The Hamaker constant is then $A_{\text{HC}} \approx 100\ k_BT$. A plot of the van der Waals interaction potential as given by Eq. (7.9) is shown later in Fig. 7.6(a). It is clear that the attraction potential can be much larger than k_BT at short distances, but decreases rapidly to be of the order of $\sim k_BT$ for a separation of $\approx$30 nm.

7.3.3. The screened Coulomb potential

The other important interaction in metallic colloid solutions is the *Coulomb or electrostatic interaction.* Metallic colloids are usually charged (typically negatively charged), and this repulsive interaction ensures the colloidal stability. In addition, in many situations of interest colloidal solutions contain additional charged species (mostly ions), which may be the products of the synthesis reaction, or even added intentionally at a later stage. The Coulomb interaction among particles is affected by these additional ions and, as we shall see, its strength is reduced. It is therefore often called the *screened* Coulomb interaction to highlight the screening effect of the ions in solution. It is the purpose of this section to give a general idea of the principles behind the screened Coulomb interaction in colloids.

Poisson–Boltzmann equation

We consider a solution with charged species (ions), each at a concentration n_i^0 [m^{-3}] and with charge q_i [C], and a single charged colloid (often called a macro-ion in this context). Because of its much larger size, the colloid dynamics is much slower than that of the ions, and it can therefore be assumed to be at a fixed position. The ions interact with themselves and with the colloid through electrostatic interactions (which result in a total electrostatic potential $\phi(\mathbf{r})$ [J]) and at the same time can move around in the solution. As a result, the density of ions $n_i(\mathbf{r})$ is position-dependent and the total density of charge is then $\rho(\mathbf{r}) = \sum_i q_i n_i(\mathbf{r})$ [C m^{-3}]. Within the electrostatic approximation, the potential can be obtained from the solution of the Poisson equation: $\nabla^2\phi(\mathbf{r}) = -\rho(\mathbf{r})/(\epsilon_0\epsilon_S)$, where ϵ_S is the *relative static dielectric constant* of the medium (water here).

This equation cannot be solved directly because the ion concentrations and therefore $\rho(\mathbf{r})$ depend on $\phi(\mathbf{r})$. Assuming the system is at thermodynamic equilibrium, we can in fact write: $n_i(\mathbf{r}) = n_i^0 \exp(-q_i\phi(\mathbf{r})/k_BT)$, from which we deduce the *non-linear Poisson–Boltzmann equation* for the electrostatic potential $\phi(\mathbf{r})$ [231]:

$$\nabla^2\phi(\mathbf{r}) = -\frac{1}{\epsilon_0\epsilon_S}\sum_i n_i^0 q_i \,\exp\left(-\frac{q_i\phi(\mathbf{r})}{k_BT}\right). \tag{7.10}$$

This transforms the problem into a non-linear equation for the potential, which mixes concepts of electromagnetic theory with thermodynamics. This is intrinsically a *mean-field* approximation that neglects spatial fluctuations in the charge distribution. This non-linear equation (with its associated boundary conditions at the colloid surface) has, in general, no analytic solutions (except in the most elementary of cases).

DLVO solution to the Poisson–Boltzmann equation

One important feature of the DLVO theory is the simplification of the Poisson–Boltzmann equation by linearizing it, i.e. by expanding the Boltzmann factor to its first order term. The zeroth-order term of the expansion is zero thanks to the assumed electro-neutrality of the solution. The linearized equation then reads:

$$\nabla^2\phi(\mathbf{r}) = \kappa^2\phi(\mathbf{r}), \tag{7.11}$$

where:

$$\boxed{\kappa^2 = \frac{1}{\epsilon_0 \epsilon_S \epsilon k_B T} \sum_i n_i^0 q_i^2 = 4\pi \lambda_B \sum_i n_i^0 z_i^2,} \tag{7.12}$$

where in the last expression for κ^2 (which is sometimes preferred for practical estimations), we have introduced the valence z_i of each ion species ($q_i = ez_i$, where e is the elementary charge), and the Bjerrum length, λ_B [m], defined as:

$$\lambda_B = \frac{e^2}{4\pi \epsilon_0 \epsilon_S k_B T}. \tag{7.13}$$

Solutions to Eq. (7.11) (for a spherical colloid) exhibit an $\exp(-\kappa r)/r$-dependence, instead of the common $1/r$-dependence of the electrostatic potential. Its strength is, accordingly, *screened* by a factor $\exp(-\kappa r)$ because of the effect of the ions. This can be viewed in simple terms as the result of the ions of opposite charge to the colloid tending to accumulate on its surface. κ^{-1} is called the *screening length* or the *Debye–Hückel screening length*, and characterizes the spatial extent of the screened Coulomb interaction. The region around the surface, whose thickness is of order κ^{-1} and where the ions reorganize as a result of interactions with the colloid, is called the *double layer* or *diffuse layer*.

The electrostatic potential for a sphere can be solved in the linear approximation from Eq. (7.11), for a given geometry and surface charge of the colloid (macro-ion). For example, for a spherical colloid of radius a with a charge valence Z (i.e. a total surface charge $Q = Ze$), the potential is:

$$\phi(\mathbf{r}) = \frac{Ze}{4\pi \epsilon_0 \epsilon_S} \frac{\mathrm{e}^{\kappa a}}{1 + \kappa a} \frac{\mathrm{e}^{-\kappa r}}{r}. \tag{7.14}$$

The r-dependence is a consequence of Eq. (7.11), while the constant arises from the boundary condition at $r = a$ (it can for example be obtained by the application of Gauss theorem on the surface $r = a$ just outside the sphere). This expression shows that the ions in the double layer, in addition to the screening effect in the r-dependence, also result in an effective (or screened) charge valence on the colloid surface ($r = a$):

$$Z_{\text{eff}} = \frac{Z}{(1 + \kappa a)}, \tag{7.15}$$

instead of Z.

Pairwise Coulomb interaction

The previous result, obtained for a single colloid, can now be extended to the pairwise interaction between two colloids. A major assumption here is that the double layer of each colloid is not affected by the presence of the other. This approximation is bound to fail for colloid separations of the order of κ^{-1}, but is often used even in this regime since no other simple approach is available. For two identical spherical colloids of radius a and separated by a distance r (center to center), we can then use the results obtained previously. The interaction potential then takes the simple form of the *screened* electrostatic interaction between two objects with effective charge $Ze/(1+\kappa a)$, i.e.

$$V_{\mathrm{Coul}}(r) = \frac{1}{4\pi\epsilon_0\epsilon_S}\left[\frac{Ze\,\mathrm{e}^{\kappa a}}{1+\kappa a}\right]^2 \frac{\mathrm{e}^{-\kappa r}}{r}. \tag{7.16}$$

It can also be written normalized with respect to the thermal energy as:

$$\boxed{\frac{V_{\mathrm{Coul}}(r)}{k_B T} = Z_{\mathrm{eff}}^2 \frac{\lambda_B \mathrm{e}^{-\kappa(r-2a)}}{r},} \tag{7.17}$$

where λ_B [m] is the Bjerrum length given in Eq. (7.13), κ^{-1} [m] the screening length given in Eq. (7.12), $r-2a$ corresponds to the gap between the colloids, and Z_{eff} [a.d.], given in Eq. (7.15), is the effective charge valence of the colloids (which determines the maximum interaction potential in the limit of contact at $r = 2a$).

The ions in the double layer, in addition to the screening effect on the r-dependence (the factor $\exp(-\kappa r)$), therefore also affect the maximum interaction potential in the limit of contact, or equivalently the effective surface charge Z_{eff} of the colloid. This latter effect can again be viewed as the effect of a layer of ions of opposite charge accumulating on the surface and partially compensating the colloid surface charge.

Note that these expressions, since they rely on the linearization of the Poisson–Boltzmann equation as described earlier, are in principle only valid when the potential is small with respect to $k_B T$. They are however widely used outside this range of validity. Numerical solutions of the non-linear equation are possible. They show that the qualitative predictions of the linear approximation are acceptable even beyond its range of validity, but a different effective charge should be used for quantitative predictions. Within DLVO theory, Ze cannot therefore be directly compared with the real surface charge of the colloid.

Example for Lee-&-Meisel Ag colloids

Let us consider again spherical silver colloids of radius $a = 30$ nm in water (representative of the Lee-&-Meisel Ag colloids) and at room temperature ($T = 290$ K). We first need to list the charged species (ions) in solution. From the synthesis reaction (see Section 7.1.2), it is clear that we have nitrate ions (NO_3^-) at a concentration of 1.04 mM and sodium ions (Na^+) at a concentration of 2.28 mM. Assuming that all the silver was reduced, there are no Ag^+ ions left. The electro-neutrality of the solution must however be ensured by the products of the silver reduction by citrate. It is unclear how this proceeds exactly, although some studies have tried to elucidate it [221]. For the sake of simplicity in this example, we will assume that three Ag^+ ions are reduced by one citrate ion $C_6H_5O_7^{3-}$, the products being only neutral species. Starting with an initial citrate concentration of 0.76 mM, $1.04/3 = 0.35$ mM react with silver ions, leaving a concentration of 0.41 mM of unreacted citrate in solution. Because citrate is a weak triple base, it could in principle transform into citric acid (and intermediate compounds). In the absence of additional pH control, the pH of the colloidal solution described above can be calculated to be around $\sim$9, in agreement with measurements. In this case, most of the 0.41 mM remaining citrate species are in the form of citrate ions (which are trivalent and negative). In summary, the charged species (ions) in solution (at pH 9) are:

- 1.04 mM of nitrate ions, NO_3^- ($z_1 = -1$),
- 2.28 mM of sodium ions, Na^+ ($z_2 = +1$),
- 0.41 mM of citrate ions, $C_6H_5O_7^{3-}$ ($z_3 = -3$),
- and negligible concentrations of H_3O^+, OH^-, and the conjugate acids of citrate.

We conclude that, in this situation, $\sum_i n_i^0 z_i^2 \approx 7$ mM.

Moreover, the static dielectric constant of water is $\epsilon_S \approx 80$ at room temperature. The Bjerrum length is therefore $\lambda_B \approx 0.72$ nm. Converting the concentrations [mM] into the volumetric densities n_i^0 [m^{-3}] and using Eq. (7.12), we deduce the screening length as $\kappa^{-1} \approx 5.1$ nm. The screening is therefore quite strong in Lee-&-Meisel Ag colloid solutions as a result of the ionic products of the reaction. This may obviously be different for other synthesis routes.

It is common under SERS conditions to add ions to the colloidal solution to introduce aggregation (as will be discussed extensively later). For a monovalent electrolyte, such as KNO_3 (which, we assume, does not interact with Ag), this changes the screening length to $\kappa^{-1} \approx 3.3$ nm for addition of 5 mM KNO_3 (final concentration), to $\kappa^{-1} \approx 2.6$ nm for 10 mM, and to $\kappa^{-1} \approx 2.2$ nm for 15 mM.

For these four examples (ionic products with 0, 5, 10, or 15 mM KNO_3), the dependence of the pairwise screened Coulomb interaction potential (normalized to $k_B T$) with separation r is shown in Fig. 7.6(a) (the surface charge of the colloid is taken to be $-1600e$ C, or $Z = -1600$). The case of an infinite screening length (i.e. no ions in solution) is also shown for comparison. The two screening effects of the ions are clearly seen in these plots:

- a more rapid decrease of the interaction with distance because of the screening effect in the double layer, and
- a slight decrease in the effective surface charge, seen as a drop in the maximum interaction potential at contact ($r = 2a$).

7.3.4. The DLVO interaction potential

Expression for the DLVO potential

In the 1940s Derjagun, Landau, Verwey and Overbeek developed (somewhat independently) what we now know as the DLVO theory to solve the electrostatic problem of an electrolyte with charged boundaries. This theory – in its simplest form – uses the linear approximation of the screened Coulomb repulsion potential discussed above, together with the (negative) interaction potential for van der Waals attraction within the framework of Hamaker theory.

The total DLVO interaction potential $U_{\text{DLVO}}(r)$ (normalized to the thermal energy $k_B T$) for two identical spheres is therefore:

$$\boxed{\frac{U_{\text{DLVO}}(r)}{k_B T} = \frac{V_{\text{vdW}}(r)}{k_B T} + \frac{V_{\text{Coulomb}}(r)}{k_B T}} , \tag{7.18}$$

where $V_{\text{vdW}}(r)$ is the van der Waals interaction given in Eq. (7.9), and V_{Coulomb} is the screened pairwise Coulomb interaction potential given in Eq. (7.16) or (7.17). This potential is the starting point for the description (at least qualitative) of many phenomena in colloid science, and in particular colloid stability and aggregation.

Note that there is, in principle, another component of the interaction among particles in colloids, which is purely hydrodynamic; i.e. the fact that moving particles produce flows that may affect other particles. This is usually ignored within DLVO theory. Moreover, the repulsion between particles may in some situations arise from *steric hindrance* rather than electrostatic repulsion. This is the case for example when long-chain molecules (like polymers) are adsorbed on the surface. This situation is also not covered by the simplest DLVO theory. We also note that the theory of interactions of charged macro-ions (colloids) is an ongoing and active field of research, and that the approximations outlined

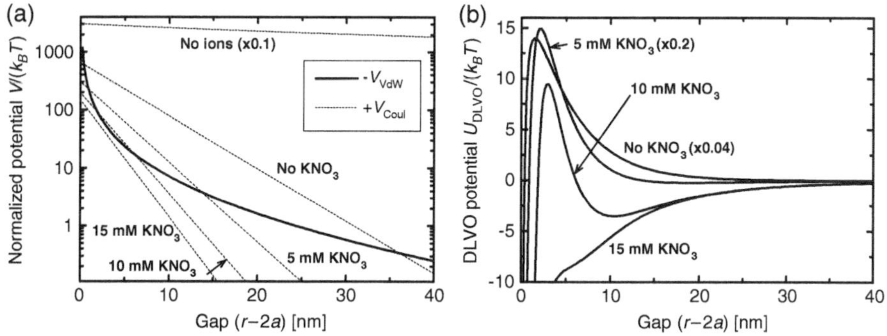

Figure 7.6. Example of application of the DLVO theory to Lee-&-Meisel Ag colloids approximated as Ag spheres of radius $a = 30$ nm. Pairwise interaction potentials are plotted as functions of the gap $r - 2a$ between two particles. (a) van der Waals interaction (shown as $-V_{\mathrm{vdW}}(r)/k_BT$, since it is negative), and screened Coulomb potential: $V_{\mathrm{Coul}}(r)/(k_BT)$ are both shown either (i) in the absence of any ions, (ii) for ionic concentrations corresponding to the reaction products of the colloid synthesis, and (iii) same as (ii) with further addition of 5, 10, or 15 mM of KNO_3. A colloid surface charge valence of $Z = -1600$ is assumed for these plots. (b) Corresponding total (Coulomb repulsion + van der Waals attraction) DLVO potential $U_{\mathrm{DLVO}}(r)/(k_BT)$. The increased screening produced by the addition of salts results in a decrease of the repulsive potential barrier that maintains stability. Two particles overcoming the positive potential hump become irreversibly bound (aggregated). At an ionic concentration of $\sim$ 15 mM the positive potential barrier has completely disappeared.

here are being revisited and reassessed all the time through different analytical and numerical methods; examples of which can be found in the literature [241–244]. Nevertheless, although many of the approximations may not always be valid, this theory provides a simple and general approach, which has been shown to agree qualitatively and semi-quantitatively with observations in many important cases.

In the following we focus on aspects of the DLVO interaction potential that are present for a wide range of chosen parameter values and form part of the basic phenomenology of colloid aggregation.

Example for Lee-&-Meisel Ag colloids

The rest of the discussion will be easier if based on an example. We therefore consider again the example of Lee-&-Meisel Ag colloids. The van der Waals and the screened Coulomb interaction have already been discussed for these. This is summarized for a few cases of interest in Fig. 7.6(a), along with the corresponding sum in Fig. 7.6(b), i.e. the total DLVO interaction potential. These curves are, of course, only indicative of the real situation which may include many complications; in particular due to the poly-dispersity in sizes and shapes. They will however serve as a useful basis for the general discussion to follow.

7.3.5. Colloid aggregation within the DLVO theory

Main features of the DLVO potential

The DLVO interaction potential in general exhibits a number of important characteristics, which are apparent in the example of Fig. 7.6(b):

- At the shortest separations, the van der Waals attraction always dominates and the DLVO potential diverges to $-\infty$. This deep potential drop is called the *primary minimum* in the jargon of colloid theory. It is responsible for irreversible aggregation. This state corresponds to the absolute minimum of the interaction potential and is therefore the normal thermodynamic equilibrium state.

- However, as clearly seen in Fig. 7.6(b), the Coulomb repulsion can strongly dominate at intermediate (but still short) gaps (at least for low ionic content). It therefore creates a large *potential barrier* preventing the system from reaching its preferred equilibrium position at $r = 2a$. In Fig. 7.6(b), the potential barrier is as high as $\approx 350\, k_B T$ for 0 mM KNO_3, and still $\approx 70\, k_B T$ for 5 mM KNO_3. The probability of overcoming such a barrier (by thermal activation) is essentially zero for all practical purposes. Although irreversible aggregation into the primary minimum is the thermodynamic equilibrium state, the potential barrier prevents the solution from reaching it, i.e. the solution is *kinetically stable or metastable.* A solution is usually considered metastable for barrier height above $\sim 15 - 20\ k_B T$. The difference between 'stability' and 'metastability' becomes then a formality for all practical purposes. The gold colloidal solutions prepared by Michael Faraday for his *Bakerian Lecture* at the Royal Society of London can still be seen today!

- The barrier height can be decreased by increasing the ionic content (i.e. by increasing the screening). Above a certain concentration, the barrier disappears completely (the van der Waals attraction then always dominates). Nothing then prevents the colloids from irreversible aggregation, forming larger and larger clusters, which eventually precipitate under gravity; this is called *coagulation.* The transition from a metastable state to a complete coagulation can be very sharp, and occurs at an ionic concentration called the *critical coagulation concentration* (CCC). In the example of Fig. 7.6(b), the CCC is just above 10 mM KNO_3.

- Just below the CCC, the potential barrier still exists but is not large enough to guarantee metastability. This is the case for example for 10 mM KNO_3 in Fig. 7.6(b), where the barrier height is of the order of $\sim 10\ k_B T$. In a first approximation, coagulation should still take place, although it may be slower. This assumes however that the interaction potential for two spheres remains a good representation of the potential for larger clusters (for example a dimer and a single sphere, etc.). If the barrier increases for larger clusters, then aggregation will stop, resulting in a metastable solution of small clusters. It has been suggested [229] that this may be the case, at least for Lee-&-Meisel colloids, at concentrations just before the CCC. The solution is then in a metastable, partially aggregated state, with small colloidal clusters. This represents a very interesting system for SERS.

- Finally, in some cases, the potential can have a shallower *secondary minimum* at larger distances after the positive hump of the potential barrier. This is seen for example in Fig. 7.6(b) at 10 mM KNO_3. The depth of this minimum is however relatively shallow (a few k_BT). This secondary minimum can be responsible for reversible binding of colloidal particles, a phenomenon called *flocculation*. This effect has not yet been evidenced for metallic colloids in a SERS context; but it is a relatively well-known phenomenon in other types of colloidal systems.

From this discussion, it is clear that DLVO theory provides a valuable tool for a qualitative description of colloid stability and aggregation. A direct comparison of the theory to experiments is not straightforward though, since some of the parameters (for example, Z) are difficult to obtain accurately[5]. Moreover, as noted earlier, the Z value that is used in the DLVO theory may not be directly equal to the real one in the linear approximation of the Poisson–Boltzmann equation. Nevertheless, it is always possible to adjust such parameters so that they predict correctly the value of the critical coagulation concentration (CCC). This is precisely what was done in the example discussed above and shown in Fig. 7.6.

Aggregation dynamics

The dynamics of coagulation/flocculation is one of the classic topics in colloid science [232], and still the source of ongoing studies [247]. We shall only sketch here the basic underpinnings of the process. Consider that there is a certain probability for two colloids to aggregate. This probability is evidently linked to – and depends strongly on – the activation energy above the repulsive potential barrier. This can be modeled as a rate constant (k_a) [s^{-1} m^3] at which the aggregation of two particles can happen, where k_a depends exponentially on the ratio of barrier height over k_BT. The initial step of any aggregation kinetics starts by forming dimers from an otherwise uniform population of single colloids. The concentration $C_1(t)$ [m^{-3}] of individual

[5] Note that Z can in principle be measured experimentally by *Zeta Potential* measurements.

colloids will decay as:

$$\frac{\mathrm{d}C_1(t)}{\mathrm{d}t} = -k_a\, C_1(t)^2. \tag{7.19}$$

The factor $C_1(t)^2$ appears because the formation of a dimer requires the presence of two colloids at the same place; if they are independent from each other this probability is proportional to $C_1(t)^2$. We assume here that this aggregation is irreversible. In fact, we can define similarly a series of concentration functions $C_n(t)$ for single colloids ($n = 1$), dimers ($n = 2$), trimers, tetramers, etc., which will have their own rate equations equivalent to (7.19). Trimers can be formed by aggregation of three colloids or aggregation of a dimer with a single colloid, etc. This forms a system of coupled differential equations for the population of clusters which can be explicitly solved [232] assuming that the aggregation rate constants k_a are independent of cluster sizes. The cluster population as a function of time then evolves according to [232]:

$$C_n(t) = C_0 \left[\frac{t}{\tau}\right]^{(n-1)} \left[1 + \frac{t}{\tau}\right]^{(-n-1)}, \tag{7.20}$$

where $n = 1$ for single colloids, $n = 2$ for dimers, $n = 3$ for trimers, etc., is the *cluster size* and $\tau = 2/(k_a C_0)$ [s] is the characteristic time for coagulation. A plot of the evolution of the populations for different cluster sizes is given in Fig. 7.7(a). As can be seen from the graph, the number of single colloids decreases rapidly, with a simultaneous increase in the number of dimers and, subsequently, in the number of trimers, tetramers, which themselves quickly disappear in favor of larger clusters. In practice, large clusters start disappearing from the kinetics of the aggregation when they cannot survive in the liquid by buoyancy, i.e. they precipitate under the effect of gravity. Overall, over a time of a few τ, the colloidal solution has fully collapsed into large clusters, often visible by eye.

The final outcome of the aggregation process depends on a variety of causes that are difficult to characterize in strict categories. A reordering of the clustered particles once they are in contact is possible in some cases. However, more often than not it is observed that colloids are kinetically trapped [232] in the original configuration thus producing fractal-like aggregation. The physical properties of fractal-like-aggregated structures of small metallic colloids have been studied in great detail for a long time with a variety of techniques [248]. An example of fractal aggregation of metallic colloids (a typical situation in dry SERS substrates) is shown in Fig. 7.7(b). Aggregated fractal-like clusters of metallic nano-particles form a class of SERS substrates in their own right (usually dried onto a planar substrate). They exhibit interesting optical properties and large EM enhancements, which have been

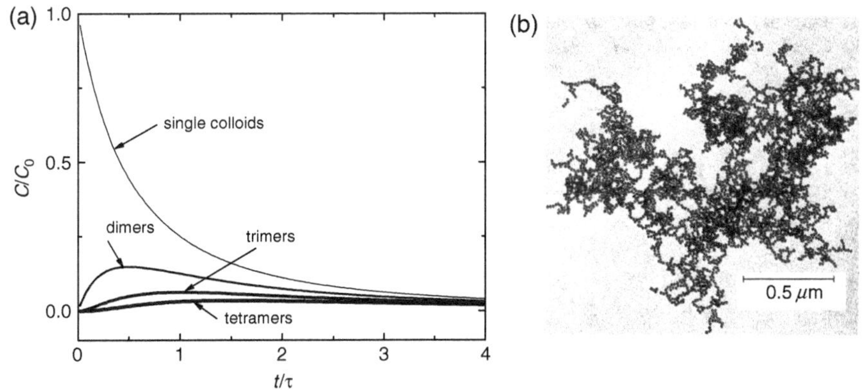

Figure 7.7. (a) Standard picture of aggregation dynamics in a colloidal suspension [232,245]. The different population of clusters evolve according to Eq. (7.20), a decrease in the population of single colloids is followed by a rapid increase in the number of dimers, and later of trimers, tetramers, etc. This eventually results in a complete coagulation, often into a fractal-like structure. This is illustrated in (b) showing the electron microscope image (from Ref. [246], reproduced with permission, copyright 1984 The American Physical Society) of the structure formed by kinetic aggregation of gold colloids in aqueous solution.

studied in the context of SERS and non-linear optical processes in some detail [249–251].

7.4. SERS WITH METALLIC COLLOIDS

Having discussed in some detail the stability of colloidal solutions, we can now tackle their use as SERS substrates. Using this as an example, it will also give us the opportunity to discuss important aspects of the problem that arise in SERS experiments with any type of substrate, like molecular adsorption or SERS fluctuations.

7.4.1. Molecular (analyte) adsorption and SERS activity

In order to profit effectively from the large SERS electromagnetic enhancements, the analytes under investigation have to be attached to or be in close proximity of the metallic surface of the substrate (by a few nm only, see Section 6.2.4). Not every molecule 'sticks' effectively to a given metal, and the *adsorption properties* of the probe under consideration can have a considerable impact on their SERS performance. This is an important basic aspect to consider when interpreting SERS experiments.

Basic principles of molecular adsorption

Adsorption is usually classified into two broad types:

- *Chemisorption*, whereby the molecule is attached chemically to the metal, for example through covalent bonding.
- *Physisorption*, when no chemical bonds are formed between the analyte and the metal, a common example being through electrostatic attractions.

Beyond this simple classification, the problem of molecular adsorption is fairly complex to approach. Firstly, the adsorption mechanisms of most molecules are still poorly understood (or not understood at all); and they are usually very difficult to study experimentally under the conditions necessary for SERS. Secondly, many parameters affect the adsorption properties, rendering systematic studies difficult. The results are therefore highly case-specific; they not only depend on the pair analyte/metal but also:

- may vary with analyte concentration (mono-layer vs multi-layer adsorption for example),
- may be affected by the presence of other species (surface charges, competition for adsorption between species, or pH-dependence), and
- may even depend on the details of the sample/substrate preparation.

As an example of the latter, consider a negatively-charged molecule in a (negatively-charged) Ag colloid solution. Any binding to the silver surface is prevented by the electrostatic repulsion and no SERS signal is observed. If now a drop of a solution of the same molecule is dried onto a planar Ag SERS substrate, then the molecule will most likely be left on the surface upon drying and a SERS signal is likely to be observed.

To add to the complexity of the problem, when a molecule is chemisorbed, it is likely that its structure is modified (or at least influenced) by the bonding to the metallic surface. It should then be viewed as a somewhat different species: a molecule/metal complex. This case was specifically discussed within the scope of the *chemical enhancement* (CE) in Section 4.8, and can have a variety of consequences: it may modify the vibrational (Raman spectrum), and electronic structure (and therefore resonance condition and Raman intensity), and may lead to additional SERS enhancement mechanisms generally engulfed in the definition of the CE (see Section 4.8 for more details).

Despite its overwhelming importance, the issue of molecular adsorption is therefore very difficult to control in general, but must at least be considered when interpreting some SERS experiments or consequences derived from them.

Charge interactions in colloidal solutions

For charge-stabilized colloidal solutions, simple electrostatic considerations can be used as a first approximation to the problem. When an analyte is

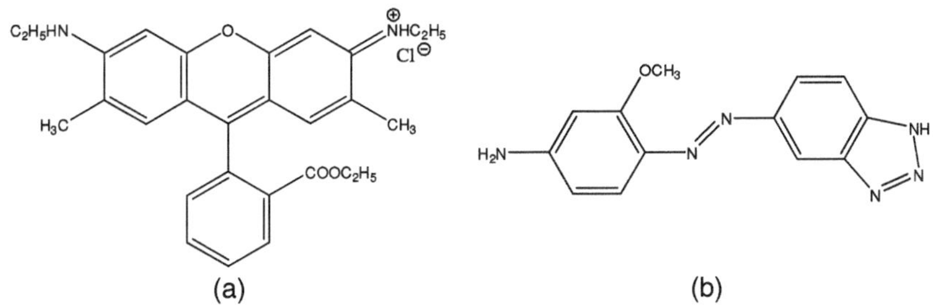

Figure 7.8. Chemical structure of (a) rhodamine 6G (RH6G), here with a counterion Cl^-, and (b) one of the benzotriazole dyes, BTZ2, developed in Ref. [89] (#2) and used in several examples in this book. The benzotriazole group is believed to bind covalently to silver (chemisorption). Further details on benzotriazole dye adsorption can also be found in Ref. [252].

added to a solution of Ag or Au colloids (assumed to be negatively charged, as in most situations) the problem is, initially, mostly electrostatic in nature:

- Negatively-charged molecular species are repelled electrostatically from the colloid surface and are unlikely to adsorb, even if they would in the absence of repulsion.
- Contrarily, positively-charged molecules are attracted to the surface, where they can for example chemisorb (by displacing some of the surface species) or simply physisorb efficiently through electrostatic bonding.
- Neutral species can also attach effectively to colloids if they have one of the many appropriate functional groups for chemical bonding to metals (such as thiol groups for attachment to gold), but the process may be slower.

In colloidal solutions, the charge problem is therefore the first consideration to take into account.

Simple examples

In fact, as a result of the importance of colloids in the development of SERS, many common SERS dyes are positively charged. This is the case for example of the widely used probe rhodamine 6G (RH6G) (shown in Fig. 7.8). Others have been specifically designed to attach covalently to silver, like the benzotriazole dyes developed in Ref. [89]. One dye of this latter family is also shown in Fig. 7.8. Other common dyes (but less common in SERS) are negatively charged and do not attach to metallic colloids. Examples include rhodamine B or fluorescein, which happens to be the dye of choice in many fluorescent-based techniques. Both dyes are similar in structure to RH6G, but are negatively charged in solution (at least at the pH of Lee-&-Meisel Ag colloids) and produce no (or a very weak) SERS signal in such solutions. This is simply a result of poor adsorption and should not be interpreted as a poor SERS enhancement factor for the substrate. In fact rhodamine B produces a SERS signal comparable to rhodamine 6G on non-charged planar substrates like the gold nano-particle arrays shown earlier in Fig. 7.4.

Effect of analyte adsorption on colloid stability

Another important aspect of molecular adsorption in the context of SERS in colloidal solutions is its potential effect on colloid stability. At analyte concentrations typical of SERS (say $< 10^{-5}$ M), the effect of the analyte on the ionic strength (and therefore on the screening length κ^{-1}) is usually negligible compared to other ions already in solution. Analyte adsorption can however strongly affect the colloid surface charge (characterized by its valence Z, assumed to be negative here) in at least two (possibly cumulative) ways:

- For positively-charged analytes, by attaching on the surface and therefore reducing the total negative charge of the colloid.
- For positively-charged or neutral analytes[6], by displacing (and replacing) the negatively-charged surface species. The adsorbate must in this case have a higher affinity to the metal than the existing surface groups.

Within the DLVO theory of colloid stability discussed earlier, this results in a smaller colloid charge valence Z, and therefore a decrease of the Coulomb repulsion. Using the example of Lee-&-Meisel Ag colloid discussed earlier, the DLVO theory predicts for example that the potential barrier (of $\approx$350 k_BT for $Z = -1600$), reduces to $\approx$8 k_BT for $Z = -600$, and entirely disappears for $Z = -400$. Once this potential barrier becomes too small, the aggregation and coagulation proceed in a similar fashion as for electrolyte-induced aggregation.

It is difficult to relate quantitatively these figures to real experiments, but semi-quantitative analysis can be performed. The analyte adsorption influence on colloid stability will depend in particular on:

- The relative concentration of analyte with respect to the colloid; or equivalently: the number of analytes per colloid. The adsorption efficiency should be taken into account in this estimate.
- The exact mechanism resulting in a change in charge (attachment only, or displacement), i.e. by how many elementary charges is the surface charge changed upon adsorption of one molecule.

---

[6] Note that it is also possible under some conditions for a negative molecule or ion to replace the negative stabilizing agent on the surface (this is the case of Cl^- discussed later), provided its affinity to the metal is much larger than that of the stabilizing agent itself. The change in surface charge then depends on the relative coverage and charge of the two species.

- The ionic strength of the solution before the analyte is added (this defines the screening length and screened charge).

As an example, a molecule like BTZ2 (see Fig. 7.8) results in coagulation of a standard Lee-&-Meisel Ag colloid solution for added concentrations of ≈ 3 μM. This corresponds to $\approx 1.7 \times 10^4$ molecules per colloid (assuming a good adsorption efficiency). The surface area of a 60 nm diameter colloid is $\approx 1.1 \times 10^4$ nm^2, and this suggests that this aggregation threshold of ≈ 3 μM must be close to mono-layer coverage of BTZ2 (≈ 1 molecule per nm^2) on the colloids. These figures suggest that BTZ2 does not bring a charge upon attachment, but does displace the citrate ions: aggregation occurs only when almost all citrate ions have been displaced (i.e. at mono-layer coverage).

As a comparison, positively-charged analytes can induce aggregation at much lower concentrations (typically $\approx$ 300 nM). In this case, every adsorbed molecule brings a positive charge, thereby canceling the negative charges of the colloids for only $\approx$1000–2000 molecules/colloid.

It is worth bearing in mind these considerations when interpreting SERS experiments. In most practical cases, one must avoid any analyte effect on the SERS substrate when possible. For colloidal solutions, it is therefore necessary to work at sufficiently low analyte concentrations to ensure that this is the case.

7.4.2. Colloid aggregation for SERS

From a SERS point of view, an unstable (aggregating) colloidal solution is undesirable because it leads to time-dependent and irreproducible results. On the other hand, the SERS enhancements obtained from small colloidal clusters are much larger than that of individual particles. The ideal system would therefore be that of a *stable* solution of small colloidal clusters (dimers would even be sufficient). The colloid aggregation step is therefore often carried out for SERS experiments, but it must be carefully controlled and studied to avoid coagulation, time-dependent results, and erroneous interpretations. This step is sometimes called *activation* in the literature, even though this term may lead to confusion.

Aggregating agents for SERS

From the previous sections, it becomes obvious that several parameters can be modified to alter the interaction and stability of the colloids. From the standpoint of SERS, there are a number of conventional routes for colloid aggregation, some of which are somewhat related to each other. These routes do not extinguish all the possibilities, but they have been frequently used in the literature and deserve some degree of attention. These are:

- The effect of the analyte itself acting as an aggregating agent by changing the surface charge Z (by attachment or displacement of charged species). This is not ideal since it is analyte-specific and concentration-dependent and should in general be avoided (if possible) by working at sufficiently small concentration, where this effect is negligible.

- Addition of a passive electrolyte such as KNO_3 (passive because the ions K^+ and NO_3^- are not thought to interact strongly with metals or any other constituents in the system). The increase in ionic concentration decreases the screened Coulomb interaction and favors aggregation. This is probably the 'cleanest' approach and its interpretation should accordingly be the simplest. One should however make sure that the resulting solution remains stable (in particular against coagulation), at least on the timescale of the experiment. Note that it is not guaranteed that there is an optimum electrolyte concentration producing partial aggregation without complete coagulation. This should ideally be assessed experimentally through careful calibration.

- Addition of an active electrolyte. In this case, the ionic strength is modified, but some of the added ions may also influence other aspects of the system. This is for example the case of KCl, which is discussed later in more detail. This is also the case of agents affecting the pH. They change the ionic strength in a non-trivial way and may affect the charge of some analytes or even the colloid surface charge (citrate for example).

- Addition of polymers or long-chain ions (for example poly(L-lysine) [221]). These may not only affect the surface charge through attachment, but also affect the stability of the sol through *steric interactions*, i.e. the colloids can no longer get close to each other because of the long-chain species on their surface. They can therefore act both as an aggregating and a stabilizing agent depending on the conditions. Under optimized conditions, one could therefore hope to obtain a stable solution of partially aggregated colloids. Such interactions can however no longer be understood within the framework of DLVO theory.

Conceptually, it is easy to understand that most of these methods of aggregation modify different parameters of the DLVO potential; i.e. either the net charge of the particles Z or the ionic concentrations n_i^0 in solution, or combinations thereof. This can be used as a basis for qualitative understanding of these effects. However, in many cases of interest, several mechanisms may operate and even interact with each other at the same time, and it is fair to say that the understanding of many aspects remains (at best) empirical. Many aspects of colloid aggregation and dynamics in liquids for SERS applications are an ongoing subject of research.

Finally, it is worth mentioning a possible alternative approach to aggregation consisting in directly synthesizing colloidal *clusters* in a controlled manner. For example, one conceptually simple approach is through chemical binding of the colloids using appropriate surface chemistry. This is however challenging and has not yet been fully explored. The use of an aggregating agent is not as clean in comparison but it is certainly much easier to carry out from a chemical point of view.

Practical approach to the problem

One of the simplest routes to colloid aggregation is the addition of salts (passive or active). It is necessary however to avoid coagulation, i.e. remain under the critical coagulation concentration (CCC), in order to avoid a complete aggregation and precipitation of the solution. Ideally, one should therefore chose a concentration just under the CCC, and ensure that: (i) the SERS signal has increased (indicating partial aggregation), and (ii) the solution remains stable, at least for the timescale envisaged for the experiment. The optimum salt concentration should therefore be determined experimentally by trial and error. This process is not always easy, mostly because of the poly-dispersity of the solution (which makes the interpretation of the results even more difficult); i.e. there is no clear transition between a stable and unstable solution. Using the SERS intensity as an indication of the aggregation state of the solution is an option but interpretations are again difficult since colloid aggregation may result in both an increased SERS signal (through the gap-plasmon effect on the SERS enhancements) and a decreased average SERS signal (because of sedimentation, i.e. falling under gravity of clusters above a certain size).

Self-limiting aggregation

For the case of the Lee-&-Meisel type of colloids, it was found [229] that a mixture of Ag colloids and 20 mM KCl in equal volume (i.e. half-concentrated colloids with a 10 mM KCl final concentration) results in a stable solution with good SERS activity.

This was interpreted using an extension of DLVO theory as a result of *self-limiting aggregation* of the colloids. Qualitatively, one has to consider the change in DLVO interaction potential for a dimer–sphere interaction compared to the standard sphere–sphere interaction. We can assume in a first approximation that the interaction potential is additive, i.e the interaction of a colloid with a dimer is simply the sum of the interactions between this colloid and each of the two colloids forming the dimer. Even within this simple approximation, the situation becomes a much more difficult many-body problem; the results of which depend (for example) on the angle of approach of the two objects. If the sphere–sphere potential is attractive (no potential barrier), the dimer–sphere potential remains attractive and aggregation proceeds toward a complete coagulation (as in the standard DLVO theory). If, however, the sphere–sphere potential has a small potential barrier of the order of say 10 k_BT as in Fig. 7.6(b) for 10 mM KNO_3, then it is conceivable that the potential barrier for the dimer–sphere interaction will be, by additivity, larger (say 20 k_BT). If this is the case, the formation of dimers is then not negligible, and single colloids are not stable. But it will become less and less probable (exponentially) to add an additional particle to the cluster because of the increased barrier height for dimer–sphere interactions and beyond. This is the principle of the self-limiting (Coulomb blocked) aggregation studied in Ref. [229]. It basically shows that partially aggregated colloids can have, under the right conditions, long term stability.

7.4.3. Focus on the 'chloride activation' of SERS signals

An important example of intertwined effects triggered by the chemical variables of the problem is the so-called 'chloride activation' of SERS signals.

There is sufficient experimental evidence to show that specific ions, like halide ions (and in particular chloride Cl^-), might be playing a special role in the 'activation' of SERS signals. For the specific case of Cl^- this is dubbed sometimes in the literature as 'chloride activation' of SERS signal[7]. Many experiments about this subject have been carried out on colloidal solutions, where the aggregating effect of ions is not always accounted for in the interpretations of the results (and in those cases where it is accounted for, it is always subject to uncertainties). There are however other experiments without colloids, where aggregation cannot confuse the results. Overall, there is a convincing case that halide ions like Cl^- or I^- play more than just one role in the explanation of the SERS signal observed experimentally. On the one hand, they contribute to the aggregation of the colloid in the form of a salt but, in addition, there may also be an 'anion-enhancement' effect [253]. We now discuss separately these two effects.

Colloid aggregation with chloride ions

There is a consensus in most of these studies that Cl^- ions attach to the metal surface efficiently (and therefore displace the stabilizing agent, for example citrate in the case of Lee-&-Meisel colloids). Any chloride-containing electrolyte should therefore be considered as an *active electrolyte*. It not only affects the ionic strength but also other components in the system, in particular the colloid surface charge.

[7] The terminology is confusing, and activation of a colloidal solution sometimes simply refers to the addition of salts to produce aggregation. The 'chloride activation' should be here understood as acting *in addition* to the conventional aggregation effects.

Let us discuss again the case of Lee-&-Meisel colloids for the sake of argument, and consider the addition of KCl to the solution. The amount of Cl^- that attaches to the Ag surface (and replaces citrate) will be at most in the sub-mM range (because of space restrictions on the colloid surface). The change in ionic strength is therefore the same as the one that would be obtained from KNO_3. By attaching to the surface, Cl^- may in addition affect the surface charge Z in a non-trivial way. In particular, it could even increase the (negative) charge. Describing in more detail the different aggregating effects of KNO_3 and KCl probably requires tools beyond the DLVO theory along with carefully designed experiments. It is however clear that KNO_3- and KCl-aggregated colloids are a different system, starting with a different surface charge, and it is therefore no surprise from this point of view that one of them (KCl) results in a more SERS active substrate; as indeed evidenced in experiments.

Additional anion enhancement

Explanations for the additional anion-enhancement effect have been discussed in some detail in the literature [253,254]. They fall into two broad categories:

- A modification of the adsorption properties of the SERS analyte as a result of Cl^- being adsorbed on the surface. This is certainly true for rhodamine 6G, for example. No SERS (or a very weak) signal is observed in the solution in the absence of Cl^-, which must therefore play a role in 'helping' the adsorption of the analyte to the surface. In addition to modifying the affinity, Cl^- ions may also modify the adsorption geometry of the analyte, which would potentially affect the SERS signal through surface selection rules.

- A modification (increase) of the chemical enhancement contribution to the SERS signal. This effect can be viewed as the creation of a surface complex involving Ag, Cl^-, and the analyte, whose electronic properties are more resonant (through an increased charge-transfer contribution, see Section 4.8), thereby resulting in a larger SERS signal.

In both cases, the anion-enhancement effect should be analyte-dependent and should therefore not be considered as part of the fundamentals of SERS. 'Chloride activation' of the SERS signal *is not a pre-requisite* for most SERS experiments, as presented too often. But it plays an unquestionable important role for specific SERS probes.

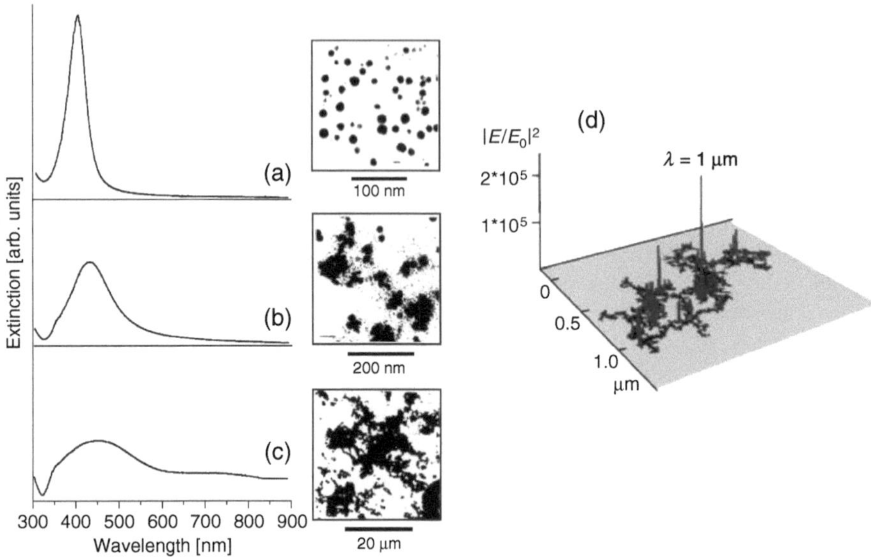

Figure 7.9. (a–c) Electron microscope images of fractal-like SERS active colloidal aggregates (Ag) (middle), with their corresponding extinction spectra (left) reproduced from Ref. [255] with permission, copyright 2002 IOP Publishing Ltd. The change in extinction spectra with contributions from coupled-LSP resonances in the clusters is evident. The relation between extinction and SERS enhancement is, however, indirect. (d) Spatial distribution of the enhancement factor in a fractal-like Ag colloid cluster [249–251] (reproduced with permission, copyright 2001 MAIK "Nauka/Interperiodica"). There are only a finite number of 'hot-spots' where the maximum electric field is highly localized. See Refs. [249–251] for further details on the numerical techniques used in these calculations.

7.4.4. SERS from 'dried' colloidal solutions

Fractal-like colloidal aggregates

As already mentioned, 'dried' colloidal solutions form a class of SERS substrate in their own right, half-way in between colloidal solutions and ordered planar substrates. They can be fabricated with ease, for example, by drying a drop of colloidal solution on a glass slide or silicon substrate. Coagulation, if not induced before with an electrolyte, will eventually occur upon drying. The analyte to be detected can also be added before drying, if desired.

Thanks to the strongly aggregated state of the colloid, large SERS enhancements, both for single-molecule EF and average EF, are typically observed on such structures. The presence of coupled-LSP plasmon resonances (gap-plasmon resonance) in these clusters can be seen in the extinction spectra, as depicted in Fig. 7.9(a–c) for different states of aggregation. Fig. 7.9(a–c) shows a clear red-shifted extinction band, compared to the position of the single-colloid plasmon resonance at $\sim$ 400 nm. This, in principle, does not say much about their SERS activity, since the connection between extinction and SERS enhancement is indirect; as pointed out before. However, as discussed in Section 6.4, gap-plasmon resonances are typically associated with highly localized regions of large EM enhancements. The nature of these 'hot-spots' in fractal-like structures has been investigated extensively theoretically [249–251]. Figure 7.9(d) shows an example of a calculated highly localized gap-plasmon resonance in a fractal-like cluster of the sort normally obtained by aggregation of metallic colloids.

These large SERS enhancements would make such structures appear as the ideal SERS substrate. Unfortunately, they tend to be extremely 'inhomogeneous', in particular: (i) the structures resulting from colloid coagulation and/or drying are very non-uniform, even on fairly large scales; this non-uniformity is magnified for the SERS signals (or enhancements), (ii) the SERS enhancements are very susceptible to small changes in the geometrical structure. Unless special care is taken to graft the colloids to the substrate, small movements of the colloids (possibly induced by the laser through heating) can substantially affect the SERS signal, and (iii) the largest EM enhancements may result in photo-bleaching or photo-destruction of the probes, possibly the source of further instability. This effect is particularly important for a fixed substrate, where the same molecules are measured for long periods of time (as long as the laser remains focused on the same spot). In short, non-uniformity and non-reproducibility render SERS experiments on these substrate very difficult to interpret at a quantitative level in many cases.

Other fixed planar SERS substrates

Fixed planar SERS substrates are nevertheless important for SERS fundamentals and applications. They offer the possibility of making repeated measurements of a well-defined spatial area, an invaluable tool for systematic studies. As a comparison, only statistical results can be obtained from SERS in solutions, because of the constant Brownian motion.

'Dried' colloids and metal-island films are the most common examples of such planar substrates but they both suffer from some the problems listed above. New types of planar SERS substrate, based on *ordered* arrays of nano-structures have been developed to remedy some of these issues. Recent approaches toward uniform arrays of well-defined nano-particles will be discussed in the next chapter. The approach of grafted colloidal clusters, discussed earlier has played an important role in many fundamental SERS studies. The principle is to use a substrate coated with a positively-charged layer, for example poly-L-lysine, and transfer a small number of colloidal clusters from a partially aggregated solution onto it. This then enables one to carry out SERS experiment on isolated individual clusters, using a

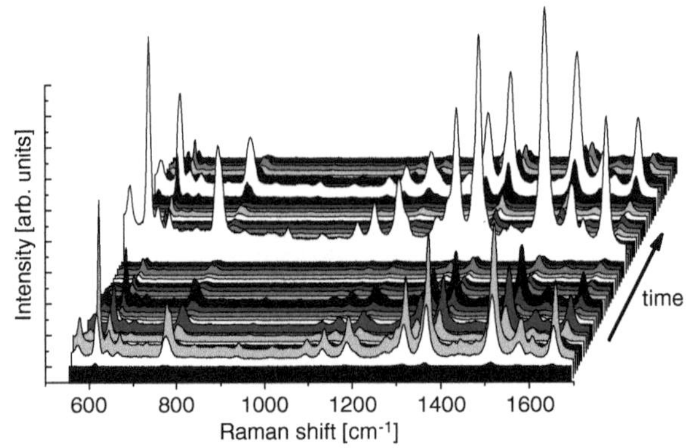

Figure 7.10. Time evolution of SERS spectra of RH6G in water. Each spectrum has been taken with 0.5 sec integration time and 3 mW at 633 nm. The sample contains Lee-&-Meisel type Ag colloids at 10 mM KCl concentration, and 100 nM RH6G. The data have been taken with a ×100 immersion objective.

high magnification objective. In this way, it is possible to study the optical properties of the individual clusters in relation to their SERS activity, and even to their morphology if the technique is combined with microscopy (for this one needs appropriate markers to match SEM images to optical spectra). The non-uniformity amongst clusters remains, but it is not an issue for single-cluster studies. Many of the most important developments in single-molecule SERS have, in fact, been achieved in substrates of this type [30,34,36].

7.4.5. SERS signal fluctuations

SERS signal fluctuations are an important feature of SERS, and have played a significant role in many of its early and current developments. It is also a critical issue for any analytical chemistry application. Moreover, it is related to claims of single-molecule sensitivity at very low dye concentration, and will be discussed in this context in Section 8.1. SERS fluctuations may refer to fluctuations in intensity or spectral shape (Raman peak positions). We discuss here some aspects of this issue from the standpoint of SERS substrates and colloidal solutions.

SERS fluctuations in liquids

Experiments in solutions are typically done with immersion objectives that are index-matched to water[8]. Figure 7.10 shows a typical time evolution of

[8] It is also possible to collect the SERS signal from the water–air interface; at the expense of losing some signal from the index of refraction mismatch at the interface and the (naturally) smaller numerical aperture typically used in these latter cases.

the signal observed experimentally at moderate dye concentration (100 nM RH6G) in colloidal solution.

The fluctuations in signal are primarily produced by the Brownian motion and diffusion of colloid clusters through the scattering volume of the objective. As discussed in 7.2.3, this occurs on a timescale of $\sim$0.1–1 s (depending on the specific objective under use) and is, therefore, easily detectable with conventional Raman systems. The SERS signal intensity should depend in this case on:

- Whether one (or more) cluster(s) is (are) present or not in the scattering volume during the integration time, together with the average diffusion coefficient.
- Where exactly in the (non-uniform) exciting laser beam (typically a Gaussian beam) the cluster is.
- The number of analyte molecules on the cluster (often related to its size and analyte concentration).

These factors are not really associated with the SERS effect itself, but rather with the *dynamics* of the underlying SERS substrate (colloidal clusters).

There are however a number of additional factors, which change from one cluster to another as a result of poly-dispersity, and are more directly related to SERS, for example:

- The plasmon resonance positions and their associated EM enhancements (single-molecule SERS EF and/or average SERS EF). Note that since the EM enhancements are wavelength dependent, these may affect the various Raman peaks differently, i.e. changes in plasmon resonance conditions from one cluster to another are associated with *both intensity and spectral shape fluctuations* [44].
- The orientation of the cluster with respect to the incident polarization. This affects the coupling of the incident light with the plasmon resonances and is (in a sense) related to the previous point [57].
- In single-molecule conditions, the orientation/position of the molecules with respect to the hot-spot(s) [162] ; see Section 8.1 for more details.

For this second group, the fluctuations induced by Brownian motion provide a simple and powerful means of sampling the SERS properties of individual clusters one at a time. This is in some ways equivalent to measurements on grafted 'dried' colloidal clusters, one at a time. In liquid, we have only one chance to probe a cluster (the time it takes for it to pass through the scattering volume). This shortcoming is counterbalanced by the fact that thousands of events (individual clusters) can be recorded in a short time (say 15 minutes). The two approaches are therefore complementary. It is also possible to affect the dynamics of Brownian motion by introducing a 'thickening agent' like glycerol, which will increase the viscosity of water at room temperature and slow down the dynamics of the colloids (see Eq. (7.6)).

SERS fluctuations on fixed substrates

Most of the fluctuations observed in liquids have their origin in the Brownian motion and poly-dispersity of the colloids and since they cannot be deconvolved from possible additional causes, they should, strictly speaking, not be called SERS fluctuations. Similar fluctuations may appear for fixed SERS substrate when moving the laser spot on the surface. Such spatial fluctuations are equivalent to the time fluctuations in liquids; they most likely originate in the non-uniformity of the substrate. For a fixed excitation on a fixed SERS substrate, one could expect these fluctuations to disappear. This is however not always the case, and some 'real' SERS fluctuations do remain. These are temporal fluctuations occurring for the SERS signal measured at a given position on a fixed substrate. The details of such SERS fluctuations are still not well understood but it is relatively easy to identify a number of possible causes:

- *Changes in the substrate morphology*
  'Fixed' planar substrates may not always be as fixed as one may think at first, especially if we look on scales of a few nanometers. For grafted colloidal clusters, minute movements (on the nanometer size scale) of one of the colloids may affect the coupled-LSP resonance condition and therefore the SERS signal. Even in the most stable substrates, large laser intensities, further magnified by the local field EM enhancements, can result in substantial heating and associated changes, for example, in the roughness of the surface. Experimental evidence for colloid heating [120] and substrate annealing [256] have recently been put forward. The understanding of the 'mechanical movements' of nano-particles induced by changes in the environment (like laser heating) is very primitive at this stage, and mostly based on indirect experimental evidence.

- *Changes in the molecular (analyte) configuration*
  These include desorption, re-orientation, diffusion, chemical transformation, or destruction of the analytes on the surface. These effects can, for example, be induced directly by the laser (photo-chemistry or photo-bleaching), or indirectly by the substrate heating produced by illumination (resulting, for example, in desorption, re-orientation, or surface diffusion). An important aspect of all these effects is that they can be 'global', i.e. affect all the molecules on the surface at the same time. They are therefore not restricted to single-molecule SERS conditions.

- *Single-molecule effects*
  Any of these effects will, however, be magnified under single-molecule SERS conditions, i.e. when the SERS signal is dominated by only one or a few of the adsorbed molecules, typically the one at an electromagnetic hot-spot (for example in a gap). For instance, if a molecule residing at a hot-spot (and contributing to the biggest fraction of the signal) photo-bleaches, then the SERS signal will disappear. In addition, *surface diffusion* of molecules in-and-out of hot-spots could, in this case, also contribute to the fluctuations. This is generally believed to occur (at least for some physisorbed analytes) but has not yet been proven conclusively. Many of these effects are most of the time hidden under the general classification of 'blinking', even though in many cases the exact origin of the fluctuations is not known or is difficult to measure. Blinking phenomena have been associated with the 'single-molecule nature' of the signal, but this connection is not necessarily strong. A change by $\sim$1 nm in the position of a colloid that was participating in a hot-spot could have dramatic consequences on the measured SERS spectrum, even if the signal comes from many molecules.

The exact origin of SERS fluctuations should therefore be assessed on a case-by-case basis, if possible experimentally (for example by varying power density or changing analytes). They may often have, in addition, more than one cause operating at the same time. Nevertheless, it is quite clear that *SERS fluctuations are not necessarily an indication of single-molecule SERS conditions*, as has been very often assumed in many reports. This is will be re-emphasized further in Section 8.1.

SERS fluctuations and applications

It is also clear that SERS fluctuations can be avoided, or at least minimized, with a suitable choice of SERS substrate, analyte concentrations, and laser power. This is critical for many applications.

The first source of fluctuations, Brownian motion in liquids or spatial non-uniformity on fixed substrates, can be avoided by improving the uniformity of the substrate itself. In addition, some degree of averaging of these fluctuations can be obtained by increasing as much as possible the integration time in liquids and/or the scattering volume (or area on a planar substrate).

Moreover, as a rule of thumb, in order to avoid the remaining sources of fluctuations, one should also choose SERS substrates that do not exhibit the largest enhancements (like those provided by gap-plasmon resonances). These are typically very sensitive to small modifications of the substrate geometry and are more likely to produce SERS signals dominated by a few molecules, which can itself be the source of further fluctuations. For applications, a SERS substrate with a relatively large average EF, and relatively small maximum single-molecule EF (i.e. no hot-spots) is the best option.

Chapter 8

Recent developments

It is fair to say that, in essence, the main principles of SERS have been well understood for 20 years or more. A quick look at some of the early reviews in the field [4] – ten years after the original discovery [1–3] – reveals that there was already a feeling that the fundamental aspects of SERS had been thoroughly explored and understood to a large degree.

Nevertheless, this chapter concentrates on important recent developments in the field of SERS. While none of these new developments requires a ground-breaking modification of the underlying principles established more than 20 years ago, they do represent milestones that set new directions and created new questions on both older and newer problems. They triggered a substantial amount of research activity and, to a large degree, invigorated the field in its quest for a deeper understanding of the basic principles themselves. Like many of the subjects treated in this book, the definition of 'important' is subjective and open to debate. We have chosen topics that appeal to the present authors for a variety of reasons, but the list is not claimed to be exhaustive. By the same token, the definition of 'recent' is also somewhat arbitrary. By *recent important results* we mean results that have appeared in the literature during (roughly) the last decade, and made considerable impact on SERS research and its potential applications. Each topic is presented in a separate self-contained section.

8.1. SINGLE-MOLECULE SERS

8.1.1. Introduction

Undoubtedly, one of the most interesting developments in SERS in the last decade has been the realization and demonstration that a single molecule can be detected with SERS. In 1997, two independent reports [30,31] claimed the observation of single-molecule emission under SERS conditions. It was concluded that SERS enhancement factors could be as large as $\sim 10^{14}$

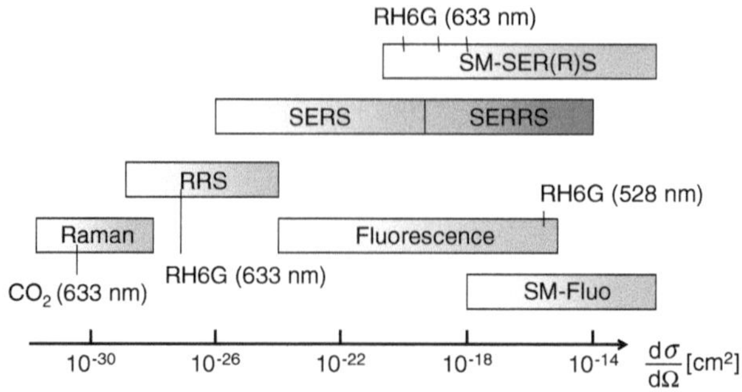

Figure 8.1. A comparative graph for the different cross-sections in different versions of Raman experiments (inspired by a similar figure in Ref. [69]). RRS=resonant Raman scattering; SE(R)RS=surface-enhanced (resonant) Raman scattering; and SM=single molecule (SERS or fluorescence). Single-molecule detection requires cross-sections which are at the top end of what is achieved in SERS and/or SERRS or fluorescence. Single-molecule detection can be more easily achieved at lower cross-sections with SERS than fluorescence, thanks to the much sharper Raman peaks.

(although this figure is misleading, and has been critically reassessed in Ref. [8]) to compensate for the intrinsically small Raman cross-sections, and that SERS probes could potentially replace fluorescent ones in several applications; for example in biology. These reports undoubtedly triggered a renewed interest in the technique in general.

The appeal of single-molecule spectroscopy is huge for many reasons, but two of them certainly include: (i) the possibility to push analytical tools to their ultimate resolution limits, and (ii) the understanding of unique single-molecule phenomena that are potentially washed out by ensemble averages. In the particular case of SERS, it was also an important argument for the intrinsic 'competition' with fluorescence spectroscopy, which had already achieved single-molecule detection at the time. As pointed out in Chapter 1, some advantages of SERS over fluorescence are its higher spectral specificity and the possibility of using infrared excitation (important in potential biological applications in living tissues, for example). It is also hoped that studies of single-molecule SERS (SM-SERS) could also lead to a better understanding of the SERS effect itself.

A comparative graph depicting the expected ranges of cross-sections for Raman processes in different situations (also including fluorescence cross-sections for comparison) is shown in Fig. 8.1. This shows that single-molecule detection is only observed under very special circumstances that require cross-sections at the top end of what is normally achievable.

Not many researchers in the field of SERS today will doubt the ability of the technique to measure (and thereby identify) single molecules in a variety of experimental situations. The path to single-molecule SERS (SM-

SERS) detection was not free of controversy and problems though; and it is still the subject of active research. SM-SERS has encountered many of the same problems found in SERS in general: fluctuations in the signals, lack of reproducibility, and a lack of understanding of the origin and conditions for SM-SERS (in terms of enhancement factors and physical–chemical variables of the problem), as well as a lack of control over the magnitudes and spatial location of hot-spots. A major problem in the early claims was the fact that single-molecule emission was inferred from indirect evidence, casting doubts over the reality of SM-SERS in the first place, and giving rise to alternative explanations and room for skepticism. This is perfectly summarized in the title of a discussion [257] by several of the authors of the first reports on SM-SERS: 'Single Molecule Raman Spectroscopy: Fact or Fiction?'. They stressed at that point that the inference of SM-SERS from their results was not straightforward, and that although much evidence supported the claim, it did not constitute a proof in the absolute sense. It was only after a large volume of evidence had been gathered over the years that claims of single-molecule detection started to be believed and understood. With this in mind, it is worth presenting the various experimental evidence of SM-SERS, stressing the virtues and limitations of the different approaches.

8.1.2. Early evidence for single-molecule detection

In order to put SM-SERS in context, we first review the evidence based on the original reports [30,31]. This presentation is largely inspired by the introductory section of Ref. [34].

By far, the largest group of evidence comes from studies of SERS on *dry silver colloidal particles* [30,32,258,259] of the Lee-&-Meisel type described in Chapter 7, and mostly with resonant or pre-resonant dyes. Silver colloids mixed with analytes at *ultra-low concentrations* are immobilized by drying or spin-coating on a suitable substrate. SERS signals from individual colloids or clusters are then collected and analyzed. Other types of SM-SERS studies were carried out in liquid colloidal solutions [31,260,261], where the SERS intensity fluctuations are then measured. The analyte concentration in both cases is chosen so that there are a *small number of analytes per colloids*, typically $\sim$0.1 molecule/colloid. The single-molecule nature of these signals can then be tentatively inferred from a number of characteristics, the most important of which being the ultra-low concentration.

Ultra-low concentrations

The low analyte concentration in such experiments suggests that, statistically speaking, there cannot be much more than one molecule per colloid [30,32,258], or one molecule in the scattering volume if several colloids are present [31]. If this is the case, then any SERS signal that is observed must originate from a single molecule.

This reasoning depends crucially on how the analyte/colloid concentrations and/or scattering volume are estimated. However, it was acknowledged early that these concentration estimates do not necessarily provide a satisfactory proof [30,257]. They are indeed prone to large errors: as commented in Chapter 7, colloid concentration is usually estimated from a knowledge of metallic mass used during preparation, for example in the citrate-reduced Lee-&-Meisel Ag type of colloid [220], and an estimate of their average size (or volume). Any non-reacted Ag, for example, or the presence of a small number of much-larger-than-average particles could lead to an overestimation of colloid concentration and therefore a dramatic underestimate of the analyte:colloid ratio. Moreover, analyte concentrations well below 1 nM (in the pM-range, as often used) require particular care to avoid contamination, wall adsorption, and dilution errors [262].

A further source of uncertainty is the fact that only a small proportion of colloid aggregates (the so-called 'hot' particles) seem to give rise to SERS signals. This means that there is a possibility that those active aggregates are the ones that have adsorbed a larger-than-average number of analytes because, for example, they have a larger effective surface area, or are composed of many individual colloids [30].

Because of these uncertainties, further statistical arguments must be used to back up the single-molecule claims. However, this is difficult in general *because the surface of the particles do not have a uniform enhancement.* Therefore, only a *very* small fraction of the molecules, and therefore at ultra-low concentration a small fraction of particles, actually emit a detectable SERS signal. This was indeed observed in Ref. [30]. As a result, the number of observable single-molecule events is a small fraction of the already small number of molecules. The resulting reliability of the statistics is extremely poor and then leaves the door open to other possible interpretations (like more selective adsorption of analytes by certain particles, or the formation of dye aggregates). The point of the *statistical reliability* is important. If the convolution of ultra-low concentrations and inhomogeneous enhancements results in the observation of events that are *extremely rare*, it is always possible to argue that these events are rare for another (rare) reason. This is particularly true for colloidal particles (like the Lee-&-Meisel colloids), where the control of size and shape is not perfect (as demonstrated in Chapter 7). There could be many reasons why some *very rare* particles show unusual behavior. The statistics then becomes unreliable and with it the claim of single-molecule detection.

The unquestionable merit of the pioneering ultra-low-concentration studies, however, was that they *did* provide the hint that SM-SERS was a real possibility. They were however not conclusive as an absolute proof of single-molecule detection, and do not constitute a practical approach to further study SM-SERS because of the poor statistical reliability.

SERS fluctuations

Another argument often put forward as evidence of SM-SERS is the fact that SERS signals exhibit strong fluctuations of various types:

- Intensity fluctuations with possible blinking (alternating on/off periods),
- Spectral shape fluctuations, in either the relative intensities of the peaks, or the peak positions (Raman shifts) and widths.

However, it is also true that such fluctuations are often observed in SERS, as discussed in Section 7.4.5, even in conditions of high analyte concentration where the signal is not believed to originate from single molecules. As a matter of fact, the intensity fluctuations can simply be attributed to changes in the underlying SERS enhancement factors and are expected for both single- and many-molecule signals. Similarly, changes in the relative intensities of the peaks are also expected if the underlying LSP resonance profile is changing [8,44,47], irrespective of the number of molecules producing the signal. Slight differences in the frequency positions and widths of the Raman peaks from one spectrum to the other are perhaps the strongest hint of SM-SERS. They are typically attributed to the slightly different chemical environments in which each molecule finds itself. But, one more time, such changes can also be observed in many-molecule conditions, for example as a result of substrate heating [120].

Particularly striking in the context of SERS fluctuations was the observation of 'sudden jumps' in the SERS spectra. Figure 8.2 shows an example from Ref. [30]. These sudden changes are often considered as a characteristic of SM emission [258,259,264], and usually attributed to surface diffusion of a single molecule in-and-out of a hot-spot or sudden photo-bleaching of the molecule [264]. Similar fluctuations have also been observed in the SERS spectra of residual amorphous carbon on the colloids, and attributed to ongoing photo-induced chemical reactions on the surface, such as photo-oxidation [10,265]. In fact, the case provided by amorphous carbon is perhaps one of the best examples to prove the existence of ongoing photo-induced processes in SERS, which can potentially include not only modifications of the molecules themselves, but also changes in the geometry of the hot-spots producing the signals.

This highlights one important aspect: SM-SERS experiments are typically carried out in conditions of large SERS enhancement factors (a necessity for single-molecule detection). Many additional effects may then arise: photo-bleaching of the dyes, photo-desorption, photo-induced surface diffusion, substrate heating [120], and possibly substrate morphology changes (through photo-oxidation for example), or even laser forces. Because many of these effects are cooperative, i.e. they are likely to affect all molecules at the same time (for example when the metal reaches a critical temperature), they could,

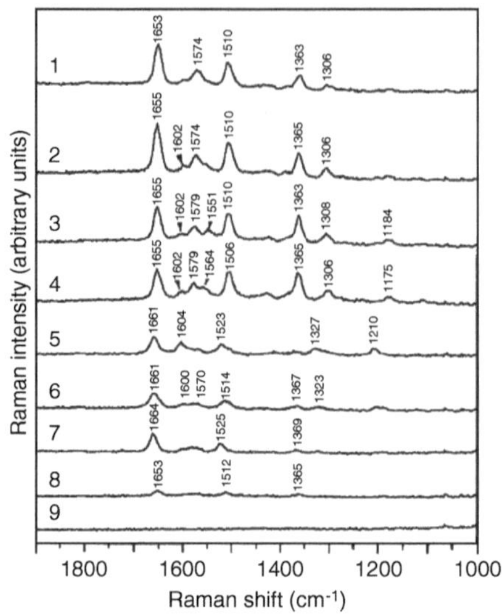

Figure 8.2. Sudden jumps in the Raman spectrum of rhodamine 6G as a function of time from Ref. [30] (reproduced with permission, copyright 1997 American Association for the Advancement of Science). Spectra have been recorded at 1 sec intervals. In this particular time sequence the intensity and frequency of the peaks changed abruptly three times, as shown in spectra 2, 5, and 8. See Ref. [30] for further details. Sudden changes in Raman intensities [263] and/or blinking/intermittent phenomena [258,259,264] have been repeatedly reported in the literature as 'signatures' of SM-SERS.

in principle, contribute to the SERS fluctuations and to blinking, even in many-molecule conditions.

Therefore SERS fluctuations of any type (intensity or spectral shape) and/or blinking, although expected and even magnified in SM-SERS conditions, do not represent by themselves a conclusive proof of SM-SERS.

Polarization studies

Other indirect evidence of single-molecule detection was put forward, for example by studying the polarization characteristics of the SERS signal (in the limit of very low concentration of dyes) and in small clusters. One of the first examples [30] of this type of study is illustrated in Fig. 8.3. In view of more recent developments in polarization studies under SERS conditions, the interpretation of such experiments was however most likely flawed. The crucial point is that the polarization effects in SERS are primarily governed by the EM coupling of the incident and emitted fields to the substrate LSP resonances and the resulting local field polarization [49]. They therefore depend mostly on the SERS substrate itself, not on the analyte(s) producing the SERS signals. Polarization studies, in their simplest form, cannot therefore constitute a proof

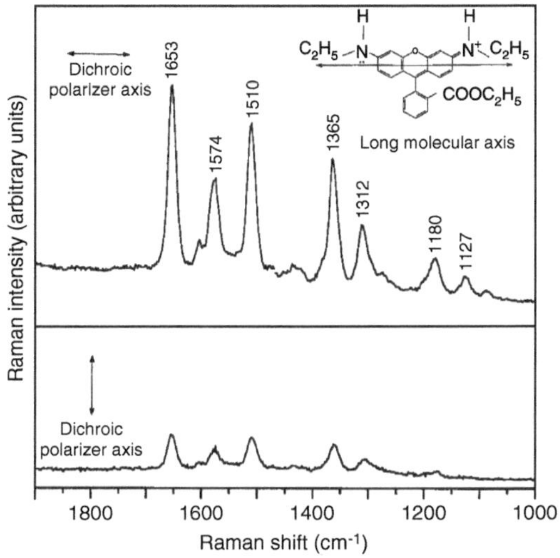

Figure 8.3. Polarized SERS spectra for two perpendicular analyzers from Ref. [30] (reproduced with permission, copyright 1997 American Association for the Advancement of Science). The presumed orientation of the molecule with respect to the incident polarization is shown explicitly. Similar polarization dependencies have also been reported in other studies [54,56] and were taken as a signature of single-molecule detection. Such an interpretation is, however, flawed because it ignores the fact that polarization effects are mostly dictated by the substrate geometry. This is by far the overriding effect for most SERS polarization studies [49], and is independent of the number of molecules.

of single-molecule detection. In fact, polarization studies in SM-SERS are at a very embryonic state still. The conditions under which some information on the orientation of single molecules can be inferred from polarization studies in SM-SERS are still being intensively debated in the literature. This is a specific area of SM-SERS that is likely to see further advancements in years to come.

Quantized SERS intensities

The original report of SM-SERS observation in liquid was further supported by the apparent observation of a Poisson (and therefore quantized) distribution of the SERS intensities [31]. The first problem with this type of argument again is that the small number of events over which the statistics is carried out (typically ~100) is not significant enough to rule out other distributions and/or interpretations. Moreover, and most importantly, as pointed out in Ref. [31], such a Poisson distribution would require a very large uniformity in the SERS signals (or enhancements), which is highly unlikely in most SERS experiments. In fact, nearly every single molecule in the scattering volume would have to be detected (and moreover be subject to the same SERS enhancement factor). These are very stringent conditions that are unlikely to be satisfied in real experiments. As pointed out recently [266], this is in contradiction with the findings of some of the other original reports of SM-SERS where only a small number of 'hot' particles gave rise to SM-SERS [30].

The existence of Poisson distributions of quantized SERS intensities, and their interpretation in terms of SM-SERS, were recently reviewed in detail [35,162]; we summarize here the main points only.

In order to have a Poisson distribution of SERS intensities, we would have to have molecules that produce exactly the same signal every time they are detected. This requires (i) to have very uniform SERS enhancements (within a factor of less than ~2), and (ii) to detect all events more or less at the same position in the scattering volume (to avoid the inhomogeneous nature of the beam from playing a role in the statistics of intensities). To these primary factors, there is a long list of additional aspects (discussed in Ref. [35]) that can conspire against a homogeneous intensity for each single molecule. But these two basic conditions are very hard to achieve experimentally for a start. Poisson distributions are therefore unlikely to be observed except, possibly, in a specifically designed and extremely challenging experiment (which has not yet been reported) where all these aforementioned factors are carefully controlled.

Moreover, there is in fact a simple explanation to the apparent observation of a Poisson distribution: the presence of 'structure' in the distribution of signals can simply be a consequence of the lack of enough sampling in the statistics, as pointed out in Ref. [162]. Based on the discussion of the EM enhancements at hot-spots in Section 6.4, we expect the SERS enhancement factors to exhibit a highly skewed long-tail distribution. When taking a small random sample of $N \sim 100$ intensities from any such distribution, a structure resembling a distribution with 'peaks' is generally observed. This is illustrated in Fig. 8.4 on a simple example. Any claim of a real Poisson distribution of SERS intensities must therefore be backed by a much larger sample statistics (~1000 events at least) to be credible.

Why do we need other approaches to SM-SERS?

It is clear from the previous discussion that most evidence put forward in the original reports of SM-SERS did not constitute a clear-cut proof and were subject to controversy or skepticism. Despite the uncertainties, this body of early evidence had led to the general acceptance that SM-SERS was a real phenomenon, in particular on dry substrates, even if no absolute proof was available. The early approaches to SM-SERS relied primarily on ultra-low analyte concentrations, which more or less ensured that only a few molecules were observed at a time. There are two major drawbacks with this approach:

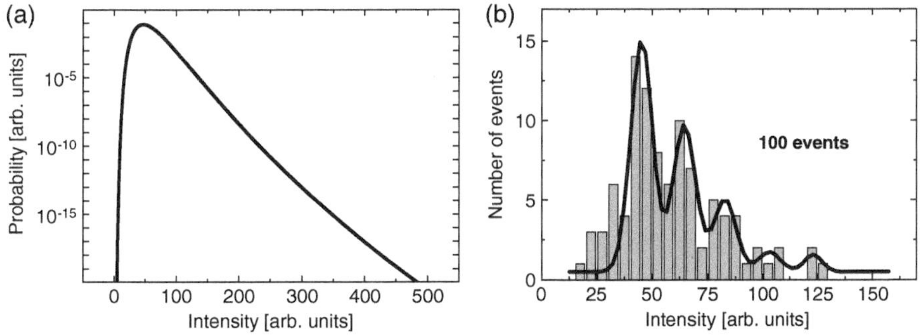

Figure 8.4. (a) A log-normal distribution with a most probable intensity at 50 and a long tail extending to high-intensity exceptional events (note the logarithmic scale for the vertical axis). Any 'long-tail' distribution function with the same characteristics will produce similar results. In (b) we show the result of taking 100 events at random and assigning them an intensity from the distribution given in (a). The lack of statistical significance in the tail of the distribution reveals itself as oscillations, i.e. an apparent 'discretization' of the signal. A tentative (and physically erroneous) fit to the data using 5 Gaussians is also shown on the histogram. Similar problems can undermine the statistical relevance of the intensity analysis in SM-SERS in liquids at ultra-low concentrations [31,260].

- Firstly, the concentration estimates may sometime be very inaccurate, casting doubts on the interpretations.
- Secondly, the probability of having a molecule at an active site is very small, leading to unreliable statistics, and making again the SM-SERS interpretation of the signals very difficult (insufficient sampling).

There was therefore a need to develop other approaches to address these two issues, two of which will be discussed in the following sub-sections.

8.1.3. Langmuir–Blodgett monolayers

An important alternative approach, which attempts to address the first point (unreliable concentration estimates), has been put forward by Aroca and coworkers in Ref. [267] and is based on the use of Langmuir–Blodgett (LB) monolayers . The technique of Langmuir–Blodgett monolayers is, in fact, quite powerful as a means to manipulate the spatial concentration of analytes in SERS experiments.

In Ref. [267], the spreading solution was calculated to achieve (on average) one molecule of the dye bis(benzimidazo)-perylene (azoPTDC) per micrometer square in the host molecule of the film, which was eicosanoic (or arachidic) acid: $C_{19}H_{39}COOH$ (AA). The LB film is then transferred onto glass slides coated with silver islands. The films are then tested for SERS signals with a Raman microscope and spatially-resolved data can be obtained. Examples of SM spectra are shown in Fig. 8.5(a) (adapted from Ref. [267]).

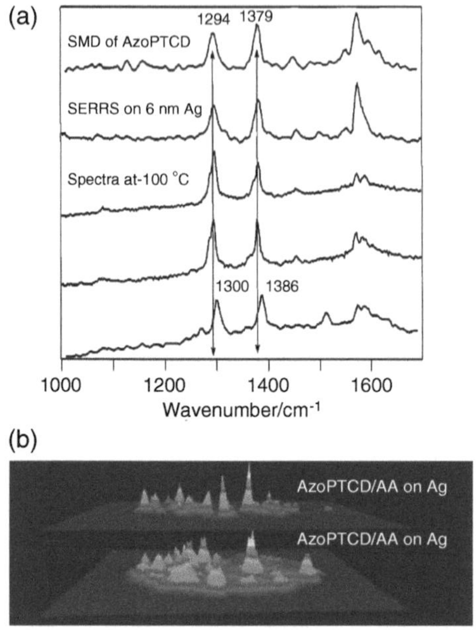

Figure 8.5. SM-SERRS on Langmuir–Blodgett films. Figure adapted from Ref. [267] (reproduced with permission, copyright 2001 Am. Chem. Soc.). (a) Individual SERRS spectra of single molecules on Langmuir–Blodgett films extracted from a mapping. The spectra are taken at −100 °C. The presence of small fluctuations in peak shapes and frequencies are attributed to signatures of the SM-detection limit. In (b) a 'global image' obtained by expanding the beam is shown. Isolated signals attributed to single molecules within the illuminated area can be seen.

It should be noted that in the specific experiment shown in Fig. 8.5, the laser is in resonance with the analyte (azoPTDC), so the experiment is formally under surface-enhanced *resonant* Raman scattering (SERRS) conditions.

The SM-SERS nature of the signals is, as before, inferred from the ultra-low-dye concentration (at most one in the scattering volume). The argument is however much stronger than in previous studies thanks to the much greater control over the dye concentration allowed by the LB deposition. In addition, these spectra show several of the characteristics normally expected for (but not constituting a proof of) SM-SERS: there are for example slight differences in the frequency positions of the peaks and their shapes from one spectrum to the other, as seen in Fig. 8.5(a). Here these can be more reliably attributed to the single-molecule character of the signal.

Another interesting insight into the nature of the signal can be obtained in this case by a global Raman-mapping technique shown in Fig. 8.5(b). In this technique, the laser beam is expanded out of focus while the signal is still coming from the focal plane. The Raman spectrum is obtained not by filtering the scattered signal with a monochromator but rather by band-pass filtering the collected image with a series of tunable edge filters. This technique only works well if Raman peaks are visible well above any type of background signal. The image is then projected onto a CCD where it is acquired. In this method, the CCD is used as an 'imaging device' in the same sense as it is used in a photographic camera. Figure 8.5(b) (adapted from Ref. [267]) shows a global Raman mapping in the Langmuir–Blodgett SERS samples. The presence of localized signals within the spot is evident from the image. It is also possible to use a standard Raman-mapping technique (rastering of a computer-controlled X-Y stage) and LB films to study the spatial localization of the Raman signals, and the breakdown of ensemble averaging when the single-molecule limit is approached as in Ref. [268].

These experiments still do not resolve the problem of the convolution between the random analyte position and the highly localized hot-spots; i.e. in order to be able to see a SM-SERS signal it is necessary for the dye to 'find the hot-spot'. Correlating the statistics of observed signals with dye concentration, potentially leads to the same problems as found in earlier studies. Still, the Langmuir–Blodgett films do provide undoubtedly a far more reliable experimental control and a much better proof of SM-SERS detection than inaccurate molecule/colloid-ratio estimates achieved by plain mixing.

8.1.4. Bi-analyte techniques

A novel approach to SM-SERS based on the use of two analytes was proposed in Ref. [34]. Its main advantage is that it addresses the second drawback of ultra-low concentration approaches (i.e. unreliable statistics). It also provides a much more direct, and therefore more reliable, proof of single-molecule detection in SERS.

Motivation

The main reasons for the unreliable statistics at ultra-low concentrations can be traced back to some very general properties of electromagnetic hot-spots (see Section 6.4), which we shall summarize briefly here. Typical SERS substrates, and in particular those with large SERS enhancement factors, exhibit a large non-uniformity in the EFs on the surface. Regions of large EFs are very localized and occupy a small fraction (typically less than $\sim 1\%$) of the metallic surface area available for molecular adsorption. If ultra-low concentrations are used (to ensure the presence of a single molecule and no more on the substrate), then the odds that the molecule adsorbs at a hot-spot (and is therefore detectable) are very small, hence the poor statistics. It is therefore highly desirable to increase the analyte concentration in order to increase this probability. One must however keep it low enough to avoid having two or more molecules adsorbed at a hot-spot (thereby losing the SM-SERS nature of the signal). Concentration estimates already have their problems, and since the exact nature (and in particular surface area) of hot-spots is not known, estimating the probability that a molecule adsorbs at a hot-spot is even less reliable. What is needed is therefore a method to assert *directly and experimentally* the single-molecule nature of the signal, not relying on concentration estimates. This is precisely what the *bi-analyte SERS* (BiASERS) method provides [34].

Principle

The method consists in measuring the SERS signal from a mixture of *two* molecular species with distinguishable SERS spectrum and (preferably) comparable SERS cross-sections (for a given laser frequency). If the concentration is such that there are many molecules at hot-spots, the SERS signal should in principle always be a mixture of these two analytes. The observation of a SERS signal of purely one type of analyte (say dye 1) is clear evidence that it comes from a very small number of molecules. For example, if the signal originates from exactly 5 random molecules, the probability of it being purely dye 1 is only 1/32, going down to 1/1024 for 10 molecules. By carrying out several BiASERS experiments at decreasing concentrations, one should in principle observe a cross-over from a regime where only 'mixed' signals are observed (many-molecule SERS) to one where only 'pure' (non-mixed) signals are observed (SM-SERS). The non-mixed signals may then be attributed to either single molecules (or a few molecules) of the same type. At intermediate concentrations, both types of signals may be observed. Such a study should allow one to determine for a given system (SERS substrate, etc.) the largest concentration where most SERS signals still originate from single molecules. It turns out that this concentration is typically much larger (by a factor $\approx$100) than that used in ultra-low concentration studies, because the majority of molecules adsorb at non-hot-spot locations and do not contribute to the SERS signal. The SM-SERS nature of the signals is nevertheless ensured by the direct experimental verification provided by BiASERS. This experimentally-determined concentration is then the optimum one for further studies of SM-SERS, since a much larger number of SM-SERS events can be recorded and analyzed.

The main advantage of this method, in addition to its simplicity as a direct proof of SM-SERS, is that it provides a route to a much better sampling of the statistics of SM-SERS events.

Experimental demonstration

The BiASERS technique was proposed and demonstrated in Ref. [34] for liquid and dry Ag colloidal solutions. The technique has since been implemented in slightly different variations [8,36,269,270], including in combination with Langmuir–Blodgett monolayers [271]. We illustrate it here, following largely Refs. [34,269].

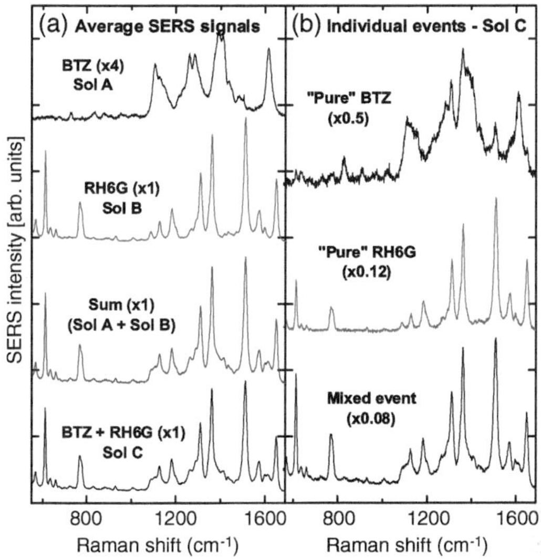

Figure 8.6. (a) Average SERS spectra from solutions A (100 nM BTZ), B (100 nM RH6G), and C (100 nM of each dye). Also shown is the sum of spectra from A and B, which is identical to the spectrum of solution C within experimental errors. (b) Representative individual spectra (integration time 0.2 s) of solution C showing a 'pure' BTZ event ($p_B = 0.91$), a 'pure' RH6G event ($p_B = 0.07$), and a mixed event ($p_B = 0.5$). The arbitrary scale is the same on both sides. The pure BTZ event still shows very small traces of RH6G due to its larger relative cross-section. Figure reproduced from Ref. [34] with permission, copyright 2006 Am. Chem. Soc.

Citrate-reduced Lee-&-Meisel silver colloids (as described in Chapter 7) are used in these experiments [220]. The two dyes under consideration here are rhodamine 6G (RH6G) and a benzotriazole dye[1] (BTZ), both shown in Fig. 7.8. Under standard conditions at 633 nm excitation, the two dyes show strong SERS spectra, which are easily distinguishable, as shown in Fig. 8.6. Of particular interest are non-overlapping peaks that allow unequivocal identification of the different dye species. Further experimental details are given in Ref. [34].

Let us briefly summarize some of the important results of these BiASERS experiments and how they can be used to support the idea of SM-SERS. We follow the main points of the discussion in Ref. [34] by focusing on a colloidal solution prepared with a mixture of equal concentration (100 nM for each dye) of RH6G and BTZ (solution C). At the same time, control samples with 100 nM of BTZ only (solution A) or RH6G only (solution B) are also

[1] The benzotriazole dye is synthesized following the procedure described in Ref. [89] (dye # 2 of this reference). It is believed to adsorb strongly (covalent bonding) to silver [89] through the triazole group.

measured for comparison. Simple estimates show that this corresponds to at least ~1000–1200 dyes of each type per colloid, a considerably larger number than anything used in the SM-SERS studies discussed earlier. A surface density of around ~0.1 dye (of each type) per nm^2 can also be estimated, indicating that steric hindrance for adsorption is not present or is negligible (typical surface area of an adsorbed dye is 1 nm^2). There is also on average between ~1 and 4 colloids (i.e. ≈1 colloid aggregate) at any given time in the scattering volume in these experiments. A series of 10^3 SERS spectra with 0.2 s integration time were collected from each solution and analyzed. The results are summarized in Fig. 8.6. The average signals of solutions A and B show that RH6G and BTZ have clearly distinguishable SERS spectra. The RH6G spectrum is stronger due to its higher SERS cross-section under these conditions. The spectrum of solution C shows a superposition of these two spectra, and is identical to the sum of the spectra from sol A and B, within experimental errors. This strongly indicates that (as expected from the low surface densities) the two dyes do not interact with each other and adsorb independently on the colloids. This 'independence' condition is an important one for a cross-talking between the two dyes would obviously result in a more complicated (correlated) statistics of the signals.

The analysis is subsequently focused on 10^3 individual spectra obtained from solution C. Most spectra exhibit a good signal, indicating that interacting colloids are on average always present in the scattering volume, and fluctuations in intensity and spectral shape are observed. These are attributed to constantly changing colloid configurations in the scattering volume produced by the unavoidable Brownian motion, and are not necessarily – by themselves – a sign of SM-SERS. More interesting are the large fluctuations observed *in the relative proportion of signals from each of the individual dyes*. For example, in Fig. 8.6(b) are shown two representative scans where the signals are composed of purely one or the other type of dye. This shows unambiguously that the SERS signal is dominated by a very small number of molecules, and represents the simplest and most direct evidence for SM-SERS sensitivity.

Similar experiments have also been carried out on dry colloidal substrates [34] and give further insights into the original SM-SERS experiments of Ref. [30]. The dye concentration is large enough to ensure population of most hot-spots by (at least) one molecule. A Raman map, as given in Ref. [34], then suggests that *all* clusters are active at these concentrations and several of them exhibit signals composed purely of the SERS spectrum of one type of dye, i.e. single- or few-molecule SERS. This is also a clear indication that SM emission is actually the norm rather than the exception in these cases. Such a conclusion could only be inferred statistically from the original low-concentration studies because of the small probability of finding a molecule at a hot-spot. However, the rarity of these events is not due to the small number of hot-spots or 'hot-particles', but to the small probability of having

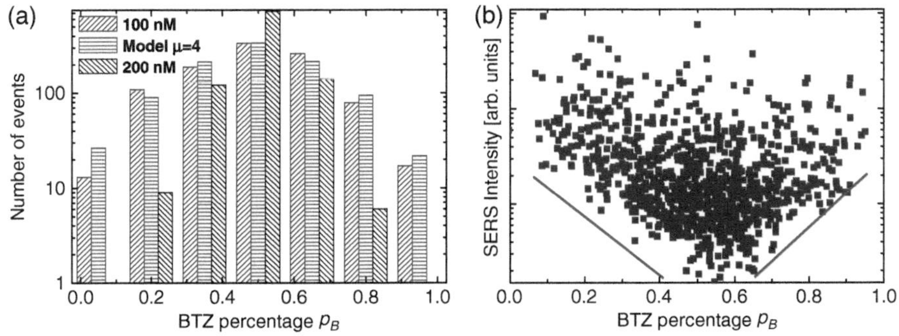

Figure 8.7. (a) Histograms of the distribution of p_B for solution C (100 nM of each dye), for a simple model with an average of $\mu = 4$ molecules per hot-spot, and for an identical solution with 200 nM of each dye. A log-scale is used to emphasize extreme events. The histograms in (A) for 200 nM lose the 'wings' of the distribution; i.e. $p_B \sim 0$ or ~ 1 (representing single-molecule events of one dye type or the other). This is, of course, expected as the concentration of both dyes increases. (b) Scatter plot of total SERS intensity vs p_B obtained from the fits. Note that spectra dominated by one type of dye ($p_B < 0.2$ or $p_B > 0.8$) are only observed for high intensity events. The figure is reproduced with permission from Ref. [34], copyright 2006 Am. Chem. Soc.

a molecule at the right place. This is easily achievable in BiASERS (due to the large concentration of dyes), while retaining at the same time the ability to distinguish 'few-molecules' from 'many-molecules' signals.

Additional benefits

Based on the previous discussion, the main advantage of the BiASERS approach is that it provides a route for improving drastically the number of measured SM-SERS events, while still ensuring their single molecule (or at least few-molecule) nature. This can be used for further analysis of SM-SERS events; for example as an experimental probe of the nature of SERS hot-spots.

In this context, the average BiASERS spectrum and the 10^3 spectra of the mixed solution (solution C) were fitted in Ref. [34] as a weighted superposition of the average spectra of BTZ and RH6G (obtained from solution A and B). The fit for the average spectrum leads to a 1:1 superposition of the average RH6G-only and BTZ-only spectra, and this is assigned to a 1:1 dye ratio. In this manner, the weighted fits enable us to extract a *single percentage* p_B characterizing the proportion of the total signal in each spectrum originating from BTZ. If the enhancement mechanism were uniform, this percentage would correspond to the proportion of molecules producing the observed BTZ signal. Accordingly, a fit of the average spectrum gives $p_B = 0.5$ (1:1 dye ratio) even if the integrated intensity is dominated by RH6G peaks (because of its higher cross-section). This simple procedure, therefore, acts as a *normalization* condition for the different cross-sections of the dyes.

The statistics of p_B is illustrated in Fig. 8.7 in two forms: (a) Histograms of the probability distribution of p_B, and (b) correlation plot of p_B with SERS intensity. p_B represents the proportion of BTZ molecules if the signals from each molecule were perfectly uniform and equal to the average signal. Because there are in excess of $\sim$1000 molecules of each type on each colloid, one would expect that $p_B \approx 0.5$, with negligible fluctuations around this value. Figures 8.6(b) and 8.7(a) show that this is clearly *not* the case, with several events where $p_B \approx 0$ or $p_B \approx 1$. The most likely explanation is that, at least for these events, the signal is dominated by a few molecules; those situated at the position of highest enhancements. As the concentration increases, the distribution of p_B should be increasingly peaked around 0.5, and the probability of extreme events ($p_B < 0.2$ or $p_B > 0.8$) should decrease drastically. This can clearly be observed experimentally as shown in Fig. 8.7(a) in the histogram for a solution identical to C, but with doubled concentrations for each dye (200 nM). It is clear that the occurrence of extreme events has virtually disappeared. Further conclusions can be obtained from such histograms, as detailed in Refs. [34,269]. For example, in Ref. [34], the hot-spot area is deduced to be only $\approx$0.17% of the total surface area. Such an estimate is almost impossible with conventional low-dye-concentration methods.

Experiments with mixtures of analytes can therefore provide further experimental evidence into the SM-SERS problem. These experiments also reveal something about the spatial inhomogeneity of the enhancement distribution, which is a key factor in the understanding and interpretation of the intensity fluctuations. The model used up to here assumes that the characteristics of the hot-spots are the same for each event, and that every molecule in the hot-spot contributes equally to the signal, which are clearly rough approximations. For example, Fig. 8.7(b) presents a clear indication that the nature of the hot-spots changes from one event to the other. If large SERS events correspond simply to instances where more molecules are present in the hot-spot area, then these events should be more likely to be of a 'mixed' type, while low intensity events should exhibit more of the extreme 'pure' dye type. The results suggest the opposite: 'pure' events only occur for high intensity events. *This suggests a strong correlation between the size of a hot-spot and its enhancement.* Such a correlation is actually predicted by all the theoretical considerations of the SERS electromagnetic enhancement treated in Chapter 6, where high enhancements are correlated with strong localization. This is another characteristic of the effect that is very difficult to prove experimentally under normal situations.

Further improvements

More sophisticated analysis tools can in fact be used for the analysis of SM-SERS events in BiASERS experiments, like those developed in Ref. [269].

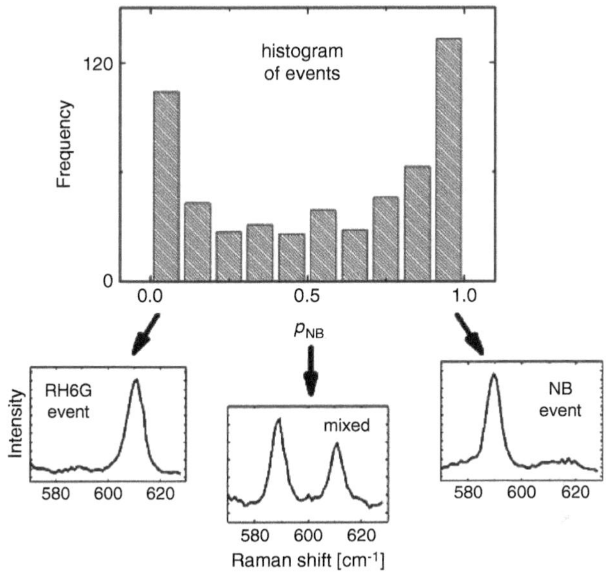

Figure 8.8. Example of bi-analyte SERS with RH6G and nile blue (NB) as BiASERS partners. The histogram (top) of relative contributions to the total signal from one of the dyes (NB in this case) shows that the largest frequency of events happens for pure single-molecule NB or RH6G events, with a smaller number being mixed events (NB+RH6G). The histogram is obtained by analyzing the intensities of the 612 cm^{-1} and 590 cm^{-1} modes of RH6G and NB, respectively, using the technique of modified principal components analysis [269]. Representative spectra contributing to different regions of the histogram are shown at the bottom for: pure NB, mixed NB + RH6G, and pure RH6G events. The figure has been adapted from Ref. [269] (reproduced with permission, copyright 2007 Am. Chem. Soc.), where further details of the analysis method and the experimental conditions can be found.

It is also advantageous in many situations to consider only two Raman peaks (each characterizing one of the two probes) that are closely spaced in energy. This avoids any influence of the dispersion (wavelength dependence) of the underlying LSP resonance. This can, for example, be implemented using rhodamine 6G (RH6G) and nile blue (NB), which have distinct SERS peaks at 612 cm^{-1} and 590 cm^{-1} respectively. This is illustrated in Fig. 8.8. The dye concentration there (2 nM) is much lower (but still larger than in ultra-low-concentration studies) and the histogram clearly shows that most signals are single-molecule SERS. These conditions correspond to the true single-molecule SERS regime, as defined in the supplementary information of Ref. [269].

Isotopic labeling of dyes

It is worth highlighting here a promising approach to improve the BiASERS method: a recent proposal and demonstration of the use of *isotopically edited probes* as BiASERS partners [36]. With isotopic substitutions, the change in

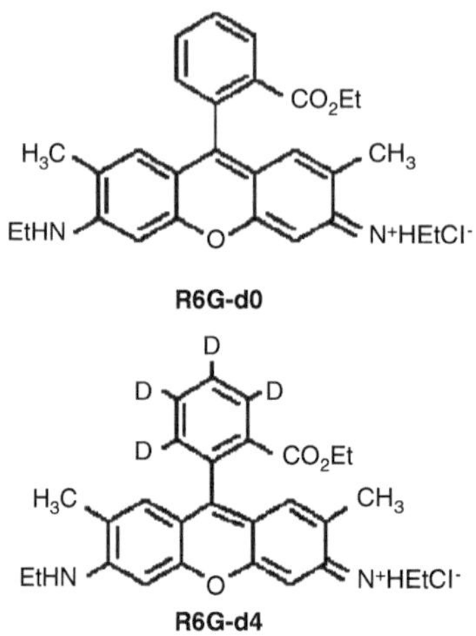

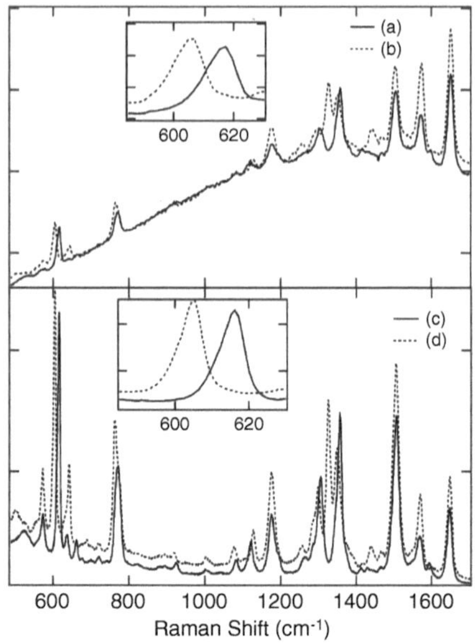

Figure 8.9. Isotopic labeling of rhodamine 6G for SERS applications, adapted from Ref. [272] (reproduced with permission, copyright 2005 Am. Chem. Soc.). The isotopic substitution of hydrogen for deuterium is not done on the main 'xanthene core' of the molecule, but rather in the phenyl moiety where only four hydrogens are replaced (left). On the right, the Raman spectra of R6G-d4 (dashed lines) and R6G-d0 (full lines) are shown for two different excitation conditions: 514 nm (*a* and *b*), and 633 nm lasers (*c* and *d*). The insets show the shifts of the $\sim$612 cm^{-1} mode of R6G-d0 upon isotopic substitution for both excitation conditions. The change of four mass units in the phenyl moiety is enough to achieve a measurable shift of key fingerprint Raman peaks. These peaks can then be used in a bi-analyte SERS experiment to study single-molecule SERS fluctuations and their statistics [36].

adsorption properties of the probe is expected to be minimal (if any at all). Still, isotopically substituted dyes can have measurable differences in their Raman spectra (in one or several fingerprint modes). Isotopic substitution therefore provides two probes with *exactly the same chemical properties but different SERS spectra*, which makes them the ideal candidates for BiASERS or multiple-analyte techniques in general.

Reference [272] provides an example of isotopic substitution in rhodamine 6G (RH6G) and its effect on SERS peaks. A substitution of the four hydrogens of the phenyl moiety by deuterium achieves a measurable change in the Raman shift of fingerprint Raman modes of RH6G, as shown in Fig. 8.9. In particular, the $\sim$612 cm^{-1} mode of RH6G, which has a strong participation of the phenyl moiety in the corresponding eigenvector [61], experiences a measurable shift that can be distinguished very easily from normal RH6G, as can be appreciated in Fig. 8.9. Note that exactly the opposite happens to the $\sim$1510 cm^{-1} Raman mode, which experiences no measurable isotopic shift, in accordance with the lack of participation of the phenyl moiety in the corresponding eigenvector [61].

With the demonstrated ability to detect single-molecule events in SERS, the idea of having probes that have exactly the same (nominal) chemical properties but different Raman spectra is quite clearly appealing, and will surely deserve more attention in the future. Isotopic substitution in molecules is a field with a long tradition in optical spectroscopy, but it has only now started to be applied in SERS-related effects and, in particular, in the framework of SM-SERS with bi-analyte methods. Refs. [36,37] demonstrate clearly its relevance in this context, and one can expect studies of this type to expand in the near future.

Bi-analyte SERS: Summary

To summarize the results of this sub-section, a mixture of two distinguishable probes circumvents many problems associated with low-concentration studies and enables the study of most SM-SERS events, instead of a very small number with unreliable statistics. This technique is simple, unambiguous, and of wide applicability to various SERS substrates. The results shown here readily demonstrate that single-molecule SERS is common, even at relatively high analyte concentration because of the hot-spot localization effect. It could further be used to study a number of outstanding issues in SERS, which we have only briefly outlined here. For example, it could shed new light into the nature of SERS hot-spots themselves, and can also be applied to determine the SERS cross-sections and enhancements with more accuracy (by increasing our confidence on events that can be truly labeled as single-molecule ones).

8.1.5. Single-molecule SERS enhancement factors

One of the most important contributions of the BiASERS method (besides increasing the reliability in the identification of single-molecule events in general) has been the possibility of estimating SERS cross sections of single molecules well beyond the level of accuracy known before. This is important for many reasons but, above all, to resolve the large discrepancies found in the literature on the question of 'how much enhancement is actually needed to see a single molecule'. This was discussed briefly in Section 4.2 and a full discussion on these topics has also been provided in Ref. [8], from where we sketch only the main points here.

Estimating single-molecule enhancement factors (SMEF), as compared to other types of EFs like the *analytical EF* (see Section 4.1.2), is in general much more complex. Amongst the problems, it is usually difficult to ensure that the SERS signal is, indeed, coming from a *single molecule*. The

BiASERS method can make a definite contribution, compared to the ultra-low-concentration approach, by allowing a sound statistics while retaining the ability to distinguish SM-SERS events.

It is important to re-emphasize an obvious point here, which is that there is no such a thing as a *unique* SMEF for a given substrate. One can see in general many different SM events with different EFs in a given substrate, reflecting the variety of enhancement situations that can be found in any typical SERS substrate. This 'spread' of SMEFs will depend on a variety of variables including: (i) the exact position (and orientation) of the molecule on the surface (and within the non-uniform exciting beam), (ii) the LSP resonance conditions with respect to the incident wavelength, (iii) incident polarization, etc. The resulting spread of SMEFs therefore reflects a characteristic range of EFs over which single-molecules can be observed. Two values, however, stand out as the most interesting ones: (i) the maximum SMEF achievable, which can be estimated from a sufficiently large pool of SM-SERS events (only possible with the BiASERS method), and (ii) the lowest EF that would still allow the observation of a SM-SERS event above the noise level, for a given experimental condition. Of particular interest (for historical reasons) is the *maximum* SMEF, for it has been the subject of a long-standing debate and the source of unrealistic claims over many years (see Section 4.2). Figure 8.10 shows two examples of SMEF determination, based on the BiASERS method with RH6G and nile blue (NB) as BiASERS partners, for two different types of substrates. The 612 and 590 cm^{-1} modes of RH6G and NB, respectively, are used for the BiASERS analysis; their respective contributions are transformed into an apparent differential cross-section by comparison with a reference compound (2B2MP). Note that these apparent cross-sections are real single-molecule cross-sections only in the cases of single-molecule events.

Determinations of the SMEFs require the measurement of the non-SERS Raman cross-section of the analyte for normalization (see Sections 2.2.8 and 4.2) and a thorough characterization of the scattering volume of the exciting/collecting optics. All these issues are discussed extensively in Ref. [8]. We shall only discuss in what follows the experimental results (and their meaning) in Fig. 8.10.

In the example of partially-aggregated colloidal Ag solutions, the results in Fig. 8.10(left) show that the maximum SMEF for the 612 cm^{-1} mode of RH6G is in the range $\sim$5–8 $\times$ 10^9, corresponding to SERS differential cross-sections of $\sim$3–5 $\times$ 10^{-18} cm^2/sr. Similar results are obtained for dry (on silicon) grafted colloidal clusters (see Chapter 7) as shown in Fig. 8.10 (right). Longer integration times ($\sim$1 s) with much smaller laser power densities are possible in this second case (compared to the liquid) because the sample does not diffuse. It is however more difficult to gain statistics in this second case, for only $\sim$10 aggregates can be seen in a typical line scan in a sparse sample, while $\sim 10^3$ spectra would be seen in an equivalent time in the liquid. Still, with this limitation in mind, the maximum SMEFs are comparable to those observed

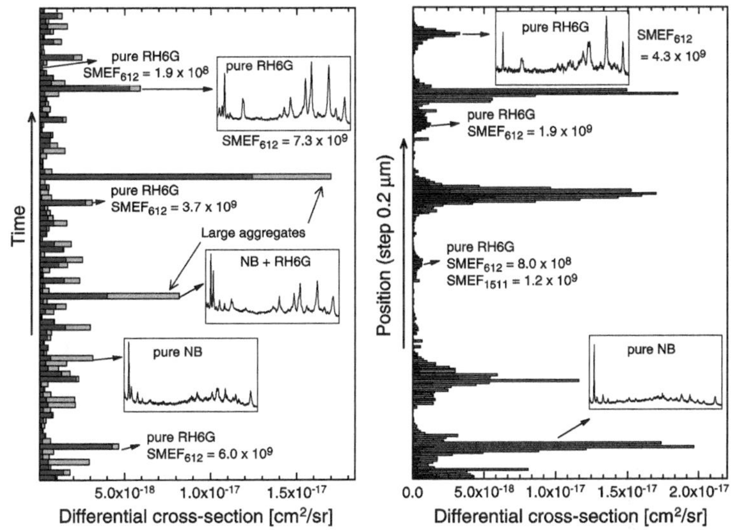

Figure 8.10. Examples of single-molecule SERS cross-section and enhancement factor measurement with the BiASERS method (using RH6G and NB as BiASERS partners); adapted from Ref. [8] (reproduced with permission, copyright 2007 Am. Chem. Soc.). Each bar corresponds to a BiASERS spectrum and the two colors in the bars represent the respective contributions of RH6G and NB to the total SERS signal. On the left, an example is shown for a colloidal solution (Ag colloids, Lee-&-Meisel type, see Chapter 7). The BiASERS spectra change as a function of *time* (vertical axis), thanks to the Brownian motion of the colloidal clusters in the liquid. On the right, a BiASERS experiment with the same dyes is performed on a *dry* sample (Ag colloids grafted on a Si wafer). In this case, the spectra change as a function of distance (vertical axis), through a linear scan using a motorized mapping stage for microscopy. This explains the different 'widths' of the single SM-SERS events in both cases. By a normalization of the signal with respect to a reference sample, the differential SERS *cross-sections* of different SM-SERS events (identified thanks to BiASERS) can be obtained. By normalizing these values with respect to the non-SERS cross-section of RH6G, the corresponding single-molecule enhancement factors (SMEFs) for the 612 and 1511 cm^{-1} peaks can be deduced.

in liquids. In addition, such experiments indicate that SMEFs of the order of 10^7–10^8 are typically sufficient to observe single-molecule SERS of resonant or pre-resonant molecules [8,273]. These are 6 or 7 orders of magnitude lower than what has sometimes been quoted in the literature!

8.1.6. Single-molecule SERS: Discussion and outlook

Pinning down a reliable demonstration of single-molecule detection in SERS is interesting for more than one reason, and this justifies the variety of approaches that have been followed in the literature. Firstly, it is the obvious academic interest of understanding the real ultimate resolution and detection capabilities of the technique but, in addition, fundamental questions like the origin of the enhancement itself rely also on it; as well as the interpretation of many experiments that depend on the assumption of single-molecule detection. Developments along those lines in SERS can only become a reality if the nature of SM-SERS is properly elucidated, without uncertainties in variables like the enhancement factors spanning ∼6–7 orders of magnitude. In a way, the story of the quest for single-molecule detection in SERS reveals the best and the worst of the advantages and limitations of the technique. Most of the inherent complexities of SERS with respect to non-enhanced situations have been thoroughly summarized and discussed in the literature [274], and the SM-SERS technique has to overcome not only these intrinsic complexities but also control certain variables to ensure reproducibility.

The current consensus, after a decade of research, is that SM-SERS is indeed possible and may even be more common than first expected. This capability can be used at first for fundamental studies, either of the SERS technique itself (in particular to investigate the optimum SM-SERS conditions), or of molecular properties that are specific to single-molecule emission.

Still, with all the recent advances in the field, SM-SERS is full of challenges ahead. The last sub-sections have highlighted the many problems encountered in the interpretation and understanding of SM-SERS experiments. Establishing that SM-SERS is indeed possible, however, does not solve by itself another major challenge, which is in fact the next 'logical' step: how can we force a single available molecule to go to the right position in order to observe its SERS signal? This will be necessary for some of the most exciting proposed applications of SM-SERS, such as single-DNA-molecule sequencing. In this particular area, the field is still in its infancy but progress might be expected in years to come. It will undoubtedly have to draw resources from other active areas of nano-science to achieve its goals.

8.2. TIP-ENHANCED RAMAN SPECTROSCOPY (TERS)

8.2.1. Introduction to TERS

Another most promising recent development in SERS has been the combination of the technique with tailor-made enhancements produced typically by a metallic tip. Many microscopic techniques like atomic force microscopy (AFM) or scanning tunneling microscopy (STM) use metallic tips. STM establishes a current through the sample and, therefore, STM tips are invariably metallic. AFM tips can be of different types, insulating or metallic (typically gold coated) depending on the exact application. A natural question at this stage is therefore whether we can bring such a metallic tip close to a surface in order to create a 'hot-spot' at a controlled position. The hot-spot characteristics can in principle be tuned by the surface/tip separation. There is the additional attraction of being able to combine SERS detection with the AFM or STM imaging capabilities; i.e. the control of the tip is not only a convenient way of producing a hot-spot with controlled properties

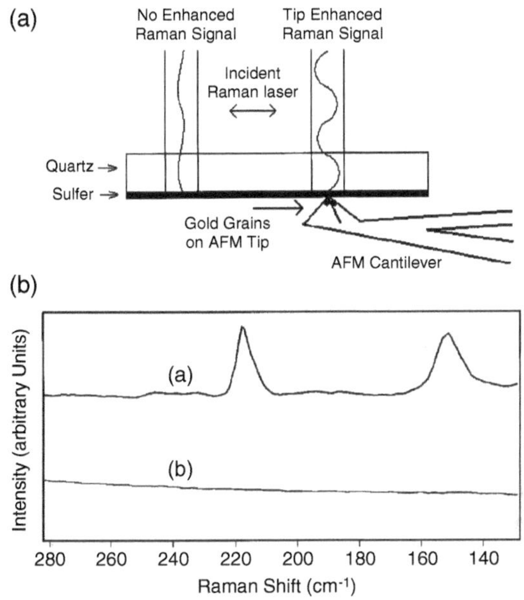

Figure 8.11. An AFM-TERS experiment showing the additional enhancement provided by a gold-coated AFM tip on the surface of a thin sulfur film ($\sim$10 μm). The figure has been adapted from Ref. [275] (reproduced with permission, copyright 2000 American Institute of Physics). (a) Schematic geometrical arrangement of the experiment. The film is illuminated and the signal collected from behind (quartz substrate). The enhanced (a) and un-enhanced (b) Raman signals coming from the sulfur film are shown in (b). The additional enhancement of the signal produced by the tip is clearly visible.

but can also provide a completely new dimension in analysis. A somewhat related aspect is the utilization of scanning near-field optical microscopy (SNOM) in combination with SERS [276], but we shall concentrate here mainly on examples which use either AFM or STM tips.

Like many emerging trends in research, what started with a few isolated reports of proof of principle rapidly proliferated into a field in its own right, and a wide variety of experimental configurations have already been proposed in the literature [68,69,275,277–279]. These configurations are distinguished mostly by: (i) the type of microscopy/tip under use, and (ii) the illumination geometry. We show here representative examples of results emphasizing the concepts that might survive future extensions and applications of the technique.

8.2.2. TERS with an atomic force microscope (AFM)

Figure 8.11 (from Ref. [275]) describes one of the first reports of tip-enhanced Raman scattering. In this particular case, a thin sulfur film is deposited on top of a quartz substrate and the laser is shone from the side

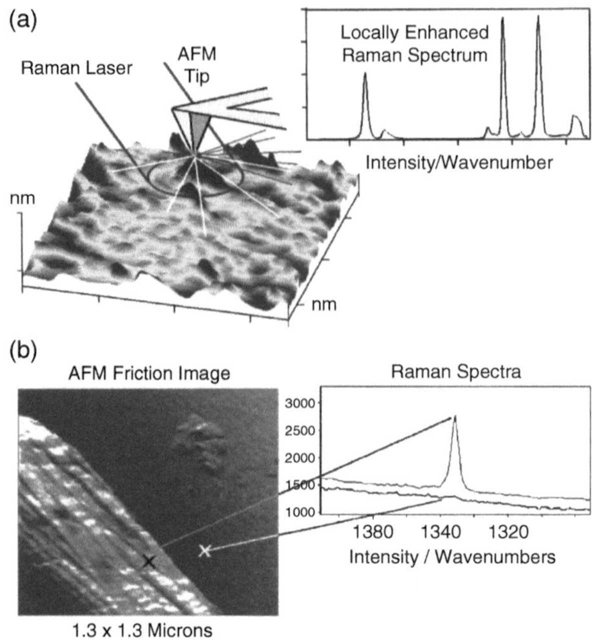

Figure 8.12. AFM-TERS scheme adapted from Ref. [277] (reproduced with permission, copyright 2002 American Institute of Physics). (a) Basic configuration of this experimental setup, in which a gold-coated AFM tip is illuminated from the side with a laser, thus achieving simultaneous AFM and SERS capabilities. (b) Example of spatially-resolved combined AFM-SERS experiment close to a diamond particle. The combination of AFM microscopy and SERS provides chemical contrast and functionality to be extrapolated down to nanometer resolution [277].

of the substrate. One can then compare the signals coming from either a free region on the film or from a spot where a gold-coated AFM tip is in close contact with the film. This is explicitly shown in Fig. 8.11(b) where the additional enhancement is obvious.

A less restrictive version of the same experiment is schematically shown in Fig. 8.12 from Ref. [277]. This configuration is less restrictive in the sense that both the AFM tip and the laser illumination come from the same side of the sample, as schematically depicted in Fig. 8.12(a). This configuration has the advantage of allowing studies on opaque substrates that cannot be accessed by the laser from underneath. The fact that the field is locally enhanced by the tip provides a valuable tool to combine spectroscopy (Raman here) and microscopy with nanometer size resolution, as demonstrated in Fig. 8.12(b).

8.2.3. TERS with a scanning tunneling microscope (STM)

Similar experiments have been carried out with scanning tunneling microscopes (STM) instead of AFMs. Figure 8.13(a) (adapted from Ref.

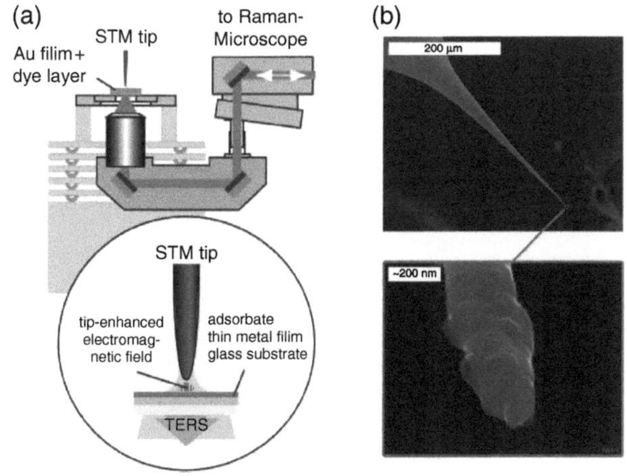

Figure 8.13. TERS with an STM setup (adapted from Ref. [69], reproduced with permission, copyright 2002 Wiley). In (a) the basic back-illumination geometry (as in Ref. [275]) is shown, with the signal being collected also from the back and sent to a Raman detection system. This setup is somewhat equivalent to the AFM one, but it allows (in addition) the use of a *tunneling current* to study a whole new range of single-molecule photo-chemical phenomena. In (b) (also from Ref. [69]) an electrochemically etched silver STM tip is shown. The knowledge of the exact geometry of the tip is essential for an estimation of the electromagnetic enhancement factors through modeling.

[69]) shows the case of a back-illuminated (as in Ref. [275]) STM-TERS setup. STM tips are obviously metallic in order to sustain the tunneling current; a typical one (obtained from electrochemical etching) is shown in Fig. 8.13(b), also from Ref. [69]. An additional difference with the AFM variant of TERS is the fact that STM-TERS needs a thin conductive film underneath (typically gold) to retrieve the tunneling current. The presence of such a metallic film has to be taken into consideration for the modeling of the enhancement, for it becomes an integral part of the electromagnetic hot-spot. This is by no means a limitation for the technique. A metal such as gold will be part of the standard choices for substrates in any case due to its chemical properties, stability, and the possibilities of binding molecules through thiol groups.

Figure 8.14 shows a clear example of TERS for the dye brilliant cresyl blue on a smooth gold film of 12 nm in thickness (from Ref. [69]). A clear enhancement of the signal when the tip is in tunneling position is easily observed. The signal for the tip far away from the sample is a factor of $\sim$16 smaller than the one for the tip in tunneling position, but this factor should not be confused with the single-molecule enhancement factor, which is much larger than this, since the signal in the presence of the tip comes from a *much smaller number of molecules* (ideally one). As pointed out in Ref.

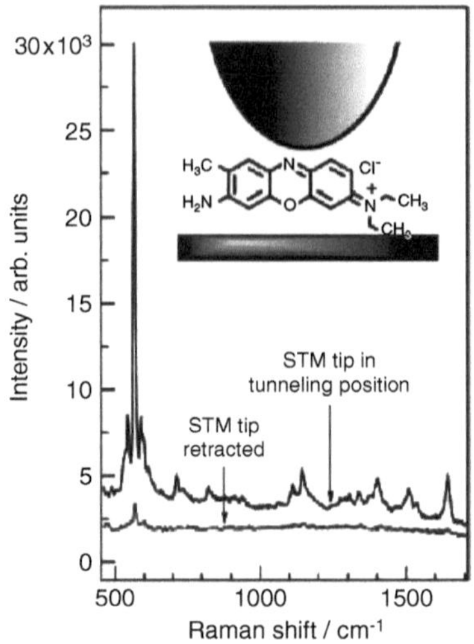

Figure 8.14. STM-TERS under resonant excitation for the dye brilliant cresyl blue on a smooth gold substrate of thickness ~12 nm (from Ref. [69], reproduced with permission, copyright 2002 Wiley). The intensity ratio (not to be confused with the enhancement factor) at ~540 cm^{-1} is a factor of ~16 between the non-enhanced signal (tip far away from the sample) and the enhanced one (tip in tunneling position). This represents a much larger enhancement factor (than a factor of ~16) if the spatial distribution of the enhancement is accounted for. See Ref. [69] for further details.

[69] the SMEF is a sensitive function of the radial distance from the center underneath the apex, as corroborated by several numerical simulations. For the specific experiment in Fig. 8.14(a) a rough estimate suggests that the signal for the tip in the tunneling position comes from approximately ~400 molecules within a radius of ~15 nm from the center of the tip, as opposed to the $\sim 1.6 \times 10^6$ molecules contributing to the non-enhanced spectrum. This is another example where the convolution between the spatial distribution of the enhancement and the number (and position) of the molecules play a crucial role in the determination of SERS EFs.

8.2.4. Theoretical calculations on tips

Due to the intrinsic attraction this technique offers, theoretical calculations of enhancement factors for model tips have been carried out. Two examples using two different methods are shown in Figs. 8.15 (from Ref. [280]) and 8.16 (from Ref. [281]).

In Fig. 8.15, the electromagnetic response of an isolated tip (without the substrate underneath) is studied numerically within the discrete dipole approximation (DDA) approach (see Section 5.3.2). The results for two in-plane polarizations but different incoming directions at a fixed wavelength

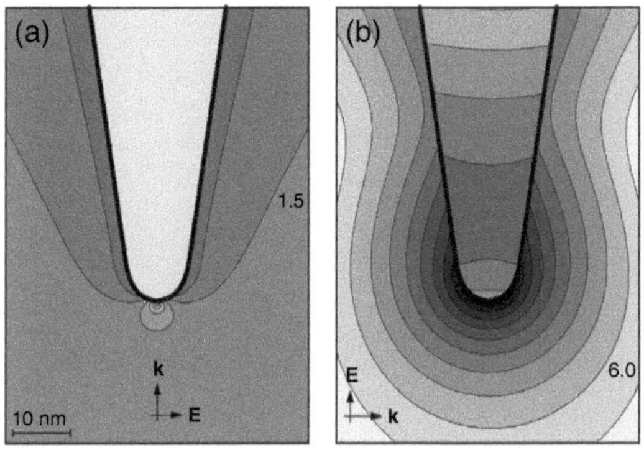

Figure 8.15. Intensity plots around a gold tip in water illuminated by a monochromatic wave at 810 nm coming from two different directions (from Ref. [280], reproduced with permission, copyright 2000 Springer-Verlag). The polarization **E** and propagation direction **k** of the incident waves are indicated in the figures. Contours of the local field enhancement factor ($|\mathbf{E}_{\mathrm{Loc}}|^2/|E_{\mathrm{Inc}}|^2$) are shown in the plots (factor of 2 between successive lines). The scaling is given by the numbers in the figures (multiples of the exciting field). No enhancement at the tip in (a); enhancement of $\sim$3000 in (b). According to Ref. [280], the local field in (b) is almost rotationally symmetric in the vicinity of the tip, despite the wave-vector of the light breaking the symmetry of the problem.

are displayed in Fig. 8.15. These calculations are not specifically aimed at SERS phenomena, but rather at exploring the possibility of optical trapping of particles produced by high-field gradients. This could be the situation in liquids for example and, in fact, the calculations in Fig. 8.15 have been performed for a tip immersed in water. One of the most obvious conclusions of these studies is that polarization plays a decisive role in the formation of a hot-spot that can be used in TERS.

Figure 8.16 shows an alternative calculation based on a finite-element numerical solution of Maxwell's equations using one of the several commercially available programs for electromagnetic finite-element analysis. The details of the calculation as well as its parameters are explained in the caption of the figure and further in Ref. [281], from where Fig. 8.16 has been adapted. Only one of the two incoming laser polarizations achieves the highest possible enhancement at the end of the tip (in agreement with the conclusions in Fig. 8.15), which is predicted to be as high as $\sim 10^{11}$ in the best possible conditions (Au substrate) for a $\sim$2 nm gap separation between tip and substrate and optimized wavelength excitation. In fact, enhancements of the order of 10^{11} are the highest enhancement factors that can be justified theoretically as coming from an electromagnetic contribution. As a matter of fact, enhancements slightly above $\sim 10^{11}$ are *only* obtained for gaps between tip and substrate that are comparable with the typical size of a dye molecule ($\sim$1 nm).

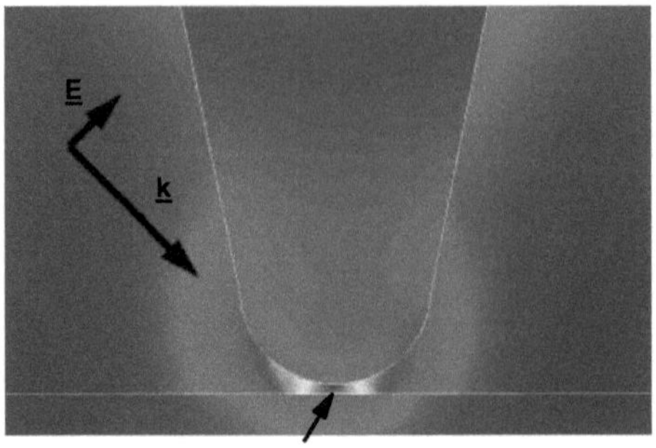

Figure 8.16. Finite-element simulation of tip-enhanced electromagnetic fields for TERS, adapted from Ref. [281] (reproduced with permisssion, copyright 2006 Am. Chem. Soc.). The simulation is done for a 20 nm radius gold tip (hemisphere) separated by a gap of 2 nm from a mica substrate in air. The hemisphere is continued by a cone of half-angle 10°; the geometry represents a model of experimentally achievable fabrication of tips by electrochemical etching [282]. The illumination configuration is similar to the experimental setup proposed in [277], and depicted explicitly in the figure. Calculations for both p- and s-polarized (parallel and perpendicular to the plane of the page, respectively) light impinging on the tip from the side at 45° are shown in Ref. [281]. Here we show the p-polarized case only for 533 nm excitation. The plot shows the spatial dependence of the SERS enhancement factor in a false color scale calculated as the fourth power of the normalized local field ($|E|^4$-approximation, see Chapter 4). The p-polarized beam shows the largest enhancement and the most centered along the axis of the tip. See Ref. [281] for further details.

8.2.5. Discussion and outlook

One important issue of TERS is that of the experimental difficulties. TERS experiments are extremely delicate, require highly specialized equipment, and very cumbersome experimental setups (with ultra-high vacuum for example). This has so far confined the technique to research labs and fundamental studies. Moreover, it is difficult in TERS experiments, as in SERS, to measure experimentally the SERS enhancement factor that is created at the tip. However, experimental results combined with theoretical predictions indicate that single-molecule SERS enhancement factors may be from moderate to high depending on several factors, but most crucially on: (i) the presence/absence of a metallic substrate underneath, and (ii) the exact details of tip shape and its type (evaporated gold on Si, chemically or electrochemically etched from bulk metal, etc.). It remains to be seen whether or not TERS hot-spots created *in a 'controlled manner'* are able to achieve SERS EFs comparable to other SERS substrates, as suggested by theory, even though the body of work at the moment goes in the direction of supporting this claim.

Controlled and reproducible enhancements could alternatively be achieved by ordered arrays of objects (normally made of gold) in which specific shapes can be arranged on a substrate in a 2D configuration. We shall treat these substrates in a separate sub-section later in this chapter, and mention here only a few aspects that serve as contrast to the tip-enhanced (TERS) approach. There are advantages and disadvantages to the ordered array approach; on the one hand we can achieve samples where the enhancement is homogeneous and well controlled, but this cannot be achieved all the way down to distances of the order of ∼1 nm (∼10 nm is the current 'optimistic' lowest limit for nano-lithography for example). Hence, this implies that the smallest gaps (highest enhancements) are still beyond reach and, in addition, the problem of positioning the analytes at the right places (hot-spots or places with high enhancements) needs to be addressed. A different approach is not to be confined to where the signal might appear, but instead to either search for it or create it locally at will. This is basically the spirit of the tip-enhanced techniques, which aims at creating a *hot-spot on demand* at a specific location on a substrate.

As for SM-SERS, TERS provides an example of a development that has opened new horizons in the field of SERS and is rapidly developing in some of the other sub-fields of the technique. One example of the latter is the increasing number of papers that are appearing on the topic of single-molecule TERS (SM-TERS) [283,284]. The field of TERS has been steadily gaining momentum in the last few years and there are good reasons to believe that the work achieved so far is to a large extent preliminary only. TERS may, for example, play a major role in achieving the goal of 'hot-spot on demand' for single-molecule detection. In any case, the possibility of combining one type of microscopy like AFM or STM with the bonus of spectroscopic information has, undoubtedly, a huge appeal and potential, and will surely unfold in decades to come.

8.3. NEW SUBSTRATES FROM NANO-TECHNOLOGY

The number of different types of SERS, and more generally plasmonics, substrates available for experimentation is increasing at high speed, with no clear end in sight to the acceleration and pace with which new alternatives are introduced and explored every year in the literature. This includes both novel nano-particles in solution as well as self-assembled or engineered structures with different levels of control over their properties. In the particular case of SERS, a tendency toward specific applications that require the technique to be more reliable and 'usable' in a wider range of cases has been the driving force to develop novel substrates. The emphasis in these cases is then placed on *optimization* (to obtain larger enhancements) and *reproducibility*. Far from attempting a complete overview of the techniques and cases being proposed in the literature (which will probably justify a book by itself), we shall arbitrarily select and discuss a few relevant examples of the current trends. These cases are more aimed at showing the underlying *principles* and concepts that are being explored at the moment, rather than showing the fine details of each case, which can always be retrieved from the original references.

8.3.1. Chemical synthesis of metallic nano-particles

Nano-particle fabrication and characterization is an active field of research within the wider field of nano-technology. Many of these nano-structures are formed by chemical reactions including 'self-organization' processes of some sort (which includes sometimes the addition of miscellaneous variables, like ultrasound [285]). Novel types of metallic nano-particles could potentially play an important role in SERS and related plasmonic effects. The most important aspects, as far as SERS is concerned, are:

- Improving the homogeneity (size distribution) through a better control during fabrication or subsequent filtering methods.
- Expanding the tunability of the associated localized-surface-plasmon resonances (through size, shape, and/or the use of composite dielectric/metal nano-particles).
- Designing nano-particles with large local field enhancement factors (either on average, or at localized positions for single-molecule detection).

Unfortunately, a large portion (but not all) of nano-particle fabrication research is carried out without specific applications to SERS or plasmonics in mind. Reports on SERS are scattered through the literature on nano-particles, but sometimes as an application mentioned in passing and without the necessary systematic studies that would be needed to establish the usefulness of the approach. One can however anticipate that many of these techniques will eventually find their way into these potential areas of applications. We discuss here two examples only (the first one for its nice pictures! and the second for its particular emphasis on SERS applications).

New shapes

As a first example, star-shaped gold nano-particles have recently been synthesized [286]. These novel nano-particles, illustrated in Fig. 8.17 have interesting LSP resonances, which are highly polarized and spatially localized. Such exotic-shaped nano-structures are being studied at present to characterize their LSP resonances and optical properties in general; i.e. many of these studies do not involve a direct application to SERS at the moment.

Nano-shells

An interesting alternative to metal particles for SERS in colloidal solutions are nano-shells, i.e. particles with a dielectric core and metallic shell. Nano-shells provide an additional degree of freedom for the *engineering* of LSP resonances in solution. This is achieved by a careful control of the thickness

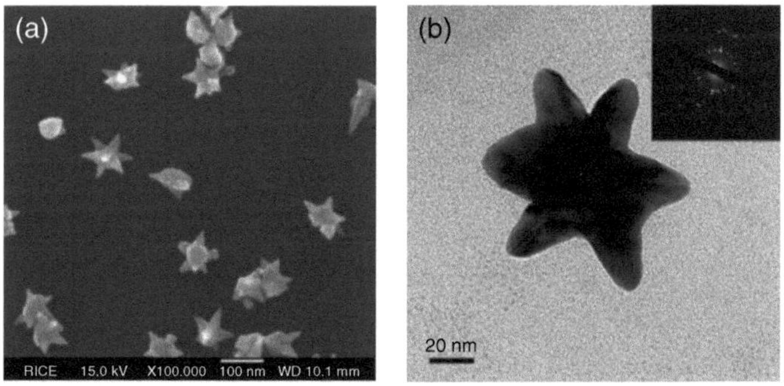

Figure 8.17. Star-shaped gold nano-particles. Figure and caption adapted from Ref. [286] (reproduced with permission, copyright 2006 Am. Chem. Soc.). SEM images (a) reveal the structure and heterogeneity, while TEM (b) and electron diffraction demonstrate that the nano-stars have defects and consist of multiple crystal domains. An extinction spectrum of the nano-star solution exhibits broad visible and NIR peaks. In addition, polarized optical scattering spectra (see Ref. [286]) show that these nano-stars have strongly anisotropic LSP resonances which could be used in plasmon-related spectroscopies, including SERS.

of the metallic shells with respect to the dielectric core, and their overall size [66,206,288]. Nano-shells have been used as efficient SERS substrates, even though they have much wider applications in plasmonics, including novel biomedical applications [289–291].

Figure 8.18(a) shows a simplified explanation of how and why the nano-shell approach is interesting from the plasmonics point of view. It provides an additional degree of tunability for the intrinsic plasmon resonance of individual particles by changing the size and the thickness of the shells. From the standpoint of SERS, nano-shells have been used as 'nano-shell colloidal solutions', in a similar fashion as typical SERS experiments on standard Ag colloids. However, rather than achieving high enhancements by *particle aggregation* as generally done in standard Ag colloidal solutions (see Chapter 7), SERS can more easily be measured from single particles with nano-shells, since the LSP resonance can be approximately *tuned* to the laser (and Raman) wavelength.

Reference [66] compared the average SERS EFs for three different modes of para-mercaptoaniline (pMA) as functions of shell thickness with the predicted enhancement from Mie theory in the $|E|^4$-approximation. This is illustrated in Fig. 8.18(b) (adapted from Ref. [66]). Despite problems with corrections due to self-absorption, the agreement between theory and experiment is quite satisfactory both quantitatively and qualitatively. The single nano-shell enhancements are probably not enough to achieve single-molecule detection. If higher enhancements are required interactions among nano-shells (either in the solution or in aggregated dry clusters) may be introduced [292]. Nevertheless, the main point of nano-shells is precisely to have the ability of

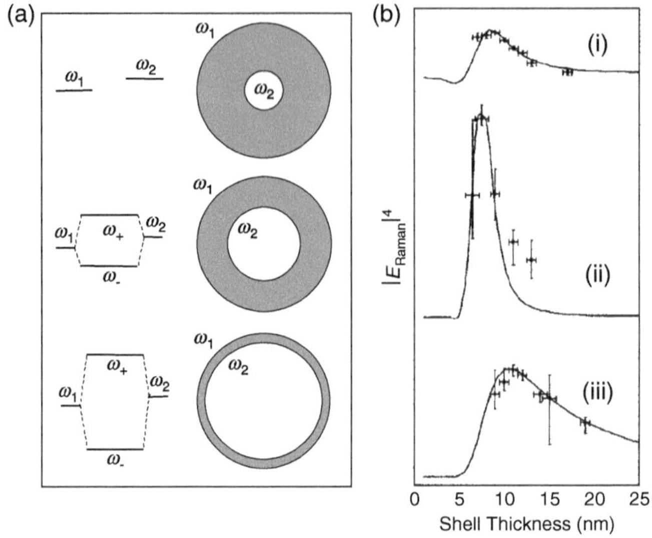

Figure 8.18. (a) Adapted from Ref. [287] (reproduced with permission, copyright 2003 American Institute of Physics): simplified 'plasmon-resonance-hybridization' picture [207] for a nano-shell. A strict treatment of LSP resonances in these spherical shells can be obtained by an extension of Mie theory (see Appendix H). A simpler approach (with some limitations) to understand why nano-shells develop a red-shifted resonance is to think of the two surfaces as carrying their own 'LSP resonance' (at different energies ω_1 and ω_2, in general), which start interacting with each other as the shell gets thinner. This creates the analog of a 'bonding' and 'anti-bonding' resonance (in analogy with atomic orbitals) that split apart in energy. This is shown schematically from top to bottom in (a). In gold nano-shells, the lowest ('bonding') resonance (ω_-) can be tuned through the visible spectrum down to the near infrared, thus providing a degree of tunability that can be exploited for SERS. (b) Figure adapted from Ref. [66] (reproduced with permission, copyright 2003 American Institute of Physics). Experimental (symbols) and predicted (lines, obtained from Mie theory in the $|E|^4$-approximation) average SERS enhancement factors for three Raman modes of pMA adsorbed on a gold-coated spherical silica cores (79 nm in diameter) with varying shell thicknesses.

controlling the LSP resonance condition of an individual colloid at will, and this provides an invaluable additional degree of freedom for the engineering of LSP resonances for specific applications.

Nano-shells are going through an intense period of innovation at the moment [293] and new varieties of nano-shells have been introduced. An example is the case of *nano-rice* [294], which possess even greater structural (geometrical) tunability than nano-shells and, by the same token, much larger enhancements produced by the elongated shape of the individual particles. These new types of nano-structures (and many others not mentioned here) are very promising for applications (one of which includes SERS) and they are still only in their infancy.

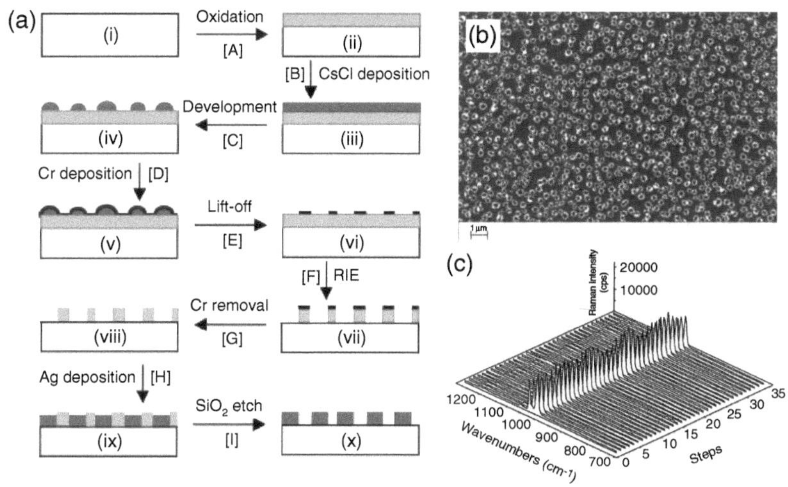

Figure 8.19. Figure and caption adapted from Ref. [64] (reproduced with permission, copyright 2003 Am. Chem. Soc.). (a) Schematic of the sequential steps (on a Si wafer) used for island lithography: (A) oxidation; (B) CsCl deposition; (C) development; (D) Cr deposition; (E) lift-off; (F) reactive ion etching; (G) Cr removal; (H) Ag deposition; and (I) SiO_2-etch. The result is a robust structure of Ag 'islands' on Si. Depending on the well diameter, the islands can grow into either pillars or torii. (b) Corresponding SEM image of silver torii on silicon after the SiO_2 was stripped away with HF. (c) Line scan of the SERS signals on a torii sample. A set of pyridine SERS spectra were taken on a torii-type structure with pyridine adsorbed (from a 0.01 M pyridine in 0.1 M NaCl aqueous solution). Spectra were taken along a 100 μm line in 3 μm steps (×20 objective) with an integration time of 1 s (633 nm laser excitation).

8.3.2. Self-organization

Self-organization is, admittedly, one of the easiest approaches to nano-fabrication. Many types of SERS substrates, including those used in early studies like aggregated colloidal clusters, are based on some form of self-organization, and new approaches along these lines are still actively pursued. We highlight here two examples only.

Island lithography

We start with island lithography [64], which is based on a series of chemical steps depicted schematically in Fig. 8.19(a). The end result is a series of Ag 'islands' which can adopt the shape of pillars or torii depending on the conditions. The topology of the array of pillars and/or torii is decided by the self-organizing patterned structure of CsCl islands that results in the development stage. The array of islands is a disordered lattice but with a large degree of homogeneity, from where these films derive (at least in part) their reproducibility. Figure 8.19(b) shows an example SEM image of a torii-type structure. Figure 8.19(c) shows, in addition, a test of the homogeneity of the substrate through a SERS line scan over a 100 μm range of pyridine dissolved in water. Estimations in Ref. [64] of the SERS enhancement factors place their values in an acceptable range to perform analytical chemistry work down to nano-molar concentrations (depending on the analyte). These types of substrates (with moderate but homogeneous enhancements) are highly desirable for applications.

Nano-sphere lithography

Another type of substrate based on self-organization is the *nano-sphere lithography* approach, developed for SERS primarily by Van Duyne and coworkers [295]. This approach has been shown to be reasonably easy to implement experimentally and fairly reproducible. Figure 8.20 describes the basic steps [295]. It basically exploits the regular patterns formed by *self-assembly* of dielectric (e.g. polystyrene) nano-spheres on a surface upon drying. The evaporation of a metal film on top of the array can be followed by the 'lift-off' of the nano-spheres themselves (in which case an array of *interstitial* sites is left on the surface) or it can be left on top of the nano-spheres to produce what has been called a 'metal-film-over-nano-sphere' (MFON) substrate [295]. This produces a wide range of possibilities and high degree of versatility for these substrates for potential applications. It has been shown, for example, that substrates with LSP resonances that can be tuned across the entire visible range can be obtained with nano-sphere lithography. In Ref. [295], it is even stated that (according to the opinion of the authors) 'the search for stable SERS substrates is completed. In order to make a significant new contribution to the fabrication of SERS substrates, one must demonstrate a surface that exceeds (the) benchmarks (of nano-sphere lithography)'. While this can be argued to be an overstatement in many respects, because it underestimates other requirements beyond those needed for simple applications, it can nevertheless be viewed as the first test to face when designing new substrates. Further details on this technique can be found in Ref. [295], and references therein.

8.3.3. Nano-lithography

Electron-beam nano-lithography

Another approach to obtain highly uniform and reproducible planar SERS substrates (with moderate enhancements) is essentially based on nano-lithography itself. In this case, the structure of the substrate is not decided through a self-organizing procedure, but rather it is carefully controlled (and can therefore be chosen) by means of conventional lithographic techniques

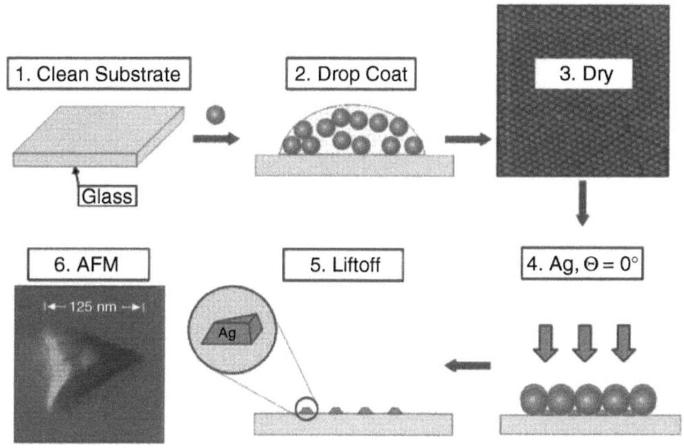

Figure 8.20. Schematic representation of the nano-sphere lithography approach for SERS substrates (reproduced from Ref. [295] with permission, copyright 2005 The Royal Society of Chemistry). Dielectric nano-spheres in solution are drop-casted onto a clean glass substrate, where they self-assemble into an array (typically an hexagonal lattice). The array provides a fairly uniform template over distances comparable to many diameters (~1–10 μm). The array is then evaporated with a metal layer (Ag in this case) of varying thicknesses (with typical values of the order of ~50 nm), and is subsequently subjected to a lift-off of the nano-spheres themselves. The final product is an array of metallic (Ag) objects defined by the *interstitial sites*. For example, an array of small *triangular* objects like the one shown in the last step can be obtained.

(of widespread use in the semiconductor industry). The benefits of the large uniformity are obvious, not only for applications, but also for fundamental studies of the SERS mechanism itself as demonstrated for example in Refs. [38,39]. In this case, *electron-beam lithography* was used to fabricate gold nano-particle arrays of well-defined size and shapes, examples of which were shown in Fig. 7.4. With the current state-of-the-art equipment, geometrical features with a resolution down to ≈10–20 nm can be achieved. This is sufficient for fabricating isolated particles with a wide variety of shapes and sizes. However, the fabrication of *interacting* particles with a precisely-controlled gap of, say 5 nm or less (required for investigating gap-plasmon effects) remains very challenging and will be surely the subject of future developments. At the moment, some sort of 'statistical approach' is followed to create gap-plasmon resonances. Many copies of closely spaced features are produced on a substrate with the hope that some of them will end up with the required gaps. There is no reason to believe though that this cannot improve in the future.

Arrays of inverted pyramids

The work in Ref. [63], illustrated in Fig. 8.21, is another example of this approach. In this case, arrays of gold-coated pits (inverted pyramids) are

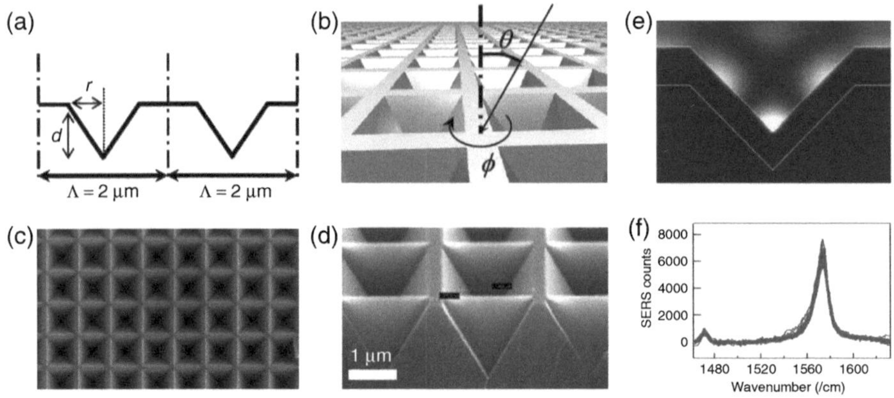

Figure 8.21. An example of SERS-active substrate of inverted pyramids fabricated by lithography on silicon (figure adapted from Ref. [63], reproduced with permission, copyright 2006 Opt. Soc. Am.). (a) 2D cross-section through the pits is shown, the pitch (Λ) is 2 μm. The aperture size (r) and, therefore, the depth (d) can be graded across the sample to study their effect. (b) Schematic representation of the sample, while (c) and (d) are top-view and cross-section SEM images of the sample before metalization with gold, respectively. (e) Intensity distribution from 2D finite-difference time-domain (FDTD) electromagnetic simulations at $\lambda = 785$ nm for $d = 1250$ nm, $\theta = 0°$ (normal incidence). The simulation shows the presence of high-field enhancement localization in the pits. (f) SERS spectra across a pit array, for the 1572 cm^{-1} line of benzenethiol in an array with $d = 1$ μm. The signals are very reproducible across the entire sample, with less than 10% residual standard deviation in the peak heights from point to point, and also in the comparison among substrates.

fabricated by means of conventional optical lithography. The samples are produced on silicon substrates by anisotropic etching using KOH, which is used to preferentially etch the (111) crystallographic planes, thus resulting in an array of inverted pyramidal square pits. These arrays are then evaporated with a thin layer of gold. According to Ref. [63], this fabrication technique is well suited to the production of plasmonic nano-structures for SERS applications because the etched surfaces are extremely reproducible, with the pyramid faces oriented at an inclination of 35.3° to the normal, and atomically smooth. While the pitch remains constant across the sample, the aperture size, r, and therefore pit depth, d, can be varied.

Figure 8.21(f) gives an idea of how homogeneous and reproducible these substrates are. Data are shown for several points on the array for the 1572 cm^{-1} line of benzenethiol using 10 s integration times with 3 mW of 633 nm laser excitation focused with a ×20 objective. Reference [63] reports that no signal could be identified on the flat gold-coated section of the sample away from the patterned area, verifying the importance of the localized surface plasmons in the pits for the SERS enhancement. The study reveals an overall reproducibility of ∼10% in both the signals from different parts of the substrate and in the comparison among substrates. Such a reproducibility is very desirable for analytical applications.

It is worth noting that arrays like the inverted pyramids potentially offer more than a mere control on the SERS enhancements. It also provides a very promising route to high-density (and high throughput with the appropriate optics) multiplexing; a highly desirable feature in biological applications. With the possibility of creating spatially-resolved samples (one in each pit) with controlled enhancements, these new substrates may have a major impact on the (bio-) analytical chemistry aspects of SERS in the near future.

Some remarks on ordered arrays

Nano-lithographic samples offer a unique opportunity to match experimental findings with theoretical predictions due to the high degree of control in the geometry. The comparison with theoretical expectations can be pushed to a substantial degree of sophistication. In Ref. [63], detailed comparisons between the plasmon resonance behavior (as a function of wavelength and incidence angle) have been provided. In addition, the spatial localization of the enhancement within the wells can be predicted. Fig. 8.21(e) shows an example of the intensity distribution inside each well.

Another important point to stress here is that the questions these ordered substrates try to address are of a fundamentally different character than those for single-molecule detection. In a way, these substrates try to 'reduce' the maximum single-molecule enhancement factors for the sake of gaining reproducibility. This should not be seen as a drawback, for these substrates produce reproducible results down to a concentration level that competes and surpasses (in many cases) other analytical tools.

8.3.4. Adaptable/Tunable SERS substrates

Another recent development has been the fabrication of SERS substrates that can be modified or tuned *a posteriori*, almost in real-time during the SERS (or plasmonics) experiments.

Adaptive silver films

Reference [65], for example, introduced the concept of *adaptive silver films* (ASF) for SERS of biomolecules and more precisely proteins. This new type of controlled self-organization is aimed at solving several problems associated with the SERS monitoring of biomolecules; specifically: the optimization of signals and the problem of denaturation. We reproduce here the explanation provided in Ref. [65]. ASFs are such that the biomaterial and the substrate act in concert. The basic idea is that while preserving properties such as conformational state and binding activity of the analyte, the analyte itself

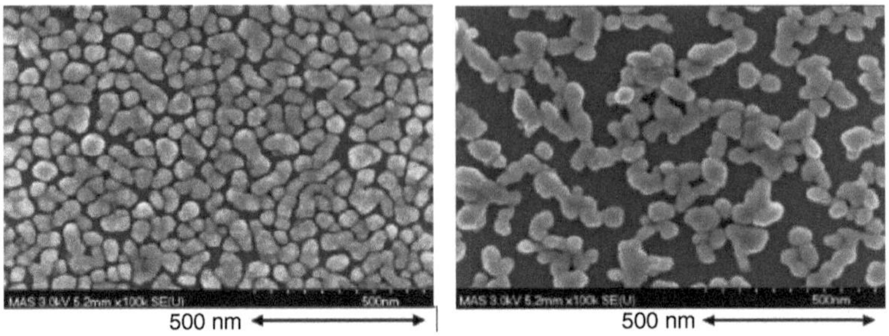

Figure 8.22. (a) From Ref. [65] (reproduced with permission, copyright 2005 Wiley): SEM images of an adaptive silver film (ASF) before (left) and after protein deposition and washing (right). The nano-scale restructuring of the film produced by the protein is used to optimize the adhesion, increase the SERS signal and improve resistance to further washing (an inevitable step in many biological applications). This method then represents a form of self-organization resulting in tailor-made nano-structures which are favorable for a specific biomolecule (in this case a protein). See Ref. [65] for further details.

produces a local restructuring of the metal surface, and achieves thereby excellent Raman enhancements. Reference [65] states that: although most SERS substrates have a predefined and static metal-dielectric nano-structure, this is not in general the best case for obtaining an optimal Raman signal from any particular analyte. Experiments reveal that a variation of evaporated metal films produces substrates with new adsorption and restructuring features that are important for SERS detection and analysis of biomolecules. According to Ref. [65], the restructuring involves competition between two processes. In the case of SERS by proteins, a protein solution can slightly dissolve the Ag particle surfaces to make them movable. At the same time, stabilization by proteins can occur, in which the protein molecules coat the silver particles and, accordingly, prevent the particles on the surface from dissolving in the buffer.

Changes in analyte deposition and restructuring can be readily observed in the SEM images in Fig. 8.22, where nano-scale restructuring is visible after protein deposition. A simple way to summarize these self-organizing SERS substrates would be to say that they achieve the most natural nano-scale structure that is compatible with the biomolecule under observation (in terms of inter-particle separations, etc.). At the same time they provide useful mechanical properties (like resistance to washing, which is a highly desirable property for biomolecules undergoing scanning processes), and tend to preserve the native structure of the probe (which is a major problem in proteins where denaturation can be easily produced by contact with surfaces). The concept of having a substrate that 'adapts' to the probe to obtain the best possible outcome in terms of signal and stability is certainly appealing and is a new idea that might find further ramifications for applications in the future. Its main drawback is the large parameter space (physical and chemical)

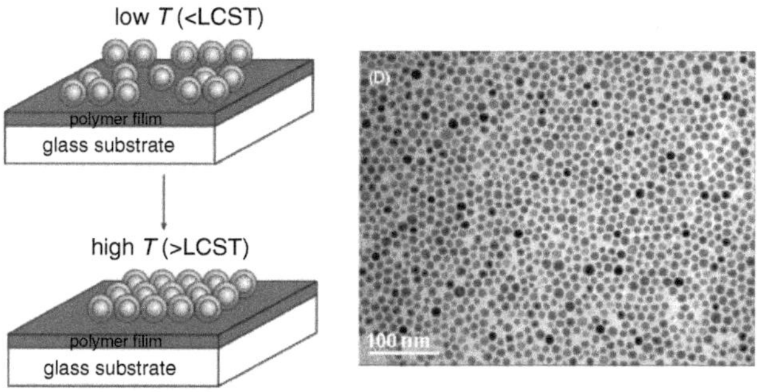

Figure 8.23. Tunable temperature-controllable SERS substrate. The figures are adapted from Ref. [296] (reproduced with permission, copyright 2005 Am. Chem. Soc.). Left: schematic illustration of a nano-particle monolayer on a thermo-responsive polymer film. Depending on temperature (over a relatively small temperature range ~10 °C) the average inter-particle distance can be controlled and, therefore, the overall (plasmon) resonant response of the layer to incoming light. Right: a TEM image showing the size and overall geometrical distribution of a polymer film with Ag nano-particles of nominal size ~20 nm. See Ref. [296] for further technical details on these substrates.

that needs to be explored for each specific probe. In that sense, ASF is more complicated to use than other more standard substrates.

Temperature-controlled substrates

Another interesting example of post-fabrication control of SERS substrates is illustrated in Fig. 8.23 from Ref. [296]. In this case, a high-density silver nano-particle film with temperature-controllable inter-particle spacing is used as SERS substrate. The technique of Langmuir–Blodgett films is used here to obtain silver nano-particles coated with a monolayer of surfactant molecules, which are then transferred to a temperature responsive polymer membrane. The (relatively) large changes in inter-particle spacing induced by the polymer membrane upon temperature changes produce different coupling conditions among the silver nano-particles. Hence, it provides a way to tune *a posteriori* the coupled-LSP resonances in these structures (and indirectly the SERS intensities).

By varying the temperature within a relatively small range (~10–15 °C), it is possible to tune the maximum of the absorption spectra by tens – and even hundreds – of nanometers. Reference [296] shows the absorption spectra moving from the near infrared (below 700 nm) and peaking in the ~600 nm region for a relatively modest change in temperature of ~10 °C. The change in the interactions among particles results, indirectly, in a change in SERS enhancements at a specific wavelength. These variations in SERS intensities can be by an order of magnitude, depending on the specific

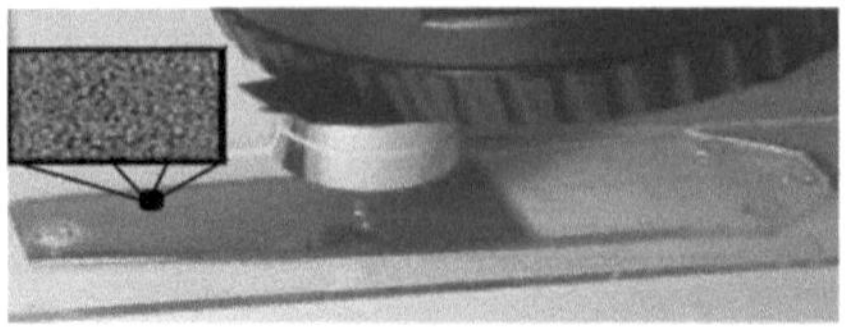

Figure 8.24. Figure and caption from Ref. [297] (reproduced with permission, copyright 2004 Elsevier). The figure shows an integrated micro-fluidic/SERS device under the $\times 10$ objective of a Raman microscope. The inset shows a SEM image of the silver–PDMS nanocomposite at approximately 10^5 magnification. See Ref. [297] for further details.

temperature. The particular example shown in Ref. [296] would not have a direct practical application *per se*; for a start the concentration of the analyte in this study (RH6G) is large (10^{-5} M) for SERS standards, and the variations in intensities by a factor of ~ 10 do not justify the use of the effect by itself at this point. However, the results in Ref. [296] should be taken as a *demonstration of principle* of a technique. The concept of tuning plasmonics nano-structures by embedding them in thermo-responsive materials (like polymers) is, undoubtedly, very interesting.

8.3.5. Micro-fluidics and SERS

The idea of combining micro-fluidic devices with embedded SERS detection capabilities has been a long-standing ambition in the field [261]. This is an area at the interface between SERS and micro-fluidics which, like some of the ideas presented in this chapter, is only beginning to take shape and is still at a 'development' level. The potential possibilities are numerous though.

An example is shown in Fig. 8.24 from Ref. [297]. The authors claim that: the widespread development of micro-fluidics has allowed the extension of efficient separations, fluid handling, and hyphenation with many detection modes to a small, portable, highly controllable physico-chemical platform. SERS offers the powerful advantage of obtaining vibrational spectroscopic information about analytes in an aqueous matrix with negligible background. According to Ref. [297], the mating of electrophoretic separations with vibrational spectroscopy on a micro-fluidic device will allow the chromatographic efficiency of capillary electrophoresis with the unequivocal analyte 'fingerprinting' capability of detailed structural information.

Most of these devices are typically created by polymer structures of PDMS (polydimethylsiloxane) which are imprinted from a matrix and completed with windows for optical access. In the case of SERS, the PDMS-devices (that creates the channels) have to be made compatible with the metallic nano-structures to produce the optical enhancement. This leads to hybrid PDMS–metal composites for micro-fluidic/SERS. It is interesting to note that, in an ideal case, we would like to produce flows (possibly containing a solvent like water and analyte molecules to be detected) through well controlled plasmonic nano-structures with very high SERS enhancements. This goal is far from sight at the moment, in terms of having the necessary control over the experimental parameters to control the signal. However, the fact that high SERS enhancements come from interactions among metallic objects separated by a few nanometers only, naturally raises the question of how fluids behave on these length scales. Micro-fluidics is, at this stage, a relatively mature field (in terms of the speed by which new areas of research develop) but its symbiosis with SERS might need an additional step into the realms of *nano-fluidics* and the way liquids (which can be binary liquids of solvents and analytes) behave on nanometer-sized channels and constrictions. The field of nano-fluidics is emerging in parallel with these developments, but can still be argued to be in a fairly undeveloped stage at present [298]. It is very likely that we will see in the next decade novel developments in this area, i.e. the symbiosis of micro/nano-fluidics and SERS, which uses a mixture of concepts from fluid dynamics, statistical physics, chemistry, surface science, electromagnetic theory, and spectroscopy.

8.4. OPTICAL FORCES

Another recent development (admittedly more on the boundaries of SERS research) has been the realization that optical forces may also play a role in some situations. Optical forces may act either on the substrate itself (for example affecting the motion of colloids in liquids) or possibly directly on the probes themselves. These aspects are only now beginning to be explored and have yet only been evidenced in a handful of situations. Laser forces have a long tradition in the field of optical manipulation of particles developed by the pioneering work of A. Ashkin in the 80s, which ultimately led to the development of laser tweezers [299]. But most of the seminal work on laser manipulation of particles has been done, primarily, on a wide range of dielectric particles, which include objects like bacteria or cells (in the $\sim$1 μm size range). The study of the effect of optical forces on metallic particles in the nanometer range, on the other hand, is relatively recent and was fueled mainly (but not only) by the interest in SERS and plasmonics.

8.4.1. A simple theory of optical forces

The theory of forces on particles produced by laser beams has a long-standing history in optics and has been summarized in many works [280]. Consider a metallic object of arbitrary shape being illuminated by an electromagnetic wave, as depicted in Fig. 8.25. If we can solve the electromagnetic scattering problem for the object (by using numerical or analytical methods, see Chapter 5 for examples), we then know formally the electric ($\mathbf{E}(\mathbf{r})$) and magnetic ($\mathbf{B}(\mathbf{r})$) fields at all positions in space. This

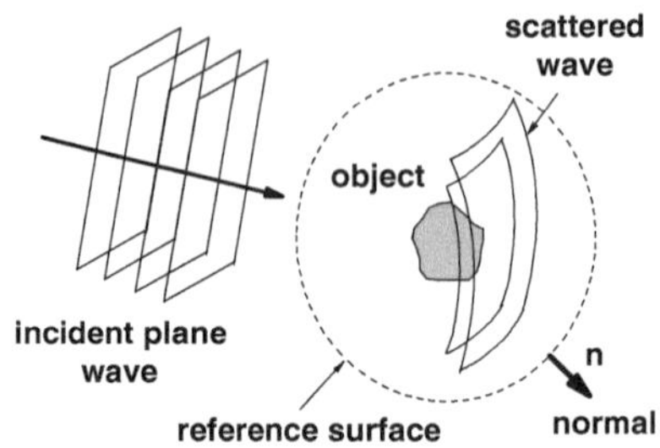

Figure 8.25. Schematic representation of the scattering of an incident electromagnetic wave by an object. If the scattering problem is solved (by any of the numerical or analytical methods available), fields at all positions $\mathbf{r}$ are then known, and the mechanical action of the beam on the particle can be evaluated by a suitable integration of Maxwell's stress tensor $\hat{T}$ (Eq. (8.1)) over a closed surface containing the object [96].

implies that we can immediately calculate the electromagnetic force exerted by the wave on the object, using one of the basic conservation laws [96] in electromagnetic theory (for momentum), which gives the force as:

$$\mathbf{F} = \oint \left\langle \hat{T}(\mathbf{r}) \cdot \mathbf{n}(\mathbf{r}) \right\rangle \mathrm{d}S, \tag{8.1}$$

where $\hat{T}(\mathbf{r})$ is *Maxwell's stress tensor* [96,280], and the integral is performed over any closed surface (with normal $\mathbf{n}$) enclosing the object; see Fig. 8.25. The brackets in this expression, mean the usual time averaging over a period, characteristic of effective magnitudes produced by harmonic fields [96].

The electromagnetic force on an arbitrary object is a fairly complicated function depending in general on polarization, shape of the incident beam, geometrical details, optical properties of the object, relative orientation of the beam with respect to the object, wavelength, etc. Only in very special circumstances can we draw simple conclusions to understand qualitatively what the laser beam does, as far as mechanical action is concerned. A big fraction of the difficulty here is to obtain the solution to the electromagnetic problem in the first place. But one situation where the analysis is greatly simplified is in the limit of very small particles compared with the wavelength [280]. The electromagnetic response of the object at $\mathbf{r}_0$ can then be approximated (in the dipolar approximation, see Section 5.1.4) by that of a dipole $\mathbf{p} = \alpha \mathbf{E}_{\mathrm{Loc}}(\mathbf{r}_0)$, where α is the dipolar polarizability (assumed to be isotropic here) of the particle. The optical force $\mathbf{F}_{\mathrm{opt}}$ in this case acquires a relatively simple form, namely [202]:

$$\mathbf{F}_{\mathrm{opt}} \propto \left[\alpha' \boldsymbol{\nabla} \left(\frac{|\mathbf{E}_{\mathrm{Loc}}(\mathbf{r})|^2}{|\mathbf{E}_0|^2} \right) + \alpha'' \frac{|\mathbf{E}_{\mathrm{Loc}}(\mathbf{r})|^2}{|\mathbf{E}_0|^2} \mathbf{k} \right]_{\mathbf{r}=\mathbf{r}_0}, \tag{8.2}$$

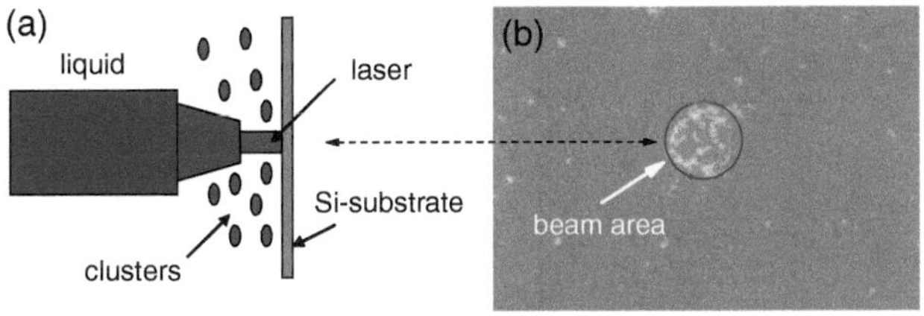

Figure 8.26. Radiation pressure in SERS-active liquids. A citrate-reduced Lee-&-Meisel Ag colloid solution with 10 mM KCl and 100 nM of RH6G is used for this particular experiment. A 633 nm laser (3 mW) is focused (with a ×100 immersion objective) onto a Si wafer standing in the *vertical* position; to avoid any effect of gravity. This is schematically represented in (a). After 5 min of shining the laser onto the wafer a clear accumulation of clusters can be seen under white light illumination in (b). The circle in (b) represents (approximately) the area where the laser is focused. The clustering is produced by radiation pressure on the colloidal clusters.

where α', α'' are the real and imaginary parts of the polarizability α, $\mathbf{E}_{\mathrm{Loc}}(\mathbf{r})$ is the local field, and $|\mathbf{E}_0|$ the incident field amplitude [202]. The main result to be highlighted here is that this expression exhibits two contributions: (i) a conservative term ($\propto \alpha'$), and (ii) a dissipative one ($\propto \alpha''$). An object without absorption can be trapped by a laser field by the action of the first term in Eq. (8.2) (the so-called 'gradient force') and this provides the standard principles of *laser tweezing* and optical manipulation of dielectric particles. The second term, on the other hand, only happens under the presence of absorption ($\alpha'' \neq 0$) and it is responsible for *radiation pressure* . One of the reasons why metallic particles are not chosen as a first option for optical tweezing is precisely because of their absorption. This effect is, moreover, expected to be magnified at wavelengths close to the localized surface plasmon resonance of the particle (where the absorption exhibits a strong resonance). A point often missed, however, is that away from this resonance, it is possible to find experimental conditions where gradient forces can overcome the effect of radiation pressure. The presence of radiation pressure cannot nevertheless be completely ignored.

8.4.2. Radiation pressure in colloidal fluids

To study the relevance of optical forces in SERS, we can study several aspects of the problem separately. For a start, metallic nano-particles used in SERS (either gold or silver) *do* have absorption, i.e. $\alpha'' \neq 0$ which implies that in liquids they will always be exposed to some amount of radiation pressure.

The existence of radiation pressure on colloids or clusters in SERS-active liquids can be readily demonstrated experimentally; Fig. 8.26 shows an example. In this case a citrate-reduced Lee-&-Meisel Ag colloid solution (as described in Chapter 7) with 10 mM KCl (to induce partial aggregation)

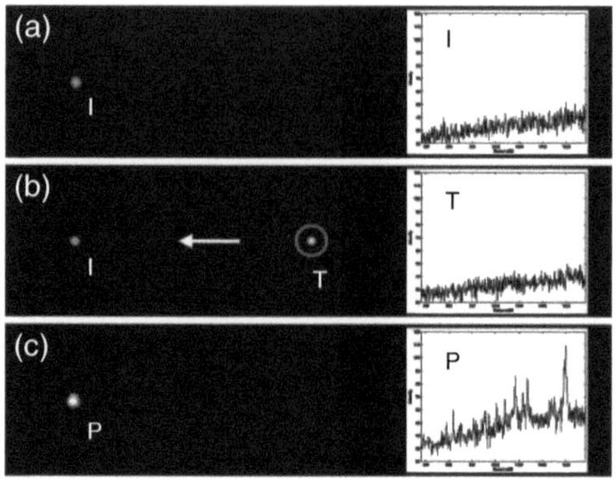

Figure 8.27. Demonstration of optical manipulation of particles in SERS (reproduced from Ref. [43] with permission, copyright 2006 Am. Chem. Soc.). The particles (silver) are visible through dark-field illumination and are covered with benzenethiol (thiophenol), which serves as the SERS probe. In (a), the SERS spectrum from a single immobilized particle (I) is shown and no SERS signal is detectable. In (b), an optically trapped particle (T) (without SERS signal, shown on the right) is moved across the optical field toward the immobilized one (by laser-trapping it). When the trapped particle is brought into near-field contact with the immobilized one, the particle pair (P) shows the signatures of the SERS signal on benzenethiol. This experiment shows the potential of combining laser manipulation of particles with SERS to achieve better control and understanding of hot-spot formation.

and 100 nM of RH6G is used, resulting in the presence of small (stable) colloidal clusters. The experiment is conducted by focusing the laser (with a horizontal immersion objective) onto a vertically-positioned Si wafer, to avoid any possible effect of gravity on the deposition of clusters on the surface. The region marked with a circle in Fig. 8.26 has been illuminated by a 3 mW HeNe laser (through a $\times 100$ immersion objective) for a period of 5 min. The laser 'pushes' (by radiation pressure) clusters that enter the scattering volume of the objective toward the surface. The obvious accumulation of clusters in the illuminated region can be readily seen from the figure, and it is a demonstration that radiation pressure in general will affect the dynamics of small clusters in liquids, even at moderate laser power. Statistical analysis of the signals in liquids should, in principle, include this effect when it might be relevant for the interpretation of results, for example in power-dependence experiments.

8.4.3. Optical trapping of metallic particles

The principles of optical manipulation of particles combined with SERS has been explicitly demonstrated in Ref. [43] and this is illustrated in Fig. 8.27. Silver particles are coated with benzenethiol (thiophenol) in this example, and the single-particle enhancement factor is not high enough to produce a measurable SERS signal. A second particle ('T' in Fig. 8.27) is trapped by the gradient force of the laser in the NIR beam and moved across the optical field to make contact with the first one. The particles are visible only because of scattering observed by dark-field illumination. Once the particles are put in contact, the SERS signatures of benzenethiol can be seen.

Like many of the results briefly reviewed in this section, such experiments are still at a development stage; but even as 'demonstrations of principle' they do show the enormous potential that the combination of laser-induced forces on metallic particles and SERS (or related techniques) can have, and surely will have, in the future.

8.4.4. Optical forces on molecules

As explained before, the effects of laser forces may not be limited to the particles themselves. References [202,300], for example, explore the possible effects of optical forces *on the probe molecules themselves*. This is something that goes well beyond the range of interests of optical manipulation of particles by lasers, but is of prime importance for the SERS problem. The individual molecules have a linear optical polarizability and, as such, they are affected by optical forces too , in particular in the very intense field gradients generated at hot-spots. We follow closely the discussion in Ref. [202] in this sub-section. A molecule on the surface experiences the optical forces created by the large field gradients produced by LSP resonances. Like any other conservative force, the conservative part of the optical force can be derived from *a scalar potential*, which depends on the 'gradient force' term in Eq. (8.2) and, hence, only on the real part of the molecular linear polarizability. Figure 8.28(b and c) shows the (normalized) calculated optical potential felt by a RH6G molecule at the hot-spots of a dimer or trimer formed by Ag colloids with characteristic sizes used in SERS. The potential is normalized in Fig. 8.28(b and c) in units of k_BT at room temperature, and per unit incident power density. For a given incident laser power, it is easy to compare from this calculation how much of an *optical trapping* potential this would produce. For large power densities (but achievable in real experiments), it is easy to show that the optical potential could be as deep as $\sim 6k_BT$; which means that the molecule can be effectively trapped and will have a relatively low probability of escaping the potential by thermal activation.

In this framework, the presence of the dissipative force cannot be ignored; i.e. the maximum force exerted by this potential will always have to be compared with the force produced by the dissipative component (absorption) to guarantee that trapping can actually occur. In Ref. [202] it was taken into account explicitly to argue in favor of an optical trapping mechanism. The dissipative force cannot be obtained from a scalar potential, but it is still possible to compare the position-dependent forces along a specific direction. This is explicitly shown in Fig. 8.28(d) for a (silver) dimer with typical dimensions and at 550 nm, where a resonance occurs in the optical potential (see Fig. 8.28(b and c)). The combined effect of conservative and dissipative forces still results in a force that has a local minimum (stable point). This implies that the trapping of a molecule may still be possible despite the presence of absorption and radiation pressure.

We note that these calculations were carried out for optimal hot-spot conditions (smallest gaps, high power, etc.), which may be rare in practical situations. Even if molecular trapping does not exist, it remains possible that molecular optical forces play a role, for example affecting possible surface diffusion of the molecules. In any case, a clear-cut direct experimental proof of optical forces on molecules (and their predicted effects on the physics and statistics of SERS signals) remains to be presented. Nonetheless, this is a current topic on the fundamental aspects of SERS that is likely to receive attention in the future and that could be linked to many details of the observed phenomenology that have been either ignored or assigned to a different cause in the past.

8.5. APPLICATIONS OF SERS

For everything that has been said so far, we still have to solve, for any particular application, the problem of actually having the molecules we want to detect at the right places! In other words, to have the analytes at those positions where they can profit from electromagnetic enhancements, firstly on the metal surface, and ideally at the hot-spots. This problem has been touched upon in several ways throughout the book but it deserves some further comments here, for it actually stands as the first major obstacle in many real practical applications of SERS. It is also a topic of many recent developments in the field. Here we close (in a way) the circle on aspects that were discussed in the introductory remarks in Chapter 1. We can revisit them now from the standpoint of having a fully developed understanding of many basic aspects of SERS.

In the spirit of the previous sections, rather than summarizing tens (possibly hundreds!) of different reports in the literature, it seems more appropriate to highlight a few principles of recent developments that might actually stand the test of time in future applications of the effect.

8.5.1. Analyte engineering and surface functionalization

We pointed out in the introduction of the book that the best SERS probes are those that actually can profit from electromagnetic enhancements by adsorbing efficiently to metals (primarily Ag or Au) and, by the same token, exhibit a large intrinsic polarizability. But we also mentioned several molecules that *do not* fall in this category. It is possible, however, to 'transform' a molecule that is not intrinsically a conventional SERS probe into one that will have useful SERS activity. Let us review a few concepts here:

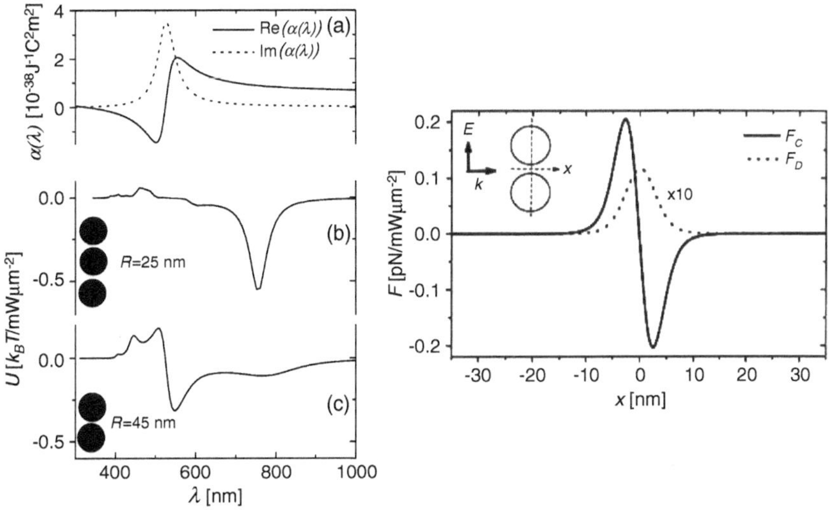

Figure 8.28. Figure adapted from Ref. [202] (reproduced with permission, copyright 2002 The American Physical Society). In (a) the optical polarizability of RH6G is simulated by an oscillator model (see Appendix D). (b) and (c) display the optical potential (conservative part) felt by a RH6G molecule at the junctions of either a dimer or trimer (formed by Ag colloids of different radii in both cases, $R = 25$ or 45 nm for the trimer and dimer, respectively, and separated by 1 nm gaps only). The optical potential is given in units of k_BT at room temperature and is normalized per unit incident power. For a large (but achievable) incident power density of $\sim$10 mW/μm^2, this produces optical potentials with a depth of $\sim 6k_BT$. (d) Comparison of dissipative (F_D) and conservative (F_C) forces along the x-axis through a silver dimer gap (see inset) (radii of the colloids are 45 nm, separation $d = 1$ nm, $\lambda = 550$ nm). The combined effect of the two forces still produces a stable point, despite the presence of a dissipative component.

- Many molecules can be attached to functional groups that will effectively bind to metals. The attachment of thiol groups is the canonical example, but it does not stop there. It is possible to do this *analyte engineering* in conjunction with *surface functionalization*. These latter two topics seem to be an active and growing area of interest in recent developments in SERS.
- The group attached to the molecules can be complementary to a group present on the functionalized surface. A typical example of this is the use of short strands of DNA, to achieve substrates that will bind only to specific complementary sequences. Another example is the functionalization of metal surfaces with anti-bodies.
- It is interesting to note that SERS has a differential advantage with respect to fluorescence in that respect. Fluorescence is another technique where analyte engineering is often required. However, for

SERS to work we only need to bring the molecule to the surface. If the molecule is in contact with the substrate, it will exhibit a SERS effect. However, it is *not* always possible to make a molecule fluoresce! (in particular small molecules). In other words: there are a lot more alternatives between *analyte engineering* and *surface functionalization* to transform a molecule into a SERS-active one, than there are possibilities to make them fluoresce! In that sense, it can be said that SERS is a lot more general as an analytical tool. In reality, it complements and feeds from the much greater experience available in fluorescence spectroscopy, and will most likely operate in parallel with it in the future.

Much effort has been put into the topic of *surface functionalization* at the moment. It is important to remind though that what matters is *surface functionalization in conjunction with controlled enhancements*; i.e. the functionalization needs to be done in the regions where the enhancement is actually present and large. Surface functionalization should certainly be a major part of many developments in terms of applications of SERS and of related surface-enhanced spectroscopies.

8.5.2. Substrate reproducibility and SERS commercialization

The issue of reproducibility and control over both the molecular adsorption (surface chemistry) and the enhancements is paramount to any application of SERS in analytical chemistry. We have mentioned them several times already in the book, and they constitute a driving force in the recent developments pertaining to the application of SERS.

It is worth highlighting in this framework that SERS substrates – specifically designed for and aimed at analytical applications – have started to appear on the market. This commercialization of SERS is an interesting development in the field, and it remains to be seen how it evolves over the next few years. An underlying theme of all these attempts at commercialization – or implementations of the technique for practical applications – seems to be the principle, already mentioned, that *very high* enhancements are 'sacrificed' in favor of homogeneity and reproducibility.

There are several review articles that have appeared lately on studies directed toward 'substrate engineering' for applications using different methods. Halas and coworkers [301], for example, present a study of tailoring of plasmonics substrates with the aim of using them for direct applications. Their approach is based on substrates created by *nano-shells* (Section 8.3) of different geometries (including ellipsoidal nano-shells, also called 'nano-rice' [301]). Furthermore, additional examples of 'rationally designed' nano-structures for SERS applications are reviewed by Mirkin and coworkers [302]. Sensitivity, stability, and reproducibility, are the three cornerstones [302] on which these studies are based. Finally, Smith [303] reviews many aspects of the practical understanding and application of the always-important case of analyte monitoring in aqueous environments. The study also focuses on the *chemical design* of surfaces and labels, highlighting one more time the present importance of surface functionalization in specific applications.

Another example of recent studies on the application side (and potential commercialization) is Ref. [17], where real-time glucose monitoring in bovine plasma is explicitly demonstrated. The study is also interesting because it combines many of the concepts already described in this chapter as possible venues for the creation of reproducible substrates, like metal-film-over-nano-sphere (Section 8.3.2) (MFON) combined with surface functionalization. Overall, these studies represent a small fraction of a much wider current trend toward reproducible substrates and applications; a trend that can only increase in years to come with better understanding and new tools from nano-technology [302], which will surely increase the potential commercialization of the technique.

8.6. EPILOGUE

The long term relevance of some of the specific examples highlighted in this chapter is surely not guaranteed. It is however fair to say that some of the broad areas, namely:

- single-molecule SERS,
- tip-enhanced Raman spectroscopy (TERS),
- development of new substrates from nano-technologies, and
- surface and analyte functionalization,

are likely to be on the agenda for a while. They played an important role in the recent renewed interest in the technique and represent a substantial part of current active research in SERS and its applications. In addition, there are many other secondary recent developments that do not fall into these main trends and tend to be more specialized (but no less interesting). This is the case for example of optical forces, discussed also in this chapter, or the case of SERS vibrational pumping [59].

Far from being complete (or even objective) the specific examples discussed in this chapter show a glimpse of some of the latest developments in SERS in the last decade in both fundamental and applied aspects. There is no reason to believe that the pace will slow down, and SERS is likely to keep expanding in key application areas, as well as improving in its microscopic understanding. On the fundamental aspects side, several outstanding issues remain unsolved; many of them related to polarization effects and surface selection rules. On the application side, the diversity of possible paths offered by combinations of technologies and methods (nano-lithography, micro/nano-fluidics, self-organization, etc.) is certainly overwhelming and sometimes even difficult to follow even for the dedicated practitioner. It is also likely that research in somewhat unrelated areas at the moment (like the creation and manipulation of particles to form patterns and arrays [304]), will start playing an increasing role in the development of SERS substrates for different applications either as intermediate (or essential) steps.

To conclude, we are not going to fall into the temptation of labeling as 'mature' the examples selected here, together with the field as a whole (in the spirit of some early reviews of the field in the 80s). Instead, we prefer to believe that the list of results presented in this chapter will be (in decades to come), not a demonstration of the maturity of the field, but rather just one more step in its historical development.

Appendix A

Density functional theory (DFT) calculations for Raman spectroscopy

The density functional theory (DFT) of the electronic structure of molecules is a vast and diverse field that spreads over hundreds of different applications and scores of different theoretical (and numerical) approaches. DFT calculations for molecules have been around for several decades now, and the field is in a high degree of development in what is usually described as *quantum chemistry*.

There are many review articles and books on the subject [305,306], as well as research articles where the principles of DFT are outlined and used for the specific prediction of Raman and IR spectra of important types of molecules [307].

It is not the objective of this appendix to review all the information on DFT available in the most specialized literature; it would not do justice to the importance and depth of the subject. We attempt to review the information here from the viewpoint of a potential *user* interested in SERS and trying to gain some insight into the Raman spectra of the probes. This brief review is therefore concentrated on the *applied* aspects of DFT for Raman calculations, in particular for the prediction of Raman spectra, cross-sections, and symmetry properties of the vibrations.

A.1. A BRIEF INTRODUCTION TO DFT

A.1.1. Computing aspects of DFT

There are several DFT packages available, some commercial and some public domain. All of them have advantages and disadvantages in their possibilities and the way they present their input/output information to the user. All of them do essentially the same, as we shall highlight later in this subsection, but they present the information differently and not all of them calculate the specific properties of interest for a given user. A judicious choice of a code depending on the specific requirements of the user is obviously the first step for a successful use of these tools. For a start, not all of them calculate the Raman or IR spectra, for example, and many codes have substantial differences in the computational approaches they use to solve the problem[1]. High performance DFT codes are unlikely at this stage to be developed by an independent isolated user; the codes are the result of decades of developments and improvements by dedicated groups in the details of numerical techniques and in the underlying principles. Some of the most widely available open source codes[2] have reached already the level of several hundred thousand programming lines over the last decade; from the point of view of a potential user, there is no option but to become familiar with the layout and input/output framework of a specific program and use them as a tool to predict useful properties of relevance to real experimental conditions. More often than not, state-of-the-art computational chemistry gives very little chances to the isolated programmer, and has been transformed into a *cooperative effort*.

Having said this, it is important (from the viewpoint of a *user*) to have a sound understanding of what the programs actually do when a calculation is performed, and the meaning of some of the basic physical principles. The raw output of the programs may be a very long and unfriendly (though more-or-less 'formatted' and 'readable') file. Commercial programs tend to develop additional software for visualization purposes of the solution, and not all of them cover the desired features. Many of the results obtained from open source DFT codes might require additional programming by the user to *visualize specific aspects of the solution*. This is particularly true for spectroscopic data. Here we report results obtained by using either Gaussian[3], a commercial software, or GAMESS (the General Atomic and Molecular Electronic Structure System), a general *ab initio* quantum chemistry package freely available[4]. Both packages can calculate the Raman spectrum of a molecule. For our purposes, all the visualization tools have been developed in our laboratory[5].

---

[1] The computational approaches differ not only in the mathematical details of the algorithms but also in their layouts. Not all of them can be 'parallelized' in several processors, for example. This might be important at the moment of calculating the vibrations, which can be 'parallelized' with massive savings in computational time.

[2] For example: ABINIT (http://www.abinit.org).

[3] http://www.gaussian.com/

[4] http://www.msg.ameslab.gov/GAMESS/

[5] Thanks are due to Matthias Meyer (Victoria University of Wellington, New Zealand) for developing the Advanced Raman Simulation Engine, a Matlab-based visualization tool for Raman tensors and spectra obtained by Gaussian.

We follow closely the presentations in Refs [305,306] to explain briefly in the following what is meant by doing a DFT calculation of the electronic structure of a molecule, as a way of highlighting the underlying principles of the method. Once this is done, the programs will be used essentially as a 'black box' to predict the characteristics of Raman spectra of specific molecules. For our purposes here, we only emphasize the conceptual aspects around DFT calculations, leaving again the more advanced details to the specialized literature in the field [305,306].

A.1.2. Principles of DFT

The aim of DFT is to compute the electronic structure of a molecule, i.e. the wave-function of all its electrons for given atomic positions. The most 'stable' geometry of the molecule – as defined by the atomic positions – is also obtained by the same token through a minimization process of the total energy as a function of geometry. If the molecular geometry (atomic positions) and the electronic structure are both known, many important physical properties of the molecule can be calculated; either directly from them or as perturbations of the ground state. Broadly speaking, DFT is 'somewhere in between' *ab initio* and semi-empirical methods. At a fundamental level, it starts by solving Schrödinger's equation for a system of coupled electrons, but some of the most difficult parts to be dealt with (like the exchange correlation) are eventually introduced through a semi-empirical parametrization of the interaction in terms of the electronic density.

Formulation of the problem

As for many other quantum chemistry approaches, the main principle is to replace a complex many-body problem (many correlated and interacting electrons) by a simpler single-electron problem, where the effect of the other electrons is treated approximately. More precisely, the DFT approach consists in solving and finding the ground state of the non-linear Schrödinger equation (Kohn–Sham equation) defined by [305]:

$$\left[-\frac{\hbar^2}{2m}\nabla^2 + V_{\mathrm{KS}}(n(\mathbf{r}),\mathbf{r})\right]\Psi(\mathbf{r}) = E\Psi(\mathbf{r}), \tag{A.1}$$

where all the symbols have the standard meaning in quantum mechanics and $V_{\mathrm{KS}}(n(\mathbf{r}),\mathbf{r})$ is the Kohn–Sham potential , which depends intrinsically on the electronic density:

$$n(\mathbf{r}) = |\Psi(\mathbf{r})|^2, \tag{A.2}$$

and it is, therefore, called a *density functional*[6]. This dependence accounts for the interaction and correlation with the other electrons. Unlike a simple problem in quantum mechanics with a standard (fixed) external potential, it is the implicit dependence of the Kohn–Sham potential on the density that transforms the problem into a non-linear problem. The solution to the Kohn–Sham equation provides a self-consistent molecular orbital wave-function for the electronic structure of the molecule.

In principle, a molecule consists of an ensemble of nuclei (in a specific molecular geometry) together with bound (inner-shell) and valence electrons interacting with both the nuclei and among themselves. The Kohn–Sham potential $V_{\mathrm{KS}}(n(\mathbf{r}),\mathbf{r})$ consists essentially of three contributions [305,306]:

$$V_{\mathrm{KS}}(n(\mathbf{r}),\mathbf{r}) = V_{\mathrm{atom}}(\mathbf{r}) + V_{\mathrm{Hartree}}(n(\mathbf{r}),\mathbf{r}) + V_{\mathrm{exchange}}(n(\mathbf{r}),\mathbf{r}), \tag{A.3}$$

where:

- $V_{\mathrm{atom}}(\mathbf{r})$ is the Coulomb interaction of the electrons with the nuclei (typically a sum of nuclear potentials centered at the atomic positions). This potential depends only on $\mathbf{r}$. In addition, depending on the specific DFT algorithm under use we can consider all the electrons and the potentials of the bare nuclei, or we can divide the electrons into *valence* and *core* (inner-shell) electrons and solve the self-consistent wave-function for the former only; i.e. the DFT calculation is performed for the valence electrons in the *screened* (by the inner-shell electrons) Coulomb interaction of the nuclei. In this latter case we talk about a *pseudo-potential* for the valence electrons.
- $V_{\mathrm{Hartree}}(n(\mathbf{r}),\mathbf{r})$ is the Coulomb repulsion among electrons. Implicit in this term is the fact that a given electron at position $\mathbf{r}$ will experience a repulsion depending only on the electron density at that point ($n(\mathbf{r})$).
- $V_{\mathrm{exchange}}(n(\mathbf{r}),\mathbf{r})$ is the *exchange potential.* The latter also depends on $n(\mathbf{r})$, and it is (in general) the one that has the largest number of different versions based on different degrees of approximation and parametrizations [306,308,309].

From a conceptual point of view, the great breakthrough that allowed the development of DFT was the realization of the fact that both $V_{\mathrm{Hartree}}(n(\mathbf{r}),\mathbf{r})$ and $V_{\mathrm{exchange}}(n(\mathbf{r}),\mathbf{r})$ could be expressed as functions of the electronic density $n(\mathbf{r})$. What DFT does is that it replaces the complexity of solving a many-body problem for the electrons by the 'simpler' problem of solving a one-

[6] In mathematical terms, a *functional* form is a function that depends on another function; i.e. one of the 'variables' of the function is a function in itself. In this particular case, the potential function depends on the *electron density*, which is itself a function (of the position), hence the name *density functional.*

electron Schrödinger equation in a non-linear potential (which depends on the electronic density).

Self-consistent solution

Solving the non-linear Kohn–Sham equation (A.1) presents challenging numerical obstacles. In almost all cases, the wave-function solution $\Psi(\mathbf{r})$ is first expressed, as a linear combination of basis functions $\phi_i(\mathbf{r})$, called the *basis set* :

$$\Psi(\mathbf{r}) = \sum_i c_i \phi_i(\mathbf{r}). \tag{A.4}$$

The equation is then replaced by a set of coupled non-linear equations on the coefficients c_i. The choice of an appropriate basis set of wave-functions is therefore a crucial step. From there, the solution normally proceeds as follows:

- An initial guess for the ground state solution is proposed.
- This initial guess defines the density $n(\mathbf{r})$, which then defines the Kohn–Sham potential (which depends on $n(\mathbf{r})$).
- The Kohn–Sham equation is solved for this fixed potential (and is therefore a linear equation). The ground state is found (as a linear combination of the same basis set of functions used at the beginning).
- The density is recalculated and compared with the density we had at the beginning. If the two are identical, convergence has been achieved and the problem is essentially solved; all calculations of physical properties follow from there, by means of suitable matrix elements of operators with the calculated wave-functions. If the initial and final densities are not the same, the Kohn–Sham equations are solved again introducing the newly calculated density in the $V_{\mathrm{KS}}(n(\mathbf{r}), \mathbf{r})$ potential and repeating the process through a new iteration, until convergence is eventually achieved. It is important to realize that even when convergence is achieved, the solution is still limited by the accuracy and degree of approximation of the original basis set with which the wave-function is constructed.

A.1.3. Important parameters

Without going deeper into the technical details of DFT, there are broadly speaking two main choices to be made at the beginning of the calculation from a mere practical (user) point of view:

- *The method*, which specifies the exact form of the Kohn–Sham potential, i.e. its dependence on $n(\mathbf{r})$, and

- The nature and type of *the basis set* to be used as wave-functions for the calculation of the self-consistent electronic field.

The method

Acronyms are typically used to refer to the different available methods. For example, the program Gaussian includes the possible use of hybrid functionals which include a mixture of Hartree–Fock exchange with DFT exchange correlation. One of the most widely used versions of this is Becke's three-parameter hybrid functional [310] with Lee–Yang–Parr [311] non-local electron correlation (commonly abbreviated as B3LYP). Full lists of acronyms for the different methods and approximations are typically available with the DFT programs, together with references to the original work where they have been developed. Accordingly, they will not be reviewed in full here.

The basis set – general considerations

The choice of the basis set depends on the nature of the problem. It is quite clear that the convergence of the self-consistent electronic field will be faster and more efficient the more the initial set of wave-functions resembles the final result. Therefore, problems involving lattices with *translational invariance* (solids or surfaces, for example) make the most efficient use of computational time by starting with a basis set of *plane waves*; which are functions that naturally respect the requisite of translational invariance. For a system with a finite size, a plane wave expansion is still possible using what is called the super-cell method, in which the finite system is periodically repeated in an artificial lattice that preserves translational invariance [305] ; this is a trick very often used for electronic calculations of surfaces. However, the most successful approach for moderately large organic molecules is based on the use of *a small number of Gaussian-like orbitals* through the methods developed in a remarkable series of ~20 papers (spanning over a decade) by J. A. Pople's group [312] and coworkers [313,314]. We only mention in passing here that the main advantage of the Gaussian expansion of orbitals is that it allows the calculation of two electron integrals (matrix elements) *analytically*, rather than the more time consuming numerical counterpart involving Slater-type orbitals.

Gaussian basis sets

There are different types of Gaussian basis sets that can be used for DFT calculations; the different basis sets add different levels of complexity and detail to the accuracy of the solutions. The addition of *polarization functions*, for example, improves the agreement with experimentally determined magnitudes (in particular when heavy atoms are involved). The different sets available are referred to by a code such as: STO-3G (minimal basis), 3-21G (small), 6-31G* (medium), or 6-31G** (large). The sets 6-31G* and 6-31G**, for example, include polarization functions for heavy atoms only, or heavy atoms *and* hydrogen, respectively. There are also 'diffuse' basis sets, for example: 6-311++G(d,s).

Choosing a basis set

Relatively fast calculations for quick checks can always be done with the small 3-21G set. It is generally accepted that research results reported in the literature have an acceptable level of accuracy at the 6-31G* (medium basis) level. The best (large) set including polarization functions for hydrogen could be unfeasible in some situations, due to the computational time involved, which scales very rapidly with the size of the molecule.

The main limitations in the size and complexity of the basis sets is most of the time based on the computational time. A rule of thumb is that the computational time required can increase exponentially with the combined total size of the basis sets and the number of atoms. This is a limit that is obviously evolving all the time, with the available computational power. Calculations on medium sized molecules with the most complete set of polarized Gaussian basis functions were impossible 10–15 years ago, except in the largest computational facilities at that time. However, they are becoming outdated now with moderate computational resources. As a general guide, the approach should be to use the largest basis set available compatible with a reasonable computation time (which has to be decided according to the size of the molecule).

A.2. APPLICATIONS OF DFT TO RAMAN

A.2.1. Principle

The most useful application of DFT calculations in a SERS context is the determination of the vibrations of the molecule (vibrational patterns), as well as the corresponding Raman shifts (vibrational energies), Raman cross-sections, and Raman tensors (see Chapter 2). The determination of the normal modes and their Raman polarizabilities and/or tensors requires a set of (normally very time consuming) steps which can be roughly summarized as follows:

- It is first necessary to find the equilibrium atomic positions (the molecular structure). This is called *geometry optimization* and can be carried out using DFT as described in the next subsection.
- The corresponding electronic structure is then known from the same DFT calculation that produces the most stable geometry of the

molecule. This enables the calculation of the force constants (which describe the interactions among atoms). The normal modes can then be computed from standard vibrational analysis as described in Section 2.7.1. The calculation of the force constants can be done directly by displacing each and every atom at a time (along three mutually perpendicular axes from their equilibrium positions) and recalculating the total energy of the configuration by a new DFT calculation. The force constants on each individual atom can be obtained by numerical differentiation of the total energy, and from here the Hessian matrix for the vibrational modes can be constructed, as explained in Section 2.7.1.

- Moreover, the linear optical polarizability tensor of the molecule can be computed using DFT by recalculating the electronic structure in the presence of an external field. This can be done with a fixed (static) external field (if the polarizability 'far from resonance' is desired) or by time-dependent DFT if the polarizability at resonance (or pre-resonance) is sought.
- In addition, the polarizability derivatives (polarizability change with respect to all individual atomic displacements) are computed using DFT by recalculating the electronic structure of the molecule with the corresponding (deformed) structure. This can be done in two ways. The first alternative is to calculate the polarizability derivatives for the entire molecule by a DFT calculation of the deformed molecule, with a deformation that follows the pattern of an eigenvector of a vibration. In that case, the Raman polarizability is obtained directly for that specific mode by direct subtraction of the polarizabilities with and without the deformation. Alternatively, the polarizability derivatives can be calculated with respect to each atomic displacement, and the corresponding Raman tensors of specific vibrations can be obtained by combining this information with the knowledge of the eigenvectors (normal mode pattern). A specific example of the latter for a simple molecule – carbon dioxide, CO_2 – will be provided in what follows.

There is a battery of technical details in these steps (that go from analytic expansions of the potential to calculate the vibrational frequencies, to direct finite-difference numerical calculations of the tensors), which will not be reviewed in detail here. The interested reader is referred to the specialized literature and the technical information of the specific DFT programs, which normally contain details of the specific numerical implementation they use.

A.2.2. Geometry optimization using DFT

As pointed out, a prerequisite for a vibrational analysis calculation is that the molecule is in its equilibrium geometrical configuration. For some standard basic molecules, a previously optimized geometry might be available in the literature but, in general, it is necessary to optimize the geometry of the structure before proceeding to the vibrational analysis. The strategy in this latter case is to start with an optimized molecular structure obtained from an *empirical force field model* , like those described in Refs [122,123], and taking these geometries as a starting point for a geometry optimization with DFT. The geometry optimization step can be extremely time consuming for moderately large molecules. It involves combining the algorithm for the Kohn–Sham self-consistent solution with a *conjugate gradient* method [315], in order to find the nearest local minimum of the total energy as a function of the positions of the atoms in the molecule. If the geometry is not well optimized, many of the vibrational frequencies might turn out to be negative at the end, which is in fact a signal for the presence of a structural instability in the original geometry. One or two small negative frequencies can sometimes be tolerated in big molecules with hundreds of vibrations and are most often the result (or drawback) of numerical accuracy rather than any real structural instability.

A.2.3. Limitations of DFT calculations for Raman

DFT is primarily used in Raman for vibrational analysis, i.e. for the determination of the normal mode pattern (including symmetry) and frequency. A known drawback of DFT determinations of vibrational spectra is the fact that they tend to overestimate slightly the vibrational frequencies [61,316]. This is normally produced by a slight overestimation of the exchange interaction and it is a well known fact that tends to be compensated, at the end, by an empirical scaling factor of the frequencies. Corrections are very rarely larger than ~10%. Except for this 'relatively minor' problem, the agreement between DFT calculations and real spectra can be remarkably good.

Moreover, the predictions of the polarizability derivatives can also be used to determine the Raman cross-sections, but again with some limitations. The standard (and easiest) Raman polarizabilities calculated by DFT derive from calculations of the static polarizability derivatives, i.e. the change in the *static* linear polarizabilities for small atomic displacements. This method is, in fact, precisely called 'static polarizability derivatives' and it has been widely used in the literature. This approach ignores any time dependence, either from the vibrational oscillations or from the incident electric field of the laser. Any electronic absorption, excited state effects, or resonance effect is therefore entirely ignored. This method can, therefore, only be valid as an estimation of the *non-resonant Raman* properties.

In order to account properly for resonance effects, and study resonant (and even pre-resonant) Raman conditions, it is necessary to generalize the approach to *time-dependent density functional theory* (TDDFT). It is however only in the last decade that TDDFT Raman calculations have appeared in the literature [60,317].

Overall, within its range of validity (non-resonant Raman), DFT Raman calculations provide an invaluable tool as an additional layer of understanding into the Raman effect and by extension into SERS. We devote the rest of this appendix to a few elementary examples of the principles underlined here, and to the practical aspects of its implementation for the determination of Raman spectra, tensors, and cross-sections.

A.3. PRACTICAL IMPLEMENTATION

One of the problems a newcomer to DFT computation of Raman spectra will find is the somewhat peculiar units and conventions used in some of the programs available for this purpose. This has resulted from the historical development of the field, together with different natural choices of units coming from chemistry and physics. To this, one has to add the confusion fueled by the sometimes improper use of the output of these programs in the literature. It is not rare to find direct comparisons of experimental Raman spectra with the *Raman activity* provided by some of the most widely used DFT programs like *Gaussian*. The Raman activity is *not* proportional to the cross-section, for the latter includes a pre-factor that depends on the vibrational frequency and even temperature (as explained in Chapter 2). It is the purpose of this section to try to clarify at least some aspects of these issues with the aim of facilitating the task of a potential reader who might want to try his/her own calculations.

We concentrate in the next few sections on some of the definitions and units appearing in the program *Gaussian* (Gaussian 03), which is one of the most commonly used DFT packages. Other programs might use the same or derived units but, nevertheless, the overview of definitions and units is quite general and can be used in more than one context.

A.3.1. Brief overview of the input and output files

We will describe in the following how a simple Raman DFT calculations of the CO_2 molecule can be carried out. Our approach is to do it in two steps, separating the geometry optimization from the Raman calculations. This is not necessary in practice but we find it convenient because the knowledge of the optimized geometry is necessary to understand the Raman output. All the information could be obtained in a single output file, but it would then be more difficult to retrieve.

Geometry optimization

Let us briefly discuss a possible input file for *Gaussian* on a specific example for the CO_2 molecule. The input file that is passed to *Gaussian* reads:

```
%Chk=co2
#T B3LYP/6-311++G(d,p) Opt

co2 opt

0 1
C        -0.4440         0.3650       0.000000
O        -0.4440         1.5250       0.000000
O        -0.4440        -0.7950       0.000000
```

The first line, starting with % is a comment. The second line is the *command line* and is arguably the most important one. This is where all the parameters of the calculation (except for the initial geometry of the molecule itself) are defined. We recognize in particular the choice of the DFT method, here `B3LYP` and the basis set `6-311++G(d,p)` (a large basis set, which is not a problem here for a small molecule). The following keywords then instruct `Gaussian` what to do. In this case `Opt` requires a *geometry optimization*. The rest of the file is the description of the molecule. The line `co2 opt` is only a title for our reference. `0 1` in the next line correspond to the charge and spin multiplicity. The other lines then list the type of atom (`C` or `O` here) and their (tentative) atomic positions (X, Y, Z) in Å.

For such an input file, *Gaussian* will carry out the geometry optimization only, and return the result, possibly in a rotated frame adapted to its internal computations. The results can be found at the end of the output file, in the following somewhat cryptic format:

```
1|1|UNPC-UNK|FOpt|RB3LYP|6-311++G(d,p)|C1O2|PCUSER|22-Nov-2007|0||#T B
3LYP/6-311++G(D,P) OPT||co2 opt||0,1|C,0.,0.,0.|O,0.,1.1607160673,0.|O
,0.,-1.1607160673,0.||Version=x86-Win32-G03RevB.03|State=1-SGG|HF=-188
.6469148|RMSD=4.258e-009|RMSF=8.509e-005|Dipole=0.,0.,0.|PG=D*H [O(C1)
,C*(O1.O1)]||@
```

The optimized geometry can be read (in Å, as required for the input file) and recast into a second input file dedicated to the Raman DFT computations. In this example, it is given as:
`|C,0.,0.,0.|O,0.,1.1607160673,0.|O,0.,-1.1607160673,0.|`

Raman DFT input file

Using this result, the second input file for Raman calculations has the form:

```
%Chk=co2
#T B3LYP/6-311++G(d,p) Freq=Raman NoSymm

co2 opt

0   1
C   0.        0.             0.
O   0.        1.1607160673   0.
O   0.       -1.1607160673   0.
```

It has the same meaning as before. We now use the keyword `Freq=Raman` to request a Raman DFT computation (vibrational analysis and polarizability derivatives). We also use `NoSymm` to prevent the program from using symmetry considerations and internal rotations of the structure. This would potentially speed up the computation in a general case, but would make the output more difficult to process (because rotations may be applied to the molecule). The `NoSymm` option is therefore easiest for a beginner, and is not an issue as long as the geometry has been previously optimized.

The output file, which may be fairly long depending on the exact options, contains toward the end the result of the Raman calculations in the following format:

```
Harmonic frequencies (cm**-1), IR intensities (KM/Mole), Raman scattering
activities (A**4/AMU), depolarization ratios for plane and unpolarized
incident light, reduced masses (AMU), force constants (mDyne/A),
and normal coordinates:
                      1                      2                      3
                     ?A                     ?A                    SGG
Frequencies --   668.3513               668.3513              1373.3652
Red. masses --    12.8774                12.8774                15.9949
Frc consts  --     3.3891                 3.3891                17.7748
IR Inten    --    33.0584                33.0584                 0.0000
Raman Activ --     0.0000                 0.0000                18.5155
Depolar (P) --     0.0000                 0.0000                 0.1639
Depolar (U) --     0.0000                 0.0000                 0.2816
Atom AN      X      Y      Z        X      Y      Z        X      Y      Z
  1   6    0.86   0.00  -0.19     0.19   0.00   0.86     0.00   0.00   0.00
  2   8   -0.32   0.00   0.07    -0.07   0.00  -0.32     0.00   0.71   0.00
  3   8   -0.32   0.00   0.07    -0.07   0.00  -0.32     0.00  -0.71   0.00
                      4
                     ?A
Frequencies --  2420.7840
Red. masses --    12.8774
Frc consts  --    44.4620
IR Inten    --   711.7854
Raman Activ --     0.0000
Depolar (P) --     0.0000
Depolar (U) --     0.0000
Atom AN      X      Y      Z
  1   6    0.00   0.88   0.00
  2   8    0.00  -0.33   0.00
  3   8    0.00  -0.33   0.00
```

We recognize the 4 normal vibrational modes of the CO_2 molecules (here modes 1 and 2 are degenerate), and their main characteristics are listed. The number of atoms here is $N = 3$ and there are $(3N - 5) = 4$ internal modes because CO_2 is a linear molecule (there would be $(3N - 6)$ otherwise, as explained in Section 2.7.1). The rest of this section is dedicated to understanding and exploiting this output.

A.3.2. Common units and definitions in Raman calculations from DFT

General units

A big fraction of the output of *Gaussian* is in *atomic units.* In particular:

- Distances are expressed in Angstroms (1 Å $= 10^{-10}$ m) or in Bohr (B hereafter):

$$1\,\mathrm{B} = 0.529\,\text{Å}. \tag{A.5}$$

- Masses are in atomic mass units (amu):

$$1\,\mathrm{amu} = 1.66 \times 10^{-27}\ \mathrm{kg}. \tag{A.6}$$

- Polarizabilities are in B^3, which is itself derived from Å^3 and is expressed in Gaussian units (not related to the name of the software!). Since $1\,\text{Å}^3 = (4\pi\epsilon_0) \times 10^{-30}$ S.I. $[\epsilon_0\ \mathrm{m}^3]$, we have

$$1\,\mathrm{B}^3 = 0.148\,\text{Å}^3 = 1.649 \times 10^{-41}\,\mathrm{S.I.} \tag{A.7}$$

One important quantity provided by the program is the Raman activity R_k $[(\epsilon_0)^2\ \mathrm{m}^4\ \mathrm{kg}^{-1}]$, which is most commonly given in $[\text{Å}^4/\mathrm{amu}]$. This a somewhat unfortunate mix of Gaussian units (in which the square of a polarizability derivative is in $[\text{Å}^4]$), and atomic units [amu] for the masses. These are related by:

$$R_k[\mathrm{S.I.}] = 7.41 \times 10^{-34}(R_k[\text{Å}^4/\mathrm{amu}]). \tag{A.8}$$

As can be seen from these examples, it is not so surprising that some of these units produce confusion among potential new users.

Vibrational modes

For a given mode, there are three important parameters that most DFT programs provide in the output for each normal mode k: the frequency ω_k $[\mathrm{s}^{-1}]$, the force constant[7] K_k $[\mathrm{kg\ s}^{-2}]$, and the reduced mass μ_k [kg] (which will be defined later). These are related through:

[7] The force constants have a different unit from those introduced in Section 2.7.1 ($f_{i,j}$ $[\mathrm{s}^{-2}]$) because they are defined here with respect to the standard coordinate [m], instead of the reduced-mass coordinates $[\mathrm{kg}^{1/2}\ \mathrm{m}]$.

$$\omega_k^2 = \frac{K_k}{\mu_k}. \tag{A.9}$$

For the specific case of the *Gaussian* output, the units are mixed and more a consequence of historical developments in chemistry, spectroscopy, and DFT: frequency is given in wave-numbers $\bar{\nu}_k = \omega_k/(2\pi c)$ in cm^{-1}, the reduced mass in amu, and force constants in mdyn/Å (1 dyn is 10^{-5} N). Expressed in these particular units, they are then linked by the relation: $K_k \equiv 5.89\times 10^{-7}\mu_k\bar{\nu}_k^2$, as can easily be checked in the output given above. The analysis of the vibrational pattern is discussed later.

Raman activity and cross-section

In addition to the normal mode mechanical properties, some programs provide some optical data, at least the *Raman activity*, R_k, and often the Raman depolarization ratio ρ_k. This is for example the case of *Gaussian*. It must be noted that the Raman activity R_k is already an *averaged* property (over the orientations of the molecule).

Two depolarization ratios, denoted P (for polarized) and U (for unpolarized) are in fact given. The one denoted P is the relevant one, ρ_k, discussed in Chapter 2. The other one ρ_k^U refers to unpolarized excitation in a Raman scattering experiment. This unpolarized depolarization ratio U comes from the time when Raman spectroscopy was performed with unpolarized atomic lines (from arc-lamps) as excitation. With the general use of lasers nowadays, this ratio is rarely relevant. Moreover, it is duplicate information since it can be inferred from ρ_k as $\rho_k^U = (2\rho_k)/(1+\rho_k)$.

From the Raman activity R_k, it is possible to calculate the absolute differential Raman cross-section using Eq. (2.86). One must however be careful with the units: the Raman activity is typically given in units of [Å^4/amu] and Eq. (A.8) must therefore be used for the conversion to S.I. units.

Denoting λ_L the wavelength of the laser (typically in nm) and $\bar{\nu}_k$ the wave-number of the vibrational mode (typically in cm^{-1}), the absolute differential Raman cross-section $\mathrm{d}\sigma_k/\mathrm{d}\Omega$ (in cm^2/sr) at room temperature (290 K) can then be obtained by re-writing Eq. (2.86) as:

$$\boxed{\begin{aligned}\frac{\mathrm{d}\sigma_k}{\mathrm{d}\Omega}[\mathrm{cm}^2\,\mathrm{sr}^{-1}] &= 5.8\times 10^{-46}\,\frac{\left[10^7/(\lambda_L\,[\mathrm{nm}]) - (\bar{\nu}_k[\mathrm{cm}^{-1}])\right]^4}{(\bar{\nu}_k[\mathrm{cm}^{-1}])}\\ &\quad\times L_M \times (R_k[\text{Å}^4/\mathrm{amu}]) \times \left(1-\exp\left(-\frac{\bar{\nu}_k[\mathrm{cm}^{-1}]}{201.56}\right)\right)^{-1}.\end{aligned}} \tag{A.10}$$

L_M, given and discussed in Section 2.4.5, is the *local-field correction*, which depends on the index of refraction of the environment (for example, it depends on whether the measurement is performed in a liquid or gas phase).

As an example, for the symmetric stretching mode of CO_2 (mode 3), we have $\bar{\nu}_k = 1373\ \text{cm}^{-1}$ and $R_k = 18.5\ \text{Å}^4/\text{amu}$. In the gas phase, we have $L_M = 1$ (the index of refraction is $n_M \approx 1$ for gases typically). The DFT calculation therefore predicts that at $\lambda_L = 633$ nm excitation:

$$\frac{\mathrm{d}\sigma_k}{\mathrm{d}\Omega} = 3.4 \times 10^{-31} \quad \text{cm}^2/\text{sr}. \tag{A.11}$$

This is in very good agreement with experimental measurements (which are probably more prone to errors than DFT on such small molecules) that yields $\mathrm{d}\sigma_k/\mathrm{d}\Omega = \approx 3 \times 10^{-31}\ \text{cm}^2/\text{sr}$ [86].

A.3.3. Normal mode patterns and Raman tensors

Vibrational analysis

The normal modes of vibration can further be characterized by the actual vibrational pattern (eigenvector) they correspond to, i.e. the relative displacements of the atoms from their respective equilibrium positions. For a given normal mode k, this can be specified either as the *real displacements* ξ_k^n [m] in Cartesian coordinates (n from 1 to $3N$ are the 3 coordinates of the N atoms in the molecule, and k is the index of the normal mode) or the *mass-weighted displacements* q_k^n [$\text{kg}^{1/2}$ m] in reduced-mass coordinates.

What is given in the *Gaussian* output[8] is the *normalized real displacements in Cartesian coordinates*, ϕ_k^n [a.d.], i.e. $(\phi_k^n)_{n=1..3N}$ is a vector of unit norm proportional to $(\xi_k^n)_{n=1..3N}$ and therefore represents the relative amplitude of the movements of the atoms in normal mode k. In mathematical terms:

$$\sum_{n=1}^{3N} (\phi_k^n)^2 = 1, \tag{A.12}$$

and the ϕ_k^n have therefore *no units* but are proportional to the real displacements ξ_k^n [m], i.e.

$$\phi_k^n = \frac{\xi_k^n}{\sqrt{\sum\limits_{p=1}^{3N} (\xi_k^p)^2}}. \tag{A.13}$$

[8] Note that we are assuming here that the molecule has not been rotated from its original geometry. Some programs (including *Gaussian*) give sometimes the vibrational eigenvectors in a different reference frame from the original input geometry; both are related by a rotation matrix which has to be checked in each specific case (called the 'Z-matrix' in most programs). Our two-step approach using the keyword `NoSymm` avoids these problems.

The ϕ_k^n are given in the output under the column `X,Y,Z` for each atom (in rows) and each normal mode k (in columns). The normal mode pattern can therefore be 'read' directly from the output. The practical visualization of these patterns is however more difficult in general (depending on the molecule), except in the simplest cases.

In our example, the CO_2 molecule is aligned along the y-axis, as specified in the second input file (for the Raman DFT). For mode 3, for example, atom 1 (C) does not move, and the O atoms (2 and 3) move in opposite directions along y (the molecular axis). This is the symmetric stretching mode (which will also be identified in Appendix B). Similarly, mode 4 can be identified as the asymmetric stretching mode, and modes 1 and 2 as the (degenerate) bending modes. The results are obviously in agreement with the patterns discussed within the framework of the bond-polarizability model in Appendix B.

There a very important point that needs to be re-emphasized here: the knowledge of the eigenvector is an essential piece of information that can, in many cases, be the missing link to have an explanation for a particular observation. Depending on the localization of the vibrational movements in the structure of the molecule, for example, one can speculate on the possible effects the *surface binding* will have on the SERS spectrum.

The linear polarizability derivatives

The prediction of Raman properties with DFT requires the computation of the *linear polarizability derivatives*, $(\partial\hat{\alpha}_L/\partial q_i)$, and these are generally also given in the output, although not always in as friendly a form as the normal mode characteristics. In *Gaussian*, for example, they are given as $\hat{\alpha}'_i = (\partial\hat{\alpha}_L/\partial\xi_i)$, i.e. the derivative with respect to the *real displacements* ξ_i and can be found at the end of the output file after the keyword '`PolarDeriv=`'. In our CO_2 example, this looks like:

```
-0.2719917|Polar=8.9809966,0.,25.9890883,0.,0.,8.9809966|PolarDeriv=0.
,0.,0.,0.,0.,0.,0.,0.,0.,0.,0.,0.,0.,0.,0.,0.,0.,0.,0.,3.8770029,0.,0.
,0.,0.,2.1700287,0.,11.5844073,0.,0.,2.1700287,0.,0.,0.,0.,3.8770029,0
.,0.,-3.8770029,0.,0.,0.,0.,-2.1700287,0.,-11.5844073,0.,0.,-2.1700287
,0.,0.,0.,0.,-3.8770029,0.|HyperPolar=0.,0.,0.,0.,0.,0.,0.,0.,0.,0.|PG
```

We have there 54 scalars following the keyword `PolarDeriv=` organized as follows:

- Let us recall first that for a given atomic coordinate i, $\hat{\alpha}'_i = (\partial\hat{\alpha}_L/\partial\xi_i)$ is a second-rank tensor, which is in addition symmetric. It can therefore be specified by 6 components, which are given in *Gaussian* in Cartesian coordinates in tetrahedral notation, i.e. in the order: $\alpha'_{ixx}, \alpha'_{ixy} = \alpha'_{iyx}, \alpha'_{iyy}, \alpha'_{ixz} = \alpha'_{izx}, \alpha'_{iyz} = \alpha'_{izy}, \alpha'_{izz}$.
- Moreover, $\hat{\alpha}'_i$ must be specified for derivation with respect to each of the $3N$ Cartesian displacements (ξ_i). This results in $18N$ components,

which for CO_2 ($N = 3$) gives 54 components. The first 6 correspond to $\hat{\alpha}'_1$, the next 6 to $\hat{\alpha}'_2$, etc.

- Moreover, ξ_1 corresponds to the x-displacement of atom 1 (x_1), ξ_2 to y_1, ξ_3 to z_1, ξ_4 to x_2, ξ_5 to y_2, etc.
- Finally, the polarizability derivative components are given by *Gaussian* in *internal units*, which here are [B^2] (a derivative of the Gaussian units [$Å^2$]).

Schematically, the polarizability derivatives are given in a linear array of length $N \times [\{x, y, z\} \times 6]$, ordered in the same hierarchy. As an example, for a three-atom molecule (e.g. CO_2) α' is given in this form:

$$\begin{aligned}&\frac{\partial\alpha_{xx}}{\partial x_1}, \frac{\partial\alpha_{xy}}{\partial x_1}, \ldots \frac{\partial\alpha_{zz}}{\partial x_1},\\ &\quad\frac{\partial\alpha_{xx}}{\partial y_1}, \frac{\partial\alpha_{xy}}{\partial y_1}, \ldots \frac{\partial\alpha_{zz}}{\partial y_1},\\ &\qquad\frac{\partial\alpha_{xx}}{\partial z_1}, \frac{\partial\alpha_{xy}}{\partial z_1}, \ldots \frac{\partial\alpha_{zz}}{\partial z_1},\\ &\qquad\quad\frac{\partial\alpha_{xx}}{\partial x_2}, \ldots\end{aligned} \tag{A.14}$$

From the output shown above, we therefore deduce for our CO_2 example (recall that atom 1 is C, while atoms 2 and 3 are O, and y is the main molecular axis):

$$\begin{aligned}&\frac{\partial\hat{\alpha}_L}{\partial x_1} = \frac{\partial\hat{\alpha}_L}{\partial y_1} = \frac{\partial\hat{\alpha}_L}{\partial z_1} = \hat{0}, \qquad \frac{\partial\hat{\alpha}_L}{\partial y_2} = -\frac{\partial\hat{\alpha}_L}{\partial y_3} = \begin{pmatrix} b & 0 & 0\\ 0 & c & 0\\ 0 & 0 & b\end{pmatrix},\\ &\frac{\partial\hat{\alpha}_L}{\partial x_2} = -\frac{\partial\hat{\alpha}_L}{\partial x_3} = \begin{pmatrix} 0 & a & 0\\ a & 0 & 0\\ 0 & 0 & 0\end{pmatrix}, \qquad \frac{\partial\hat{\alpha}_L}{\partial z_2} = -\frac{\partial\hat{\alpha}_L}{\partial z_3} = \begin{pmatrix} 0 & 0 & 0\\ 0 & 0 & a\\ 0 & a & 0\end{pmatrix},\end{aligned} \tag{A.15}$$

$$\begin{aligned}\text{with}\quad & a = 3.870029\ \mathrm{B}^2, \quad b = 2.1700287\ \mathrm{B}^2, \quad c = 11.5844073\ \mathrm{B}^2\\ \text{or}\quad & a = 1.082993\ \text{Å}^2, \quad b = 0.607263\ \text{Å}^2, \quad c = 3.241792\ \text{Å}^2.\end{aligned} \tag{A.16}$$

A note on the rotation matrix

It should be noted, however, that the polarizability derivatives α'_i are given with respect to the rotated Z-matrix coordinate frame, which *Gaussian* uses by default internally. However, *this does not apply to the relative atomic displacements* ϕ_k^n describing the normal modes – these are given in the original coordinate frame. These two potentially different coordinate frames can be the source of many confusions. One way to overcome this problem is to rotate the vibration vectors into the same coordinate frame as the polarizability derivatives. The matrix A that *Gaussian* uses to transform the coordinate frame can be found in the output (labeled as *rotation matrix*), if the keyword `IOP(2/33)=1` is appended to the input file. An even easier option, which is the one used here, is to carry out the calculations in two steps as explained above. No rotation is then performed in the Raman calculation, thereby avoiding these issues altogether.

Raman tensors

The Raman tensors of the normal modes are directly related to the polarizability derivatives, and can be obtained from an expression similar to Eq. (2.67) of Section 2.5.3. We give details below of such a derivation in the case of *Gaussian*. To this end, we must relate the derivative with respect to the normal coordinate Q_k in the definition of the Raman tensor to derivatives with respect to the real displacements ξ^n (which are given in the output).

Using the notations of Section 2.7.1, the normalized eigenvector $\mathbf{A}_k$ (adimensional because it is a normalized vector), which defines the normal mode pattern in mass-weighted coordinates is proportional to the vector $(\sqrt{m^n}\phi_k^n)_{n=1..3N}$, where ϕ_k^n are the normalized displacements given in the output of *Gaussian*. To express this concisely, it is convenient to introduce and define the *reduced mass* μ_k [kg] of a normal mode k as:

$$\mu_k = \sum_{n=1}^{3N} m^n (\phi_k^n)^2. \tag{A.17}$$

μ_k is in fact conveniently also given in the output file. The eigenvector $\mathbf{A}_k$ (which must be a unit vector) then takes the form:

$$\mathbf{A}_k = \frac{1}{\sqrt{\mu_k}} \begin{pmatrix} \sqrt{m^1}\phi_k^1 \\ \vdots \\ \sqrt{m^n}\phi_k^n \\ \vdots \\ \sqrt{m^{3N}}\phi_k^{3N} \end{pmatrix}. \tag{A.18}$$

Moreover, by the definition of Q_k, we have $\mathbf{q} = \sum_k Q_k \mathbf{A}_k$, from which we deduce:

$$\xi^n = \frac{q^n}{\sqrt{m^n}} = \sum_{k=1}^{3N} Q_k \frac{A_k^n}{\sqrt{m^n}} = \sum_{k=1}^{3N} Q_k \frac{\phi_k^n}{\sqrt{\mu_k}}. \tag{A.19}$$

Using this expression, the Raman tensor can then be expressed in terms of the polarizability derivatives with respect to the real displacements as:

$$\boxed{\hat{R}_k = \left(\frac{\partial \hat{\alpha}_L}{\partial Q_k}\right) = \frac{1}{\sqrt{\mu_k}} \sum_{n=1}^{3N} \phi_k^n \left(\frac{\partial \hat{\alpha}_L}{\partial \xi^n}\right).} \tag{A.20}$$

Let us illustrate this final formula again on the CO_2 example. For mode $k = 3$ (the bond-stretching mode) only the displacements corresponding to y_2 ($\phi_3^5 = 0.71$) and y_3 ($\phi_3^8 = -0.71$) are non-zero and the reduced mass is $\mu_3 = 15.9949$ amu (see output). We therefore have:

$$\begin{aligned}\hat{R}_3 &= \frac{1}{\sqrt{\mu_3}} \left[0.71 \left(\frac{\partial \hat{\alpha}_L}{\partial y_2}\right) - 0.71 \left(\frac{\partial \hat{\alpha}_L}{\partial y_3}\right)\right] \\ &= \begin{pmatrix} 0.215 & 0 & 0 \\ 0 & 1.146 & 0 \\ 0 & 0 & 0.215 \end{pmatrix} \text{Å}^2 \text{ amu}^{-1/2}.\end{aligned} \tag{A.21}$$

One can in fact check the self-consistency of this derivation by calculating the Raman activity R_3 and depolarization ratio ρ_3 of this mode directly from its Raman tensor as given above. To do so, we first calculate the two invariants of the Raman tensor using Eqs (2.56) and (2.57): $(\bar{\alpha}_3')^2 = 0.2758$ Å^4/amu and $(\bar{\gamma}_3')^2 = 0.8676$ Å^4/amu. We then use Eqs (2.83) and (2.84) to find $R_3 = 18.49$ Å^4/amu and $\rho_3 = 0.164$, in agreement with the figures provided directly by *Gaussian* in the output.

The other modes are in fact Raman inactive and Eq. (A.20) indeed results in a zero Raman tensor because of cancellations between the terms in the sum. This is the same cancellation that occurs in more basic treatments of the Raman effect like the bond-polarizability model (see Appendix B).

Raman polarizability tensor

In some instances, including SERS, knowing the Raman polarizability tensor may be more convenient than the Raman tensor itself. This can be achieved using Eq. (2.81) together with Eq. (2.77) to obtain the zero-point amplitude.

For the bond-stretching mode of CO_2, the zero-point amplitude is $b_3 = 4.5 \times 10^{-25}$ kg$^{1/2}$ m or equivalently $b_3 = 0.11$ amu$^{1/2}$ Å. The Raman polarizability tensor of this mode is therefore (ignoring the temperature term, negligible here at room temperature):

$$\hat{\alpha}_3 = b_3\hat{R}_3 = \begin{pmatrix} 0.024 & 0 & 0 \\ 0 & 0.126 & 0 \\ 0 & 0 & 0.024 \end{pmatrix} \text{Å}^3. \tag{A.22}$$

This can also be expressed in S.I. units as:

$$\hat{\alpha}_3 = \begin{pmatrix} 0.26 & 0 & 0 \\ 0 & 1.4 & 0 \\ 0 & 0 & 0.26 \end{pmatrix} \times 10^{-41}\text{S.I.} \tag{A.23}$$

The Raman polarizability tensor can then be directly used in the phenomenological description of Raman scattering presented in Section 2.4 and its extensions to SERS. In fact, we can again check the self-consistency of our calculations by estimating the Raman cross-section directly from the Raman polarizability tensor using Eq. (2.60).

General remarks

Other available DFT programs (like *GAMESS*, for example) can produce *directly* the Raman tensors in the output if requested. There are several reasons, however, why we have chosen to explain the origin of how to obtain the Raman tensors from the *Gaussian* output; the most important of which is that it shows explicitly the interconnection between the polarizability derivatives, the eigenvectors of the modes, and the Raman polarizability/tensor. Knowing the polarizability derivatives with respect to individual atomic movements is a useful information in any case that tends to be hidden when we look at the Raman polarizability of the overall molecule. There are several possible situations when one would like to actually backtrack on the steps that lead to the Raman polarizability/tensor. One simple example is that of a Raman *inactive* mode that results from the cancellation of symmetric atomic movements that have (individually) large polarizability derivatives. This is a mode that is likely to show *activation* upon small deformations of the molecular structure, which can happen on surfaces (and therefore in SERS).

The eigenvectors of the vibrations have, in addition, important clues on whether (and how) the vibrations are going to be perturbed or not, depending primarily on how the molecule binds to the surface. If the eigenvector of a mode is 'localized' in a region of the molecular structure that is not binding to the surface, it will be relatively less perturbed than another vibration that has the binding moiety as part of its eigenvector. Some specific Raman active vibrations can sometimes even disappear from the spectrum when the molecule is anchored to the surface, and we shall show a specific example of this for benzenethiol under SERS conditions later.

A.4. EXAMPLES OF DFT CALCULATIONS FOR SERS APPLICATIONS

We shall only present here a few results of DFT calculations applied to SERS from our own research, without claiming them to be the best possible examples.

A.4.1. Validation of absolute Raman cross-sections of reference compounds

An important (and one could argue fundamental) task in this field is to understand the origin and magnitude of SERS enhancements [8] and to be able to measure and quantify them properly. It is therefore very important to have *reference compounds* that are reliable, easily accessible (and preferably non-toxic), and with known Raman cross-sections for characteristic modes. Good measurements of absolute Raman cross-section is a very difficult task in general [86], and require careful experimental conditions. Whenever possible, experimental values should be preferred over calculations. But the reliability of experimental values (with possible intrinsic systematic errors) is something that should always be assessed carefully depending on the origin of the experiment. DFT calculations come here as an aid, to double-check the absolute differential Raman cross-sections of reference compounds. The relative cross-sections of other compounds are then easily measured by comparison to the references (see Section 2.2.8). It can never be seen as a substitute for the experiment, but it adds an additional level of confidence to the results.

Such an approach was discussed in detail in Ref. [8], for example. Some of the results were illustrated in Table 2.2.

A.4.2. Raman tensor and vibrational pattern visualizations

Anything to do with the symmetries of the Raman tensors and the characteristics of the eigenvectors is normally hidden from experimental access in SERS (or Raman in general). Some of the information regarding the symmetry of the tensor is of course contained in the depolarization ratios (Chapter 2), but the possibility of visualizing both the Raman tensors of the modes and their corresponding eigenvectors (vibrational pattern) is, arguably, one of the biggest advantages of DFT Raman calculations.

Raman tensors can then either be visualized as matrices or 3D-graphical representations of them; both methods have advantages and disadvantages.

One more time, it is difficult to be very prescriptive with respect to existing software, because their capabilities keep changing all the time. But a rule of thumb at the moment is that most visualization tools have to be entirely developed by the user[9].

Example

Figure A.1 shows an explicit example of both Raman tensor and eigenvalue visualization for benzenethiol (thiophenol), which is a molecule that has been used many times in a SERS context. Benzenethiol is a standard probe on gold substrates, due to the advantage of a thiol group that will bind effectively to the surface. It is also a relatively simple molecule (basically, a benzene ring with a thiol replaced at one of the hydrogen positions), which means that it can be calculated by DFT within a reasonable time and with modest computational resources. Moreover, its adsorption properties on surfaces are more likely to be understood than other more complicated molecules. Figure A.1 shows the calculated (DFT) Raman spectrum, as well as the experimental Raman (pure benzenethiol) and SERS spectra. The DFT calculation provides the Raman cross-sections for the different modes, but it does not give the broadening of the peaks[10], which are introduced manually ($\sim$15 cm^{-1} for all peaks) to plot the spectrum in Fig. A.1. The relative peak intensities cannot therefore be readily compared (only the integrated intensities). At the top of Fig. A.1, the vibrational patterns and tensors of several selected vibrations are shown.

The eigenvectors of the different vibrations are represented by the standard method [61] of representing a pattern of vectors at each atomic position. All these modes are *in-plane* modes and, hence, we chose to represent the pattern of displacements as seen from the 'top', with the molecule lying on a plane. For the Raman tensors ($\hat{R}$), on the other hand, we do not have an easy way of representing them graphically. One possibility (often used) is to draw a polar plot of the scalar quantity:

$$I = \left| \mathbf{e} \cdot \hat{R} \cdot \mathbf{e} \right|^2 , \tag{A.24}$$

where $\mathbf{e}$ is a rotating unit vector. The polar plot then represents I as a function of the direction of $\mathbf{e}$ and conveys a 'visual' impression of the main characteristics of the tensor (uni-axial, isotropic, with many lobes, etc.). This

[9] The figures for this subsection have been kindly provided by Matthias Meyer, Victoria University of Wellington, New Zealand.

[10] The intrinsic widths of the peaks can be, in principle, obtained in DFT if an *anharmonic coupling analysis* of the modes is carried out. This is, however, very time consuming and only very rarely justified. An estimated intrinsic broadening in the range 10–20 cm^{-1} for all the peaks is normally enough to reproduce the experimental spectra rather well.

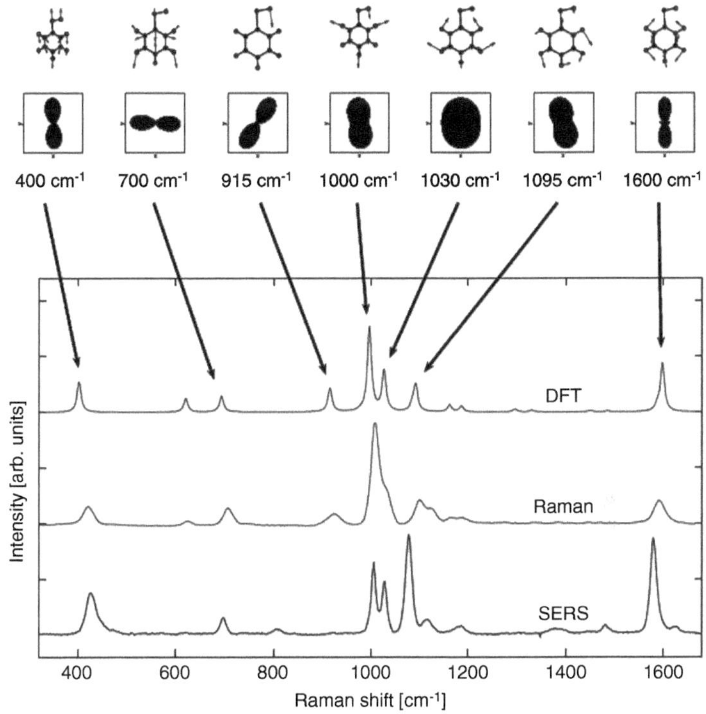

Figure A.1. DFT, normal Raman, and SERS spectra of benzenethiol (thiophenol). The labels with the frequencies should be taken as approximate values, because the exact frequency depends on whether we refer to the normal Raman, SERS, or DFT spectrum. All these modes are *in-plane* modes, which means that the eigenvectors can be plotted by looking at the molecule flat on a plane. The *polar plots* representing the corresponding Raman tensors (from Eq. (A.24)) are also seen in a projection on the same plane. The mode at $\sim$915 cm^{-1} is absent from the SERS spectrum. Its vibrational pattern involves a large movement of the H atom on the thiol group, indicating this must be the binding site of the molecule to the metal surface, as expected.

particular scalar representation of the tensor is very easy to plot as projections along the main planes ($x - y$, $x - z$, and $y - z$), and it gives an idea of the symmetry of the tensor by a simple visual inspection. An alternative way of explaining it, would be to say that Eq. (A.24) gives basically the intensity that would be measured in a Raman experiment with *parallel* polarization detection, if the signal of a fixed single molecule could indeed be measured (without SERS enhancement).

Figure A.1 represents the projections of tensors in the $x - y$ plane only, but the full representation is also possible by looking at the different projections along $x - y$, $y - z$, or $x - z$ planes; a specific example for only one vibration is explicitly shown in Fig. A.2.

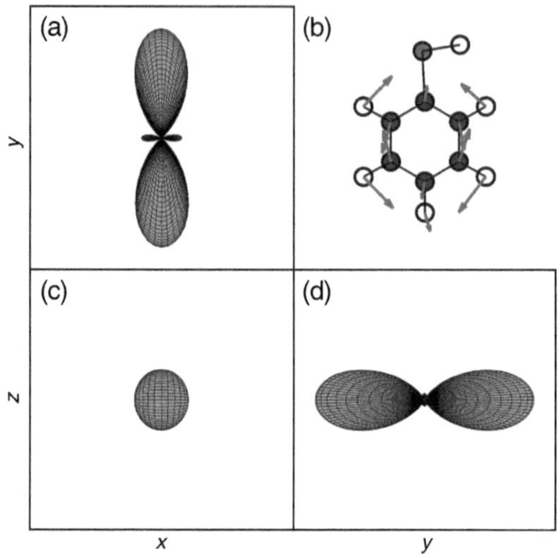

Figure A.2. Different projections of the tensor for the ~1600 cm^{-1} vibration in benzenethiol. The pattern of atom displacements of the eigenvector is also shown as in Fig. A.1. The different views show the projection of the scalar magnitude defined by Eq. (A.24) along the $x-y$, $y-z$, and $x-z$ planes, respectively. As can be seen from the different projections, this particular Raman tensor is very uni-axial and mainly confined along the y-axis (i.e. along the direction defined by the C–S bond).

Discussion

Note that there is a wealth of information that can be discussed when we have simultaneous access to a DFT calculation, the normal Raman spectrum, and the SERS spectrum. We only discuss some very basic aspects here to convey a general idea. Note, for example, that the mode at ~915 cm^{-1} completely disappears in the SERS spectrum, despite being predicted in DFT and observed in the normal molecule in the (pure) liquid. The reason for this can be tracked down to the eigenvector. This mode corresponds to a bending of the thiol group as a whole, and accordingly, does not exist when the molecule is bound to a surface through the thiol group. Without going into much detail here, we also note that there is activation of modes in the SERS spectrum (like the one at ~1480 cm^{-1}). The reason for the activation of these modes can most of the time be tracked down to perturbation of the corresponding eigenvectors (not shown here) by the presence of the surface.

In general, a more complicated information to be analyzed is that regarding the changes in relative intensities for Raman active modes. There can be more than one reason for this. One obvious suspect is *surface selection rules* (see Chapter 4 and Ref. [48]). Note that the spectrum does not necessarily need to be single-molecule to observe the effects of surface selection rules; if all the molecules attach to the surface in the same way, these effects are not washed

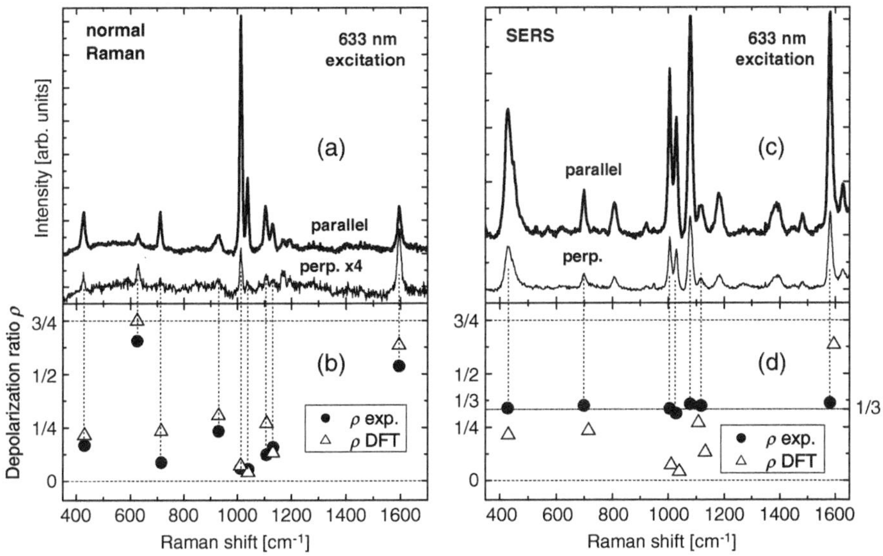

Figure A.3. (a) Normal Raman spectra of pure benzenethiol measured in back-scattering (in a cell) for parallel and perpendicular scattering configurations from Ref. [48] (reproduced with permission. Copyright 2008 Wiley). In (b) the depolarization ratios of different modes are shown together with the prediction for the same modes from DFT calculations. In (c) and (d) we show the corresponding situation under SERS conditions in a diluted colloidal solution. The theoretical DFT values are reproduced again in (d) for clarity. The experimental values for ρ under SERS conditions are all $\sim 1/3$, thus demonstrating the overriding effect of local-field polarization at hot-spots, see Ref. [57] for further details.

out by averaging over many molecules. More often for small molecules with strong binding, however, the effect of the surface has to be included explicitly in the DFT calculation if a detailed understanding of the symmetry of the modes is sought.

A.4.3. Depolarization ratio breakdown under SERS conditions

As a final brief glimpse into the use and application of DFT calculations for molecules that are relevant to SERS, we show in Fig. A.3 another aspect of the benzenethiol DFT calculation which, one more time, adds a different level of understanding in a SERS context; the example is again taken from our own research in Ref. [48]. We compare here the depolarization ratios of the modes predicted by DFT with the experimentally observed values at 633 nm excitation. The overall agreement between the observed and measured depolarization ratios in non-SERS conditions is excellent. This is a strong indication that the DFT predictions are correct and we can truly claim to understand the symmetry and Raman tensors in the bare molecule. However, the equivalent situation for benzenethiol under SERS conditions is quite different. A common characteristic – seen many times under SERS conditions – is that the SERS depolarization ratios of the modes are different from those of the bare molecule. As can be easily appreciated from Fig. A.3(d), all modes have a depolarization ratio of $\sim 1/3$ under SERS conditions. Without straying too much into the details of Ref. [48], we only mention in passing here that this breakdown of the normal depolarization ratios for modes with different symmetries is a natural consequence of the presence of highly uni-axial electromagnetic local fields at the hot-spots producing the SERS signals. Any attempt to use polarization effects in SERS (including the possibility of extending the technique to SERS optical activity [57]) cannot ignore the overriding effect of the local field [49], and these effects can only be understood if the depolarization ratios and symmetries of the Raman modes of the bare molecules are available and understood. The change in depolarization ratio under SERS conditions is a natural consequence of the more general problem of 'surface selection rules' as described in Section 4.5 (and in the original work by Moskovits in Ref. [52]), but for randomly occurring hot-spots rather than flat surfaces; which is the topic of Ref. [48].

Appendix B

The bond-polarizability model

The so-called *bond-polarizability model* provides an alternative approximate (empirical) method to evaluate the Raman polarizability tensor for a given known normal mode. In the following, we review briefly the basic principles of the method which, in addition, provides a good pedagogical illustration of the microscopic link between Raman polarizability and vibrations (and in particular their symmetries and selection rules).

B.1. PRINCIPLE AND IMPLEMENTATION

B.1.1. Principle

Given two atoms linked by a covalent chemical bond, it is a well established fact that the electronic properties of the bond are transferable (to a large extent) to another molecule where the same bond occurs. The basic idea of the bond-polarizability model is to assume that the total Raman polarizability for a given normal mode can be obtained from the *sum of the changes in linear polarizability of the individual bonds* participating in the oscillation. It essentially recasts the classical formulation (in terms of a single normal coordinate oscillation Q_k) into a sum over individual bond movements. The bond polarizabilities have to be added as tensors (matrices) to account for their relative orientations[1].

[1] Chemical bonds are highly 'uni-axial' objects, and their response to external fields are typically tensorial in character. The optical properties of a single bond can be described (phenomenologically) by a tensor in a fixed system of axes, which normally includes the direction of the bond as one of the principal axes. If the contributions of two or more tensors are to be added to obtain the total contribution from a molecule, the matrix representation of the tensors has to be rotated to the same reference system of axes before summation.

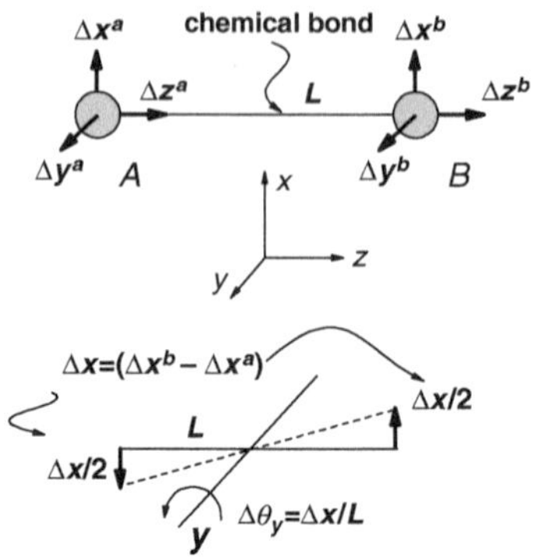

Figure B.1. Effect of atomic displacements on the bond polarizability between two atoms A and B. The (fixed) system of reference is such that the z-axis is along the bond. Small displacements along the x-axis are equivalent (up to a translation) to a relative displacement of $\Delta x = \Delta x^b - \Delta x^a$ of B with respect to A, and therefore to an opposite displacement of $\Delta x/2$ each. This can then be viewed as a rotation of the bond by an angle: $\Delta\theta_y \sim \Delta x/L$, where L is the length of the bond. A similar conclusion holds for displacements along y. Relative displacements along z results in a simple elongation (or contraction) of the bond along its main axis.

This model was originally developed by Volkenstein [125] to account for the optical properties of complex macromolecules. Implicit in this model is the idea that there are no local field corrections for a given bond due to the presence of others, and that there are no 'retardation effects'; i.e. all the bonds experience the same electric field with the same phase. Both approximations are extremely good for small and medium-sized molecules. Volkenstein [125] developed the method to understand the (seemingly) intractable problem of vibrations in complex polymer conglomerates and biomolecules (like protein globules).

When a molecule is vibrating according to a given normal mode pattern (at frequency ω_k), each individual bond of the molecule changes periodically (also at frequency ω_k) its *length* and/or its *orientation* with respect to the incident field. This will, in turn, modulate the linear polarizability of the bond. The total change in polarizability, obtained by adding the contributions from all the bonds, constitutes the Raman polarizability of the mode. The challenge is to add the contributions of all the individual bonds and in particular to decide which ones add up or cancel out.

B.1.2. Calculation of bond polarizabilities

Consider a chemical bond between two atoms A and B as in Fig. B.1. The linear polarizability tensor is a symmetric tensor which, in the appropriate system of coordinates (shown in Fig. B.1), can be represented by a diagonal matrix of the form:

$$\hat{\alpha}_{\text{bond}} = \begin{pmatrix} \alpha & 0 & 0 \\ 0 & \beta & 0 \\ 0 & 0 & \gamma \end{pmatrix}. \tag{B.1}$$

We have allowed for the possibility of the bond having two different polarizabilities along x and y, even though in most cases (by rotational symmetry) we will have $\alpha = \beta \neq \gamma$.

As far as these two atoms are concerned, any atomic movement can always be expressed in terms of the six basic displacements along the axes depicted in Fig. B.1. In a similar fashion as for the normal modes of a full molecule, these six degrees of freedom can be described as 6 characteristic patterns. For example, if $\Delta x^a = \Delta x^b$, the corresponding movement (along the x-axis) is a translation. Such a translation has no effect on the linear polarizability of the bond, and the same applies to translation along y and z. What matters, then, are the relative displacements of the two atoms: $\Delta x = \Delta x^b - \Delta x^a$, $\Delta y = \Delta y^b - \Delta y^a$, and $\Delta z = \Delta z^b - \Delta z^a$.

Let us now consider the effect of these differential movements on the polarizability.

- Along z, the opposite displacements result in a small stretching (elongation or contraction) of the bond.

- Along x, the relative displacements can be viewed, as illustrated in Fig. B.1, as a small rotation of the bond around the y-axis by an angle:

  $$\Delta\theta_y \sim \Delta x / L, \tag{B.2}$$

  where L is the length of the bond.

- Similarly the movement along y can be viewed as a small rotation of the bond around the x-axis, by an angle:

  $$\Delta\theta_x \sim -\Delta y / L. \tag{B.3}$$

We can now evaluate the effect of all these perturbations on the linear polarizability. For bond rotations first, the linear polarizability tensor of the bond is not modified, but only its orientation with respect to the incident field. To express it in matrix form in the original coordinate frame, it must therefore be rotated. For a relative displacement Δx along the x-axis, corresponding to a small rotation by $\Delta\theta_y$ around the y-axis, the resulting polarizability matrix is therefore:

$$\hat{\alpha}(\Delta x) = [R_y(\Delta\theta_y)]^{-1} \cdot \begin{pmatrix} \alpha & 0 & 0 \\ 0 & \beta & 0 \\ 0 & 0 & \gamma \end{pmatrix} \cdot R_y(\Delta\theta_y)$$
$$\approx \begin{pmatrix} \alpha & 0 & (\gamma-\alpha)\Delta x/L \\ 0 & \beta & 0 \\ (\gamma-\alpha)\Delta x/L & 0 & \gamma \end{pmatrix}, \qquad \text{(B.4)}$$

where $R_y(\Delta\theta_y)$ denotes the rotation matrix by an angle $\Delta\theta_y$ around the y-axis:

$$R_y(\Delta\theta_y) = \begin{pmatrix} \cos\Delta\theta_y & 0 & \sin\Delta\theta_y \\ 0 & 1 & 0 \\ -\sin\Delta\theta_y & 0 & \cos\Delta\theta_y \end{pmatrix} \approx \begin{pmatrix} 1 & 0 & \Delta\theta_y \\ 0 & 1 & 0 \\ -\Delta\theta_y & 0 & 1 \end{pmatrix}, \qquad \text{(B.5)}$$

where the terms of order $\Delta\theta_y^2$ were discarded in the last expression ($\Delta\theta_y \ll 1$).

The differential change in polarizability is then:

$$\frac{\partial\hat{\alpha}}{\partial\Delta x} = \begin{pmatrix} 0 & 0 & (\gamma-\alpha)/L \\ 0 & 0 & 0 \\ (\gamma-\alpha)/L & 0 & 0 \end{pmatrix}. \qquad \text{(B.6)}$$

Using the same procedure for displacements along y, we obtain:

$$\frac{\partial\hat{\alpha}}{\partial\Delta y} = \begin{pmatrix} 0 & 0 & 0 \\ 0 & 0 & (\beta-\gamma)/L \\ 0 & (\beta-\gamma)/L & 0 \end{pmatrix}. \qquad \text{(B.7)}$$

For movements along z (stretching), the tensor itself changes with bond length, and the change can only be described as:

$$\frac{\partial\hat{\alpha}}{\partial\Delta z} = \begin{pmatrix} \alpha' & 0 & 0 \\ 0 & \beta' & 0 \\ 0 & 0 & \gamma' \end{pmatrix}, \qquad \text{(B.8)}$$

where $\alpha' = \mathrm{d}\alpha/\mathrm{d}L$, $\beta' = \mathrm{d}\beta/\mathrm{d}L$, and $\gamma = \mathrm{d}\gamma/\mathrm{d}L$. Finally, the total change in $\hat{\alpha}$ produced by an arbitrary displacement of one atom with respect to the other can always be calculated as:

$$\delta\hat{\alpha} = \frac{\partial\hat{\alpha}}{\partial\Delta x}\Delta x + \frac{\partial\hat{\alpha}}{\partial\Delta y}\Delta y + \frac{\partial\hat{\alpha}}{\partial\Delta z}\Delta z. \qquad \text{(B.9)}$$

Table B.1 Experimental polarizability parameters for the C–H bond in two different molecules CH_4 and C_2H_6 from Refs [74,318]. The same transferability of parameters (within small differences) holds for other standard chemical bonds found in molecules, thus justifying the utilization of the bond-polarizability model as a sum of contributions from individual bonds under non-resonant conditions. Note that in this particular case the C–H bond parameters are truly uni-axial with $\gamma - \alpha = \gamma - \beta$, and $\alpha' = \beta'$. The parameters are given in Gaussian units.

	C–H bond	
Parameter	CH_4	C_2H_6
$\gamma - \alpha$ [Å^3]	0.276	0.229
$\gamma - \beta$ [Å^3]	0.276	0.229
α' [Å^2]	0.449	0.392
β' [Å^2]	0.449	0.392
γ' [Å^2]	2.491	2.506

B.1.3. Practical implementation

The strategy of the bond-polarizability model, then, is quite simple. Still, it obviously requires first the knowledge of the normal modes and of the empirical bond polarizability parameters. Then, for a given normal mode vibration, the relative movements of two chemically-bound atoms in the structure of the molecule are found and Δx, Δy, and Δz are derived. From these and the parameters of the bond $(\gamma - \alpha)$, $(\beta - \gamma)$, α', β', γ', the net contribution of this bond to the Raman process can be assessed. In many cases, $\alpha = \beta$ and $\alpha' = \beta'$ by symmetry, leaving only 3 empirical parameters for the bond polarizability. The same procedure is repeated for all possible bonds in the molecule for a given normal mode, and the final result is the (tensorial) sum of all these contributions, taking into account the different orientations of the bonds and the relative amplitudes of the movements.

The bond-polarizability model is one of the most successful semi-empirical (and phenomenological) models that account for most experimental observations in vibrational Raman scattering of molecules. Its validity is justified not only by its simplicity but also by the experimental fact indicating that the parameters of bonds are *transferable* to a large extent among different molecules. An explicit example is shown in Table B.1, where the C–H bond parameters entering the bond-polarizability model are compared for CH_4 and C_2H_6 (in Gaussian units[2]). The numbers given in the table are those that reproduce the corresponding experimental (non-resonant) Raman data of the compounds. Similar values are obtained for other compounds, thus providing an experimental justification to the approximations of the model.

[2] These polarizability values can be converted to the SI system ([ϵ_0 m^3] for γ, α, and β, and [ϵ_0 m^2] for γ', α', and β') by multiplying them by $4\pi\epsilon_0$.

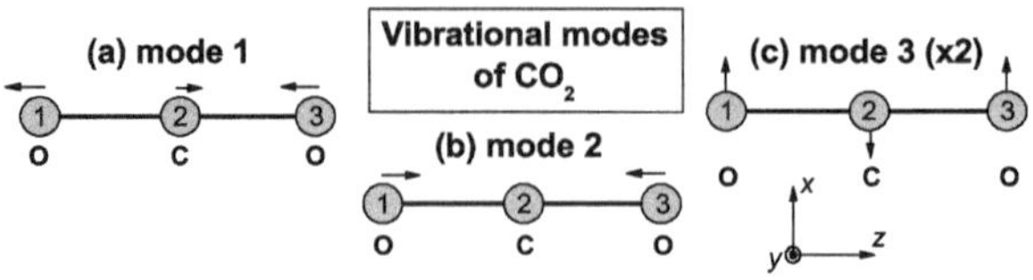

Figure B.2. The 4 normal modes ($3N - 5 = 4$) of a linear symmetric triatomic molecule like carbon dioxide (CO_2). The third eigenvector in (c) is doubly degenerate by symmetry (it can happen on the plane of the page or perpendicular to it). The eigenvector in (b) is a *symmetric* stretch of the bonds where the central atom remains at rest, while (a) is an *asymmetric* vibration with the atoms on both ends moving in phase in the same direction and by the same amount, while the atom at the center moves in the opposite direction by a different amount (to keep the center of mass at rest). The relative magnitude of the displacements depends on the relative masses of the atoms.

The bond-polarizability model has strong limitations, of course, when it comes to resonance phenomena (in particular in the transferability of parameters). But it otherwise represents a very useful phenomenological tool to explain basic facts about selection rules and the microscopic origin of the Raman effect, as illustrated in the following.

B.2. A SIMPLE EXAMPLE IN DETAIL

We now present one of the simplest possible examples of some of the concepts outlined so far: the case of a linear triatomic molecule ($N = 3$), carbon dioxide (CO_2) for example, depicted in Fig. B.2. We will use this example to illustrate the connection between normal mode symmetry and Raman selection rules. The example also links with the DFT calculation of the same molecule in Appendix A, thus providing a phenomenological approach to the same problem.

B.2.1. Bond-polarizability analysis

The normal mode analysis of the linear triatomic molecule is a textbook example; it is solved, for example, in one of the problems about small oscillations in Ref. [104]. Here we are interested in going one step further and analyze the Raman polarizability of the molecule; we therefore skip the details of the vibrational analysis (also illustrated in Appendix A) and simply use the final result. Essentially, this molecule has $3N - 5 = 4$ internal vibrations, which are shown in Fig. B.2. The first two are an *asymmetric* and a *symmetric* stretch of the bonds with respect to the central atom, respectively. The other two are bond bending vibrations and are degenerate (same frequency) because the vibration can happen along two mutually independent and perpendicular planes going through the axis of the molecule.

We are interested in the effect of these vibrations on the Raman spectrum, in light of the bond-polarizability model. Let us denote $\alpha' = \beta'$, γ', and $\gamma - \alpha = -(\beta - \gamma)$ the empirical parameters of the C=O bond, where C is atom A and O is atom B using the earlier definitions, i.e. the relative displacement must be defined as atom B (O) minus atom A (C). The atoms are indexed as in Fig. B.2 and all tensors will be expressed in the coordinate frame in which the z direction is along the main axis of the molecule, x is on the plane of the page and perpendicular to z, and y is perpendicular to the plane of the page (and therefore to z and x); see Fig. B.2. The first C=O bond (1–2) is denoted L (for left) and the second (2–3) is denoted R (for right). Let us consider this second bond since it is in exactly the same configuration as in the definitions. The relative displacement along z is $\Delta Z_R = \Delta z_3 - \Delta z_2$ and the corresponding contribution to the Raman polarizability is therefore:

$$\frac{\partial \hat{\alpha}^R}{\partial \Delta Z_R} \Delta Z_R = \begin{pmatrix} \alpha' & 0 & 0 \\ 0 & \alpha' & 0 \\ 0 & 0 & \gamma' \end{pmatrix} \Delta Z_R. \tag{B.10}$$

Almost the same applies to the first C=O bond, but its orientation is opposite and the analysis must be done in this frame with $z' = -z$, and the resulting tensor must be symmetrically inversed (which leaves it unchanged, a general property of second-rank tensors). We therefore have $\Delta Z'_L = \Delta z'_1 - \Delta z'_2 = \Delta z_2 - \Delta z_1$, and:

$$\frac{\partial \hat{\alpha}^L}{\partial \Delta Z_L} \Delta Z_L = \begin{pmatrix} \alpha' & 0 & 0 \\ 0 & \alpha' & 0 \\ 0 & 0 & \gamma' \end{pmatrix} \Delta Z_L. \tag{B.11}$$

B.2.2. Raman polarizabilities

Mode 1 in Fig. B.2(a) is an asymmetric stretch, where one bond is contracting by the same amount as the other is expanding. We therefore have $\Delta z_1 = \Delta z_3$, which implies $\Delta Z_R = -\Delta Z_L$ in the previous two equations. Accordingly, the two contributions cancel each other exactly when they are added; i.e. this mode is *Raman inactive.*

Let us now analyze mode 2 in Fig. B.2(b). This is a symmetric stretch of the bonds along the molecular axis (z-axis here), and we have $\Delta z_2 = 0$ and $\Delta z_3 = -\Delta z_1$, which implies $\Delta Z_R = \Delta Z_L$. The total polarizability change of the molecule is then given by (the sum of the two bonds' contributions):

$$\delta \hat{\alpha}_1 = 2\Delta z_3 \begin{pmatrix} \alpha' & 0 & 0 \\ 0 & \alpha' & 0 \\ 0 & 0 & \gamma' \end{pmatrix}. \tag{B.12}$$

This mode will therefore be *Raman active* and will have a Raman tensor with the symmetry of the general form given by (B.12).

Finally, mode 3 in Fig. B.2(c) is a bond bending mode and corresponds to $\Delta x_1 = \Delta x_2$, and therefore $\Delta X_R = -\Delta X_L$. The same analysis as for mode 2 then reveals that this mode (in reality two modes due to degeneracy) is also Raman inactive.

Accordingly, of the four internal vibrations of the symmetric linear triatomic molecule only one (the symmetric stretch vibration) turns out to be Raman active and will have a Raman tensor of the general form given by (B.12). The same result can be confirmed by a DFT calculation as discussed in Appendix A. The Raman inactivity is a simple consequence, within the bond-polarizability model, of the cancellation of the contributions of different bonds because of symmetry.

B.2.3. A brief comment on the symmetry

We could have obtained this result by a straightforward symmetry analysis. As mentioned in Chapter 2, it is for simple molecules with well defined symmetries that the power of group theory can be really applied. The linear triatomic molecule has a center of inversion; as mentioned in Section 2.7.2, there is a mutual exclusion between Raman and IR active modes in molecules with a center of inversion. The key issue is the behavior of the normal mode under the 'inversion operation'. If we invert the whole pattern of displacements with the operation $\mathbf{r} \to -\mathbf{r}$ in Fig. B.2(b), we obtain basically the same pattern. This eigenvector remains therefore unchanged by inversion and it is said to be *even* (Gerade)[3]. The patterns in Figs B.2(a) and (c), on the contrary, change to their *negatives* under the operation $\mathbf{r} \to -\mathbf{r}$; i.e. all vectors point in the opposite direction. This is then an 'odd' (Ungerade) mode. It turns out that, for molecules with a center of inversion, only *even* modes are Raman active while *odd* modes are IR active. The bond-polarizability model has, however, the big advantage of showing the microscopic origin of selection rules, as a direct consequence of cancellations or additions of contributions of different bonds to the total Raman polarizability depending on the specific eigenvectors of different modes. In a real (bigger) molecule (like typical SERS analytes), it is unlikely that these cancellations or additions can be evaluated manually, and we have to rely on more sophisticated approaches like DFT to get the final answer. Still, a simple phenomenological approach like the one presented here contributes to the basic understanding of the topic.

[3] The language of group theory uses (for historical reasons) the German words *Gerade* (even) or *Ungerade* (odd), to classify the inversion symmetries.

Appendix C

A brief overview of Maxwell's equations in media

In this appendix, we review briefly some basic results of classical electromagnetic theory that are most relevant to SERS and plasmonics. The aim is obviously not to give a detailed account of classical electromagnetism, to which some entire textbooks are dedicated [96]. Accordingly, we focus on aspects that are directly relevant to electromagnetic effects in SERS, namely harmonic electromagnetic waves in media. We will also constrain to the simplest (and most common in SERS) case of non-magnetic media. This appendix is also intended to define the notations and conventions chosen for electromagnetism in this book. This is particularly important here because of the various definitions and units encountered in the literature discussing electromagnetism. We follow here again the choice made for the rest of this book to use S.I. units or their derivatives.

C.1. MAXWELL'S EQUATIONS IN VACUUM

C.1.1. The equations

The electromagnetic field is commonly described by two vector fields: the electric field $\mathbf{E}$ [V m^{-1}, or m kg s^{-3} A^{-1}] and the magnetic induction $\mathbf{B}$ [T for Tesla, or kg s^{-2} A^{-1}]. These fields are created (and affected) by charges either static, represented by the charge density ρ [C m^{-3}, or m^{-3} s A], or moving, represented by the current density $\mathbf{j}$ [A m^{-2}]. Charges and currents are therefore called *sources* of the electromagnetic field. Quantities, X, are formally defined for every time t [s] at every point $\mathbf{r}$ [m] in space: $X(\mathbf{r}, t)$.

The equations governing electromagnetic phenomena, which relate these quantities and their spatial and temporal dependence, are the *Maxwell's equations*:

$$\nabla \cdot \mathbf{B} = 0 \tag{C.1}$$

$$\nabla \times \mathbf{E} + \frac{\partial \mathbf{B}}{\partial t} = 0 \tag{C.2}$$

$$\nabla \cdot \mathbf{E} = \frac{\rho}{\epsilon_0} \tag{C.3}$$

$$\nabla \times \mathbf{B} - \frac{1}{c^2}\frac{\partial \mathbf{E}}{\partial t} = \mu_0 \mathbf{j}. \tag{C.4}$$

Note that these equations can be separated into two groups: the first two equations on the one hand, where only the fields appear, and the last two on the other hand, which contain the source terms (ρ or $\mathbf{j}$). ϵ_0 and μ_0 are fundamental constants called the *electric permittivity* and *magnetic permeability* of vacuum, respectively. c is the speed of light and is related to these by $c = 1/\sqrt{\epsilon_0\mu_0}$. The usual convention is to define c and μ_0 and derive the value of ϵ_0 from these. In S.I. units, we have:

$$\begin{array}{lll} c = 2.997\,924\,58 \times 10^{8} & & \mathrm{m\ s^{-1}} \\ \mu_0 = 4\pi \times 10^{-7} & \approx 1.257 \times 10^{-6} & \mathrm{m\ kg\ s^{-2}\ A^{-2}} \\ \epsilon_0 = 10^7/(4\pi c^2) & \approx 8.854 \times 10^{-12} & \mathrm{m^{-3}\ kg^{-1}\ s^4\ A^2}. \end{array} \tag{C.5}$$

Combining the last two equations (for the sources), one obtains the continuity equation for charges:

$$\frac{\partial \rho}{\partial t} + \nabla \cdot \mathbf{j} = 0. \tag{C.6}$$

The conservation of electric charge is therefore implicit in Maxwell's equations. Note that these equations govern the dynamics of the electromagnetic fields as a function of a *given* charge and current distribution. But charges and currents are themselves affected by the electromagnetic fields. In order to study the motion of charges itself, Maxwell's equations have to be complemented with the expression for the force created by the fields on a point charge q [C or s A] moving with a velocity $\mathbf{v}$ [m s^{-1}], called the Lorentz force [96]:

$$\mathbf{F} = q(\mathbf{E} + \mathbf{v} \times \mathbf{B}). \tag{C.7}$$

Maxwell's equations as stated above are valid *in vacuum* in a wide range of situations and can in fact be directly justified from the quantum field theory of electrodynamics.

C.1.2. Maxwell's equations for harmonic fields in vacuum

Harmonic fields, which are fields whose time evolution is harmonic (oscillating as a sine), are very important in electromagnetic theory. One reason is that every field evolution can be viewed as a coherent superposition of harmonic fields, using Fourier analysis. This is the basis for the quantum description of the fields. The whole theory of electromagnetism could in fact be presented in terms of harmonic fields, with special cases like electrostatics and magnetostatics being limiting cases. Moreover, as we will see later, the electromagnetic response of media (and linear response theory in general) is more easily described by Fourier analysis, in both time and space. Finally, from a more practical point of view, monochromatic light is a special case of a harmonic electromagnetic field, and any experiments related to optics, and in particular to SERS, will be mainly concerned with harmonic electromagnetic fields (or possibly superpositions of them).

Complex notations for harmonic fields

Electromagnetic radiation, including light, consists of electromagnetic waves oscillating in time at a frequency ν, or angular frequency $\omega = 2\pi\nu$. If a system is excited by such an external electromagnetic wave, it imposes an oscillating behavior at the same frequency for all physical quantities. All quantities then have the so-called *harmonic* time dependence of the type $\cos(\omega t + \phi)$ (with possibly a different phase ϕ for each quantity).

It is common in electromagnetism (and in many other areas) to introduce complex notations when discussing harmonic time dependencies. Each real quantity $X(\mathbf{r}, t)$ (scalar or vector), with a harmonic time dependence at frequency ω, can be described by a complex quantity $\underline{X}(\mathbf{r}, \omega)$ with the following convention[1]:

$$X(\mathbf{r}, t) = \mathrm{Re}\left(\underline{X}(\mathbf{r}, \omega)\mathrm{e}^{-\mathrm{i}\omega t}\right). \tag{C.8}$$

The exponential factor automatically accounts for the harmonic time dependence while the complex amplitude $\underline{X}(\mathbf{r}, \omega)$ describes the spatial dependence of the amplitude and phase. The time derivatives are then simply obtained from:

$$\frac{\partial X}{\partial t} = \mathrm{Re}\left(-\mathrm{i}\omega\underline{X}\mathrm{e}^{-\mathrm{i}\omega t}\right) \quad \text{and} \quad \frac{\partial^2 X}{\partial t^2} = \mathrm{Re}\left(-\omega^2\underline{X}\mathrm{e}^{-\mathrm{i}\omega t}\right). \tag{C.9}$$

[1] An alternative convention is to use $\exp(+\mathrm{i}\omega t)$. The complex quantities are then complex conjugates, but it leads to exactly the same physical results for real quantities. We have used the same convention ($\exp(-\mathrm{i}\omega t)$) throughout this book.

The advantage of complex notation is that it takes automatically into account the harmonic time dependence. All physical relations can then be rewritten and solved for the spatial dependence of the complex quantities, ignoring time dependence. The above relations can then be used to derive the real quantities from their complex notations. It is common practice to denote the complex quantity corresponding to a given real quantity X by the same notation X, or sometimes $X(\omega)$; we will adopt this convention in this book, except in this appendix, where we will use explicitly $\underline{X}(\mathbf{r},\omega)$ for pedagogical purposes.

Finally, it is worth noting that this complex notation approach is more general than the simple case of a harmonic dependence of the fields. For a general time dependence, the complex quantity $\underline{X}(\mathbf{r},\omega)$ in fact corresponds (up to a constant factor) to the temporal Fourier transform of $X(\mathbf{r},t)$. Defining $\underline{X}(\mathbf{r},-\omega) = \underline{X}(\mathbf{r},\omega)^*$, the general time dependence of $X(\mathbf{r},t)$ can be inferred from its Fourier components $\underline{X}(\mathbf{r},\omega)$:

$$X(\mathbf{r},t) = \frac{1}{2}\int_{-\infty}^{\infty} \mathrm{d}\omega \underline{X}(\mathbf{r},\omega)\mathrm{e}^{-\mathrm{i}\omega t}, \tag{C.10}$$

which is a generalization of Eq. (C.8) to non-harmonic time dependence. Note that the Fourier components can then be defined as:

$$\underline{X}(\mathbf{r},\omega) = \frac{1}{\pi}\int_{-\infty}^{\infty} \mathrm{d}t X(\mathbf{r},t)\mathrm{e}^{+\mathrm{i}\omega t}. \tag{C.11}$$

The complex notations are therefore simply a special case of temporal Fourier transform for a harmonic time dependence. It is often more convenient to think in terms of complex notations (rather than Fourier transforms) for electromagnetic problems with harmonic dependence (which is the case for most optical problems, and in particular for SERS).

Maxwell's equations in vacuum in complex notations

Using these definitions, Maxwell's equations in vacuum in complex notations (in terms of $\underline{\mathbf{E}}(\mathbf{r},\omega)$ and $\underline{\mathbf{B}}(\mathbf{r},\omega)$) then read:

$$\nabla \cdot \underline{\mathbf{B}} = 0 \tag{C.12}$$

$$\nabla \times \underline{\mathbf{E}} - \mathrm{i}\omega\underline{\mathbf{B}} = 0 \tag{C.13}$$

$$\epsilon_0 \nabla \cdot \underline{\mathbf{E}} = \underline{\rho} \tag{C.14}$$

$$\nabla \times \underline{\mathbf{B}} + \frac{\mathrm{i}\omega}{c^2}\underline{\mathbf{E}} = \mu_0 \underline{\mathbf{j}}. \tag{C.15}$$

C.1.3. Plane wave solutions in free-space

When studying harmonic fields with complex notations, we mentioned that this was equivalent to studying the components of the temporal Fourier transform of the fields $\underline{X}(\mathbf{r}, \omega)$ (instead of $X(\mathbf{r}, t)$). The Fourier analysis of Maxwell's equations can be extended to space by considering the spatial and temporal (4 dimensional) Fourier transforms of the fields $\underline{X}(\mathbf{k}, \omega)$. This analysis is for example crucial for the quantization of the electromagnetic field. We will not dwell on this here, since it is not directly relevant to SERS. A simple approach to this Fourier analysis is to consider solutions where the complex fields have a dependence $\underline{X}(\mathbf{r}, \omega) = X_0 \exp(\mathrm{i}\mathbf{k} \cdot \mathbf{r})$, where X_0 is a (complex) constant in space and time, and $\mathbf{k}$ is called the wave-vector. This corresponds for the real quantity to $X(\mathbf{r}, t) = |X_0| \cos(\mathbf{k} \cdot \mathbf{r} - \omega t + \phi)$. As we have seen before, the time derivatives are then $\partial \underline{X}/\partial t = -\mathrm{i}\omega \underline{X}$. The spatial derivatives simplify in a similar manner to give for a vector field: $\nabla \cdot \underline{\mathbf{X}} = \mathrm{i}\mathbf{k} \cdot \underline{\mathbf{X}}$ and $\nabla \times \underline{\mathbf{X}} = \mathrm{i}\mathbf{k} \times \underline{\mathbf{X}}$.

If we look for such solutions for the free-space electromagnetic field in vacuum, i.e. in the absence of charges and currents, it derives directly from the above Maxwell's equations (C.12)–(C.15) that:

- $\mathbf{E}_0$ and $\mathbf{B}_0$ are perpendicular to $\mathbf{k}$ (from Eqs (C.12) and (C.14)), and also perpendicular to each other (from the other two equations).
- For Eqs (C.13) and (C.15) to be compatible, and therefore for a solution to exist, ω and $\mathbf{k}$ must be related by $\omega = c|\mathbf{k}|$. This is called the *dispersion relation.* We moreover need the condition $\mathbf{B}_0 = (-\mathrm{i}/\omega)\mathbf{k} \times \mathbf{E}_0$, which in particular means that $|\mathbf{E}_0| = c|\mathbf{B}_0|$.
- Such solutions are simply *plane electromagnetic waves*, propagating with a wave-vector $\mathbf{k}$ [m^{-1}] (i.e. with a spatial period $2\pi/|\mathbf{k}|$ [m]), and oscillating in time at angular frequency ω [rad s^{-1}], or frequency $\nu = \omega/(2\pi)$ [s^{-1} or Hz], or temporal period $2\pi/\omega$ [s].

C.2. MAXWELL'S EQUATIONS IN MEDIA

C.2.1. Microscopic and macroscopic fields

In order to generalize Maxwell's equations to the case of electromagnetic fields in matter, the first step is to define the *macroscopic fields.* The reason for this is simple. Consider for example a crystal (periodic structure of atoms). At a microscopic (atomic) level, the charge distribution is highly non-uniform and non-trivial (positive charges confined to the nucleus and negative charges in the electron cloud). The details of the electromagnetic fields at this level are therefore likely to be very complicated. Moreover, we will only rarely have access experimentally to such fine details. It is therefore natural to consider average electric and magnetic fields on a length scale larger than the details of the atomic structure. These average fields are called *macroscopic fields*, while the 'real' fields are called *microscopic fields*. Moreover it can be shown that macroscopic fields are governed by the same Maxwell's equations as the microscopic field [96], only with an average (macroscopic) charge and current density. Note however that, fundamentally, these equations are only exact for the microscopic fields. For macroscopic fields, they are nevertheless an excellent approximation at length scales larger than the averaging length scale[2].

The distinction between microscopic and macroscopic fields is rarely highlighted and they are typically represented by exactly the same notation. In the absence of further information, it is in most cases the macroscopic fields that are studied. This will be the case in the following and in the rest of this book. The microscopic field can however be relevant when relating the macroscopic optical properties to the microscopic structure of matter.

C.2.2. The electromagnetic response of the medium

We have so far seen how the microscopic details of the fields in matter could be averaged into a macroscopic field, which is the relevant physical quantity for electromagnetic effects in media. Moreover, *the electromagnetic response of a medium is governed by the interaction of the fields with internal charges and their dynamics*. It is therefore important to separate the contribution from these internal charges to that of possible external charges. The charge density can then be written as $\rho = \rho_{\text{int}} + \rho_{\text{ext}}$, with a similar relation for the current density. We have 'control' over the external sources only. The internal charges and their electromagnetic response are intrinsic to the medium. Writing Maxwell's equations for a medium then consists in expressing the response of the internal charges to the macroscopic electromagnetic excitation, i.e. finding a relation of the form $\rho_{\text{int}} = f(\mathbf{E}, \mathbf{B})$, and $\mathbf{j}_{\text{int}} = f(\mathbf{E}, \mathbf{B})$. Substituting these expressions leads to the so-called *Maxwell's equations in media*, which model the dynamics of the *macroscopic* fields $\mathbf{E}$ and $\mathbf{B}$ as a function of *external* sources ρ_{ext} and $\mathbf{j}_{\text{ext}}$.

The response of the internal charges to an electromagnetic field can however be very complicated to model in detail. For this reason, Maxwell's equations in media can take many different forms depending on the chosen approximations and their intended range of validity. It is common in textbooks to treat the problem separately for dielectric media in the electrostatic approximation,

[2] The averaging procedure is difficult to define rigorously, see for example Ref. [96]. The length scale for averaging needs to be larger than the lattice constant for a crystal, for example, to be meaningful; say a few nanometers. The notion of macroscopic field should therefore in principle fail at length scales of the order of 10 nm, and a full microscopic study would be necessary. In practice, macroscopic fields are used down to length scales of nm, or even Å. The main reason is that it gives reasonable results and that there are simply no easy general alternatives.

then for conducting media (for dc currents), and then for magnetic media in the magnetostatics approximation (stationary currents). The three treatments are then combined and generalized to apply to oscillating (harmonic) fields. Such an approach has some merits for the understanding of the physical origin of the electromagnetic response, but the generalization to the case of harmonic fields is not always rigorous and can be the source of confusion[3].

We have chosen here to introduce the electromagnetic response of media directly from the general framework of linear response theory. This approach is more directly related, we believe, to the cases of interest for SERS and plasmonics: that of harmonic fields. It also allows a clear separation of the two aspects of the problem:

- The first one is to provide a unified framework in which the electromagnetic response of most media can be described. We will see in the next two sections that under certain approximations, most media of interest to SERS can be fully described by a single function: the dielectric function $\epsilon(\omega)$. This function is introduced here empirically, and its use is justified simply because it agrees well with experiments.
- A separate subject is to actually relate the electromagnetic response (for example $\epsilon(\omega)$) to the microscopic properties of the medium of interest. When possible, such a treatment can formally justify the empirical introduction of the dielectric function and its range of validity. It also provides a more direct understanding of the physical origin of the electromagnetic response and can sometimes lead to approximate analytical expressions for the dielectric function. This is however a very difficult undertaking in general, and a separate treatment is required for each type of medium under consideration (metal, dielectric, etc.). Because of this complexity, we will therefore mostly ignore this aspect here and focus on the more general empirical approach, which is sufficient in most optical studies.

C.2.3. Electric polarization and magnetization

The first step is to define the macroscopic quantities that we will use to represent the response of the medium to an electromagnetic field. The most obvious would be to use the macroscopic charge density ρ_{int} and current density $\mathbf{j}_{\text{int}}$ and express them as functions of $\mathbf{E}$ and $\mathbf{B}$. However, this approach does not facilitate the physical interpretations and has seldom been used. The standard approach, which draws its origin from electrostatics and magnetostatics, attempts to treat separately the effects of the electric

[3] One typical example of confusion is the problem associated with the conductivity. The dc conductivity for conducting media 'becomes' an optical conductivity for the harmonic fields, whose physical significance is not always clear; see the discussion later in this appendix.

and magnetic fields on the internal charges. To this end, it is first necessary to rewrite Maxwell's equations in a form where electric and magnetic response are more clearly separated and characterized.

The electric field $\mathbf{E}$, first, modifies the internal charge and current density distribution. This can be modeled as a polarization charge and current density: ρ_{pol} and $\mathbf{j}_{\text{pol}}$, linked by the relation for charge conservation (continuity equation): $\partial\rho_{\text{pol}}/\partial t + \nabla \cdot \mathbf{j}_{\text{pol}} = 0$. To represent these polarization charges and current distributions, it is common and more physical to introduce a new macroscopic vector field $\mathbf{P}$ [C m^{-2} or m^{-2} s A] called *electric polarization*, defined by:

$$\nabla \cdot \mathbf{P} = -\rho_{\text{pol}} \quad \text{and} \quad \frac{\partial \mathbf{P}}{\partial t} = \mathbf{j}_{\text{pol}}. \tag{C.16}$$

It is always possible to find such a $\mathbf{P}$ although it is not unique (any time-independent field $\mathbf{P}_0$ satisfying $\nabla \cdot \mathbf{P}_0 = 0$ can be added to it). $\mathbf{P}$ represents physically a continuous distribution of electric dipoles (and its amplitude therefore corresponds to a dipole moment per unit volume). The advantage of working with $\mathbf{P}$ instead of ρ_{pol} and $\mathbf{j}_{\text{pol}}$ is two-fold: firstly, as we will see, it enables us to rewrite Maxwell's equations for media in a form nearly identical as for vacuum; and secondly, it is much easier to physically understand the response of the medium to an electric field in terms of an induced electric polarization (induced dipole per unit volume) rather than induced charge distribution and currents.

Similarly, for a magnetic medium, the magnetic induction $\mathbf{B}$ modifies the internal current density distribution, creating magnetic currents $\mathbf{j}_{\text{mag}}$. It is assumed that the magnetic field does not affect the charge density distribution, and therefore: $\rho_{\text{mag}} = 0$ and $\nabla \cdot \mathbf{j}_{\text{mag}} = 0$, from charge conservation. Instead of working with $\mathbf{j}_{\text{mag}}$, it is also possible to define a new macroscopic vector field $\mathbf{M}$ [m^{-1} A] called *magnetization* as:

$$\nabla \times \mathbf{M} = \mathbf{j}_{\text{mag}}. \tag{C.17}$$

$\mathbf{M}$ represents physically a continuous distribution of magnetic moments (magnetic moment per unit volume).

The total internal current density therefore has the form $\mathbf{j}_{\text{int}} = \mathbf{j}_{\text{pol}} + \mathbf{j}_{\text{mag}}$, and the internal charge density is simply $\rho_{\text{int}} = \rho_{\text{pol}}$. Using these definitions, the two Maxwell's equations (C.3) and (C.4) with source terms can be written for the macroscopic fields in a medium as:

$$\epsilon_0 \nabla \cdot \mathbf{E} = \rho_{\text{tot}} = -\nabla \cdot \mathbf{P} + \rho_{\text{ext}} \tag{C.18}$$

$$\frac{1}{\mu_0}\nabla \times \mathbf{B} - \epsilon_0 \frac{\partial \mathbf{E}}{\partial t} = \mathbf{j}_{\text{tot}} = \frac{\partial \mathbf{P}}{\partial t} + \nabla \times \mathbf{M}. \tag{C.19}$$

These equations are usually simplified by introducing two macroscopic fields, the *electric displacement* $\mathbf{D}$ [C m^{-2} or m^{-2} s A]

$$\mathbf{D} = \epsilon_0 \mathbf{E} + \mathbf{P}, \tag{C.20}$$

and the magnetic field[4] $\mathbf{H}$ [m^{-1} A]

$$\mathbf{H} = \frac{\mathbf{B}}{\mu_0} - \mathbf{M}. \tag{C.21}$$

Combined with the two other Maxwell's equations (C.1) and (C.2), we obtain the full set of Maxwell's equations in media (for macroscopic fields):

$$\nabla \cdot \mathbf{B} = 0 \tag{C.22}$$

$$\nabla \times \mathbf{E} + \frac{\partial \mathbf{B}}{\partial t} = 0 \tag{C.23}$$

$$\nabla \cdot \mathbf{D} = \rho_{\text{ext}} \tag{C.24}$$

$$\nabla \times \mathbf{H} - \frac{\partial \mathbf{D}}{\partial t} = \mathbf{j}_{\text{ext}}. \tag{C.25}$$

In these equations, the contributions from internal and external charges have been clearly separated. The external charges appear as source terms ρ_{ext} and $\mathbf{j}_{\text{ext}}$ in a similar way as for the microscopic Maxwell's equations (C.1)–(C.4). The internal charge contribution is hidden in the electric polarization $\mathbf{P}$ (contained in $\mathbf{D}$), and magnetization $\mathbf{M}$ (contained in $\mathbf{H}$). These quantities represent the electromagnetic response of the medium to the electromagnetic field.

Note that these equations are strictly equivalent to the macroscopic Maxwell's equations written in terms of $\rho_{\text{int}} + \rho_{\text{ext}}$ and $\mathbf{j}_{\text{int}} + \mathbf{j}_{\text{ext}}$. They are simply rewritten in terms of the new macroscopic fields $\mathbf{D}$ and $\mathbf{H}$, which gives an equivalent representation of the internal charges and currents. The reason why these equations are more important is that it is easier to relate physically $\mathbf{D}$ and $\mathbf{H}$ (or $\mathbf{P}$ and $\mathbf{M}$), the response, to $\mathbf{E}$ and $\mathbf{B}$, the excitation. Note also that the separation of the currents into polarization and magnetic currents is entirely artificial at this stage. It is simply in anticipation of the next step where $\mathbf{P}$ will be related to $\mathbf{E}$ and $\mathbf{M}$ to $\mathbf{B}$. One could have equally defined $\mathbf{P}$ in terms of total internal currents instead of polarization currents, and taken $\mathbf{M} = 0$. The equations would be equally valid, but the description of the response of a magnetic medium would then require to have $\mathbf{P}$ related to both

[4] The exact denominations of $\mathbf{B}$ and $\mathbf{H}$ are not always used carefully, and $\mathbf{B}$ is sometimes called the magnetic field. Strictly speaking, $\mathbf{H}$ represents the magnetic field (magnetic field strength to be precise) and $\mathbf{B}$ is the magnetic induction or magnetic flux density.

E and **B**. The dependence on **B** would then be fairly complicated because **P** is not adapted to represent magnetic moment distributions. The definitions in this section should therefore be viewed as a mathematical re-arrangement of Maxwell's equations, guided by physical considerations. It 'prepares' the equations in a form that lends itself easily to the actual modeling of the electromagnetic response of the medium.

C.2.4. Constitutive relations

For these equations to be useful, it is now necessary to model explicitly the electromagnetic response of the medium. Instead of finding the relations $\rho_{\text{int}} = f(\mathbf{E}, \mathbf{B})$ and $\mathbf{j}_{\text{int}} = f(\mathbf{E}, \mathbf{B})$, we will look for the more physical relations:

$$\mathbf{P} \text{ or } \mathbf{D} = f(\mathbf{E}, \mathbf{B}), \quad \text{and} \quad \mathbf{M} \text{ or } \mathbf{H} = f(\mathbf{E}, \mathbf{B}), \tag{C.26}$$

called the *constitutive relations.*

The first approximation, which was already hinted at when we defined **P** and **M**, is to assume that **P** is affected only by **E**, while **M** is affected only by **B**. Moreover, for simplicity, we will also restrict from now on to non-magnetic media, which encompass most cases relevant to SERS. This means that $\mathbf{M} = 0$ and $\mathbf{B} = \mu_0 \mathbf{H}$. We therefore only need to relate **P** or **D** to **E**. We will also constrain ourselves to isotropic media, for which the mathematical treatment is much simpler.

The linear approximation

From there, the simplest (and most used) approximation is the *linear approximation* in which the response is assumed to depend linearly on excitation. By its very nature, the linear approximation works well for small-amplitude fields and should fail for large excitation field intensities (where *non-linear effects* become important [99]). These non-linear effects (if at all present) are most of the time neglected in a SERS context and we will therefore ignore them in this book.

In its simplest form, the linear approximation would lead to a relation of the form: $\mathbf{P}(\mathbf{r}, t) = \epsilon_0 \chi \mathbf{E}(\mathbf{r}, t)$, where χ is a non-dimensional constant characterizing the response of the medium. Such a relation is said to be *local,* both in space and time, because it relates the values of the quantities at a given point in space and time only to values at exactly the same point in space and time. This is for example the case of Maxwell's equations in vacuum, which relate the fields and charges at the same point **r** and time t. This assumption is however very restrictive, and it is reasonable to assume that in general, the internal charge and current distribution (and therefore **P**) in a medium at a given point **r** and time t could also depend on the values of the fields **E** and **B** at different points around **r** (non-local in space) and at times preceding t (non-local in time).

The EM response functions

A more general relation (still within the linear approximation) is:

$$\mathbf{P}(\mathbf{r},t) = \epsilon_0 \int \mathrm{d}\mathbf{r}' \int_{-\infty}^{t} \mathrm{d}t' \chi(\mathbf{r},\mathbf{r}',t,t')\mathbf{E}(\mathbf{r}',t'), \tag{C.27}$$

where $\mathbf{P}(\mathbf{r},t)$ now depends linearly on the electric field around $\mathbf{r}$ and before time t.

Such an expression is however impractical, and the fields are in general decomposed onto a plane wave basis (Fourier analysis of Maxwell's equations) as[5]:

$$\mathbf{E}(\mathbf{r},t) = \frac{1}{2} \int \mathrm{d}\mathbf{q} \int \mathrm{d}\omega \underline{\mathbf{E}}(\mathbf{q},\omega) \exp(\mathrm{i}\mathbf{q}\cdot\mathbf{r} - \mathrm{i}\omega t). \tag{C.28}$$

$\underline{\mathbf{E}}(\mathbf{q},\omega)$ is then the Fourier transform of $\mathbf{E}(\mathbf{r},t)$ both in space ($\mathbf{q}$) and time (ω). One can show that the optical response can then be described within the linear approximation by the simpler expression:

$$\underline{\mathbf{P}}(\mathbf{q},\omega) = \epsilon_0 \chi(\mathbf{q},\omega)\underline{\mathbf{E}}(\mathbf{q},\omega). \tag{C.29}$$

$\chi(\mathbf{q},\omega)$ is called the complex relative electric susceptibility of the medium. It is a *response function* relating the Fourier components of the response, $\mathbf{P}$, to the excitation, $\mathbf{E}$. The above equation should therefore be viewed as the definition of $\chi(\mathbf{q},\omega)$. In many electromagnetic problems, it is common to use $\epsilon(\mathbf{q},\omega) = 1 + \chi(\mathbf{q},\omega)$ instead, to characterize the electromagnetic response of the medium. $\epsilon(\mathbf{q},\omega)$ is called the *complex relative dielectric function*, and it follows easily that:

$$\underline{\mathbf{D}}(\mathbf{q},\omega) = \epsilon_0 \epsilon(\mathbf{q},\omega)\underline{\mathbf{E}}(\mathbf{q},\omega). \tag{C.30}$$

The dependence of $\epsilon(\mathbf{q},\omega)$ on $\mathbf{q}$ is termed *spatial dispersion* and represents the fact that the response at a given point depends on the excitation at other points in space. The dependence on ω is called *temporal dispersion*, and indicates that the response at a given time may depend on the excitation at previous times.

Because the electromagnetic response is defined for the Fourier components of the fields (through the constitutive relation (C.30)), Maxwell's equations in media (Eqs (C.22)–(C.25)) also take a simple form in Fourier space. They

---

[5] The factor 1/2 here is not important and is chosen so that the Fourier components are equal to the complex notations of the fields introduced earlier for harmonic fields.

are written below in terms of **E** and **H**:

$$\mu(\mathbf{q},\omega)\mathbf{q}\cdot\underline{\mathbf{H}}(\mathbf{q},\omega)=0 \tag{C.31}$$

$$\mathbf{q}\times\underline{\mathbf{E}}(\mathbf{q},\omega)-\omega\mu_0\mu(\mathbf{q},\omega)\underline{\mathbf{H}}(\mathbf{q},\omega)=0 \tag{C.32}$$

$$\mathrm{i}\epsilon_0\epsilon(\mathbf{q},\omega)\mathbf{q}\cdot\underline{\mathbf{E}}(\mathbf{q},\omega)=\underline{\rho}_{\text{ext}}(\mathbf{q},\omega) \tag{C.33}$$

$$\mathbf{q}\times\underline{\mathbf{H}}(\mathbf{q},\omega)+\omega\epsilon_0\epsilon(\mathbf{q},\omega)\underline{\mathbf{E}}(\mathbf{q},\omega)=\underline{\mathbf{j}}_{\text{ext}}(\mathbf{q},\omega). \tag{C.34}$$

The magnetic response was included here for reference, using the response function $\mu(\mathbf{q},\omega)$ defined in a similar way as $\epsilon(\mathbf{q},\omega)$, i.e. $\mathbf{H}(\mathbf{q},\omega) = \mathbf{B}(\mathbf{q},\omega)/(\mu_0\mu(\mathbf{q},\omega))$. For non-magnetic media, $\mu(\mathbf{q},\omega)=1$.

The actual form of $\epsilon(\mathbf{q},\omega)$ (and $\mu(\mathbf{q},\omega)$) for a given medium depends on the microscopic details of the medium under consideration. Modeling it can be extremely complicated and requires making additional approximations appropriate to the case being studied. Provided these response functions are known, it is then relatively easy to solve the above equations for the Fourier components. This leads to plane wave solutions for electromagnetic waves in the medium. However, things can become extremely difficult when there are boundaries between two media (which is arguably always the case in real systems). Because the boundaries are defined in real space, Fourier analysis is not adapted to model them, except possibly in the simplest cases such as planar interfaces.

The local approximation

Fortunately, in many situations (and in many cases for SERS), the spatial dispersion (**q**-dependence) can be neglected. Physically, this means that the response at a given point only depends on the excitation at the *same point* at previous times (but not on the excitation at other points). This results in a dependence on ω only, e.g. $\epsilon(\omega)$, and is called the *local approximation* (meaning local in space, the response in time is still 'non-local' and depends on previous times). This approximation can be grossly justified when the typical length scales of the problem are much larger than the inter-atomic distance in the medium because only wave-vectors with $|\mathbf{q}|\approx 0$ then contribute to the Fourier integrals. The local approximation is also commonly used in situations where it cannot really be justified rigorously (including in SERS). It is still a useful approximation (provided one realizes its limitations) since solving the problem with spatial dispersion is in general intractable even for the simplest cases.

We will use the local approximation in this book, and the frequency-dependent relative dielectric function is then defined by:

$$\underline{\mathbf{D}}(\mathbf{r},\omega)=\epsilon_0\epsilon(\omega)\underline{\mathbf{E}}(\mathbf{r},\omega). \tag{C.35}$$

$\epsilon(\omega)$ is called in this context the *local* relative dielectric function. Note that because $\epsilon(\omega)$ is defined in terms of Fourier components, it may be complex. Because $\chi(\omega) = \epsilon(\omega) - 1$ is a causal linear response function, it must satisfy some general properties called Kramers–Kronig relations[6] [319]. *For a non-magnetic, isotropic medium in the linear and local approximation,* $\epsilon(\omega)$ *fully characterizes the electromagnetic response of the medium.*

As was the case for $\epsilon(\mathbf{q}, \omega)$, linking $\epsilon(\omega)$ to the microscopic properties of the medium can still be a challenge, even without spatial dispersion. However, because of its direct link with the optical properties of the medium (see later), $\epsilon(\omega)$ can be measured experimentally over a wide range of frequencies of interest (by the experimental technique of *ellipsometry* [88], for example). It can therefore simply be taken as a well-defined parameter characterizing the electromagnetic response of the medium.

In the local approximation, Maxwell's equations can be written for fields that have the temporal Fourier components (i.e. they depend on ω), but are not Fourier-transformed for the spatial variables, (i.e. they still depend on $\mathbf{r}$). In other words, the field amplitudes will look like: $\underline{X}(\mathbf{r}, \omega)$. For a linear isotropic non-magnetic medium, in the absence of external charges and currents, this leads to the following Maxwell's equations:

$$\nabla \cdot \underline{\mathbf{H}}(\mathbf{r}, \omega) = 0 \tag{C.36}$$

$$\nabla \times \underline{\mathbf{E}}(\mathbf{r}, \omega) - \mathrm{i}\omega\mu_0 \underline{\mathbf{H}}(\mathbf{r}, \omega) = 0 \tag{C.37}$$

$$\nabla \cdot \underline{\mathbf{D}}(\mathbf{r}, \omega) = 0 \tag{C.38}$$

$$\nabla \times \underline{\mathbf{H}}(\mathbf{r}, \omega) + \mathrm{i}\omega \underline{\mathbf{D}}(\mathbf{r}, \omega) = 0 \tag{C.39}$$

which should be used in conjunction with the *constitutive relation*:

$$\underline{\mathbf{D}}(\mathbf{r}, \omega) = \epsilon_0 \epsilon(\omega) \underline{\mathbf{E}}(\mathbf{r}, \omega). \tag{C.40}$$

Note that these equations reduce to that in vacuum by taking $\epsilon(\omega) = 1$. This form of Maxwell's equations is the basis for most electromagnetic studies in this book. The components of the fields in these equations can be viewed as the temporal Fourier components. These equations then apply to any time-dependent problem, and the time dependence can be recovered by the inverse Fourier transform (which requires solving the equations for all ω). However, they are most often applied to problems with harmonic fields oscillating at a given frequency ω. This is the case of most linear optical studies, including SERS, where excitation with monochromatic light is assumed. In this case, it is easier to view the components of the fields appearing in the equations as their complex notations, as defined in Section C.1.2.

---

[6] Kramers–Kronig relations come from the analytic properties of the response function to satisfy *causality*, i.e. the fact that (in the time domain) the 'response' cannot come before the 'cause' producing it. See Ref. [319] for more details.

Internal charges in the local approximation

One simple consequence of the above equations is that if $\epsilon(\omega) \neq 0$, then from Eqs (C.40) and (C.38) we have: $\nabla \cdot \underline{\mathbf{D}}(\omega) = \nabla \cdot \underline{\mathbf{E}}(\omega) = \nabla \cdot \underline{\mathbf{P}}(\omega) = 0$. From the definition of **P** in Section C.2.3, this means that the macroscopic internal charge density ρ_{int} is zero everywhere. Note that this does not mean that the microscopic charges do not respond to excitation, they do respond in the form of an electric polarization, which can still be non-zero, but must have a zero divergence.

C.2.5. Boundary conditions between two media

We have so far considered electromagnetic fields in infinite media. In most cases of interest, the media are not infinite, but are bounded by interfaces with other media or with vacuum. Maxwell's equations are valid everywhere within a given medium, but to solve a general electromagnetic problem, it is necessary to relate the electromagnetic fields on either side of an interface between two media. These relations are called the *boundary conditions*.

The boundary conditions can be directly inferred from the form of Maxwell's equations (C.22)–(C.25) or (C.36)–(C.39). The standard approach consists in integrating the equations on the surface of an imaginary (infinitesimally small) cylinder using Gauss' theorem (or along the contour of a small square loop using Stokes' theorem) around the interface. It can be found in most standard textbooks [96] and we give here the results without further details. At an interface between two media (denoted 1 and 2), we have:

- From Eq. (C.22), the normal component of **H** (**B** for magnetic media) is continuous at the interface.
- From Eq. (C.23), the tangential component of **E** is continuous.
- From Eq. (C.24), the normal component of **D** is continuous.
- From Eq. (C.25), the tangential component of **H** is continuous.

These expressions are fairly general and are true, in particular, for the value of the fields in space and time $\mathbf{X}(\mathbf{r}, t)$ (from Eqs (C.22)–(C.25)), or for their temporal Fourier components (or complex notations) $\underline{\mathbf{X}}(\mathbf{r}, \omega)$ (from Eqs (C.36)–(C.39)). They do not apply however to the spatial Fourier components $\underline{X}(\mathbf{q}, \omega)$ that need additional considerations regarding the spatial dispersion.

It is worth noting that in the local approximation, the first and second conditions are redundant because of the second Maxwell's equation (C.37). Similarly, the third and fourth conditions are also equivalent because of the fourth Maxwell's equation (C.39). For problems where the electric field plays a more important role – like SERS – we therefore usually use only the second and third conditions. Denoting $\mathbf{n}$ the unit normal vector at the interface, these two boundary conditions can be written in a concise form as:

$$(\mathbf{E}_1 - \mathbf{E}_2) \times \mathbf{n} = 0 \tag{C.41}$$

$$(\mathbf{D}_1 - \mathbf{D}_2) \cdot \mathbf{n} = 0. \tag{C.42}$$

Finally, the continuity of the normal component of $\mathbf{D}$ implies a discontinuity of the normal component of $\mathbf{E}$. This discontinuity is artificial and only due to the introduction of infinitely sharp interfaces to model the boundary between two media. Note that such discontinuity appears for the macroscopic field, but not for the microscopic field. This discontinuity breaks down the condition $\nabla\cdot\mathbf{E} = \nabla\cdot\mathbf{P} = 0$ locally at the interface, which means that there can be a local macroscopic internal charge density at the surface. This is usually represented as an *internal surface charge density*, σ_{int}, and it can be shown that:

$$\sigma_{\text{int}} = \epsilon_0(\mathbf{E}_2 - \mathbf{E}_1) \cdot \mathbf{n}, \tag{C.43}$$

where $\mathbf{n}$ is the normal directed from medium 1 to 2. All of this is standard material of electromagnetic theory [96].

C.3. OTHER ASPECTS RELEVANT TO SERS AND PLASMONICS

C.3.1. The microscopic field

The relation between microscopic and macroscopic field

Maxwell's equations for media, together with the constitutive relations and boundary conditions, can be used to model and predict the macroscopic electromagnetic fields in most situations of interest to SERS and plasmonics. However, from the point of a view of a molecule or an atom, what matters are the microscopic fields at the molecule position, not the average macroscopic field around its position. In vacuum, the two are equal, but in a medium the microscopic field can be affected by the EM response of the medium in the local environment of the molecule. This *local field*[7] problem is discussed in most standard textbooks on electromagnetic theory [96,97]. It results in the introduction of the *local field correction factor*, of particular importance in the description of Raman scattering in liquids (see

[7] The denomination 'local field' in this context refers to the microscopic field (as opposed to macroscopic). In a SERS context, the local field also refers to the macroscopic field at the molecule position. To try to avoid confusion, we will call the former the microscopic field here, except in the common expression 'local field correction', which refers to the correction introduced by the difference between microscopic and macroscopic fields.

Section 2.4.5). Here we only sketch the derivation and provide the final result (see p. 160 in Ref. [96] for details).

The field felt by a reference molecule in a medium described by its dielectric constant ϵ_M can be expressed as coming from two main contributions: that from the external charges $\mathbf{E}_{\text{ext}}$ and that from the internal charges in the medium $\mathbf{E}_{\text{int}}$. Both are taken into account in the macroscopic field $\mathbf{E}$ derived from Maxwell's equations, but $\mathbf{E}_{\text{int}}$ is not correctly described in the vicinity of the molecule because of the averaging process. To obtain the microscopic field ($\mathbf{E}_{\text{micro}}$), we therefore need to subtract this macroscopic internal contribution $\mathbf{E}_{\text{int}}$ and replace it by the microscopic field created by the local environment (few molecules or atoms in the immediate vicinity) ($\mathbf{E}_{\text{close}}$), i.e.:

$$\mathbf{E}_{\text{micro}} = \mathbf{E} - \mathbf{E}_{\text{int}} + \mathbf{E}_{\text{close}}. \tag{C.44}$$

The macroscopic internal contribution can be obtained by considering the field created at the molecule position by a macroscopic electric polarization $\mathbf{P}$ in the medium and turns out to be [96]:

$$\mathbf{E}_{\text{int}} = -\frac{1}{3\epsilon_0}\mathbf{P}. \tag{C.45}$$

$\mathbf{E}_{\text{close}}$ is more difficult to estimate and depends on the symmetry of the local environment of the reference molecule or atom. In a cubic lattice environment $\mathbf{E}_{\text{close}}$ vanishes due to symmetry, as shown explicitly in Ref. [96]. In a completely random environment like a liquid, we also have $\mathbf{E}_{\text{close}} \sim 0$.

Using the constitutive relation for the macroscopic quantities in the medium, i.e. $\mathbf{P} = \epsilon_0(\epsilon_M - 1)\mathbf{E}$, and Eq. (C.45), we therefore obtain:

$$\mathbf{E}_{\text{micro}} = \frac{\epsilon_M + 2}{3}\mathbf{E}. \tag{C.46}$$

This expression links the microscopic field, really felt by a molecule embedded in a medium of dielectric constant ϵ_M, to the macroscopic field described by Maxwell's equations for media.

This relation has important consequences when relating microscopic atomic or molecular properties to their macroscopic EM response.

Local field correction factor

One such aspect relates to molecular polarizabilities (linear or Raman), discussed in Chapter 2. These are microscopic properties, and therefore relate the molecular response to the microscopic field felt by the molecule. When these are used to predict macroscopic properties (such as cross-section), which are defined with respect to macroscopic fields, an additional factor arises from the application of Eq. (C.46). This *local field correction factor*, L_M, is defined as:

$$(L_M)^{1/4} = \frac{\epsilon_M + 2}{3} \tag{C.47}$$

and simply reflects the ratio between microscopic and macroscopic fields. See also Section 2.4.5 for further discussion and for an explanation of the choice of the 1/4 exponent.

Clausius–Mossotti relation

A similar effect arises when relating the (microscopic) atomic optical polarizabilities (denoted α [ϵ_0 m^3] here) in a crystal to their macroscopic response, usually characterized by the dielectric function ϵ [a.d.]. The electric polarization can be expressed as the sum of the contribution from the induced dipole in each atom (with density n [m^{-3}]).

$$\mathbf{P} = n\alpha\mathbf{E}_{\text{micro}} = n\alpha\frac{\epsilon + 2}{3}\mathbf{E}. \tag{C.48}$$

From the definition of ϵ in the constitutive relation, we also have $\mathbf{P} = \epsilon_0(\epsilon - 1)\mathbf{E}$. We therefore deduce:

$$\epsilon = 1 + \frac{n\alpha/\epsilon_0}{1 - \frac{1}{3}n\alpha/\epsilon_0}. \tag{C.49}$$

or the inverse relation:

$$\alpha = \frac{3\epsilon_0}{n}\left[\frac{\epsilon - 1}{\epsilon + 2}\right]. \tag{C.50}$$

Both expressions link the *microscopic* (molecular) polarizability with the *macroscopic* response characterized by the dielectric function. An example of use of this link is given in Appendix D. Equation (C.50) is known as the *Clausius–Mossotti* equation.

C.3.2. Plane waves in media

Dispersion relation in media

As was the case for the fields in vacuum, plane wave solutions also play an important role in media. For a linear, isotropic, non-magnetic medium in the local approximation, Maxwell's equations (C.36)–(C.39) take a form similar to that in vacuum (C.12)–(C.15). The difference is the fact that $\epsilon(\omega) \neq 1$, and the fact that it can be a complex number. We consider again solutions where the complex fields have a dependence $\underline{X} = X_0 \exp(\mathrm{i}\mathbf{k} \cdot \mathbf{r})$, where X_0 is a constant in space and time, and here we allow $\mathbf{k}$ to be a complex vector. Furthermore, we also assume that $\epsilon(\omega) \neq 0$ (this special case is discussed in more detail in Chapter 3 in the context of plasmons). It derives directly from Maxwell's equations in media (C.36)–(C.39) that the plane wave solutions in media have essentially the same characteristics as in vacuum, except for the dispersion relation (relating ω and $\mathbf{k}$), which now reads:

$$\epsilon(\omega)\omega^2 = c^2 \mathbf{k} \cdot \mathbf{k}. \tag{C.51}$$

Note that $\mathbf{k} \cdot \mathbf{k} = |\mathbf{k}|^2$ is only true if $\mathbf{k}$ is real. If $\mathbf{k}$ is not real, then the field amplitudes decay exponentially as $\exp(-\mathrm{Im}(\mathbf{k}) \cdot \mathbf{r})$. Such waves are then called *evanescent* in the directions where $\mathrm{Im}(\mathbf{k})$ is non-zero. From the above dispersion relation, one can see that $\mathbf{k}$ can be real only if $\epsilon(\omega)$ is real and positive. This is then called a non-absorbing medium (at frequency ω) since waves can propagate without decay of the field amplitudes. The opposite is not true: evanescent waves can indeed exist when $\epsilon(\omega)$ is real positive! But they are associated with a boundary in that case; i.e. they appear because of a boundary condition between two media. Plane waves in media and in the presence of planar boundaries are discussed extensively in Appendix F. Here we only use them to highlight the link between the dielectric function and more 'conventional' optical properties.

Link between dielectric function and optical properties

Plane wave solutions represent electromagnetic waves as we know them 'in everyday life', and in particular in optics. The optical properties of media at a given frequency are often characterized, not by its dielectric function, but by its *refractive index* n [a.d.] (or index of refraction) and its *absorption coefficient* κ [a.d.]. The two representations are equivalent and related by:

$$\epsilon(\omega) = (n(\omega) + \mathrm{i}\kappa(\omega))^2, \tag{C.52}$$

where $n + \mathrm{i}\kappa$ is sometimes called the complex refractive index.

This representation is simpler for plane waves because of the form of the dispersion relation (C.51). Let us consider a plane wave propagating along a fixed direction (say Ox). In vacuum, its wave-vector would be $k_0 = \omega/c$. In a medium, its wave-vector, $\mathbf{k} = k\mathbf{e}_x$, and frequency ω are related by the dispersion relation (C.51): $k = k_0\sqrt{\epsilon(\omega)} = k_0(n(\omega) + \mathrm{i}\kappa(\omega))$. The refractive index n then represents the effect of the medium on *propagation*, through $\mathrm{Re}(k) = k_0 n$, while κ represents the *absorption* of the medium. The field amplitudes decay exponentially as $\exp(-\text{Im}(k)x)$, which is equivalent to a decay of the field intensity as $\exp(-\alpha x)$, where $\alpha = 2k_0\kappa$ [m^{-1}] is also called the *absorption coefficient* of the medium.

The representation in terms of $n(\omega)$ and $\kappa(\omega)$ is therefore more adapted to problems with plane electromagnetic waves, i.e. to ray optics. The use of $\epsilon(\omega)$ is usually more suited to electromagnetic problems with non-planar boundary conditions (and therefore to most EM problems encountered in SERS). The two representations are equivalent, and one can derive one from the other using Eq. (C.52), or the inverse relations. Writing $\epsilon = \epsilon' + \mathrm{i}\epsilon''$, we have:

$$\epsilon' = n^2 - \kappa^2 \quad \text{and} \quad \epsilon'' = 2n\kappa \tag{C.53}$$

$$n = \sqrt{\frac{\epsilon' + |\epsilon|}{2}} \quad \text{and} \quad \kappa = \frac{\epsilon''}{\sqrt{2}\sqrt{\epsilon' + |\epsilon|}}. \tag{C.54}$$

Note that for non-active media (i.e. media without optical gain, where no energy can be created), we always have $\kappa \geq 0$ and therefore $\epsilon'' \geq 0$, and $\kappa = \epsilon'' = 0$ *only* for non-absorbing media. Moreover, we can see that we always have $n \geq 0$, and $n \approx 0$ for $\epsilon' < 0$ and $|\epsilon'| \gg |\epsilon''|$ (these conditions are met typically for metals at visible frequencies). $n(\omega)$ and $k(\omega)$ can be measured experimentally in optical experiments, such as reflectance spectroscopy or ellipsometry, from which values of $\epsilon(\omega)$ can be obtained.

Note also that the above relations are valid only for a non-magnetic medium ($\mu = 1$). For a magnetic medium, one needs to replace ϵ by the product $\epsilon\mu$. It is then possible to have $n < 0$ (the so-called negative refraction) when both ϵ and μ are negative. This condition has never been encountered in natural media, but can be realized in some 'artificial' materials. It has recently been the subject of intense research for its possible application as a 'perfect lens' and for other exotic electromagnetic effects [133].

C.3.3. Electromagnetic problems in SERS

We have in the preceding sections introduced the electromagnetic response of media within linear response theory. This is not the conventional approach but we tried (as we did with Chapter 2 on Raman spectroscopy) to present a 'different' approach to the subject, rather than just repeating the information that is available in a multitude of classic books in the field (like Ref. [96]). This approach, although a bit more abstract than the conventional approach, presents in our opinion several advantages:

- It is very general (and applies for example to dielectric and conductors within the same framework) and particularly suited to harmonic fields problems, in particular in the local approximation ($\epsilon = \epsilon(\omega)$).

- It provides a rigorous definition of $\epsilon(\omega)$ as a response function, and highlights the fact this function contains all information required for most electromagnetic problems (in the local approximation).
- It avoids the need to relate the macroscopic response, characterized by $\epsilon(\omega)$, to its microscopic origin. This microscopic origin is a problem of great interest in itself and it is briefly outlined in Section D.2, but is not crucial to the understanding of electromagnetic phenomena in SERS. We can therefore view $\epsilon(\omega)$ as an 'empirical parameter' describing the optical response of the media.

In the next section, we will connect this approach to a more standard one from electrostatics. But before, it is useful to summarize the main results from a SERS point of view: many electromagnetic problems in SERS consists of a collection of one or more electromagnetic media, typically dielectric or metals, assembled in a more or less complex configuration. The electromagnetic excitation can consist of a monochromatic (frequency ω) laser beam coming from infinity, (represented either by a plane wave or a more complex beam, such as a Gaussian beam), or of a localized source (represented by a dipolar source).

- The electromagnetic response of each medium can be represented in the local and linear approximation by $\epsilon(\omega)$ and taken as a known parameter of the problem. $\epsilon(\omega)$ is well characterized for many dielectrics and metals of interest to SERS (see Appendix E for some examples).
- The problem then consists of solving Maxwell's equations as given in Eqs (C.36)–(C.39), with the constitutive relation (C.40) for each medium, and the appropriate boundary conditions (C.41)–(C.42) at interfaces. The excitation is represented by conditions at infinity (for beam excitation) or by source terms (for localized excitation). More details on these aspects are given in Chapter 5.

C.3.4. Link with the static approach

Before closing this presentation of Maxwell's equations in media, we will now try to link the linear response approach to the more conventional one, which starts from electrostatics and magnetostatics and is then generalized to oscillating fields. We ignore the magnetic response again for simplicity. We use the term electrostatics in its general sense here, meaning that it can include stationary electric currents, or slowly varying fields (compared to the timescale of the electromagnetic response).

Dielectrics and conductors

In electrostatics, it is common to separate the electric response of a medium into two contributions, with different physical origins.

The first is associated with the response of bound electrons. Under the influence of a constant electric field, a given atom remains electrically neutral, but the motion of its bound electrons is slightly affected resulting in a first approximation in the creation of an induced dipole moment. Macroscopically, this leads to the appearance of a continuous distribution of dipole moments, i.e. a static electric polarization $\mathbf{P}_s$. The electric response of an isotropic dielectric medium can then be described in the linear approximation as:

$$\mathbf{P}_s = \epsilon_0 \chi_s \mathbf{E}, \tag{C.55}$$

where χ_s [a.d.] is called the *static* relative electric (linear) susceptibility. Note that χ_s must be a *real number*, because it relates two real fields. This is different from the response function $\chi(\omega)$, which can be complex, since it relates two Fourier components of the fields. If $\mathbf{P}_s$ is varying slowly in time, this results in a polarization current $\mathbf{j}_{\text{pol}} = \partial \mathbf{P}_s/\partial t$. As shown in Section C.2.3, the polarization charges and currents can be included in Maxwell's equations by defining the *static* electric displacement $\mathbf{D}_s = \epsilon_0 \mathbf{E} + \mathbf{P}_s$. $\mathbf{D}_s$ can be directly related to $\mathbf{E}$ by:

$$\mathbf{D}_s = \epsilon_0 \epsilon_s \mathbf{E}, \tag{C.56}$$

where $\epsilon_s = 1 + \chi_s$ [a.d.] is the *static* relative dielectric constant (the term dielectric constant is usually kept for the static case, while dielectric function is used for the frequency-dependent response function $\epsilon(\omega)$). One feature of this approach is that, if the fields do not vary in time, then the polarization current is zero. But we also know that physically, a constant field may induce currents in a conducting medium.

The response of conducting charges must therefore be treated separately, and in the linear approximation, the electric field in a conductor will therefore create a conduction current:

$$\mathbf{j}_{\text{cond}} = \sigma_{\text{dc}} \mathbf{E}, \tag{C.57}$$

where σ_{dc} is the dc conductivity. Note that the above expression is simply Ohm's law for a (resistive) conductor.

In a non-conducting dielectric, there are only bound charges, and the situation is simply described by $\mathbf{D}_s$. In a conducting medium, there are both bound and conducting charges and the two electromagnetic responses, $\mathbf{D}_s$ and $\mathbf{j}_{\text{cond}}$ must be taken into account.

Low-frequency limit

This distinction between polarization and conduction currents seems perfectly reasonable on physical grounds for electrostatic problems. The total internal current can then, in general, be expressed as:

$$\mathbf{j}_{\text{int}} = \mathbf{j}_{\text{pol}} + \mathbf{j}_{\text{cond}} = \epsilon_0 \chi_s \frac{\partial \mathbf{E}}{\partial t} + \sigma_{\text{dc}} \mathbf{E}. \tag{C.58}$$

Internal currents are therefore induced both by $\mathbf{E}$ and its temporal variations, $\partial \mathbf{E}/\partial t$. This distinction is therefore relevant in situations where $\mathbf{E}$ and its time derivative are conceptually very different, for example in 'electrostatic' problems with slowly varying fields and stationary currents.

However, when dealing with harmonic (oscillating) fields, this distinction is no longer necessary because the time derivative of a field is directly related to the field (with a factor $-\mathrm{i}\omega$ proportionality in complex notation). This factor introduces some temporal dispersion, which can therefore automatically take into account the effect of the time derivatives of the field. This is why the concept of conductivity is not necessary for harmonic fields. The oscillating electric polarization $\underline{\mathbf{P}}(\omega)$ can model currents, not only proportional to $\partial \mathbf{E}/\partial t$ as is the case in electrostatics, but also to $\mathbf{E}$ simply by introducing some temporal dispersion in the response function $\epsilon(\omega)$. The value of the dc conductivity must therefore be contained in $\epsilon(\omega)$.

For a dielectric conducting medium, the two descriptions must be equivalent. The electrostatic description should apply to harmonic fields in the limit of low frequencies $\omega \to 0$. We can then rewrite Eq. (C.58) for harmonic fields in this limit:

$$\underline{\mathbf{j}}_{\text{int}}(\omega) = (-\mathrm{i}\omega\epsilon_0\chi_s + \sigma_{\text{dc}})\underline{\mathbf{E}}(\omega). \tag{C.59}$$

Now recall that the electric polarization $\mathbf{P}$ was defined in Section C.2.3 by Eq. (C.16), which implies $-\mathrm{i}\omega\underline{\mathbf{P}}(\omega) = \underline{\mathbf{j}}_{\text{int}}(\omega)$. We can therefore derive the low-frequency response functions, $\chi(\omega)$, and therefore $\epsilon(\omega)$ in terms of the static dielectric constant and dc conductivity:

$$\epsilon(\omega) = \epsilon_s + \frac{\mathrm{i}}{\omega\epsilon_0}\sigma_{\text{dc}}. \tag{C.60}$$

The above expression is a requirement for the equivalence of the electrostatics and low-frequency linear response approaches. As a matter of fact, it provides a simple expression for $\epsilon(\omega)$ at low frequencies, as a function of two well-characterized properties of a dielectric conducting medium: its static dielectric constant and its dc conductivity.

Likewise, it highlights one important point, which is often the source of confusion: the static electric polarization $\mathbf{P}_s$ of electrostatics (or static electric displacement $\mathbf{D}_s$), and the electric polarization $\mathbf{P}$ (or electric displacement $\mathbf{D}$) appearing in the general Maxwell's equations for media as defined in Section C.2.3 are (in general) two different physical quantities for conducting media. In particular, they are not even equal in the limiting case of low frequency ω. In fact, for a slowly varying uniform harmonic excitation of the form: $\mathbf{E}(t) = \mathbf{E}_0 \cos(\omega t)$, which in complex notation corresponds to $\underline{\mathbf{E}}(\omega) = \mathbf{E}_0$, we have from Eq. (C.56):

$$\mathbf{D}_s(t) = \epsilon_s \mathbf{E}_0 \cos(\omega t), \tag{C.61}$$

while from Eqs (C.40) and (C.60):

$$\mathbf{D}(t) = \mathrm{Re}(\underline{\mathbf{D}}(\omega)\mathrm{e}^{-\mathrm{i}\omega t}) = \mathrm{Re}(\epsilon(\omega)\mathbf{E}_0\mathrm{e}^{-\mathrm{i}\omega t}) = \mathbf{D}_s(t) + \frac{\sigma_{\mathrm{dc}}\mathbf{E}_0}{\omega}\sin(\omega t). \tag{C.62}$$

It is clear that the difference between $\mathbf{D}(t)$ and $\mathbf{D}_s(t)$ is due to the conduction currents. It can be seen from their respective definitions that they are related by:

$$\frac{\partial \mathbf{D}}{\partial t} = \frac{\partial \mathbf{D}_s}{\partial t} + \mathbf{j}_{\mathrm{cond}}. \tag{C.63}$$

The optical conductivity

In an attempt to generalize the electrostatic approach to the case of harmonic fields at a frequency ω, it is common to introduce two response functions, $\epsilon_s(\omega)$ and $\sigma_s(\omega)$, which by analogy with the electrostatic ones are defined as:

$$\underline{\mathbf{D}}_s(\omega) = \epsilon_0 \epsilon_s(\omega)\underline{\mathbf{E}}(\omega) \tag{C.64}$$
$$\underline{\mathbf{j}}_{\mathrm{cond}}(\omega) = \sigma_s(\omega)\underline{\mathbf{E}}(\omega). \tag{C.65}$$

$\sigma_s(\omega)$ is then called the *optical conductivity*, and $\epsilon_s(\omega)$ is usually called the frequency-dependent relative dielectric function (like $\epsilon(\omega)$, but unfortunately it is a different response function because $\mathbf{D}_s$ is different from $\mathbf{D}$!). Because they are defined with respect to the Fourier components (or complex quantities) of the fields, note that these two quantities are in general complex, which was not the case for their static counterparts, ϵ_s and σ_{dc}. From Eq. (C.62), it is easy to see that the 'standard' relative dielectric function $\epsilon(\omega)$ can then be obtained by:

$$\epsilon(\omega) = \epsilon_s(\omega) + \frac{\mathrm{i}}{\omega\epsilon_0}\sigma_s(\omega). \tag{C.66}$$

Discussion of the optical conductivity

The latter approach is in our opinion very confusing, and in many cases un-physical. We discuss it here only for reference because it appears regularly in the literature.

We have shown in the previous section that $\epsilon(\omega)$ was sufficient to fully characterize the optical response of the medium. It has a direct physical meaning since it can be directly measured in optical experiments. Its real part corresponds physically to optical scattering, which affects propagation, while its imaginary part to optical absorption by the medium. It also makes sense to define the static dielectric constant ϵ_s and dc conductivity σ_{dc} because they can be measured experimentally and correspond to a different physical reality: induced current proportional to a variation of $\mathbf{E}$ with time for the former, and proportional to the field $\mathbf{E}$ itself for the latter.

This is not the case in general for $\epsilon_s(\omega)$ and $\sigma_s(\omega)$. By defining them, we implicitly assume that they represent different physical realities, in fact the response of bound charges as opposed to that of conduction charges. But these different responses cannot be distinguished from the electromagnetic point of view since both are merged in Maxwell's equations into a single response function $\epsilon(\omega)$. It is true that the distinction could in principle be made theoretically, for example by modeling separately the response of bound charges ($\epsilon_s(\omega)$) and that of free charges ($\sigma_s(\omega)$). But since both quantities are in general complex, it is not possible from an experimental measurement of $\epsilon(\omega)$ to distinguish their respective contributions in Eq. (C.66).

The best approach, we believe, is therefore to avoid unnecessary complications, and simply work with the 'standard' relative dielectric function $\epsilon(\omega)$, which can be measured experimentally without further assumptions.

Appendix D

Lorentz model of the atomic/molecular polarizability

The different types of spectroscopies introduced in Chapter 2 have all been drawn (ultimately) from the concept of *optical polarizability*. Due to the importance of *polarizability theory* in optical spectroscopy as a whole, and for the Raman effect in particular, we provide here a few details of one of the most used (and simplest) models for the optical response of molecules and condensed matter: the Lorentz model. Furthermore, the model has connections with several concepts treated throughout the book, including the dielectric response of metals (Drude model) treated in Appendix E for the special cases of silver and gold, and molecular polarizability models of some important SERS probes.

D.1. THE LORENTZ OSCILLATOR

D.1.1. Principle

The linear optical polarizability $\alpha_L(\omega)$ of a molecule has been introduced in Section 2.4.3 from a purely phenomenological point of view. There it was taken as an empirical parameter. In reality, the linear optical polarizability is entirely determined by (and linked to) the electronic structure of the molecule. Given the electronic structure of a molecule, the linear optical polarizability can be, in principle, obtained from first principles starting from the electronic orbitals of the molecule and their coupling to electromagnetic radiation. Such a calculation requires the tools of quantum mechanics and perturbation theory. Luckily, some of the basic results to understand qualitatively the physics involved in the problem can be obtained from a classical treatment, first introduced by H.A. Lorentz almost a century ago (hereafter called the Lorentz model). This classical treatment then only requires a reinterpretation of the meaning of the terms in order to be translated into the quantum-mechanical result. As mentioned also in Section 2.7.1, it is a unique (and very fortunate!) property of harmonic oscillators that some of the results from the quantum-mechanical approach have direct links to similar expressions in the classical treatment.

The Lorentz model consists in treating the interaction of an electromagnetic wave with a specific electronic state as a classical (damped) harmonic oscillator with a natural frequency ω_0 [rad s^{-1}]. This approximation turns out to be extremely successful (and ultimately justified) because the action of the field is a small perturbation to the real interactions felt by the electrons. The equation of motion for the coordinate $\mathbf{r}$ [m] representing a small perturbation of the electrons from the ground state in an external electric field $\mathbf{E}$ [V m^{-1}] is [104]:

$$m\left(\frac{\mathrm{d}^2\mathbf{r}}{\mathrm{d}t^2} + \omega_0^2\mathbf{r} + \Gamma\frac{\mathrm{d}\mathbf{r}}{\mathrm{d}t}\right) = -e\mathbf{E}, \tag{D.1}$$

where Γ [s^{-1}] is a 'dissipation coefficient' which models phenomenologically all the internal and external interactions of the electronic cloud with everything but the field. Both $-e$ [C] and m [kg] may be taken as an *effective* charge and mass, when relevant. Equation (D.1) is that of a classical forced harmonic oscillator with damping. This can be solved easily using complex notations for harmonic fields. Taking $\mathbf{E} = \mathrm{Re}(\mathbf{E}_0 \exp(-\mathrm{i}\omega t))$, we seek a solution of the form $\mathbf{r} = \mathrm{Re}(\mathbf{r}_0 \exp(-\mathrm{i}\omega t))$ and obtain:

$$\mathbf{r}_0 = \frac{-e/m}{(\omega_0^2 - \omega^2 - \mathrm{i}\Gamma\omega)}\mathbf{E}_0. \tag{D.2}$$

The (complex) induced electric dipole moment is then given by $\mathbf{p}_0 = -e\mathbf{r}_0$ [C m]. Since by definition of the linear optical polarizability, $\mathbf{p}_0 = \alpha_L(\omega)\mathbf{E}_0$, we deduce[1]:

$$\alpha_L(\omega) = \frac{e^2/m}{(\omega_0^2 - \omega^2 - \mathrm{i}\Gamma\omega)}, \tag{D.3}$$

which is the *classical linear optical polarizability* [ϵ_0 m^3]. Note that $\alpha_L(\omega)$ is in general *complex*, therefore introducing a phase-shift between the incident field and the induced dipole. In quantum mechanics, the numerator of this

[1] It is worth remembering that the sign of the imaginary parts of all these expressions depends on the convention used for the oscillating fields; here we assume an $\exp(-\mathrm{i}\omega t)$ dependence resulting in a positive imaginary part.

expression is proportional to the matrix element of the dipolar transition [84] ; it is a measure of the oscillator strength. In addition, ω_0 is the energy difference between two electronic levels that are connected by a dipole-allowed transition by symmetry, and Γ represents all dissipative terms coming from higher order interactions with the environment and radiation damping.

D.1.2. Multiple transitions (multiple resonances)

A molecule can have more than one electronic resonance (different dipole-allowed transitions in the electronic structure). If that is the case, the linear optical polarizability can be generalized to a sum of the form:

$$\alpha_L(\omega) = (e^2/m) \sum_i \frac{f_i}{[\omega_i^2 - \omega^2 - \mathrm{i}\Gamma_i \omega]}, \tag{D.4}$$

where f_i is the *oscillator strength* of the transition at ω_i with damping constant Γ_i. The oscillator strengths satisfy the sum rule $\sum_i f_i = Z$, where Z is the number of electrons of the molecule [96].

D.1.3. Example: linear optical polarizability of rhodamine 6G

The most direct experimental manifestation of the linear optical polarizability of a molecule is through the measurement of its optical absorption. The free-space absorption cross-section in a medium of refractive index n_M is given by Eq. (4.79):

$$\sigma_{\mathrm{Abs}}^M = \frac{P_{\mathrm{Abs}}}{S_{\mathrm{Inc}}} = \frac{(L_M^{1/2})\omega}{n_M \epsilon_0 c} \frac{\mathrm{Im}\,(\alpha_L(\omega))}{3}, \tag{D.5}$$

where L_M is the local field correction factor (see Section 2.4.5). For a fluorophore like rhodamine 6G (RH6G), this can be measured easily by UV/Vis absorption (see Section 2.3.1). The linear optical polarizability of RH6G can therefore be partly inferred from a fit of the UV/Vis absorption spectrum of RH6G shown in Fig. 2.7. In a first approximation, we consider that only *one* transition dominates the optical properties of RH6G, and gather the effects of all other possible transitions at much higher energies as part of a background resulting in a real frequency-independent polarizability α_∞. With the additional definitions of: $\mu_0 = 2\pi c/\Gamma_0$, $\omega = 2\pi c/\lambda$, and $\omega_0 = 2\pi c/\lambda_0$, we can therefore write:

$$\alpha_L(\lambda) = \frac{\alpha_0}{[1 - (\lambda_0/\lambda)^2 - \mathrm{i}\lambda_0^2/(\mu_0 \lambda)]} + \alpha_\infty. \tag{D.6}$$

Using Eq. (2.21) for the UV/Vis absorbance and Eq. (D.5) for the absorption cross-section, the parameters that best fit the UV/Vis spectrum of RH6G in Fig. 2.7 are:

$$\lambda_0 = 526 \text{ nm}, \quad \mu_0 = 8000 \text{ nm}, \quad \alpha_0 = 3.9 \times 10^{-39} \text{ S.I.} \tag{D.7}$$

These values are similar to those provided in Ref. [202] where a similar model was used. Using in addition the fact that $\alpha_0 + \alpha_\infty$ should correspond to the static polarizability [202], one deduces $\alpha_\infty = 2.2 \times 10^{-39}$ S.I.

Note that μ_0 is basically the damping constant but expressed as a wavelength. In that way, we achieve an easier expression where all the units in the denominator of the first term in Eq. (D.6) cancel out, and we do not have to use the speed of light explicitly in the expression (it is absorbed in the definition of μ_0). We shall use a similar definition for the "damping constant" in Appendix E dealing with the dielectric functions of Ag and Au.

D.2. LINK WITH MACROSCOPIC PROPERTIES

D.2.1. Dielectric function in a dilute medium

In a dilute medium, the mutual interactions among atoms or molecules can be neglected and, therefore, we can ignore the microscopic local field correction treated in C.3.1. This is equivalent to assuming that the dielectric function (and refractive index) is close to 1, $\epsilon(\omega) \approx 1$ (i.e. the medium is not strongly optically active). In this case, the individual contribution of each molecule simply adds up. If there are n molecules per unit volume, the total induced polarization is $\mathbf{P} = n\alpha_L\mathbf{E}$, and we have simply:

$$\epsilon(\omega) = 1 + n\alpha_L(\omega)/\epsilon_0 = 1 + \frac{ne^2/m\epsilon_0}{(\omega_0^2 - \omega^2 - \mathrm{i}\Gamma\omega)}. \tag{D.8}$$

A plot of the corresponding $\mathrm{Re}\,[\epsilon(\omega)]$ and $\mathrm{Im}\,[\epsilon(\omega)]$ as a function of the *reduced frequency* ω/ω_0 is shown in Fig. D.1. The caption of Fig. D.1 summarizes the main aspects.

D.2.2. Dielectric function in solids

This last expression represents the contribution of one or several *discrete* transitions to the dielectric function $\epsilon(\omega)$ and this is typical of molecules. In solids, however, the electronic levels can form a *continuum* of transitions and, therefore, the contributions have to be suitably integrated (rather than summed) with a weight function proportional to the density of electronic states. The end result is a dielectric function that does not look like a sum of Lorentzians (as in Eq. (D.8)), but has more complicated frequency

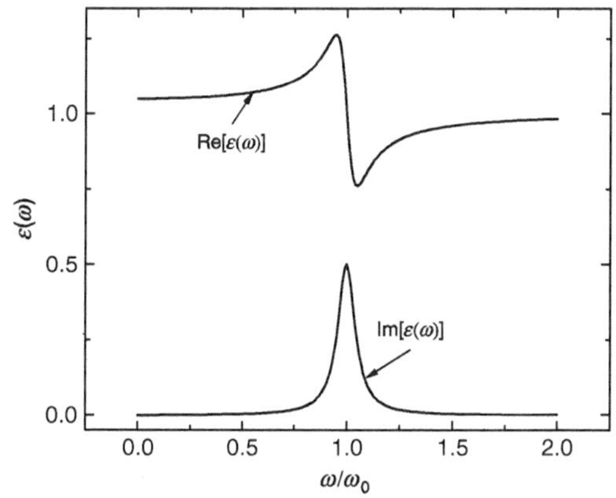

Figure D.1. Real and imaginary parts of $\epsilon(\omega)$ in and around the resonance frequency ω_0. The imaginary part peaks at $\omega \sim \omega_0$, slightly shifting the maximum to lower frequencies for larger Γ's. The plot has been generated for the dimensionless quantities $(ne^2/m\epsilon_0\omega_0^2) = 0.05$ (i.e. a "dilute system") and $(\Gamma^2/\omega_0^2) = 0.1$. The real part increases toward the resonance on the low energy side, $(\omega/\omega_0) < 1$; and can go negative for $(\omega/\omega_0) > 1$ for much denser media (depending on the strength of the resonance, measured here by the parameter $ne^2/m\epsilon_0\omega_0^2$). The region around ω_0 where $\text{Re}[\epsilon(\omega)]$ has a negative slope is the region known in classical optics as the *anomalous dispersion* region [96].

dependencies that reflect the integration process over the density of states. Without going into further details here, we only mention that this leads to the theory of *critical points* in the dielectric function, of widespread use in semiconductors [75,320,321]. We come back to this point when addressing the dielectric function of gold in Appendix E. Note also that the local field correction may be important in solids.

D.2.3. The metallic limit

From the point of view of SERS, it is interesting also to study a special case of expression (D.8). Electrons in a metal can be moved by the action of an external electric field without having a restoring force (in the equivalent oscillator picture); the maximum speed achieved by the electrons being only limited by scattering (losses) processes. Accordingly, if a certain fraction of the electrons (with density n_c) can move freely – as it would happen for the conduction electrons in metals – we can consider these electrons to have $\omega_0 = 0$ in Eq. (D.8), thus resulting in [97]:

$$\boxed{\epsilon(\omega) = \epsilon_b(\omega) - \frac{n_c e^2}{m\epsilon_0(\omega^2 + i\omega\Gamma)},} \tag{D.9}$$

where $\epsilon_b(\omega)$ is the *background* contribution from all other transitions. The term $\epsilon_b(\omega)$ will represent the so-called *inter-band transitions* that can coexist in a metal with the free electron contribution modeled by the second term.

As we shall see in Appendix E, silver can be relatively well modeled by the contribution of the conduction electrons only ($\omega_0 = 0$) (Drude model), while gold requires the addition of at least two critical points, which come from inter-band transitions. Both can be considered from a very general point of view to be "extensions" of the Lorentz model for an oscillator.

D.3. SUMMARY

The power of the phenomenology described by the Lorentz oscillator should not be underestimated. It can basically describe in simple terms the optical response of atoms and molecules, and with a few extensions to account for band effects and inter-band transitions in solids, it can also be extended to metals and semiconductors [75,320,321]. The next appendix continues with the same underlying subject, but from a more practical point of view, and specifically focusing on the optical properties of silver and gold; arguably two of the most important materials for SERS and plasmonics.

Appendix E

Dielectric function of gold and silver

Silver (Ag) and gold (Au) are by far the most used materials for SERS and plasmonics, and their optical properties are crucial to theoretical predictions and comparisons with experiments. It is therefore useful to develop further the arguments given in Section 3.2 and provide an analytic model for their respective optical properties. A detailed measurement of the optical properties of Ag and Au was carried out by Johnson and Christy [137]. This was based on the measurement of reflection and transmission on carefully prepared surfaces to avoid contributions from scattering. Another similar measurement was reported later by Palik [139]. In addition, another set of experiments used a different technique, based on surface plasmon resonances [150]. These three sets of data are consistent overall but not always in the details. In the following, we will provide an analytic model dielectric function for Ag and Au that is broadly consistent with the three aforementioned sets of experimental data, but do not exactly follow any one of these sets. These models are guided by the previous physical considerations but their aim is primarily to simplify the modeling of EM problems relevant to SERS and plasmonics. These analytical models are further discussed in Ref. [322].

E.1. MODEL DIELECTRIC FUNCTION FOR SILVER

E.1.1. Analytical expression

A good parametrization (based on the Drude model, see Section 3.2) of the optical properties of silver in the visible, near-IR, and near-UV range is given

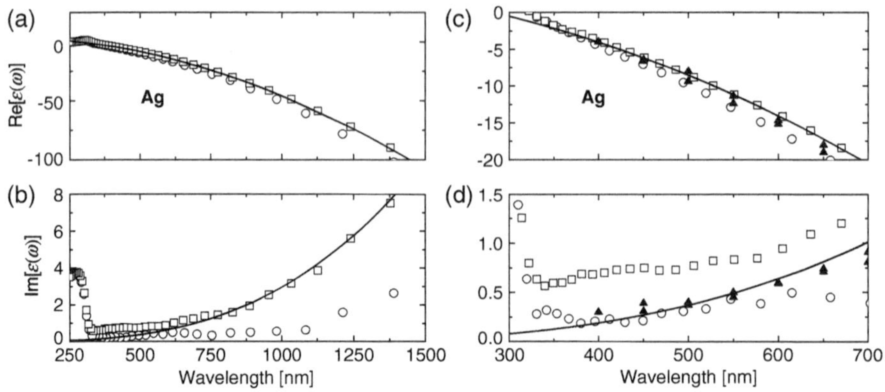

Figure E.1. Real (a and c) and imaginary (b and d) parts of the dielectric function of Ag from the near-UV to the far-IR range (a and b) and in the 'plasmonics' region (c and d). The solid lines are the model dielectric function of Ag given by Eq. (E.1). The experimental data have been taken from different sources: Ref. [137] (circles), Ref. [139] (squares), and Ref. [150] (solid triangles).

by [139,165,322]:

$$\epsilon_{\mathrm{Ag}}(\lambda) = \epsilon_\infty \left(1 - \frac{1}{\lambda_p^2\left(\frac{1}{\lambda^2} + \frac{i}{\mu_p \lambda}\right)}\right), \qquad \text{(E.1)}$$
$$\text{where} \quad \epsilon_\infty = 4, \quad \lambda_p = 282 \text{ nm}, \quad \mu_p = 17\,000 \text{ nm},$$

where λ is the wavelength (in nm).

Note that this is the same expression as that given for a Drude model in Eq. (3.6), but expressed in terms of wavelengths instead of frequency: $\lambda = 2\pi c/\omega$ and $\mu_p = 2\pi c/\gamma_0$ (which corresponds to $(2\pi)/\gamma_0 = 57$ fs). These parameters provide a good Drude fit for the real optical properties of Ag in the region of interest. This fit provides a value for ϵ_∞ from which the plasma frequency for Ag is derived: $\hbar\omega_p \approx 4.4$ eV, or $\lambda_p = 282$ nm.

E.1.2. Comparison to experimental results

The comparison between the experimental dielectric function of Ag and the analytical model given by Eq. (E.1) is shown in Fig. E.1. It is important to stress again that experimental measurements from different sources do not necessarily agree with each other. There are several possible reasons for this. Firstly, the quality/purity of the material may influence its optical properties.

Secondly, various techniques can be used to measure the optical properties: reflection/transmission measurements, as in Refs [137,139], ellipsometry, or the study of surface plasmon resonances as in Ref. [150]. For metals, the first two methods may be affected by plasmon-related effects in the 'plasmonics' region, while the third one automatically takes them into account (and even uses them for the measurement). Any attempts at fitting the optical properties should therefore be a compromise between these conflicting experimental results. The analytical model proposed here in Eq. (E.1) is one possible compromise.

Except for a few imperfections in the range 200–300 nm (where plasmon-related effects, possibly activated by surface imperfections, are expected) and in absorption in the far-IR tail, one could argue that the overall shape of $\epsilon(\omega)$ for Ag can be accounted for by a simple Drude model, as in Eq. (E.1) with a constant contribution ϵ_∞ from inter-band transitions .

E.2. MODEL DIELECTRIC FUNCTION OF GOLD

Gold is the second most important metal for SERS; widely preferred in many applications over silver for its relatively easier surface chemistry, good bio-compatibility, and stability under atmospheric conditions (oxidation).

E.2.1. Analytical expression

The optical properties of Au are slightly more difficult to represent with an analytical model. One reason is the more important role played by inter-band transitions in the visible/near-UV region. Gold has at least two inter-band transitions at $\lambda \sim 470$ and 325 nm that do play a role in $\epsilon(\omega)$ and have to be included explicitly. Their line-shapes are not very well accounted for by a simple Lorentz peak as in a molecular transition because of band effects (see Appendix D). A different type of analytic model for these two transitions in the near-UV has to be included to achieve a reasonable representation of $\epsilon(\omega)$. We shall not dwell on the details here, but only mention that it is possible to include a family of analytical models for transitions in solids which satisfy a set of minimum requirements (like Kramers–Krönig consistency) and reproduce most of the line-shapes in $\epsilon(\omega)$ observed experimentally. These are the so-called *critical points*, of common use in semiconductors. The optical properties of Au can therefore be represented relatively well (as functions of

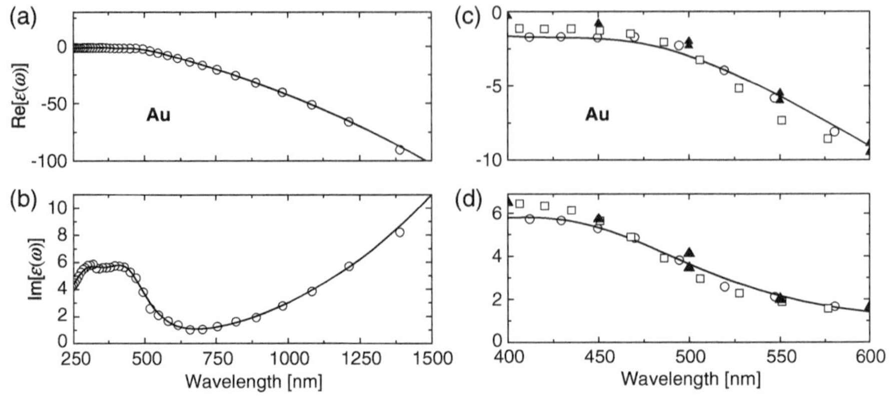

Figure E.2. Real (a and c) and imaginary (b and d) parts of the dielectric function of Ag from the near-UV to the far-IR range (a and b) and in the 'plasmonics' region (c and d). The solid lines are the model dielectric function of Au given by Eq. (E.2). The experimental data have been taken from different sources: Ref. [137] (circles), Ref. [139] (squares), and Ref. [150] (solid triangles).

$\lambda = 2\pi/\omega$) by the expression [322]:

$$\begin{aligned}\epsilon_{\mathrm{Au}}(\lambda) &= \epsilon_\infty \left(1 - \frac{1}{\lambda_p^2\left(\frac{1}{\lambda^2} + \frac{i}{\mu_p \lambda}\right)}\right) \\ &+ \sum_{n=1,2} \frac{A_n}{\lambda_n}\left[\frac{e^{i\phi_n}}{\frac{1}{\lambda_n} - \frac{1}{\lambda} - \frac{i}{\mu_n}} + \frac{e^{-i\phi_n}}{\frac{1}{\lambda_n} + \frac{1}{\lambda} + \frac{i}{\mu_n}}\right],\end{aligned} \tag{E.2}$$

where $\epsilon_\infty = 1.54$, $\lambda_p = 177.5\,\mathrm{nm}$, $\mu_p = 14\,500\,\mathrm{nm}$,
$A_1 = 1.27$, $\phi_1 = -\pi/4$, $\lambda_1 = 470\,\mathrm{nm}$, $\mu_1 = 1900\,\mathrm{nm}$,
$A_2 = 1.10$, $\phi_2 = -\pi/4$, $\lambda_2 = 325\,\mathrm{nm}$, $\mu_2 = 1060\,\mathrm{nm}$.

Note that in this case, the plasma frequency λ_p does not have a direct physical meaning, since plasma oscillations are affected by the presence of inter-band transitions and we do not have $\epsilon(\lambda_p) = 0$ as for the Drude model[1].

E.2.2. Comparison to experimental results

The result of this parametrization of the dielectric function is shown in Fig. E.2 and compared to experimental results. As in the case of

[1] Note that the definition (and value) of λ_p is slightly different to that used in Ref. [322] because the constant ϵ_∞ is factored out in the Drude model expression of (E.2).

Ag, except for a few unavoidable imperfections – which are a mixture of experimental problems in the determination of $\epsilon(\omega)$ in real samples and natural shortcomings of the analytical expressions used in the fit – the agreement between experiment and model is quite satisfactory. The first term in Eq. (E.2) accounts for the contribution of free electrons in Au, while the two other terms in the sum account for the two inter-band transitions at ~470 and 325 nm.

E.3. REMARKS ON THE MODEL DIELECTRIC FUNCTIONS

These parametrizations of $\epsilon(\omega)$ for Ag and Au can be used for theoretical predictions of the SERS enhancement as a very good starting point, with some provisos in mind which we now discuss.

E.3.1. Limitations of the models

Silver and gold are such crucial metals for SERS that it makes sense to expand the explanation of their optical properties. Equations like (E.1) are directly used in the modeling of metallic nano-structures used in SERS. It is important to understand not only its origins but also its limitations. In particular, we can worry about how relevant the imperfections of the simple Drude model with respect to the experimental data are for any type of conclusion derived from the modeling. The imperfections from a pure Drude model in metals have typically two possible causes. One of them is the presence of inter-band transitions on top of the contribution coming from free electrons. In practice, it means that the contribution from one or many additional oscillators in $\epsilon_b(\omega)$ in Eq. (3.6) must be taken into account, in addition to the free-electron term. This can produce departures from an ideal Drude behavior and this is the case of gold, as shown already. In addition, for reasons that are further explained in Chapter 3, the determination of optical constants is particularly complicated around the region where $\text{Re}[\epsilon(\omega)] \sim -1$. In this region, surface plasmon excitations that would not be activated in an ideal, atomically flat, surface become accessible through imperfections. This explains, at least partially, the anomalies seen in the optical constants of Ag at ~300 nm. This can produce 'experimental imperfections' in the data that should not be taken into account as real transitions. A combination of theoretical (band structure) and experimental data suggests that the optical properties of Ag are entirely determined by a Drude term, and that departures from that around ~300 nm are purely due to experimental imperfections. Then the approach for Ag is to try fitting the 'tails' of both $\text{Re}[\epsilon(\omega)]$ and $\text{Im}[\epsilon(\omega)]$ for $\lambda > 400$ nm and extrapolate the Drude behavior throughout. This is the meaning of Eq. (E.1). This approach is, we believe, no less valuable (if not better) than relying on a single set of experimental values, which may be affected by the experimental problems discussed above. Most of the

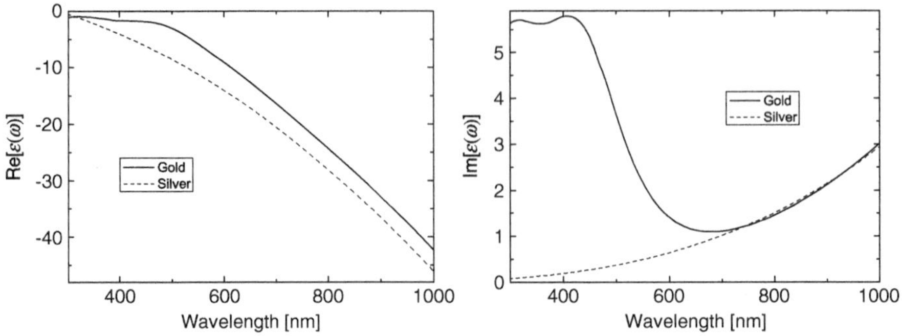

Figure E.3. Comparison of the dielectric functions, real part (left) and imaginary part (right) of silver and gold over the (extended) visible range.

calculations presented in this book will therefore rely on Eq. (E.1) (for silver) and Eq. (E.2) (for gold) for computing the dielectric functions.

E.3.2. Comparison between Ag and Au

The differences between silver and gold in terms of their plasmonic properties have been highlighted in several instances in this book. We re-emphasize here these differences by direct inspection of their respective dielectric function as given above. The comparison between their dielectric functions is shown explicitly in Fig. E.3 in the wavelength range of interest.

The presence of inter-band transitions for gold is the main origin of the differences between gold and silver, as far as the optical response is concerned. It increases $\mathrm{Re}(\epsilon)$ in the region where plasmon resonances should have occurred, thereby shifting them toward longer wavelength compared to silver. More importantly, it increases dramatically the optical absorption ($\propto \mathrm{Im}[\epsilon(\omega)]$) in this region, to make it much larger for Au than for Ag in the region where $\mathrm{Re}[\epsilon(\omega)] < 0$. This conveys completely different properties to both metals as far as plasmons are concerned. This was discussed in several places. We re-emphasize the fact that the higher absorption for Au at $\lambda < 600$ nm produces more 'lossy' plasmon resonances that result in less enhancement in general. It is equally important to remark that this is no longer true at longer wavelengths, typically $\lambda > 600$–650 nm (see Fig. E.3). Gold should therefore exhibit (and does exhibit) comparable plasmonic properties as silver (in particular in terms of local field enhancements) in this longer wavelength range.

Ag is largely preferred for its higher SERS enhancements, but it has the drawback of poorer stability and lower bio-compatibility compared to Au. Real applications will in practice find a tradeoff between these properties depending on the case. What Fig. E.3 demonstrates is that working at longer wavelengths (for example 633 nm and upwards) is a simple solution to the problem, provided appropriate structures can be designed to work in this range (which is indeed the case, see Chapter 6 for examples).

Appendix F

Plane waves and planar interfaces

In this appendix, we review some relevant results and summarize important expressions pertaining to plane waves and planar interfaces. These results can be found in many textbooks, but the emphasis is often on 'common' plane waves propagating in dielectrics. We attempt here to emphasize the problems associated with plane waves in absorbing media, such as metals, for their relevance to SERS and plasmonic effects. Particular attention is given to the 'commonly-swept-under-the-carpet' details, such as conventions pertaining to square roots of complex numbers or signs, for classic problems involving absorbing media, and in particular for the determination of surface modes such as surface plasmon–polaritons.

F.1. THE PLANE WAVE ELECTROMAGNETIC FIELDS

We recall here briefly the most important formulas pertaining to plane waves. We only consider monochromatic harmonic waves at frequency ω (corresponding to a free-space wave-vector $k_0 = \omega/c$), and all the fields refer to complex fields (with the $\exp(-\mathrm{i}\omega t)$ time-dependence convention). We consider a plane wave in a medium of relative dielectric function ϵ (possibly frequency dependent, and possibly complex). We exclude the possibility of an active medium (i.e. with gain), and the $\exp(-\mathrm{i}\omega t)$ time-dependence convention then implies that $\mathrm{Im}(\epsilon) \geq 0$.

F.1.1. General expressions

A general plane wave then corresponds to an electromagnetic field given as:

$$\boxed{\begin{cases} \mathbf{E}(\mathbf{r}) = \mathbf{E}_0 \exp(\mathrm{i}\mathbf{k}\cdot\mathbf{r}), \\ \mathbf{H}(\mathbf{r}) = \mathbf{H}_0 \exp(\mathrm{i}\mathbf{k}\cdot\mathbf{r}), \end{cases}} \tag{F.1}$$

where $\mathbf{k}$ is the wave-vector (possibly complex). The dispersion relation (which is imposed by Maxwell's equations, see Appendix C) can be written in terms of the $\mathbf{k}$ vector as:

$$\boxed{\epsilon(\omega)\frac{\omega^2}{c^2} = \epsilon(\omega)k_0^2 = \mathbf{k}\cdot\mathbf{k} = k_x^2 + k_y^2 + k_z^2,} \tag{F.2}$$

where $k_0 = \omega/c$ is the free-space wave-vector. Note that $\mathbf{k}\cdot\mathbf{k} \neq |\mathbf{k}|^2$ in general, if $\mathbf{k}$ is a complex vector. It is convenient in some situations to write the wave-vector as $\mathbf{k} = k\mathbf{n}$, where $\mathbf{n}$ is a vector (possibly complex) satisfying $\mathbf{n}\cdot\mathbf{n} = 1$ (and is therefore unique up to a sign, but is not necessarily a *unit vector*), and k is the wave-vector amplitude, possibly complex and also defined up to a sign by $k^2 = \mathbf{k}\cdot\mathbf{k} = k_0^2\epsilon$.

Moreover, $\mathbf{E}_0$ (complex vector) must satisfy $\mathbf{n}\cdot\mathbf{E}_0 = 0$ (transverse field). $\mathbf{H}_0$, is then fully determined by:

$$\boxed{\mathbf{H}_0 = \epsilon_0 c\frac{k}{k_0}\mathbf{n}\times\mathbf{E}_0,} \tag{F.3}$$

which further ensures that $\mathbf{n}\cdot\mathbf{H}_0 = 0$ (transverse field again). This is also equivalent to defining $\mathbf{H}_0$ and deriving:

$$\boxed{\mathbf{E}_0 = -\frac{1}{\epsilon(\epsilon_0 c)}\frac{k}{k_0}\mathbf{n}\times\mathbf{H}_0.} \tag{F.4}$$

These two relations are also sometimes expressed in terms of the vacuum impedance defined as $Z_0 = (\epsilon_0 c)^{-1} \approx 376.73$ ohms. Moreover, they can usually be simplified using $k/k_0 = \pm\sqrt{\epsilon}$ (where the sign depends on the problem at hand and the conventions chosen). They further imply that $\mathbf{E}_0\cdot\mathbf{H}_0 = 0$ (note that this does not necessarily imply that $\mathbf{E}$ and $\mathbf{H}$ are perpendicular when $\mathbf{E}_0$ and $\mathbf{H}_0$ are general complex numbers). Finally, note that $\mathbf{E}_0$ and $\mathbf{H}_0$ are the vectorial complex amplitudes of the fields, i.e. they characterize their amplitude, relative phase, and polarization. They should *not* be confused with the corresponding real amplitudes $|\mathbf{E}_0|$ and $|\mathbf{H}_0|$, which only characterize the amplitude of the waves.

The above equations and considerations are very general. In particular, they allow for the possibility of a complex k and/or complex $\mathbf{n}$. For these waves to exist in the infinite region, one must however make sure that the fields do not diverge at infinity, i.e., only waves with $\mathrm{Im}\,(\mathbf{k}\cdot\mathbf{r}) \geq 0$ at infinity can represent real physical waves.

F.1.2. Propagating plane waves

Most physical waves have $\mathbf{n}$ real, and $\mathbf{n}$ is then a real unit vector corresponding to the direction of propagation. These waves are then called *homogeneous plane waves.*

If k is also real, then $\mathbf{k}$ is a real vector, and the wave is a *propagating plane wave*, or propagating homogeneous plane wave. Note that from Eq. (F.2), ϵ is necessarily real positive in this case. This is the most intuitive situation and the one used to approximate light beams. In that case (and only in that case), the real amplitudes of the fields are directly related through:

$$|\mathbf{H}_0| = \sqrt{\epsilon}(\epsilon_0 c)|\mathbf{E}_0|. \tag{F.5}$$

The complex Poynting vector for such a wave is then given by (this is not valid in general for non-real k or $\mathbf{n}$):

$$\boxed{\mathbf{S} = \frac{1}{2}\mathbf{E} \times \mathbf{H}^* = S_0\mathbf{n} = \sqrt{\epsilon}\frac{\epsilon_0 c}{2}|\mathbf{E}_0|^2\mathbf{n}.} \tag{F.6}$$

S_0 then corresponds physically to the excitation density $[\mathrm{W/m^2}]$ of the plane wave and, therefore, relates this measurable physical quantity to the amplitude $E_0 = |\mathbf{E}_0|$. Let us take for example a laser beam of 1 mW, which we assume for the sake of argument of uniform intensity over an area of 1 mm^2, i.e. an excitation density of $S_0 = 1000\ \mathrm{W/m^2}$. The electric field amplitude of such a beam in air ($\epsilon = 1$) is then[1] $E_0 \approx 870$ V/m.

F.1.3. Evanescent plane waves

We now focus on cases where k is not real, but $\mathbf{n}$ is still real. The wave then propagates along $\mathbf{n}$, but with a propagation wave-vector $\mathrm{Re}(k)\mathbf{n}$ (which is real). In addition, from Eq. (F.1), the field amplitudes decay exponentially along the propagation direction (along $\mathbf{n}$) as $\exp(-\mathrm{Im}(k)\mathbf{n}\cdot\mathbf{r})$. The wave is *evanescent,* and is an evanescent homogeneous plane wave[2].

One can further qualify such a wave depending on the relative magnitude of $\mathrm{Re}(k)$ and $\mathrm{Im}(k)$. In particular, if $|\mathrm{Im}(k)| \ll |\mathrm{Re}(k)|$ (in a low absorption medium for example), then the decay length is much larger than the spatial

---

[1] It is tempting to estimate the parameters needed to attain $E_0 \approx 3 \times 10^6$ V/m, i.e. the threshold for dielectric breakdown (ionization) in air: A 12 mW beam on a 1 $\mu\mathrm{m}^2$ area would achieve this condition. However, dielectric breakdown is a DC phenomenon, i.e. it occurs for a *constant* electric field, whereas our fields are oscillating and the actual breakdown may therefore occur at a completely different threshold.

[2] The term homogeneous is sometimes reserved for propagating plane waves, i.e. with k and $\mathbf{n}$ real, see the discussion of inhomogeneous plane waves in the next section.

wavelength of propagation. The wave can therefore propagate over relatively long distances before its amplitude has substantially decayed. Such a wave can be called *pseudo-propagating*, even if it is strictly speaking, an evanescent wave. In other cases, the wave is truly evanescent.

Note that evanescent waves cannot exist by themselves in the whole of space, as is the case for propagating plane waves. They would indeed be of infinite amplitude in the direction opposite to $\mathbf{n}$, so they must be bounded by an interface along there. They must therefore be excited by an incident wave on this interface. For example, such an evanescent wave can be created inside an absorbing material when an incident wave impinges onto its surface (see later).

Finally, for both cases of homogeneous plane waves, propagating or evanescent, one can show that the electric and magnetic fields are always perpendicular to each other and to the direction of propagation. A general homogeneous plane wave can then always be decomposed as *a (complex) superposition of two linearly polarized homogeneous plane waves* (i.e. with $\mathbf{E}_0$ real vector up to a complex phase). For more details on this specific point, see for example Refs [96,151] and Section F.2.1.

F.1.4. Inhomogeneous plane waves: hybrid propagating/evanescent waves

If $\mathbf{n}$ is not real, the wave is then called an *inhomogeneous plane wave.* This term refers to the fact that the (plane) surfaces of equal amplitudes are not the same as those of equal phase (as was the case in the two previous situations of propagating and evanescent homogeneous plane waves). To put it differently, the wave amplitude decays along a direction that is different from that of propagation.

To understand the meaning of this, let us consider the case where k is real (which requires ϵ real positive, for example a non-absorbing dielectric). It is then possible to show that the wave propagates along the direction of $\mathrm{Re}(\mathbf{n})$, but decays along that of $\mathrm{Im}(\mathbf{n})$ (note that $\mathrm{Re}(\mathbf{n}) \perp \mathrm{Im}(\mathbf{n})$ because of the condition $\mathbf{n} \cdot \mathbf{n} = 1$). Moreover, the electric field polarization $\mathbf{E}_0$, may have non-zero components along both directions (those of propagation and decay). Such a wave may be viewed as a hybrid propagating/evanescent plane wave.

In the case of a complex k, the situation is even more complicated, since the wave will then be evanescent along the two perpendicular directions, $\mathrm{Re}(\mathbf{n})$ and $\mathrm{Im}(\mathbf{n})$, but with different propagation wave-vectors and decay lengths. See for example p. 298 in Ref. [96] for more details.

Inhomogeneous plane waves could be considered at first as mathematical oddities, but they actually appear in many electromagnetic problems at planar interfaces. For example, the wave created inside a dielectric upon total internal reflection at an interface with another dielectric (of higher refractive index) is a inhomogeneous plane wave: it propagates along the surface, but decays perpendicular to it. More relevant to our subject here,

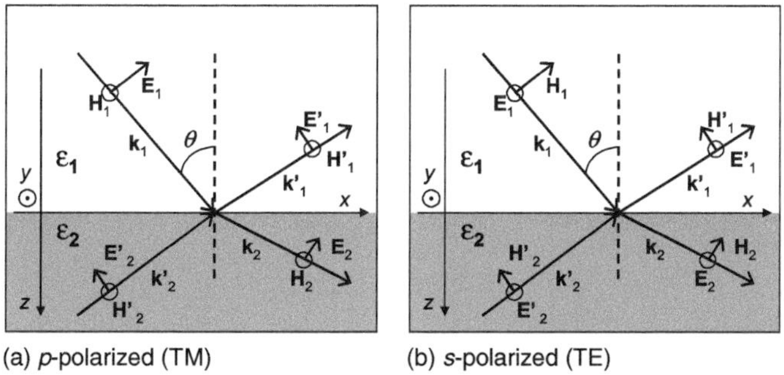

Figure F.1. Schematic illustrating the two types of polarization for an homogeneous plane wave (characterized by wave-vector $\mathbf{k}_1$ and electric and magnetic fields $\mathbf{E}_1$ and $\mathbf{H}_1$) impinging on a planar interface. The figure is drawn in the plane of incidence (xOz). Also shown are the other waves involved in the general solution of the problem, discussed in Section F.2.2.

surface plasmon–polaritons (SPPs) at planar interfaces are also examples of inhomogeneous plane waves: they are composed of one inside the metal, and another one in the dielectric.

F.2. PLANE WAVES AT A SINGLE PLANAR INTERFACE

In this section, we study the scattering of a plane wave by an infinite plane interface between two materials (as illustrated in Fig. F.1). This problem is treated in numerous textbooks [96,151] from the viewpoint of reflection/refraction of waves at interfaces. Rather than repeating these, we adopt here a somewhat different and more general approach, which emphasizes some of the issues relevant to plasmonics. We therefore start from the most general problem (even if its physical relevance does not justify it), and specialize gradually its solution toward the classic problem of reflection/refraction. This is probably not the easiest approach to reflection/refraction, but it has the benefit of deriving the existence of surface plasmon–polariton modes along the way, thereby placing them in the same general framework as reflection/refraction at interfaces. It also extends itself naturally to the case of multi-layer interfaces; very important in plasmonics, and treated in the next section.

F.2.1. Plane wave polarization at an interface

We avoid here a general discussion of plane wave polarization parameters (Stokes parameters), which can be found e.g. in Refs [96,151], and focus on the relevant case of an incident (homogeneous) plane wave impinging on an infinite planar surface, as illustrated in Fig. F.1. It is customary to define the *plane of incidence*, containing the two vectors $\mathbf{n}$ (real) and $\mathbf{e}_z$, normal to the planar surface (which is parallel to xOy in our conventions, see Fig. F.1). It can then be shown that any (homogeneous) plane wave polarization can be represented as a linear (possibly complex) superposition of two linearly polarized waves (with polarization mutually perpendicular and also perpendicular to the direction of propagation $\mathbf{n}$). It therefore suffices to study these two cases:

- s-polarized wave (from 'senkrecht' for perpendicular, in German), for a wave with linear polarization perpendicular to the plane of incidence, i.e. $\mathbf{E}_0 \perp \mathbf{e}_z, \mathbf{n}$. Such waves are also called *Transverse Electric* (TE) waves.
- p-polarized wave (from 'parallel' for parallel in both German and English!), for a wave with linear polarization parallel to the plane of incidence. In this case, the magnetic field must be perpendicular to the plane of incidence, $\mathbf{H}_0 \perp \mathbf{e}_z, \mathbf{n}$. Such waves are also called *Transverse Magnetic* (TM) waves.

This separation is convenient since the nature of the wave polarization (s or p) *is conserved at planar interfaces.* It is therefore possible to study independently these two cases, and any general case can then be inferred from an appropriate superposition of the solutions of these two problems.

The study of the polarization of inhomogeneous plane wave is outside the scope of this appendix. Let us just say that inhomogeneous plane waves cannot, in the most general case, be represented as the superposition of s- and p-polarized waves. However, in all cases here, the inhomogeneous plane waves that we may encounter will always have been created from a homogeneous plane wave (impinging on a surface, for example). Since the nature (s or p) of the polarization is conserved, the homogeneous plane wave imposes its polarization on the others (including any inhomogeneous plane waves that may arise). It is therefore sufficient for our purpose to consider only inhomogeneous plane waves with s- or p- polarization.

F.2.2. General solution for plane waves at a planar interface

General considerations, definitions, and conventions

We now derive important relations resulting from the boundary conditions at a planar interface, restricting ourselves to s- or p-polarized waves. The interface is at $z = 0$, and it delimits two half-spaces characterized by their dielectric functions ϵ_1 for $z < 0$, and ϵ_2 for $z > 0$. We assume for generality that there are several plane waves in each half-space (we will show that two are sufficient), characterized by their electric fields and wave-vectors: $\mathbf{E}_1$, $\mathbf{k}_1$, $\mathbf{E}'_1$, $\mathbf{k}'_1$, $\mathbf{E}_2$, $\mathbf{k}_2$, $\mathbf{E}'_2$, $\mathbf{k}'_2$ (see Fig. F.1). The derivation will be carried out in the most general case (homogeneous or inhomogeneous plane waves). All quantities are hence *a priori* complex numbers. The aim here is to find the relations between these that are imposed by the interface. The results will then be specialized to physical cases of interest such as reflection/refraction at single or multi-layer interfaces, and a physical meaning will then be assigned to these waves.

We assume (without loss of generality) that the plane of incidence is xOz and therefore $k_{1y} = k'_{1y} = k_{2y} = k'_{2y} = 0$. The most general plane wave (say in region 1) within these assumptions can be characterized by its wave-vector:

$$\mathbf{k}_1 = k_{1x}\mathbf{e}_x + k_{1z}\mathbf{e}_z, \tag{F.7}$$

where k_{1x} and k_{1z} may be complex.

The electromagnetic boundary conditions at $z = 0$ will necessarily contain factors of the type $\exp(\mathrm{i}k_{1x}x)$, and its equivalent for all the other plane waves. Since the boundary conditions must be valid for all x, it implies that:

$$k_{1x} = k'_{1x} = k_{2x} = k'_{2x} = k_x, \tag{F.8}$$

which defines a common k_x for all waves[3]. This is not a surprise since the translational invariance of the problem along x automatically implies conservation of k_x.

The dispersion relations then imply that:

$$\epsilon_1 k_0^2 = k_x^2 + k_{1z}^2, \tag{F.9}$$

and the same relations for the other waves, simply changing the index and/or adding the prime ($'$). This means that $\mathbf{k}_1$ is entirely determined by the knowledge of k_x, up to a sign for k_{1z}; this is why there are only two plane waves in each region in the most general case and, moreover, we have $k'_{1z} = -k_{1z}$ (if they were equal, both waves could be combined into one) and more precisely:

$$k_{1z} = -k'_{1z} = \pm\sqrt{\epsilon_1 k_0^2 - k_x^2}. \tag{F.10}$$

The choice of sign will depend on the physical meaning of the waves represented by $\mathbf{k}_1$ and $\mathbf{k}'_1$ in the problem under consideration. The same arguments apply to region 2: $k'_{2z} = -k_{2z}$, with the same sign indetermination. All wave-vectors are therefore entirely determined (up to signs) by the knowledge of k_x. The choice of the signs is rarely discussed, since it is generally obvious in most common situations (problems with a propagating incident wave). This is no longer so obvious for problems with metals and in relation to surface modes. It will therefore be discussed in detail in the following.

[3] We will see later that k_x conservation, combined with the dispersion relations, simply leads to Snell's law (or Descartes' law) for the reflection and refraction angles at planar interfaces.

We now treat the cases of p-polarized and s-polarized waves separately.

p-polarized or TM waves

TM waves are characterized by a magnetic field perpendicular to the plane of incidence, i.e. along $\mathbf{e}_y$. We can therefore, using Eqs (F.1) and (F.4), write their electromagnetic field as:

$$\textbf{TM}: \quad \begin{cases} \mathbf{H}_1 = (H_{1y}\mathbf{e}_y)\exp(ik_{1x}x + ik_{1z}z) \\ \mathbf{E}_1 = \dfrac{H_{1y}}{\omega\epsilon_0\epsilon_1}(k_{1z}\mathbf{e}_x - k_{1x}\mathbf{e}_z)\exp(ik_{1x}x + ik_{1z}z). \end{cases} \tag{F.11}$$

The same expression exists for the other three waves, simply changing the index and/or adding the prime ($'$).

We then write explicitly the two (independent) boundary conditions at $z = 0$: the continuity of the tangential magnetic field:

$$H_{1y} + H'_{1y} = H_{2y} + H'_{2y}, \tag{F.12}$$

and the continuity of the tangential electric field (Recall that $k'_{1z} = -k_{1z}$ and $k'_{2z} = -k_{2z}$):

$$\frac{H_{1y}k_{1z}}{\epsilon_1} - \frac{H'_{1y}k_{1z}}{\epsilon_1} = \frac{H_{2y}k_{2z}}{\epsilon_2} - \frac{H'_{2y}k_{2z}}{\epsilon_2}. \tag{F.13}$$

These equations can be rewritten in matrix form as:

$$\begin{pmatrix} H_{1y} \\ H'_{1y} \end{pmatrix} = \frac{1}{2}\begin{pmatrix} 1+K^p & 1-K^p \\ 1-K^p & 1+K^p \end{pmatrix}\begin{pmatrix} H_{2y} \\ H'_{2y} \end{pmatrix}, \tag{F.14}$$

or by inversion as:

$$\begin{pmatrix} H_{2y} \\ H'_{2y} \end{pmatrix} = \frac{1}{2}\begin{pmatrix} 1+(K^p)^{-1} & 1-(K^p)^{-1} \\ 1-(K^p)^{-1} & 1+(K^p)^{-1} \end{pmatrix}\begin{pmatrix} H_{1y} \\ H'_{1y} \end{pmatrix}, \tag{F.15}$$

where

$$K^p = \frac{\epsilon_1 k_{2z}}{\epsilon_2 k_{1z}}. \tag{F.16}$$

These matrix equations are simply equivalent to the electromagnetic boundary conditions at $z = 0$, for the four TM polarized waves.

Finally, we note that the discontinuity of the normal electric field E_z results in the presence of a surface charge density wave on the interface ($z = 0$) given by:

$$\rho^p_{\text{Surf}} = \frac{k_x}{\omega}(H_{1y} + H'_{1y}) \left[\frac{1}{\epsilon_1} - \frac{1}{\epsilon_2}\right] \exp(ik_x x). \tag{F.17}$$

This does not occur for TE waves since the electric field is parallel to the plane.

s-polarized or TE waves

TE waves are characterized by an electric field perpendicular to the plane of incidence, i.e. along $\mathbf{e}_y$. We can therefore, using Eqs (F.1) and (F.3), write their electromagnetic field as:

$$\textbf{TE}: \begin{cases} \mathbf{E}_1 = (E_{1y}\mathbf{e}_y)\exp(ik_{1x}x + ik_{1z}z) \\ \mathbf{H}_1 = \dfrac{E_{1y}}{\omega\mu_0}(-k_{1z}\mathbf{e}_x + k_{1x}\mathbf{e}_z)\exp(ik_{1x}x + ik_{1z}z). \end{cases} \tag{F.18}$$

The same expression exists for the three other waves, again by simply changing the index and/or adding the prime (′).

We write again explicitly the two (independent) boundary conditions at $z = 0$ for the continuity of the tangential fields. In fact, we obtain similar relations as before for TM waves, only replacing H_y/ϵ by E_y:

$$E_{1y} + E'_{1y} = E_{2y} + E'_{2y}, \tag{F.19}$$

and

$$E_{1y}k_{1z} - E'_{1y}k_{1z} = E_{2y}k_{2z} - E'_{2y}k_{2z}. \tag{F.20}$$

These equations can be rewritten in matrix form as before:

$$\begin{pmatrix} E_{1y} \\ E'_{1y} \end{pmatrix} = \frac{1}{2}\begin{pmatrix} 1+K^s & 1-K^s \\ 1-K^s & 1+K^s \end{pmatrix}\begin{pmatrix} E_{2y} \\ E'_{2y} \end{pmatrix}, \tag{F.21}$$

where:

$$K^s = \frac{k_{2z}}{k_{1z}} = \frac{\epsilon_2}{\epsilon_1}K^p. \tag{F.22}$$

These general matrix expressions can now be used for the solution of more specific problems.

A note on k_x conservation

The previous arguments have shown that for a general planar interface, the wave-vector along the directions of translational invariance (k_x in this case) is conserved and, therefore, it is the same for all waves. In theory, and for perfect translational invariance, k_x must be real or otherwise the field amplitude would diverge at infinity (either at $x \to +\infty$ or $x \to -\infty$). In practical experiments also, the system is excited by an external wave, usually modeled as a plane wave propagating in a non-absorbing dielectric, which therefore imposes again that k_x is real.

It is nevertheless convenient to allow k_x to be complex for several reasons:

- We study here the electromagnetic response of the interface to an *external* excitation. This external source is *a priori* not constrained by translational invariance (for example for an emitting dipole) and could in principle be decomposed as a sum of plane waves with any (possibly complex) k_x.
- When studying the response of our planar system in terms of electromagnetic modes, the modes will be characterized by dispersion relations linking k_x and ω. If the system is excited by a propagating plane wave, then k_x and ω are real. This external excitation should couple fully to a mode where both k_x and ω are real (and match the external excitation parameters). For modes where k_x and ω cannot be both real, the coupling cannot be perfect, but is still possible and therefore worth studying. One can always recover the case where k_x is real by assuming that ω is complex. The modes are then viewed as *virtual modes* (with a finite lifetime). The other point of view, where ω is real and k_x complex is equivalent and more adapted to situations with *continuous wave* (CW) excitations (which are common). The fact that k_x is complex then simply reflects the fact that the mode decays spatially away from the region where the external excitation was applied (which is never infinite in practice and therefore automatically breaks the translational invariance). It is then an *evanescent wave*. These points of view were also discussed in Section 3.3.4.

F.2.3. Physical waves in a semi-infinite region

In the preceding section, we have made no assumption on the physical reality of the waves in region 1 and 2, and which source created them in the first place. This aspect needs to be addressed to model practical situations. Let us therefore consider two semi-infinite regions, region 1 with $z < 0$ and region 2 with $z > 0$, as in Fig. F.1.

Physical waves in non-absorbing regions

In a 'classic' reflection/refraction problem in its simplest form, both regions are non-absorbing dielectrics (both ϵ_1 and ϵ_2 are real positive). This is the most commonly studied electromagnetic boundary condition problem. A wave propagates in region 1 from $z \to -\infty$ toward the surface (i.e. with $k_{1z} > 0$, and k_x also real); it is usually called the *incident wave.* This wave is reflected, therefore resulting in a second wave in region 1 with $k'_{1z} = -k_{1z} < 0$ propagating away from the surface; the *reflected or scattered wave.* A wave is also created in region 2 and is called the *refracted or transmitted wave.* If this wave is propagating, it cannot propagate from $z \to +\infty$, where no source is present, and must therefore propagate toward $z \to +\infty$, i.e. $k_{2z} > 0$. If the transmitted wave is evanescent, then one must impose $\mathrm{Im}(k_{2z}) > 0$ to avoid divergence of the field. The choice of sign for k_{2z} is therefore always determined by one of these two conditions.

Physical waves in the general case

We would like to generalize the preceding consideration to a general case where all quantities may be complex. Two waves may in principle exist in each region for a given k_x (as considered in the general case), and with opposite z-wave-vectors $k_{1z} = -k'_{1z}$ (possibly complex) and $k_{2z} = -k'_{2z}$. One then needs to distinguish between two situations in each region. For region 1 first:

- If k_{1z} is real ($\mathrm{Im}(k_{1z}) = 0$), then both waves are propagating, in opposite directions along z (and may or may not be evanescent in the x-direction in the general case). We can choose, by convention, to take $k_{1z} > 0$. The wave with wave-vector $\mathbf{k}_1$ then represents an *incident wave.* This automatically implies that $k'_{1z} < 0$ and the corresponding wave is therefore a *reflected or scattered wave.* Both are *a priori* acceptable physical waves. However, the incident wave (with $k_{1z} > 0$) originates from $z \to -\infty$ and therefore only exists in physical problems where such a source is present at $z \to -\infty$. This is the case in many common problems, where an exciting propagating wave impinges on the surface from $z \to -\infty$.

- We must now define the choice of sign when k_{1z} is not real. If $\mathrm{Im}(k_{1z}) \neq 0$, then both waves are evanescent (along z). However, for a wave to exist at $z \to -\infty$, one must have $\mathrm{Im}(k_z) < 0$ to avoid a divergence of the field amplitude. Since $k_{1z} = -k'_{1z}$, this can only be fulfilled by one of the two waves, which we choose *by convention* to be the wave represented by $\mathbf{E}'_1$ and $\mathbf{k}'_1$. We must therefore have: $\mathrm{Im}(k'_{1z}) < 0$ (or equivalently $\mathrm{Im}(k_{1z}) > 0$) and $\mathbf{E}_1 = \mathbf{0}$. We will moreover denote this wave again as the *scattered wave.* The reason for this choice is that it cannot be an incident wave in the same sense as understood before, since an incident evanescent wave coming from $z \to -\infty$ would never physically reach the interface (because its amplitude decays).

Within these conventions, we can therefore summarize our terminology and choice of sign in the most general case:

$$\text{If} \quad \text{Im}(k_{1z}) = 0, \quad \begin{array}{l} \text{then} \;\; \text{Re}(k_{1z}) > 0, \\ \text{else} \;\; \text{Im}(k_{1z}) > 0 \quad \text{and} \quad \mathbf{E}_1 = \mathbf{0}. \end{array} \tag{F.23}$$

The wave represented by $\mathbf{k}_1$ will be called the *incident wave* and can only exist if $\text{Im}(k_{1z}) = 0$ and if a source is present at $z \to -\infty$, The wave represented by $\mathbf{k}'_1$ will be called the *reflected or scattered wave.*

For consistency, similar conventions are chosen to decide the signs of k_{2z} and k'_{2z} in region 2 ($z > 0$):

$$\text{If} \quad \text{Im}(k_{2z}) = 0, \quad \begin{array}{l} \text{then} \;\; \text{Re}(k_{2z}) > 0 \\ \text{else} \;\; \text{Im}(k_{2z}) > 0 \quad \text{and} \quad \mathbf{E}'_2 = \mathbf{0}. \end{array} \tag{F.24}$$

These conventions can be discussed by distinguishing again between the two situations:

- If k_{2z} is real ($\text{Im}(k_{2z}) = 0$), then both waves are propagating, in opposite direction along z (and may or may not be evanescent in the x-direction in the general case). Again, both are *a priori* acceptable physical waves. Since by convention $k'_{2z} < 0$, the wave with $\mathbf{k}'_2$ originates from $z \to +\infty$, and only exists in physical problems where such a source is present at $z \to +\infty$. It could be called the 'second incident wave', but in most problems of interest, there is no such source and this wave is not present: $\mathbf{E}'_2 = \mathbf{0}$. The other wave in region 2, the one with wave-vector $\mathbf{k}_2$ such as $k_{2z} > 0$ propagates toward $z \to +\infty$, and will be called the *transmitted wave.*

- If $\text{Im}(k_{2z}) \neq 0$, then both waves are evanescent (along z). Moreover, by convention, $\text{Im}(k_{2z}) > 0$ and $\text{Im}(k'_{2z}) < 0$. The second condition is in fact unphysical, since it leads to a divergence of the field amplitude at $z \to +\infty$. We must therefore impose $\mathbf{E}'_2 = \mathbf{0}$ in this case. Only one wave can then exist in medium 2, the one represented by $\mathbf{k}_2$, for which $\text{Im}(k_{2z}) > 0$. By analogy with the previous case, we will also call this wave the *transmitted wave.*

In summary, the sign conventions given in Eqs (F.23) and (F.24) ensure that the following assignments are always physically correct:

- The wave represented by ($\mathbf{k}_1$,$\mathbf{E}_1$) is the incident wave. It exists only if $\text{Im}(k_{1z}) = 0$ and if a source is present at $z \to -\infty$.

- The wave represented by ($\mathbf{k}'_1$,$\mathbf{E}'_1$) is the reflected or scattered wave.
- The wave represented by ($\mathbf{k}_2$,$\mathbf{E}_2$) is the transmitted wave.
- The wave represented by ($\mathbf{k}'_2$,$\mathbf{E}'_2$) is the 'second incident wave' and is absent in most problems of interest.

Finally, note that the preceding arguments are no longer valid if region 1 (or region 2) does not extend to $z \to -\infty$ ($z \to +\infty$), for example if there is another interface at $z < 0$ ($z > 0$). The two waves are then physically acceptable in all cases (but it is then a different EM problem).

Which way is the scattered wave going?

One may find the previous discussion rather excessive. After all, simply looking at Fig. F.1, the incident wave should be the one going down, and the scattered wave the one going up. This is in some way true, but this naive view breaks down, when complex wave-vector replaces real ones! This situation in fact only arises when deriving the surface modes and can be (and has been) ignored in most 'classic' problems, but surface modes are precisely the ones important to *plasmonics*. This issue will therefore be discussed in more detail later in relation to surface modes.

Here, we only highlight a few aspects of the problem. Consider a situation where k_{1z} is not real. The scattered wave must therefore satisfy $\text{Im}(k'_{1z}) < 0$ (and there cannot be any incident wave). This simply tells us that the field amplitude of the scattered wave decays as one moves away from the interface into region 1, as intuitively expected for an evanescent wave. Intuition also suggests that phase propagation should also go away from the interface, i.e. $\text{Re}(k'_{1z}) < 0$. In fact, we have:

$$2\text{Im}(k'_{1z})\text{Re}(k'_{1z}) = \text{Im}((k'_{1z})^2) = k_0^2\text{Im}(\epsilon_1) - \text{Im}(k_x^2). \tag{F.25}$$

It is, therefore, possible that some solutions for which $\text{Im}(k_x^2) > 0$ may have the real and imaginary parts of k'_{1z} of opposite sign, i.e. $\text{Re}(k'_{1z}) > 0$. In this case, the scattered wave propagates *toward* the interface, while its amplitude decays *away* from the interface. This situation is similar in some ways with what happens in *negative refractive index* materials. Although counterintuitive, this solution is entirely acceptable and this will be further argued in the context of surface modes. Should this wave be called an incident or scattered wave? We will argue that it should be called a scattered wave, despite the fact that it propagates toward the surface. This denomination is motivated by the fact that such a wave does not require the presence of an EM source at $z \to -\infty$ (this will be shown in Section F.2.5), as would be the case for a real incident wave. These considerations highlight the difficulties that arise when complex numbers are involved and that standard definitions should not be taken for granted. In particular, the incident and scattered waves cannot be defined as $\mathrm{Re}(k_{1z}) > 0$ and $\mathrm{Re}(k'_{1z}) < 0$, as intuition would suggest. The rigorous definitions are the one given in the preceding section. In fact, if the wrong definitions are used, surface plasmon–polariton modes would only exist for ideal (non-absorbing) metal/dielectric interfaces. Finally (and fortunately), both definitions turn out to be equivalent in the most common cases where k_x is real. These intrinsic complexities of surface waves also explain why they are normally 'swept under the carpet' in most conventional treatments of the problem.

We make from now on the common assumption that there are no waves coming from $z \to +\infty$, and only one wave is then present in region 2; the transmitted wave. We therefore have $\mathbf{E}'_2 = \mathbf{H}'_2 = \mathbf{0}$. In region 1, we may have two waves, the incident wave (if it has a physical origin) and the scattered wave.

F.2.4. The Fresnel coefficients

The amplitude of the fields for the incident, scattered, and transmitted waves are then related by the matrix expressions given in Eqs (F.14) and (F.21), simply taking $H'_{2y} = 0$ or $E'_{2y} = 0$. For both TM and TE cases, this can be written as:

$$\begin{pmatrix} F_{1y} \\ F'_{1y} \end{pmatrix} = \frac{1}{2} \begin{pmatrix} 1+K & 1-K \\ 1-K & 1+K \end{pmatrix} \begin{pmatrix} F_{2y} \\ 0 \end{pmatrix}, \tag{F.26}$$

where $F \equiv H$ and $K \equiv K^p$ for TM waves, while $F \equiv E$ and $K \equiv K^s$ for TE waves.

Rather than this matrix expression, it is often more convenient to define two Fresnel coefficients (which are complex) relating the field amplitudes of the scattered and transmitted waves to that of the incident wave. The *Fresnel reflection coefficient* is defined as:

$$r = \frac{F'_{1y}}{F_{1y}} = \frac{1-K}{1+K}. \tag{F.27}$$

The *Fresnel transmission coefficient* is defined as:

$$t = \frac{F_{2y}}{F_{1y}} = \frac{2}{1+K}. \tag{F.28}$$

Although the definitions take a common form for TM and TE waves, the corresponding coefficients, r^p and t^p for TM waves, r^s and t^s for TE waves are different: firstly K is different (K^p in Eq. (F.16) vs K^s in Eq. (F.22)), and secondly they relate different field amplitudes (the magnetic field for TM waves, the electric field for TE waves). These coefficients, using Eqs (F.16) and (F.22), and the conventions in Eqs (F.23) and (F.24), can in fact be viewed as functions of k_x (which may be complex). More specific expressions of r^p, t^p, r^s, and t^s will be given in Sections F.3.3 and F.3.4.

F.2.5. Surface modes

Description of the surface modes

In Chapter 3, we have discussed the existence of electromagnetic modes called bound or surface modes, which may exist in the absence of any incident wave. They, accordingly, correspond to solutions for which $\mathbf{E}_1 = \mathbf{H}_1 = \mathbf{0}$, or in terms of the Fresnel coefficients to $r \to \infty$ and $t \to \infty$. It is straightforward to show, using either Eq. (F.14) or directly the general expressions of r and t, that the only possible solutions correspond to the condition $K = -1$ ($K^p = -1$ for TM waves or $K^s = -1$ for TE waves).

For TE waves, the condition $(K_s)^2 = 1$ implies $\epsilon_1 = \epsilon_2$ (i.e. the interface does not exist), which we exclude, and the condition $K^s = -1$ cannot therefore be met. We conclude that there are *no surface modes for a single interface with TE polarization.*

For TM waves, the condition $K^p = -1$ is equivalent to $\epsilon_1 k_{2z} = -\epsilon_2 k_{1z}$ or to $\epsilon_1 k_{2z} = \epsilon_2 k'_{1z}$ since $k_{1z} = -k'_{1z}$ (here the wave in region 1 is the scattered wave represented by $\mathbf{k}'_1$). In fact, using the dispersion relations, the more general condition $(K^p)^2 = 1$ can be shown to be equivalent to:

$$\boxed{k_x^2 = k_0^2 \frac{\epsilon_1 \epsilon_2}{\epsilon_1 + \epsilon_2}.} \tag{F.29}$$

This expression relates k_x to ω (through the ω-dependence of k_0, ϵ_1 and/or ϵ_2) and is therefore the *dispersion relation for the surface modes.* Note also that for all the surface modes, the condition $K^p = -1$ further implies that $H_{2y} = H'_{1y}$ (from Eq. (F.14)).

Rigorous derivation

To complete the proof, we need however to find the additional requirement for having $K^p = -1$ (and not $K^p = +1$). This is not as straightforward as it may seem, and requires taking extra care in the handling of the complex square root, as we show now. Using the expression for k_x^2 (Eq. (F.29)), the dispersion relations then imply that (see Eq. (F.10)):

$$k'_{1z} = -k_{1z} = \pm k_0 \epsilon_1 \sqrt{\frac{1}{\epsilon_1 + \epsilon_2}} \quad \text{and} \quad k_{2z} = \pm k_0 \epsilon_2 \sqrt{\frac{1}{\epsilon_1 + \epsilon_2}}. \tag{F.30}$$

In all these expressions, there are in principle two choices for the square root of a complex number. When writing $\sqrt{z}$ for a complex number as above, we however explicitly choose one of these two signs. We use in this book the principal ('standard') complex square root convention, i.e. the one with a positive real part (or if it is zero, the one with the positive imaginary part). More explicitly: for $-\pi < \phi \le \pi$, $\sqrt{r\exp(\mathrm{i}\phi)} = \sqrt{r}\exp(\mathrm{i}\phi/2)$. We then always have $\mathrm{Re}(\sqrt{z}) \ge 0$, and $\mathrm{Im}(\sqrt{z})$ has the same sign as $\mathrm{Im}(z)$. One must take extra care in manipulating these complex square roots, because the function is not continuous on the negative real axis. This implies for example that $\sqrt{z^2} \ne z$ and $\sqrt{z_1 z_2} \ne \sqrt{z_1}\sqrt{z_2}$ in the general case (these equalities are only true up to a sign).

To determine the appropriate choice of sign for k'_{1z} and k_{2z} (and therefore find out if $K^p = +1$ or $K^p = -1$), we need to apply the physical considerations discussed in Section F.2.3. We will assume in the following that ϵ_1 is real positive (non-absorbing dielectric) to try to keep the discussion simple (it is also the case of most interest to practical situations). Because of the discontinuity of the complex square root, we need to differentiate between several possible situations for region 2:

- If ϵ_2 is real positive (non-absorbing medium), then k'_{1z} and k_{2z} are both real. The wave in region 1 is the scattered wave represented by $\mathbf{k}'_1$, and from our conventions (Eq. (F.23)) must therefore satisfy $k'_{1z} < 0$. Similarly, the wave in region 2 must satisfy $k_{2z} > 0$ to be physically acceptable. These two conditions, together with $\epsilon_2 > 0$, imply that $K^p = +1$ (and not -1), and we therefore conclude that *there are no surface modes if both ϵ_1 and ϵ_2 are real positive* . Note however that the solution with $k'_{1z} > 0$ is physically acceptable but represents an incident wave, a possibility that we have excluded here in the context of surface modes. This solution will however be recovered as a particular case of incident wave mode (with no scattered wave, since we considered only one wave in region 1). These are called the Brewster modes and will be discussed later in the context of reflection/refraction at the interface.

- If ϵ_2 is not real, i.e. $\mathrm{Im}(\epsilon_2) > 0$, then both k_{1z} and k_{2z} are non-real. The wave in region 1 (scattered wave), must then satisfy $\mathrm{Im}(k'_{1z}) < 0$ to be physically acceptable. Using $\mathrm{Im}(\epsilon_2) > 0$, one can easily deduce that $\mathrm{Im}(\sqrt{1/(\epsilon_1+\epsilon_2)}) < 0$ and therefore:

$$k'_{1z} = -k_{1z} = +k_0\epsilon_1\sqrt{\frac{1}{\epsilon_1+\epsilon_2}}. \tag{F.31}$$

  Similarly, the wave in region 2 must satisfy $\mathrm{Im}(k_{2z}) > 0$ to be physically acceptable. Elementary (but tedious) complex calculus can be used to show that the sign for k_{2z} must then be chosen in the most general case as:

$$k_{2z} = +k_0\epsilon_2\sqrt{\frac{1}{\epsilon_1+\epsilon_2}}. \tag{F.32}$$

Note that this is not equivalent to $k_{2z} = +k_0\sqrt{(\epsilon_2)^2/(\epsilon_1+\epsilon_2)}$. We conclude that we have in this case $\epsilon_1 k_{2z} = -\epsilon_2 k_{1z}$, i.e. $K_p = -1$. *The corresponding solution is a surface mode.* Note that k_x is complex in this case, and these are therefore *evanescent* surface modes.

- In addition, we also need to treat separately the case where ϵ_2 is real, but negative (this cannot happen in reality, but is the case of a perfect non-absorbing metal, for example). If $\epsilon_2 < -\epsilon_1$ first, then both k'_{1z} and k_{2z} are pure imaginary. $\mathrm{Im}(k'_{1z}) < 0$ and $\mathrm{Im}(k_{2z}) > 0$ then imply that k_{1z} and k_{2z} are almost given by Eqs (F.31) and (F.32), but both with opposite signs. The condition $K^p = -1$ remains true, and the corresponding solution is again a surface mode. Moreover, k_x is real for these modes, which are therefore truly *propagating surface modes.* The fact that we have to differentiate this case with the previous one is in fact a mathematical artifact introduced by the discontinuity of the complex square root. In fact, one can show that in both cases (ϵ_2 non-real and $\epsilon_2 < -\epsilon_1$ real), the wave-vectors can be expressed by a unified expression[4] as:

$$k'_{1z} = -k_{1z} = -\mathrm{i}k_0\epsilon_1\sqrt{\frac{-1}{\epsilon_1+\epsilon_2}}, \tag{F.33}$$

$$k_{2z} = -\mathrm{i}k_0\epsilon_2\sqrt{\frac{-1}{\epsilon_1+\epsilon_2}}, \tag{F.34}$$

which always ensure $\mathrm{Im}(k'_{1z}) < 0$ and $\mathrm{Im}(k_{2z}) > 0$ when $\epsilon_2 < -\epsilon_1$ (real) or ϵ_2 is non-real. In these two cases, $\mathrm{Im}(k'_{1z}) \neq 0$, and the scattered wave is evanescent. *The surface modes are therefore non-radiative.*

- Finally, if ϵ_2 is real negative but with $-\epsilon_1 < \epsilon_2 < 0$, then both k_{1z} and k_{2z} are real. As for the case of a non-absorbing medium with $\epsilon_2 > 0$, we must have $k'_{1z} < 0$ and $k_{2z} > 0$. But since $\epsilon_2 < 0$ here, this now implies that $K^p = -1$ and the corresponding solutions are also *surface modes.* We have for these modes:

[4] We use here a mathematical 'trick' to move the singularity (discontinuity) to the $\mathrm{Re}(\epsilon_2) > 0$ side where it does not matter.

$$k'_{1z} = -k_{1z} = -k_0\epsilon_1\sqrt{\frac{1}{\epsilon_1+\epsilon_2}}, \tag{F.35}$$

$$k_{2z} = -k_0\epsilon_2\sqrt{\frac{1}{\epsilon_1+\epsilon_2}}. \tag{F.36}$$

Furthermore, these modes are characterized by a k_x that is *pure imaginary*. These are, therefore, not propagating at all along the surface, they are *localized* modes and are also *radiative* modes since the scattered wave is characterized by k'_{1z} real.

The nature of these surface modes, along with the correct expressions for k'_{1z} and k_{2z}, are summarized in Fig. 3.7of Chapter 3.

Energy considerations

It is interesting to pause briefly and analyze the different types of modes identified above in terms of energy and energy propagation considerations.

One can show from the above expressions that when $\mathrm{Im}(\epsilon_2) > 0$ (cases (e) and (f) in Fig. 3.7), we have $\mathrm{Re}(k'_{1z}) > 0$ for the scattered wave, i.e. it propagates toward the interface (but is evanescent away from it). A similar situation arises in some cases for the transmitted wave in region 2:

$$\begin{aligned}&\text{if} \quad -\epsilon_1 - \sqrt{\epsilon_1^2 + (\mathrm{Im}(\epsilon_2))^2} < \mathrm{Re}(\epsilon_2) < -\epsilon_1 + \sqrt{\epsilon_1^2 + (\mathrm{Im}(\epsilon_2))^2},\\ &\text{then} \quad \mathrm{Re}(k_{2z}) < 0,\end{aligned} \tag{F.37}$$

i.e. the transmitted wave comes toward the interface. Similarly, the modes for $-\epsilon_1 < \epsilon_2 < 0$ real (case (b) in Fig. 3.7) correspond to propagating scattered and transmitted waves, but with no incident wave. One may wonder how this is possible in terms of energy conservation.

The answer to these apparent contradictions lie in the fact that k_x is not real in these cases. This first means that the corresponding surface mode can be viewed either as a virtual mode or an evanescent wave. Taking this latter point of view and assuming $\mathrm{Im}(k_x) > 0$ without loss of generality, the mode cannot exist everywhere or it would diverge in the direction opposite to where it is evanescent. It must therefore be bounded on the $x \to -\infty$ side, say at $x = 0$. Whatever source maintains this surface mode excitation may then input any required amount of energy through this boundary, therefore solving the global energy conservation issue.

The question still remains of whether the energy is conserved locally in the case where $\mathrm{Im}(\epsilon_2) > 0$ (cases (e) and (f) in Fig. 3.7). The scattered wave appears to increase in amplitude as it propagates toward the surface; is it compatible with energy conservation? Let us therefore consider a small box in region 1, as depicted in Fig. F.2. The real part of the complex Poynting

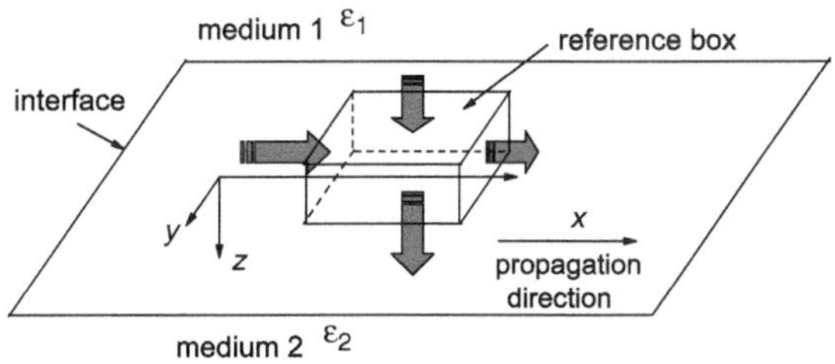

Figure F.2. Schematic diagram for energy conservation considerations of surface modes. The balance of energy is performed in a 'reference box' in medium 1 (which is non-dissipative) and the energy flux is schematically represented by block arrows. In this particular case, the wave seems to *increase* its amplitude as it propagates toward the surface in the z-direction. A careful analysis shows that this is achieved at the expense of attenuating its amplitude along the propagation direction x. Energy conservation is (fortunately) not violated.

vector for the scattered wave in region 1 takes the form:

$$\mathrm{Re}(\mathbf{S}_1) = \frac{|H'_{1y}|^2}{2\epsilon_1\epsilon_0 c k_0} \mathrm{e}^{-2\mathrm{Im}(k_x)x} \mathrm{e}^{-2\mathrm{Im}(k'_{1z})z} \left[\mathrm{Re}(k_x)\mathbf{e}_x + \mathrm{Re}(k'_{1z})\mathbf{e}_z\right]. \qquad (\mathrm{F}.38)$$

Along the z-direction, the net energy exiting the box is therefore:

$$W_z = \frac{\mathrm{d}(\mathrm{Re}(\mathbf{S}_1 \cdot \mathbf{e}_z))}{\mathrm{d}z} \Delta x \Delta y \Delta z = -2C\mathrm{Im}(k'_{1z})\mathrm{Re}(k'_{1z}), \qquad (\mathrm{F}.39)$$

where $C > 0$ derives from (F.38). Since $\mathrm{Im}(k'_{1z}) < 0$ and $\mathrm{Re}(k'_{1z}) > 0$, we deduce that $W_z > 0$, meaning that the wave 'gains energy by going through the box' along z. This is indeed possible if this gained energy is balanced by lost energy through the other faces of the box. In fact, along the x-direction, the net energy exiting the box is:

$$W_x = \frac{\mathrm{d}(\mathrm{Re}(S_x))}{\mathrm{d}x} \Delta x \Delta y \Delta z = -2C\mathrm{Im}(k_x)\mathrm{Re}(k_x), \qquad (\mathrm{F}.40)$$

where $C > 0$ is the same as above. From Eq. (F.29), one can show that the real and imaginary parts of k_x are necessarily of the same sign: i.e. the surface wave propagates along the surface in the same direction as it decays. This implies that $W_x < 0$: the wave 'loses energy by going through the box' along x. This energy is in fact transferred to the z-direction. Since the medium in region 1 is non-dissipative, energy conservation would require $W_x + W_z = 0$. This is in fact the case from the above expressions. To see this equality, one can for example take the imaginary part of the dispersion relation $(k'_{1z})^2 + k_x^2 = \epsilon_1 k_0^2$ which implies $\mathrm{Im}(k'_{1z})\mathrm{Re}(k'_{1z}) + \mathrm{Im}(k_x)\mathrm{Re}(k_x) = 0$.

Final remarks

Among the modes we discussed, the most interesting ones are those arising when $\text{Re}(\epsilon_2) < 0$. For a metal for example, they correspond to the surface plasmon–polaritons (SPPs) of the planar metal/dielectric interface, and are discussed in detail in Section 3.4 of Chapter 3.

We discuss briefly here the other family of surface modes, i.e. when $\text{Re}(\epsilon_2) > 0$ (cases (a) and (c) of Fig. 3.7), since they are rarely mentioned. As shown above, there are no surface modes with $\text{Re}(\epsilon_2) > 0$ and $\text{Im}(\epsilon_2) = 0$. It seems therefore strange that such modes would exist for $\text{Im}(\epsilon_2) > 0$ (since it can in principle be as small as we want whilst being not zero). In fact, as $\text{Im}(\epsilon_2)$ approaches 0, the solution for the surface modes approaches the solution found earlier for Brewster modes (which are incident wave modes with no scattered wave). In particular, the scattered wave with $\text{Im}(k'_{1z}) < 0$ and $\text{Re}(k'_{1z}) > 0$ converges to a wave with $\text{Im}(k'_{1z}) = 0$ and $\text{Re}(k'_{1z}) > 0$. This problem highlights the limitations of the classification of electromagnetic modes in terms of surface modes and incident wave modes (and of the distinction between incident and scattered waves). The surface modes (no incident wave, only a scattered wave) obtained when $\text{Re}(\epsilon_2) > 0$ and $\text{Im}(\epsilon_2) \neq 0$ are similar in essence to the Brewster modes (incident wave modes with no scattered wave) and could be interpreted as modified (because evanescent) incident wave modes. These subtleties are (in most cases) irrelevant, and disappear when applied to practical problems.

F.2.6. Incident wave modes

For all the other electromagnetic modes of the system, there must be an incident wave (characterized by k_{1z} real positive) and therefore an electromagnetic source in region 1. The corresponding electromagnetic mode is then entirely determined by the knowledge of k_x (which may be complex) and the incident wave field amplitudes (using the Fresnel coefficients to determine the reflected and transmitted wave amplitudes).

The case of most relevance experimentally is when the incident wave is a homogeneous plane wave propagating from $z \to -\infty$, which we will now consider. Our approach therefore rejoins at this stage, the more conventional treatment of the reflection/refraction of a plane wave at a planar interface found in many textbooks [96,151].

F.3. REFLECTION/REFRACTION AT A PLANAR INTERFACE

The problem of a plane wave incident on a planar interface is a classic electromagnetic problem of great importance for many applications. It is accordingly treated in detail in most textbooks (see for example [96,151]). We summarize here the main results, firstly for reference and completeness, and secondly to emphasize often neglected aspects more relevant to plasmonics (such as the local field enhancements). The problem of a plane wave incident on a single planar interface has essentially been solved in the previous section when we defined the Fresnel coefficients in Section F.2.4. We only expand here on this solution and specialize it for the problem at hand. The treatment of TM and TE waves is again separated, but we note that TM waves are the ones that are most relevant in many plasmonics problems.

F.3.1. Incident, reflected, and transmitted waves

In a physical context, a propagating (incident) plane wave can only exist in a non-absorbing medium, i.e. with a real positive dielectric constant. We therefore impose that $\epsilon_1 > 0$ (and real) in region 1 and consider a propagating plane wave in this region, i.e. with a wave-vector $\mathbf{k}_1$ real, i.e. k_x and k_{1z} are both real and $k_x^2 + k_{1z}^2 = k_0^2\epsilon_1$ (which implies that $k_x \leq k_0\sqrt{\epsilon_1}$). Within the conventions defined in Eq. (F.23), the incident wave corresponds to $\mathbf{E}_1$, $\mathbf{k}_1$, with $k_{1z} > 0$, and represents a wave propagating from $z \to -\infty$ toward the interface (see also Fig. F.1). We then have:

$$k_{1z} = -k'_{1z} = +\sqrt{\epsilon_1 k_0^2 - k_x^2}. \tag{F.41}$$

In this context it is also common to characterize the incident plane wave by its *angle of incidence* θ (angle between $\mathbf{k}_1$, real, and the plane normal $\mathbf{e}_z$, see Fig. F.1). Defining $k_1 = k_0\sqrt{\epsilon_1}$ (real) as the wave-vector amplitude, we have:

$$k_{1x} = k_1 \sin\theta \quad \text{and} \quad k_{1z} = k_1 \cos\theta. \tag{F.42}$$

The other wave in region 1 with $\mathbf{E}'_1$, $\mathbf{k}'_1$, whose wave-vector along z is opposite to $\mathbf{k}_1$, is a wave propagating away from the interface and corresponds to the *reflected or scattered wave.*

In the second half-space $z > 0$, we have shown that only one wave may physically exist in the absence of sources at $z \to +\infty$. It is the *transmitted wave* (also called the *refracted wave,* in particular when it is a propagating wave), and is represented by $\mathbf{E}_2$ and $\mathbf{k}_2$. We have shown that k_{2z} must satisfy the conditions in Eq. (F.24). Because k_x is real here, one can show in a similar fashion as discussed in Section F.2.3 for the scattered wave, that the real and imaginary parts of k_{2z} must be of the same sign. The conditions in Eq. (F.24) are then equivalent to the simple condition: $\text{Re}(k_{2z}) > 0$, which is what is intuitively expected for the transmitted wave. The choice of sign for k_{2z} is then fully determined and is given by:

$$k_{2z} = +\sqrt{\epsilon_2 k_0^2 - k_x^2}. \tag{F.43}$$

F.3.2. Snell's law

Although it is usually not used in the context of plasmonics, we recall briefly Snell's law (or Descartes' law) for completeness. As mentioned earlier, it derives automatically from the conservation of k_x, together with the dispersion relations.

For the reflected wave first, one can define a reflection angle, θ_r in a similar fashion as the angle of incidence for the incident wave: $k'_{1x} = k_1 \sin\theta_r$ and $k'_{1z} = k_1 \sin\theta_r$, i.e. it is the angle between $\mathbf{k}'_1$ and the normal to the plane. Since $k'_{1x} = k_{1x}$ and $k'_{1z} = -k_{1z}$, it is clear that $\theta_r = -\theta$.

Similarly, an angle of refraction, θ_t, can be defined for the transmitted (or refracted) wave when it is propagating. Defining $k_2 = k_0\sqrt{\epsilon_2}$, the dispersion relation reads $k_{2x}^2 + k_{2z}^2 = k_2^2$, which can be written as: $k_{2x} = k_2 \sin\theta_t$ and $k_{2z} = k_2 \sin\theta_t$. From $k_{2x} = k_x$, we then deduce Snell's law:

$$n_1 \sin\theta = n_2 \sin\theta_t, \tag{F.44}$$

where the refractive indices $n_1 = \sqrt{\epsilon_1}$ and $n_2 = \sqrt{\epsilon_2}$ have been used instead of the dielectric constants. The above considerations are only valid if ϵ_2 is real positive and $n_2 > n_1 \sin\theta$, which ensures that the refracted wave is propagating. In other cases, it is evanescent and the angle of refraction loses its meaning.

This highlights the limitations of Snell's law, which is in general used only in the context of *geometrical optics*, where all waves are propagating. A more general approach from an electromagnetic point of view, is to characterize the waves by their wave-vectors, thereby allowing them to be evanescent or even inhomogeneous. Indeed, we focus on this approach hereafter.

F.3.3. TM or *p*-polarized waves

Fresnel coefficients

The Fresnel coefficients for reflection, r^p, and transmission, t^p, for a *p*-polarized wave can be written as (see Eqs (F.27) and (F.28)):

$$\boxed{r^p = \frac{H'_{1y}}{H_{1y}} = \frac{1-K^p}{1+K^p} = \frac{\epsilon_2 k_{1z} - \epsilon_1 k_{2z}}{\epsilon_2 k_{1z} + \epsilon_1 k_{2z}},} \tag{F.45}$$

and

$$\boxed{t^p = \frac{H_{2y}}{H_{1y}} = \frac{2}{1+K^p} = \frac{2\epsilon_2 k_{1z}}{\epsilon_2 k_{1z} + \epsilon_1 k_{2z}}.} \tag{F.46}$$

Both coefficients are in general complex. These definitions are sometimes given in terms of the electric field amplitude but this leads to problems for p-polarized waves since the electric field has two components (along z and x), and this should be avoided.

These formulae (and K^p) can be viewed as functions of k_x, using the expressions (F.41) and (F.43). For propagating incident waves ($k_x < k_1$), equivalent expressions can also be given in terms of the angle of incidence θ (more common in optics), for example:

$$K^p = \frac{\epsilon_1 k_{2z}}{\epsilon_2 k_{1z}} = \frac{n_1\sqrt{1 - \frac{n_1^2}{n_2^2}\sin^2\theta}}{n_2\cos\theta}. \tag{F.47}$$

Reflected field

These expressions can further be used to write similar relations for the electric field amplitudes. For the reflected field, we have:

$$\frac{E'_{1x}}{E_{1x}} = -r^p, \qquad \frac{E'_{1z}}{E_{1z}} = r^p. \tag{F.48}$$

Using the (real) amplitude of the incident field, $E_1 = |\mathbf{E}_1|$, and the fact that $|E_{1x}| = E_1\cos\theta$ and $|E_{1z}| = E_1\sin\theta$, the (real) amplitude of the reflected field, $E'_1 = |\mathbf{E}'_1|$, can be written in short as $E'_1 = |r^p|E_1$. It is then easy to deduce, using Eq. (F.6), that the reflected power density S'_1 is related simply to the incident power density S_1 as:

$$\frac{S'_1}{S_1} = \frac{(E'_1)^2}{(E_1)^2} = |r^p|^2 = R^p, \tag{F.49}$$

where the last equality defines R^p, which is called the reflection coefficient (not to be confused with the Fresnel reflection coefficient r^p). R^p is real with $0 \leq R^p \leq 1$ and corresponds to the proportion of the incident energy that is reflected at the interface. Conservation of energy implies that the proportion of energy transmitted through the interface is:

$$T^p = 1 - R^p = \frac{4\mathrm{Re}(K^p)}{|1 + K^p|^2}, \tag{F.50}$$

where T^p is the transmission coefficient (not to be confused with the Fresnel transmission coefficient t^p). If medium 2 is absorbing, this energy is absorbed (in the form of heat) in the medium. If not, this energy is propagated by the transmitted wave. In this latter case, T^p could also be inferred from the calculation of $\mathbf{S}_2$ (projected along z) using Eq. (F.6).

Finally, these expressions can be rewritten in terms of field enhancement factors at the interface. The local field intensity enhancement factor (LFIEF) can be decomposed into its perpendicular ($\perp$) and tangential ($\|$) field contributions. The perpendicular LFIEF at the surface in region 1 (outside) is:

$$M^p_{\perp 1} = \frac{|E_{1z} + E'_{1z}|^2}{|E_1|^2} = |1 + r^p|^2 \sin^2\theta = \frac{4\sin^2\theta}{|1 + K^p|^2}, \tag{F.51}$$

and the parallel LFIEF is:

$$M^p_{\|1} = \frac{|E_{1x} + E'_{1x}|^2}{|E_1|^2} = |1 - r^p|^2 \cos^2\theta = \frac{4|K^p|^2\cos^2\theta}{|1 + K^p|^2}. \tag{F.52}$$

The total LFIEF is therefore:

$$M^p_1 = \frac{|\mathbf{E}_1 + \mathbf{E}'_1|^2}{|E_1|^2} = M^p_{\perp 1} + M^p_{\|1} = \frac{4(|K^p|^2\cos^2\theta + \sin^2\theta)}{|1 + K^p|^2}. \tag{F.53}$$

Transmitted field

For the transmitted field, we have:

$$\frac{E_{2x}}{E_{1x}} = K^p t^p, \qquad \frac{E_{2z}}{E_{1z}} = \frac{\epsilon_1}{\epsilon_2} t^p, \tag{F.54}$$

or equivalently:

$$\frac{E_{2x}}{E_1} = K^p t^p \cos\theta, \qquad \frac{E_{2z}}{E_1} = -\frac{\epsilon_1}{\epsilon_2} t^p \sin\theta. \tag{F.55}$$

The general relation for the (real) electric field amplitude $E_2 = |\mathbf{E}_2|$ is therefore more complicated. Only if the transmitted wave is propagating ($\mathbf{k}_2$ real), we can use Eq. (F.5) for the incident and transmitted wave and deduce:

$$\frac{E_2}{E_1} = \sqrt{\frac{\epsilon_1}{\epsilon_2}}|t^p|. \tag{F.56}$$

In the general case, the best we can do is to use Eq. (F.5) for the incident wave only, and deduce:

$$\frac{E_2}{E_1} = |t^p|\sqrt{|K^p|^2\cos^2\theta + \frac{\epsilon_1^2}{|\epsilon_2|^2}\sin^2\theta}$$

$$= \sqrt{\frac{\epsilon_1}{|\epsilon_2|}}|t^p|\sqrt{\left|1 - \frac{\epsilon_1}{\epsilon_2}\sin^2\theta\right| + \frac{\epsilon_1}{|\epsilon_2|}\sin^2\theta}. \tag{F.57}$$

This general expression is equivalent to the previous one only when ϵ_2 is real positive and $\epsilon_2 > \epsilon_1 \sin^2\theta$, i.e. when the transmitted wave is propagating.

The local field intensity enhancement factors at the surface in region 2 (inside) also follow. For the perpendicular case, we have:

$$M^p_{\perp 2} = \frac{|E_{2z}|^2}{|E_1|^2} = \frac{\epsilon_1^2}{|\epsilon_2|^2}\frac{4\sin^2\theta}{|1+K^p|^2}. \tag{F.58}$$

Moreover, because of continuity, we have for the parallel case:

$$M^p_{\parallel 2} = M^p_{\parallel 1} = \frac{4|K^p|^2\cos^2\theta}{|1+K^p|^2}, \tag{F.59}$$

and we deduce the total field enhancement factor as:

$$\begin{aligned} M^p_2 = \frac{|\mathbf{E}_2|^2}{|E_1|^2} &= M_{\perp 2} + M_{\parallel 2} \\ &= \frac{4}{|1+K^p|^2}\left[\left|1 - \frac{\epsilon_1}{\epsilon_2}\sin^2\theta\right| + \frac{\epsilon_1}{|\epsilon_2|}\sin^2\theta\right]. \end{aligned} \tag{F.60}$$

This final expression can be simplified, but only when the transmitted wave is propagating, to:

$$M^p_2 = \frac{|\mathbf{E}_2|^2}{|E_1|^2} = M^p_{\perp 2} + M^p_{\parallel 2} = \frac{4}{|1+K^p|^2}\frac{\epsilon_1}{|\epsilon_2|}. \tag{F.61}$$

F.3.4. TE or *s*-polarized waves

We now briefly go through the similar relations that are obtained for *s*-polarized waves.

Fresnel coefficients

The Fresnel coefficients for reflection, r^s, and transmission, t^s, for TE waves are given by:

$$\boxed{r^s = \frac{E'_{1y}}{E_{1y}} = \frac{1-K^s}{1+K^s} = \frac{k_{1z}-k_{2z}}{k_{1z}+k_{2z}},} \tag{F.62}$$

and

$$t^s = \frac{E_{2y}}{E_{1y}} = \frac{2}{1 + K^s} = \frac{2k_{1z}}{k_{1z} + k_{2z}}, \tag{F.63}$$

where K^s can also be expressed in terms of the angle of incidence as:

$$K^s = \frac{k_{2z}}{k_{1z}} = \frac{n_2\sqrt{1 - \frac{n_1^2}{n_2^2}\sin^2\theta}}{n_1\cos\theta}. \tag{F.64}$$

Reflected field

The electric field (which is simpler here compared to p-waves since it has only one component) is already fully characterized by the Fresnel coefficients. The (real) amplitude of the reflected field, $E_1' = |\mathbf{E}_1'|$, can be directly expressed in short as $E_1' = |r^s|E_1$. Also, the reflected power density S_1' is related to the incident power density S_1 as:

$$\frac{S_1'}{S_1} = R^s = |r^s|^2, \tag{F.65}$$

and it follows that:

$$T^s = 1 - R^s = \frac{4\text{Re}(K^s)}{|1 + K^s|^2}, \tag{F.66}$$

where R^s and T^s are the reflection and transmission coefficients.

There is only one type of local field intensity enhancement factor here, since the electric field is parallel to the interface. At the surface in region 1 (outside) it is:

$$M_1^s = M_{\|1}^s = \frac{|E_{1y} + E_{1y}'|^2}{|E_1|^2} = |1 + r^s|^2 = \frac{4}{|1 + K^s|^2}. \tag{F.67}$$

Transmitted field

For the transmitted field, we simply have for the real amplitude $E_2 = |t^s|E_1$. The local field enhancement factor at the surface in region 2 (inside) is:

$$M_2^s = M_{\|2}^s = M_{\|1}^s = \frac{4}{|1 + K^s|^2}. \tag{F.68}$$

F.3.5. Special cases

We briefly discuss a few special cases of interest. These arise when K^p (or K^s) takes special values:

- If $K^p = 1$, then $r^p = 0$: there is *no reflected wave* and all the power is therefore transmitted ($T^p = 1$) through the interface. Note that the condition $K^s = 1$ cannot be met (otherwise $\epsilon_1 = \epsilon_2$). The condition $K^p = 1$ is similar in some ways to the one encountered for surface modes ($K^p = -1$) in Section F.2.5, and has already been discussed there. Both imply $(K_p)^2 = 1$, which is equivalent to:

$$k_x^2 = k_0^2 \frac{\epsilon_1 \epsilon_2}{\epsilon_1 + \epsilon_2}. \tag{F.69}$$

  To ensure $K_p = +1$, one must in addition have both ϵ_1 and ϵ_2 real positive. The condition is then equivalent to *Brewster's angle* condition [96] for the angle of incidence: $\tan\theta = n_2/n_1$, where $n_i = \sqrt{\epsilon_i}$ are the refractive indices. These solutions are non-radiative incident wave modes (since there is no scattered wave) and can be called the *Brewster modes*, with the dispersion relation given above. As seen by the similarity in the dispersion relation, the Brewster modes are dual to the surface modes introduced earlier in the sense that there is only one wave in each region. The physical meaning of the wave in region 1 is, however, different (incident wave vs scattered wave).

- If $K^p = 0$, then $r^p = 1$ (and therefore $R^p = 1$ and $T^p = 0$): all power is therefore reflected, but note that there is still a transmitted wave since $t^p \neq 0$ (and it must then be evanescent along z to ensure energy conservation). A similar but more general situation is when K^p is pure imaginary, which again implies that $R^p = 1$ and $T^p = 0$. It is the same situation as for $K^p = 0$, but the reflected wave incurs an additional phase shift since $r^p \neq 1$. These cases correspond to *total internal reflection* (TIR) . They may occur for both p- or s-polarization. For a propagating incident wave, this is only possible if both ϵ_1 and ϵ_2 are real and $\epsilon_2 < \epsilon_1$. If ϵ_2 is real negative, then TIR always occurs. If ϵ_2 is real positive, then the condition for TIR can be shown to be:

$$\boxed{\sin\theta \geq n_2/n_1,} \tag{F.70}$$

  which is only possible if $n_2 < n_1$. The same expression can in fact be deduced simply from Snell's law. Total internal reflection plays an important role in plasmonics as a method for creating evanescent waves for which the range of k_x values is extended compared to propagating waves. This is a 'trick' often used to excite surface plasmon–polariton modes; see Chapter 3 for more details.

- Finally, if $k_{1z} = 0$, then the incident wave impinges parallel to the plane surface (with perfectly grazing incidence). It is easy to see that such a beam would in practice never 'touch' the surface. Within our treatment, we have $|K^p| \to \infty$, $r^p = -1$, and $t^p = 0$. This means that there is no transmitted field and that the reflected field cancels exactly the incident field. The only solution is therefore a zero field everywhere. This non-physical result is simply a consequence of representing the incident wave by a plane wave (which therefore extends to infinity in the transverse direction).

F.4. MULTI-LAYER INTERFACES

We now extend the results to the case of a multi-layer interface, which is a common situation in most plasmonics problems. We give here only the results relevant to the numerical implementation of the electromagnetic solution.

F.4.1. Principle

We consider the more general case of N planar interfaces, i.e. $N+1$ regions (see Fig. F.3). The planar interfaces are at planes $z = Z_1 = 0$ for the first one, and $z = Z_i (i = 1 \dots N)$ for the others. Alternatively, the thickness of each finite region is $L_i = Z_i - Z_{i-1}$, with the convention $L_1 = 0$. In a similar fashion as before, in each region $i = 1 \dots N+1$ characterized by ϵ_i, there are two counter-propagating plane waves with fields $\mathbf{E}_i$, $\mathbf{H}_i$,and $\mathbf{E}'_i$, $\mathbf{H}'_i$. In each region i, the wave-vector is given as $\mathbf{k}_i = k_{ix}\mathbf{e}_x + k_{iz}\mathbf{e}_z$ and the dispersion relation gives $k_{ix}^2 + k_{iz}^2 = k_0^2\epsilon_i$. As before, k_x is conserved through each interface, and the other component of $\mathbf{k}$ can therefore be obtained from the dispersion relation, up to a sign which we assign, using the same convention as for the single interface (for k_x real):

$$k_{iz} = +\sqrt{\epsilon_i k_0^2 - k_x^2} \quad \text{and} \quad k'_{iz} = -k_{iz} = -\sqrt{\epsilon_i k_0^2 - k_x^2}. \tag{F.71}$$

Within this convention, $\mathbf{E}_1$ is the incident wave, $\mathbf{E}'_1$ the reflected wave, and $\mathbf{E}_{N+1}$ is the transmitted wave (and as before $\mathbf{E}'_{N+1} = \mathbf{0}$). Two counter-propagating waves may exist in intermediate regions, and both are physically acceptable since the region is bounded on both sides along z. Region 1 is non-absorbing and $k_1 = k_0\sqrt{\epsilon_1}$ is real with $k_{1z} = k_1 \cos\theta$ and $k_{1x} = k_1 \sin\theta$.

Finally, note that if $k_{jz} = 0$ for a given region j, then the corresponding wave in region j propagates along x (all power is reflected at this interface between region j and region $j+1$). This does not excite any waves in the subsequent

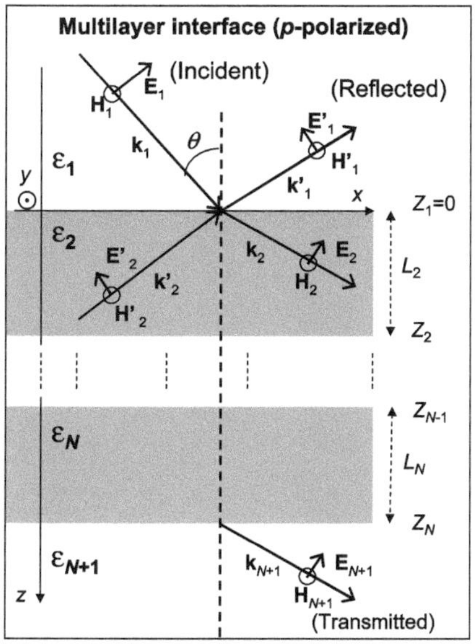

Figure F.3. Schematic diagram illustrating the problem of a p-polarized (TM) propagating plane wave impinging on a multi-layer planar interface. The figure is drawn in the plane of incidence ($x\mathrm{O}z$).

regions $i > j$ and the problem can therefore be reduced to a problem with j regions ($j-1$ interfaces). We can therefore consider that $k_{iz} \neq 0$, except possibly in the final layer $k_{(N+1)z} = 0$.

F.4.2. p-polarized or TM waves

We start with the case of p-waves, which is more relevant to plasmonics.

We can compute for each interface $i = 1 \dots N$ between region i and $i+1$, K_i^p (equivalent to K^p for a p-wave on a single interface):

$$K_i^p = \frac{\epsilon_i k_{(i+1)z}}{\epsilon_{i+1} k_{iz}}. \tag{F.72}$$

We now consider the boundary conditions at interface i. The problem has, in fact, been treated earlier. The only modification is that, since the interface is not at $z = 0$, but $z = Z_i$, the amplitudes carry an additional phase, which leads to:

$$\begin{pmatrix} H_{iy} \exp(\mathrm{i}k_{iz} Z_i) \\ H'_{iy} \exp(-\mathrm{i}k_{iz} Z_i) \end{pmatrix} = \frac{1}{2} \begin{pmatrix} 1 + K_i^p & 1 - K_i^p \\ 1 - K_i^p & 1 + K_i^p \end{pmatrix} \begin{pmatrix} H_{(i+1)y} \exp(\mathrm{i}k_{(i+1)z} Z_i) \\ H'_{(i+1)y} \exp(-\mathrm{i}k_{(i+1)z} Z_i) \end{pmatrix}. \tag{F.73}$$

There is a potential problem with this approach: since k_{iz} may be imaginary and Z_i could be large, then the exponential factors could be almost zero (and lead to floating-point precision problems in numerical implementations). To avoid this problem, it is better to express the fields in each region $i > 1$ with a different origin of the z-axis at $z = Z_{i-1}$, i.e. (taking $Z_0 = 0$):

$$\bar{H}_{iy} = H_{iy}\exp(ik_{iz}Z_{i-1}) \quad \text{and} \quad \bar{H}'_{iy} = H'_{iy}\exp(-ik_{iz}Z_{i-1}), \tag{F.74}$$

and similar expressions for the electric field components. The boundary condition relation can then be written as:

$$\begin{pmatrix} \bar{H}_{iy} \\ \bar{H}'_{iy} \end{pmatrix} = M_i \begin{pmatrix} \bar{H}_{(i+1)y} \\ \bar{H}'_{(i+1)y} \end{pmatrix}, \tag{F.75}$$

with

$$M_i = \frac{1}{2}\begin{pmatrix} (1+K_i^p)\mathrm{e}^{-ik_{iz}L_i} & (1-K_i^p)\mathrm{e}^{-ik_{iz}L_i} \\ (1-K_i^p)\mathrm{e}^{+ik_{iz}L_i} & (1+K_i^p)\mathrm{e}^{+ik_{iz}L_i} \end{pmatrix}. \tag{F.76}$$

Since $\bar{H}'_{(N+1)y} = 0$, we can now deduce:

$$\begin{pmatrix} \bar{H}_{1y} \\ \bar{H}'_{1y} \end{pmatrix} = M \begin{pmatrix} \bar{H}_{(N+1)y} \\ 0 \end{pmatrix}, \tag{F.77}$$

with

$$M = \left(\prod_{i=1}^{i=N} M_i\right) = \begin{pmatrix} M_{11} & M_{12} \\ M_{21} & M_{22} \end{pmatrix}. \tag{F.78}$$

Note that the matrix product must be carried out in the right order, i.e. $M = M_1 \times M_2 \times \cdots$.

The reflection and transmission Fresnel coefficients of the multi-layer can therefore be directly computed from the above relation, i.e.

$$r^p = \frac{H'_{1y}}{H_{1y}} = \frac{\bar{H}'_{1y}}{\bar{H}_{1y}} = \frac{M_{21}}{M_{11}}, \tag{F.79}$$

and

$$t^p = \frac{\bar{H}_{(N+1)y}}{H_{1y}} = \frac{1}{M_{11}}. \tag{F.80}$$

Note that we could have used $H_{(N+1)y}$ instead of $\bar{H}_{(N+1)y}$ in this final expression. The results would then have differed by a phase only. This choice is only a matter of convention (and of the definition of t^p for a multi-layer interface), and one must only be careful to remain consistent with this choice. This issue does not exist for the reflection coefficient.

The fields in other regions can then be deduced for example by computing recurrently $\bar{H}_{iy}/H_{1y}$ and $\bar{H}'_{iy}/H_{1y}$.

Finally, it is worth noting that the local field enhancement factors in region 1 can be expressed in terms of r^p, exactly as given in Section F.3.3 for the single interface problem. In other regions, they must be deduced from a computation of the field amplitudes.

F.4.3. *s*-polarized or TE waves

The calculations can be carried out in a similar fashion for TE waves, only replacing H's by E's and K_i^p by the corresponding definition of K_i^s following Eq. (F.22). The expressions for r^s and t^s in terms of M_{ij} are then exactly the same as given above for r^p and t^p.

F.4.4. Particular cases of interest

The case of two interfaces (three regions) is common and the results can still be reasonably expressed analytically. Denoting r_1^p, r_2^p, t_1^p, t_2^p, the Fresnel coefficients of the first two interfaces, and $L = L_2$ their separation, we have for the combination of the two:

$$r^p = \frac{r_1^p + r_2^p e^{2ik_{2z}L}}{1 + r_1^p r_2^p e^{2ik_{2z}L}}, \tag{F.81}$$

and

$$t^p = \frac{t_1^p t_2^p e^{ik_{2z}L}}{1 + r_1^p r_2^p e^{2ik_{2z}L}}. \tag{F.82}$$

The same relations apply for the corresponding coefficients for *s*-polarized waves.

From there one could follow the treatment of Section F.2.5 to study the surface modes of, for example, a metal slab using the above expressions (with a great deal of effort).

F.4.5. Implementation in Matlab

The previous expressions for a general multi-layer can easily be implemented in Matlab and we provide here an example. Our implementation (available from the book website) allows the calculation of all relevant properties for a propagating plane wave (either *p*- (TM) or *s*- (TE) polarized) impinging on a general multi-layer interface, as a function of both the wavelength λ and/or the angle of incidence θ.

Parameters

The parameters necessary for the implementation are defined in the following variables:

- `nNbSurf [1 x 1]`: Number of interfaces (N), corresponding to $N+1$ regions.
- `lambda [L x 1]`: Column vector with wavelengths (in nm).
- `Cepsilon` $\{$`[L x 1]`$\}_{N+1}$: Cell containing $N+1$ column vectors corresponding to the wavelength-dependent relative dielectric function ϵ_i in each region $i = 1 \dots N+1$. For silver and gold, these column vectors can be defined as `eAg = EpsAg(lambda)` and `eAu = EpsAu(lambda)`. For a dielectric with constant ϵ_M such as water ($\epsilon_M = 1.77$), one can use: `eWater = 0*lambda + 1.77`.
- `CL` $\{$`[1 x 1]`$\}_N$: Cell containing N scalars with the region thicknesses L_i. Note that `CL{1}=0` is automatically enforced since $L_1 = 0$ in our definitions.
- `aideg [1 x A]`: Row vector with angles of incidences (in degrees).
- `[d=0] [1 x 1]`: Scalar with distance (in nm) from interfaces where local field enhancements will be estimated. $d = 0$ by default.

Function description

The main Matlab function is:
`stResMulti [struct] = MultiRef (sPol [string], nNbSurf [1 x 1], lambda [L x 1], Cepsilon` $\{$`[L x 1]`$\}_{N+1}$, `CL` $\{$`[1 x 1]`$\}_N$, `aideg [1 x A], [d=0] [1 x 1])`

It calculates the Fresnel coefficients and field enhancements for a multi-layer structure excited with either a *p*-polarized (TM) wave (`sPol='TM'`) or an *s*-polarized (TE) wave (`sPol='TE'`). Note that this function does not deal with the special case where $k_{iz} = 0$ (which can in practice always be avoided, see the discussion earlier). The function returns a `stResMulti` structure (see the details below) with the results.

The `stResMulti` structure contains as fields all the arguments of the `MultiRef` function for future reference. Moreover, it contains the following additional fields containing the results. Firstly, the Fresnel coefficients:

- `rP [L x A]` (or `rS [L x A]`): Matrix of Fresnel reflection coefficients, $r^p(\lambda, \theta)$ for each wavelength and angle of incidence.
- `tP [L x A]` (or `rS [L x A]`): Same for the Fresnel transmission coefficients.

Secondly, the amplitude of the fields in each region, from which the fields at any point in space can be computed. In the TM polarization case, they are:

- `HiyOvH1y {[L x A]}` $_{N+1}$: Cell of $i = 1 \dots N+1$ matrices with magnetic field amplitude of the first wave in region i, $\bar{H}_{iy}/H_{1y}$ (see the definition earlier).
- `HpiyOvH1y {[L x A]}` $_{N+1}$: Cell of $i = 1 \dots N+1$ matrices with magnetic field amplitude of the second wave in region i, $\bar{H}'_{iy}/H_{1y}$.
- The electric field amplitudes are returned in a similar fashion, using as a reference amplitude: $E_1 = (H_{1y}/|H_{1y}|)\,|\mathbf{E}_1| = H_{1y}/[(\epsilon_0 c)\sqrt{\epsilon_1}]$, i.e. E_1 is a complex amplitude such as $|E_1| = |\mathbf{E}_1|$, but with the same phase as H_{1y}. The corresponding entries in the structure are:
  - `EixOvE1 {[L x A]}` $_{N+1}$ for $\bar{E}_{ix}/E_1$,
  - `EizOvE1 {[L x A]}` $_{N+1}$ for $\bar{E}_{iz}/E_1$,
  - `EpixOvE1 {[L x A]}` $_{N+1}$ for $\bar{E}'_{ix}/E_1$,
  - `EpizOvE1 {[L x A]}` $_{N+1}$ for $\bar{E}'_{iz}/E_1$.

Similar names are used for TE polarization, only replacing E by H and vice versa.

In addition, the local field enhancements (with respect to the incident field) are computed for parallel and perpendicular orientations with respect to the surface, at distances d on either side of each interface ($i = 1 \dots N$), i.e. at positions $z = Z_i - d$ ('outside') and $z = Z_i + d$ ('inside'). We assume implicitly that d is small enough for 'outside' to be in region i and 'inside' in region $i+1$. The local field enhancements are $M^{\perp}_{\text{Loc}} = |E_z(z)|^2/|E_1|^2$ and $M^{\parallel}_{\text{Loc}} = |E_x(z)|^2/|E_1|^2$. The corresponding fields in the structure are:

- `MoutPerp {[L x A]}` $_N$ for $M^{\perp}_{\text{Loc}}$ 'outside' interface i (in region i).
- `MoutPara {[L x A]}` $_N$ for $M^{\parallel}_{\text{Loc}}$ 'outside' interface i (in region i).
- `MinPerp {[L x A]}` $_N$ for $M^{\perp}_{\text{Loc}}$ 'inside' interface i (in region $i+1$).
- `MinPara {[L x A]}` $_N$ for $M^{\parallel}_{\text{Loc}}$ 'inside' interface i (in region $i+1$).

The same applies to TE waves, but $M^{\perp}_{\text{Loc}}$ is always zero.

Example

As an example, we copy below the Matlab script used to produce Fig. 3.14(a), showing the angle dependence of the reflectivity for excitation of propagating SPPs in the Otto configuration (see Fig. 3.14 for details):

```
nM=1.33; % Water - Dielectric for PSPPs
nP=1.766; % Sapphire - Prism

Lgap=500; % in nm, good for 633nm, nM=1.33

% lambda (in nm) is column vector
lambda=[514 633 1000]';

% aideg is row vector; angle of incidence in degrees
aideg=[48:0.005:60]; % Otto, 633nm, nM=1.33

% Epsilon Metal (column vector)
epsMetal=epsAg(lambda); % Analytical expression for Ag

% Defines Otto configuration, 2 interfaces, 3 layers
nNbSurf=2;
Cepsilon{1}=nP^2+0*lambda; % Prism
Cepsilon{2}=nM^2+0*lambda; % Dielectric for PSPP
Cepsilon{3}=epsMetal; % Metal
CL{1}=0; % interfaces positions
CL{2}=Lgap;

% solves the problem
resMulti=MultiRef ('TM',nNbSurf, lambda, Cepsilon, CL, aideg);

%Reflectivity (reflection coefficient $R^p$)
% R is matrix length(lambda) x length(aideg)
R=abs(resMulti.rP).^2;

% plot results
plot(aideg, R);
```

This script can be found on the book website (http://www.victoria.ac.nz/raman/book) and similar scripts are provided for the Kretschmann configuration and the Kretschmann configuration with an additional adsorbed layer.

F.5. DIPOLE EMISSION CLOSE TO A PLANAR INTERFACE

Finally, the problem of a dipole emitting close to a planar interface (or even a multi-layer) can also be solved analytically, see for example Refs [12,169].

Unfortunately, the resulting expressions are not particularly friendly, and involve integrals that can only be evaluated numerically. In this section, we simply give a brief overview of the results and their physical significance.

We consider a dipole at a distance d from the surface (at $z = 0$), i.e. located at $z = -d$, in the half-space $z < 0$ with ϵ_1 real and positive. The cases of a dipole parallel and perpendicular to the surface are treated separately and the general case can be deduced from these. In all the expressions, $k_1 = k_0\sqrt{\epsilon_1}$, and $s = k_x/k_1$ is used as the variable for integration, for simplicity. The Fresnel reflection coefficients r^p and r^s are those of the interface (Eqs (F.45) and (F.62) for a single interface, Eq. (F.79) for multi-layers). They are calculated for $k_x = sk_1$ and are therefore functions of s. For $0 \leq s \leq 1$, this corresponds to the problem of an incidence wave with angle of incidence $\sin\theta = s$. For $s > 1$, $k_x > k_1$ and $k_{1z} = k_1\sqrt{1-s^2}$ must therefore be pure imaginary; the wave in region 1 is inhomogeneous and evanescent along z. The choice of $\mathrm{Im}(k_{1z}) > 0$ is implicit in the previous expression. This is necessary for compatibility with the expressions of r^p and r^s given earlier in Eqs (F.45) and (F.62), and is taken into account in the expressions below. The physical evanescent wave in region 1, however, would be the one with $-k_{1z}$ (i.e. what we called earlier the reflected wave).

F.5.1. Total decay rates

The *total* EM decay rate enhancement factors for perpendicular and parallel dipoles are then given by [12]:

$$M_{\mathrm{Tot}}^{\perp} = 1 + \frac{3}{2}\int_0^{\infty} \mathrm{Re}\left\{\frac{s^3}{\sqrt{1-s^2}} r^p \mathrm{e}^{2\mathrm{i}k_1 d\sqrt{1-s^2}}\right\} \mathrm{d}s, \tag{F.83}$$

and

$$M_{\mathrm{Tot}}^{\parallel} = 1 + \frac{3}{4}\int_0^{\infty} \mathrm{Re}\left\{\left[\frac{r^s}{\sqrt{1-s^2}} - r^p\sqrt{1-s^2}\right] s\mathrm{e}^{2\mathrm{i}k_1 d\sqrt{1-s^2}}\right\} \mathrm{d}s. \tag{F.84}$$

These integrals must in general be evaluated numerically.

F.5.2. Radiative decay rates

The radiative decay rate enhancement is usually separated into two contributions: that emitted radiatively in the upper ($z < 0$) or lower ($z > 0$) half-space. If the lower region is absorbing (like a metal), then the second contribution is zero. These expressions are usually derived from the full solution of the electromagnetic problem. It is however simpler (this is left as an exercise to the reader) to obtain them from the application of the optical reciprocity theorem (ORT), as described in Section 4.4.6.

We focus here only on the upper half-space, where we have [12]:

$$M^{\perp}_{\mathrm{Rad},z<0} = \frac{1}{2} + \frac{3}{4}\int_0^1 \frac{s^3}{\sqrt{1-s^2}}|r^p|^2 \mathrm{d}s + \\ + \frac{3}{2}\int_0^1 \mathrm{Re}\left\{\frac{s^3}{\sqrt{1-s^2}} r^p \mathrm{e}^{2\mathrm{i}k_1 d\sqrt{1-s^2}}\right\} \mathrm{d}s. \tag{F.85}$$

This can also be rewritten as:

$$M^{\perp}_{\mathrm{Rad},z<0} = \frac{3}{4}\int_0^{\pi/2} \left|1 + r^p \mathrm{e}^{2\mathrm{i}k_1 d\cos\theta}\right|^2 \sin^3\theta \mathrm{d}\theta, \tag{F.86}$$

which actually results immediately from the application of the ORT. Similarly

$$M^{\parallel}_{\mathrm{Rad},z<0} = \frac{1}{2} + \frac{3}{8}\int_0^1 \left[\frac{|r^s|^2}{\sqrt{1-s^2}} + |r^p|^2\sqrt{1-s^2}\right] s\mathrm{d}s + \\ + \frac{3}{4}\int_0^1 \mathrm{Re}\left\{\left[\frac{r^s}{\sqrt{1-s^2}} - r^p\sqrt{1-s^2}\right] s\mathrm{e}^{2\mathrm{i}k_1 d\sqrt{1-s^2}}\right\} \mathrm{d}s. \tag{F.87}$$

This can also be rewritten as:

$$M^{\parallel}_{\mathrm{Rad},z<0} = \frac{3}{8}\int_0^{\pi/2} \left[\left|1 + r^s \mathrm{e}^{2ik_1 d\cos\theta}\right|^2 \right. \\ \left. + \left|1 - r^p \mathrm{e}^{2ik_1 d\cos\theta}\right|^2 \cos^2\theta\right] \sin\theta \mathrm{d}\theta, \tag{F.88}$$

which also results immediately from the application of the ORT.

Moreover, the application of the ORT has the advantage that it automatically yields the radiation profile (which can be read directly in the expressions above in the integrands as a function of θ, angle of incidence).

Appendix G

Ellipsoids in the electrostatic approximation

In this appendix, we provide a number of analytic expressions in relation with the problem of the *ellipsoid in the electrostatic approximation* (see Section 5.1.4). Particular attention is given to the special cases of spheroids (oblate and prolate) for which expressions in closed form can be obtained. These can be used in a fairly straightforward manner to implement numerically most calculations related to this problem. Expansions in terms of the aspect ratio are also given to study the limiting cases of a 'flat' oblate spheroid and a needle-like prolate spheroid. Finally, we also give a brief description of a possible Matlab implementation of these results in the context of SERS modeling.

A discussion of some of these aspects, including additional references, can be found in electromagnetic textbooks, e.g. [95,149]. For a direct connection with SERS, see review articles like Refs. [4,218].

G.1. GENERAL CASE

We do not give here the details of the derivation, which can for example be found in Ref. [149], but only the most important results, which are then used to discuss some of the key EM indicators of interest for SERS.

G.1.1. Some definitions

A general ellipsoid can be described, in the appropriate coordinate frame, by the equation:

$$\frac{x^2}{a^2}+\frac{y^2}{b^2}+\frac{z^2}{c^2}=1, \tag{G.1}$$

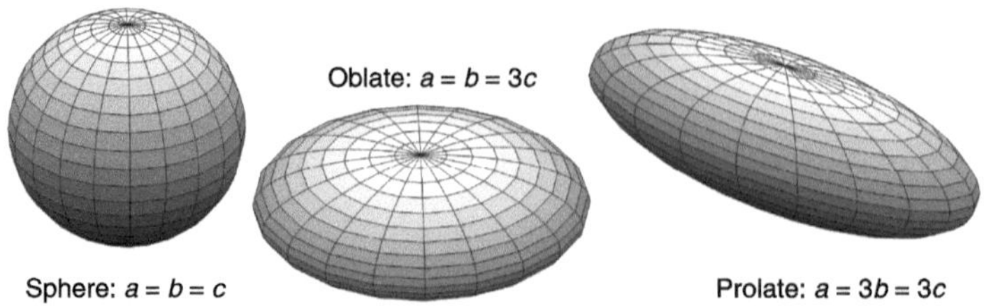

Figure G.1. Schematic representation of special cases of ellipsoids: a sphere, an oblate spheroid and a prolate spheroid (both with an aspect ratio of $a/c = 3$).

where $a \geq b \geq c$ (by convention) are the semi-axis lengths. Several special cases can be identified (see Fig. G.1):

- If $a = b = c$, the ellipsoid has maximum symmetry and is a *sphere* of radius a.
- If $a = b > c$, we have an ellipsoid of revolution (around the z-axis). It can be viewed as a sphere squashed along one direction (pumpkin-like), and it is called an *oblate spheroid.*
- If $a > b = c$, we have again an ellipsoid of revolution (around the x-axis). It can be viewed as a sphere elongated along one direction (rugby-ball-like), and it is called a *prolate spheroid* [1].
- If $a > b > c$, the ellipsoid may be called a scalene ellipsoid.

The two special cases of oblate and prolate spheroids are usually sufficient to approximate many particles of interest. They also yield simpler analytical expressions. They can moreover be conveniently characterized (up to a global scaling of dimensions) by a single quantity: their *aspect ratio* defined here as $h = a/c$.

G.1.2. Ellipsoidal coordinates

Unfortunately, in order to solve the Laplace equation for an ellipsoid using separation of variables, it is necessary to use ellipsoidal coordinates, which are not as 'nice' as spherical coordinates, and with which most people are not very familiar. There is an infinite family of ellipsoidal coordinates, each being the natural coordinate system of an ellipsoid with given aspect ratios (semi-axis ratios). The expressions are in fact different for oblate and prolate spheroids, and we therefore exclude this case for the moment and consider

[1] It seems more appropriate from New Zealand to call this shape 'rugby ball' instead of 'football', as it would be surely called in the US.

in the following: an ellipsoid with $a > b > c$ (the spheroid cases can then be obtained *a posteriori* by taking the appropriate limit).

The corresponding ellipsoidal coordinates are denoted (ξ, η, ζ) and are defined as the three solutions of the following equation (where u is the unknown):

$$\frac{x^2}{a^2+u} + \frac{y^2}{b^2+u} + \frac{z^2}{b^2+u} = 1, \tag{G.2}$$

with

$$-a^2 < \zeta < -b^2, \qquad -b^2 < \eta < -c^2, \qquad -c^2 < \xi < \infty. \tag{G.3}$$

The ellipsoid boundary equation (Eq. (G.1)) therefore corresponds simply to the surface $\xi = 0$. One can show that to a given point with Cartesian coordinates (x, y, z) corresponds one set of ellipsoidal coordinates (ξ, η, ζ). The opposite is not exactly true: to a given (ξ, η, ζ), corresponds a unique triplet (x^2, y^2, z^2), and therefore 8 points related by symmetries around the axes. More specifically, we have:

$$\begin{cases} x^2 &= \dfrac{(a^2+\xi)(a^2+\eta)(a^2+\zeta)}{(b^2-a^2)(c^2-a^2)} \\ y^2 &= \dfrac{(b^2+\xi)(b^2+\eta)(b^2+\zeta)}{(a^2-b^2)(c^2-b^2)} \\ z^2 &= \dfrac{(c^2+\xi)(c^2+\eta)(c^2+\zeta)}{(a^2-c^2)(b^2-c^2)}. \end{cases} \tag{G.4}$$

G.1.3. The electrostatic solution

Solution for the potential

We consider an ellipsoid, with dielectric constant $\epsilon(\omega)$, embedded in a dielectric medium with dielectric constant ϵ_M, and placed in a constant and uniform external electric field $\mathbf{E}_0 = E_0\mathbf{e}_z = -\nabla\phi_0$. Here the field polarization is chosen along one of the main axes; the general case can be obtained by superposition and will be discussed later. The electric potential solutions inside and outside are given in ellipsoidal coordinates as [149]:

$$\phi_{\text{in}} = \frac{\phi_0}{1 + L_3\frac{\epsilon-\epsilon_M}{\epsilon_M}} = \frac{3\epsilon_M}{3L_3\epsilon + \epsilon_M(3-3L_3)}\phi_0, \tag{G.5}$$

and

$$\phi_{\text{out}} = \phi_0 - \phi_0 3\beta_3 F(c,\xi). \tag{G.6}$$

As we shall see, β_3 is the non-dimensional polarizability (for excitation along the z-axis) and is analogous to β_S for the sphere (Eq. (6.15)):

$$\boxed{\beta_3(\omega) = \frac{\epsilon(\omega) - \epsilon_M}{3L_3\epsilon(\omega) + \epsilon_M(3 - 3L_3)}.} \tag{G.7}$$

$F(d,\xi)$ is an auxiliary function given by:

$$F(d,\xi) = abc \int_{\xi}^{\infty} \frac{dq}{2(d^2+q)f(q)}, \tag{G.8}$$

where

$$f(q) = \sqrt{(a^2+q)(b^2+q)(c^2+q)}. \tag{G.9}$$

L_3 is a geometrical factor (discussed later) given as:

$$\boxed{L_3 = F(c,0) = abc \int_0^{\infty} \frac{dq}{2(c^2+q)f(q)}.} \tag{G.10}$$

Field solution and polarizability

The electric field can then be derived from the electric potential. The internal field is constant and aligned with the incident field, as was the case for a sphere:

$$\mathbf{E}_{\text{in}} = \frac{3\epsilon_M}{3L_3\epsilon + \epsilon_M(3 - 3L_3)}\mathbf{E}_0 = (1 - 3\beta_3 L_3)\mathbf{E}_0, \tag{G.11}$$

from which we derive a similar relation for the electric polarization inside the ellipsoid (see Section 6.2.1):

$$\mathbf{P}_M = 3\epsilon_0\epsilon_M\beta_3\mathbf{E}_0. \tag{G.12}$$

As for the sphere, the polarization is uniform across the volume ($V_E = (4/3)\pi abc$) of the ellipsoid, and equivalent to a dipole moment $\mathbf{p}_M = \alpha_3\mathbf{E}_0$, where the dipolar polarizability (for excitation along the z-axis) is:

$$\boxed{\alpha_3 = 3\epsilon_0\epsilon_M V_E \beta_3 = 4\pi\epsilon_0\epsilon_M abc \frac{\epsilon - \epsilon_M}{3L_3\epsilon + \epsilon_M(3 - 3L_3)}.} \quad \text{(G.13)}$$

The outside field is the sum of the external field $\mathbf{E}_0$ and the scattered field. After some manipulation, it can be expressed as:

$$\mathbf{E}_{\text{out}} = \mathbf{E}_0 + \mathbf{E}_{\text{sca}} = E_0 \left(1 - 3\beta_3 F(c,\xi)\right) \mathbf{e}_z + 3\beta_3 E_0 \frac{abc}{f(\xi)} \left(\mathbf{e}_\xi \cdot \mathbf{e}_z\right) \mathbf{e}_\xi, \quad \text{(G.14)}$$

where a mixture of ellipsoidal and Cartesian coordinates was used for simplicity. Expressed in this form, the similarities with the case of a perfect sphere (Section 6.2.3) are easily noted. The unit vector $\mathbf{e}_\xi$ is the normal to the ellipsoid surface for points on this surface (i.e. when $\xi = 0$), and can otherwise be expressed as the general expression:

$$\mathbf{e}_\xi = \frac{1}{\sqrt{\frac{x^2}{(a^2+\xi)^2} + \frac{y^2}{(b^2+\xi)^2} + \frac{z^2}{(c^2+\xi)^2}}} \left[\frac{x}{a^2+\xi}\mathbf{e}_x + \frac{y}{b^2+\xi}\mathbf{e}_y + \frac{z}{c^2+\xi}\mathbf{e}_z\right]. \quad \text{(G.15)}$$

Effect of incident polarization

The previous expressions were obtained for an exciting field polarized along z. Because the ellipsoid is not fully symmetric like the sphere, the results should depend on the incident polarization. The previous treatment in fact remains valid for polarization along x or y, only replacing z by x or y, L_3 by $L_1 = F(a, 0)$ or $L_2 = F(b, 0)$, and β_3 (and α_3) by the corresponding β_1 (and α_1) or β_2 (and α_2).

The L_i's are in general called *geometrical factors* or *depolarization factors* and are amongst the most important parameters for the optical properties of the ellipsoids. They can, in fact, be interpreted as depolarization factors [149], but only when $\epsilon_M = 1$ or when one considers the polarization $\mathbf{P}_M$ with respect to the embedding medium. For further discussion of this aspect, see Ref. [149]. We have in addition the important properties:

$$\boxed{L_1 + L_2 + L_3 = 1 \quad \text{and} \quad 0 \leq L_1 \leq L_2 \leq L_3 \leq 1,} \quad \text{(G.16)}$$

where the latter inequalities arise from the convention $a \geq b \geq c$. Note that it is possible to obtain 'simpler' analytical expressions for the L_i's in the special case of oblate or prolate spheroids as given in Sections G.2 and G.3.

Since the L_i's are in general distinct (except for the special case of the sphere where $L_1 = L_2 = L_3 = 1/3$), the dipolar polarizabilities are also different depending on the incident field polarization. For a general incident polarization, the solution is simply the sum of the solutions for each of the three components along the main ellipsoid axes. If $\mathbf{E}_0 = E_{0x}\mathbf{e}_x + E_{0y}\mathbf{e}_y + E_{0z}\mathbf{e}_z$, the induced dipole, for example, is:

$$\mathbf{p}_M = \alpha_1 E_{0x}\mathbf{e}_x + \alpha_2 E_{0y}\mathbf{e}_y + \alpha_3 E_{0z}\mathbf{e}_z. \tag{G.17}$$

The induced dipole is therefore not necessarily aligned with the incident polarization (except when it is aligned with one of the main axes of the ellipsoid). In the following discussion, we will focus again on the special case of incident polarization along one of the main axes, z here, but the results can easily be extended to the other two axes, or to a general polarization (by superposition).

G.1.4. Some important EM indicators for ellipsoids

Far-field properties

Following the treatment of the sphere and using again the dipolar approximation, it is straightforward to obtain the absorption (approximately equal to extinction in this approximation) and scattering cross-sections of the ellipsoid in the ES approximation for incident polarization along one of the main axes:

$$\sigma_{\text{Ext}} \approx \sigma_{\text{Abs}} = 4\pi\ k_M\ abc\ \text{Im}(\beta_i(\omega)), \tag{G.18}$$

and

$$\sigma_{\text{Sca}} = \frac{8\pi}{3}\ (k_M)^4\ (abc)^2\ |\beta_i(\omega)|^2, \tag{G.19}$$

where $k_M = n_M\omega/c$ and $i = 1, 2, 3$ depending on the axis of the incident polarization.

Local fields in the ES approximation

Of particular interest to us is the electric field just outside the ellipsoid at the surface, i.e. at $\xi = 0$. Using $F(c, 0) = L_3$ and $f(0) = abc$, we have:

$$\boxed{\mathbf{E}_{\text{out}}(\xi = 0) = E_0\left[(1 - 3L_3\beta_3)\mathbf{e}_z + 3\beta_3\left(\mathbf{e}_\xi \cdot \mathbf{e}_z\right)\mathbf{e}_\xi\right].} \tag{G.20}$$

The similarity with the expression for the sphere (Eq. (6.18)) is again noted, and it can easily be recovered here by taking $L_3 = 1/3$.

We can now write, as for the sphere, a local field intensity enhancement factor (LFIEF), normal ($\perp$) and parallel ($//$) to the ellipsoid surface, at any point $\mathbf{r}$ on the surface ($\xi = 0$):

$$\boxed{\begin{cases} M_{\text{Loc}}^{\perp}(\mathbf{r},\omega) = \dfrac{|E_\xi|^2}{E_0^2} = A_3^{\perp}(\omega)\,|\mathbf{e}_\xi \cdot \mathbf{e}_z|^2 \\ \text{where} \quad A_3^{\perp}(\omega) = |1 + (3 - 3L_3)\beta_3(\omega)|^2\,; \end{cases}} \tag{G.21}$$

and

$$\boxed{\begin{cases} M_{\text{Loc}}^{||}(\mathbf{r},\omega) = \dfrac{|E_\eta|^2 + |E_\varsigma|^2}{E_0^2} = A_3^{||}(\omega)[1 - |\mathbf{e}_\xi \cdot \mathbf{e}_z|^2] \\ \text{where} \quad A_3^{||}(\omega) = |1 - 3L_3\beta_3(\omega)|^2. \end{cases}} \tag{G.22}$$

We can deduce the LFIEF on the surface as:

$$\boxed{M_{\text{Loc}}(\mathbf{r},\omega) = A_3^{||}(\omega) + (A_3^{\perp}(\omega) - A_3^{||}(\omega))\,|\mathbf{e}_\xi \cdot \mathbf{e}_z|^2\,,} \tag{G.23}$$

and the SERS EF derives from:

$$F_{E4}^{0} = (M_{\text{Loc}})^2. \tag{G.24}$$

Note that the scalar product can for example be obtained as a function of the Cartesian coordinates of a point on the surface (i.e. for $\xi = 0$ only) as:

$$|\mathbf{e}_\xi \cdot \mathbf{e}_z|^2 = \frac{z^2}{c^4\left(x^2/a^4 + y^2/b^4 + z^2/c^4\right)}. \tag{G.25}$$

Finally, we have the same useful expression as for the sphere:

$$\frac{A_3^{||}(\omega)}{A_3^{\perp}(\omega)} = \frac{(\epsilon_M)^2}{|\epsilon(\omega)|^2}. \tag{G.26}$$

See Section 6.2.3 for a discussion of this relation.

Average enhancement factors in the ES approximation

The average enhancement factors cannot be written here in closed form, but can be expressed in terms of the surface averages $\langle|\mathbf{e}_\xi \cdot \mathbf{e}_z|^2\rangle$ and $\langle|\mathbf{e}_\xi \cdot \mathbf{e}_z|^4\rangle$. We have for example:

$$\langle M^{\perp}_{\mathrm{Loc}}(\omega)\rangle = A^{\perp}_3(\omega)\langle|\mathbf{e}_\xi \cdot \mathbf{e}_z|^2\rangle, \tag{G.27}$$

$$\langle M_{\mathrm{Loc}}(\omega)\rangle = A^{\parallel}_3(\omega) + (A^{\perp}_3(\omega) - A^{\parallel}_3(\omega))\langle|\mathbf{e}_\xi \cdot \mathbf{e}_z|^2\rangle, \tag{G.28}$$

and,

$$\langle F^0_{E4}(\omega_L)\rangle = \left[A^{\parallel}_3(\omega_L)\right]^2 + 2A^{\parallel}_3(\omega_L)\left[A^{\perp}_3(\omega_L) - A^{\parallel}_3(\omega_L)\right]\langle|\mathbf{e}_\xi \cdot \mathbf{e}_z|^2\rangle + \left[A^{\perp}_3(\omega_L) - A^{\parallel}_3(\omega_L)\right]^2\langle|\mathbf{e}_\xi \cdot \mathbf{e}_z|^4\rangle. \tag{G.29}$$

The actual calculation of the surface averages $\langle|\mathbf{e}_\xi \cdot \mathbf{e}_z|^2\rangle$ and $\langle|\mathbf{e}_\xi \cdot \mathbf{e}_z|^4\rangle$ is not an easy task. In fact, in the case of a general ellipsoid, even the surface area cannot be expressed analytically in a simple form. It is however possible to calculate these averages analytically for the special cases of spheroids. The results are given for reference in Sections G.2 and G.3 along with a number of other analytic expressions relevant to spheroids. Most of these expressions are usually sufficiently complicated by themselves, and their utility could be questioned. They are, however, very useful for numerical calculations of the spheroid optical properties in the ES approximation and are provided mostly to this end.

Depolarization and radiative corrections for spheroids

It was proposed [218,219] that the corrections to the polarizability of the sphere discussed in Section 6.2.1 could be generalized to the case of the ellipsoid and it was argued that such corrections agree well with exact results. However, as already pointed out in Section 6.2.1, these corrections are already inadequate for the sphere, and they are therefore unlikely to perform any better for the ellipsoid. These corrections are therefore no better *a priori* than the ESA itself and are in fact identical to it in the limit of small $k_M a$. If they appear to agree with exact results, it is most likely because the ESA would also agree, and the additional complications of these corrections are then unnecessary. As for the sphere, more investigations are needed in this area to clarify the situation.

G.1.5. Some aspects of the numerical implementation

Geometrical factors

The geometrical (or depolarization) factors can, by definition, be obtained from the integrals:

$$L_1 = F(a,0) = \frac{abc}{2}\int_0^{\infty} \frac{\mathrm{d}q}{(a^2+q)^{3/2}(b^2+q)^{1/2}(c^2+q)^{1/2}}, \tag{G.30}$$

$$L_2 = F(b,0) = \frac{abc}{2}\int_0^{\infty} \frac{\mathrm{d}q}{(a^2+q)^{1/2}(b^2+q)^{3/2}(c^2+q)^{1/2}}, \tag{G.31}$$

$$L_3 = F(c,0) = \frac{abc}{2}\int_0^{\infty} \frac{\mathrm{d}q}{(a^2+q)^{1/2}(b^2+q)^{1/2}(c^2+q)^{3/2}}, \tag{G.32}$$

and from $a \geq b \geq c$, we recall that:

$$L_1 \leq L_2 \leq L_3 \qquad \text{and} \qquad L_1 + L_2 + L_3 = 1. \tag{G.33}$$

One can also show that these factors can be expressed as *surface integrals* on the surface of the ellipsoid (S) as:

$$L_3 = \frac{1}{4\pi}\int\int_S \frac{z}{r^3}(\mathbf{e}_\xi \cdot \mathbf{e}_z)\mathrm{d}S, \tag{G.34}$$

where $r = |\mathbf{r}|$, and with equivalent relations for L_1 and L_2.

Other important surface integrals

Also of interest are the integrals defining the surface area S of the ellipsoid and the surface averages $\langle|\mathbf{e}_\xi \cdot \mathbf{e}_z|^2\rangle$ and $\langle|\mathbf{e}_\xi \cdot \mathbf{e}_z|^4\rangle$, and their counterparts for the x and y axes. We have:

$$S = \int\int_S \mathrm{d}S, \tag{G.35}$$

$$\langle|\mathbf{e}_\xi \cdot \mathbf{e}_z|^2\rangle = \frac{1}{S}\int\int_S |\mathbf{e}_\xi \cdot \mathbf{e}_z|^2\,\mathrm{d}S, \tag{G.36}$$

and similar expressions for the other averages.

All these surface integrals (including the ones to calculate the L_i's) can in principle be written as integrals on 2 angles, for example, by carrying out the following change of variables:

$$\begin{cases} x &= a\sin\theta\cos\phi \\ y &= b\sin\theta\sin\phi \\ z &= c\cos\theta \end{cases} \quad \text{with} \quad \begin{matrix} 0 \le \theta \le \pi \\ 0 \le \phi < 2\pi. \end{matrix} \tag{G.37}$$

Note that in this context, θ and ϕ are not the usual angles of the spherical coordinates. The infinitesimal surface element on the surface of the ellipsoid then takes the form:

$$\mathrm{d}S = \sin\theta\sqrt{s}\mathrm{d}\theta\mathrm{d}\phi, \tag{G.38}$$

where

$$\begin{aligned} s &= (abc)^2\left[\frac{x^2}{a^4}+\frac{y^2}{b^4}+\frac{z^2}{c^4}\right] \\ &= b^2c^2\sin^2\theta\cos^2\phi + a^2c^2\sin^2\theta\sin^2\phi + a^2b^2\cos^2\theta. \end{aligned} \tag{G.39}$$

Note that for a point on the surface ($\xi = 0$), we moreover have the simple relation:

$$\mathbf{e}_\xi = \frac{abc}{\sqrt{s}}\left[\frac{x}{a^2}\mathbf{e}_x + \frac{y}{b^2}\mathbf{e}_y + \frac{z}{c^2}\mathbf{e}_z\right]. \tag{G.40}$$

Example of numerical implementation in Matlab

We give below examples of some of these integrals using this change of variables.

$$S = \int_0^{2\pi}\int_0^{\pi} \sin\theta\sqrt{s(\theta,\phi)}\mathrm{d}\theta\mathrm{d}\phi, \tag{G.41}$$

$$\langle|\mathbf{e}_\xi\cdot\mathbf{e}_z|^2\rangle = \frac{1}{S}\int_0^{2\pi}\int_0^{\pi}\frac{a^2b^2\sin\theta\cos^2\theta}{\sqrt{s(\theta,\phi)}}\mathrm{d}\theta\mathrm{d}\phi, \tag{G.42}$$

and

$$L_3 = \frac{1}{4\pi}\int_0^{2\pi}\int_0^{\pi}\frac{abc\cos^2\theta\sin\theta}{(a^2\sin^2\theta\cos^2\phi + b^2\sin^2\theta\sin^2\phi + c^2\cos^2\theta)^{3/2}}\mathrm{d}\theta\mathrm{d}\phi. \tag{G.43}$$

These (horrible) expressions can be rewritten using special functions called incomplete elliptic integrals, but are probably best left for numerical estimation.

As an example, the surface area of the ellipsoid can be simply computed in Matlab with the following (assuming, a, b, and c have been defined):

```
Sdiff = @(t,f) sin(t).*(b^2*c^2*sin(t).^2.*cos(f).^2+ ...
    a^2*c^2*sin(t).^2.*sin(f).^2+a^2*b^2*cos(t).^2).^(1/2);
S = dblquad(Sdiff,0,pi,0,2*pi)
```

The first command defines the function of θ (t) and ϕ (f) to integrate and the second simply computes the integral. All the important properties of the ellipsoid (including the L_i's) can be computed as surface integrals in this way. Many key EM indicators then derive easily. These are implemented in a few ready-to-use Matlab scripts that can be found on the book website (www.victoria.ac.nz/raman/book).

Finally, in the special case of spheroids, these surface integrals can in fact be computed and closed-form analytical expressions can be obtained. These may be more convenient for some studies and can also be used to understand the limiting cases. Matlab scripts using these direct expressions are also provided. The relevant formulas for the two possible cases of spheroids, oblate and prolate, are listed for reference without further justification in the following two sections.

G.2. OBLATE SPHEROID (PUMPKIN)

We will go rapidly in this section through the most important expressions that can be derived for oblate spheroids (see Fig. G.1).

G.2.1. Geometrical factors

For an oblate spheroid ($a = b > c$), the ellipsoid has symmetry of revolution around the z-axis, its aspect ratio is $h = a/c$, and the *eccentricity* is by definition:

$$e_o = 1 - c^2/a^2 \qquad (0 \le e_o < 1). \tag{G.44}$$

We moreover have $L_1 = L_2 < 1/3 < L_3$ and:

$$\begin{aligned} L_1 = L_2 &= \frac{1}{2e_o^2}\left[\frac{\sqrt{1-e_o^2}}{e_o}\arcsin(e_o) - (1-e_o^2)\right], \\ L_3 = 1 - 2L_1 &= \frac{1}{e_o^2}\left[1 - \frac{\sqrt{1-e_o^2}}{e_o}\arcsin(e_o)\right]. \end{aligned} \tag{G.45}$$

This reduces to $L_1 = L_2 = L_3 = 1/3$ for $e_o \to 0$ (sphere), as expected.

G.2.2. Surface averages

The surface averages can be written conveniently using the auxiliary function f_o defined as:

$$f_o = \frac{1-e_o^2}{2e_o} \ln\left(\frac{1+e_o}{1-e_o}\right). \tag{G.46}$$

The surface area is then:

$$S = 2\pi a^2 \left[1 + f_o\right]. \tag{G.47}$$

The surface averages involved in the calculations of local field enhancement factors are:

$$\langle |\mathbf{e}_\xi \cdot \mathbf{e}_z|^2 \rangle = \frac{1}{e_o^2} \cdot \frac{1-f_o}{1+f_o}, \tag{G.48}$$

from which we derive by symmetry:

$$\begin{aligned}\langle |\mathbf{e}_\xi \cdot \mathbf{e}_x|^2 \rangle = \langle |\mathbf{e}_\xi \cdot \mathbf{e}_y|^2 \rangle &= \frac{1}{2}(1 - \langle |\mathbf{e}_\xi \cdot \mathbf{e}_z|^2 \rangle) \\ &= \frac{1}{2e_o^2} \cdot \frac{(1+e_o^2)f_o - (1-e_o^2)}{1+f_o}.\end{aligned} \tag{G.49}$$

Moreover,

$$\langle |\mathbf{e}_\xi \cdot \mathbf{e}_z|^4 \rangle = \frac{1}{e_o^4} \cdot \frac{3 - 2e_o^2 - 3f_o}{1+f_o}, \tag{G.50}$$

from which we derive by symmetry:

$$\begin{aligned}\langle |\mathbf{e}_\xi \cdot \mathbf{e}_x|^4 \rangle = \langle |\mathbf{e}_\xi \cdot \mathbf{e}_y|^4 \rangle &= \frac{3}{8}(1 - 2\langle |\mathbf{e}_\xi \cdot \mathbf{e}_z|^2 \rangle + \langle |\mathbf{e}_\xi \cdot \mathbf{e}_z|^4 \rangle) \\ &= \frac{3(1-e_o^2)}{8e_o^4} \cdot \frac{3 - e_o^2 - (3+e_o^2)f_o}{1+f_o}.\end{aligned} \tag{G.51}$$

Finally, in the limit of $e_o \to 0$ (sphere), we have:

$$f_o \approx 1 - \frac{2}{3}e_o^2 - \frac{2}{15}e_o^4 + O(e_o^6), \tag{G.52}$$

from which we recover the sphere results: $\langle |\mathbf{e}_\xi \cdot \mathbf{e}_{x,y,z}|^2 \rangle = 1/3$ and $\langle |\mathbf{e}_\xi \cdot \mathbf{e}_{x,y,z}|^4 \rangle = 1/5$, as expected.

G.2.3. Limit of large aspect ratio

Also of interest is the case where $e_o \to 1$, i.e. a very flat, disk-like, pumpkin, with a large aspect ratio ($h = a/c \gg 1$). We can then expand the previous results in terms of $1/h = c/a = \sqrt{1 - e_o^2} \to 0$, and we obtain:

$$L_1 = L_2 \approx \frac{\pi}{4} \frac{c}{a} - \frac{c^2}{a^2} + O\left(\frac{c^3}{a^3}\right) \quad \text{and} \quad L_3 \approx 1. \tag{G.53}$$

Moreover $f_o \to 0$ and, more precisely:

$$f_o \approx \frac{c^2}{a^2} \ln\left(\frac{2a}{c}\right) + O\left(\frac{c^4}{a^4} \ln\left(\frac{2a}{c}\right)\right), \tag{G.54}$$

which implies:

$$S \approx 2\pi a^2, \tag{G.55}$$

$$\langle |\mathbf{e}_\xi \cdot \mathbf{e}_z|^2 \rangle \approx 1, \tag{G.56}$$

$$\langle |\mathbf{e}_\xi \cdot \mathbf{e}_x|^2 \rangle = \langle |\mathbf{e}_\xi \cdot \mathbf{e}_y|^2 \rangle \approx \frac{c^2}{a^2} \left(\ln\left(\frac{2a}{c}\right) - \frac{1}{2}\right) + O\left(\frac{c^4}{a^4} \ln\left(\frac{2a}{c}\right)\right), \tag{G.57}$$

$$\langle |\mathbf{e}_\xi \cdot \mathbf{e}_z|^4 \rangle \approx 1, \tag{G.58}$$

and

$$\langle |\mathbf{e}_\xi \cdot \mathbf{e}_x|^4 \rangle = \langle |\mathbf{e}_\xi \cdot \mathbf{e}_y|^4 \rangle \approx \frac{3c^2}{4a^2} + O\left(\frac{c^4}{a^4} \ln\left(\frac{2a}{c}\right)\right). \tag{G.59}$$

G.3. PROLATE SPHEROID (RUGBY BALL)

We again go rapidly in this section through the most important expressions that can be derived for prolate spheroids (see Fig. G.1).

G.3.1. Geometrical factors

For a prolate spheroid ($b = c < a$), the ellipsoid has symmetry of revolution around the x-axis, its aspect ratio is $h = a/b$, and the *eccentricity* is by definition:

$$e_p = 1 - b^2/a^2 \qquad (0 \le e_p < 1). \tag{G.60}$$

In addition, we have $L_1 < 1/3 < L_2 = L_3$ and:

$$L_1 = \frac{1-e_p^2}{e_p^2}\left[-1 + \frac{1}{2e_p}\ln\left(\frac{1+e_p}{1-e_p}\right)\right],$$
$$L_3 = L_2 = (1-L_1)/2 = \frac{1}{2e_p^2}\left[1 - \frac{1-e_p^2}{2e_p}\ln\left(\frac{1+e_p}{1-e_p}\right)\right]. \tag{G.61}$$

This reduces to $L_1 = L_2 = L_3 = 1/3$ for $e_p \to 0$ (limit of a sphere), as expected.

G.3.2. Surface averages

The surface averages can be written conveniently using the auxiliary function f_p defined as:

$$\text{with} \quad f_p = \frac{1}{e_p\sqrt{1-e_p^2}} \arcsin(e_p). \tag{G.62}$$

The surface area is then

$$S = 2\pi b^2 [1 + f_p]. \tag{G.63}$$

The surface averages involved in the calculations of local field enhancement factors are:

$$\langle |\mathbf{e}_\xi \cdot \mathbf{e}_x|^2 \rangle = \frac{1-e_p^2}{e_p^2} \cdot \frac{f_p - 1}{1 + f_p}, \tag{G.64}$$

and

$$\langle |\mathbf{e}_\xi \cdot \mathbf{e}_y|^2 \rangle = \langle |\mathbf{e}_\xi \cdot \mathbf{e}_z|^2 \rangle = \frac{1}{2e_p^2} \cdot \frac{1 - (1 - 2e_p^2) f_p}{1 + f_p}. \tag{G.65}$$

Furthermore:

$$\langle |\mathbf{e}_\xi \cdot \mathbf{e}_y|^4 \rangle = \langle |\mathbf{e}_\xi \cdot \mathbf{e}_z|^4 \rangle = \frac{3}{8e_p^4} \cdot \frac{3 - 2e_p^2 - (3 - 4e^2) f_p}{1 + f_p}, \tag{G.66}$$

and

$$\langle|\mathbf{e}_\xi\cdot\mathbf{e}_x|^4\rangle = 1 - 4\langle|\mathbf{e}_\xi\cdot\mathbf{e}_z|^2\rangle + \frac{8}{3}\langle|\mathbf{e}_\xi\cdot\mathbf{e}_z|^4\rangle$$
$$= \frac{1-e_p^2}{e_p^4}\cdot\frac{3-e_p^2-3(1-e_p^2)f_p}{1+f_p}. \tag{G.67}$$

Finally, in the limit of $e_p \to 0$ (sphere), we have:

$$f_p \approx 1 + \frac{2}{3}e_p^2 + \frac{8}{15}e_p^4 + O(e_p^6), \tag{G.68}$$

from which we recover again the sphere results, as expected.

G.3.3. Limit of large aspect ratio

Also of interest is the case where $e_p \to 1$, i.e. a 'very elongated rugby ball', more like a Cuban cigar, with a large aspect ratio ($h = a/b \gg 1$). We can then expand the previous results in terms of $1/h = b/a = \sqrt{1-e_p^2} \to 0$, and we obtain:

$$L_1 \approx \frac{b^2}{a^2}\left(\ln\left(\frac{2a}{b}\right) - 1\right) + O\left(\frac{b^4}{a^4}\ln\left(\frac{2a}{b}\right)\right) \text{ and } L_2 = L_3 \approx 1/2. \tag{G.69}$$

Moreover $f_p \to \infty$ and, more precisely:

$$f_p \approx \frac{\pi}{2}\frac{a}{b} - 1 + \frac{\pi}{4}\frac{b}{a} - \frac{2}{3}\frac{b^2}{a^2} + O\left(\frac{b^3}{a^3}\right), \tag{G.70}$$

which implies:

$$S \approx \pi^2 ab, \tag{G.71}$$

$$\langle|\mathbf{e}_\xi\cdot\mathbf{e}_y|^2\rangle = \langle|\mathbf{e}_\xi\cdot\mathbf{e}_z|^2\rangle \approx \frac{1}{2}, \tag{G.72}$$

$$\langle|\mathbf{e}_\xi\cdot\mathbf{e}_x|^2\rangle \approx \frac{b^2}{a^2} + O\left(\frac{b^2}{a^2}\right), \tag{G.73}$$

$$\langle|\mathbf{e}_\xi\cdot\mathbf{e}_y|^4\rangle = \langle|\mathbf{e}_\xi\cdot\mathbf{e}_z|^4\rangle \approx \frac{3}{8}, \tag{G.74}$$

and

$$\langle|\mathbf{e}_\xi\cdot\mathbf{e}_x|^4\rangle \approx \frac{4}{\pi}\frac{b^3}{a^3} - 3\frac{b^4}{a^4} + O\left(\frac{b^4}{a^4}\right). \tag{G.75}$$

Appendix H

Mie theory and its implementation

H.1. INTRODUCTION

H.1.1. Motivation

Mie theory is concerned with the description of the scattering properties of a sphere (dielectric or metallic, for example) of arbitrary size. It can easily be extended to coated spheres (or spherical multi-layers) and even (although less easily) to multiple spheres. The theory has one great advantage in that it is one of the few available exact solutions of the electromagnetic problem. One disadvantage is that it is not very user-friendly since it makes intensive use of some special functions (spherical Bessel functions and spherical harmonics). This is further complicated by the fact that several definitions and conventions co-exist in the literature, which makes it difficult to compile different aspects of the theory from different sources.

Mie theory [158] is simple conceptually, but can be cumbersome to implement. The aim of this appendix is therefore three-fold:

- To describe first the simple concepts underlying Mie theory, without dwelling too much on its technical (mathematical) details.
- To compile a consistent list of the most useful expressions of Mie theory and its derivatives. This should enable the interested reader to adapt, expand, or implement the theory toward his/her own specific needs. We also emphasize aspects of Mie theory such as dipolar emission that are not usually discussed extensively in textbooks, but are very relevant to SERS and plasmonics.
- To provide an easy-to-use implementation, in the form of Matlab codes, available from the book website (www.victoria.ac.nz/raman/book). The aim of this appendix is therefore to provide sufficient understanding for the reader to use the codes knowledgeably and efficiently.

Understanding Mie theory is certainly *not necessary* to understand SERS and other plasmon-related effects. It does however provide a well-tested and reliable framework to understand many aspects of the problems, as shown for example in Section 6.2, without having to worry about the approximations that are made.

H.1.2. Overview of this appendix

There are a number of excellent textbooks on Mie theory. Chapter 4 of Ref. [149] by Bohren and Huffman is, for example, a clear and detailed presentation of Mie theory in its simplest form: for plane wave excitation (PWE). Ref. [196] also provides a detailed description with an extensive discussion of the influence of the various parameters. Because there is no need to repeat what has already been done well by others, we attempt in this appendix (following the pattern of the previous appendices) to provide a slightly different approach from these textbooks.

Because we are aware that some readers may be easily put off by the apparent complexity of the equations involved in Mie theory, we have attempted to adopt a 'graded' approach to this presentation. We therefore begin in Section H.2 by introducing the concepts of Mie theory, while keeping the technical details to a bare minimum. Then, in Section H.3 we re-analyze these concepts in detail and provide the most important mathematical expressions. We realize that this two-level approach may appear more cumbersome to those readers who were prepared to tackle directly the whole lot, but we hope this choice will make this appendix more appealing (and more useful) to a broader readership.

We then focus in Section H.4, on the specific case of plane wave excitation (PWE) of a sphere. This case is often considered as *being* Mie theory, while other aspects are classified as extensions of Mie theory. Such extensions, in particular for dipole emission and coated spheres are discussed in Section H.5. Finally, Section H.6 is devoted to the numerical implementation of these concepts in Matlab. This section is meant as a brief introduction of the Matlab codes for Mie theory that can be downloaded from the book website (www.victoria.ac.nz/raman/book).

This organization, we believe, should achieve the aims of this appendix to provide (i) a simple introduction to Mie theory, and (ii) a consistent list of the important formulas to be used as a reference and in relation to our specific Matlab implementation.

H.2. THE CONCEPTS OF MIE THEORY

H.2.1. The electromagnetic equations

For simplicity, we confine ourselves to homogeneous, isotropic, *non-magnetic* materials, whose optical properties are characterized by a *local* (possibly frequency-dependent) dielectric function $\epsilon(\omega)$. We also restrict ourselves to harmonic time dependence, and all the fields are represented in complex notation (see Appendix C). Most EM problems related to SERS and plasmonics can then be reduced (see Chapter 5) to that of finding the divergence-less electric field solution of the Helmholtz equation or *vectorial wave equation*:

$$\boxed{\nabla^2 \mathbf{E} + k^2 \mathbf{E} = 0 \quad \text{and} \quad \nabla \cdot \mathbf{E} = 0,} \tag{H.1}$$

where the *wave vector* is given by:

$$k^2 = \frac{\omega^2}{c^2} \epsilon(\omega). \tag{H.2}$$

The solution must in addition satisfy the appropriate *boundary conditions at interfaces and/or infinity.* The magnetic field is then easily derived from:

$$\mathbf{H} = \frac{1}{\mathrm{i}\omega\mu_0} \nabla \times \mathbf{E}. \tag{H.3}$$

Finding the divergence-less solutions of the vectorial wave equation in a given volume of space is not in itself difficult. The tricky part is to 'stitch' these together, i.e. to find the ones that satisfy the correct boundary conditions at all interfaces. The most common approach to this is to express the solutions in a coordinate system where all interfaces are surfaces of constant coordinates. For example, spheres are simply defined as $r = a$ in spherical coordinates. This obviously limits enormously the variety of geometries of EM problems that can be handled. The most common are those in either Cartesian coordinates (rectangular objects), with spherical or cylindrical symmetries, or possibly in elliptic/parabolic/hyperbolic/spheroidal coordinates.

Mie theory is concerned with *spherical objects* and therefore simply consists in expressing the field solutions in spherical coordinates and finding the appropriate stitching conditions.

H.2.2. The vectorial wave equation in spherical coordinates

It can be shown that a general divergence-less solution of the vector wave equations can be written in spherical coordinates as an infinite series of a complete set of vector spherical harmonics as:

$$\mathbf{E}(\mathbf{r}) = E_0 \sum_{n=0}^{\infty} \sum_{m=-n}^{m=n} a_{nm} \mathbf{M}_{nm}^{(i)}(k, \mathbf{r}) + b_{nm} \mathbf{N}_{nm}^{(i)}(k, \mathbf{r}), \tag{H.4}$$

where k was defined earlier in Eq. (H.2), and depends on the optical properties of the medium under consideration through the relative dielectric function ϵ. a_{nm} and b_{nm} can be any *complex* coefficients. E_0 is a (possibly complex) electric field amplitude (arbitrary at this stage) to make these coefficients non-dimensional. Note that the sum is over two indices, n a positive integer, and m that is restrained to be $|m| \leq n$. Physically, n can be interpreted as a total angular momentum, while m is the projection of this momentum along the z-axis. This interpretation is a direct consequence of the spherical symmetry and is completely analogous to the same classification done in quantum mechanics for potentials with spherical symmetry.

The vectors $\mathbf{M}_{nm}^{(i)}(k, \mathbf{r})$ and $\mathbf{N}_{nm}^{(i)}(k, \mathbf{r})$ are called *vector spherical harmonics* (VSHs) or *multipole fields*. They physically correspond to the electric and magnetic field created by a given multipolar distribution of sources at the origin. For example, $\mathbf{N}_{1m}^{(i)}(k, \mathbf{r})$ corresponds to the electric field of an electric dipole, and $\mathbf{M}_{1m}^{(i)}(k, \mathbf{r})$ to that of a magnetic dipole. For $n > 1$, they correspond to the electric fields of higher order electric and magnetic multipoles (quadrupole, etc.). This interpretation is not necessary for the implementation of Mie theory, but is often mentioned in the literature, in expressions such as the 'electric dipole component' or 'quadrupolar component' of the field.

The VSHs are (arguably) fairly complicated functions of r, θ, and ϕ. In fact, many mathematical difficulties of Mie theory come from handling these functions. To their credit, the r-dependence and angular dependence (θ and ϕ) can be decoupled. Their angular dependence is described in terms of *spherical harmonics*, while their r-dependence involves *spherical Bessel functions*. Both groups of functions are fairly common in physics and relatively well supported in most commonly used mathematical software (like Matlab). There are, in fact, four types of VSHs and they are differentiated by the index $i = 1 \ldots 4$. They only differ in the type of the spherical Bessel function used for the r-dependence and the index to be used is determined by the nature of the electric field at the origin (singular/infinite or not) and at infinity (plane wave or outgoing/ingoing spherical waves). The details of the mathematical properties and actual calculations of the vector spherical harmonics are not necessary to understand Mie theory (only to implement it), and they are therefore left for later.

One important property is that $\mathbf{M}$ and $\mathbf{N}$ are related in a way that enables one to simply derive the expansion of the magnetic field associated with the electric field in Eq. (H.4):

$$\mathbf{H}(\mathbf{r}) = H_0 \sum_{n=0}^{\infty} \sum_{m=-n}^{m=n} b_{nm} \mathbf{M}_{nm}^{(i)}(k, \mathbf{r}) + a_{nm} \mathbf{N}_{nm}^{(i)}(k, \mathbf{r}), \tag{H.5}$$

where

$$H_0 = \frac{kE_0}{\mathrm{i}\omega\mu_0}. \tag{H.6}$$

The other important property is that the set of VSHs $\mathbf{M}^{(i)}_{nm}(k, \mathbf{r})$ and $\mathbf{N}^{(i)}_{nm}(k, \mathbf{r})$ for $n = 0 \dots \infty, |m| \leq n$ forms a complete set of *orthogonal* angular-dependent functions. The orthogonality relations will be defined more clearly later. If we consider two fields that are equal for all θ and ϕ, which in terms of their expansion correspond to a single equality of the form $\sum_{n,m} \cdots = \sum_{n,m} \cdots$, then the orthogonality of the VSHs implies that this single equality is equivalent to the equality of the expansion coefficients for each pair (n, m). A complicated single equality of two series can therefore be recast into many simple equalities on the coefficients of the expansions. This property is what makes tractable the solution of EM problems with a spherical symmetry.

H.2.3. Scattering by a sphere

Expansion of field solutions in VSHs

We can now use this expansion formalism to solve the EM problem of a single sphere excited by an incident wave with a given (known) electric field $\mathbf{E}_{\mathrm{Inc}}(\mathbf{r})$ at frequency ω (free-space wavelength $\lambda = 2\pi c/\omega$). The sphere is centered at the origin O and its radius is a. Its dielectric function is denoted by ϵ_{in} (possibly complex and frequency-dependent), and that of the surrounding medium is ϵ_M. The corresponding wave vectors, inside the sphere, and in the embedding medium, are given as:

$$k_{\mathrm{in}} = \sqrt{\epsilon_{\mathrm{in}}}\frac{\omega}{c} \quad \text{and} \quad k_M = \sqrt{\epsilon_M}\frac{\omega}{c}. \tag{H.7}$$

Note that for consistency, $\sqrt{\epsilon_{\mathrm{in}}}$ must be chosen such that $\mathrm{Im}(k_{\mathrm{in}}) \geq 0$.

In scattering problems where an *incident field* $\mathbf{E}_{\mathrm{Inc}}$ is imposed externally, it is convenient to decompose the field outside the sphere as the sum of the incident field and a *scattered field*, i.e.

$$\mathbf{E}_{\mathrm{out}} = \mathbf{E}_{\mathrm{Inc}} + \mathbf{E}_{\mathrm{Sca}}. \tag{H.8}$$

The field inside the sphere is denoted $\mathbf{E}_{\mathrm{in}}$.

In order to apply the appropriate boundary conditions, it is necessary to express the field solutions in spherical coordinates. To do so, and because the fields must also be solutions of the vector wave equation, we follow the treatment of the previous section and expand these fields in sums of vector spherical harmonics. For reasons that will be explained later, the scattered field must be expanded in terms of $\mathbf{M}_{nm}^{(3)}$ and $\mathbf{N}_{nm}^{(3)}$ and is therefore given by:

$$\boxed{\mathbf{E}_{\text{Sca}}(\mathbf{r}) = E_0 \sum_{n,m} c_{nm}\mathbf{M}_{nm}^{(3)}(k_M, \mathbf{r}) + d_{nm}\mathbf{N}_{nm}^{(3)}(k_M, \mathbf{r}),} \tag{H.9}$$

where c_{nm} and d_{nm} are *unknown* complex coefficients to be determined. The corresponding magnetic field then automatically derives from Eq. (H.5), with $H_0 = k_M E_0/(\mathrm{i}\mu_0\omega)$.

The incident field can usually be expanded (at least for the region containing the surface $r = a$ where the boundary conditions will be applied) in terms of $\mathbf{M}_{nm}^{(1)}$ and $\mathbf{N}_{nm}^{(1)}$:

$$\boxed{\mathbf{E}_{\text{Inc}}(\mathbf{r}) = E_0 \sum_{n,m} a_{nm}\mathbf{M}_{nm}^{(1)}(k_M, \mathbf{r}) + b_{nm}\mathbf{N}_{nm}^{(1)}(k_M, \mathbf{r}),} \tag{H.10}$$

where a_{nm} and b_{nm} are *known* complex coefficients, entirely determined by the form of the incident field.

A similar exercise can be carried out for the field inside the sphere. Because the interior of the sphere contains the origin, the only possible expansion is in terms of $\mathbf{M}_{nm}^{(1)}$ and $\mathbf{N}_{nm}^{(1)}$:

$$\boxed{\mathbf{E}_{\text{in}}(\mathbf{r}) = E_0 \sum_{n,m} \alpha_{nm}\mathbf{M}_{nm}^{(1)}(k_{\text{in}}, \mathbf{r}) + \beta_{nm}\mathbf{N}_{nm}^{(1)}(k_{\text{in}}, \mathbf{r}),} \tag{H.11}$$

where α_{nm} and β_{nm} are *unknown* complex coefficients to be determined. Note that the magnetic field in this case derives from Eq. (H.5), but by replacing H_0 with $H_{0,\text{in}} = k_{\text{in}} E_0/(\mathrm{i}\mu_0\omega)$.

Matching of the boundary conditions

The unknown coefficients must then be determined by imposing two independent EM boundary conditions at the sphere surface ($r = a$), for example continuity of both the tangential electric and tangential magnetic fields. This is where the whole apparatus of the vector spherical harmonics becomes crucial. Each boundary condition takes the form of a single equality of the form $\sum_{n,m} \cdots = \sum_{n,m} \cdots$, which must be satisfied for all θ and ϕ. Thanks to the orthogonality relations of the spherical harmonics, this can be recast in the form of an equality for each value of the pair (n, m). One can show that for each (n, m), the boundary conditions result in a set a four independent expressions involving the r-dependent part of the VSHs (spherical Bessel functions) at $r = a$, the four unknown coefficients c_{nm}, d_{nm}, α_{nm}, and β_{nm}, and the known coefficients a_{nm}, b_{nm}. Moreover, these four equations can in fact be split into two independent groups of *two equations with two unknowns* containing respectively α_{nm}, a_{nm}, c_{nm}, and β_{nm}, b_{nm}, d_{nm}. These two systems of two equations are *linear* in the coefficients, and solving them is therefore a straightforward matter.

Solution of the EM problem

For a given pair (n, m), the linearity of the obtained system of two equations implies that the two unknown coefficients α_{nm} and c_{nm} are simply proportional to the known coefficient a_{nm}. A similar conclusion applies for β_{nm} and d_{nm}, which are proportional to b_{nm}. Obtaining and solving these equations require looking in more detail at the structure of the VSHs, and will be dealt with later. One can show that the proportionality coefficients are in fact independent of m and we can therefore write for the scattered field:

$$\boxed{c_{nm} = \Gamma_n a_{nm} \quad \text{and} \quad d_{nm} = \Delta_n b_{nm},} \tag{H.12}$$

and for the internal field:

$$\boxed{\alpha_{nm} = A_n a_{nm} \quad \text{and} \quad \beta_{nm} = B_n b_{nm}.} \tag{H.13}$$

The exact expressions for Γ_n, Δ_n, A_n, and B_n are given later.

To summarize this latest development, the coefficients for the expansion in VSHs of the scattered and internal fields are 'simply' proportional to the coefficients of expansion of the incident wave. Therefore, Γ_n and Δ_n characterize the optical response of the sphere for the scattered field. They are sometimes called *magnetic and electric susceptibilities of the sphere*, respectively, since they refer to the magnetic and electric multipole fields. Similarly, A_n and B_n characterize the internal optical response. All these coefficients are in general frequency-dependent (especially if ϵ_{in} is frequency-dependent, as for metals). Their frequency dependence describes all aspects of the wavelength-dependent optical response of the sphere, i.e. its *optical resonances*. This will be discussed in more detail later.

These four expressions complete the solution of the EM problem. Indeed the electric and magnetic fields inside and outside the sphere are now fully determined by the knowledge of the coefficients of their expansion in VSHs.

Expansion coefficients of the incident wave

It is clear from the previous arguments that the expansion coefficients a_{nm} and b_{nm} of the incident wave are important input parameters, since all the coefficients of the field solutions are deduced from them. a_{nm} and b_{nm} can be quite complex for a general incident field, for example for Gaussian beam excitation. However, they simplify substantially for the two most common (and useful) cases of incident field: plane wave excitation and dipolar emission. In these cases, with a few assumptions that do not affect the generality of the solution thanks to symmetries, a_{nm} and b_{nm} are zero for all m except $|m| = 1$ or $m = 0$. This property is then automatically transferred to the other expansion coefficients. This means that all the double sums on (n, m) in fact become simple sums on n.

In the next section (where detailed expressions will be given) we shall keep the double sums for generality, but it is recomforting to remember that they reduce to simple sums on n for the two important cases of plane wave excitation and dipolar emission.

Summary

We can summarize the Mie theory approach to the EM problem of scattering by a sphere.

- The incident field is expanded as a sum of VSHs, i.e. a_{nm} and b_{nm} are determined.
- The sphere optical response is calculated, i.e. the coefficients Γ_n and Δ_n for the scattered field, and A_n and B_n if the internal field is sought.
- From there, the coefficients of the scattered field, c_{nm} and d_{nm} are derived, and possibly those of the internal field α_{nm} and β_{nm}.
- The full EM solution for the electric and magnetic fields is then defined analytically in terms of series of VSHs with the previously determined coefficients.

Although the solution has then been found, it remains a fairly complex expression that may be difficult to interpret physically. In order to do so, it is therefore interesting to (i) be able to implement the previous theory numerically to extract some useful information from it, and (ii) study analytically some of the most interesting properties. These form the subject of the following sections.

H.2.4. Optical resonances of the sphere

As already mentioned, the optical response of the sphere is fully determined by its susceptibilities. For the scattered field, these are Γ_n and Δ_n, and are in general complex and frequency-dependent. Resonances, i.e. a large increase in the optical response, may appear when Γ_n and Δ_n are very large, i.e. for frequencies where: $\Gamma_n^{-1}(\omega) \approx 0$ or $\Delta_n^{-1}(\omega) \approx 0$. The response would be 'infinite' if the equality were perfect, and in practice, the inverse of the susceptibilities may approach ~ 0, but not be exactly 0. When they exist, these resonances are sometimes labeled with respect to the multipole field (VSHs) they affect. For example, $\Gamma_1^{-1} \approx 0$ corresponds to a dipolar magnetic resonance of the sphere, $\Delta_1^{-1} \approx 0$ to a dipolar electric resonance, $\Delta_2^{-1} \approx 0$ to a quadrupolar electric resonance, etc.

Such resonances may appear in different contexts. For example, for a non-absorbing dielectric sphere (a latex sphere for example) of medium size (≈ 5 μm), strong and sharp resonances occur for large n, typically $n = 10$–20. These are sometimes called Mie resonances of the sphere and are associated with whispering gallery modes.

Another important case to us here is that of small (sub-wavelength) metallic spheres. In this case, there are no resonances of the magnetic type ($\Gamma_n^{-1} \approx 0$), and there exists one resonance (at one specific frequency) of the electric type ($\Delta_n^{-1} \approx 0$) for each multipolar order n. These correspond to the *localized surface plasmon (LSP) resonances* of the sphere (see Section 6.2). As before, the dipolar LSP resonance corresponds to $\Delta_1^{-1} \approx 0$, quadrupolar to $\Delta_2^{-1} \approx 0$. We see here how a basic knowledge of Mie theory becomes useful in understanding the common terminology of dipolar LSP resonance, etc. We also mention here that the resonance frequencies of the LSP resonances increase with n tending toward the limiting value defined by $\epsilon(\omega) = -\epsilon_M$. We will discuss these in more detail once we have obtained the expressions for the susceptibilities in the next section.

Finally, note that these optical resonances will appear only if they are 'excited', i.e. if the corresponding coefficient in the expansion of the incident wave is non-zero (and not too small). For example, we will show later that for a dipolar source on the z-axis and oriented along (Oz), the incident field has no magnetic component ($a_{nm} = 0$). Then no effect of the magnetic-type resonances (even if they exist for the sphere in question) will be observed for such an incident field. We also emphasize that in actual experiments the resonances will be apparent only if they are sharp (or well separated) enough to be resolved.

H.2.5. Some aspects of the practical implementation of Mie theory

We have so far explained the concepts that lead to the formal analytical solution of the scattering of an EM wave by a sphere. We now discuss how this theory may be used in practice. Because of the complexity of the functions appearing in the various expressions, most practical uses of the theory must ultimately involve a numerical estimation of these expressions. The technicalities of such an implementation will be discussed later on, and we focus here only on some general aspects.

Truncation of the series

Firstly, it is impossible to compute the summations for an arbitrary large number of terms. The summations must therefore be truncated: i.e. $n \le N$. Fortunately, for most applications, the convergence of the series is quite good, which means that relatively small N's are sufficient typically to achieve convergence (depending on the dielectric function). It remains necessary, however, to ensure that N is sufficiently large. This can be done for example by making sure that the results are the same for larger N's. Typically, for spheres with sub-wavelength size, $N \approx 30$–60 is more than sufficient, which is well within desktop computer capabilities. However, larger N's are usually necessary for larger spheres. Also, as will be discussed later, larger N's may be required for specific properties where the convergence is not as fast, such as for the study of dipolar emission in the vicinity of the sphere surface. Finally, one has to be aware of the possible numerical problems when handling large N's. The numbers involved may then be very large or very small and may lead to overflow, significant loss of precision, or cancellation problems.

Determination of the known coefficients

We have assumed so far that the coefficients of expansion of the incident wave were known. This is formally true, but in practice, they must be determined from the form of the incident field and this step is not usually straightforward. We will study these expansions for two common useful cases: plane wave excitation (PWE) and dipole emission.

Derivation of EM properties

Finally, the previous treatment has focused mainly on finding the field solutions at any given point in space. In practice, we may want to study additional properties of the EM problem, which depend on the field solutions, but may not be easily derived from it. In particular, average properties, such as average local field enhancement factor on the surface or total radiated power, can only be obtained by integrating functions of the electric field. Although the integration could be carried out numerically, it would require computing the electric field at a large number of points, a time-consuming exercise. It is therefore useful to carry out such integrations analytically. An important part of Mie theory therefore also provides analytical expressions for the most useful EM properties.

H.3. BASIC FORMULAS OF MIE THEORY

In this section, we now focus on the actual mathematical implementation of the concepts described previously. The aim is to compile all the necessary definitions, conventions, and useful expressions in a single self-consistent body. This may be used as a future reference to develop the theory further or as a guide to the Matlab implementation that can be found on the book website. We also try to provide the physical origin of the expressions when possible but do not give details of the mathematical derivations.

H.3.1. Conventions

The basics of Mie theory – in its simplest from – is very well treated by Bohren and Huffman in Ref. [149] (Chapter 4), for example. Many other books and scientific papers also deal with Mie theory [95,196] and its extensions. One annoying issue is that the conventions and notations may vary from one treatment to another, making it cumbersome to follow one approach when we are familiar with another one. The conventions chosen in Ref. [149] are probably the most physical and most adequate to the treatment of plane wave excitation (PWE) of a single sphere. However, many extensions of the theory (in particular generalized Mie theory) tend to prefer a different convention and we therefore adopt here the conventions of Gérardy and Ausloos [201]. The relation with other conventions is discussed in Ref. [196].

H.3.2. Spherical coordinates: A brief reminder

Several definitions or conventions can be used for spherical coordinates and we list here the main results for the convention we choose. A point M is represented by (r, θ, ϕ) with:

- $r \geq 0$ is the distance from origin O.
- $0 \leq \theta \leq \pi$ is the co-latitude, angle between $\mathbf{e}_z$ and $\mathbf{OM}$.
- $0 \leq \phi \leq 2\pi$ is the longitude, angle between $\mathbf{e}_x$ and the projection of $\mathbf{OM}$ on (xOy).

The spherical coordinates are therefore related to the Cartesian coordinates by:

$$\begin{cases} x &= r\sin\theta\cos\phi \\ y &= r\sin\theta\sin\phi \\ z &= r\cos\theta. \end{cases} \tag{H.14}$$

Moreover, the unit base vectors in Cartesian and spherical coordinates are related through:

$$\begin{cases} \mathbf{e}_r &= \sin\theta\cos\phi \;\; \mathbf{e}_x + \sin\theta\sin\phi \;\; \mathbf{e}_y + \cos\theta \;\; \mathbf{e}_z \\ \mathbf{e}_\theta &= \cos\theta\cos\phi \;\; \mathbf{e}_x + \cos\theta\sin\phi \;\; \mathbf{e}_y - \sin\theta \;\; \mathbf{e}_z \\ \mathbf{e}_\phi &= -\sin\phi \;\; \mathbf{e}_x + \cos\phi \;\; \mathbf{e}_y. \end{cases} \tag{H.15}$$

The inverse relations are:

$$\begin{cases} \mathbf{e}_x &= \sin\theta\cos\phi \ \mathbf{e}_r + \cos\theta\cos\phi \ \mathbf{e}_\theta - \sin\phi \ \mathbf{e}_\phi \\ \mathbf{e}_y &= \sin\theta\sin\phi \ \mathbf{e}_r + \cos\theta\sin\phi \ \mathbf{e}_\theta + \cos\phi \ \mathbf{e}_\phi \\ \mathbf{e}_z &= \cos\theta \ \mathbf{e}_r - \sin\theta \ \mathbf{e}_\theta. \end{cases} \tag{H.16}$$

H.3.3. Definition and properties of the vector spherical harmonics

Definition

The vector spherical harmonics are divergence-less solutions of the vector wave equations, i.e. solutions of Eq. (H.1). Within the conventions that we shall use here [201], the VSHs are defined (and derived) as follows. We first define *scalar* functions $\zeta_{nm}(k, \mathbf{r})$ as:

$$\zeta_{nm}(k, \mathbf{r}) = \frac{1}{\sqrt{n(n+1)}} z_n(kr) Y_{nm}(\theta, \phi), \tag{H.17}$$

where z_n is a spherical Bessel functions, and $Y_{nm}(\theta, \phi)$ are the spherical harmonics with the following standard convention (for normalization upon integration):

$$Y_{nm}(\theta, \phi) = \sqrt{\frac{2n+1}{4\pi}\frac{(n-m)!}{(n+m)!}} P_n^m(\cos(\theta)) \mathrm{e}^{\mathrm{i}m\phi}. \tag{H.18}$$

$P_n^m(\cos(\theta))$ are the associated Legendre functions, defined using the Condon–Shortley phase (this phase is not always included in other treatments), i.e.

$$P_n^m(x) = \frac{(-1)^m}{2^n n!}(1-x^2)^{m/2}\frac{\mathrm{d}^{n+m}}{\mathrm{d}x^{n+m}}(x^2-1)^n \quad (m \geq 0), \tag{H.19}$$

and

$$P_n^{(-m)} = (-1)^m \frac{(n-m)!}{(n+m)!} P_n^m. \tag{H.20}$$

We therefore have the following property:

$$Y_{n,-m}(\theta, \phi) = (-1)^m Y_{n,m}^*(\theta, \phi) = (-1)^m Y_{n,m}(\theta, -\phi). \tag{H.21}$$

From there, $\mathbf{M}_{nm}(k, \mathbf{r})$ and $\mathbf{N}_{nm}(k, \mathbf{r})$ are defined as

$$\mathbf{M}_{nm}(k, \mathbf{r}) = \nabla \times (\zeta_{nm}(k, \mathbf{r})\,\mathbf{r}), \tag{H.22}$$

and

$$\mathbf{N}_{nm}(k, \mathbf{r}) = \frac{1}{k}\boldsymbol{\nabla} \times \mathbf{M}_{nm}(k, \mathbf{r}). \tag{H.23}$$

Finally, four types of vector spherical harmonics can in principle be defined, depending on which spherical Bessel function is used for the r-dependence: the spherical Bessel function of the first kind (denoted j or $z^{(1)}$), of the second kind (y or $z^{(2)}$), or the spherical Hankel function of the first kind ($h^{(1)} = j + \mathrm{i}y$ or $z^{(3)}$) or of the second kind ($h^{(2)} = j - \mathrm{i}y$ or $z^{(4)}$). Only j is regular at 0, and the corresponding vector spherical harmonics $\mathbf{M}_{nm}^{(1)}$ and $\mathbf{N}_{nm}^{(1)}$ must be chosen to expand a field that is defined and finite at the origin. Moreover, it can be shown that $h^{(1)}$ corresponds to an outgoing spherical wave (radiation field) at infinity, and the corresponding vector spherical harmonics $\mathbf{M}_{nm}^{(3)}$ and $\mathbf{N}_{nm}^{(3)}$ must be chosen to expand the scattered (radiated) field at infinity. Finally for regions that are neither at infinity nor containing the origin, a superposition of both $\mathbf{M}_{nm}^{(1)}$, $\mathbf{N}_{nm}^{(1)}$, and $\mathbf{M}_{nm}^{(3)}$, $\mathbf{N}_{nm}^{(3)}$ is the most general solution. By convention, z_n with no superscript denotes any spherical Bessel function and the same convention applies to $\mathbf{M}_{nm}$ and $\mathbf{N}_{nm}$.

Some important properties

There are a few simple and useful properties of vector spherical harmonics. First they are related by:

$$k\mathbf{N}_{nm}(k, \mathbf{r}) = \boldsymbol{\nabla} \times \mathbf{M}_{nm}(k, \mathbf{r}), \tag{H.24}$$

and

$$k\mathbf{M}_{nm}(k, \mathbf{r}) = \boldsymbol{\nabla} \times \mathbf{N}_{nm}(k, \mathbf{r}). \tag{H.25}$$

This enables one to derive simply the magnetic field expansion in Eq. (H.5) from the electric field expansion in Eq. (H.4).

Secondly, they form a *complete orthogonal basis* for the divergence-less solutions of the wave vector equation (and therefore of Maxwell's equations in the absence of sources). The orthogonality is related to that of spherical harmonics and concerns the angular dependence only. More specifically, we have:

$$\int\int \mathbf{M}^{(i)}_{nm}(k,r,\Omega)\cdot\mathbf{N}^{(j)*}_{n'm'}(k,r,\Omega)\mathrm{d}\Omega = 0 \quad \forall(m,m',n,n';(i,j)=1\ldots 4), \tag{H.26}$$

where $*$ denotes the complex conjugate, $\Omega = (\theta,\phi)$, $\mathrm{d}\Omega = \sin\theta\mathrm{d}\theta\mathrm{d}\phi$ and $\int\int = \int_{\theta=0}^{\theta=\pi}\int_{\phi=0}^{\phi=2\pi}$. There are equivalent relations for all pairs $(\mathbf{M}_{nm}, \mathbf{M}_{n'm'})$ and $(\mathbf{N}_{nm}, \mathbf{N}_{n'm'})$, but these are not valid when $n = n'$ and $m = m'$ (a vector spherical harmonic is not orthogonal to itself). We then have:

$$\int\int \mathbf{M}^{(i)}_{nm}(k,r,\Omega)\cdot\mathbf{M}^{(j)*}_{n'm'}(k,r,\Omega)\mathrm{d}\Omega = \delta_{n,n'}\delta_{m,m'}z_n^{(i)}(kr)\left[z_n^{(j)}(kr)\right]^*, \tag{H.27}$$

and

$$\begin{aligned}\int\int \mathbf{N}^{(i)}_{nm}(k,r,\Omega)\cdot\mathbf{N}^{(j)*}_{n'm'}(k,r,\Omega)\mathrm{d}\Omega &= \delta_{n,n'}\delta_{m,m'} \\ \times \Bigg[n(n+1)\frac{z_n^{(i)}(kr)[z_n^{(j)}(kr)]^*}{|kr|^2} \\ + \frac{[z_n^{(i)}(kr)+krz_n'^{(i)}(kr)][z_n^{(j)}(kr)+krz_n'^{(j)}(kr)]^*}{|kr|^2}\Bigg].\end{aligned} \tag{H.28}$$

These orthogonality relations are important, and for example useful for determining the coefficients of the expansion of a given electric field in a sum of vector spherical harmonics.

Other useful expressions

It follows directly from standard properties of spherical harmonics that if k is real, then:

$$\mathbf{M}^{(1)}_{n,-m}(k,\mathbf{r}) = (-1)^m\left(\mathbf{M}^{(1)}_{n,m}(k,\mathbf{r})\right)^* \tag{H.29}$$

as well as the equivalent expression for $\mathbf{N}^{(1)}$. Note that this expression is valid only when k is real *and* the spherical Bessel function is j, since it requires it to be real.

Moreover, the following additional expressions may be useful when manipulating VSHs:

$$\iint \left(\mathbf{M}^{(i)}_{nm}(k,r,\Omega) \times \mathbf{M}^{(j)*}_{n'm'}(k,r,\Omega) \right) \cdot \mathbf{e}_r \, \mathrm{d}\Omega = 0, \tag{H.30}$$

$$\iint \left(\mathbf{N}^{(i)}_{nm}(k,r,\Omega) \times \mathbf{N}^{(j)*}_{n'm'}(k,r,\Omega) \right) \cdot \mathbf{e}_r \, \mathrm{d}\Omega = 0, \tag{H.31}$$

and

$$\iint \left(\mathbf{M}^{(i)}_{nm}(k,r,\Omega) \times \mathbf{N}^{(j)*}_{n'm'}(k,r,\Omega) \right) \cdot \mathbf{e}_r \, \mathrm{d}\Omega$$
$$= \delta_{n,n'} \delta_{m,m'} \frac{z_n^{(i)}(kr)}{kr} \left[z_n^{(j)}(kr) + kr z_n'^{(j)}(kr) \right]^* . \tag{H.32}$$

VSHs components

From the previous definitions, it is possible to write down expressions for the components (coordinates) of the VSHs. These expressions are separable with respect to the three spherical coordinates r, θ, and ϕ:

$$\begin{cases} \mathbf{M}_{nm}(k,\mathbf{r}) \cdot \mathbf{e}_r &= 0 \\ \mathbf{M}_{nm}(k,\mathbf{r}) \cdot \mathbf{e}_\theta &= \mathrm{i} Z_n^0(kr) T^1_{nm}(\theta) \mathrm{e}^{\mathrm{i}m\phi} \\ \mathbf{M}_{nm}(k,\mathbf{r}) \cdot \mathbf{e}_\phi &= -Z_n^0(kr) T^3_{nm}(\theta) \mathrm{e}^{\mathrm{i}m\phi} \end{cases} \tag{H.33}$$

$$\begin{cases} \mathbf{N}_{nm}(k,\mathbf{r}) \cdot \mathbf{e}_r &= Z_n^1(kr) T^2_{nm}(\theta) \mathrm{e}^{\mathrm{i}m\phi} \\ \mathbf{N}_{nm}(k,\mathbf{r}) \cdot \mathbf{e}_\theta &= Z_n^2(kr) T^3_{nm}(\theta) \mathrm{e}^{\mathrm{i}m\phi} \\ \mathbf{N}_{nm}(k,\mathbf{r}) \cdot \mathbf{e}_\phi &= \mathrm{i} Z_n^2(kr) T^1_{nm}(\theta) \mathrm{e}^{\mathrm{i}m\phi} \end{cases} \tag{H.34}$$

where we have defined the r-dependence auxiliary functions:

$$\begin{aligned} Z_n^0(x) &= z_n(x), \\ Z_n^1(x) &= z_n(x)/x, \\ Z_n^2(x) &= [x z_n(x)]'/x, \end{aligned} \tag{H.35}$$

and the θ-dependent auxiliary functions:

$$\begin{aligned} T^1_{nm}(\theta) &= m \frac{T_{nm}(\theta)}{\sin\theta}, \\ T^2_{nm}(\theta) &= n(n+1) T_{nm}(\theta), \\ T^3_{nm}(\theta) &= \frac{\partial T_{nm}}{\partial \theta}(\theta), \end{aligned} \tag{H.36}$$

where

$$T_{nm}(\theta) = \frac{1}{\sqrt{n(n+1)}} Y_{nm}(\theta, \phi = 0). \tag{H.37}$$

Note that the T functions are *real* and have the following properties:

$$\begin{aligned}
T_{n,-m} &= (-1)^m \, T_{n,m}, \\
T^1_{n,-m} &= (-1)^{m+1} \, T^1_{n,m}, \\
T^2_{n,-m} &= (-1)^m \, T^2_{n,m}, \\
T^3_{n,-m} &= (-1)^m \, T^3_{n,m}.
\end{aligned} \tag{H.38}$$

Asymptotic forms of VSHs

The asymptotic behavior of spherical Bessel functions is known and can therefore be used to find the asymptotic behavior of VSHs.

At infinity ($r \to \infty$), the relevant VSHs are $\mathbf{M}^{(3)}$ and $\mathbf{N}^{(3)}$, and their asymptotic expressions are:

$$\mathbf{M}^{(3)}_{nm}(k, \mathbf{r}) \approx (-\mathrm{i})^{n+1} \frac{\mathrm{e}^{\mathrm{i}kr}}{kr} \mathrm{e}^{\mathrm{i}m\phi} \left[\mathrm{i}T^1_{nm}(\theta)\mathbf{e}_\theta - T^3_{nm}(\theta)\mathbf{e}_\phi \right], \tag{H.39}$$

$$\mathbf{N}^{(3)}_{nm}(k, \mathbf{r}) \approx (-\mathrm{i})^{n} \frac{\mathrm{e}^{\mathrm{i}kr}}{kr} \mathrm{e}^{\mathrm{i}m\phi} \left[T^3_{nm}(\theta)\mathbf{e}_\theta + \mathrm{i}T^1_{nm}(\theta)\mathbf{e}_\phi \right]. \tag{H.40}$$

They represent radiation electromagnetic fields, as expected, with an $\mathrm{e}^{\mathrm{i}kr}/kr$ dependence and no component along $\mathbf{e}_r$ (i.e. transverse field).

Physical interpretation of VSHs

The VSHs can be interpreted as multipole fields. For a detailed discussion of this aspect, see for example Ref. [96], Chapter 9. We will only illustrate here this important aspect on a few simple examples. Let us for example consider $\mathbf{M}^{(3)}_{1,0}$ and $\mathbf{N}^{(3)}_{1,0}$. One can show that they take the form:

$$\mathbf{M}^{(3)}_{1,0}(k, \mathbf{r}) = -\frac{\mathrm{e}^{\mathrm{i}kr}}{kr} \left[1 + \frac{\mathrm{i}}{kr} \right] \sqrt{\frac{3}{8\pi}} \sin\theta \mathbf{e}_\phi, \tag{H.41}$$

and

$$\mathbf{N}^{(3)}_{1,0}(k, \mathbf{r}) = \mathrm{i} \frac{\mathrm{e}^{\mathrm{i}kr}}{kr} \left\{ \left[\frac{\mathrm{i}}{kr} - \frac{1}{(kr)^2} \right] \sqrt{\frac{3}{2\pi}} \cos\theta \mathbf{e}_r \right.$$

$$+\left[1+\frac{\mathrm{i}}{kr}-\frac{1}{(kr)^2}\right]\sqrt{\frac{3}{8\pi}}\sin\theta\mathbf{e}_\theta\Bigg\}. \tag{H.42}$$

One may recognize in the latter expression the electric field created by an electric dipole at the origin and oriented along (Oz): $\mathbf{p} = p\mathbf{e}_z$. More precisely, the electric field of such an electric dipole is:

$$\mathbf{E}(\mathbf{r}) = \frac{\mathrm{i}k^3 p}{\sqrt{6\pi}\epsilon_0\epsilon_M}\mathbf{N}^{(3)}_{1,0}(k,\mathbf{r}). \tag{H.43}$$

It results immediately that $\mathbf{M}^{(3)}_{1,0}$ corresponds to the magnetic field created by the same electric dipole.

Another view is to recognize that $\mathbf{M}^{(3)}_{1,0}$ also represents the electric field created by a magnetic dipole at the origin and oriented along (Oz): $\boldsymbol{\mu} = \mu\mathbf{e}_z$. More precisely, the electric field of such a magnetic dipole is:

$$\mathbf{E}(\mathbf{r}) = \frac{-k^3\mu}{\sqrt{6\pi}\epsilon_0\epsilon_M c}\mathbf{M}^{(3)}_{1,0}(k,\mathbf{r}). \tag{H.44}$$

These considerations can be generalized to other m values: $\mathbf{M}^{(3)}_{1,\pm1}$ (and $\mathbf{N}^{(3)}_{1,\pm1}$) represent the electric field of a magnetic (electric) dipole perpendicular to the z-axis or the magnetic field of a similar electric (magnetic) dipole.

The generalization is possible also for $n \geq 2$. The only change is the order of the multipole, which is given by n, i.e. $\mathbf{N}^{(3)}_{2,m}$ for $m = -2..2$ represent the electric fields of the 5 independent quadrupoles that can be formed at the origin, etc. In fact these VSHs could be considered as a formal definition of radiative multipolar sources in electromagnetic theory [96].

H.3.4. Expressions for the susceptibilities

As discussed previously, the susceptibilities are proportionality factors that express the (unknown) expansion coefficients of the scattered and internal fields in terms of those (known) of the incident field; see Eqs (H.12)–(H.13). They arise from the matching of the boundary conditions at the spherical interface. We skip the technical details here and only give the final result. First, we define the adimensional quantities:

$$\boxed{x = k_M a = 2\pi\sqrt{\epsilon_M}\frac{a}{\lambda} \quad \text{and} \quad s = \frac{k_{\text{in}}}{k_M} = \frac{\sqrt{\epsilon_{\text{in}}}}{\sqrt{\epsilon_M}}.} \tag{H.45}$$

x and sx then appear as the arguments of the Bessel functions (r-dependence at $r = a$). Note that both depend on frequency ω (or wavelength λ).

Scattered field susceptibilities

We then have for the magnetic susceptibility related to the scattered field:

$$\boxed{\Gamma_n = \frac{s\psi_n(x)\psi_n'(sx) - \psi_n(sx)\psi_n'(x)}{\psi_n(sx)\xi_n'(x) - s\xi_n(x)\psi_n'(sx)},} \tag{H.46}$$

and for the corresponding electric susceptibility:

$$\boxed{\Delta_n = \frac{\psi_n(x)\psi_n'(sx) - s\psi_n(sx)\psi_n'(x)}{s\psi_n(sx)\xi_n'(x) - \xi_n(x)\psi_n'(sx)}.} \tag{H.47}$$

These two expressions are probably the most important and most used of Mie theory. We have used for convenience the Riccati–Bessel functions defined as:

$$\boxed{\psi_n(\rho) = \rho j_n(\rho) \qquad \xi_n(\rho) = \rho h_n^{(1)}(\rho).} \tag{H.48}$$

These functions are in many cases more practical to use than the spherical Bessel functions in the context of Mie theory.

Internal field susceptibilities

Similar expressions can be obtained for the magnetic and electric susceptibilities related to the internal field:

$$A_n = \frac{s\psi_n(x)\xi_n'(x) - s\xi_n(x)\psi_n'(x)}{\psi_n(sx)\xi_n'(x) - s\xi_n(x)\psi_n'(sx)}, \tag{H.49}$$

and

$$B_n = \frac{s\psi_n(x)\xi_n'(x) - s\xi_n(x)\psi_n'(x)}{s\psi_n(sx)\xi_n'(x) - \xi_n(x)\psi_n'(sx)}. \tag{H.50}$$

In fact, one can show using the Wronskian of the Riccati–Bessel functions, that the numerator of these two expressions simply reduces to is, which results in the simpler expressions:

$$A_n = \frac{\mathrm{i}s}{\psi_n(sx)\xi_n'(x) - s\xi_n(x)\psi_n'(sx)}, \tag{H.51}$$

and

$$B_n = \frac{\mathrm{i}s}{s\psi_n(sx)\xi_n'(x) - \xi_n(x)\psi_n'(sx)}. \tag{H.52}$$

As already mentioned, the susceptibilities of the sphere only depend on n and not on m. Their formal expression is complicated but they can be evaluated numerically.

Useful expansions for small spheres

Alternatively, approximate expressions may be used, for example in the case of a small sphere (i.e. small x). The following expansions of the lowest order susceptibilities for small sphere may be useful in this context. Keeping terms up to order x^6, we have:

$$\Delta_1 \approx \frac{2\mathrm{i}}{3}x^3\frac{s^2-1}{s^2+2}\left[1 - \frac{3}{5}x^2\frac{s^2-2}{s^2+2} - \frac{2}{3}\mathrm{i}x^3\frac{s^2-1}{s^2+2}\right]^{-1}, \tag{H.53}$$

and up to order x^8:

$$\Gamma_1 \approx \frac{\mathrm{i}}{45}x^5(s^2-1)\left[1 - \frac{1}{21}x^2(2s^2-5)\right]^{-1}, \tag{H.54}$$

$$\Delta_2 \approx \frac{-\mathrm{i}}{30}x^5\frac{s^2-1}{s^2+3/2}\left[1 + \frac{5}{14}x^2\frac{1}{s^2+3/2}\right]^{-1}. \tag{H.55}$$

H.3.5. More on optical resonances

As discussed previously, optical resonances (i.e. large optical response) occur at frequencies where the inverse of a susceptibility approaches ~0. In the previous expressions, this corresponds to when the denominator becomes ~0. The resonance condition for the magnetic susceptibilities Γ_n and A_n is the same and reduces to:

$$\frac{\xi_n'(x)}{\xi_n(x)} \approx s\,\frac{\psi_n'(sx)}{\psi_n(sx)}, \tag{H.56}$$

while for the electric susceptibilities Δ_n and B_n, we obtain:

$$\frac{\xi_n'(x)}{\xi_n(x)} \approx \frac{1}{s} \frac{\psi_n'(sx)}{\psi_n(sx)}. \tag{H.57}$$

These resonances have been discussed in various works (for example Ref. [95]) and they correspond to the so-called normal modes of the sphere. These equations may be fulfilled exactly, but only for *complex* frequencies and the corresponding solution are then *virtual modes* (see the discussion in Chapter 3 on 'quasi-particles'). Alternatively, approximate solutions can be sought for *real* frequencies. These are then the resonance frequencies, and they appear as increased optical response in various EM properties.

For metals (with a Drude-like optical response), it can be shown that in the limit of vanishing x (small spheres), there is no solution of the first type (magnetic), and there is one resonance frequency of the electric type for each multipole order n. These are the *localized surface plasmon resonances* of the sphere.

Let us for example have a closer look at the special case of the dipolar electric resonance $n = 1$: For small x, we have:

$$\Delta_1 = \mathrm{i}\frac{2}{3}x^3\frac{s^2-1}{s^2+2} + O(x^5) = \mathrm{i}\frac{2}{3}k_M^3 a^3 \frac{\epsilon_{\text{in}} - \epsilon_M}{\epsilon_{\text{in}} + 2\epsilon_M} + O(x^5). \tag{H.58}$$

The resonance condition in this limit is therefore the usual condition $\epsilon_{\text{in}} = -2\epsilon_M$ for the dipolar LSP resonance in the electrostatic approximation (see Section 6.2). Outside the range of this approximation, we usually have to resort to numerical calculations of Δ_1.

We now focus in the rest of this section on some important 'average' EM properties of the problem. Since we can in principle compute numerically the field at any point in space, average properties could be derived by suitable integrations. Nonetheless, this would be a very cumbersome and time-consuming approach and it is, therefore, important to continue the analytical approach as far as possible. In fact, a big advantage of Mie theory is that it can still produce analytical expressions for cases of interest, with a controllable degree of approximation from an otherwise *exact* solution.

H.3.6. Absorption, scattering, and extinction for an incident beam

The EM characteristics that are most often considered in scattering problems are the absorption, scattering, and extinction cross-sections or coefficients. These concern mostly the 'classical scattering configuration', where the incident wave is a beam coming from infinity. We therefore focus on this case here, and will treat separately the case of excitation by a localized source in the next subsection.

Definitions

These basic concepts of general scattering theory (see, for example, Section 3.4 in Ref. [149] for details) can be defined in terms of the following quantities:

- Scattered power: P_{Sca} [W] is the power radiated by the scattered electromagnetic field ($\mathbf{E}_{\mathrm{Sca}}$).
- Absorbed power: P_{Abs} [W] is the power absorbed in the dielectric/metallic objects during the scattering process.
- Extinguished power: $P_{\mathrm{Ext}} = P_{\mathrm{Sca}} + P_{\mathrm{Abs}}$ [W] corresponds to the power extracted from the incident wave.

The following expressions can be used to calculate these quantities:

$$P_{\mathrm{Sca}} = \int\int_S \mathrm{Re}\left(\frac{1}{2}(\mathbf{E}_{\mathrm{Sca}} \times \mathbf{H}^*_{\mathrm{Sca}}) \cdot \mathbf{n}\right) r^2 \mathrm{d}S, \tag{H.59}$$

where S is a surface enclosing all dielectric/metallic objects, and $\mathbf{n}$ is the unit vector normal to the surface (and pointing outwards). Similarly, we have:

$$P_{\mathrm{Abs}} = -\int\int_S \mathrm{Re}\left(\frac{1}{2}(\mathbf{E} \times \mathbf{H}^*) \cdot \mathbf{n}\right) r^2 \mathrm{d}S. \tag{H.60}$$

Using $\mathbf{E}_{\mathrm{out}} = \mathbf{E}_{\mathrm{Inc}} + \mathbf{E}_{\mathrm{Sca}}$ and $P_{\mathrm{Ext}} = P_{\mathrm{Abs}} + P_{\mathrm{Sca}}$, one can show that:

$$P_{\mathrm{Ext}} = -\int\int_S \mathrm{Re}\left(\frac{1}{2}(\mathbf{E}_{\mathrm{Inc}} \times \mathbf{H}^*_{\mathrm{Sca}} + \mathbf{E}_{\mathrm{Sca}} \times \mathbf{H}^*_{\mathrm{Inc}}) \cdot \mathbf{n}\right) r^2 \mathrm{d}S. \tag{H.61}$$

Cross-sections [m^2] are then defined with respect to the incident field intensity (power density) S_{Inc} [$\mathrm{W/m}^2$] as:

$$\sigma_{\mathrm{Sca}} = P_{\mathrm{Sca}}/S_{\mathrm{Inc}}, \tag{H.62}$$

with similar definitions for σ_{Abs} and σ_{Ext}.

Finally, for a simple object like a sphere, the single-particle scattering coefficient [a.d.] can be defined by normalizing with the geometric cross-section σ_{geom} [m^2]:

$$Q^{\mathrm{NP}}_{\mathrm{Sca}} = \sigma_{\mathrm{Sca}}/\sigma_{\mathrm{geom}}. \tag{H.63}$$

For a sphere of radius a, for example, $\sigma_{\mathrm{geom}} = \pi a^2$.

Expressions within Mie theory

These expressions can now be evaluated within the framework of Mie theory, i.e. in terms of the expansion coefficients. For this, S is chosen to be a sphere containing all the objects (the single sphere here but it can be generalized to multiple spheres). One must use the expressions for integrals of the VSHs given earlier along with the following property of the Riccati-Bessel functions: $\mathrm{Re}(\mathrm{i}\xi_n(x)\xi_n'(x)^*) = \mathrm{Re}(-\mathrm{i}\xi_n(x)^*\xi_n'(x)) = 1$. Within our conventions, the incident intensity of the plane wave is $S_{\mathrm{Inc}} = |E_0 H_0|/2 = \epsilon_0 c\sqrt{\epsilon_M}|E_0|^2/2$. We also use the fact that $S_0 = (E_0 H_0^*)/2 = \mathrm{i}S_{\mathrm{Inc}}$. The *scattering cross-section* (derived from the calculation of the scattered power) then takes the form:

$$\begin{aligned}\sigma_{\mathrm{Sca}} &= \frac{1}{k_M^2}\sum_{n,m}\left[|c_{nm}|^2 + |d_{nm}|^2\right] \\ &= \frac{1}{k_M^2}\sum_{n,m}\left[|a_{nm}|^2|\Gamma_n|^2 + |b_{nm}|^2|\Delta_n^2|\right]. \end{aligned} \tag{H.64}$$

The extinguished power can be expressed in a similar fashion in terms of expanded incident and scattered fields. After some manipulation, one obtains the *extinction cross-section* for a general incident wave as:

$$\begin{aligned}\sigma_{\mathrm{Ext}} &= \frac{-1}{k_M^2}\sum_{n,m}\left[\mathrm{Re}(a_{nm}c_{nm}^*) + \mathrm{Re}(b_{nm}d_{nm}^*)\right] \\ &= \frac{-1}{k_M^2}\sum_{n,m}\left[|a_{nm}|^2\mathrm{Re}(\Gamma_n) + |b_{nm}|^2\mathrm{Re}(\Delta_n)\right]. \end{aligned} \tag{H.65}$$

The absorbed power can then be inferred from energy conservation, $P_{\mathrm{Abs}} = P_{\mathrm{Ext}} - P_{\mathrm{Sca}}$, resulting in the *absorption cross-section*:

$$\sigma_{\mathrm{Abs}} = \frac{-1}{k_M^2}\sum_{n,m}\left[|a_{nm}|^2\left(|\Gamma_n|^2 + \mathrm{Re}(\Gamma_n)\right) + |b_{nm}|^2\left(|\Delta_n|^2 + \mathrm{Re}(\Delta_n)\right)\right]. \tag{H.66}$$

It could also be derived analytically (but not without effort) from the internal field VSHs expansions. It is also interesting to consider as a consistency check the case where the sphere is non-absorbing, i.e. ϵ_{in} is real positive and therefore s is real. One can then show that $\mathrm{Re}(\Gamma_n^{-1}) = -1$ and, hence, $\mathrm{Re}(\Gamma_n) = -|\Gamma_n|^2$, with the same expressions for Δ_n. This implies that $\sigma_{\mathrm{Abs}} = 0$, as expected from physical arguments.

H.3.7. Absorption and radiation for a localized source

For a localized source, the approach differs from that of a beam, because the source creates a radiation field, even in the absence of any objects. This means that the expansion of the incident field in terms of $\mathbf{M}^{(1)}$ and $\mathbf{N}^{(1)}$ given in Eq. (H.10) cannot be valid in the far field. It remains valid in the vicinity of the sphere surface (and the previous treatment therefore remains correct), but for far-field properties, one must in addition carry out the expansion of the incident field in terms of $\mathbf{M}^{(3)}$ and $\mathbf{N}^{(3)}$, i.e.:

$$\mathbf{E}_{\text{Inc-Far}}(\mathbf{r}) = E_0 \sum_{n,m} e_{nm}\mathbf{M}_{nm}^{(3)}(k_M, \mathbf{r}) + f_{nm}\mathbf{N}_{nm}^{(3)}(k_M, \mathbf{r}), \tag{H.67}$$

where e_{nm} and f_{nm} are *known* complex coefficients, which derive from the nature of the localized source.

The radiation field in the presence of objects is then the sum of the incident field (from the source) and the scattered field, and its far-field intensity is a result of the interference between these two fields.

Definitions

P_{Abs}, P_{Ext}, and P_{Sca} can be defined as for an incident beam, provided that the volume delimited by the surface S does *not* contain the localized source. The expressions from the previous subsection therefore remain valid, but only P_{Abs} retains is physical meaning. There are however other quantities of interest here, namely:

- the power radiated by the source in the absence of object: P_0
- the power radiated in the scattering problem (with the objects): P_{Rad}
- the total power extracted from the source (with the objects): $P_{\text{Tot}} = P_{\text{Abs}} + P_{\text{Rad}}$.

Note that for a localized source, P_{Abs} is also denoted P_{NR}, since it corresponds to the power emitted non-radiatively.

Expressions within Mie theory

One can compute integrals similar to that for P_{Sca} and deduce:

$$P_0 = \frac{|E_0 H_0|}{2} \frac{1}{k_M^2} \sum_{n,m} \left[|e_{nm}|^2 + |f_{nm}|^2 \right]. \tag{H.68}$$

Note however, that this expression is not useful in general since P_0 is usually known (for example, power radiated by a dipole). More important is the corresponding expression for the power radiated in the presence of the scatterer (sphere):

$$P_{\mathrm{Rad}} = \frac{|E_0 H_0|}{2} \frac{1}{k_M^2} \sum_{n,m} \left[|c_{nm} + e_{nm}|^2 + |d_{nm} + f_{nm}|^2 \right]. \tag{H.69}$$

Moreover, P_{Tot} can be derived from $P_{\mathrm{Tot}} = P_{\mathrm{Abs}} + P_{\mathrm{Rad}}$, where $P_{\mathrm{Abs}} = P_{\mathrm{NR}}$ is obtained from Eq. (H.66).

H.3.8. Far-field radiation profile

The scattered field takes the form of a *radiation field* far away from the object and decays as $\mathrm{e}^{\mathrm{i}k_M r}/r$. The radial dependence of the VSHs can therefore be removed by taking their asymptotic value, and one can show that the *radiation profile* can then be estimated from:

$$\begin{aligned}\frac{\mathrm{d}P}{\mathrm{d}\Omega}(\theta,\phi) &= \frac{\sqrt{\epsilon_M}\epsilon_0 c|E_0|^2}{2} \\ &\times \frac{1}{k_M^2}\left[\left|\sum_{n,m}\frac{(-\mathrm{i})^n}{\sqrt{n(n+1)}}\left((d_{nm}+f_{nm})\frac{\partial Y_{nm}}{\partial\theta} + (c_{nm}+e_{nm})\frac{mY_{nm}}{\sin\theta}\right)\right|^2\right. \\ &+ \left.\left|\sum_{n,m}\frac{(-\mathrm{i})^n}{\sqrt{n(n+1)}}\left((c_{nm}+e_{nm})\frac{\partial Y_{nm}}{\partial\theta} + (d_{nm}+f_{nm})\frac{mY_{nm}}{\sin\theta}\right)\right|^2\right].\end{aligned} \tag{H.70}$$

This expression can be used to compute numerically the radiation profile. It is valid for excitation by a localized source or an incident beam, the coefficients e_{nm} and f_{nm} being zero in the latter case.

H.3.9. The local field at the surface

The local field can be calculated at any point outside the sphere from:

$$\begin{aligned}\mathbf{E}_{\mathrm{out}}(\mathbf{r}) &= \mathbf{E}_{\mathrm{Inc}}(\mathbf{r}) + \mathbf{E}_{\mathrm{Sca}}(\mathbf{r}) \\ &= E_0 \sum_{n,m} a_{nm}\mathbf{M}_{nm}^{(1)}(k_M,\mathbf{r}) + b_{nm}\mathbf{N}_{nm}^{(1)}(k_M,\mathbf{r}) \\ &\quad + c_{nm}\mathbf{M}_{nm}^{(3)}(k_M,\mathbf{r}) + d_{nm}\mathbf{N}_{nm}^{(3)}(k_M,\mathbf{r}).\end{aligned} \tag{H.71}$$

This expression can be computed numerically at any point of interest. From it, important properties can be deduced, such as $M_{\mathrm{Loc}}(\mathbf{r})$ and $F_{E4}^0(\mathbf{r})$ (of relevance to SERS).

However, it can be cumbersome or even impractical to compute it at a large number of points. In particular for average properties at the surface, it may be faster to carry out the averaging analytically rather than numerically. This is possible for example for $\langle M_{\mathrm{Loc}} \rangle$ (the average local field intensity enhancement factor at the surface of the sphere). We restrict ourselves here to the (most important) case of excitation by an incident beam and assume that the parameter E_0 in Mie theory is the incident field amplitude. The result is:

$$\langle M_{\mathrm{Loc}} \rangle = \frac{1}{4\pi x^4} \sum_{n,m} \Big[|a_{nm}|^2 x^2 |\psi_n(x) + \Gamma_n \xi_n(x)|^2 + \\ + |b_{nm}|^2 x^2 |\psi_n'(x) + \Delta_n \xi_n'(x)|^2 \\ + |b_{nm}|^2 n(n+1) |\psi_n(x) + \Delta_n \xi_n(x)|^2 \Big] . \quad \text{(H.72)}$$

H.4. PLANE WAVE EXCITATION OF A SPHERE: THE 'ORIGINAL MIE THEORY'

The general expressions obtained in the previous two sections can now be rewritten (and simplified) for the particular case of plane wave excitation (PWE) of a sphere. This case is often considered as *being* Mie theory, while any further developments are considered as extensions of Mie theory. This is, indeed, the case originally solved by Mie [158].

H.4.1. Expansion of a plane wave in vector spherical harmonics

To complete the solution of the EM problem in practice, it remains to find the expansion in VSHs of the incident wave, i.e. find a_{nm} and b_{nm}. In standard Mie theory, this is usually done for a plane wave, but can in principle be carried out for any type of excitation solution of Maxwell's equations (for example, for a Gaussian-like beam).

We leave aside one more time the technical aspects of the derivation and only give the final result. In the context of Mie theory, it is customary (and easier!) to choose a plane wave propagating along $\mathbf{e}_z$ and polarized along $\mathbf{e}_x$ (all cases are equivalent to this one by spherical symmetry). The incident field is then:

$$\mathbf{E}_{\mathrm{Inc}}(\mathbf{r}) = e^{ik_M z} E_{\mathrm{Inc}} \mathbf{e}_x, \quad \text{(H.73)}$$

where E_{Inc} denotes its amplitude.

It is therefore natural to take $E_0 = E_{\mathrm{Inc}}$ as a reference amplitude for our expansions. The power of the incident wave is then $P_{\mathrm{Inc}} = |S_0| = |E_0 H_0|/2 = \sqrt{\epsilon_M} \epsilon_0 c |E_{\mathrm{Inc}}|^2/2$.

The expansion of such a field in VSHs is then given by Eq. (H.10) with coefficients:

$$\begin{cases} a_{nm} &= b_{nm} = 0 \quad \forall |m| \neq 1 \\ a_{n1} &= K_n \\ a_{n,-1} &= K_n \\ b_{n1} &= K_n \\ b_{n,-1} &= -K_n, \end{cases} \tag{H.74}$$

with

$$K_n = \mathrm{i}^{n+1}\sqrt{\pi(2n+1)}. \tag{H.75}$$

Only indices with $|m| = 1$ appear in the expansion, which implies that for PWE only these indices will appear in the expansion of the scattered and internal field. This, together with the fact that $|a_{n,1}|^2 = |a_{n,-1}|^2 = |b_{n,1}|^2 = |b_{n,-1}|^2 = |K_n|^2 = \pi(2n+1)$, simplifies considerably the general expressions obtained in the previous section, as we show now.

H.4.2. Extinction, scattering, and absorption for plane wave excitation

In the context of PWE of a single sphere, the extinction, scattering, and absorption cross-sections are often characterized in terms of the corresponding (adimensional) coefficients defined in Eq. (H.63). From Eq. (H.64), we obtain the scattering coefficient:

$$\boxed{Q_{\mathrm{Sca}}^{\mathrm{NP}} = \sigma_{\mathrm{Sca}}/(\pi a^2) = \frac{2}{x^2}\sum_{n=1}^{\infty}(2n+1)\left(|\Gamma_n|^2 + |\Delta_n|^2\right).} \tag{H.76}$$

From Eq. (H.65), we obtain the extinction coefficient:

$$\boxed{Q_{\mathrm{Ext}}^{\mathrm{NP}} = \sigma_{\mathrm{Ext}}/(\pi a^2) = \frac{-2}{x^2}\sum_{n=1}^{\infty}(2n+1)\left[\mathrm{Re}(\Gamma_n) + \mathrm{Re}(\Delta_n)\right].} \tag{H.77}$$

The absorption coefficient then results simply from:

$$\boxed{Q_{\mathrm{Abs}}^{\mathrm{NP}} = Q_{\mathrm{Ext}}^{\mathrm{NP}} - Q_{\mathrm{Sca}}^{\mathrm{NP}}.} \tag{H.78}$$

H.4.3. Average local field at the surface

The average local field enhancement factor for PWE also reduces to:

$$\langle M_{\text{Loc}} \rangle = \frac{1}{2x^4} \sum_n (2n+1) \left[x^2 |\psi_n(x) + \Gamma_n \xi_n(x)|^2 \right.$$
$$\left. + x^2 |\psi_n'(x) + \Delta_n \xi_n'(x)|^2 + n(n+1) |\psi_n(x) + \Delta_n \xi_n(x)|^2 \right] . \quad \text{(H.79)}$$

H.4.4. Useful expansions for plane wave excitation

The expansions given earlier for the susceptibilities are still valid and we can write similar expressions for the far-field coefficients. Let us denote:

$$\beta_S = \frac{s^2 - 1}{s^2 + 2}, \quad \text{(H.80)}$$

which is the non-dimensional dipolar polarizability of the sphere in the ES approximation (see Section 6.2.1). We consider only the lowest order term, and corrections of the order up to x^3 to this term. For the scattering coefficient, for example, only Δ_1 (electric dipole) then contributes to these terms, and we have:

$$Q_{\text{Sca}}^{\text{NP}} \approx \frac{8}{3} x^4 |\beta_S|^2 \left[1 + \frac{6}{5} x^2 \text{Re} \left(\frac{s^2 - 2}{s^2 + 2} \right) - \frac{4}{3} x^3 \text{Im}(\beta_S) \right] . \quad \text{(H.81)}$$

Similar expressions can be obtained for the other coefficients.

H.5. EXTENSIONS OF MIE THEORY

H.5.1. Emitter close to a sphere

In the context of SEF (and to some extent of SERS), it is in general as important to study the emission process as the excitation process. It is therefore interesting to extend Mie theory's result to the case of a sphere excited by a dipole emitting in its vicinity. This problem may seem more complicated, but can be solved relatively easily using the tools already developed. The only difference lies in the nature of the incident field, which must now be taken as being the field of a dipole. The only technical difficulty is therefore to expand the dipole field in terms of VSHs.

Expansion of a dipole field in VSHs

We give here the final result without proof. We consider a dipole $\mathbf{p} = p\mathbf{e}_p$ at a position defined by $\mathbf{R}_p$. We assume that the dipole is in the non-absorbing embedding medium of dielectric constant ϵ_M (i.e. it cannot be inside the sphere). The dipolar field (which acts as the incident field here), in the region of the sphere surface, can then be expanded in terms of $\mathbf{M}^{(1)}$ and $\mathbf{N}^{(1)}$ as:

$$\mathbf{E}_{\rm Inc}(\mathbf{r}) = E_{p0} \sum_{n=0}^{\infty} \sum_{m=-n}^{m=n} a_{nm} \mathbf{M}_{nm}^{(1)}(k_M, \mathbf{r}) + b_{nm} \mathbf{N}_{nm}^{(1)}(k_M, \mathbf{r}), \tag{H.82}$$

with

$$E_{p0} = \frac{\mathrm{i} k_M^3 p}{\sqrt{6\pi \epsilon_0 \epsilon_M}}, \tag{H.83}$$

and

$$\begin{cases} a_{nm} &= \sqrt{6\pi}(-1)^m \mathbf{e}_p \cdot \mathbf{M}_{n,-m}^{(3)}(k_M, \mathbf{R}_p) \\ b_{nm} &= \sqrt{6\pi}(-1)^m \mathbf{e}_p \cdot \mathbf{N}_{n,-m}^{(3)}(k_M, \mathbf{R}_p). \end{cases} \tag{H.84}$$

These expressions are not exactly 'user-friendly', but they can be computed numerically when required. From these coefficients, it is easy to determine the scattering coefficients c_{nm} and d_{nm}, and all the required EM properties as before.

It is important to note that this expansion is only valid for $r < R_p$. This is not a problem since the surface at which the boundary condition is applied is fully contained within this volume. However, this means that it does not apply to the incident field in the far-field region, i.e. to the field radiated by the dipole in the absence of the sphere. In this case a similar expansion is possible in terms of $\mathbf{M}^{(3)}$ and $\mathbf{N}^{(3)}$, namely:

$$\mathbf{E}_{\rm Inc}(\mathbf{r}) = E_{p0} \sum_{n=0}^{\infty} \sum_{m=-n}^{m=n} e_{nm} \mathbf{M}_{nm}^{(3)}(k_M, \mathbf{r}) + f_{nm} \mathbf{N}_{nm}^{(3)}(k_M, \mathbf{r}), \tag{H.85}$$

with

$$\begin{cases} e_{nm} = \sqrt{6\pi}(-1)^m \mathbf{e}_p \cdot \mathbf{M}_{n,-m}^{(1)}(k_M, \mathbf{R}_p) = \sqrt{6\pi} \mathbf{e}_p \cdot \left(\mathbf{M}_{n,m}^{(1)}(k_M, \mathbf{R}_p) \right)^* \\ f_{nm} = \sqrt{6\pi}(-1)^m \mathbf{e}_p \cdot \mathbf{N}_{n,-m}^{(1)}(k_M, \mathbf{R}_p) = \sqrt{6\pi} \mathbf{e}_p \cdot \left(\mathbf{N}_{n,m}^{(1)}(k_M, \mathbf{R}_p) \right)^* . \end{cases} \tag{H.86}$$

The choice of E_{p0} may appear a bit strange (or arbitrary). In fact, it was chosen such that the expansion of the field of a dipole at the origin and along $(\mathrm{O}z)$ is simply $e_{1,0} = 1$ with all the other coefficients zero. This convention is convenient since the quantity $|S_{p0}| = |E_{p0} H_{p0}|/2$ is then simply related to the power P_0 radiated by the dipole in the absence of objects:

$$\frac{|E_{p0}H_{p0}|}{2} = k_M^2 P_0 \quad \text{with} \quad P_0 = \frac{\omega^4 |p|^2 \sqrt{\epsilon_M}}{12\pi\epsilon_0 c^3}. \tag{H.87}$$

Simplification for spherical symmetry

These expressions are fairly general and valid for a dipole at any position outside the sphere. They were included here since they can be useful for generalizing Mie theory to multiple spheres. However, for a single sphere or a sphere with multi-layers, the treatment can be greatly simplified owing to the spherical symmetry of the problem. In particular, we can choose without loss of generality a specific angular position for the dipole. A convenient choice is to take the dipole on the positive z-axis, i.e. $\theta = \phi = 0$, and at a distance $R_p > a$ from the center. We will also consider two cases: dipole perpendicular ($\perp$) to the surface, i.e. along (Oz) $\mathbf{e}_p = \mathbf{e}_z = \mathbf{e}_r$ (since $\theta = \phi = 0$), and dipole parallel ($\|$) to the surface, i.e. $\mathbf{e}_p \perp \mathbf{e}_r$. The coefficients of the expansion then simplify substantially. We define the adimensional coefficient $x_p = k_M R_p$.

For the *perpendicular dipole*, only $m = 0$ coefficients are non-zero, and we have:

$$\begin{cases} a_{nm} &= 0 \quad \forall (n,m) \\ b_{nm} &= 0 \quad \forall m \neq 0 \\ b_{n0} &= \sqrt{\frac{3}{2}(2n+1)(n+1)n}\frac{\xi_n(x_p)}{x_p^2}. \end{cases} \tag{H.88}$$

and similarly:

$$\begin{cases} e_{nm} &= 0 \quad \forall (n,m) \\ f_{nm} &= 0 \quad \forall m \neq 0, \\ f_{n0} &= \sqrt{\frac{3}{2}(2n+1)(n+1)n}\frac{\psi_n(x_p)}{x_p^2}. \end{cases} \tag{H.89}$$

For the *parallel dipole*, only $|m| = 1$ coefficients are non-zero, and we obtain:

$$\begin{cases} a_{nm} &= b_{nm} = 0 \quad \forall |m| \neq 1, \\ a_{n,1} &= \sqrt{\frac{3}{8}(2n+1)}\frac{\xi_n(x_p)}{x_p}\mathrm{i}\mathbf{e}_p \cdot (\mathbf{e}_\theta - \mathrm{i}\mathbf{e}_\phi), \\ a_{n,-1} &= \sqrt{\frac{3}{8}(2n+1)}\frac{\xi_n(x_p)}{x_p}\mathrm{i}\mathbf{e}_p \cdot (\mathbf{e}_\theta + \mathrm{i}\mathbf{e}_\phi), \\ b_{n,1} &= \sqrt{\frac{3}{8}(2n+1)}\frac{\xi_n'(x_p)}{x_p}\mathbf{e}_p \cdot (-\mathbf{e}_\theta + \mathrm{i}\mathbf{e}_\phi), \\ b_{n,-1} &= \sqrt{\frac{3}{8}(2n+1)}\frac{\xi_n'(x_p)}{x_p}\mathbf{e}_p \cdot (\mathbf{e}_\theta + \mathrm{i}\mathbf{e}_\phi), \end{cases} \tag{H.90}$$

and the same expressions for e_{nm} and f_{nm} replacing ξ_n by ψ_n.

Furthermore, we note that for Eqs (H.68) and (H.87) to be compatible, one must have:

$$\sum_{n,m} |e_{nm}|^2 + |f_{nm}|^2 = 1. \tag{H.91}$$

It is possible to verify that this is indeed the case for the expansion coefficients above, using the following properties of Riccati–Bessel functions [177]:

$$\sum_{n=1}^{\infty}(2n+1)(n+1)n\psi_n(x)^2 = 2x^4/3, \tag{H.92}$$

$$\sum_{n=1}^{\infty}(2n+1)\psi_n(x)^2 = x^2 - \psi_0(x)^2, \tag{H.93}$$

$$\sum_{n=1}^{\infty}(2n+1)\psi_n'(x)^2 = x^2/3 + \psi_0(x)^2. \tag{H.94}$$

Important EM characteristics

The power radiated by the dipole in the absence of any objects is known for standard EM theory and was given in Eq. (H.87). Using the expansion coefficients of the previous section, and the definitions in Eqs (H.66) and (H.69), we can now give the expressions of P_{Rad} and P_{Abs} for a dipolar source in the presence of a sphere [177]. We focus here on the corresponding enhancement factors, $M_{\text{Rad}} = P_{\text{Rad}}/P_0$ and $M_{\text{NR}} = P_{\text{Abs}}/P_0$, which are more relevant to SEF and plasmonics (see Section 4.4.3). We again consider the perpendicular and parallel dipole cases:

$$M_{\text{Rad}}^{\perp} = \frac{3}{2x_p^4}\sum_{n=1}^{\infty} n(n+1)(2n+1)|\psi_n(x_p) + \Delta_n\xi_n(x_p)|^2, \tag{H.95}$$

$$M_{\text{Rad}}^{\parallel} = \frac{3}{4x_p^2}\sum_{n=1}^{\infty}(2n+1)\left[|\psi_n(x_p) + \Gamma_n\xi_n(x_p)|^2 + |\psi_n'(x_p) + \Delta_n\xi_n'(x_p)|^2\right], \tag{H.96}$$

$$M_{\text{NR}}^{\perp} = \frac{-3}{2x_p^4}\sum_{n=1}^{\infty} n(n+1)(2n+1)|\xi_n(x_p)|^2(\text{Re}(\Delta_n) + |\Delta_n|^2), \tag{H.97}$$

$$M_{\text{NR}}^{\parallel} = \frac{-3}{4x_p^2}\sum_{n=1}^{\infty}(2n+1)\left[|\xi_n(x_p)|^2(\text{Re}(\Gamma_n) + |\Gamma_n|^2) + |\xi_n'(x_p)|^2(\text{Re}(\Delta_n) + |\Delta_n|^2)\right]. \tag{H.98}$$

From these expressions, one can deduce the total EM decay rate enhancement factors:

$$M^{\perp}_{\mathrm{Tot}} = M^{\perp}_{\mathrm{Rad}} + M^{\perp}_{\mathrm{NR}}, \quad \text{and} \quad M^{\|}_{\mathrm{Tot}} = M^{\|}_{\mathrm{Rad}} + M^{\|}_{\mathrm{NR}}. \tag{H.99}$$

In addition, the modified radiation profile $\mathrm{d}P/\mathrm{d}\Omega$ of the dipole can also be deduced from Eq. (H.70), and thus the corresponding differential radiative enhancement factors (see Section 4.4.3) $M^{d}_{\mathrm{Rad}} = (\mathrm{d}P/\mathrm{d}\Omega)/(3P_0/8\pi)$. For example, for the perpendicular dipole, we have:

$$M^{d\,\perp}_{\mathrm{Rad}}(\theta,\phi) = \frac{4\pi}{x_p^4}\left|\sum_{n=1}^{\infty}(-i)^n\sqrt{2n+1}\,(\psi_n(x_p)+\Delta_n\xi_n(x_p))\,\frac{\partial Y_{n0}}{\partial\theta}\right|^2. \tag{H.100}$$

We note again for self-consistency that taking $\Delta_n = 0$ results (after some manipulations) in $M^{d\,\perp}_{\mathrm{Rad}}(\theta,\phi) = \sin^2\theta$ as expected. Also, integration over all angles leads to the above result for $M^{d\,\perp}_{\mathrm{Rad}}$.

Dipole self-reaction

There is an alternative approach to obtaining M_{Tot}, which must be consistent with this one. It uses the self-reaction approach to dipolar emission and was discussed in Section 4.4. M_{Tot} can be obtained from the value of the scattered field at the dipole position (self-reaction field) as:

$$M_{\mathrm{Tot}} = 1 + \frac{6\pi\epsilon_0\epsilon_M}{k_M^3}\mathrm{Im}\left(\mathbf{e}_d^*\cdot\mathrm{E}_{\mathrm{Sca}}(\mathbf{r}_p)\right). \tag{H.101}$$

Using the expansion of E_{Sca} in terms of c_{nm} and d_{nm}, one can therefore deduce M_{Tot} directly from this expression. We obtain for the perpendicular and parallel dipoles:

$$M^{\perp}_{\mathrm{Tot}} = 1 + \frac{3}{2x_p^4}\sum_{n=1}^{\infty} n(n+1)(2n+1)\mathrm{Re}\left[\Delta_n\xi_n(x_p)^2\right], \tag{H.102}$$

$$M^{\|}_{\mathrm{Tot}} = 1 + \frac{3}{4x_p^2}\sum_{n=1}^{\infty}(2n+1)\mathrm{Re}\left[\Gamma_n\xi_n(x_p)^2 + \Delta_n\xi_n'(x_p)^2\right]. \tag{H.103}$$

Again, it is possible (but not easy) to show that these expressions are mathematically equivalent to those obtained previously [177].

Convergence of the total decay rate series

Although it may not be obvious from a direct inspection of the previous expressions, the series for M_{Tot} suffer from *slow-convergence problems* for dipoles very close to the surface. This problem is in fact already present when solving the same problem in the electrostatic approximation, as discussed in Section 6.2.1. The total angular momentum was there called l, instead of n, and we shall therefore use l for this discussion. The reason for this slow convergence is that from the point of view of the dipole, the sphere surface looks flat (like a plane) from a short distance, and the spherical harmonics expansion is not adapted to the planar geometry (which would formally correspond to $l \to \infty$).

We can illustrate this slow convergence by plotting the magnitude of the terms in the series as a function of order l for the special case of a dipole at a distance $d = 1$ nm from a silver sphere of radius 50 nm in water, as shown in Fig. H.1(a and b). In this specific example, it is clear that terms up to $l \approx 400$ are necessary to ensure convergence. Unfortunately, numerical computation of the exact Mie terms in our implementation fails at much lower orders ($l \approx 60$) because the special functions (spherical Bessel) then lie outside our floating-point precision. Exact computations of M_{Tot} using the Mie theory expressions would then require sophisticated numerical implementations with high precision and a large number of terms in the series.

Fortunately, there is a much simpler (and almost as accurate approach), which consists in using the ES approximation (ESA) to estimate the higher order terms in the series. This approach is valid because the exact Mie terms in the series converge toward the ESA results, as evident in Fig. H.1(a and b).

We therefore write here (without justification) the ESA result for a perpendicular and a parallel dipole at a distance d from the sphere, which read:

$$M_{\text{Tot}}^{\perp} = 1 + \sum_{l=1}^{\infty} \frac{3}{2(k_M(a+d))^3}(l+1)^2 \left(\frac{a}{a+d}\right)^{2l+1} \times \operatorname{Im}\left(\frac{\epsilon - \epsilon_M}{\epsilon + \frac{l+1}{l}\epsilon_M}\right), \tag{H.104}$$

$$M_{\text{Tot}}^{\parallel} = 1 + \sum_{l=1}^{\infty} \frac{3}{4(k_M(a+d))^3}l(l+1) \left(\frac{a}{a+d}\right)^{2l+1} \times \operatorname{Im}\left(\frac{\epsilon - \epsilon_M}{\epsilon + \frac{l+1}{l}\epsilon_M}\right). \tag{H.105}$$

To estimate M_{Tot} in practice, one should therefore use Eqs (H.102)–(H.103) up to a given (reasonable) order (typically the same as used for the other

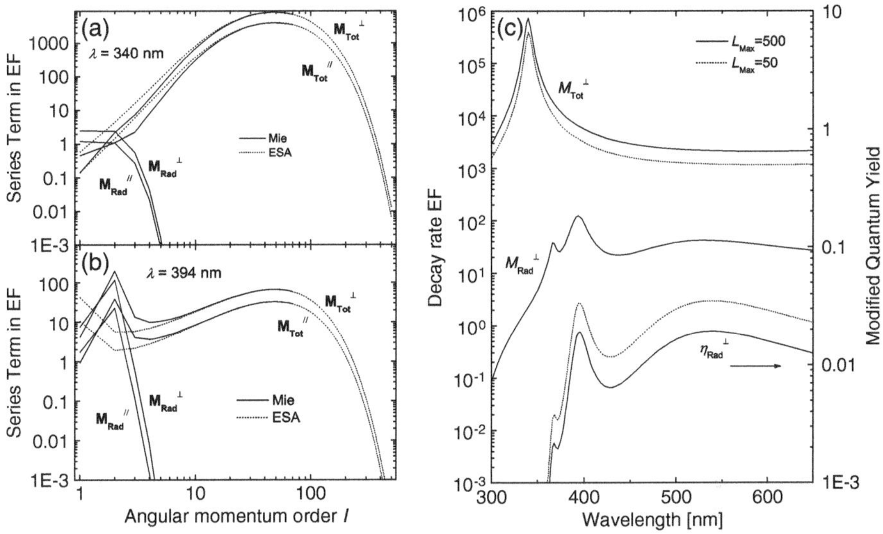

Figure H.1. This figure illustrates the importance of higher order terms for the calculation of M_{Tot}, in this example for a dipole at $d = 1$ nm from the surface of a $a = 50$ nm radius silver sphere in water. The l-dependence of the terms in the series for the total and radiative decay rates is shown in (a) and (b) for two wavelengths: (a) $\lambda = 340$ nm, corresponding to the non-radiative resonance, and (b) $\lambda = 394$ nm corresponding to the quadrupolar radiative resonance (the strongest in this case). The convergence of the series for M_{Rad} is very fast, but quite the opposite for M_{Tot}. The Mie computation of the coefficient is in fact no longer possible in our implementation above $l \approx 60$. However, the Mie results then become identical to those in the electrostatic approximation (ESA), and the ESA result should then be used for computations. The importance of the inclusion of high order terms for M_{Tot} (and therefore for $\eta_{\mathrm{Rad}}^{\mathrm{EM}}$) is also illustrated in the full wavelength dependence shown in (c) where the results of calculations with a series truncation at $L_{\mathrm{Max}} = 50$ and $L_{\mathrm{Max}} = 500$ are compared. The calculations of terms for $L_{\mathrm{Max}} = 500$ are carried out in the ESA for $l > 50$.

series) and then add the remaining higher orders up to a large $l \approx 500$, calculated within the ESA, according to Eqs (H.104)–(H.105).

Large errors would occur if these higher order terms are not included, especially for the smallest separation between dipole and surface, as shown in Fig. H.1.

H.5.2. Coated spheres

Principle

It is also fairly straightforward to adapt the theory to multi-layers of spheres, i.e. several (instead of one) concentric interfaces, the simplest example of which is a coated sphere, or a spherical void.

We therefore consider $k = 1 \ldots K$ spherical surfaces with radii a_k. These therefore delimit $K+1$ volumes V_k for $k = 0 \ldots K$ as $a_k \le r \le a_{k+1}$ ($a_0 = 0$ and $a_{K+1} = \infty$). V_0 is a sphere, and V_K is the outside (embedding) medium as for the single sphere. The other volumes are spherical shells. Each volume V_k is described by a dielectric function $\epsilon_k(\omega)$, with $\epsilon_K = \epsilon_M$ real, as before. The wave vector in each region is also defined as $k_k = \sqrt{\epsilon_k(\omega)}\omega/c$. This system is excited by an incident wave $\mathbf{E}_{\text{Inc}}$ in the embedding medium (V_K), also as before.

Similarly to the case of a single sphere, the spherical symmetry of the problem suggests that we should expand the electric field solution in each volume in terms of VSHs. Because most volumes do not contain the origin or infinity, the most general expansion is in terms of both $\mathbf{M}^{(1)}/\mathbf{N}^{(1)}$ and $\mathbf{M}^{(3)}/\mathbf{N}^{(3)}$. In volume V_k, we therefore have:

$$\mathbf{E}(\mathbf{r}) = E_0 \sum_{n,m} \alpha^k_{nm}\mathbf{M}^{(1)}_{nm}(k_k, \mathbf{r}) + \beta^k_{nm}\mathbf{N}^{(1)}_{nm}(k_k, \mathbf{r})$$
$$+ \gamma^k_{nm}\mathbf{M}^{(3)}_{nm}(k_k, \mathbf{r}) + \delta^k_{nm}\mathbf{N}^{(3)}_{nm}(k_k, \mathbf{r}). \qquad \text{(H.106)}$$

Following the discussion of the single sphere case, we moreover have for V_0 (central sphere) $\gamma^0_{nm} = \delta^0_{nm} = 0$. Also for the outer region V_K, $\alpha^K_{nm} = a_{nm}$ and $\beta^K_{nm} = b_{nm}$ correspond to the coefficients of the incident wave, and are therefore known, and $\gamma^K_{nm} = c_{nm}$ and $\delta^K_{nm} = d_{nm}$ correspond to the coefficients of the scattered wave and need to be determined.

It now remains to apply the appropriate boundary conditions at each surface $r = a_k$. This is very similar to what was done for the single sphere case, and can be done here recursively, starting with $r = a_1$. For this first case, $\gamma^0_{nm} = \delta^0_{nm} = 0$ inside, and it is therefore exactly the same situation as for the single sphere case. We therefore deduce that:

$$\gamma^1_{nm} = \Gamma_n \alpha^1_{nm} \quad \text{and} \quad \delta^1_{nm} = \Delta_n \beta^1_{nm}, \qquad \text{(H.107)}$$

where Γ_n and Δ_n are the susceptibilities of the sphere defined in Eqs (H.46) and (H.47). but replacing the arguments of the functions: x by $k_1 a_1$, and sx by $k_0 a_1$.

For the next interface, $r = a_2$, the situation is more complicated because we have 4 coefficients defining the field on each side (instead of $2+4$ in the first case). However, the above expression shows that γ^1_{nm} is proportional to α^1_{nm} and δ^1_{nm} is proportional to β^1_{nm}, which therefore reduces the problem to $2+4$ as before. We can therefore deduce again a proportionality relation between γ^2_{nm} and α^2_{nm}, and this can be carried on recursively for all interfaces.

Recursive implementation: Scattered field

Without going further into the technicalities, we simply give the results necessary to define the full electromagnetic solution. We define first for $k = 1 \dots K$, $x_k = k_k a_k$ and $s_k = k_{k-1}/k_k$ so that $s_k x_k = k_{k-1} a_k$. For all $k = 1 \dots K$, we have:

$$\gamma_{nm}^k = \Gamma_n^k \alpha_{nm}^k, \quad \text{and} \quad \delta_{nm}^k = \Delta_n^k \beta_{nm}^k, \tag{H.108}$$

where the proportionality coefficients (which represent the susceptibilities of each interface) are defined as:

$$\Gamma_n^0 = \Delta_n^0 = 0, \tag{H.109}$$

and by recurrence on $k = 1 \dots K$:

$$\Gamma_n^k = N_n^k(\Gamma)/D_n^k(\Gamma), \tag{H.110}$$

with:

$$\begin{aligned} N_n^k(\Gamma) = & \left[\psi_n(s_k x_k) + \Gamma_n^{k-1} \xi_n(s_k x_k)\right] \psi_n'(x_k) \\ & - s_k \psi_n(x_k) \left[\psi_n'(s_k x_k) + \Gamma_n^{k-1} \xi_n'(s_k x_k)\right], \end{aligned} \tag{H.111}$$

and

$$\begin{aligned} D_n^k(\Gamma) = & s_k \xi_n(x_k) \left[\psi_n'(s_k x_k) + \Gamma_n^{k-1} \xi_n'(s_k x_k)\right] \\ & - \left[\psi_n(s_k x_k) + \Gamma_n^{k-1} \xi_n(s_k x_k)\right] \xi_n'(x_k), \end{aligned} \tag{H.112}$$

while

$$\Delta_n^k = N_n^k(\Delta)/D_n^k(\Delta), \tag{H.113}$$

with

$$\begin{aligned} N_n^k(\Delta) = & \psi_n(x_k) \left[\psi_n'(s_k x_k) + \Delta_n^{k-1} \xi_n'(s_k x_k)\right] \\ & - s_k \left[\psi_n(s_k x_k) + \Delta_n^{k-1} \xi_n(s_k x_k)\right] \psi_n'(x_k), \end{aligned} \tag{H.114}$$

and

$$D_n^k(\Delta) = s_k \left[\psi_n(s_k x_k) + \Delta_n^{k-1} \xi_n(s_k x_k)\right] \xi_n'(x_k)$$

$$- \xi_n(x_k) \left[\psi'_n(s_k x_k) + \Delta_n^{k-1} \xi'_n(s_k x_k)\right]. \tag{H.115}$$

Recursive implementation: Internal fields

Since the coefficients of the incident wave are known, the coefficients of the scattered wave are then obtained from:

$$c_{nm} = \gamma_{nm}^K = \Gamma_n^K a_{nm} \quad \text{and} \quad d_{nm} = \delta_{nm}^K = \Delta_n^K b_{nm}. \tag{H.116}$$

This entirely defines the electric field in the outer region.

If necessary, the coefficients describing the field inside the spherical shells can also be derived by downward recurrence for $k = K \dots 1$ by generalizing Eqs (H.51) and (H.52):

$$\alpha_{nm}^{k-1} = A_n^k \alpha_{nm}^k, \quad \text{and} \quad \beta_{nm}^{k-1} = B_n^k \beta_{nm}^k, \tag{H.117}$$

with

$$A_n^k = N_n^k(A)/D_n^k(A), \tag{H.118}$$

where

$$N_n^k(A) = s_k \psi_n(x_k) \xi'_n(x_k) - s_k \xi_n(x_k) \psi'_n(x_k) = i s_k, \tag{H.119}$$

and

$$D_n^k(A) = \left[\psi_n(s_k x_k) + \Gamma_n^{k-1} \xi_n(s_k x_k)\right] \xi'_n(x_k)$$
$$- s_k \xi_n(x_k) \left[\psi'_n(s_k x_k) + \Gamma_n^{k-1} \xi'_n(s_k x_k)\right], \tag{H.120}$$

while, on the other hand:

$$B_n^k = N_n^k(B)/D_n^k(B), \tag{H.121}$$

with

$$N_n^k(B) = s_k \psi_n(x_k) \xi'_n(x_k) - s_k \xi_n(x_k) \psi'_n(x_k) = i s_k, \tag{H.122}$$

and

$$D_n^k(B) = s_k \left[\psi_n(s_k x_k) + \Delta_n^{k-1} \xi_n(s_k x_k)\right] \xi'_n(x_k)$$
$$- \xi_n(x_k) \left[\psi'_n(s_k x_k) + \Delta_n^{k-1} \xi'_n(s_k x_k)\right]. \tag{H.123}$$

These coefficients complete the solution of the electromagnetic problem.

Further properties

From them, one can in principle calculate the field at any point in space, by using the appropriate expansion. In particular, all the EM properties related to the outside field (in the medium) can be derived from the same formulas as for Mie theory, but using Γ_n^K, Δ_n^K, and x_K instead of Γ_n, Δ_n, and x. This applies for example to far-field properties and local field at the surface of the outer sphere. It can also be applied for plane wave excitation or dipolar emission.

Needless to say, this is where the analytical work finds its boundary of applicability or 'usefulness'. Anything beyond this level of description becomes considerably cumbersome to handle analytically, and we might as well solve the problem numerically. These expressions are already hard enough to write!, and it will be difficult to manipulate them further. However, once they are written, they lend themselves easily to a numerical implementation. We provide such an implementation with Matlab on the book website, and it should be easy for the interested reader to implement it on a different platform too. This is, in fact, the main reason why we have listed all these expressions here.

H.5.3. Multiple spheres and generalized Mie theory (GMT)

The extension of Mie theory to cases with *more* than one sphere is called generalized Mie theory (GMT). The problem becomes (obviously) substantially more complicated. We shall not dwell on the details of GMT here, and we only make a few comments in passing for completeness.

When several spheres are present, each of them contributes to a scattering wave that is seen by the others. In other words, the electromagnetic response of the spheres interferes with each other through their respective scattered fields. The scattered field has now contribution from spherical waves (VSHs) emanating from all the individual spheres. For a system of N spheres, for example, the scattered field has to be now expanded as:

$$\mathbf{E}_{\text{Sca}}(\mathbf{r}) = E_0 \sum_{i=1}^{N} \sum_{n,m} c_{nm}^i \mathbf{M}_{nm}^{(3)}(k_M, \mathbf{r} - \mathbf{r}_i) + d_{nm}^i \mathbf{N}_{nm}^{(3)}(k_M, \mathbf{r} - \mathbf{r}_i), \tag{H.124}$$

which is the 'equivalent' of Eq. (H.9) but for many spheres centered at $\mathbf{r}_i$, with $i = 1 \ldots N$. Note that the scattered fields now are a superposition of VSHs originating from the centers of the different spheres, and that the coefficients

c^i_{nm}'s and d^i_{nm}'s pick up an additional index 'i' that identifies which sphere they are representing.

The 'stitching' problem of the coefficients at the boundaries of the spheres becomes altogether a more difficult problem, because the condition $r_i = a$ is natural for only one spherical coordinate system (that of sphere i). One must therefore express all VSHs of a sphere as expansion onto the VSHs basis of the others. A lot of the additional complexity comes from this step. Without going into further details, this eventually results in a coupled system of linear equations with as many unknowns as coefficients in the expansions $(N \times (N_{\mathrm{Max}} \times (N_{\mathrm{Max}} + 1))$. The corresponding interaction matrix for the coefficients must then be inverted to obtain a solution.

For two spheres, there are still a few simplifications possible in the interaction of the respective VSHs, because we can always choose to align them along z. For more than two spheres, the symmetry is lost and the problem has to be solved in full.

H.6. EXAMPLE OF IMPLEMENTATION OF MIE THEORY WITH MATLAB

In this section, we discuss some aspects of the implementation of Mie theory, with a focus on using Matlab as a numerical platform. The use of mathematical softwares such as Matlab simplifies greatly the implementation thanks to their 'easy' handling of *complex numbers*, and of some of the *special functions* appearing in Mie theory. In addition, Matlab is particularly well suited for handling arrays.

Various Mie theory codes are already available on the web, including some for Matlab. Nevertheless, we feel it will be useful to provide codes that implement the theory, exactly as it was presented in the previous sections, i.e. with the same conventions, same notations, and same formulas. This will hopefully enable the user to use the codes knowledgeably and efficiently, and even to adapt or extend them toward more specific needs.

H.6.1. Common problems

Here are some of the common problems that may be encountered when implementing Mie theory:

- Handling of complex numbers:
  This is not an issue in Matlab.
- Calculation of special functions:
  Bessel functions are already implemented in Matlab, from which spherical Bessel (or Riccati–Bessel) functions are easily derived. Otherwise, algorithms exist to calculate these 'from scratch', see for example Section 3.6 of Ref. [196]. Associated Legendre functions are also implemented in Matlab.

- Overflow and precision issues:
  The magnitude of the numbers involved can sometimes be very small or very large; especially for large order n. One has to make sure that such numbers are properly handled, or precision may be lost. In Matlab, numbers as small as 10^{-300} and as large as 10^{300} can be handled fairly reliably in floating-point calculations. This is enough accuracy for multiplication/division, but problems may occur when subtracting/adding two large numbers with similar magnitudes that *compensate* each other. The result may then be zero for a subtraction! because of the limited precision, while it should have been in reality finite and non-negligible. This type of problem can (in principle) be avoided by choosing the appropriate algorithm.

H.6.2. Other issues specific to Matlab

There are at least two potential problems one has to be aware when implementing Mie theory in Matlab.

Firstly, as a rule of thumb one should only use the first and second order Bessel functions `besselj` and `bessely`, and derive Hankel functions from them. There are errors (at least in the current versions) with the implementation of the Hankel function `besselh` (in Matlab).

Another potential source of errors in Matlab codes of Mie theory is the use of the Hermitian transpose operator (`'`) instead of the normal transpose operator (`.'`). In Matlab, most variables are either vector or matrices, and it is often necessary to transpose them. We will always use the command `transpose` to do so, instead of the commonly used shortcut (`'`). Both are equivalent for *real* matrices, but for complex matrices, (`'`) carries out a Hermitian transpose, i.e. transpose *and* complex conjugation. Because most matrices for Mie theory are complex, it is important to apply the relevant operator. We will therefore avoid using the operators (`.'`) and (`'`), and use the commands `transpose` for normal transpose (`.'`), and `ctranspose` for Hermitian transpose (`'`).

H.6.3. Some aspects of our implementation

Matlab has been designed to work with data organized in vectors or matrices, even for problems that are not related to linear algebra. Computations are therefore faster when carried out in matrix form, rather than using loops over scalar elements of an array. In our implementation, we attempt to make use of this (powerful) particularity of Matlab.

Our codes have been developed with the aim of performing fast wavelength-dependence calculations, i.e. it solves the EM problem for one single geometry, but $L \geq 1$ wavelengths. The wavelength dependence will appear in the rows of the object (first row = first wavelength, etc.).

These codes, along with further description of their use and examples, are available from the book website: www.victoria.ac.nz/raman/book, and additional information or news will be updated when appropriate.

References

[1] M. Fleischmann, P. J. Hendra, and A. J. McQuillan. Raman spectra of pyridine adsorbed at a silver electrode. *Chem. Phys. Lett.*, 26:163–166, 1974.

[2] D. L. Jeanmaire and R. P. Van Duyne. Surface Raman spectroelectrochemistry Part I. Heterocyclic, aromatic, and aliphatic amines adsorbed on the anodized silver electrode. *J. Electroanal. Chem.*, 84:1–20, 1977.

[3] M. G. Albrecht and J. A. Creighton. Anomalously intense Raman-spectra of pyridine at a silver electrode. *J. Am. Chem. Soc.*, 99: 5215–5217, 1977.

[4] M. Moskovits. Surface-enhanced spectroscopy. *Rev. Modern Phys.*, 57: 783–826, 1985.

[5] A. Otto. Surface-enhanced Raman scattering: 'classical' and 'chemical' origins. In M. Cardona and G. Güntherodt, editors, *Light Scattering in Solids IV*, volume 4 of *Top. Appl. Phys.*, Springer-Verlag, Berlin, 1984.

[6] A. Campion and P. Kambhampati. Surface-enhanced Raman scattering. *Chem. Soc. Rev.*, 27:241–250, 1998.

[7] R. Aroca. *Surface-enhanced vibrational spectroscopy*. John Wiley & Sons, Chichester, 2006.

[8] E. C. Le Ru, E. Blackie, M. Meyer, and P. G. Etchegoin. SERS enhancement factors: a comprehensive study. *J. Phys. Chem. C*, 111: 13794–13803, 2007.

[9] E. C. Le Ru, P. G. Etchegoin, J. Grand, *et al.* The mechanisms of spectral profile modification in surface enhanced fluorescence. *J. Phys. Chem. C*, 111:16076–16079, 2007.

[10] E. C. Le Ru and P. G. Etchegoin. Sub-wavelength localization of hot-spots in SERS. *Chem. Phys. Lett.*, 396:393–397, 2004.

[11] W. L. Barnes, A. Dereux, and T. W. Ebbesen. Surface plasmon subwavelength optics. *Nature*, 424:824–830, 2003.

[12] L. Novotny and B. Hecht. *Principles of nano-optics*. Cambridge University Press, Cambridge, 2006.

[13] C. L. Haynes, A. D. McFarland, and R. P. Van Duyne. Surface-enhanced Raman spectroscopy. *Anal. Chem.*, 77:338A–346A, 2005.

[14] R. Mukhopadhyay. Raman flexes its muscles. *Anal. Chem.*, 79: 3265–3270, 2007.

[15] M. J. Pelletier. Quantitative analysis using Raman spectrometry. *Appl. Spectrosc.*, 57:20A–42A, 2003.

[16] G. Sabatte, R. Keir, M. Lawlor, *et al.* Comparison of surface-enhanced resonance Raman scattering and fluorescence for detection of a labeled antibody. *Anal. Chem.*, 80:2351–2356, 2008.

[17] O. Lyandres, N. C. Shah, C. R. Yonzon, *et al.* Real-time glucose sensing by surface-enhanced Raman spectroscopy in bovine plasma facilitated by a mixed decanethiol/mercaptohexanol partition layer. *Anal. Chem.*, 77:6134–6139, 2005.

[18] M. Larsson and J. Lindgren. Analysis of glutathione and immunoglobulin G inside chromatographic beads using surface-enhanced Raman scattering spectroscopy. *J. Raman Spectrosc.*, 36:394–399, 2005.

[19] A. Rasmussen and V. Deckert. Surface- and tip-enhanced Raman scattering of DNA components. *J. Raman Spectrosc.*, 37:311–317, 2006.

[20] S. Cinta-Pinzaru, N. Peica, B. Kustner, *et al.* FT-Raman and NIR-SERS characterization of the antimalarial drugs chloroquine and mefloquine and their interaction with hematin. *J. Raman Spectrosc.*, 37:326–334, 2006.

[21] J. Binoy, I. H. Joe, V. S. Jayakumar, *et al.* DFT based relaxed PES scan studies and SERS of anti-cancer drug combretastatin A-4. *Laser Phys. Lett.*, 2:544–550, 2005.

[22] A. V. Szeghalmi, L. Leopold, S. Pinzaru, *et al.* Adsorption of 6-mercaptopurine and 6-mercaptopurine-ribosideon silver colloid: A pH-dependent surface-enhanced Raman spectroscopy and density functional theory study. II. 6-mercaptopurine-riboside. *Biopolymers*, 78:298–310, 2005.

[23] J. M. Sylvia, J. A. Janni, J. D. Klein, and K. M. Spencer. Surface-enhanced Raman detection of 2,4-dinitrotoluene impurity vapor as a marker to locate landmines. *Anal. Chem.*, 72:5834–5840, 2000.

[24] K. Chen, K. C. Vo-Dinh, F. Yan, *et al.* Direct identification of alizarin and lac-dye on painting fragments using surface-enhanced Raman scattering. *Anal. Chim. Acta*, 569:234–240, 2006.

[25] K. Chen, M. Leona, K. C. Vo-Dinh, *et al.* Application of surface-enhanced Raman scattering (SERS) for the identification of anthraquinone dyes used in works of art. *J. Raman Spectrosc.*, 37: 520–527, 2006.

[26] R. J. H. Clark. Pigment identification on medieval manuscripts by Raman microscopy. *J. Mol. Struct.*, 347:417–444, 1995.

[27] C. L. Haynes, C. R. Yonzon, X. Zhang, and R. P. Van Duyne. Surface enhanced Raman sensors: early history and the development of sensors for quantitative biowarfare agent and glucose detection. *J. Raman Spectrosc.*, 36:471–484, 2005.

[28] Z. Q. Tian. Surface enhanced Raman spectroscopy: advancements and applications. *J. Raman Spectrosc.*, 36:466–470, 2005.

[29] M. Moskovits. Surface-enhanced Raman spectroscopy: a brief retrospective. *J. Raman Spectrosc.*, 36:485–496, 2005.

[30] S. Nie and S. R. Emory. Probing single molecules and single nanoparticles by surface-enhanced Raman scattering. *Science*, 275: 1102–1106, 1997.

[31] K. Kneipp, Y. Wang, H. Kneipp, *et al.* Single molecule detection using surface-enhanced Raman scattering (SERS). *Phys. Rev. Lett.*, 78: 1667–1670, 1997.

[32] H. Xu, E. J. Bjerneld, M. Käll, and L. Börjesson. Spectroscopy of single hemoglobin molecules by surface enhanced Raman scattering. *Phys. Rev. Lett.*, 83:4357–4360, 1999.

[33] A. Otto. What is observed in single molecule SERS, and why?. *J. Raman Spectrosc.*, 33:593–598, 2002.

[34] E. C. Le Ru, M. Meyer, and P. G. Etchegoin. Proof of single-molecule sensitivity in surface enhanced Raman scattering (SERS) by means of a two-analyte technique. *J. Phys. Chem. B*, 110:1944–1948, 2006.

[35] P. G. Etchegoin, M. Meyer, and E. C. Le Ru. Statistics of single molecule SERS signals: is there a Poisson distribution of intensities?. *Phys. Chem. Chem. Phys.*, 9:2007 3606–3010.

[36] J. A. Dieringer, R. B. Lettan II, K. A. Scheidt, and R. P. Van Duyne. A frequency domain existence proof of single-molecule surface-enhanced Raman spectroscopy. *J. Am. Chem. Soc.*, 129:16249–16256, 2007.

[37] E. Blackie, E. C. Le Ru, M. Meyer, M. Timmer, B. Burkett, P. Northcote, and P. G. Etchegoin. Bi-analyte SERS with isotopically edited dyes. *Phys. Chem. Chem. Phys.*, 10:4147–4153, 2008.

[38] N. Félidj, J. Aubard, G. Lévi, *et al.* Controlling the optical response of regular arrays of gold particles for surface-enhanced Raman scattering. *Phys. Rev. B*, 65:2002 075419–1–9.

[39] N. Félidj, J. Aubard, G. Lévi, *et al.* Optimized surface-enhanced Raman scattering on gold nanoparticle arrays. *Appl. Phys. Lett.*, 82:3095–3097, 2003.

[40] C. L. Haynes and R. P. Van Duyne. Plasmon-sampled surface-enhanced Raman excitation spectroscopy. *J. Phys. Chem. B*, 107:7426–7433, 2003.

[41] A. D. McFarland, M. A. Young, J. A. Dieringer, and R. P. Van Duyne. Wavelength-scanned surface-enhanced Raman excitation spectroscopy. *J. Phys. Chem. B*, 109:11279–11285, 2005.

[42] S. Lal, N. K. Grady, G. P. Goodrich, and N. J. Halas. Profiling the near field of a plasmonic nanoparticle with Raman-based molecular rulers. *Nano Lett.*, 6:2338–2343, 2006.

[43] F. Svedberg, Z. Li, H. Xu, and M. Käll. Creating hot nanoparticle pairs for surface-enhanced Raman spectroscopy through optical manipulation. *Nano Lett.*, 6:2639–2641, 2006.

[44] E. C. Le Ru, M. Dalley, and P. G. Etchegoin. Plasmon resonances of silver colloids studied by surface enhanced Raman spectroscopy. *Cur. Appl. Phys.*, 6:411–414, 2006.

[45] E. C. Le Ru and P. G. Etchegoin. Rigorous justification of the $|E|^4$ enhancement factor in surface enhanced Raman spectroscopy. *Chem. Phys. Lett.*, 423:63–66, 2006.

[46] S. J. Lee, Z. Guan, H. Xu, and M. Moskovits. Surface-enhanced Raman spectroscopy and nano-geometry: the plasmonic origin of SERS. *J. Phys. Chem. C*, 111:17985–17988, 2007.

[47] E. C. Le Ru, P. G. Etchegoin, J. Grand, *et al.* Surface enhanced Raman spectroscopy on nanolithography-prepared substrates. *Cur. Appl. Phys.*, 8:467–470, 2008.

[48] E. C. Le Ru, M. Meyer, E. Blackie, and P. G. Etchegoin. Advanced aspects of electromagnetic SERS enhancement factors at a hot-spot. *J. Raman Spectrosc.*, 39:1127–1134, 2008.

[49] E. C. Le Ru, J. Grand, N. Félidj, *et al.* Experimental verification of the SERS electromagnetic model beyond the $|E|^4$ approximation: Polarization Effects. *J. Phys. Chem. C*, 112:8117–8121, 2008.

[50] A. Otto and M. Futamata. Electronic mechanisms of SERS. In *Surface-Enhanced Raman Scattering - Physics and Applications*, volume 103 of *Top. Appl. Phys.*, Springer-Verlag, Berlin, pp. 147–182, 2006.

[51] J. R. Lombardi and R. L. Birke. Time-dependent picture of the charge-transfer contributions to surface enhanced Raman spectroscopy. *J. Chem. Phys.*, 126:2007 244709–1.

[52] M. Moskovits. Surface selection rules. *J. Chem. Phys.*, 77:4408–4416, 1982.

[53] A. M. Michaels, J. Jiang, and L. Brus. Ag nanocrystal junctions as the site for surface-enhanced Raman scattering of single rhodamine 6G molecules. *J. Phys. Chem. B*, 104:11965–11971, 2000.

[54] J. Jiang, K. Bosnick, M. Maillard, and L. Brus. Single molecule Raman spectroscopy at the junctions of large Ag nanocrystals. *J. Phys. Chem. B*, 107:9964–9972, 2003.

[55] H. X. Xu and M. Käll. Polarization-dependent surface-enhanced Raman spectroscopy of isolated silver nanoaggregates. *Chem. Phys. Chem.*, 4: 1001–1005, 2003.

[56] T. O. Shegai and G. Haran. Probing the Raman scattering tensors of individual molecules. *J. Phys. Chem. B*, 110:2459–2461, 2006.

[57] P. G. Etchegoin, C. Galloway, and E. C. Le Ru. Polarization-dependent effects in surface-enhanced Raman scattering (SERS). *Phys. Chem. Chem. Phys.*, 8:2624–2628, 2006.

[58] K. Kneipp, Y. Wang, H. Kneipp, *et al.* Population pumping of excited vibrational states by spontaneous surface-enhanced Raman scattering. *Phys. Rev. Lett.*, 76:2444–2447, 1996.

[59] R. C. Maher, C. M. Galloway, E. C. Le Ru, *et al.* Vibrational pumping in surface enhanced Raman scattering (SERS). *Chem. Soc. Rev.*, 37: 965–979, 2008.

[60] L. Jensen and G. C. Schatz. Resonance Raman scattering of rhodamine 6G as calculated using time-dependent density functional theory. *J. Phys. Chem. A*, 110:5973, 2006.

[61] H. Watanabe, N. Hayazawa, Y. Inouye, and S. Kawata. DFT vibrational calculations of rhodamine 6G adsorbed on silver: analysis of tip-enhanced Raman spectroscopy. *J. Phys. Chem. B*, 109:5012–5020, 2005.

[62] G. A. Baker and D. S. Moore. Progress in plasmonic engineering of surface-enhanced Raman-scattering substrates toward ultra-trace analysis. *Anal. Bioanal. Chem.*, 382:1751–1770, 2005.

[63] N. M. B. Perney, J. J. Baumberg, M. E. Zoorob, *et al.* Tuning localized plasmons in nanostructured substrates for surface-enhanced Raman scattering. *Opt. Express*, 14:847–857, 2006.

[64] M. Green and F. M. Liu. SERS substrates fabricated by island lithography: the silver/pyridine system. *J. Phys. Chem. B*, 107: 13015–13021, 2003.

[65] V. P. Drachev, M. D. Thoreson, V. Nashine, *et al.* Adaptive silver films for surface-enhanced Raman spectroscopy of biomolecules. *J. Raman Spectrosc.*, 36:648–656, 2005.

[66] J. B. Jackson, S. L. Westcott, L. R. Hirsch, *et al.* Controlling the surface enhanced Raman effect via the nanoshell geometry. *Appl. Phys. Lett.*, 82:257–259, 2003.

[67] M. J. Natan. Concluding remarks: surface enhanced Raman scattering. *Faraday Discuss.*, 132:321–328, 2006.

[68] N. Hayazawa, Y. Inouye, Z. Sekkat, and S. Kawata. Near-field Raman scattering enhanced by a metallized tip. *Chem. Phys. Lett.*, 335:369–374, 2001.

[69] B. Pettinger, G. Picardi, R. Schuster, and G. Ertl. Surface-enhanced and STM-tip-enhanced Raman spectroscopy at metal surfaces. *Single Mol.*, 5–6:285–294, 2002.

[70] D. A. Long. *The Raman effect, a unified treatment of the theory of Raman scattering by molecules*. John Wiley & Sons Ltd., Chichester, 2002.

[71] W. Demtröder. *Laser spectroscopy*. Springer, Berlin, 2002.

[72] P. Y. Yu and M. Cardona. *Fundamentals of semiconductors: physics and materials properties*. Springer, Berlin, 2004.

[73] A. Yariv. *Quantum electronics*. Wiley, New York, 1989.

[74] D. Bermejo. *Introducción a la espectroscopía Raman*. The General Secretariat of the Organization of American States, Washington, 1988. (in Spanish).

[75] M. Cardona. *Light scattering in solids II*, volume 50 of *Top. Appl. Phys.*, Springer-Verlag, Berlin, Heidelberg, New York, 1982.

[76] C. V. Raman and K. S. Krishnan. A new type of secondary radiation. *Nature*, 121:501–502, 1921.

[77] R. S. Krishnan and R. K. Shankar. Raman effect: history of the discovery. *J. Raman Spectrosc.*, 10:1981.

[78] W. Kiefer. Recent advances in linear and nonlinear Raman spectroscopy I. *J. Raman Spectrosc.*, 38:1538–1553, 2007.

[79] B. Valeur. *Molecular fluorescence. Principles and applications*. Wiley-VCH, Weinheim, 2002.

[80] H. Haken, H. C. Wolf, and W. D. Brewer. *Molecular physics and elements of quantum chemistry: introduction to experiments and theory*. Springer-Verlag, Berlin, 2004.

[81] C. Eggeling, J. Widengren, R. Rigler, and C. A. M. Seidel. Photobleaching of fluorescent dyes under conditions used for single-molecule detection: evidence of two-step photolysis. *Anal. Chem.*, 70: 2651–2659, 1998.

[82] C. Eggeling, A. Volkmer, and C. A. M. Seidel. Molecular photobleaching kinetics of rhodamine 6G by one- and two-photon induced confocal fluorescence microscopy. *Chem. Phys. Chem.*, 6:791–804, 2005.

[83] C. Julien, A. Débarre, D. Nutarelli, *et al.* Single molecule study of perylene orange photobleaching in thin sol-gel films. *J. Phys. Chem. B*, 109:23145–23153, 2005.

[84] L. D. Landau and E. M. Lifshitz. *Quantum mechanics*. 3rd edition, Elsevier, Amsterdam, 2003.

[85] B. Schrader and D. S. Moore. Laser-based molecular spectroscopy for chemical analysis - Raman scattering processes (IUPAC Recommendations 1997). *Pure & Appl. Chem.*, 69:1451–1468, 1997.

[86] H. W. Schrötter and H. W. Klöckner. *Raman scattering cross sections in gases and liquids*. Springer, Berlin, 1979, pp. 123–166.

[87] S. Shim, C. M. Stuart, and R. A. Mathies. Resonance Raman cross-sections and vibronic analysis of rhodamine 6G from broadband stimulated Raman spectroscopy. *Chem. Phys. Chem.*, 9:697–699, 2008.

[88] R. M. A. Azzam and N. M. Bashara. *Ellipsometry and polarized light*. North-Holland, Amsterdam, 1977.

[89] D. Graham, C. McLaughlin, G. McAnally, *et al.* Synthesis of novel monoazo benzotriazole dyes specifically for surface enhanced resonance Raman scattering. *Chem. Commun.*, 1998:1187–1188, 1998.

[90] J. Widengren, Ü. Mets, and R. Rigler. Fluorescence spectroscopy of triplet states in solution: a theoretical and experimental study. *J. Phys. Chem.*, 99:13368–13379, 1995.

[91] J. R. Lakowicz. *Principles of fluorescence spectroscopy.* 3rd edition, Springer, New York, 2006.

[92] D. Magde, R. Wong, and P. G. Seybold. Fluorescence quantum yields and their relation to lifetimes of rhodamine 6G and fluorescein in nine solvents: improved absolute standards for quantum yields. *Photochem. Photobiol.*, 75:327–334, 2002.

[93] M. P. Gordon, T. Ha, and P. R. Selvin. Single-molecule high-resolution imaging with photobleaching. *Proc. Nat. Acad. Sci.*, 101:6462–6465, 2004.

[94] J. Widengren and R. Rigler. Mechanisms of photobleaching investigated by fluorescence correlation spectroscopy. *Bioimaging*, 4:149–157, 1996.

[95] J. A. Stratton. *Electromagnetic theory.* McGraw-Hill, New York, 1941.

[96] J. D. Jackson. *Classical electrodynamics.* 3rd edition, Wiley, New York, 1998.

[97] A. López Dávalos and D. Zanette. *Fundamentals of electromagnetism.* Springer, Berlin, 1999.

[98] E. M. Purcell and C. R. Pennypacker. Scattering and absorption of light by nonspherical dielectric grains. *Astrophys. J.*, 186:705–714, 1973.

[99] Y. R. Shen. *The principles of non-linear optics.* Wiley, New York, 1984.

[100] D. L. Rousseau, R. P. Bauman, and S. P. S. Porto. Normal mode determination in crystals. *J. Raman Spectrosc.*, 10:253–290, 1981.

[101] W. Hayes and R. Loudon. *Scattering of light by crystals.* Wiley, New York, 1975.

[102] Y. Öhrn. *Elements of molecular symmetry.* Wiley, New York, 2000.

[103] M. Tinkham. *Group theory and quantum mechanics.* McGraw-Hill, New York, 1964.

[104] L. D. Landau and E. M. Lifshitz. *Mechanics.* 3rd edition, Elsevier, Amsterdam, 2004.

[105] W.-H. Yang, J. Hulteen, G. C. Schat, and R. P. Van Duyne. A surface-enhanced hyper-Raman and surface-enhanced Raman scattering study of trans-1,2-bis(4-pyridyl)ethylene adsorbed onto silver film over nanosphere electrodes. Vibrational assignments: experiment and theory. *J. Chem. Phys.*, 104:4313–4323, 1996.

[106] J. T. Golab, J. R. Sprague, K. T. Carron, *et al.* A surface enhanced hyper-Raman scattering study of pyridine adsorbed onto silver: experiment and theory. *J. Chem. Phys.*, 88:7942–7951, 1988.

[107] C. K. Johnson and S. A. Soper. Nonlinear surface-enhanced spectroscopy of silver colloids and pyridine: hyper-Raman and second-harmonic scattering. *J. Phys. Chem.*, 93:7281–7285, 1989.

[108] T. Itoh, Y. Ozaki, H. Yoshikawa, *et al.* Hyper-Rayleigh scattering and hyper-Raman scattering of dye-adsorbed silver nanoparticles induced by a focused continuous-wave near-infrared laser. *Appl. Phys. Lett.*, 88: 084102–084104, 2006.

[109] R. Eisberg and R. Resnick. *Quantum physics of atoms, molecules, solids, nuclei, and particles*. John Wiley & Sons, New York, 1974.

[110] G. Placzek. Rayleigh-Streuung und Raman-Effekt. In E. Marx, editor, *Handbuch der Radiologie*, volume 6 of *Handbuch der Radiologie*, Academische Verlag, Leipzig, pp. 205–374, 1934.

[111] P. Brüsch, editor. *Phonons: theory and experiments II*. Springer-Verlag, Berlin, 1986.

[112] S. K. Ma. *Statistical mechanics*. World Scientific, Philadelphia, 1985.

[113] P. M. Morse. *Thermal physics*. W. A. Benjamin, Inc., New York, 1969.

[114] F. Reif. *Fundamentals of statistical and thermal physics*. McGraw-Hill, New York, 1965.

[115] R. C. Maher, L. F. Cohen, J. C. Gallop, *et al.* Temperature-dependent anti-Stokes/Stokes ratios under surface-enhanced Raman scattering conditions. *J. Phys. Chem. B*, 110:6797–6803, 2006.

[116] D. Steele. *Theory of vibrational spectroscopy*. Saunders, Philadelphia, 1976.

[117] L. A. Woodward. *Introduction to the theory of molecular vibrations and vibrational spectroscopy*. Oxford University Press, Oxford, 1972.

[118] S. P. McGlynn, L. G. Vanquickenborne, M. Kinoshita, and D. G. Carroll. *Introduction to applied quantum chemistry*. Holt, Rinehart, and Winston, Inc., New York, 1972.

[119] R. C. Maher, L. F. Cohen, E. C. Le Ru, and P. G. Etchegoin. On the experimental estimation of surface enhanced Raman scattering (SERS) cross sections by vibrational pumping. *J. Phys. Chem. B*, 110: 19469–19478, 2006.

[120] E. C. Le Ru and P. G. Etchegoin. Vibrational pumping and heating under SERS conditions: fact or myth?. *Faraday Discuss.*, 132:63–75, 2006.

[121] H. Poulet and J. P. Mathieu. *Vibration spectra and symmetry of crystals*. Gordon and Breach, Paris, 1976.

[122] U. Berkert and N. L. Allinger. *Molecular mechanics (American Chemical Society Monograph 177)*. American Chemical Society, Washington, 1982.

[123] C. L. Brooks III, M. Karplus, and B. M. Pettitt. *Proteins: a theoretical perspective of dynamics, structure, and thermodynamics*, volume 71 of *Adv. Chem. Phys.*, John Wiley & Sons, New York, 1988.

[124] P. J. Robinson and K. A. Holbrook. *Unimolecular reactions*. Wiley, New York, 1972.

[125] M. V. Volkenstein. *Molecular biophysics*. Academic Press, New York, 1977.

[126] A. J. Haes and R. P. Van Duyne. A unified view of propagating and localized surface plasmon resonance biosensors. *Anal. Bioanal. Chem.*, 379:920–930, 2005.

[127] X. Liu, D. Song, Q. Zhang, *et al.* Wavelength-modulation surface plasmon resonance sensor. *Trends Anal. Chem.*, 24:887–893, 2005.

[128] S. A. Maier and H. A. Atwater. Plasmonics: localization and guiding of electromagnetic energy in metal/dielectric structures. *J. Appl. Phys.*, 98:2005 011101–1–10.

[129] J. M. Pitarke, V. M. silkin, E. V. Chulkov, and P. M. Echenique. Surface plasmons in metallic structures. *J. Opt. A: Pure Appl. Opt.*, 7:S73–S84, 2005.

[130] E. Ozbay. Plasmonics: merging photonics and electronics at nanoscale dimensions. *Science*, 311:189–193, 2006.

[131] Bill Barnes and Roy Sambles. Metal surfaces light up. *Phys. World*, 19: 17–21, 2006.

[132] A. V. Zayatsa, I. I. Smolyaninov, and A. A. Maradudin. Nano-optics of surface plasmon polaritons. *Phys. Rep.*, 408:131–314, 2005.

[133] S. A. Maier. *Plasmonics: Fundamentals and applications*. Springer-Verlag, New York, 2007.

[134] E. Prodan, C. Radloff, N. J. Halas, and P. Nordlander. A hybridization model for the plasmon response of complex nanostructures. *Science*, 302:419–422, 2003.

[135] P. M. Platzman and P. A. Wolf. *Waves and interactions in solid state plasmas (Solid State Physics, Suppl. No 13)*. Academic Press, New York, London, 1973.

[136] M. Dressel and G. Grüner. *Electrodynamics of solids: optical properties of electrons in matter*. Cambridge University Press, Cambridge, 2002.

[137] P. B. Johnson and R. W. Christy. Optical constants of noble metals. *Phys. Rev. B*, 6:4370–4379, 1972.

[138] G. W. Ford and W. H. Weber. Electromagnetic interactions of molecules with metal surfaces. *Phys. Rep.*, 113:195–287, 1984.

[139] E. D. Palik, editor. *Handbook of optical constants of solids III*. Academic Press, New York, 1998.

[140] F. Wang and Y. R. Shen. General properties of local plasmons in metal nanostructures. *Phys. Rev. Lett.*, 97:2006 206806–1–4.

[141] D. Pines. Collective energy losses in solids. *Rev. Mod. Phys.*, 28:184–199, 1956.

[142] D. L. Mills and E. Burstein. Polaritons: the electromagnetic modes of media. *Rep. Prog. Phys.*, 37:817–926, 1974.

[143] R. H. Ritchie. Plasma losses by fast electrons in thin films. *Phys. Rev.*, 106:874–881, 1957.

[144] C. J. Powell and J. B. Swan. Origin of the characteristic electron energy losses in aluminum. *Phys. Rev.*, 115:869–875, 1959.

[145] E. A. Stern and R. A. Ferrell. Surface plasma oscillations of a degenerate electron gas. *Phys. Rev.*, 120:130–136, 1960.

[146] A. D. Boardman. Chapter 1 - Hydrodynamic theory of plasmon-polaritons on plane surfaces. In *Electromagnetic surface modes*. Wiley, Chichester, 1982, pp. 1–76.

[147] H. Kawamura. *Polaritons*. Pergamon, New York, 1974, p. 247.

[148] E. N. Economou and K. L. Ngai. Surface plasma oscillations and related surface effects in solids. *Adv. Chem. Phys.*, 27:265–354, 1974.

[149] C. F. Bohren and D. R. Huffman. *Absorption and scattering of light by small particles*. John Wiley & Sons Inc., New York, 1983.

[150] J. R. Sambles, G. W. Bradbery, and F. Yang. Optical excitation of surface plasmons: an introduction. *Contemp. Phys.*, 32:173–183, 1991.

[151] M. Born and E. Wolf. *Principles of optics.* 7th (expanded) edition, Cambridge University Press, Cambridge, 1999.

[152] E. T. Arakawa, M. W. Williams, R. N. Hamm, and R. H. Ritchie. Effect of damping on surface plasmon dispersion. *Phys. Rev. Lett.*, 31: 1127–1129, 1973.

[153] R. W. Alexander, G. S. Kovener, and R. J. Bell. Dispersion curves for surface electromagnetic waves with damping. *Phys. Rev. Lett.*, 32: 154–157, 1974.

[154] A. Otto. Excitation of nonradiative surface plasma waves in silver by method of frustrated total reflection. *Z. Phys.*, 216:398–410, 1968.

[155] E. Kretschmann and H. Raether. Radiative decay of non radiative surface plasmons excited by light. *Z. Naturf. A*, 23:2135, 1968.

[156] E. Kretschmann. Determination of the optical constants of metals by excitation of surface plasmons. *Z. Phys.*, 241:313–324, 1971.

[157] B. Hecht, H. Bielefeldt, L. Novotny, *et al.* Local excitation, scattering, and interference of surface plasmons. *Phys. Rev. Lett.*, 77:1889–1892, 1996.

[158] G. Mie. Articles on the optical characteristics of turbid tubes, especially colloidal metal solutions. *Ann. Phys.*, 25:377–445, 1908.

[159] R. Fuchs and K. L. Kliewer. Optical modes of vibration in an ionic crystal sphere. *J. Opt. Soc. Am.*, 58:319–330, 1968.

[160] H. Wang, D. W. Brandl, P. Nordlander, and N. J. Halas. Plasmonic nanostructures: Artificial molecules. *Acc. Chem. Res.*, 40:53–62, 2007.

[161] J. R. Lombardi, R. L. Birke, T. Lu, and J. Xu. Charge transfer theory of surface enhanced Raman spectroscopy; Herzberg-Teller contributions. *J. Chem. Phys.*, 84:4174–4180, 1986.

[162] E. C. Le Ru, P. G. Etchegoin, and M. Meyer. Enhancement factor distribution around a single surface-enhanced Raman scattering hot spot and its relation to single molecule detection. *J. Chem. Phys.*, 125: 204701–204714, 2006.

[163] W. B. Cai, B. Ren, X. Q. Li, *et al.* Investigation of surface-enhanced Raman scattering from platinum electrodes using a confocal Raman microscope: dependence of surface roughening pretreatment. *Surf. Sci.*, 406:9–22, 1998.

[164] M. Kerker, D.-S. Wang, and H. Chew. Surface enhanced Raman-scattering (SERS) by molecules adsorbed at spherical-particles. *Appl. Opt.*, 19:4159–4174, 1980.

[165] R. Rojas and F. Claro. Theory of surface enhanced Raman-scattering in colloids. *J. Chem. Phys.*, 98:998–1006, 1993.

[166] E. M. Purcell. Spontaneous emission probabilities at radio frequencies. *Phys. Rev.*, 69:681, 1946.

[167] K. H. Drexhage, H. Kuhn, and F. P. Scäfer. Variation of fluorescence decay time of a molecule in front of a mirror. *Ber. Bunsenges. Phys. Chem.*, 72:329, 1968.

[168] P. Goy, J. M. Raymond, M. Gross, and S. Haroche. Observation of cavity-enhanced single-atom spontaneous emission. *Phys. Rev. Lett.*, 50: 1903–1906, 1983.

[169] R. R. Chance, A. Prock, and R. Silbey. Molecular fluorescence and energy transfer near interfaces. *Adv. Chem. Phys.*, 37:1–65, 1978.

[170] H. Metiu. Surface enhanced spectroscopy. *Prog. Surf. Sci.*, 17:153–320, 1984.

[171] J. I. Gersten and A. Nitzan. Photophysics and photochemistry near surfaces and small particles. *Surf. Sci.*, 158:165–189, 1985.

[172] E. Dulkeith, A. C. Morteani, T. Niedereichholz, *et al.* Fluorescence quenching of dye molecules near gold nanoparticles: radiative and nonradiative effects. *Phys. Rev. Lett.*, 89:2002 203002–1–4.

[173] S. M. Barnett, B. Huttner, and R. Loudon. Spontaneous emission in absorbing dielectric media. *Phys. Rev. Lett.*, 68:3698–3701, 1992.

[174] J. Dalibard, J. Dupont-Roc, and C. Cohen-Tannoudji. Vacuum fluctuations and radiative reaction: identification of their respective contributions. *J. Phys.*, 43:1617–1638, 1982.

[175] L. A. Blanco and F. J. García de Abajo. Spontaneous light emission in complex nanostructures. *Phys. Rev. B*, 69:2004 205414–1–12.

[176] C. Cohen-Tannoudji, J. Dupont-Roc, and G. Grynberg. *Photons and atoms: introduction to quantum electrodynamics*. Wiley, New York, 1989.

[177] H. Chew. Transition rates of atoms near spherical surfaces. *J. Chem. Phys.*, 87:1355–1360, 1987.

[178] L. D. Landau, E. M. Lifshitz, and L. P. Pitaevskii. *Electrodynamics of continuous media*. 2nd edition, Pergamon, Oxford, 1984.

[179] J.-Y. Courtois, J.-M. Courty, and J. C. Mertz. Internal dynamics of multilevel atoms near a vacuum-dielectric interface. *Phys. Rev. A*, 53: 1862–1878, 1996.

[180] M. Kahl and E. Voges. Analysis of plasmon resonance and surface-enhanced Raman scattering on periodic silver structures. *Phys. Rev. B*, 61:14078–14088, 2000.

[181] E. J. Bjerneld, F. Svedberg, P. Johansson, and M. Käll. Direct observation of heterogeneous photochemistry on aggregated Ag nanocrystals using Raman spectroscopy: the case of photoinduced degradation of aromatic amino acids. *J. Phys. Chem. A*, 108:4187–4193, 2004.

[182] B. Pettinger, B. Ren, G. Picardi, *et al.* Tip-enhanced Raman spectroscopy (TERS) of malachite green isothiocyanate at Au(111): bleaching behavior under the influence of high electromagnetic fields. *J. Raman Spectrosc.*, 36:541–550, 2005.

[183] A. Campion, J. E. Ivanecky, C. M. Child, and M. Foster. On the mechanism of chemical enhancement in surface-enhanced Raman scattering. *J. Am. Chem. Soc.*, 117:11807–11808, 1995.

[184] T. W. Koo, S. Chan, L. Sun, *et al.* Specific chemical effects on surface-enhanced Raman spectroscopy for ultra-sensitive detection of biological molecules. *Appl. Spectrosc.*, 58:1401–1407, 2004.

[185] W. E. Doering and S. M. Nie. Single-molecule and single nanoparticle SERS: Examining the roles of surface active sites and chemical enhancement. *J. Phys. Chem. B*, 106:311–317, 2002.

[186] E. J. Liang and W. Kiefer. Chemical effect of SERS with near-infrared excitation. *J. Raman Spectrosc.*, 27:879–885, 1996.

[187] Z. Q. Tian. General discussions section. *Faraday Discuss.*, 132:309–319, 2006.

[188] D.-Y. Wu, S. Duan, B. Ren, and Z.-Q. Tian. Density functional theory study of surface-enhanced Raman scattering spectra of pyridine adsorbed on noble and transition metal surfaces. *J. Raman Spectrosc.*, 36:533–540, 2005.

[189] J. A. Creighton. The resonance Raman contribution to SERS: pyridine on copper or silver in aqueous media. *Surf. Sci.*, 173:665–672, 1986.

[190] J. I. Gersten, R. L. Birke, and J. R. Lombardi. Theory of enhance light scattering from molecules adsorbed at the metal-solution interface. *Phys. Rev. Lett.*, 43:147–150, 1979.

[191] Y. Xie, D. Y. Wu, G. K. Liu, *et al.* Adsorption and photon-driven charge transfer of pyridine on a cobalt electrode analyzed by surface enhanced Raman spectroscopy and relevant theories. *J. Electroanal. Chem.*, 554: 417–425, 2003.

[192] M. Pagliai, L. Bellucci, M. Muniz-Miranda, *et al.* A combined Raman, DFT and MD study of the solvation dynamics and the adsorption process of pyridine in silver hydrosols. *Phys. Chem. Chem. Phys.*, 8: 171–178, 2006.

[193] F. W. King, R. P. Van Duyne, and G. C. Schatz. Theory of Raman-scattering by molecules adsorbed on electrode surfaces. *J. Chem. Phys.*, 69:4472–4481, 1978.

[194] S. Efrima and H. Metiu. Classical theory of light-scattering by an adsorbed molecule.1. Theory. *J. Chem. Phys.*, 70:1602–1613, 1979.

[195] W. H. Weber and G. W. Ford. Enhanced Raman-scattering by adsorbates including the nonlocal response of the metal and the excitation of nonradiative modes. *Phys. Rev. Lett.*, 44:1774–1777, 1980.

[196] M. Kerker. *The scattering of light and other electromagnetic radiation.* Academic Press, New York, 1969.

[197] A. F. Stevenson. Solution of electromagnetic scattering problems as power series in the ratio (dimension of scatterer)/(wavelength). *J. Appl. Phys.*, 24:1134–1142, 1953.

[198] J. Van Bladel. *Electromagnetic fields*. McGraw-Hill, New York, 1964.

[199] A. T. Young. Rayleigh scattering. *Appl. Opt.*, 20:533–535, 1981.

[200] R. E. Kleinman and T. B. A. Senior. *Rayleigh scattering*. Elsevier, Amsterdam, 1986, pp. 1–70.

[201] J. M. Gérardy and M. Ausloos. Absorption spectrum of clusters of spheres from the general solution of Maxwell's equations. II. Optical properties of aggregated metal spheres. *Phys. Rev. B*, 25:4204–4229, 1982.

[202] H. Xu and M. Käll. Surface-plasmon-enhanced optical forces in silver nanoaggregates. *Phys. Rev. Lett.*, 89:246802–246805, 2002.

[203] P. Johansson, H. Xu, and M. Käll. Surface-enhanced Raman scattering and fluorescence near metal nanoparticles. *Phys. Rev. B*, 72:2005 035427–1–17.

[204] E. C. Le Ru, C. Galloway, and P. G. Etchegoin. On the connection between optical absorption/extinction and SERS enhancements. *Phys. Chem. Chem. Phys.*, 8:3083–3087, 2006.

[205] S. Asano and G. Yamamoto. Light-scattering by a spheroidal particle. *Appl. Opt.*, 14:29–49, 1975.

[206] E. Prodan, P. Nordlander, and N. J. Halas. Electronic structure and optical properties of gold nanoshells. *Nano Lett.*, 3:1411–1415, 2003.

[207] P. Nordlander and E. Prodan. Plasmon hybridization in nanoparticles near metallic surfaces. *Nano Lett.*, 4:2209–2213, 2004.

[208] W.-H. Yang, G. C. Schatz, and R. P. Van Duyne. Discrete dipole approximation for calculating extinction and Raman intensities for small particles with arbitrary shapes. *J. Chem. Phys.*, 103:869–875, 1995.

[209] M. A. Yurkin and A. G. Hoekstra. The discrete dipole approximation: an overview and recent developments. *J. Quant. Spectrosc. Rad. Transf.*, 106:558–589, 2007.

[210] B. T. Draine and P. J. Flatau. User guide to the discrete dipole approximation code DDSCAT 6.1. http://arxiv.org/abs/astro-ph/0409262v2 arXiv, 2004.

[211] D. W. Ward and K. A. Nelson. Finite-difference time-domain (FDTD) simulations of electromagnetic wave propagation using a spreadsheet. *Comput. Appl. Eng. Educ.*, 13:213–221, 2005.

[212] K. Kunz and R. Luebbers. *The finite difference time domain method for electromagnetics*. CRC press, Boca Raton, FL, 1993.

[213] A. Taflove and S. Hagness. *Computational electrodynamics: the finite-difference time-domain method*. Artech House, Boston, 2000.

[214] J. Jin. *The finite element method in electromagnetics*. 2nd edition, Wiley-IEEE Press, New York, 2002.

[215] M. I. Mishchenko, L. D. Travis, and A. A. Lacis. *Scattering, absorption, and emission of light by small particles*. Cambridge University Press, Cambridge, 2002.

[216] A. Wokaun, J. P. Gordon, and P. F. Liao. Radiation damping in surface-enhanced Raman scattering. *Phys. Rev. Lett.*, 48:957–960, 1982.

[217] M. Meier and A. Wokaun. Enhanced fields on large metal particles: dynamic depolarization. *Opt. Lett.*, 8:581–583, 1983.

[218] K. L. Kelly, E. Coronado, L. L. Zhao, and G. C. Schatz. The optical properties of metal nanoparticles: the influence of size, shape, and dielectric environment. *J. Phys. Chem. B*, 107:668–677, 2003.

[219] E. J. Zeman and G. C. Schatz. An accurate electromagnetic theory study of surface enhancement factors for Ag, Au, Cu, Li, Na, Al, Ga, In, Zn, and Cd. *J. Phys. Chem.*, 91:634–643, 1987.

[220] P. C. Lee and D. Meisel. Adsorption and surface-enhanced Raman of dyes on silver and gold sols. *J. Phys. Chem.*, 86:3391–3395, 1982.

[221] C. H. Munro, W. E. Smith, M. Garner, *et al.* Characterization of the surface of a citrate-reduced colloid optimized for use as a substrate for surface-enhanced resonance Raman scattering. *Langmuir*, 11:3712–3720, 1995.

[222] S. Schneider, P. Halbig, H. Grau, and Ulrich Nickel. Reproducible preparation of silver sols with uniform particle size for applications in surface-enhanced Raman spectroscopy. *Photochem. Photobiol.*, 60: 605–610, 1994.

[223] U. Nickel, K. Mansyreff, and S. Schneider. Production of monodisperse silver colloids by reduction with hydrazine: the effect of chloride and aggregation on SER(R)S signal intensity. *J. Raman Spectrosc.*, 35: 101–110, 2004.

[224] J. A. Creighton, C. G. Blatchford, and M. G. Albrecht. Plasma resonance enhancement of Raman scattering by pyridine adsorbed on silver or gold sol particles of size comparable to the excitation wavelength. *J. Chem. Soc.: Faraday Trans.*, 2:790–798, 1979.

[225] D. L. Van Hyning and C. F. Zukoski. Formation mechanisms and aggregation behavior of borohydride reduced silver particles. *Langmuir*, 14:7034–7046, 1998.

[226] J. J. Mock, M. Barbic, D. R. Smith, *et al.* Shape effects in plasmon resonance of individual colloidal silver nanoparticles. *J. Chem. Phys.*, 116:6755–6759, 2002.

[227] H. S. Kim, J. H. Ryu, B. Jose, *et al.* Formation of silver nanoparticles induced by poly(2,6-dimethyl-1,4-phenylene oxide). *Langmuir*, 17: 5817–5820, 2001.

[228] Z. S. Pillai and P. V. Kamat. What factors control the size and shape of silver nanoparticles in the citrate-ion reduction method?. *J. Phys. Chem. B*, 108:945–951, 2004.

[229] M. Meyer, E. C. Le Ru, and P. G. Etchegoin. Self-limiting aggregation leads to long-lived metastable clusters in colloidal solutions. *J. Phys. Chem. B*, 110:6040–6047, 2006.

[230] Z. Tai, J. Zhang, J. Gao, and G. Xue. Surface-enhanced Raman scattering study of the surface coordination of porphyrins adsorbed on silver. *J. Mat. Chem.*, 3:417–420, 1993.

[231] R. J. Hunter. *Foundations of colloid science*, volume I and II of Oxford University Press, Oxford, 1989.

[232] D. F. Evans and H. Wenneerström. *The colloidal domain: where physics, chemistry, biology, and technology meet*. Wiley-VHC, New York, 1999.

[233] B. J. Berne and R. Pecora. *Dynamic light scattering (with applications to chemistry, biology, and physics)*. Dover Publications Inc., New York, 2000.

[234] K. Varis, M. Mattila, S. Arpiainen, *et al.* Reflection of focused beams from opal photonic crystals. *Optics Express*, 13:2653–2667, 2005.

[235] Y. Han and D. G. Grier. Confinement-induced colloidal attractions in equilibrium. *Phys. Rev. Lett.*, 91:038302–038306, 2003.

[236] B. V. Derjaguin and L. D. Landau. Theory of the stability of strongly charged lyophobic sols and of the adhesion of strongly charged particles in solutions of electrolytes. *Acta Phys. (USSR)*, 14:633–662, 1941.

[237] E. J. Verwey and J. Overbeek. *Theory of the stability of lyophobic colloids*. Elsevier, Amsterdam, 1948.

[238] A. S. Davydov. *Quantum mechanics*. Pergamon Press, Oxford, 1965.

[239] J. O. Hirschfelder, C. F. Curtis, and R. B. Bird. *Molecular theory of gases and liquids*. Wiley, New York, 1954.

[240] L. D. Landau and E. M. Lifshitz. *Statistical physics, part II*. Elsevier, Amsterdam, 2003.

[241] I. D'Amico and H. Löwen. Effective forces between macroions: a Monte-Carlo study. *Physica A*, 237:25–30, 1997.

[242] M. J. Stevens, M. L. Falk, and M. O. Robbins. Interactions between charged spherical macroions. *J. Chem. Phys.*, 104:5209–5219, 1996.

[243] H. Löwen, E. Allahyarov, C. N. Likos, *et al.* Charged colloids, polyelectrolytes and biomolecules viewed as strongly coupled Coulomb systems. *J. Phys. A: Math. Gen.*, 36:5827–5834, 2003.

[244] J. C. Crocker and D. G. Grier. When like charges attract: the effects of geometrical confinement on long-range colloidal interactions. *Phys. Rev. Lett.*, 77:1897–1900, 1996.

[245] H.R. Kruyt, editor. *Colloid science*. Elsevier, New York, 1952.

[246] D. A. Weitz and M. Oliveria. Fractal structures formed by kinetic aggregation of aqueous gold colloids. *Phys. Rev. Lett.*, 52:1433–1436, 1984.

[247] S. Bastea. Aggregation kinetics in a model colloidal suspension. *Phys. Rev. Lett.*, 96:028305–028308, 2006.

[248] D. A. Weitz, M. Y. Lin, and J. S. Huang. *Physics of complex and supramolecular fluids*. Wiley, New York, 1987.

[249] V. M. Shalaev. Electromagnetic properties of small-particle composites. *Phys. Rep.*, 272:61–137, 1996.

[250] A. K. Sarychev and V. M. Shalaev. Electromagnetic field fluctuations and optical nonlinearities in metal-dielectric composites. *Phys. Rep.*, 335:275–371, 2000.

[251] V. M. Shalaev. *Nonlinear optics of random media: Fractal composites and metal-dielectric films*. Springer, Berlin, 2000.

[252] C. McLaughlin, D. Graham, and W. E. Smith. Comparison of resonant and non-resonant conditions on the concentration dependence of surface enhanced Raman scattering from a dye adsorbed on silver colloid. *J. Phys. Chem. B*, 106:5408–5412, 2002.

[253] A. Otto, A. Bruckbauer, and Y. X. Chen. On the chloride activation in SERS and single molecule SERS. *J. Mol. Struct. (Theochem)*, 661–662: 501–514, 2003.

[254] W. Grochala, A. Kudelski, and J. Bukowska. Anion-induced charge-transfer enhancement in SERS and SERRS spectra of rhodamine 6G on a silver electrode: how important is it?. *J. Raman Spectrosc.*, 29: 681–685, 1998.

[255] K. Kneipp, H. Kneipp, I. Itzkan, *et al.* Surface-enhanced Raman scattering and biophysics. *J. Phys.: Cond. Matt.*, 14:R597–R624, 2002.

[256] W. Zhang, T. Schmid, B.-S. Yeo, and R. Zenobi. Near-field heating, annealing, and signal loss in tip-enhanced Raman spectroscopy. *J. Phys. Chem. C*, 112:2104–2108, 2008.

[257] S. R. Emory, S. Nie, K. Kneipp, and G. R. Harrison. Single-molecule Raman spectroscopy - fact or fiction?. *Chimia*, 53:35–37, 1999.

[258] A. M. Michaels, M. Nirmal, and L. E. Brus. Surface enhanced Raman spectroscopy of individual rhodamine 6G molecules on large Ag nanocrystals. *J. Am. Chem. Soc.*, 121:9932–9939, 1999.

[259] A. Weiss and G. Haran. Time-dependent single-molecule Raman scattering as a probe of surface dynamics. *J. Phys. Chem. B*, 105: 12348–12354, 2001.

[260] K. Kenipp, H. Kneipp, G. Deinum, *et al.* Single-Molecule detection of cyanine dye in silver colloidal solution using near-infrared surface-enhanced Raman scattering. *Appl. Spectrosc.*, 52:175–178, 1998.

[261] P. Etchegoin, R. C. Maher, L. F. Cohen, *et al.* New limits in ultrasensitive trace detection by surface enhanced Raman scattering (SERS). *Chem. Phys. Lett.*, 375:84–90, 2003.

[262] P. Hildebrandt and M. Stockburger. Surface-enhanced resonance Raman spectroscopy of rhodamine 6G adsorbed on colloidal silver. *J. Phys. Chem.*, 88:5935–5944, 1984.

[263] R. C. Maher, L. F. Cohen, and P. G. Etchegoin. Single molecule photobleaching observed by surface enhanced resonant Raman scattering (SERRS). *Chem. Phys. Lett.*, 352:378–384, 2002.

[264] Y. Maruyama, M. Ishikawa, and M. Futamata. Thermal activation of blinking in SERS signals. *J. Phys. Chem. B*, 108:673–678, 2004.

[265] A. Kudelski and B. Pettinger. SERS on carbon chain segments: monitoring locally surface chemistry. *Chem. Phys. Lett.*, 321:356–362, 2000.

[266] M. Moskovits, L. L. Tay, J. Yang, and T. Haslett. SERS and the single molecule. In *Optical properties of nanostructured random media*, volume 82 of *Top. Appl. Phys.*, Springer-Verlag, Berlin, pp. 215–226, 2002.

[267] C. J. L. Constantino, T. Lemma, P. A. Antunes, and R. Aroca. Single-molecule detection using surface-enhanced resonance Raman scattering and Langmuir–Blodgett monolayers. *Anal. Chem.*, 73:3674–3678, 2001.

[268] P. J. G. Goulet, N. P. W. Pieczonka, and R. F. Aroca. Mapping single-molecule SERRS from Langmuir-Blodgett monolayers on nanostructured silver island films. *J. Raman Spectrosc.*, 36:574–580, 2005.

[269] P. G. Etchegoin, M. Meyer, E. Blackie, and E. C. Le Ru. Statistics of single-molecule surface enhanced Raman scattering signals: fluctuation analysis with multiple analyte techniques. *Anal. Chem.*, 79:8411–8515, 2007.

[270] Y. Sawai, B. Takimoto, H. Nabika, *et al.* Observation of a small number of molecules at a metal nanogap arrayed on a solid surface using surface-enhanced Raman scattering. *J. Am. Chem. Soc.*, 129:1658–1662, 2007.

[271] P. J. G. Goulet and R. F. Aroca. Distinguishing individual vibrational fingerprints: single-molecule surface-enhanced resonance Raman scattering from one-to-one binary mixtures in Langmuir-Blodgett monolayers. *Anal. Chem.*, 79:2728–2734, 2007.

[272] D. Zhang, Y. Xie, S. K. Deb, *et al.* Isotope edited internal standard method for quantitative surface-enhanced Raman spectroscopy. *Anal. Chem.*, 77:3563–3569, 2005.

[273] N. P. W. Pieczonka and R. F. Aroca. Single molecule analysis by surfaced-enhanced Raman scattering. *Chem. Soc. Rev.*, 37:946–954, 2008.

[274] N. P. W. Pieczonka and R. F. Aroca. Inherent complexities of trace detection by surface-enhanced Raman scattering. *Chem. Phys. Chem.*, 6:2473–2484, 2005.

[275] M. S. Anderson. Locally enhanced Raman spectroscopy with an atomic force microscope. *Appl. Phys. Lett.*, 76:3130–3132, 2000.

[276] D. Zeisel, V. Deckert, R. Zenobi, and T. Vo-Dinh. Near-field surface-enhanced Raman spectroscopy of dye molecules adsorbed on silver island films. *Chem. Phys. Lett.*, 283:381–385, 1998.

[277] M. S. Anderson and W. T. Pike. A Raman-atomic force microscope for apertureless-near-field spectroscopy and optical trapping. *Rev. Scientific Instruments*, 73:1198–1203, 2002.

[278] D. S. Bulgarevich and M. Futamata. Apertureless tip-enhanced Raman microscopy with confocal epi-illumination/collection optics. *Appl. Spectrosc.*, 58:757–761, 2004.

[279] A. Hartschuh, N. Anderson, and L. Novotny. Near-field Raman spectroscopy using a sharp metal tip. *J. Microsc.*, 210:234–240, 2002.

[280] L. Novotny. *Near-field optics and surface plasmon polaritons*, volume 81 of *Top. Appl. Phys.*, Springer-Verlag, Berlin, 2000.

[281] A. Downes, D. Salter, and A. Elfick. Finite element simulations of tip-enhanced Raman and fluorescence spectroscopy. *J. Phys. Chem. B*, 110: 6692–6698, 2006.

[282] L. Billot, L. Berguiga, M. L. de la Chapelle, *et al.* Production of gold tips for tip-enhanced near-field optical microscopy and spectroscopy:

analysis of the etching parameters. *Eur. Phys. J. Appl. Phys.*, 31: 139–146, 2006.

[283] K. F. Domke and B. Pettinger. Comment on scanning-probe Raman spectroscopy with single-molecule sensitivity. *Phys. Rev. B*, 75: 236401–236403, 2007.

[284] C. C. Neacsu, J. Dreyer, N. Behr, and M. B. Raschke. Reply to Comment on 'scanning-probe Raman spectroscopy with single-molecule sensitivity'. *Phys. Rev. B*, 75:236402–236405, 2007.

[285] C. Li, W. Cai, Y. Li, *et al.* Ultrasonically induced Au nanoprisms and their size manipulation based on aging. *J. Phys. Chem. B*, 110: 1546–1552, 2006.

[286] C. L. Nehl, H. Liao, and J. H. Hafner. Optical properties of star-shaped gold nanoparticles. *Nano Lett.*, 6:683–688, 2006.

[287] P. Etchegoin, L. F. Cohen, H. Hartigan, *et al.* Electromagnetic contribution to surface enhanced Raman scattering revisited. *J. Chem. Phys.*, 119:5281–5289, 2003.

[288] F. Tam, C. Moran, and N. Halas. Geometrical parameters controlling sensitivity of nanoshell plasmon resonances to changes in dielectric environment. *J. Phys. Chem. B*, 108:17290–17294, 2004.

[289] A. M. Gobin, D. P. O'Neal, N. J. Halas, *et al.* Near infrared laser tissue welding using nanoshells as an exogenous absorber. *Lasers Surg. Med.*, 37:123–129, 2005.

[290] C. Loo, L. R. Hirsch, M.-H. Lee, *et al.* Gold nanoshell bioconjugates for molecular imaging in living cells. *Optics Lett.*, 30:1012–1014, 2005.

[291] D. P. O'Neal, L. R. Hirsch, N. J. Halas, *et al.* Photo-thermal tumor ablation in mice using near infrared-absorbing nanoparticles. *Cancer Lett.*, 109:171–176, 2004.

[292] H. Wang, G. Goodrich, F. Tam, *et al.* Controlled texturing modifies the surface topography and plasmonic properties of Au nanoshells. *J. Phys. Chem. B*, 109:11083–11087, 2005.

[293] N. Halas. Playing with plasmons: tuning the optical resonant properties of metallic nanoshells. *MRS Bull.*, 30:362–367, 2005.

[294] H. Wang, D. W. Brandl, F. Le, P. Nordlander, and N. J. Halas. Nanorice: a hybrid plasmonic nanostructure. *Nano Lett.*, 6:827–832, 2006.

[295] J. A. Dieringer, A. D. McFarland, N. C. Shah, *et al.* Surface enhanced Raman spectroscopy: new materials, concepts, characterization tools, and applications. *Faraday Discuss.*, 132:9–26, 2006.

[296] Y. Lu, G. L. Liu, and L. P. Lee. High-density silver nanoparticle film with temperature-controllable inter-particle spacing for a tunable surface enhanced Raman scattering substrate. *Nano Lett.*, 5:5–9, 2005.

[297] R. M. Connatser, L. A. Riddle, and M. J. Sepaniak. Metal-polymer nanocomposites for integrated microfluidic separations and surface enhanced Raman spectroscopic detection. *J. Separation Science*, 27: 1545–1550, 2004.

[298] S. C. Hendy, M. Jasperse, and J. Burnell. Effect of patterned slip on micro- and nanofluidic flows. *Phys. Rev. E*, 73:016303–016311, 2005.

[299] A. Ashkin. Optical trapping and manipulation of neutral particles using lasers. *Proc. Nat. Acad. Sci.*, 94:4853–4860, 1997.

[300] F. Svedberg and M. Käll. On the importance of optical forces in surface-enhanced Raman scattering (SERS). *Faraday Discuss.*, 132:35–44, 2006.

[301] S. Lal, N. K. Grady, J. Kundu, *et al.* Tailoring plasmonics substrates for surface enhanced spectroscopies. *Chem. Soc. Rev.*, 37:898–911, 2008.

[302] M. J. Banholzer, J. E. Millstone, L. Qin, and C. A. Mirkin. Rationally designed nanostructures for surface-enhanced Raman spectroscopy. *Chem. Soc. Rev.*, 37:885–897, 2008.

[303] W. E. Smith. Practical understanding and use of surface enhanced Raman scattering/surface enhanced resonance Raman scattering in chemical and biological analysis. *Chem. Soc. Rev.*, 37:955–964, 2008.

[304] A. van Blaaderen, J. P. Hoogenboom, D. L. J. Vossen, *et al.* Colloidal epitaxy: playing with the boundary conditions of colloidal crystallization. *Faraday Discuss.*, 123:107–119, 2003.

[305] F. Nogueira, A. Castro, and M. A. L. Marques. *A tutorial on density functional theory*, volume 620 of *Lecture Notes in Physics*, Springer-Verlag, Berlin, pp. 218–256, 2003.

[306] R. M. Dreizler and E. K. U. Gross. *Density functional theory: an approach to the quantum many-body problem*. Springer-Verlag, Berlin, 1990.

[307] J. Neugebauer, M. Reiher, C. Kind, and B. A. Hess. Quantum chemical calculations of vibrational spectra of large molecules- Raman and IR spectra for buckminsterfullerene. *J. Comput. Chem.*, 23:895–910, 2002.

[308] J. P. Perdew and A. Zunger. Self-interaction correction to density-functional approximations for many-electron systems. *Phys. Rev. B*, 23: 5048–5079, 1981.

[309] J. P. Perdew and Y. Wang. Accurate and simple analytic representation of the electron-gas correlation energy. *Phys. Rev. B*, 45:13244–13249, 1992.

[310] A. D. Becke. Density-functional thermochemistry. III. The role of exact exchange. *J. Chem. Phys.*, 98:5648–5652, 1993.

[311] C. Lee, W. Yang, and R. G. Parr. Development of the Colle–Salvetti correlation-energy formula into a functional of the electron density. *Phys. Rev. B*, 37:785–789, 1988.

[312] J. A. Pople. Nobel lecture: quantum chemical models. *Rev. Mod. Phys.*, 71:1267–1274, 1999.

[313] W. J. Hehre, R. F. Stewart, and J. A. Pople. Self-consistent molecular-orbital methods. I. Use of Gaussian expansions of Slater-type atomic orbitals. *J. Chem. Phys.*, 51:2657–2664, 1969.

[314] R. Krishnan, J. S. Binkley, R. Seeger, and J. A. Pople. Self-consistent molecular orbital. XX. A basis set for correlated wave functions. *J. Chem. Phys.*, 72:650–654, 1980.

[315] J. Dongarra, I. Duff, D. Sorensen, and H. van der Vorst. *Solving linear systems on vector and shared memory computers*. SIAM, Philadelphia, 1991.

[316] A. A. El-Azhary and H. U. Suter. Correlated ab-initio force fields and vibrational analysis of the spectra of isoxazole and isothiazole. *J. Phys. Chem.*, 99:12751–12758, 1995.

[317] C. Van Caillie and R. D. Amos. Raman intensities using time-dependent density functional theory. *Phys. Chem. Chem. Phys.*, 2:2123–2129, 2000.

[318] F. Orduna, C. Domingo, S. Montero, and W. F. Murphy. Gas phase Raman intensities of C_2H_2, C_2HD, and C_2D_2. *Mol. Phys.*, 45:65–75, 1982.

[319] L. D. Landau, E. M. Lifshitz, and L. P. Pitaevskiĭ. *Electrodynamics of continuous media*. 2nd edition, Elsevier, Amsterdam, 2004.

[320] M. Campoy-Quiles, G. Heliotis, R. Xia, *et al.* Ellipsometric characterization of the optical constants of polyfluorene gain media. *Adv. Functional Mater.*, 15:925–933, 2005.

[321] M. Campoy-Quiles, P. G. Etchegoin, and D. D. C. Bradley. On the optical anisotropy of conjugated polymer thin films. *Phys. Rev. B*, 72: 045209–045225, 2005.

[322] P. G. Etchegoin, E. C. Le Ru, and M. Meyer. An analytic model for the optical properties of gold. *J. Chem. Phys.*, 125:2006 164705–1–3.

Index

Kolibri 360 GmbH
Ferdinand-Jühlke-Straße 7
99095 Erfurt, Deutschland
produktsicherheit@kolibri360.de